HIGHLIGHTS OF ASTRONOMY

INTERNATIONAL ASTRONOMICAL UNION

UNION ASTRONOMIQUE INTERNATIONALE

HIGHLIGHTS OF ASTRONOMY

VOLUME 11A

AS PRESENTED AT THE XXIIIrd GENERAL ASSEMBLY OF THE IAU, 1997

EDITED BY

J. ANDERSEN

General Secretary of the Union

KLUWER ACADEMIC PUBLISHERS

DORDRECHT / BOSTON / LONDON

The Library of Congress Cataloged this serial publications as follows:
71–159657

ISBN 0-7923-5341-2

ISBN 0-7923-5617-9 (set)

Published on behalf of
the International Astronomical Union
by
Kluwer Academic Publishers, P.O. Box 17, 3300 AA Dordrecht, The Netherlands.

Sold and distributed in North, Central and South America
by Kluwer Academic Publishers,
101 Philip Drive, Norwell, MA 02061, U.S.A.

In all other countries, sold and distributed
by Kluwer Academic Publishers,
P.O. Box 322, 3300 AH Dordrecht, The Netherlands.

Printed on acid-free paper

Printed in the Netherlands

SUMMARY OF CONTENTS

Volume 11A: pages 1-588, Volume 11B: pages 589-1159.

TABLE OF CONTENTS

VOLUME 11A

VOLUME 11B

PREFACE

The XXIIIrd IAU General Assembly took place in Kyoto, Japan, in August 1997. It featured an exceptionally rich scientific programme: Between three Invited Discourses, six Symposia, 23 Joint Discussions, three Special Scientific Sessions, and well over 100 Commission and Working Group meetings, a total of over 800 oral and 1100 poster papers were presented. On behalf of the Union, I thank my predecessor, Immo Appenzeller, for the overall planning of this programme, as well as the many scientific and local organisers whose hard work assured the success of the individual events.

As a result, the present volume of the Highlights of Astronomy, which presents the Invited Discourses and the proceedings or summaries of the Joint Discussions and Special Scientific Sessions, has been divided into two separate sub-volumes (Vol. 11A+B) in order to retain a manageable book size. In addition, the Proceedings of the six Symposia (nos. 183-188) will be published separately in the regular IAU Symposium Series.

I extend my best thanks to all authors and Editors of the reports on the individual events who submitted their contributions by the specified deadline. Still, a number of key contributions were only received after long delays, and for three events no manuscript was received at all. Together with some confusion over LaTeX style file versions, a number of health issues at the Secretariat, and unexpected other commitments of the writer's time, this has delayed the completion of this volume considerably from the original schedule. I thank Ms. Monique Léger-Orine for assistance in tracking and retrieving the contributions, and especially Copenhagen astronomy students Michael Cramer Andersen, Margit Dühring, and Brian Lindgren Jensen for their dedicated, vigorous, and painstaking efforts to produce the final camera-ready manuscript; without their help, this volume would have been even further delayed and less uniform in presentation.

Finally, I thank Drs. G. Marcy and M. Mayor for providing the data from which the cover diagram of Vol. 11A was prepared, and Mr. Eckhard Slawik of Waldenburg, Germany, for the cover photo of Vol. 11B; I hope that these figures may in part compensate for the missing contributions concerning just those topics.

Copenhagen and Paris, August 1998
Johannes Andersen
General Secretary

INVITED DISCOURSES

THE HUBBLE DEEP FIELD

ROBERT WILLIAMS
Space Telescope Science Institute
Baltimore, MD, USA

1. Introduction

Great progress has been made in recent years in understanding the large-scale structure of the universe. Recall that it is only within this century that we have even come to realize the existence of other galaxies as separate entities. Until the 1920's the universe did not extend beyond the Milky Way. Novae and Cepheid variables changed that picture, and for the past 75 years extragalactic astronomy has been one of the most active and fertile areas of science. The concept of an expanding universe and its beginning in a Big Bang all derive from the discovery of external galaxies.

Distant galaxies are faint and have small angular sizes, therefore their study has remained the province of the largest telescopes. Before the launch of Hubble Space Telescope, ground-based telescopes had succeeded in detecting distant galaxies out to redshifts of $z \sim 1$, and in establishing certain of their characteristics. But, there was uncertainty as to how much further HST could push the study of distant galaxies given its modest 2.4m diameter mirror and the fact that the surface brightnesses of cosmologically distant objects decrease as $(1+z)^4$.

For this reason, one of the early observations that was scheduled immediately after the first servicing mission of HST in December 1993 to repair spherical aberration was the re-imaging of the cluster of galaxies 0939+4713 at $z = 0.4$ that had been observed previously by Dressler et al. (1994) from both the ground and with the aberrated HST. The ten-orbit WFPC2 image demonstrated HST's ability to resolve structure in distant galaxies, showing spiral and elliptical galaxies with a clarity approaching that achieved for the Coma cluster from the ground. Spirals are seen to be relatively abundant in 0939+4713, although they generally show an anomalous morphology.

The 0939+4713 observation was followed some months later by a GO program of Dickinson (1995) to image for 32 orbits the strong radio galaxy 3C 324 at redshift $z = 1.2$. The final combined image of 3C 324 and associated galaxies was very interesting in that it revealed numerous ellipticals, but unlike 0939+4713, no spiral galaxies. Furthermore, the E's were measured to have a $r^{1/4}$ radial light distribution characteristic of dynamically relaxed stellar systems. The ellipticals thus showed evidence of already being old at this higher redshift, whereas spiral galaxies were not yet present. The extent to which this difference in the two clusters indicated basic evolutionary properties of E's and S's was difficult to ascertain because of the small sample involved of only two clusters. However, it did demonstrate the power of HST to discern galaxy structure at high redshift.

2. The HDF Campaign

Following informal discussions among the Institute scientific staff on what additional observations HST might undertake that would make an important contribution to our understanding of galaxy evolution, the decision was taken to devote a substantial fraction of Director's Discretionary time in Cycle 4 to this topic. Certain desired characteristics of any deep image were agreed upon by us, and these included (1) selecting fields with low extinction, known to include distant galaxies, (2) observing in the HST Continuous Viewing Zone to maximize observing time, (3) obtaining color information by imaging in at least 2 wavelength bands, (4) setting the total exposure time so the background noise would be reduced below CCD systematic noise levels, (5) dithering the exposures to recover sub-pixel spatial resolution, (6) imaging the flanking fields around any primary field to provide an HST image that matched the field size of ground-based telescopes, and (7) making the

J. Andersen (ed.), Highlights of Astronomy, Volume 11A, 3–15.

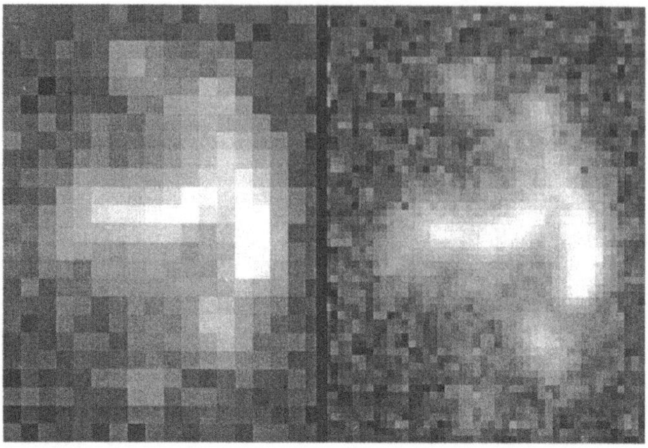

Figure 1. Comparison of two combined images of an object in the HDF, produced from the same dataset using different procedures. The left image is the result of the shift-and-add technique, and the right image is the result of 'drizzling' (from Fruchter & Hook 1997).

data non-proprietary as soon as it could be properly reduced so it would be available immediately for broad community study and follow-up.

What our discussions did not satisfactorily resolve was which fields should be imaged. An external advisory committee was therefore convened to define appropriate programs that could be executed. They agreed with most of the desirable characteristics that had been set forth, but strong discussion focused on what field(s) should be studied. The majority favored selection of a few targeted fields with known high-redshift galaxies, as opposed to several committee members who insisted that a deep study of one or two random, representative fields would yield the most significant results. Arguments pro and con were intensively debated throughout the day until the inability of the majority to agree upon which fields to target took its toll! At this point, the minority argument of simply selecting a random field took on greater credibility, and eventually was what was decided upon.

A group of scientists at ST ScI formed a team to identify a field and execute a deep image, which we called the Hubble Deep Field (HDF). The field eventually selected was a random one that was chosen in the northern CVZ that fit all the above criteria, and whose specific coordinates were dictated by the presence of independent pairs of HST guide stars. The observations were carried out over a two week period in December 1995, with 342 dithered exposures obtained over 150 consecutive orbits in the four WFPC2 broad-band filters F300W, F450W, F606W, and F814W (Williams et al. 1996). The flanking fields were also imaged in the I band for 1–2 orbits each.

The individual exposures in each passband were cleaned and combined by a process that was developed for this observing procedure by Fruchter & Hook (1997) called Variable Pixel Linear Reconstruction (VPLR), or 'drizzling'. VPLR maps the detector input pixels, whose sizes are arbitrarily shrunken to a smaller size commensurate with the dither pattern, onto an output grid whose pixel size is chosen to match that of the telescope resolution and the dither scheme. The flux assigned to each output pixel is the sum of the fluxes of the various input pixels that are mapped onto that output pixel, weighted by the fractions of the area of each input pixel that overlap that output pixel. It is as if the input flux drizzled down onto the output grid, hence the name 'drizzling' given to this technique. It is an excellent procedure for registering dithered data in the presence of geometrical distortion of the detector, and it conserves both flux and surface brightness. Drizzling does not treat background noise optimally, but it is an excellent method to recover spatial resolution lost by detector undersampling. Fig. 1 shows a comparison of a feature from the HDF dataset that has been combined by two techniques, shift-and-add vs. drizzling, and the improved resolution yielded by drizzling is apparent.

A gray-scale image of the final HDF produced by combining the exposures from the B, V,

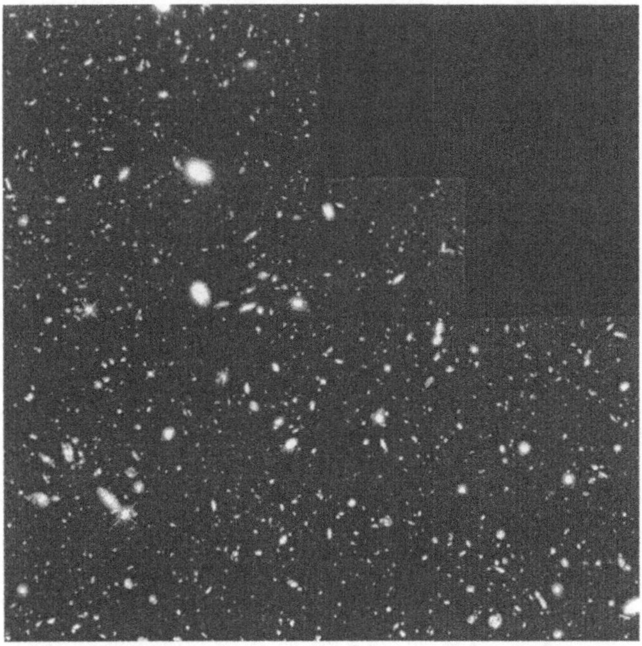

Figure 2. Gray-scale composite drizzled image of the HDF, obtained from combining the majority of individual dithered exposures in B, V, and I bands

and I bands is shown in Fig. 2 on a logarithmic intensity scale to enhance the detail visible at all brightness levels. The faintest sources have brightnesses of about $V = 29^{\mathrm{m}}$, and there are more than 2,000 identifiable objects in the image, most of which are extragalactic. No solar system objects appear in the combined HDF image since any such object would display motion during the observing sequence, and the cosmic ray cleaning algorithm therefore treats them as artifacts and rejects them. The objects in the HDF thus consist of galactic stars and galaxies, and the distinction between these two classes can generally be made from their image sizes, compared with the point spread function (PSF). Distance determinations for the sources are the most important information necessary for interpretation of the data, and these require follow-up spectroscopy, but some conclusions can be drawn without knowledge of the specific object distances.

3. The Stellar Component

Study of the stars in the HDF provides constraints on the stellar luminosity function at faint magnitudes, and several groups have worked on this problem beginning with the identification of the galactic stars. Elson, Santiago, & Gilmore (1996=ESG) and Gould, Bahcall, & Flynn (1997=GBF) each determine the image size for all objects in the HDF down to $V = 28^{\mathrm{m}}$, the limit at which noise makes discrimination of a stellar from a non-stellar object problematical. A color-magnitude diagram of all stellar-like objects in the HDF is shown in Fig. 3, taken from ESG. There are two grouping of objects: one of which is blue and consists of 50 sources, and another which is red, consisting of about 10 objects. The blue group is too blue to be anything but white dwarfs if they are indeed stars, and their brightnesses then require them to all have distances of about 10 kpc. Any realistic model of the Galaxy leads one to expect even more WD's at closer distances, but such objects do not appear in the CM diagram. The hypothesis that these objects are WD's must therefore be rejected, and the blue objects, which are all fainter than 26th magnitude, must reasonably be considered to be point-like extragalactic objects. This leaves the 9 reddish objects as the only likely stars in the HDF down to $V = 28^{\mathrm{m}}$.

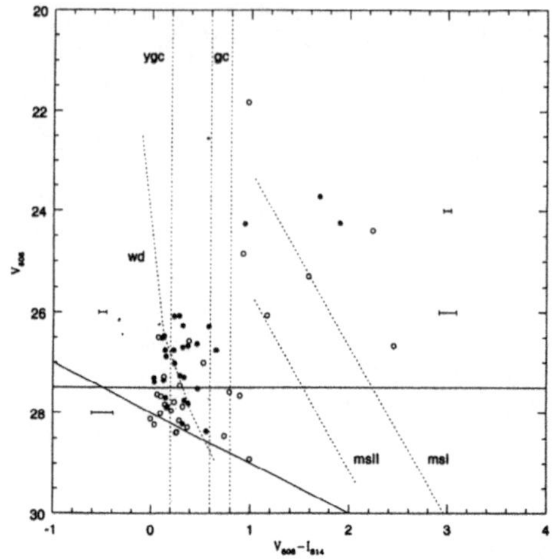

Figure 3. Color-magnitude diagram of all stellar-like objects in the HDF brighter than $V = 28^m$ (from ESG 1996).

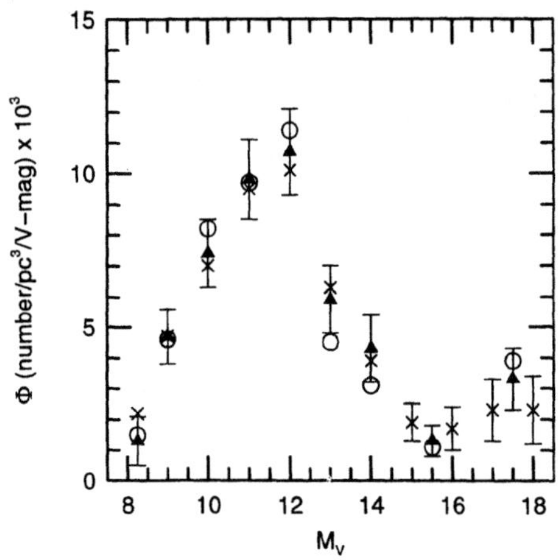

Figure 4. Stellar luminosity function derived from HST data including the Groth Strip. Note the pronounced turn down for luminosities fainter than $M_v = 12$ (from GBF 1997).

Three different groups (ESG; GBF; Mendez et al. 1996) have taken the HDF stellar data and fitted them to models of the Galaxy in order to determine properties of the disk and halo. The specifics of their procedures and conclusions differ, but they agree on the basic finding concerning the stellar luminosity function: it must turn down for magnitudes fainter than $M_v = 12$, as can be seen in Fig. 4, from GBF—a result of some significance.

4. Galaxy Number Counts

Turning our attention to the galaxies in the HDF, one of the fundamental relations that can be determined is that of galaxy numbers per unit magnitude interval. Number counts have been obtained straightforwardly for the HDF from photometry of the sources in the different passbands, and the results for the B and I bands are shown in Fig. 5. The HDF results are seen to agree

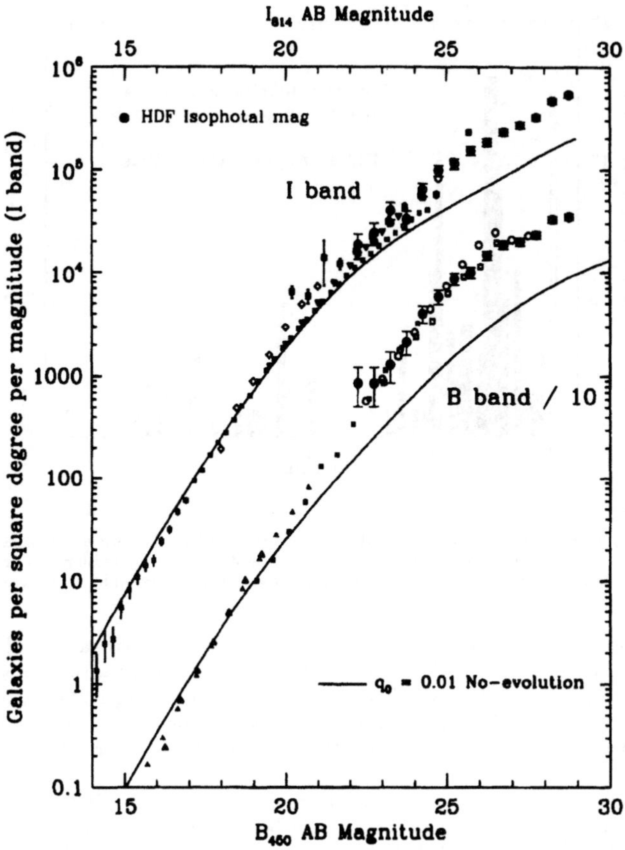

Figure 5. Galaxy number counts in the HDF from the B and I bands, compared with ground-based results and non-evolving model predictions (from Ferguson & Babul 1997).

with previous ground-based results at brighter magnitudes where there is overlap, and to smoothly extend beyond them at fainter magnitudes. Of interest is the comparison of the data with the relationship predicted for a non-evolving galaxy luminosity function, shown by the solid line in Fig. 5. The actual data exceed the expected counts at fainter magnitudes in both B and I, with the discrepancy being greatest in the blue. This is the origin of the well known 'faint blue galaxy' excess that has prompted certain scenarios for the recent evolution of galaxies such as the existence of dwarf galaxy bursts at epochs $0.5 < z < 1$ (Ferguson & Babul 1997). Clearly, the data require either the numbers or the luminosities of galaxies to have changed in relatively recent times although the number counts alone do not dictate which alternative is correct.

5. HDF at other Wavelengths

The HDF has now been imaged in wavelength regions other than the optical, and sources have been identified with the optical galaxies. Roughly 100 hours of radio mapping of the HDF and flanking fields has been done at 8 GHz using the VLA by Kellermann, Fomalont, and colleagues, in which they achieved a detection threshold of 10 μJy and a spatial resolution of 4 arcsec (Fomalont et al. 1997). They find seven sources, of which six can be identified with galaxies visible in the HDF image. Three of the radio sources are associated with red elliptical galaxies of redshift $z \sim 1$ and

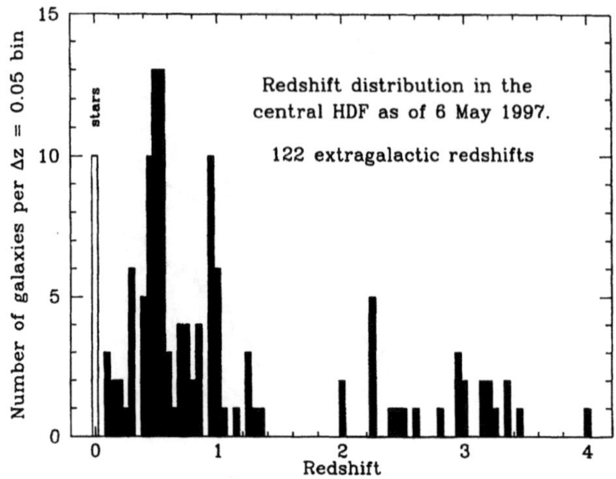

Figure 6. Histogram of spectroscopic redshifts for HDF objects measured by the various consortia with the Keck 10m + LRIS. Smaller redshifts are from the magnitude-limited samples, and larger redshifts are from the U- and B-band dropout samples. (Compiled from available data in the literature and websites by M. Dickinson).

having normal appearance. The other three sources are identified with disk galaxies of intermediate redshift, $z \sim 0.5$, and in each case there are indications of interactions with nearby galaxies, as if radio emission in disk systems is stimulated by interactions and mergers. One remaining weak radio source has no corresponding optical counterpart and may represent a galaxy enshrouded in dust or having such a high redshift ($z > 6$) that the Lyman limit is longward of the I band, i.e., an I-band dropout.

ESA's ISO observatory has also devoted 12.5 hours of time to image the HDF with its ISO-CAM at 6.7μ and 15μ. Two groups, from Imperial College London and Saclay, are analyzing the data (Serjeant et al. 1997; Cesarsky et al. 1997), and their studies have identified sources down to a $100\mu Jy$ detection threshold with a 6–12 arcsec resolution. They find a strong correlation between the sources at each of the IR wavelengths and also between the ISO and the VLA radio sources. They interpret the IR emission to be due to dust (primarily PAH's) and to stellar winds associated with an old stellar population.

6. Redshift Distribution

The excellent spatial resolution and depth of the HDF have caused this field to be chosen for intensive ground-based follow-up study. In particular, groups having access to the Keck 10m telescopes have committed considerable time and resources to obtaining redshifts for as many of the HDF sources as possible, to aid in the interpretation of the HST imaging. Furthermore, the different research groups have coordinated their spectroscopic programs in order to eliminate duplication and they have made most of the redshifts available to the community before publication. At the present time, after two observing seasons, 122 galaxy redshifts have been obtained within the HDF, and their distribution is shown in Fig. 6.

The histogram of redshifts consists basically of two distinct samples. The first is the approximate magnitude-limited sample of the Keck consortia, which have obtained spectra of HDF objects down to $R = 24^{\rm m}$, and the second is the group of U-band dropouts that were selected on the basis of their likely having high redshift. Almost all of the objects with $z < 1.5$ belong to the magnitude-limited sample, and the $z > 1.5$ objects belong to the second, color-selected sample. The non-uniformities, or peaks, in the redshift distribution are statistically significant, and the median redshift of $< z > \sim 0.5$ for the magnitude-limited sample is consistent with the findings of ground-based studies such as the CFRS (Lilly et al. 1996).

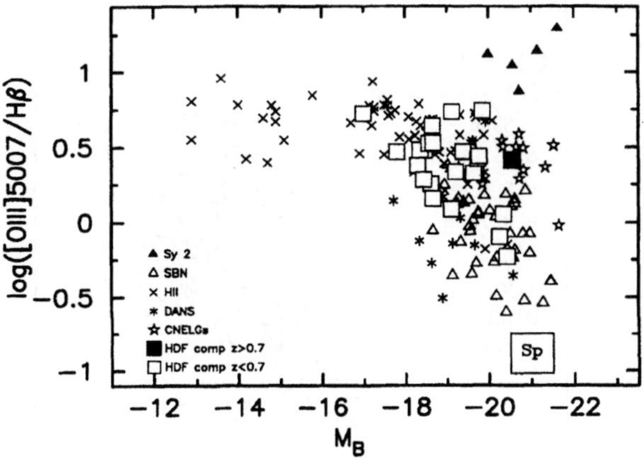

Figure 7. Relation between [O III]/$H\beta$ flux vs. absolute blue magnitude for compact galaxies in the HDF. The galaxies have intermediate redshifts and moderately strong emission lines. They occupy the same region of the diagram as starburst galaxies, as opposed to other types of active galaxies (from Guzman et al. 1997).

The morphologies of the objects in the HDF are related to their redshifts. Normal Hubble sequence galaxies are seen primarily at low redshifts, whereas a very high fraction of objects with $z > 1$ have disturbed morphologies and evidence for interactions (Abraham et al. 1996). This is especially borne out by the appearances of the U-band dropout sample, which are all irregular and disturbed (Steidel et al. 1996). Colley et al. (1996) have determined the 2-point correlation function for HDF galaxies, and they find that the function increases for object separations of 2 arcsec or less, especially for the U-band dropout sample, which are primarily at high redshifts. This indicates either a high rate of interactions and mergers or the fact that closely separated HDF objects may in fact be simply the visible, higher surface brightness components of an underlying galaxy.

7. Spectroscopic Follow-Up Results

The 'DEEP' collaboration at Lick/UCSC has studied three separate samples of galaxies in the HDF and flanking fields, for which they have obtained Keck 10m LRIS spectroscopy. The first sample consists of the brighter ($I < 23.7^{\rm m}$), higher surface brightness, compact (half-light radius < 0.5 arcsec) objects in the flanking fields. They identify 61 such galaxies, having a range in redshifts of roughly $0.4 < z < 1$ (Guzman et al. 1997). For these objects they find that the spectra consistently show strong emission lines, particularly [O II] 3727, with an H&K break and a blue continuum. The [O II] strengths point to star formation rates of 0.1–10 $M_\odot$/yr, and the line widths indicate that these compact systems are more massive than present day dwarf galaxies. The relationship between [O III] 5007/$H\beta$ vs. absolute blue magnitude for these compact galaxies, shown in Fig. 7, is identical to that of present day starburst galaxies, as opposed to Liners, Seyferts, or other types of active galaxies.

The second sample consists of isolated disk galaxies brighter than $I = 22.5^{\rm m}$ and with inclinations $i > 30^{\rm o}$. There are 8 such systems which have redshifts in the range $0.15 < z < 0.75$, and Keck spectra of them were taken with the slit oriented along the major axes so that rotation curves were obtained for the galaxies (Vogt et al. 1997). The velocity turnovers were observed for most of the galaxies, and the results modeled in order to determine the mass distributions and M/L ratios. The corresponding Tully-Fisher relation was derived for this intermediate redshift sample, and it was found to have the same slope as the local T-F relation, but shifted 0.4 mag brighter. Thus, either the luminosities of the sample are higher or the masses of these galaxies lower than local disk galaxies.

The third DEEP galaxy sample consists of the U-band dropouts in the primary HDF that have confirmed redshifts $z > 2.2$, of which there are 12 galaxies (Lowenthal et al. 1997). The median

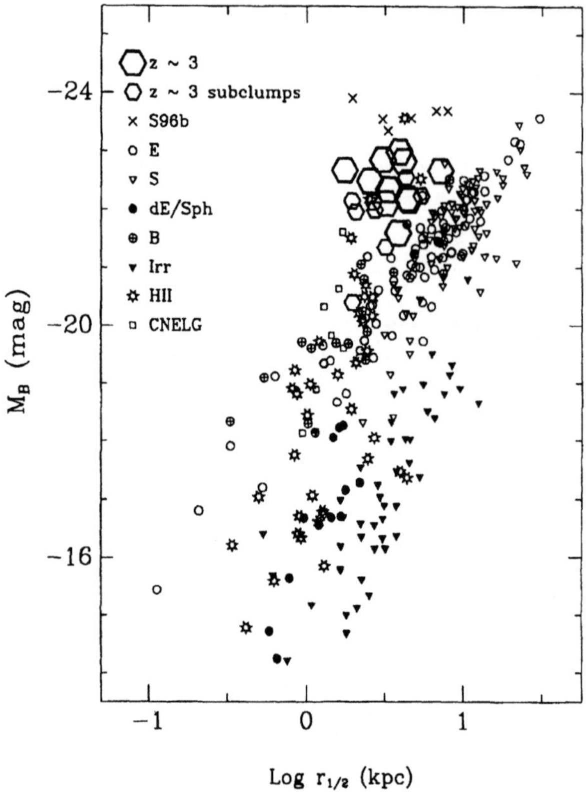

Figure 8. Half-light radii of the high-redshift sample of HDF galaxies (hexagons) vs. their blue luminosities, compared with various morphological types of galaxies at the present epoch. The high-z HDF objects have a mean radius of 3 kpc, and are more compact for their luminosities than any other types of galaxies (from Lowenthal et al. 1997).

optical luminosity of these galaxies is $2L_*$, thus they are not overly luminous, and their half-light radii are smaller than present epoch galaxies of their luminosity, as can be seen in Fig. 8. The median half-light radius of the sample is less than 4 kpc, so the luminous component of these galaxies is quite small and compact. The median redshift of the sample is $< z >= 3.0$, and their averaged spectrum shows a blue continuum with characteristic ISM absorption lines, but quite weak Ly-α emission. The strengths of the UV absorption lines indicate low metallicity, and the combination of weak Ly-α with the very blue continuum suggests a moderate star formation rate coupled with dust.

What these U-band dropout systems are is difficult to say. They could be the early spheroids of present day massive elliptical galaxies, or they could represent the bulges of spirals. Perhaps they are sub-galactic clumps that will eventually merge to form the more massive systems of the present epoch. Whatever their evolutionary phase, they must be present today in some other form than we now observe them at high redshift because such compact systems of those luminosities are not observed at the present epoch.

8. Photometric Redshifts

The determination of the distances, i.e., redshifts, to the galaxies in the HDF is crucial to the interpretation of the data and to our understanding of the evolution of galaxies. The Keck 10m+LRIS is capable of obtaining the spectra of galaxies down to about $R = 26^{\mathrm{m}}$ with some effort, and therefore the faintest 3 magnitudes of galaxies in the HDF are beyond the limit at which we can reasonably expect to be able to measure redshifts spectroscopically in the coming decades. The possibility that approximate redshifts might be determined from broad-band photometry of the galaxies has therefore been explored by various groups, as has the identification of probable high-redshift galaxies based upon their broad-band colors.

The passage of the Lyman break through longer wavelength filter passbands for increasing redshifts produces a color signature in galaxies that is easy to identify. A sudden change in flux between two successive passbands is likely to be due to their placement on either side of the Lyman or Balmer discontinuities. Madau et al. (1996) have demonstrated that one can define regions in a 2-color diagram which are occupied by galaxies within a specific redshift range, and this procedure can be used to isolate galaxies within a redshift interval. Similarly, the Lyman break is redward of the U band for redshifts $z > 2.2$, and Steidel and colleagues (Steidel & Hamilton 1992; Steidel et al. 1996) have made use of this fact to define a sample of high redshift candidates which they have studied from the ground. This procedure has been used by various groups to define intermediate and high redshift samples of galaxies in the HDF.

One can carry this method further and use the observed fluxes of different types of galaxies, for which existing data are primarily confined to the present epoch, to determine what their broad band fluxes should be over a range of wavelengths and redshifts. A knowledge of the UV spectra of galaxies is necessary for this procedure since UV wavelengths are shifted into the optical at higher redshifts; therefore the observed IUE spectra of galaxies are essential data. Galaxy evolution models that predict how galaxy spectra evolve with time may be used to make corrections to local epoch spectra, and the effects of dust absorption may also be modeled and corrected for. With this information, templates of galaxy broad-band fluxes are constructed for different epochs, i.e., redshifts, and compared with the colors of individual galaxies measured from the HDF images in the four passbands. A redshift is assigned to each galaxy from the template that produces the smallest flux residuals, and a photometric redshift can be determined for any object having measurable broad-band fluxes.

A number of investigators have undertaken to refine the process of determining redshifts based upon broad-band colors or fluxes, and most of them utilize templates of galaxy spectra which they compare with the measured fluxes of the sample of galaxies (Gwyn & Hartwick 1996; Lanzetta et al. 1996; Mobasher et al. 1996; Sawicki, Lin, & Yee 1996). An alternative approach has been employed by Connolly et al. (1995), who fit polynomials to the observed spectra of galaxies of different redshift, using the measured spectra as a basis for the comparison with the sample of galaxies. An advantage of this method is that it does not depend on the assumption of template spectra for the basis set. A disadvantage is that it requires a calibrating set of galaxies with observed spectra, and very few exist in the regime $z > 0.5$.

A comparison of photometric redshifts determined for objects in the HDF with their spectroscopic redshifts is shown in Fig. 9 for two cases, viz., with and without an IR broad-band flux measurement. The inclusion of a flux point longward of 1μ is important to the accuracy of photometric redshifts because cool stars can mimic the spectral distribution of galaxies at moderate redshift if passbands in only the optical are used. The addition of one IR passband permits detection of the Balmer break at intermediate redshift, and resolves this degeneracy. One can see that the accuracy of photometric redshifts is greatly improved by the addition of an IR flux point. Hogg et al. (1998) conducted a blind test of predicted photometric redshifts vs. their Keck-measured spectroscopic redshifts for selected HDF galaxies, and their results are shown in Fig. 10. For redshifts in the range $z < 1$ the results are surprisingly good, and they demonstrate the validity of photometric redshifts to 10% accuracy.

9. Star Formation Rates

The use of photometric redshifts, which can be used to determine the approximate redshifts of even the faintest objects in the HDF, is central to the interpretation of the data and it opens up the HDF to important analysis of galaxy evolution. The luminosities of all objects imaged in the HDF are directly determinable for each of the passbands, as is the conversion of angular size to linear dimension. And, the variation with time, or epoch, of any quantity measurable from the

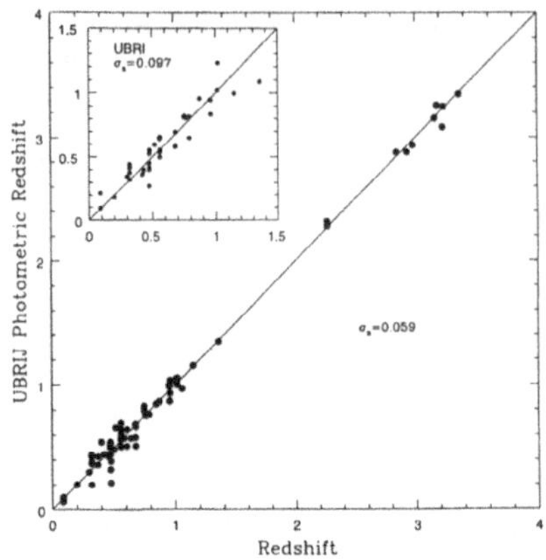

Figure 9. Keck-measured spectroscopic redshifts for galaxies in the HDF vs. photometric redshifts determined by Connolly et al. (1997).

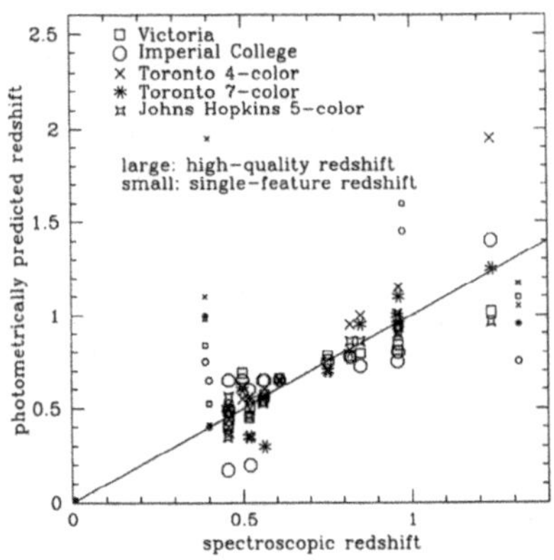

Figure 10. Blind test comparison of Keck 10m spectroscopic redshifts vs. photometric redshifts determined by different groups (from Hogg et al. 1998).

HDF can be determined, including the galaxy luminosity function, morphologies, space densities, interaction/merger rates, and star formation rates.

An example of fundamental information that can be obtained with photometric redshifts from the HDF is the derivation by Sawicki et al. (1996) of the galaxy luminosity function at different redshifts. By measuring the total brightness in the F450W filter of each galaxy they binned the data by redshift and derived the resulting luminosity function shown in Fig. 11. The bright end of the LF increases and the faint end steepens back to the epoch $z = 3$, after which the LF fades back to its low redshift value for $z = 4$.

Star formation rates (SFR) can also be determined as a function of redshift from the HDF, and this has produced one of the notable results from the Deep Field. Active star formation manifests itself by the presence of short-lived hot, luminous stars whose signatures are supernovae, H II regions, and a strong UV continuum. The rate of star formation is directly related via the initial mass function to the SNe rate, the $H\alpha$ and [O II] luminosities from H II regions, and the UV luminosity of the galaxy. All of these indicators have been calibrated for galaxies, and observation of any of them will yield a star formation rate. The HDF images in the various passbands provide

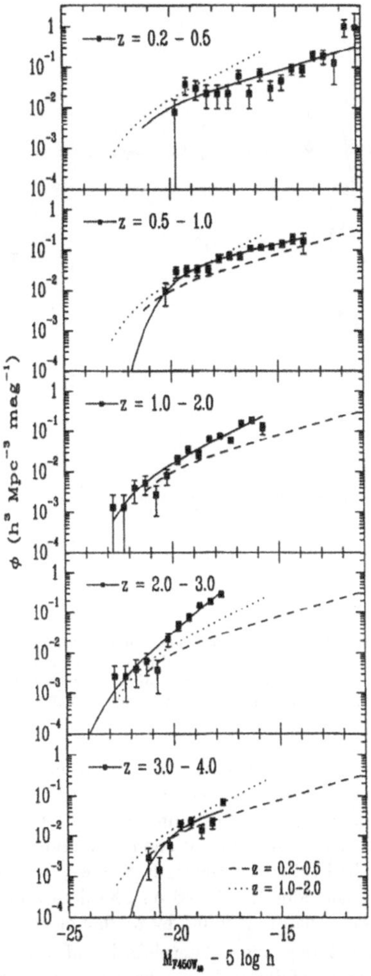

Figure 11. Evolution of the galaxy luminosity function with redshift for HDF objects, using photometric redshifts (from Sawicki, Lin, & Yee 1996).

the UV luminosities of galaxies for different redshifts, and these can be converted directly into the corresponding SFR at earlier epochs.

The local SFR at the present epoch has been determined from an $H\alpha$ objective prism survey of galaxies (Gallego et al. 1995), and an evaluation of star formation at intermediate redshifts has come from the ground-based Canada-France redshift survey (Lilly et al. 1996), whose magnitude-limited sample shows a steep rise in SFR with redshift out to $z = 1$. Madau et al. (1996) have used the HDF to probe the SFR for earlier epochs at $z > 1$ by using 2-color criteria to define U-, B-, and V-band dropouts, corresponding to galaxies with mean redshifts of 2.75, 4.0, and 5.5, respectively. They measured the mean UV luminosity of galaxies in each of these redshift intervals to derive the average SFR for that epoch. The results of all the above investigations enable the history of star formation to be pieced together, and is shown in Fig. 12.

A definite evolution of the SFR is apparent in which star formation begins modestly at high redshifts of $z > 4$, and increases to a maximum at epochs of $z \sim 1 - 2$. It then declines rapidly to

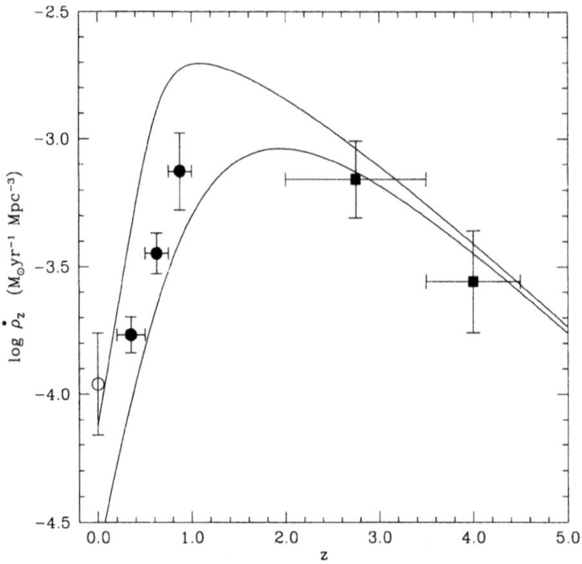

Figure 12. Global star formation rate (SFR) vs. redshift as determined from galaxy emission line strengths and UV continuum flux. The SFR's for $z < 1$ are derived from ground-based surveys, while $z > 1$ values are determined by Madau et al. (1996) from the HDF, and are uncertain because of the unknown effects of dust absorption. The solid curves are independent determinations from quasar absorption system abundances by Pei & Fall (1995).

the present time, such that the global SFR is now an order of magnitude lower than it was at its peak. This result is consistent with the Keck spectra of galaxies in the HDF, which show stronger [O II] and H I emission lines at intermediate redshifts of $z \sim 1$ than at higher or lower redshifts (cf. Guzman et al. 1997; Lowenthal et al. 1997). It should be noted that a separate determination of the SFR can be made from the IR fluxes of the HDF galaxies, caused by re-radiation of UV light by dust, and Rowan-Robinson has done this based on ISO data and concludes that the SFR for $z > 2$ may be substantially higher than that found by Madau et al.

It is significant that a totally independent determination of SFR history can be deduced from the abundances of the absorbing gas of damped *Ly-α* systems in quasars. Metal lines are detectable in the DLA systems and these should originate in the ISM of galaxies, so they track the SFR since heavy elements are created and injected into the ISM by the winds and explosions of massive stars. Heavy element abundances have been determined for the DLA's over a wide range in redshift, and the SFR is directly related to the change in abundance with redshift. Pei & Fall (1995) have analyzed the DLA abundances from models and have deduced the corresponding SFR with redshift. As their curves show in Fig. 12, their results are in quite good agreement with the SFR derived from the HDF. It thus appears that the HDF includes objects back at an epoch before most of the stars and galaxies now in existence had yet formed.

10. Summary

The Hubble Deep Field is a rich source of data encompassing both stellar and extragalactic astronomy. As further research on the HDF is performed using new techniques, new facts will emerge and some of the above results will need to be modified. The real significance of the HDF is its demonstration that by combining both space and ground-based telescopes we can now seriously study galaxies out to redshift $z > 3$ and down to 29th magnitude. After centuries of having to confine study of the universe to local objects, it falls to us to live at a time when technology has developed the tools that enable us to address the most fundamental questions of our science: how the stars and galaxies now around us developed and evolved from the very first inhomogeneities that formed out of the early cosmic sea.

References

Abraham, R.G. et al. 1996, MNRAS, **279**, L47
Cesarsky, C. et al. 1997, in preparation
Colley, W.N., Rhoads, J.E., Ostriker, J.P.,& Spergel, D.N. 1996, ApJ, **473**, L63
Connolly, A.J. et al. 1995, AJ, **110**, 2655
Connolly, A.J., Szalay, A.S., Dickinson, M., Subbarso, M.U., & Brunner, R.J. 1997, ApJ, **486**, L11
Dickinson, M. 1995, in Fresh Views of Elliptical Galaxies, ed. A. Buzzoni, A. Renzini, A. Serrano, ASP Conf. Ser. No. 86, p. 283
Dressler, A., Oemler, A.J., Sparks, W.B.,& Lucas, R.A. 1994, ApJ, **435**, 23
Elson, R.A.W., Santiago, B.X.,& Gilmore, G.F. 1996, New Astron, **1**, 1
Ferguson, H.C.& Babul, A. 1997, MNRAS, in press
Fomalont, E.B., Kellermann, K.I., Richards, E.A., Windhorst, R.A.,& Partridge, R.B. 1997, ApJ, **475**, L5
Fruchter, A.S.,& Hook, R.N. 1997, in Applications of Digital Image Processing, ed. A. Tescher, Proc. SPIE, vol. 3164, in press
Gallego, J., Zamorano, J., Aragon-Salamanca, A.,& Rego, M. 1995, ApJ, **455**, L1
Gould, A., Bahcall, J.N.,& Flynn, C. 1997, ApJ, **482**, 913
Guzman, R. et al. 1997, ApJ, **489**, 559
Gwyn, S.D.J.,& Hartwick, F.D.A. 1996, ApJ, **468**, L77
Hogg, D.W. et al. 1998, AJ, in press
Lanzetta, K.M., Yahil, A.,& Fernandez-Soto, A. 1996, Nature, **381**, 759
Lilly, S.J., Le Fevre, O., Hammer, F.,& Crampton, D. 1996, ApJ, **460**, L1
Lowenthal, J.D. et al. 1997, ApJ, **481**, 673
Madau, P. et al. 1996, MNRAS, **283**, 1388
Mendez, R.A., Minniti, D., De Marchi, G., Baker, A.,& Couch, W.J. 1996, MNRAS, **283**, 666
Mobasher, B., Rowan-Robinson, M., Georgakakis, A.,& Eaton, N. 1996, MNRAS, **282**, L7
Pei, Y.C.& Fall, S.M. 1995, ApJ, **454**, 69
Sawicki, M.M., Lin, H.,& Yee, H.D.C. 1996, AJ, **113**, 1
Serjeant, S.B.G. et al. 1997, MNRAS, **289**, 457
Steidel, C.C. & Hamilton, D. 1992, AJ, **104**, 941
Steidel, C.C., Giavalisco, M., Dickinson, M.,& Adelberger, K.L. 1996, AJ, **112**, 352
Vogt, N.P. et al. 1997, ApJ, **479**, L121
Williams, R.E. et al. 1996, AJ, **112**, 1335

CATACLYSMIC VARIABLE STARS

BRIAN WARNER
Department of Astronomy, University of Cape Town, Rondebosch 7700, South Africa

1. Introduction

The evolution of single stars on and away from the main sequence is well understood. A degenerate core is formed in a star as the star leaves the main sequence and expands to a giant with a radius typically 50 - 500 $R_\odot$. Observationally it is known that most stars are members of binary systems, and among these many have orbital periods less than 100 y. It can happen, therefore, that the expanding envelope of the primary of a binary system can reach to the secondary. As this happens, the primary fills its Roche tidal lobe and transfers matter to the secondary; if the primary has a radiative envelope the rate at which this occurs exceeds the Eddington limit of the secondary, which therefore repels the incoming gas, forming a common envelope around the two stars. Friction within the envelope causes the stars to spiral towards each other until the energy and angular momentum extracted from the binary orbit and transferred to the envelope are sufficient to eject the common envelope as a planetary nebula, leaving a short period binary ($P_{orb} \lesssim$ 1d) comprising a white dwarf and a main sequence star.

This mechanism of producing short period binaries containing white dwarfs, proposed by Ostriker and by Paczynski (1976), is the probable origin of the class of objects known as Cataclysmic Variable Stars (CVs), which encompass the classical novae, dwarf novae, nova-like variables and a variety of related objects. Evidence has been accumulating for forty years (Crawford & Kraft 1956, Warner 1995a) that every CV consists of a secondary star (usually a dwarf, but a few systems contain giants) filling its Roche lobe and transferring mass to a white dwarf primary. In systems of normal chemical composition the orbital periods lie between 75 mins and ~250 d, with the majority having $P_{orb} \lesssim$ 8 h. A few hydrogen-free systems are known for which 17 mins < P_{orb} < 50 mins. It should be noted that CVs are very compact binary systems: for $P_{orb} \lesssim$ 2.5 h such a binary would fit inside the Sun.

Cataclysmic variables, interesting as they are in their own right, provide insight into the physics and phenomenology of a range of other objects. In particular, the gas flowing from the secondary creates an accretion disc around the primary, and it is in the CVs that quasi-stable and time-dependent accretion discs are observationally easiest to study, which informs interpretations of discs around young stars, in X-ray binaries and in active galactic nuclei. The low mass X-ray binaries, with their neutron star and black hole primaries, are higher energy analogues of the CVs.

The history of CV research follows closely the development of astronomical technology. The rapidity with which the luminosities and spectra of CVs vary around an orbital cycle has always put strong demands on instrumentation. Early photographic photometry contributed only to the long term light curves, revealing the large amplitude variations of dwarf novae and classical novae. It was the pioneering photoelectric photometry of A.P. Linnell and M.F. Walker in the 1950s that showed that CVs vary on very short time scales, typically with variations of 0.01 to 0.2 mag within minutes. This flickering is now recognised as a characteristic of the mass transfer process, having its origins in the shocked region where the mass transfer stream impacts onto the outer rim of the accretion disc, and in the inner regions of the disc itself. Walker's discovery of low amplitude coherent 71 s variations in the remnant of Nova Herculis 1934 (DQ Her) (Walker 1956) was the most spectacular discovery of this era of CV research.

The later development of photon counting photometry, and its compatibility with on-line application of computers (Nather & Warner 1971, Warner 1988a), reinvigorated high time resolution photometry of CVs, leading to discovery of more examples of rapid coherent

J. Andersen (ed.), Highlights of Astronomy, Volume 11A, 16–27.

brightness variations and a range of quasi-periodic variations during the outbursts of dwarf novae (Warner & Robinson 1972a).

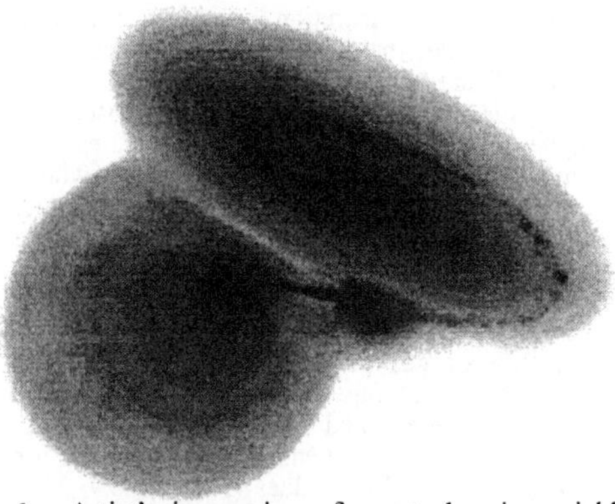

Fig. 1. Artist's impression of a cataclysmic variable star. For a colour version see the front cover of PASP Vol. 105, February 1993.

The requirement for high time resolution spectra initially limited CV spectrography to possessors of very large telescopes. The pioneering work of R.P. Kraft (1961, 1962, 1964), obtaining trailed spectra with the prime focus spectrograph of the 200 in reflector, proved that all CVs are binaries and established that the characteristic emission line spectra are also a signature of the mass transfer process.

The deep potential well into which the accreting gas falls leads to high energy phenomena in the vicinity of the primary, with the result that CVs are luminous objects in the far ultraviolet and X-ray regions, observed extensively by satellites such as IUE, Einstein, ROSAT, EXOSAT, ASCA and the HST. Perhaps the most exciting result has been the discovery of TeV gamma rays from the nova-like variable AE Aqr, modulated with the 33s rotation period of the primary (Meintjies *et al.* 1994). The same object, and a few other CVs, have been detected at radio wavelengths.

Since the discovery of strong linear and circular optical polarization in AM Her in 1976 (Tapia 1977) a further dimension has been added to CV research. Large magnetic fields (10^6 - 10^8 G) are known among field white dwarfs. The presence of strong fields in CV primaries, first detected through optical polarization, has a dramatic effect on the mass transfer process. If the field is sufficiently strong ($\gtrsim 10^7$ G) the stream, ionized by hard radiation from the vicinity of the primary, threads onto the field lines and falls directly to the primary without forming an accretion disc. Furthermore, the interaction between the field of the primary and that of the secondary is sufficient to synchronize the primary's rotation with the orbital revolution of the secondary. Such systems are known as polars, or AM Her stars. The infalling gas passes through a standoff shock before settling onto the white dwarf. This shock-heated gas is a source of ~10 keV bremsstrahlung and polarized optical and infrared cyclotron emission. Reprocessed radiation from the surface of the white dwarf adds a ~40 eV component.

If the primary's field is weaker, synchronization generally does not happen, and an accretion disc can form, but its inner edge is determined by the region where the magnetosphere controls the final mass flow onto the primary. These objects are known as intermediate polars (IPs); DQ Her and AE Aqr are examples, but in most IPs the rotation period of the primary is typically ~20 mins.

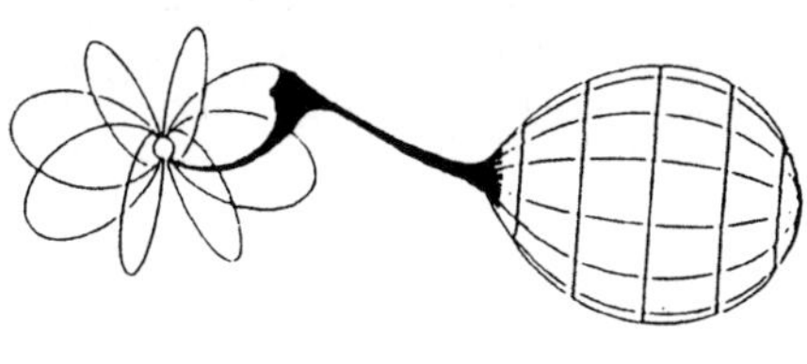

Fig. 2. Schematic view of a polar, showing the gas stream becoming attached to field lines of the primary. From Cropper (1990).

To these basic structures of CVs yet another dimension of behaviour is seen in their long term behaviour. The rate of mass transfer from the secondary may vary on a range of time scales. The stream itself is structured, probably through turbulent or magnetically controlled flow out of the secondary; this is seen in rapid luminosity variations, particularly in polars (Wood, Imamura & Wolff 1992). Some quasiperiodic variations, on time scales of years, have been detected through mean brightness variations, which are probably caused by magnetic cycles in the secondary, analogous to the sunspot cycle in the Sun (Applegate 1992; Warner 1988b). The most dramatic effect is seen when mass transfer from the secondary drops by a very large amount, sometimes for years, leading to a fall in mean brightness by several magnitudes as the accretion luminosity ceases. This is seen commonly in the polars, and has the advantage that the absorption spectrum of the primary can then be observed, and its magnetic field strength determined from Zeeman effect (these fields agree with those derived from the spacing of cyclotron harmonics observed in the spectra at high transfer rates: Wickramasinghe 1993). The cause of these low states is not yet understood. Similar low states are seen among some of the nova-like variables, where they have been attributed to deterministic chaos inherent in the feedback mechanism caused by heating of the secondary's atmosphere by the accretion luminosity (Wu, Wickramasinghe & Warner 1995a).

In a certain range of mass transfer rate the accretion discs of non-magnetic systems have thermal instabilities. This is the result of there being two different equilibrium structures for an accretion disc - a consequence of the opacity of gas peaking near $T \sim 10^4$ K so the requirements on opacity can be satisfied either in a high temperature convective disc or a low temperature disc. As the temperature of a disc is determined by the rate at which it must radiate the energy released by mass passing through it there are equilibrium structures for high and for low rates of mass transfer, but not for intermediate values (Hoshi 1979; Meyer & Meyer-Hofmeister 1981). If the secondary provides such an intermediate value ($\sim 10^{13}$ - 5×10^{16} gm s^{-1}) no equilibrium is possible and the disc cycles between high temperature and low temperature states. This is the explanation of dwarf nova outbursts, where discs increase in luminosity by factors of 10-100 for a few days mostly on time scales of 10-100 d (with a few outbursting on time scales of 1-10 y).

Finally we must consider the effect of accumulation of hydrogen-rich material onto the surface of the primary. It has long been realised that this will lead to a thermonuclear runaway at the boundary of the degenerate core producing the classical nova eruption (Mestel 1952). For long it was thought that dwarf novae and classical novae are separate kinds of CVs, but it should be clear from the above discussion that all CVs must pass through the nova eruption stage, and that the present state and classification of a CV depends largely on its rate of mass transfer (and magnetic state). Indeed, in recent years connections between the various subclasses have been observed. After a nova eruption the system remains bright for decades, probably because of enhanced mass transfer caused by heating of the secondary's atmosphere by the white dwarf heated to $\sim 10^5$K by the eruption. As the nova remnant cools so mass transfer from the secondary is reduced and the stable accretion disc changes to an unstable one - such dwarf nova behaviour has been seen long after eruptions in, e.g., GK Per V1017 Sgr and V446 Her (Honeycutt *et al.* 1998).

Nova eruptions in polars have a different consequence: interaction between the ejected shell and the secondary, exchanging orbital and shell angular momentum, with subsequent collapse of part of the temporary common envelope onto the primary, can break synchronicity of the primary's rotation. Nova Cyg 1975 (V1500 Cyg) has an orbital period of 0.1396 d and a primary rotation period (found from modulation of optical polarization) of 0.1372 d. The latter is found to be increasing at a rate which will restore synchronism in ~185 y (Schmidt, Liebert & Stockman 1995). This is very short compared with the intervals between eruptions (probably 10^3 - 10^5 y), so V1500 Cyg spends most of its life as a polar. Other slightly desynchronized polars are BY Cam and RXJ 1940-1025, which we may deduce were prehistoric novae.

There must be a number of bright nova remnants unrecognised among the known nova-like variables: in this century several novae have arisen from stars found from archival plates to have had pre-nova brightnesses of ~12 - 13 mag, brightnesses to which they eventually returned after a decade or so. Among the present known nova-likes, therefore, there must be nova remnants from past centuries, as well as some of the novae to occur in the next century. Surprisingly few novae were recognised and announced prior to 1850: there were only Nova Vulpecula 1670 and Nova Ophiuchi 1848 (some supernovae of course were observed, and some bright stars were observed and their positions measured without the observer being aware at the time that they were novae; Nova Sagittae 1783 was noticed but not proclaimed).

I will now select a few topics from the CV field in which recent advances in understanding have been achieved. These will illustrate the wide range of physical principles and processes that are brought to bear in this arena of astrophysical research.

2. The Short-period Dwarf Novae: SU UMa stars

A well defined subclass of dwarf novae, named after SU UMa, the first to be recognised, is observed to have a bimodal distribution of outburst behaviour: outbursts lasting ~ 2 days and less frequent ($\gtrsim$ 150 d) outbursts that last for ~20 days and are about one magnitude brighter at maximum. The latter are known as superoutbursts. In a few SU UMa stars only superoutbursts occur; these are very infrequent, with recurrence times greater than a year (e.g. WZ Sge, which had superoutbursts in 1913, 1946 and 1978). There is strong observational evidence that a superoutburst is 'triggered' by an ordinary outburst - in the sense that an ordinary outburst is underway when the additional rise to supermaximum occurs.

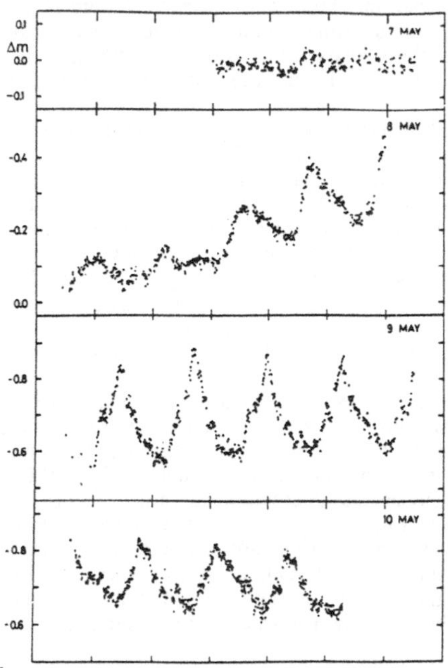

F[illegible] the growth of superhumps during a superoutburst From Semeniuk (1980).

A crucial clue to the understanding of the superoutburst phenomenon was the discovery of 'superhumps' during the maximal phase of the light curve (Vogt 1974; Warner 1975): all superoutbursts show brightness modulations with periods typically a few percent greater than the orbital period. The beat period between the superhump and orbital periods is usually ~2 d. After many attempts to explain this behaviour the solution emerged unexpectedly from multiple particle simulations of accretion discs (Whitehurst 1988): for sufficiently small mass ratio ($q = M_2/M_1 \lesssim 0.25$) accretion discs with large radii are perturbed into an elliptical form, the major axis of which precesses with the ~2 d period. More complete simulations are able to reproduce the observed superhump modulations, which arise in tidal stresses in the disc, caused by passage of the secondary past the disc particles near apastron (Murray 1996).

From the geometry of Roche-lobe filling (semi-detached) binaries with main sequence secondaries, the requirement $q \lesssim 0.25$ translates into a condition on the orbital period: $P_{orb} \lesssim 3$ h. This is compatible with the observation that all such short P_{orb} dwarf novae are SU UMa stars.

During normal outbursts of dwarf novae the accretion disc expands in order to accept the angular momentum of gas falling onto the primary. For $q \lesssim 0.25$ the 3:1 resonance of particle orbits with P_{orb} is accessible to the outer edge of the disc. In normal outbursts of SU UMa stars only a small fraction of the disc mass is accreted on the primary, and the disc radius does not increase to the resonant radius. The gradual increase of mass and radius eventually results in a normal outburst in which the outer disc does reach resonance, and it is this that drives the disc into its elliptical shape and the consequent large tidal stress maintains the disc in a state of high viscosity until about half of its mass has drained onto the primary (Ichhikawa, Hirose & Osaki 1989; Lubow 1991).

Recently a group of stars has been recognised as representing an extremum of the SU UMa type: the stars ER UMa, V1159 Ori, RZ Leo and DI UMa show normal outbursts at intervals of ~4 d and superoutbursts at ~50 d. These must have mass transfer rates just marginally below what would result in stable discs (Warner 1995b). In the dwarf novae of longer orbital period, such high rates lead to the Z Cam phenomenon, in which a slight increase of mass transfer out of the secondary causes a stable disc (i.e. cessation of dwarf nova outbursts), keeping the star in an apparent outburst state for weeks or months. We may hope to see such a transition in one of these newly appreciated objects (they have been known for long as irregular variables, but ironically the rapidity with which they change state obscured their true nature. They are in a state of constant change - one normal outburst is not completed before the next begins, as a result they are of lower outburst range with no true quiescent state attained).

3. Ultrashort-period Systems: the AM CVn Stars

Among the variety of possible interactions of the stars in the initial binary system there are channels that lead (e.g. after two common envelope phases) to two helium cores revolving around each other. Two such degenerate stars can only be in a semi-detached state if their orbital period is very short. A few helium-rich stars showing brightness modulations on time scales of tens of minutes have been known for thirty years, but it is only recently that their light curves have been understood as ultrashort-period analogues of the more familiar hydrogen-rich CVs.

AM CVn (HZ 29) was found by Smak (1967) to have a double humped light curve with a period of ~17½ min. High speed flickering indicated a mass-transferring binary (Warner & Robinson 1972b). The modulation period is not strictly constant, which is now interpreted as being a superhump phenomenon: at such a short P_{orb} it is inevitable that $q << 0.25$ (the secondary mass $M_2 \sim 0.09\ M_\odot$) and as the rate of mass transfer in AM CVn is quite high the accretion disc is maintained at a large radius and continually precessing (a few short P_{orb} systems with low amplitude permanent superhumps are also known among the H-rich CVs). The orbital modulation with a constant period of 1028.8 s has recently been detected (Harvey *et al.* 1998) and the disc precession period of 13.4 h has been observed spectroscopically from the asymmetry of line profiles that occur in a precessing elliptical disc (Patterson, Halpern & Shambrook 1993).

Despite the unusual nature of the secondary (a very low mass degenerate helium star) the range of CV behaviour is shown among the AM CVn stars: AM CVn and EC15330-1403 are nova-like, with permanently high mass transfer; V803 Cen and CP Eri are nova-likes possessing states of low transfer rate; CR Boo is possibly a dwarf nova similar to V1159 Ori; and GP Com is in a low state which may eventually have a superoutburst (Warner 1995c).

The AM CVn stars found so far are bright ($m_v \sim$ 13-15) and mostly at high galactic latitude (where surveys for UV-rich stars have been made). There are therefore many more to be found. With their short P_{orb} and relative proximity ($\lesssim$ 100 pc) the AM CVn stars should be apparently strong sources of gravitational radiation. They constitute an evolutionary channel leading ultimately to single DB white dwarfs. No strongly magnetic helium systems have yet been identified.

4. Secular and Cyclical Evolution of CVs

Emerging from common envelope systems as detached binaries, evolving to become semi-detached interacting binaries, continually reducing their secondaries' masses, expelling mass via nova eruptions (and stellar winds from the high state discs) - CVs clearly require a mechanism that extracts angular momentum from their orbits in order to drive the mass transfer process (for $q \lesssim 1$, which is the case for almost all CVs, mass transfer from the secondary to the primary increases their orbital separation, so mass transfer is not a self-sustaining process).

For the shortest period systems emission of gravitational radiation is an inevitable, and possibly the principal, angular momentum sink. At longer periods ($P_{orb} \gtrsim 2$ h) an additional process is required to explain the observed rates of transfer of 10^{-9} - 10^{-8} $M_{\odot}$ y^{-1}. This has been identified as magnetic braking, in which the wind from the secondary, coupled to large radii onto its magnetic field lines, extracts angular momentum from the secondary which, being tidally locked in synchronism with the orbital revolution, ultimately extracts angular momentum from the orbit (Verbunt & Zwaan 1981). Quantitative estimates of the strength of this process show that it is able to generate the observed mass transfer rates (Mestel & Spruit 1987; McDermott & Taam 1989; Tout & Pringle 1992a).

The observed low space density of CVs having $2 \lesssim P_{orb} \lesssim 3$ h (the so-called orbital period gap) has been thought to be the result of a reduction of magnetic braking when the secondary is so reduced in mass that it becomes fully convective, changing the strength or topology of its magnetic field (Robinson *et al.* 1981) but it has recently been suggested (Clemens *et al.* 1998) that the expected change in shape of the mass-radius relation of mass sequence stars in the vicinity of 0.3 $M_{\odot}$ may suffice to explain the period gap: CVs race through the 2-3 h P_{orb} range at a high transfer rate. Interestingly, CVs may as a result become the best objects for studying the internal structure of low mass main sequence stars.

When the mass of the secondary in H-rich systems falls to ~0.08 $M_{\odot}$ the secondary becomes degenerate and expands in radius as it further decreases in mass. The combination of mass transfer and orbital angular momentum loss then causes the orbital period to increase. For H-rich systems this occurs at $P_{orb} \cong 75$ min, which agrees well with the minimum P_{orb} for such CVs. For helium systems the minimum is at $P_{orb} \sim 5$ mins; no helium CVs have yet been found with such short periods - they will have high mass transfer rates and consequently short lifetimes near the minimum period.

Magnetic braking combined with gravitational radiation predicts that the mass transfer rate should be largely determined by P_{orb}. This is strongly at variance with observation: at a given P_{orb} there is a wide range of transfer rate (factors as large as 10^3 or more). This manifests itself in the variety of CV subtypes that exist at any P_{orb} : non-magnetic systems range from dwarf novae with very infrequent outbursts to highly luminous nova-likes with stable discs. It is clear that there is at least one other parameter that determines the current rate of transfer. It has been suggested (Livio & Shara 1987) that nova eruptions heat the secondary and cause excess mass transfer for ~100 y, with the result that the system must fall into a low state for long periods of time in order to maintain the secular mean transfer rate determined by the angular momentum sink. This would account for the apparent absence of CVs identifiable with ancient (Chinese and Korean) records of bright naked eye novae. The slow recovery from a low state (of 'hibernation') brings a CV up through the dwarf nova state to a stable disc before the next nova eruption. Such cyclical evolution (on time scales short compared with orbital evolution) may well happen, but the cause is not yet clear. Chaos inherent in the mass transfer irradiation-enhanced process is a possible mechanism (Wu, Wickramasinghe & Warner 1995b).

5. Supersoft X-ray Sources

A new type of CV has been observed, which fills a gap in the theoretical range of mass transfer possibilities. If a binary with $q > 1$ reduces its separation (through magnetic braking) to the point where its secondary fills its Roche lobe, and if the secondary has a radiative envelope, then a stable high rate of mass transfer (~10^{-7} $M_{\odot}$ y^{-1}) ensues. At such high rates of accretion stable nuclear burning occurs in the outer layers of a white dwarf, with a luminosity ~10^3 $L_{\odot}$ and a flux peaking at an energy of ~40 eV. A number of such 'supersoft' X-ray sources are known, and have been explained in the above manner (van den Heuvel *et al.* 1992; Kahabka & van den Heuvel 1997). Steady nuclear burning also occurs in some post novae for up to a decade after the nova eruption (e.g. V1974 Cyg and GQ Mus).

6. Properties of Accretion Discs

In high state accretion discs - found in nova-likes, post nova eruption systems ('nova remnants') and dwarf novae near maximum of outburst - the observed flux comes from an almost two dimensional object. The (apparent) absolute magnitude is therefore sensitive to orbital inclination. This is most directly demonstrated in the nova remnants, where distances and hence absolute magnitudes are obtained from the eruption behaviour (the 'absolute magnitude - rate of decline relationship'): (apparent) absolute magnitudes are correlated with inclination in the way expected for a limb-darkened flat disc (Warner 1986a); this is confirmed by independent measurements of UV luminosities (Selvelli *et al.* 1990). This effect explains the existence of a few novae with anomalously large eruption amplitudes (e.g. DQ Her and T Aur: Payne-Gaposchkin 1957): these are systems of high inclination. Using this effect, DY Pup was predicted (Warner 1995a) and confirmed (Warner & Van Zyl 1998) to be an eclipsing nova remnant.

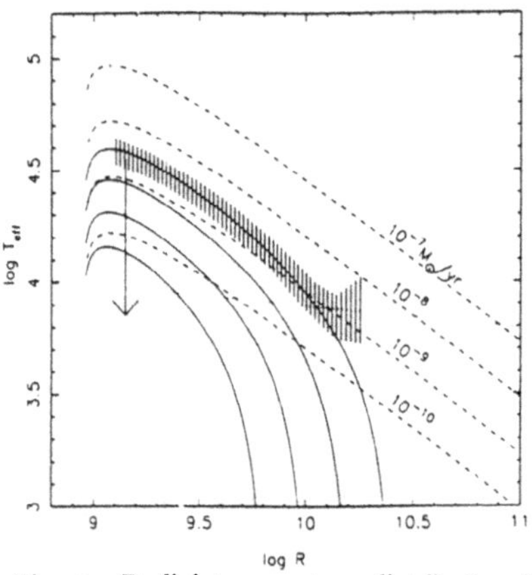

Fig. 4. Radial temperature distribution in the accretion disc of a dwarf nova during outburst. Dashed lines are theoretical distributions for non-steady discs at various rates of mass transfer. The hatched region is the observed distribution for Z Cha. From Mineshige (1991).

In eclipsing CVs the tomographic technique introduced by Horne (1985) can be employed to map the intensity distribution over the surface of the accretion disc. Repeated in different wavelength bands a colour temperature distribution as a function of radial distance from the primary is deduced, which can be compared with the theoretical temperature distribution for steady state (Shakura & Sunyaev 1973) or non-equilibrium (e.g. Mineshige & Wood 1989) discs. There is general agreement for high state (e.g. Horne & Cook 1985; Rutten *et al.* 1992) and low state discs (Mineshige & Wood 1989). These analyses lead directly to measurement of the mass transfer rates (through the discs): typically 10^{15} - 10^{16} gm s^{-1} in dwarf novae at

Using a tomographic restoration method on the emission line profiles of accretion disc spectra Marsh & Horne (1988) show how to deduce the velocity distribution in the disc. From these Doppler maps Horne and Saar (1991) find an average intensity distribution $I(r) \propto r^{-3/2}$ as a function of radial distance r from the primary This relationship is what would be expected if the chromosphere of a disc is heated by process similar to those in cool stars, i,.e. the effects of a self-sustaining dynamo in the differentially rotating disc.

The identification of the mechanism by which sufficient effective viscosity is generated in a disc to account for the transition time scales observed in dwarf nova outbursts has long been a problem. Currently the most popular model is a dynamical instability that requires only very small magnetic fields initially to be present in the disc (Balbus & Hawley 1991). This is capable of producing an unsteady self-sustaining dynamo with sufficient magnetically-generated viscosity (Tout & Pringle 1992b).

Half of the potential energy of accreting gas is radiated away in the disc; in a simple model with a non-rotating primary the remaining half would be expected to be lost from the boundary layer (BL) that connects the inner edge of the disc to the primary. For low accretion rates this layer will be optically thin and BL shock temperatures $T \sim 10^8$ K, emitting X-rays at ~10 keV, should result. At high accretion rates the BL is optically thick and the thermalized radiation would emerge at $T \sim 2 \times 10^5$ K (Pringle & Savonije 1979). In fact, although hard X-rays are observed, and high mass transfer systems do have lower energy emission (e.g. Patterson & Raymond 1985), the luminosities and temperatures of the latter systems are too low. From the ionization structure of winds from CV discs Hoare & Drew (1991) find BL temperatures in the range 5×10^4 - 1×10^5 K. It remains uncertain why the BLs are of lower temperature and luminosity than the simple energy balance argument predicts.

7. Dwarf Nova Oscillations

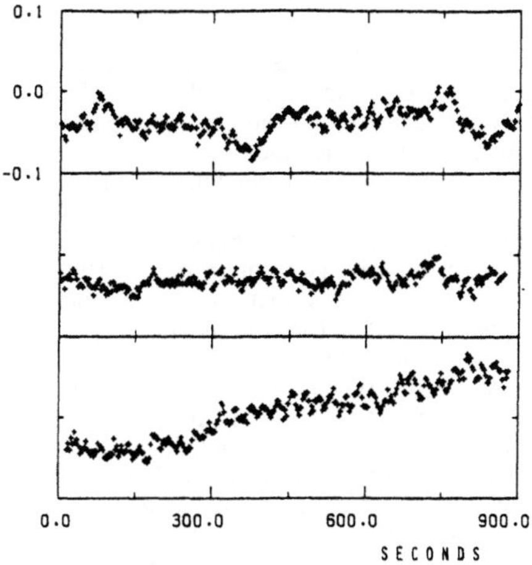

Fig. 5. Dwarf nova oscillations seen in the outburst light curve of TY PsA. From Warner, O'Donoghue & Wargau (1989).

As an example of the unexpected or exotic phenomena that arise in CVs we can consider the short time scale periodic or quasiperiodic brightness modulations that occur in most dwarf novae during outburst, and spasmodically in some nova-likes. Discovered at very low amplitudes (seen rarely in the light curve with amplitudes ~1%; more generally only detectable by Fourier analysis with amplitudes as low as 0.02%), dwarf nova oscillations (DNOs) have periods typically in the range 10-30 s (Warner and Robinson 1972a; Warner 1995a) with each star showing a small range. The periods change systematically during outburst, passing through a minimum value at maximum luminosity. The coherence of the oscillations also varies during outburst, being largest near maximum light but often showing good coherence for intervals of 1-2 h with abrupt small changes of period between. The DNOs are observed in SS Cyg, U Gem and VW Hyi in soft X-rays (0.04 - 0.5 keV) (e.g. Watson, King & Heise 1985); these three CVs are relatively near to the Earth, so soft X-ray emission is not totally absorbed by interstellar gas. DNOs have also been observed in the far UV (Mauche 1998).

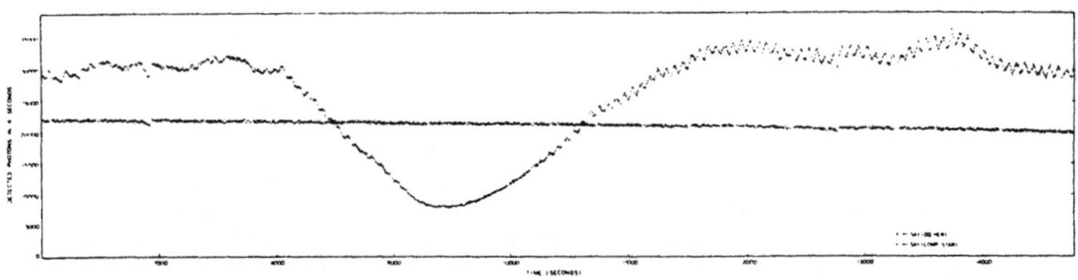

Fig. 6. Light curve of DQ Her at 4s time resolution showing eclipse of the 71s oscillations. The horizontal light curve is that of a comparison star. From Nather (1978).

Less coherent modulations, known as quasi periodic oscillations (QPOs), are also observed in dwarf novae during outburst (Patterson, Robinson & Nather 1977). These have periods a few times larger than the DNOs.

The origin of DNOs is not yet certain. A promising model invokes magnetically controlled accretion by relatively weak fields of the white dwarf primaries, the surface layers of which are spun up by enhanced accretion during dwarf nova outburst (Warner 1995d). The QPOs may be caused by non-radial oscillations in the accretion disc (Carroll *et al.* 1985).

An interesting practical use for DNOs arises from measurement of their phase variations during eclipse. As in the case of the 71s oscillations in DQ Her, where the very high stability of the pulse enables phase shifts to be made with precision (Warner *et al.* 1972, Robinson 1998), phase shifts in HT Cas and UX UMa (Nather & Robinson 1974; Patterson 1981) demonstrate that the whole accretion disc is emitting the DNO optical radiation: this is deduced to be reprocessed modulated hard radiation from the vicinity of the primary.

8. Intermediate Polars

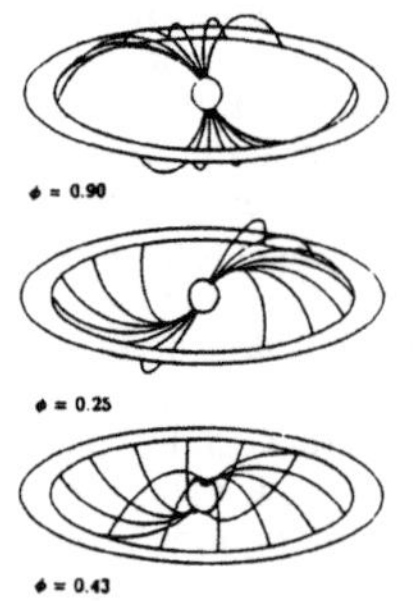

Fig. 7. Schematic view of accretion in an intermediate polar, viewed from three different rotational phases. From Hellier, Cropper & Mason (1991).

The interaction of accretion disc plasma with the rotating magnetosphere of a white dwarf is currently beyond the realm of realistic computation. There is a wealth of observed phenomena awaiting quantitative interpretation; partial progress has been made in understanding some of these.

Most IPs are probably close to a spin equilibrium where the torques acting on the primary through the field lines that connect it to parts of the disc that spin more rapidly or less rapidly (plus the material torque as accreting gas hits the primary) are in balance. From a theoretical evaluation of these torques (Ghosh & Lamb 1979; Lamb 1989) it is possible to deduce the magnetic moment of the primary from its observed spin period and the accretion rate (found from estimating the total accretion luminosity, which is predominantly emitted at X-ray wavelengths). Such an approach gives surface field strengths ~0.5 - 5 $\times$ 10^6 G for the primaries in IPs. The large unpolarized optical flux from the truncated accretion disc dilutes any cyclotron emission from the material accreting along field lines to the primary; nevertheless, circular polarization has been detected in the infrared of three IPs - presumably those with the largest field strengths - which indicate fields in the range 2 - 5 $\times$ 10^6 G (Wickramasinghe, Wu & Ferrario 1991; Buckley *et al.* 1997).

In the X-ray region IPs are both luminous and modulated: most IPs have been discovered as X-ray sources. In general the modulation period is the spin period of the primary, caused by the X-ray emitting accretion regions being carried around as the primary rotates. There are usually two accretion regions, near the magnetic poles of the primary, rather like auroral arcs. The magnetic field axis is inclined to the rotation axis in all IPs so far studied; magnetic and rotational alignment would largely eliminate the spin modulation, but such an IP would still be a strong X-ray emitter.

Although there is strong evidence that the majority of IPs possess (truncated) accretion discs, it is possible to have a primary with field strong enough to prevent formation of a disc but too weak to ensure synchronous rotation. Then a process of 'discless accretion' occurs, with the accretion stream impacting directly onto the asynchronously rotating magnetosphere of the primary. This impact zone is another source of X-ray emission, but its modulation period is the synodic rather than the sidereal period of the primary. This situation may, however, be confused with that where stream material partially continues over the face of the disc and impacts onto the rotating magnetosphere. These cases may be distinguished by differences in modulation amplitude. Hellier (1992) finds most IPs to have discs, but Buckley *et al.* (1997) make a strong case for discless accretion in RXJ1712.6-2414.

The anisotropic hard radiation emitted by the asynchronously rotating primary periodically intercepts various parts of the binary that are revolving at the orbital period. These features may be the secondary star, the accretion stream or the disc thickening where it is impacted by the stream. Each of these features reprocesses the X-rays/UV into optical radiation, whose brightness may be modulated (through geometric or optical depth effects) on P_{orb}. This can lead not only to optical modulation at sidereal and synodic spin periods but also at other orbital sidebands of these spin periods (Warner 1986b). In some IPs the full set of predicted modulations is seen (e.g. FO Aqr: de Martino *et al.* 1994). The relative phases of these modulations can help to identify which component of the CV is reprocessing the hard radiation.

Changes of spin period on time scales of ~10^6 y are observed, both spin-ups and spin-downs occurring. These are to be expected of systems not perfectly in spin equilibrium - the present rotation period of the primary being determined by the average mass transfer rate over the past 10^5 years, which is unlikely to match the current transfer rate from the secondary. (This is also the case in DQ Her, whose 71s spin period is decreasing on a time scale ~10^5 y as it attempts to find equilibrium with the high mass transfer rate of a recent post nova).

9. **Polars**

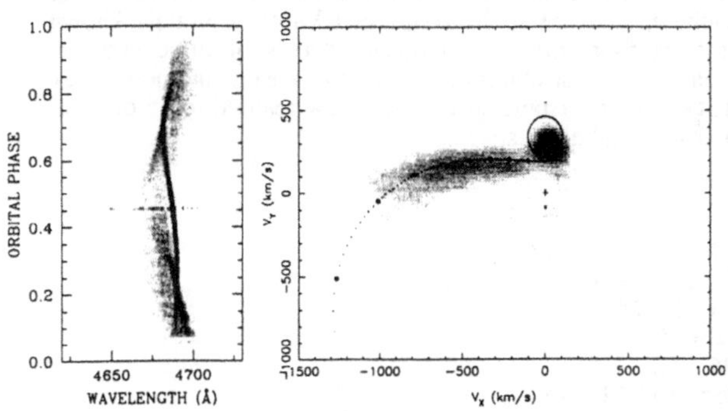

Fig. 8. Time resolved spectrum (left) and doppler tomogram (right) of the polar HU Aqr. The computed outline of the secondary star and the stream velocity trajectory are for a chosen set of parameters. From Schwope *et al.* (1995).

The power of doppler tomography to delineate the stream flow in strongly magnetic CVs is shown in Figure 8. The initial time-resolved spectrum maps onto velocity coordinates and shows where the line emission is located - in HU Aqr clearly it is from the inter-star stream and from the side of the secondary pointing towards (and hence heated by) the primary and its accretion column (Schwope *et al.* 1995). The stream trajectory is a function of the mass ratio q; in the case of HU Aqr this leads to $q = 0.25$. Then from the observed duration of eclipse in this system, the inclination is reliably determined at 85°. In such eclipsing polars, therefore, some system parameters can be very well determined.

10. **Nova Eruptions**

The accumulation of hydrogen-rich material on the surface of a white dwarf leads ultimately to the pressure and temperature at the base of the layer reaching the point where nuclear reactions commence. As this is at a level where the gas is semi-degenerate a thermonuclear runaway ensues which ejects the gas accumulated since the previous eruption, and can also excavate part of the core of the white dwarf.

The duration between eruptions depends sensitively on the mass of the white dwarf: for masses approaching the Chandrasekhar limit the ejected mass is ~10^{-6} $M_\odot$ and can occur on time scales of ~100 y (for mean mass transfer rates ~10^{-8} $M_\odot$ y^{-1}). This agrees with observations of recurrent novae. In general the recurrence time is ~10^3 - 10^5 y and ~10^{-4} $M_\odot$ is ejected. The greater frequency and energy of eruption in CVs of high mass biases mass determinations towards the high end of the range.

Total ejected shell masses are best determined from radio observations at 3-90 GHz, where the shell remains optically thin until 100-500 d after eruption (Hjellming 1990). Optical and infrared spectra of the shells typically show overabundances of CNO and sometimes Ne (Livio & Truran 1992). This is caused either by mixing between the core and the bottom of the H-rich layer in the interval between eruptions, or to direct dredge-up during eruption (Kovetz & Prialnik 1985).

Imaging of novae in the earliest stages of eruption has become possible in the radio region through the use of very long baseline interferometry (e.g. QU Vul: Taylor *et al.* 1998) and at optical and UV wavelengths through the use of HST (e.g. Nova Cyg 1992: Paresce *et al.* 1995). The high resolution of HST also reveals remarkable detail in the dispersing shells of old novae (e.g. recurrent nova T Pyx: Shara *et al.* 1997).

11. Prospects

Cataclysmic Variable stars, of which I have here given only a small example of their behaviour, will remain a growth industry for the foreseeable future. They are bright and populous in the X-ray region, and so will be prominent in the surveys of the next generation of X-ray satellites. The brightest quiescent CVs in the Magellanic Clouds are within reach of the new very large telescopes in the southern hemisphere; study of such groups of CVs at accurately known distances will reveal more of the multi-parameter evolutionary details of these systems. Furthermore, the brightest CVs remain the best objects for studies of steady and non-steady accretion discs, which will aid progess in understanding the discs now realised to be of great importance in a variety of other astrophysical objects.

References

Applegate, J.H., 1992. *ApJ*, **385**, 621.
Balbus, S.A. & Hawley, J.F., 1991. *ApJ*, **376**, 214.
Buckley, D.A.H., *et al.*, 1997. *MNRAS*, **287**, 117.
Carroll, B.W., *et al.*, 1985. *ApJ*, **266**, 529.
Clemens, C. *et al.*, 1998. *ApJ*, **496**, 352.
Crawford, J.A. & Kraft, R.P., 1956. *ApJ*, **123**, 44.
Cropper, M.S., 1990. *Sp. Sci. Rev.* **54**, 195.
de Martino, D., *et al.*, 1994. *Astr. Astrophys.*, **284**, 125.
Ghosh, P. & Lamb, F.K., 1979. *ApJ*, **234**, 296.
Harvey, D.A. *et al.*, 1998. *ApJ Letts.*, **493**, L105.
Hellier, C., 1992. *MNRAS*, **258**, 578.
Hellier, C., Cropper, M.S. & Mason, K.O., 1991. *MNRAS*, **248**, 233.
Hjellming, R.M., 1990. *Lect. Notes Phys.* **364**, 169.
Hoare, M.G. & Drew, J.E. 1991. *MNRAS*, **249**, 452.
Honeycutt, R.K. *et al.*, 1998. *ApJ*, **495**, 933.
Horne, K., 1985. *MNRAS*, **213** , 129.
Horne, K. & Cook, M.C., 1985. *MNRAS*, **214**, 307.
Horne, K. & Saar, S.H., 1991. *ApJ*, **374**, L55.
Hoshi, R., 1979. *Prog. Theor. Phys.*, **61**, 1307.
Ichikawa, S., Hirose, M. & Osaki, Y., 1993. *Pub. astr. Soc. Japan*, **45**, 243.
Kahabka, P. & van den Heuvel, E.P.J., 1997. *Ann. Rev. Astr. Astrophys.*, **35**, 69.
Kovetz, A. & Prialnik, D., 1985. *ApJ*, **291**, 812.
Kraft, R.P., 1961. *ApJ*, **134**, 171.
Kraft, R.P., 1962. *ApJ*, **135**, 408.
Kraft, R.P., 1964. *ApJ*, **139**, 457.
Lamb, F.K., 1989, in *Timing Neutron Stars*, eds. H. Ogelman & E.P.J. van den Heuvel, Kluwer, p. 649.
Livio, M. & Truran, J.W., 1992. *ApJ*, **389**, 695.
Lubow, S., 1991. *ApJ*, **381**, 259.
Marsh, T.R. & Horne, K., 1988. *MNRAS*, **235**, 269.
Mauche, C., 1998. *ApJ* In press.
McDermott, P.N. & Taam, R.E., 1989. *ApJ*, **342**, 1019.
Meintjies, P.J. *et al.*, 1994. *ApJ*, **434**, 292.
Mestel, L., 1952. *MNRAS*, **112**, 598.
Mestel, L. & Spruit, H.C., 1987. *MNRAS*, **226**, 57.
Meyer, F. & Meyer-Hofmeister, E., 1981. *Astr. Astrophys.*, **104**, L10.
Mineshige, S., 1991. *MNRAS*, **250**, 253.
Mineshige, S. & Wood, J., 1989. *MNRAS*, **241**, 259.
Murray, J. 1996. *MNRAS*, **279**, 402.
Nather, R.E., 1978. *PASP*, **90**, 477.
Nather, R.E. & Robinson, E.L., 1974. *ApJ*, **190**, 637.
Nather, R.E. & Warner, B., 1971. *MNRAS*, **152**, 209.
Paczynski, B., 1976. *Int. Astr. Union Symp. No. 73*, p. 75.

Paresce, F. *et al.*, 1995. *Astr. Astrophys.*, **299**, 823.
Patterson, J., 1981. *ApJ Suppl.*, **45**, 517.
Patterson, J. & Raymond, J.C., 1985. *ApJ*, **292**, 535.
Patterson, J., Halpern, J.P. & Shambrook, A., 1993. *ApJ*, **419**, 83.
Patterson, J., Robinson, E.L. & Nather, R.E., 1977. *ApJ*, **214**, 144.
Payne-Gaposchkin, C., 1957. *The Galactic Novae*, North-Holland, Amsterdam.
Robinson, E.L., *et al.*, 1981. *ApJ*, **251**, 611.
Robinson, E.L. *et al.*, 1998. *ApJ*, In press.
Rutten, R.G.M. *et al.*, 1992. *Astr. Astrophys.*, **265**, 159.
Schmidt, G.D., Liebert, J. & Stockman, H.S., 1995. *ApJ*, **441**, 414.
Schwope, A.D. *et al.*, 1995. *Astr. Soc. Pacific Conf. Ser.*, **85**, 166.
Selvelli, P.L. *et al.*, 1990. *IAU Colloq. No.* 122, p. 65.
Semeniuk, I., 1980. *Astr. Astrophys. Suppl.*, **39**, 29.
Shakura, N.I. & Sunyaev, R.A., 1973. *Astr. Astrophys.*, **24**, 337.
Shara, M.M. & Prialnik, D., 1994. *Astr. J.*, **107**, 1542.
Shara, M.M. *et al.*, 1997. *Astr. J.*, **114**, 258.
Smak, J., 1967. *Acta Astr.*, **17**, 255.
Tapia, S., 1977. *ApJ*, **212**, L125.
Taylor, A.R. *et al.*, 1998. *Nature*, **335**, 235.
Tout, C.A. & Pringle, J.E., 1992. *MNRAS*, **256**, 269.
Tout, C.A. & Pringle, J., 1992b. *MNRAS*, **259**, 604.
van den Heuvel, E.P.J. *et al.*, 1992. *Astr. Astrophys.*, **262**, 97.
Verbunt, F. & Zwaan, C., 1981. *Astr. Astrophys.*, **100**, L7.
Vogt, N., 1974. *Astr. Astrophys.*, **36**, 369.
Walker, M.F., 1956. *ApJ*, **123**, 68.
Warner, B., 1975. *MNRAS*, **170**, 219.
Warner, B., 1986a. *MNRAS*, **222**, 11.
Warner, B., 1986b. *MNRAS*, **219**, 347.
Warner, B., 1988a. *High Speed Astronomical Photometry*, Cambridge University Press.
Warner, B., 1988b. *Nature*, **336**, 129.
Warner, B., 1995a. *Cataclysmic Variable Stars*, Cambridge University Press.
Warner, B., 1995b. *Astrophys. Sp. Sci.*, **226**, 187.
Warner, B., 1995c. *Astrophys. Sp. Sci.*, **225**, 249.
Warner, B., 1995d. *Astr. Soc. Pacific Conf. Ser.*, **85**, 343.
Warner, B. & Robinson, E.L., 1972a. *Nature Phys. Sci.*, **239**, 2.
Warner, B. & Robinson, E.L., 1972b. *MNRAS*, **159**, 101.
Warner, B. & van Zyl, L., 1998. In preparation.
Warner, B., O'Donoghue, D. & Wargau, W., 1989. *MNRAS*, **238**, 73.
Warner, B. *et al.*, 1972. *MNRAS*, **159**, 321.
Watson, M.G., King, A.R. & Heise, J., 1985. *Sp. Sci. Rev.*, **40**, 127.
Whitehurst, R., 1988. *MNRAS*, **232**, 35.
Wickramasinghe, D.T., 1993. in *Cataclysmic Variables and Related Physics*, eds. O. Regev & G. Shaviv, Inst. Phys. Publ., Bristol, p. 213.
Wickramasinghe, D.T., Wu, K. & Ferrario, L., 1991. *MNRAS*, **249**, 460.
Wood, K.S., Imamura, J.N. & Wolff, M.T., 1992. *ApJ*, **398**, 593.
Wu, K., Wickramasinghe, D.T. & Warner, B. 1995a. *Astrophys. & Sp. Sci. Library* **205**, 315.
Wu, K., Wickramasinghe, D.T. & Warner, B., 1995b. *Proc. Astr. Soc. Australia*, **12**, 60.

BLACK HOLES IN THE UNIVERSE

I. D. NOVIKOV
Theoretical Astrophysics Center, Juliane Maries Vej 30, DK-2100 Copenhagen Ø, Denmark;
University Observatory, Juliane Maries Vej 30, DK-2100 Copenhagen Ø, Denmark;
Astro Space Centre of P.N. Lebedev Physical Institute,
Profsoyuznaya 84/32, Moscow, 117810, Russia;
Nordita, Blegdamsvej 17, DK-2100 Copenhagen Ø, Denmark.

1. Introduction

Some 30 years ago very few scientists thought that black holes may really exist. Attention focussed on the black hole hypothesis after neutron stars had been discovered. It was rather surprising that astrophysicists immediately 'welcomed' black holes. They found their place not only in the remnants of supernova explosions but also in the nuclei of galaxies and quasars.

Of all conceptions of the human mind perhaps the most fantastic is the black hole. Black holes are neither bodies nor radiation. They are clots of gravity. The study of black hole physics extends our knowledge of the fundamental properties of space and time. Quantum processes occur in the neighborhood of black holes, so that the most intricate structure of the physical vacuum is revealed. Even more powerful (catastrophically powerful) quantum processes take place inside black holes (in the vicinity of the singularity).

It may be said that black holes are a door to a new, very wide field of study of the physical world.

In this paper I will give a brief review of some problems of physics and astrophysics of black holes. For systematic discussion of the problems see the books: Thorne *et al.* (1986), Novikov and Frolov (1989), section "Black Holes" in the book Kawaler *et al.* (1997), and Frolov and Novikov (1998).

Let us start from physics.

A black hole is region in space-time from which no signal can escape to an external observer. A black hole's boundary is a so called event horizon.

After the gravitational collapse of a celestial body and formation of a black hole the external gravitational field of it asymptotically approaches a standard equilibrium configuration known as the Kerr-Newman field and characterized by just three numbers: mass, angular momentum and charge.

2. Physics outside a black hole

Black holes reside in curved space. If a black hole has nonzero angular momentum then anything near a black hole will be dragged along by the vortex gravitational field. In this section I will consider a black hole without electric charge (Kerr black hole). The horizon's surface area can be written in terms of its mass M and angular momentum $J = aM$, where a is an angular momentum per unit mass ($c = 1$, $G = 1$):

$$A = 4\pi(r_H^2 + a^2), \tag{1}$$

$$r_H = M + \sqrt{M^2 - a^2}. \tag{2}$$

The rotational energy, or corresponding mass M_{rot}, of a Kerr black hole is the following

$$M_{rot} = M - \left[\frac{1}{2}M\left(M + \sqrt{M^2 - a^2}\right)\right]^{1/2} \tag{3}$$

J. Andersen (ed.), Highlights of Astronomy, Volume 11A, 28–42.

This rotational energy (energy of the vortex gravitational field) can be extracted (in principle) from a black hole.

The black hole is a clot of gravity, there is not any real matter on the horizon. In spite of this fact the horizon looks for an external observer (outside the black hole) and behaves as a physical membrane which is made from a two-dimensional viscous fluid with definite mechanical, electrical and thermodynamic properties.

This remarkable viewpoint is known as the membrane paradigm (see Thorne *et al.* (1986) for a review). According to this paradigm the interaction of the horizon with the external universe is described in terms of familiar laws for the horizon fluid, e.g. the Navier-Stokes equation, Maxwell's equations, a tidal force equation, and the equations of thermodynamics. It is very important to emphasize that the membrane paradigm is not an approximation method or some analogy. It is an exact formalism which gives exactly the same results as the standard formalism of the General Relativity. Because the laws governing the horizon's behavior have familiar forms, they are powerful for understanding intuitively and computing quantitatively the interaction of black holes with complex environments.

In subsequent parts of this section we will consider some manifestations of the physical properties of the black hole's membrane, that resided in the three dimensional space.

2.1. MECHANICAL PROPERTIES OF THE HORIZON'S MEMBRANE

According to the membrane formalism, from the point of view of an external observer the hole's membrane has definite surface mass density and the surface pressure and viscosity.

The formula for the mass density is

$$\sum = -\frac{1}{8\pi}\theta^H, \qquad \theta^H \equiv \frac{d(\Delta A)}{\Delta A dt}, \tag{4}$$

where θ^H is a fractional change of area of a surface element per unit time of an observer at infinity. One can see that for the case of a black hole in equilibrium (for example, a nonrotating (Schwarzschild) or a Kerr hole in the empty space) $\sum = 0$. The value of θ^H is always non-negative, consequently $\sum$ is always non-positive.

There is surface pressure p^H in the membrane. For a Schwarzschild hole it is:

$$p^H = \frac{1}{32\pi M} \approx \left(10^{42}\frac{\text{dyne}}{\text{cm}}\right)\left(\frac{M_\odot}{M}\right). \tag{5}$$

From the point of view of the membrane formalism the gravity of a black hole in equilibrium is produced by p^H.

The horizon's shear viscosity η^H and the horizon's bulk viscosity ζ^H are correspondingly:

$$\eta^H = \frac{1}{16\pi} \approx 10^{37}\frac{\text{g}}{\text{sec}}, \tag{6}$$

$$\zeta^H = -\frac{1}{16\pi} = -10^{37}\frac{\text{g}}{\text{sec}}. \tag{7}$$

Because the membrane paradigm regards a black hole as a two dimensional membrane with familiar mechanical properties it is rather easier to understand intuitively and compute quantitatively what happens with a black hole under some definite conditions. Let us consider a few examples.

If a black hole occurs as a result of gravitational collapse of an asymmetric celestial body (without rotation), then a nonspherical hole arises at the first moment. The hole's membrane is deformed and there is no balance between the surface pressure of the membrane and its gravity. So the membrane vibrates and radiates gravitational waves. The waves carry away the energy of the membrane deformation. This together with the membrane viscosity makes the horizon settle down into an absolutely spherical equilibrium shape.

Another example is a shape of the membrane of a rotating black hole. Centrifugal forces make a hole's membrane bulge out at its equator. The balance between the surface pressure, gravity and centrifugal forces determines the shape of the horizon's membrane.

Let us consider one very unusual property of the horizon's membrane. We emphasized above that the differential equations which describe the interaction of the horizon with the external universe are familiar physical laws (e.g. the Navier-Stokes equation and so on). But the solutions of the equations are determined also by the boundary conditions. In the case of standard physics the boundary conditions must be imposed at some initial moment or in the infinite past. That is not so for the hole's horizon! The point is that the horizon is a boundary between light-speed signals that can and those that cannot ever escape to spatial infinity. But this fact depends on the processes in the future, not in the past.

Whether a signal can escape depends on the region of spacetime to the future of the signal's source. It means that the motion of the horizon at any moment of time depends not on what has happened to the horizon in the past but what will happen to the horizon in the future.

This property can be illustrated by the problem of a free fall of a thin spherical shell of a matter of mass ΔM into a Schwarzschild hole with mass M. The spacetime geometry is that of Schwarzschild both interior to the shell and outside it. In the interior the Schwarzschild mass is M and outside it is $M + \Delta M$. Now the light-speed signals with world lines at $r = 2M$ cannot be the boundary of the non-escape region because these signals and outgoing signals just outside $r = 2M$ will get caught and pulled into the hole by the added gravity of the shell when in the future the shell passes through them. The real boundary is generated by light-speed signals world lines of which are just outside of the surface $r = 2M$. In the past, long before the shell arrives at the horizon this surface practically coincides with $r = 2M$. Then, as the shell nears it the surface (which is the real boundary, meaning the real horizon) starts to expand. This is because the world lines of its generators go farther an farther from $r = 2M$. This is their property in the Schwarzschild spacetime, and it does not depend on the approaching shell. When the shell finally passes through it, the added shell's gravity starts influence the motions of the generators of the surface, the horizon suddenly stops expanding and freezes at $r = 2(M + \Delta M)$. These behaviors of the horizon are dictated by the properties of propagation of the light-speed signals which generate the horizon and which have the property to propagate at $r = 2(M + \Delta M)$ **after** crossing with the shell. Thus, this behavior of the horizon **before** crossing with the shell (its expansion) depends on the events in the future (the crossing with the shell).

One refers to this dependence of future events as the "teleological" nature of the horizon (see Thorne *et al.* 1986). I would like to emphasize that these behaviors looks as if the hole's membrane lives in time which flows in the opposite direction: from the future into the past. Indeed in this case the change of the size of the horizon looks very natural and causal. If we accept this point of view, we should consider the extraction of the shell from the hole, and just after this extraction of the shell from the membrane at $r = 2(M + \Delta M)$, the horizon starts contracting and settles down to $r = 2M$. We will see in Section 3 that this unusual property, namely, "feeling" information from the infinite future of the external observer, is a characteristic property not only of the horizon but also of the interior of a black hole.

2.2. BLACK-HOLE ELECTRODYNAMICS

A black hole horizon behaves as an electrically conducting sphere. To understand this let us ask what could be the external manifestation of the electric conductivity of a body in a flat spacetime. The simplest manifestation is the following. If one brings a positive electric charge close to a metal sphere then free electrons on the sphere's metal surface will be displaced with respect to the ions by the Coulomb electric forces. It polarizes the sphere. As a result, the electric field lines form a characteristic configuration in the space around the sphere. Now if one moves the charge parallel to the surface of the sphere from one position to another one, the characteristic configuration of the electric lines comes to a new place with some delay. This delay is determined by the resistivity of the sphere's metal surface. It turns out that if one brings a charge close to a non-spinning black hole, there is a similarity between the picture of the field lines in the vicinity of the black hole and the analogous picture in the vicinity of a metal sphere in a flat spacetime. Now the curvature of spacetime distorts the field line rather than displacement of real charges on the horizon. Nevertheless, it looks like the field of the charge polarizes the horizon.

If one moves the charge parallel to the hole's horizon to another position, then the configuration of the electric lines will settle down at the new place with some delay. Now it is determined by the finite time of propagation of electromagnetic signals. Nevertheless one can interpret it as an effective resistivity of the horizon.

In general one can say that a horizon's membrane behaves as a metal sphere with a surface resistivity equal to $R_H = 4\pi \approx 377$ohms.

The membrane paradigm gives insight into possible behaviors of rotating black holes in interaction with magnetized plasma. We will draw an analogy with a dynamo. In its rotor the motion of wire coils in a magnetic field produces an electromotive force compelling the charges to flow through the conductor. A black hole is also a special dynamo of great size. If a spinning black hole is immersed in an external magnetic field, a powerful electric field will also develop in its vicinity. The magnetic field is created by the interstellar gas flowing into a black hole. The magnetic field lines will tend to rotate along with the spinning black hole. The motion of any magnetic field generates an electric field. In the case of a rapidly rotating, magnetized black hole, the electric field generated near its edges can produce an enormous voltage difference between the poles of the hole and its equatorial region:

$$\Delta V \approx (10^{20}\text{volts}) \left(\frac{a}{M}\right) \left(\frac{M}{10^9 M_\odot}\right) \left(\frac{B}{10^4 G}\right), \tag{8}$$

where B is the magnetic field in the vicinity of the black hole. It is as though the spinning black hole was a huge battery. The electric field is responsible for accelerating the charged particles of the plasma and causing them to move along the magnetic lines of force. The total power output is

$$P \approx \left(10^{45}\frac{\text{erg}}{\text{sec}}\right) \left(\frac{a}{M}\right) \left(\frac{M}{10^9 M_\odot}\right) \left(\frac{B}{10^4 G}\right), \tag{9}$$

Probably this mechanism is the main "engine" of the active galactic nuclei.

2.3. THERMODYNAMICS OF BLACK HOLES

From many aspects of the thermodynamics of black holes, I will discuss the problem of the black hole's thermal quantum radiation and the related problem of the thermal atmosphere of a black hole.

S.Hawking (1974) claimed that a black hole should emit thermal radiation with temperature

$$T_H = \frac{\hbar}{8\pi k} M^{-1} \approx (10^{-7} K) \left(\frac{M_\odot}{M}\right). \tag{10}$$

How, in simple physical terms, could one understand that a black hole behaves like an ordinary body with temperature T_H. A key insight into thermal emission from a hole come from theoretical discoveries in the mid-1970s (see Unruh, 1976). The crucial point is the existence of the event horizon for some definite classes of observers. For example, an accelerated observer in an empty spacetime has a horizon. This observer cannot receive information from the region beyond the horizon. The virtual particles' vacuum fluctuation waves are not confined solely to the region above the horizon; part of each fluctuation wave is beyond the horizon and part is within the region which the observer can see. According to quantum mechanics this principle lack of information about vacuum fluctuation waves leads to the conclusion (for an accelerated observer) that they are real waves. As a result, this observer is bathed in a perfect bath of thermal radiation with temperature $T = \hbar a/(2\pi k)$, where a is the observer's acceleration. Since a static observer just above a Schwarzschild horizon can be viewed as analogous to an accelerated observer in flat spacetime with acceleration $a = c^2/z$, where z is the distance from the horizon, such an observer should feel himself bathed in thermal radiation with local temperature $T = \hbar/(2\pi kz)$. This thermal radiation forms a thermal atmosphere of the hole. The radiation, climbing up through the hole's gravitational field, would be redshifted by a factor $\left(1 - \frac{2M}{r}\right)^{1/2}$. It will emerge with temperature T_H. Most of the photons and other particles fly upward a short distance and are then pulled back down by the hole's enormous gravity. A few of the particles manage to escape the hole's gravitational grip and evaporate into space. These particles form the Hawking radiation.

Note that a free falling observer does not feel this thermal atmosphere. He "sees" only vacuum fluctuations to consist of pairs of virtual particles.

The process of the Hawking quantum evaporation is very slow. The total lifetime is proportional to the cube of M. For a 20 solar mass black hole it is 10^{70}years.

In principle, the interactions of a black hole with the external Universe can change the process of extraction of the thermal energy from a black hole atmosphere drastically (see the review in referenced books).

3. Physics inside a black hole

3.1. BLACK HOLE'S INTERIOR

What can one say about the interior of a black hole?

This problem was the subject of a very active investigation last decades. There is a great progress in these researches. We know some important properties of the realistic black hole's interior, but some details and crucial problems are still the subject of much debate.

A very important point for understanding the problem of the black hole's interior is the fact that the path into the gravitational abyss of the interior of a black hole is a progression in time. We recall that inside a spherical hole, for example, the radial coordinate is timelike. It means that the problem of the black hole interior is an *evolutionary problem*. In this sense it is completely different from a problem of an internal structure of other celestial bodies, stars for example.

In principle, if we know the conditions on the border of a black hole (on the event horizon), we can integrate the Einstein equations in time and learn the structure of the progressively more and more deep layers inside the black hole. Conceptually it looks simple, but there are two types of principal difficulties which prevent realizing this idea consistently.

The first difficulty is the following. The internal structure of a general black hole, which rotates, even soon after its creation depends crucially on the conditions on the event horizon at tremendous future of the external observer (formally at the infinite future). It is because the light-like signal can propagate inside a black hole from the tremendous future to those regions if the regions are deep enough in the hole. The limiting light like signals which propagate from (formally) infinite future of the external observer form a border inside a black hole which is called a Cauchy horizon.

Thus, the structure of the regions inside a black hole depends crucially on the fate of the black hole at infinite future of an external observer. For example, it depends on the final state of the black hole evaporation, on possible collisions of the black hole with another black hole, and it depends on the fate of the Universe itself. It is clear that theoretitions feel themselves uncomfortable under such circumstances.

The second serious problem is related with existing of a singularity inside a black hole. Close to the singularity, where the curvatures of the spacetime approach Planck levels, the Classical General Relativity is not applicable. We have not the final version of the quantum theory of gravity yet, thus any extension of the discussion of physics in this region would be highly speculative. Fortunately, as we shall see, these singular regions are deep enough in the black hole interior and they are *in the future* with respect to overlying and *preceding* layers of the black hole where curvatures are not so high and which can be described by well-established theory.

The first attempts to investigate the interior of a Schwarzschild black hole have been made in the late 70's. It has been demonstrated that the parts of the interior of a black hole long after its formation and subsequent perturbations are virtually free of perturbations because the gravitational radiation from aspherical initial perturbations becomes infinitely diluted as it reaches these regions. But this result breaks down in general case when the angular momentum or the electric charge is not equal to zero. The reason for that is related with the fact that the topology of the interior of the rotating or/and charged black holes solution differs drastically from the Schwarzschild one. The key point is that the interior of this black hole possess a Cauchy horizon. This is a surface of infinite blueshift. Infalling gravitational radiation propagates inside the black hole along paths approaching the generators of the Cauchy horizon, and the energy density of this radiation will suffer an infinite blueshift as it approaches the Cauchy horizon.

In general the evolution with time into the black hole deeps looks like the following. There is a weak flux of gravitational radiation into a black hole through the horizon because of small perturbations outside of it. When this radiation approaches the Cauchy horizon it suffers an infinite blueshift. This infinite blueshift radiation together with the radiation scattered on the curvature of spacetime inside the black hole leads to formation of the curvature singularity of the spacetime along the Cauchy horizon. The infinite tidal gravitational forces arrise here.

A detail discussion one can find in the following works: Poisson and Israel (1990), Ori (1992), Bonanno *et al.* (1995), Israel (1997) and references therein.

3.2. QUANTUM EFFECTS

In previous discussions we emphasized that the internal structure of black holes is a problem of the evolution in time starting from boundary conditions on the event horizon for all moment of time up to the infinite future of the external observer.

It is very essential to know boundary conditions up to infinity because we observed that the essential events: mass inflation and singularity formation, happened along the Cauchy horizon which brought information from the infinite future of the external spacetime. However, even an isolated black hole in the asymptotically flat spacetime cannot exist forever. It will be evaporated by Hawking quantum radiation. So far we discussed the problem without taking into account this ultimate fate of black holes. Even without going into details it is clear that quantum evaporation of the black holes is crucial for the whole problem.

What can we say about general picture of the black hole interior accounted quantum evaporation? To account quantum evaporation we have to change the boundary conditions on the event horizon as compared to the boundary conditions discussed above. Now they should include the flux of negative energy across the horizon which is related with the quantum evaporation. The last stage of the quantum evaporation when the mass of the black hole becomes comparable to the Planck mass, $m_{pl} = (\hbar/6)^{1/2} \approx 2.2 \times 10^{-5}$g, is unknown. At this stage the spacetime curvature near the horizon reaches l_{pl}^{-2}, where l_{pl} is the Planckian length, $l_{pl} = (G\hbar/c^3)^{1/2} \approx 1.6 \times 10^{-33}$cm. It means that from the point of view of the semi-classical physics a singularity arises here. Probably at this stage the black hole has characteristics of an extreme black hole, when an external event horizon and internal Cauchy horizon coincide.

As for the processes inside a true singularity in the black hole interior, they can be treated only in the framework of an unified quantum theory incorporating gravitation which is unknown.

4. Astrophysics of black holes

Do black holes exist in the Universe or are they only an abstract concept of the human mind? In principle, a black hole could be built artificially. However, this meets such grandiose technical difficulties that it looks impossible, at least in the immediate future. In fact, the artificial building of a black hole looks even more problematic than an artificial creation of a star. Thus we have to conclude that the physics of black holes, as well as the physics of stars, is the physics of celestial bodies. Stars definitely exist, but what may one say about the existence of astrophysical black holes?

Modern astrophysics deals with two types of black holes in the Universe:
1) *stellar black holes*, that is black holes of stellar masses, that were born when massive stars died;
2) *supermassive black holes* with masses up to $10^9 M_\odot$ and greater at centers of galaxies ($M_\odot = 2 \times 10^{33}$g is the mass of the Sun).

These two types of black holes have been discovered. The third possible type of astrophysical black holes - primordial black holes will be discussed in subsection 4.5. Our main attention in Section 4 is focused on the possible observational manifestation of black holes.

4.1. THE ORIGIN OF STELLAR BLACK HOLES

"When all the thermo-nuclear sources of energy are exhausted a sufficiently heavy star will collapse" – this is the first phrase of the abstract of a remarkable paper by Oppenheimer and Snyder (1939). Every statement of this paper accords with ideas that remain valid today. The authors conclude the abstract by the following sentence: "... an external observer sees the star shrinking to its gravitational radius." This is the modern prediction of the formation of black holes when massive stars die.

How heavy should a star be to turn into a black hole? The answer is not simple. A star that is not massive enough ends up either as a white dwarf or a neutron star. There are upper limits on the masses of both these types of celestial bodies. For white dwarfs it is the *Chandrasekhar limit*, which is about $(1.2 - 1.4) \times M_\odot$. For neutron stars it is the *Oppenheimer-Volkoff limit.* The exact value of this limit depends on the equation of state at matter density higher than the density of nuclear matter $\rho_0 = 2.8 \times 10^{14}\text{g cm}^{-3}$. The modern theory gives for the maximal mass of a non-rotating neutron star the estimate $(2-3) \times M_\odot$. Rotation can increase maximal mass of a non-rotating neutron star only slightly up to 25%. Thus one can believe that the upper mass limit for neutron stars should not be greater than $M_0 \approx 3M_\odot$. If a star at the very end of its evolution has mass greater than M_0 it must turn into a black hole. However this does not mean that all normal stars (on the "main sequence" of the Herzsprung-Russell diagram) with masses $M > M_0$ are black hole progenitors. The point is that the final stages of evolution of massive stars are poorly understood. Steady mass loss, catastrophic mass ejection and even disruption at supernovae explosions are possible. These processes can considerably reduce mass of a star at the end of its evolution. Thus the initial mass of black hole progenitors could be essentially greater than M_0.

There are different estimates for the minimal mass M_* of a progenitor star that still forms a black hole. Uncertainty is $M_* \approx (10-40)M_\odot$ and even more. Note that the evolution of stars in close binary systems differs from the evolution of sole stars because of mass transfer from one star to another. The conclusions about masses of black hole progenitors in this case could be essentially different.

One can try to estimate how many black holes have been created by stellar collapse in our Galaxy during its existence. The estimates give the number of the order 10^9.

4.2. DISK ACCRETION ONTO BLACK HOLES

For the purpose of finding and investigating black holes, two specific cases of accretion are of a particular importance: accretion in binary systems and accretion onto supermassive black holes that probably reside at the centers of galaxies. In both cases the accreting gas has big specific angular momentum. As a result the gas elements circle around the black hole in Keplerian orbits, forming a disk or a torus around it. Viscosity plays a crucial role for the accretion. It removes angular momentum from each gas element, permitting it to gradually spiral inward toward the black hole. At the same time the viscosity heats the gas, causing it to radiate. Probable sources of viscosity are turbulence in the gas disk and random magnetic fields. Unfortunately, we are not near to a good physical understanding of the effective viscosity. Large-scale magnetic fields can also play an important role in the physics of accretion.

The properties of the accreting disk are determined by the rate of gas accretion. An important measure of any accretion luminosity of a black hole is provided by the Eddington critical luminosity

$$L_E = 4\pi G M_h \mu m_p c/\sigma_T = \left(1.3 \times 10^{38}\frac{\text{erg}}{\text{s}}\right)\mu\left(\frac{M_h}{M_\odot}\right). \tag{4.1}$$

Here M_h is the mass of a black hole, μ is the molecular weight of electron, m_p is the rest mass of the proton, and σ_T is the Thomson cross section. It is the luminosity at which the radiation pressure just balances the gravitational force of the mass M_h for a fully ionized plasma.

A useful measure of the accretion rate $\dot{M}$ is the so-called "critical accretion rate":

$$\dot{M}_E = L_E c^{-2}, \tag{4.2}$$

where L_E is given by equation (4.1). We shall use the dimensionless expression $\dot{m} \equiv \dot{M}/\dot{M}_E$.

The first models of the disk accretion were rather simple. They focused on the case of moderate rate of accretion $\dot{m} < 1$. Subsequently theories for $\dot{m} \sim 1$ and $\dot{m} \gg 1$ were developed. They take into account complex processes in radiative plasma and various types of instabilities.

The source of luminosity for disk accretion is the gravitational energy that is released when gas elements in the disk spiral down. Most of the gravitational energy is released, generating most of the luminosity, from the inner parts of the disk. According to the theory for these simplest models the total luminosity of the disk is the following

$$L \approx \begin{cases} \left(3 \times 10^{36}\frac{\text{erg}}{\text{s}}\right)\left(\frac{\dot{M}}{10^{-9}M_\odot/\text{yr}}\right) & \text{, for a nonrotating hole ,} \\ \left(3 \times 10^{37}\frac{\text{erg}}{\text{s}}\right)\left(\frac{\dot{M}}{10^{-9}M_\odot/\text{yr}}\right) & \text{, for a maximally rotating hole .} \end{cases} \tag{4.3}$$

The accretion rate $\dot{M}$ is an arbitrary external parameter, which is determined by the source of gas (for example, by the flux of gas from the upper atmosphere of the companion star in a binary system). We normalized $\dot{M}$ by the value $M_0 = 10^{-9} M_\odot/\text{yr}$ because this is probably the typical rate at which a normal star is dumping gas onto a companion black hole. In this model the accretion gas is assumed to be relatively cool, with its temperature much less than the virial temperature corresponding to the potential energy in the gravitational field. As estimates show, a *geometrically thin disk* (with heights $h \ll r$) might be formed under these conditions. This is the so-called *standard disk model.* In this model the electron and ion temperatures are equal, and the disk is effectively optically thick. The temperature of the gas in the inner parts of the disk reaches $T \approx 10^7 - 10^8$K. In this region electron scattering opacity modifies the emitted spectrum so that it is no longer the blackbody spectrum. Instead, the total spectrum of the disk radiation is a power law $F \sim \omega^{1/3}$ with an exponential cut off at high frequencies. The innermost regions of such "standard" disks are probably unstable. The thin accretion disk model is unable to explain the hard spectra observed in accretion flows around black holes in many observable cases.

A few types of hot accretion flow models have been proposed. Among them a model with a hot corona above a standard thin accretion disk. In another model the ions in the inner region are hot $T_i \approx 10^{11}$K but the electrons are considerably cooler $T_e \approx 10^9$K. This inner disk is thicker than in the "standard" model and produces most of the X-ray emission. The models with hot ions and cooler electrons are optically thin.

Further development of the theory of disk accretion led to more sophisticated models. It have been demonstrated that when the luminosity reaches the critical one (corresponding to $\dot{m} \equiv \dot{M}/\dot{M}_E$ of the order of unity) radiation pressure in the inner parts of the disk dominates the gas pressure and the disk is thermally and viscously unstable. For especially big $\dot{m} > 80$ the essential part of the energy of the plasmais lost by advection into the black hole horizon because the radiation is trapped in the accretion gas and is unable to escape. This process stabilizes the gas flow against perturbations. Advection can also be important for smaller $\dot{m}$. For high mass accretion rates the height of the accretion disk becomes comparable to its radius. In modern models the radial pressure gradients and the motion of gas elements along radius are taken into account. In the innermost parts of the disk and down to the black hole the flow of gas is supersonic.

Recently, a new class of optically thin hot disk solutions has been discovered. In this model the most of the viscously dissipated energy is advected with the accreting gas, with only a small fraction of the energy being radiated. It is because the gas density is so low that the radiative efficiency is very poor. These models were named advection-dominated. They have been applied successfully to a few concrete celestial objects.

In conclusion we note that in some models of disk accretion electron-positron pair production can be important. We believe that new models involving recent achievements of plasma physics will play a key role in the modern astrophysics of black holes.

4.3. EVIDENCE FOR BLACK HOLES IN STELLAR BINARY SYSTEMS

Probably the best evidence that black holes exist comes from studies of X-ray binaries. The arguments that are used to prove that an X-ray binary contains a black hole, are as follows:

1. The X-ray emitting object in a binary system is very compact, and therefore cannot be an ordinary star. Thus it is either a neutron star or a black hole. This argument comes mainly from analysis of the features of emitted X-rays.
2. Analysis of the observational data allows one to determine the orbital motion in the binary system makes it possible to obtain the mass of the compact object. The data on the observed velocity of the optical companion star is of the most importance. Note that the Newtonian theory is always sufficient for the analysis. The technique of weighing stars in binaries is well known in astronomy. If the mass of the compact component is greater than the maximal possible mass of neutron stars $M_0 \approx 3M_\odot$ (see Section 4.1), then it is a black hole.

It is worth noting that this evidence is somewhat indirect because it does not confront us with the specific relativistic effects that occur near black holes and which are peculiar to black holes alone. However, it is the best that modern astronomy has proposed so far. In spite of these circumstances, we believe that the logic of the arguments is reliable enough.

According to the generally accepted interpretation, we have the necessary observational confirmation only for a few systems at the present time. For these systems, we have strong reasons to

TABLE 1. Black-hole candidates in binary systems [Cherepashchuk (1996)].

System	Spectral type of the optical companion	Orbital period (days)	Mass of the compact companion (in $M_\odot$)	Mass of the optical companion (in$M_\odot$)	X-ray luminosity (erg/sec)
Cyg X-1 (V 1357 Cyg)	O9.7Iab	5.6	7-18	20-30	$\sim 8 \times 10^{37}$
LMC X-3	B(3-6)II-III	1.7	7-11	3-6	$\sim 4 \times 10^{38}$
LMC X-1	O(7-9)III	4.2	4-10	18-25	$\sim 2 \times 10^{38}$
A0620-00 (V616 Mon)	K(5-7)V	0.3	5-17	~ 0.7	$\leq 10^{38}$
GS 2023+338 (V 404 Cyg)	K0IV	6.5	10-15	0.5-1.0	$\leq 6 \times 10^{38}$
GRS 1121-68 (XN Mus 1991)	K(3-5)V	0.4	9-16	0.7-0.8	$\leq 10^{38}$
GS 2000+25 (QZ Vul)	K(3-7)V	0.3	5.3-8.2	~ 0.7	$\leq 10^{38}$
GRO J0422+32 (XN Per 1992= =V518 Per)	M(0-4)V	0.2	2.5-5.0	~ 0.4	$\leq 10^{38}$
GRO J1655-40 (XN Sco 1994)	F5IV	2.6	4-6	~ 2.3	$\leq 10^{38}$
XN Oph 1977	K3	0.7	5-7	~ 0.8	$\leq 10^{38}$

believe that the compact X-ray emitting companions are black holes. Some characteristics of these leading black hole candidates are summarized in Table 1 [according to Cherepashchuk (1996)].

The most plausible masses of compact objects in these systems are considerably larger than $M_0 \approx 3M_\odot$. The strongest candidate are those which have a *dynamical lower limit* of the mass of the compact object (or so-called *mass function* $f(M)$) greater than $3M_\odot$. ¿From this point of view the strongest candidates are GS 2033+338 (f(M)=$6.5M_\odot$), GS 2000+25 (f(M)=$5M_\odot$), and XN Oph 1977 (f(M)=$4M_\odot$).

The total number of systems that are frequently mentioned as possible candidates for black holes of stellar mass is about 20. All seriously discussed candidates are X-ray sources in binary systems. Some of them are persistent, other are transient. Begelman and Rees (1996) summarize the present status as follows: "There is also overwhelming evidence for black holes in our own galaxy, formed when ordinary massive stars die, each weighting a few times as much as the Sun". Most of experts now agree with this unambiguous conclusion.

During the more than 25 years since the discovery of the first black hole candidate Cyg X-1 only a few new candidates have been added. This is in contrast to the rapid increase of the number of identified neutron stars. At present many hundreds of neutron stars have been identified in the Galaxy. About 100 of them are in binary systems. One might conclude that black holes in binary systems are exceedingly rare objects. This is not necessarily true, however. The small number of identified black hole candidates may as well be related to the specific conditions which are necessary for their observable manifestation.

According to estimations the evolutionary stage when a black hole binary continuously radiates X-ray may last only 10^4 years. We can thus detect it only during this short period. In effect, the population of black-hole binaries may be much larger than what we can presently observe. Such systems may be as common as neutron star binaries.

4.4. SUPERMASSIVE BLACK HOLES IN GALACTIC CENTERS

Since the middle of this century astronomers have come across many violent or even catastrophic processes associated with galaxies. These processes are accompanied by powerful releases of energy and are fast not only by astronomic but also by earthly standards. They may last only a few days or even minutes. Most such processes occur in the central parts of galaxies, the galactic nuclei.

About one percent of all galactic nuclei eject radio-emitting plasma and gas clouds, and are themselves powerful sources of radiation in the radio, infrared, and especially, the "hard" (short wavelength) ultraviolet, X-ray and gamma regions of the spectrum. The full luminosity of the nucleus is in some cases $L \approx 10^{47}$erg/s, and millions of times the luminosity of the nuclei of more quiet galaxies, such as ours. These objects were called *active galactic nuclei* (AGN). Practically all the energy of activity and of the giant jets released by galaxies originates from the centers of their nuclei.

Quasars form a special subclass of AGN. Their characteristic property is that their total energy release is hundreds of times greater than the combined radiation of all the stars in a large galaxy. At the same time the average linear dimensions of the radiating regions are small: a mere one-hundred-millionth of the linear size of a galaxy. Quasars are the most powerful energy sources registered in the Universe to date. What processes are responsible for the extraordinary outbursts of energy from AGN and quasars?

Learning about the nature of these objects involves measuring their sizes and masses. This is not easy at all. The central emitting regions of AGN and quasars are so small that telescope view reveals them just as point sources of light. Fortunately quite soon after the discovery of the quasar 3C 273 it was shown that its brightness changed. Sometimes it changes very rapidly, in less than a week. After this discovery, even faster variability (at timescale of a few hours or less) were detected in other galactic nuclei. From these variations one could estimate the dimensions of the central parts of the nuclei that are responsible for radiation. The conclusion was that these regions were not more than a few light-hours in diameter. That is, they are comparable to the solar system in size.

In spite of the rather small linear dimensions of quasars and many galactic nuclei, their masses turned out to be enormous. They were first estimated using formula (4.1). For quasistatic objects the luminosity cannot be essentially greater than L_E. A comparison of the observed luminosity with the expression (4.1) gives an estimate of the lower limit of the central mass. In some quasars this limit is $M \approx (1-10^2) \times 10^7 M_{\odot}$. These estimates are supported by data on the velocities within the galactic nuclei of stars, and gas clouds accelerated in the gravitational fields of the center of the nuclei. We will discuss this in the end of this Section.

Great mass but small linear dimensions prompt the guess that there could be a black hole. This would account for all the extraordinary qualities of these objects. Now it is generally accepted that in AGN there are supermassive black holes with accretion gas (and maybe also dust) disks. One of the most important facts implied by observations, especially by means of radio telescopes, is the existence of directed jets from the nuclei of some active galaxies. For some of the objects there are evidence that radio components move away from the nucleus at ultrarelativistic velocities. The existence of an axis of ejection strongly suggests the presence of some stable compact gyroscope, probably a rotating black hole. In some cases one can observe evidence that there is also precession of this gyroscope. An essential role in the physics of processes in the centers of AGN is probably played by black-hole electrodynamics.

In the model of a supermassive black hole with an accretion disk for AGN one requires sources of fuel – gas or dust. The following sources have been discussed: gas from nearby galactic companion (the result of interaction between the host galaxy and the companion), interstellar gas of the host galaxy, disruption of stars by high velocity collisions in the vicinity of a black hole, disruption of stars by the tidal field of the black hole and some others.

Clearly, the processes taking place in quasars and other galactic nuclei are still a mystery in many respects. But the suggestion that we are witnessing the work of a supermassive black hole with an accretion disk seems rather plausible. Rees (1990) advocates a hypothesis that the massive black holes are not only in the active galactic nuclei but, also in the centers of "normal" galaxies (including nearby galaxies and our own Milky Way). They are quiescent because they are now starved of fuel (gas). Observations show that galactic nuclei were more active in the past. Thus, "dead quasars" (massive black holes without fuel) should be common at the present epoch.

How can these black holes be detected? It has been pointed out that black holes produce cusp-like gravitational potentials and hence they should produce cuspy-like density distributions of the stars in the central regions of galaxies. Some authors have argued that the brightness profiles of the central regions of particular galaxies imply that they contain black holes. However the arguments based only on surface brightness profiles are inconclusive. The point is that a high central number density of stars in a core with small radius can be the consequence of dissipation, and a cusp-like profile can be the result of anisotropy of the velocity dispersion of stars. Thus these properties taken alone are not sufficient evidence for the presence of a black hole.

The reliable way to detect black holes in the galactic nuclei is analogous to the case of black holes in binaries. Namely, one must prove that there is a large dark mass in a small volume, and that it can be nothing other than a black hole. In order to obtain such a proof we can use arguments based on both stellar kinematics and surface photometry of the galactic nuclei.

If the distribution of the mass M and the luminosity L as functions of the radius are known we can consider the mass-to-light ratio M/L (in solar units) as a function of radius. This ratio is well known for different types of stellar populations. As a rule this ratio is between 1 and 10 for elliptical galaxies and globular clusters (old stellar population dominates there). If for some galaxy the ratio M/L is almost constant at rather large radii (and has a"normal" value between 1 and 10) but rises rapidly (toward values much larger than 10) as one approaches the galactic centre, then there is evidence for a central dark object (probably a black hole).

As an example consider galaxy NGC 3115 which is at a distance of 9.2Mpc from us [Kormendy and Richstone (1992)]. For this galaxy $M/L \approx 4$ and almost constant over a large range of radii $r > 4''$ (in angular units). This value is normal for a bulge of this type of galaxy. At radii $r < 2''$ the ratio M/L rises rapidly up to $M/L \approx 40$. If this is due to a central dark mass added to a stellar distribution with constant M/L, then $M_H = 10^{9.2\pm0.5} M_\odot$.

Is it possible to give another explanation of the large mass-to-light ratio in the central region of a galaxy? We cannot exclude the possibility that a galaxy contains a central compact cluster of dim stars. But it is unlikely. The central density of stars in the galaxy NGC 3115 is not peculiar. It is the same as in the centers of globular clusters. The direct observational data (spectra and colors) of this galaxy do not give any evidence of a dramatic population gradient near the center. Thus, the most plausible conclusion is that there is a central massive black hole.

Unfortunately, it is difficult to detect massive black holes in giant elliptical galaxies with active nuclei, where we are almost sure black holes must exist because we observe their active manifestation [Kormendy (1993)]. The reason for this is a fundamental difference between giant elliptical galaxies (the nuclei of some of them are among the most extreme examples of AGN), dwarf elliptical galaxies and spiral galaxies. Dwarf ellipticals rotate rapidly and star velocity dispersions are nearly isotropic. Giant elliptical galaxies do not rotate significantly and they have the anisotropic velocities. It is not so easy to model these dispersions. Furthermore, giant elliptical galaxies have large cores and shallow brightness profiles. Consequently, the projected spectra are dominated by light from large radii, where a black hole has no effect.

The technique described above has been used to search for black holes in galactic nuclei. Another possibility is to observe rotational velocities of gas in the vicinity of the galactic center. So far (the middle of 1997) black hole detections have been reported for the following galaxies: M32, M31, NGC 3377, NGC 4594, Milky Way, NGC3115, M87, and NGC 4258 [for review see Kormendy (1993), see also Miyoshi *et al.* (1995) Kormendy and Richstone (1992)]. Some evidence for a supermassive black hole in NGC 4468B was reported by Kormendy *et al.* (1997).

Special investigations were performed in the case of the galaxy M87 [see Dressler (1989) for review of earlier works and Lauer *et al.* 1992]. This is a giant elliptical galaxy with active nucleus and a jet from the center. At present there is secure stellar-dynamical evidence for a black hole with mass $M \approx 3 \cdot 10^9 M_\odot$ in this galaxy. The Hubble Space Telescope has revealed a rotating

TABLE 2. Estimated masses of black holes in galactic nuclei.

Galaxy	Mass of black hole (in $M_\odot$)
M31	2×10^7
M32	$(2-5) \times 10^6$
Milky Way	3×10^6
NGC 4594	10^9
NGC 3115	10^9
NGC 3377	10^8
M87	2.4×10^9
NGC 4258(M106)	3.6×10^7

disk of gas orbiting the central object in the galaxy [Ford *et al.* (1994), Harms *et al.* (1994)]. The estimated mass of the central object is $M = 2.4 \times 10^9 M_\odot$. The presence of a black hole in M87 is especially important for our understanding of the nature of the central regions of galaxies because in this case we observe also the activity of the "central engine".

Radio-astronomical observations of the nucleus of the galaxy NGC 4258 are of special interest [Miyoshi *et al.* (1995)]. Using radio interferometry technique of observation of maser lines of molecules of water in gas clouds orbiting in the close vicinity of the nucleus, the observers obtained the angular resolution 100 times better that in the case of observations by the Hubble Space Telescope. The spectral resolution is 100 times better as well. According to the interpretation of the observations the center of NGC 4258 harbors a thin disk which was measured on scales of less than one light-year. The mass of the central object is $3.6 \times 10^7 M_\odot$. According to the opinion of Begelman and Rees (1996): "It represents truly overwhelming evidence for a black hole... NGC 4258 is the system for which it is hardest to envisage that the mass comprises anything but a single black hole".

In conclusion we list in Table 2 estimates of masses of black holes in the nuclei of some galaxies [Kormendy (1993), van der Marel (1995), Miyoshi *et al.* (1995)].

Progress in this field is very rapid and in the nearest future our knowledge about evidence of supermassive black holes in galactic nuclei will be more profound.

4.5. PRIMORDIAL BLACK HOLES

Modern astrophysics considers also the third possible type of black holes in the Universe - primordial black holes. These black holes might appear from inhomogeneities at the very beginning of the expansion of the Universe. Their masses can be arbitrary, but primordial black holes with $M \leq 10^{15}$g would have radiated away their mass by the Hawking quantum process in a time $t \leq 10^{10}$years (the age of the Universe) Only primordial black holes with mass $M > 10^{15}$g could exist in the contemporary Universe.

Searches for PBHs attempt to detect a diffuse photon (or another particle) background from a distribution of PBHs or to search directly for the final emission stage of individual black holes. Using the theoretical spectra of particles and radiation emitting by evaporating black holes of different masses, one can calculate the theoretical backgrounds of photons and other particles produced by a distribution of PBHs emitting over the lifetime of the Universe. The level of this background depends on the integrated density of PBHs with initial masses in the considered range.

A comparison of the theoretical estimates with the observational cosmic ray and γ-ray backgrounds place an upper limit on the integrated density of PBHs with initial masses in this range. According to estimates of MacGibbon and Carr (1991),, this limit corresponds to $\approx 10^{-6}$ of the integrated mass density of the visible matter in the Universe (matter in the visible galaxies). The comparison of the theory with other observational data gives more weak limits (for review see Halzen *et al.* 1991, Coyne 1993).

The search for high energy gamma-ray bursts as direct manifestation of the final emission of the evaporating (exploding) individual PBHs has continued for more than 20 years. No positive evidence for existence of PBHs has been reported (see Cline and Hong (1992, 1994).

A population of PBHs whose influence is small today may have been more important in the earlier epochs of the evolution of the Universe. Radiation from PBHs could perturb the usual picture of cosmological nucleosynthesis, distort the microwave background and produced too much entropy in relation to the matter density of the Universe. Limits on the density of PBHs, now or at earlier times, can be used to provide information on the homogeneity and isotropy of the very early Universe, when they were formed. For review see Novikov *et al.* (1979), Carr *et al.* (1994).

The final state of the black hole evaporation is still unclear. There is a possibility that the endpoint of the black hole evaporation is a stable relic. The possible role of such relics in cosmology was first discussed by MacGibbon (1987), for the recent review see Barrow *et al.* (1992).

5. Conclusions

Black holes are absolutely unusual objects. In spite of all the progress the nature of space and time in the black holes remains to a large extent a mystery for us. Some aspects of the problem still look as scientific toys interesting only for specialists.

As for the practical realization of new ideas, I would like to conclude the paper with the remaining that in the middle of the last century even such a practical (now!) thing as electricity looked as a scientific toy. When at that period Mr. Gladstone asked Faraday about the practical worth of electricity, M.Faraday answered: "One day, Sir, you may tax it".

I am an optimist and believe in enormous promise of the new field of research concerned physics and astrophysics of black holes.

6. References

Barrow, I.D., Copeland E.J. and Liddle A.R. (1982) *Phys. Rev.* **D51**, 1741.
Begelman, M.C. and Rees M.I. (1996) *Gravity's Fatal Attraction: Black holes in the Universe* (Scientific American Library) W.H. Freeman.
Bonanno, A., Droz. S., Israel, W. & Morsink, S. (1995) *Proc. Roy. Soc* **A450** 553.
Carr, B.J., Gilbert J. H., and J. E. Lidsey J. E. (1994) *Phys. Rev.* **D50**, 4853.
Cherepashchuk, A.M. (1996) *Russian Physics - Uspekhi*, **39**, 759.
Cline, D.B. and Hong W. (1992) *Astrophys, J*, **401**, L57.
Cline, D.B. and Hong W.(1994) *BAAS* **185**, 116.
Coyne, D.G. (1993) in *International symposium on black holes, membranes, wormholes and superstrings.* (Singapore, World Scientific Publishing Co.), ed. by S. Kalara and D. V. Nanapoulos, p. 159.
Dresler, A. (1989) in D. E. Osterbrock and J. S. Miller (eds), *IAU Symposium 134, Active Galactic Nuclei,* Kluwer, Dordrecht, p.217.
Ford, H. C., Harms R. J., Tsvetanov Z. I., Hartig G. F., Dressel L. L., Kriss G. A., Bohlin R. C., Davidsen A. F., Margon B., and Kochhar A. K. (1994) *Astrophys, J*, **435**, L27.
Frolov, V.P. andNovikov I.D. (1998) Black Holes Physics: Basic concept and New Developments, Kluwer Acad. Publ., in preparation.
Halzen, F., Zas E., MacGibbon J. H., and Weekes T. C. (1991) *Nature* **353**, 807.
Harms, R. J., Ford H. C., Tsvetanov Z. I., Hartig G. F., Dressel L. L., Kriss G. A., Bohlin R. C., Davidsen A. F., Margon B., and Kochhar A. K. (1994) *Astrophys, J*, **435**, L35.
Hawking, S.W. (1974) *Nature*, **248**, 30.
Israel, W. (1997) in the book Relativistic Astrophysics ed. Jones B. and Markovic D., Cambridge University Press, p.173.
Kawaler, S.D., Novikov I.,Srinivasan G. (1997) *Stellar Remnants,* Saas-Fee Advanced Course 25, Springer, p. 237.
Kormendy, J. (1993) In *The Nearest Active Galaxies*, Eds. J.E. Beckman, H. Netzer, and L. Colina (Consejo Superior de Investigaciones Cientificas, Madrid), p. 197.
Kormendy, J. and Richstone D. (1992) *Astrophys. J.* **393**, 559.
Kormendy, J., Bender R., Maggorian J., Tremaine S., Gebhardt K., Richstone D., Dressler A., Faber S. M., Grillmair C., and Lauer T. R. (1997) astro-ph/9703188, 27Mar97.
Lauer, T. R. *et al.* (1992) *Astronomical J.* **103**, 703.
MacGibbon, J.H. (1987) *Nature* **329**, 308.
MacGibbon, J.H. andCarr B.J. (1991) *Astrophys. J.* **371**, 447.
Miyoshi, M., Moran J., Herrnstein J., Greenhill L., Nakai N., Diamond P., and M. Inoue M. (1995) *Nature*, **373**, 127.
Novikov, I.D. and Frolov V.P. (1989) Physics of Black Holes. Kluwer Acad. Publ. (Russian version (19 86, Nauka, Moscow).

Novikov, I. D., A. G. Polnarev A. G., Starobinsky A. A. and Zeldovich Ya. B. (1979) *Astron. and Astrophys.* **80**, 104.
Oppenheimer, J.R. and Snyder H. (1939) *Phys. Rev.* **56**, 455.
Ori, A. (1992) *Phys. Rev. Lett.* **68**, 2117.
Poisson, E., & Israel, W. (1990) *Phys. Rev.* **D41**, 1796.
Rees, M. (1990) Science **247**, N4944, 16 February, p.817.
ndent Thorne, K.S., Price R.H. and Macdonald D.A. (1986) Black Holes: The Membrane Paradigm, Yale Univ. Press, New Haven.
van der Marel, R.P. (1995), In I.Appenzeller (ed.), *Highlights of Astronomy*, **10**, p.527.

JOINT DISCUSSIONS

JD1

Abundance Ratios in the Oldest Stars

Chairperson: B. Barbuy

Editors: B. Barbuy & M. S. Bessel

JOINT DISCUSSION 1: ABUNDANCE RATIOS IN THE OLDEST STARS

BEATRIZ BARBUY
Universidade de São Paulo
C. P. 3368, São Paulo 01060-970, Brazil

AND

MICHAEL S. BESSELL
Mt. Stromlo Observatory
Private Bag, Weston Creek PO, 2611 Canberra, ACT, Australia

1. Introduction

Joint Discussion 1 was supported by Division IV (Stars) and Commission 29 (Stellar Spectra), and co-supported by Commissions 28 (Galaxies), 36 (Theory of Stellar Atmospheres) and 37 (Stellar Clusters and Associations). Members of the scientific organizing committee were: N. Arimoto (Japan), B. Barbuy (Brazil), T. Beers (USA), J. Bergeron (Germany), M. Bessell (Australia), R. Cayrel (France), G. Gilmore (UK), B. Gustafsson (Sweden), F. Matteucci (Italy), P. Nissen (Denmark), and M. Rich (USA).

The inspiration for this meeting was the growing overlap and connections between previously separate areas of astrophysical research, namely, studies of stellar abundances, the bulges of galaxies, the gaseous components of nearby galaxies and the clouds (some of which may be primordial) responsible for the narrow absorption lines in quasars.

The signature of the early chemical evolution of our Galaxy is imprinted in the abundance ratios of the oldest stars. We recall that element ratios are determined by a mix of the relative rates of different types of supernovae, the stellar IMF, and the relative histories of star formation rates and gaseous flows, and thus encapsulate much of the history of star formation and ISM evolution in galaxies. Hence, abundance ratios in stars are a primary probe for testing theories of galaxy formation and evolution.

We do not know how the Galaxy formed: both the Eggen, Lynden-Bell & Sandage (1962) and the Searle & Zinn (1978) scenarios may be accommodated in the recent proposal of van den Bergh (1993) where the inner Galaxy follows ELS, whereas the outer Galaxy formation conforms to the Searle-Zinn proposition. A combination of abundance ratios, ages derived from colour-magnitude diagrams, and kinematical properties, can give us the required information to trace the past history of our Galaxy. We note here, that although stellar evolution and model atmospheres are not discussed in the proceedings both topics are of fundamental underlying importance. Model atmospheres are used to derive temperatures, colors and bolometric corrections of stars that are used not only in abundance analyses but also in deriving the ages of stars by comparing CM diagrams with HR diagrams. This process is under close scrutiny because of the apparent difference between the ages of the oldest stars and the expansion age of the universe.

In the present JD, abundance ratios in the halo and bulge, which are the oldest components of our Galaxy, are reviewed. In particular, the abundances of CNO, α-elements, r- and s-elements relative to Fe in the halo and bulge as compared to the disk. Let us note that, due to improved model atmospheres on the one hand, and selections of stellar samples based on kinematical properties on the other, the most recent abundance results, such as that of Nissen & Schuster (1997), show unexpected features.

A comparison between the abundances found in elliptical galaxies and from QSO absorption lines and those predicted by chemical evolution models could help us understand the early chemical evolution of ellipticals and the intergalactic gas.

J. Andersen (ed.), Highlights of Astronomy, Volume 11A, 47–48.

Elliptical galaxies are currently interpreted in essentially three ways with regard to their stellar populations: (i) an old stellar population, (ii) a combination of an old plus a young stellar component, (iii) an originally old stellar population, strongly modified by merging processes with other galaxies - the main post-merger characteristic is an intermediate age stellar population formed during the merging process. Despite the different evolutionary scenarios suggested by these interpretations, it is important to note how homogeneous most of these objects are, a large percentage of them following tight relations in the fundamental plane (Burstein et al. 1986; Djorgovski & Davis 1987). In particular, how does one explain the tight correlation between the metallicity indicator Mg_2 versus velocity dispersion and other quantities, in view of a possible variation of [Mg/Fe] in those galaxies ?

The determination of abundances in intergalactic clouds at increasing distances in the line-of-sight of quasars enables the direct study of the chemical evolution with time towards the past. A comparison between these abundances and those found in Galactic metal-poor stars of similar metallicities may lead to a better understanding of chemical evolution as a whole in different parts of our Universe.

Finally, it is important to point out that this is the last General Assembly which will precede the general availability of telescopes of 8-12m class. As analysis of the elemental abundances in the oldest stars, quasar absorption lines and abundances deduced from population synthesis in the bulges of ellipticals/spirals are stated goals of essentially all these telescopes, 1997 is a particularly important time to take stock of our current knowledge in these fields. Future GAs will discuss the progress made by the years 2000 or 2003 following the technological leaps in faint object spectroscopy.

References

Burstein, D., Dressler, A., Faber, S.M., Davies, R., Lynden-Bell, D., Terlevich, R., Wegner, G.: 1986, in *Galaxy Distances and Deviations from Universal Expansion*, eds. B. Madore, B. Tully, Reidel, p. 123

Djorgovski, S., Davis, M.: 1986, in *Galaxy Distances and Deviations from Universal Expansion*, eds. B. Madore, B. Tully, Reidel, p. 135

Eggen, O.J., Lynden-Bell, D., Sandage, A. 1962, *ApJ*, **136**, 748

Nissen, P.E., Schuster, W.J. 1997, *A&A*, **326**, 751

Searle, L., Zinn, R. 1978, *ApJ*, **225**, 357

van den Bergh, S. 1993, *ApJ*, **411**, 178

CHEMICAL EVOLUTION AND EXTREMELY METAL-POOR STARS

ANDREW MCWILLIAM
Observatories of the Carnegie Institution of Washington
813 Santa Barbara St., Pasadena, CA. 91101, U.S.A.
andy@marmite.ociw.edu

1. Introduction

Early abundance studies (e.g. Pagel 1968) showed that neutron-capture heavy elements ($Z > 30$) are present in halo stars, but deficient relative iron. Truran (1981) argued that at low [Fe/H] the chemical enrichment time scale was shorter than the lifetime of low-mass AGB progenitors, which are the main source of solar system heavy elements. He proposed that in the halo the heavy elements were produced by high mass stars, in type II supernova events (SNII), by rapid neutron capture nucleosynthesis (the r-process).

Spite & Spite (1978) investigated the trend of heavy element abundances with metallicity, from a small sample of halo stars. They found that at [Fe/H]~ -1.5 the halo [heavy element/Fe] ratio is approximately solar; but at lower [Fe/H] there is a roughly linear decrease of [heavy element/Fe] with declining [Fe/H]. Subsequent observations confirmed the general trend of heavy elements in the halo: [M/Fe]$\sim$0 down to [Fe/H]~ -2, followed by a linear decline in [M/Fe] to lower [Fe/H] (e.g. Gilroy et al 1988, Lambert 1987).

Additional evidence for the role of SNII in halo heavy element synthesis comes from the trend of [Eu/Fe] with [Fe/H]. Europium is an almost pure r-process element (Käppeler et al. 1989) and its abundance trend with metallicity is similar to the α element trend (e.g. O and Mg made in massive stars). The element ratios show an increase in [M/Fe] as [Fe/H] decreases from 0 to -1; below this point [Eu/Fe] and [α/Fe] remain constant at $\sim$+0.3 dex. For α elements this behavior is thought to be due to the change in the relative contributions from type II SN and type Ia SN in the disk and halo (Tinsley 1979). The trend for Eu also indicates production by massive stars (e.g. SNII). Near [Fe/H]~ -2.5 Eu appears to decline relative to [Fe/H] (like other heavy elements, but unlike the α elements). This abundance trend has been used to constrain the numerous proposed astrophysical sites of the r-process (e.g. Mathews & Cowan 1990).

Evidence for a dispersion in heavy element abundances at low metallicity was first uncovered by Griffin et al. (1982), who found one star, HD 115444, with a huge excess of Eu. Gilroy et al (1988) claimed a dispersion in [heavy element/Fe] for metal-poor halo giants, as did Gratton & Sneden (1990, 1991) and Ryan et al (1991). However, this was not universally accepted: Baraffe & Takahashi (1993) compared the results for individual stars from different studies, and found that the observed dispersion was consistent with the measurement errors. Also, Gratton & Sneden (1994) found no dispersion in [Sr/Fe] for their study of halo stars with [Fe/H]$= -3$ to -0.5 dex, and concluded that measurement errors in the earlier works were responsible for the apparent dispersion.

2. Extremely Metal-Poor Stars

2.1. HEAVY ELEMENT ABUNDANCES

In our study (McWilliam et al 1995, MPSS95) we analyzed spectra of 33 halo red giants from the sample of Beers et al (1992, BPS), with [Fe/H] ranging from -4 to -2. The measurement uncertainty of the abundance analysis was evaluated by use of a Taylor series expansion, propagated from the S/N of the spectra. We found a 2.5 dex range in [Sr/Fe] (Figure 1), and a typical abundance uncertainty near ±0.2 dex; this confirmed previous claims of an intrinsic abundance dispersion, and was consistent with the early heavy element abundance trends.

J. Andersen (ed.), Highlights of Astronomy, Volume 11A, 49–52.

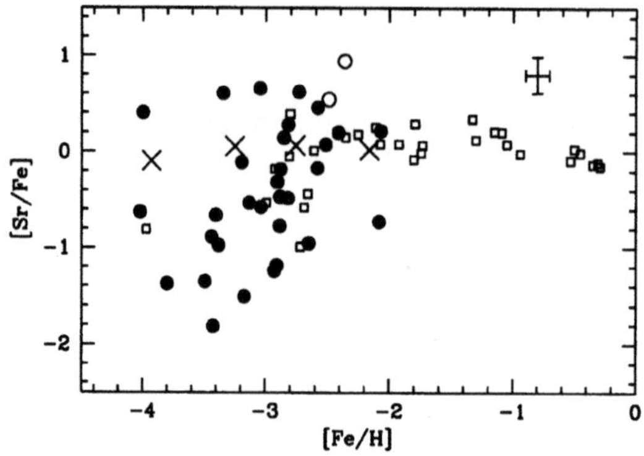

Figure 1. [Sr/Fe] as a function of [Fe/H]. Filled circles from MPSS95, open squares from Gratton & Sneden (1990, 1991). Open circles indicate CH stars, which do not reflect their original composition. Crosses show the mean [Sr/Fe] values for bins 0.5 dex wide in [Fe/H]; note that the average [Sr/Fe] ratio is approximately independent of [Fe/H].

The intrinsic heavy element dispersion cannot be due to AGB nucleosynthesis, such as seen in population I barium stars. The most compelling reason for this is that the heavy elements in these stars exhibit a nearly pure r-process abundance pattern (Figure 2), whereas AGB nucleosynthesis is achieved via the s-process. Also, the frequency of population I barium stars is too low to explain the halo [Sr/Fe] dispersion. The primordial origin of the dispersion is also supported by a similar [Sr/Fe] dispersion found in halo dwarfs (Ryan et al. 1996).

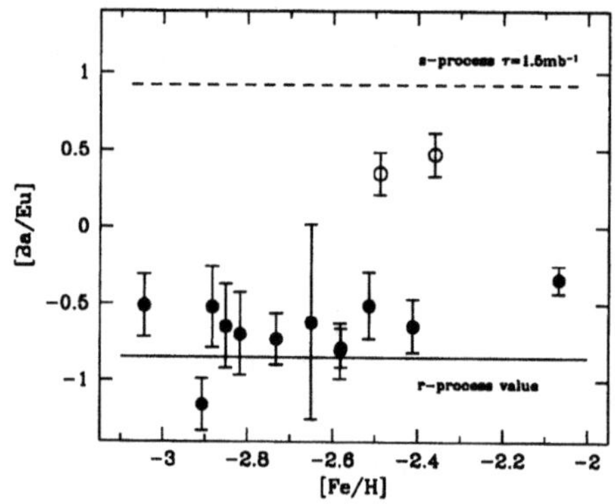

Figure 2. [Ba/Eu] for stars in the MPSS95 sample (McWilliam 1997). Open circles indicate CH stars, which are affected by AGB nucleosynthesis and do not reflect their primordial composition.

The observed 300-fold range in [Sr/Fe] seen in Figure 1 sets the minimum range of Sr/Fe yield ratios from the progenitor SN. The observed dispersion also requires that homogenization processes at these low metallicities were insufficient to mask the abundance patterns from individual SN events. Furthermore, since the largest [heavy element/Fe] ratios are approximately 20 times the asymptotic halo values, the SN producing these large ratios must be correspondingly rare. The dispersion in Figure 1 shows that homogenization did occur by [Fe/H]~ -2.5 dex.

These observations suggests a stochastic model of chemical evolution, in which stars were born from isolated regions. In each region the SN events were drawn from a random sampling of the SN yield function. As the number of SN events increased the composition approached the average value, with complete homogenization when the full range of SN yields was sampled. The stochastic model is supported by a number of observations:

1. The trend of decreasing dispersion in Sr/Fe with increasing [Fe/H].
2. The average [Sr/Fe] ratio is roughly constant; although the number of stars with high Sr/Fe ratio is small.

3. The presence of stars with highly unusual chemical composition: CS 22892-052 with a large excess of heavy elements, CS 22949-037 which is apparently deficient in iron-peak elements by ~ 1 dex, and CS 22897–008 which has an unusually high [Sr/Ba] ratio.
4. The shape of the observed [Sr/Fe] function between [Fe/H]=−2.5 and −3.0 can be matched with a stochastic model (Searle & McWilliam 1997).

For a very simple-minded stochastic chemical evolution model, in which the iron yield is constant for all SN types and mixes with the same mass of ISM, the dispersion in the Sr/Fe number ratio is given by:

$$\sigma_n = \sigma_1/\sqrt{n} \tag{1}$$

Where n is the number of SN events contributing to the composition, and σ_n is the dispersion of the gas after n SN events. Of course σ_1 represents the Sr/Fe yield function from individual SN events, which the observations show is not Gaussian. Equation 1 indicates that if stars near [Fe/H]=−3.5 are composed of material from many SN events then the dispersion in supernova Sr/Fe yield ratios must be even larger than the presently observed values.

The model can be used to estimate the mass of the regions, which formed the stars with [Fe/H] between −4 and −2.5: Because the largest observed heavy-element/Fe ratio is ~20 times the asymptotic value, the frequency of the progenitor SN events can be no more than 1/20 of the total population. Since homogenization is observed to occur near [Fe/H]=−2.5 the full range of SN yields is approximately sampled by this metallicity, which occurs in at least 20 SN events. If each SNII produces ~ 0.1 $M_\odot$ of Fe, then in 20 events 2.0$M_\odot$ of Fe is produced; to dilute this to [Fe/H]=−2.5 requires 10^5–10^6 $M_\odot$ of hydrogen. Curiously, this is similar to the mass of giant molecular clouds seen in the Galaxy today. In this case a single SN event should result in material with [Fe/H]~ −3.8, which is close to the value of [Fe/H]=−4 predicted by Audouze & Silk (1995) from the physics of SN ejecta.

2.2. IRON PEAK ELEMENT ABUNDANCES

For Iron-peak elements MPSS95 found three new trends: a decline in [Cr/Fe] and [Mn/Fe] ratios and an increase in [Co/Fe] below [Fe/H]~ −2.5. This is additional evidence that chemical evolution below [Fe/H]~−2.5 was significantly different from the main population II component, and could be interpreted as evidence of population III or an early population II. Because the iron-peak ratios at low metallicity are not seen at higher metallicity the composition of stars in the MPSS95 sample cannot be due to a dilution of population II composition with pure hydrogen.

The unusual composition shows that SN iron-peak yield ratios must have a range, at least as large as the observations.

A viable chemical evolution model must simultaneously explain the tight correlation of the [Co/Cr] ratios (Figure 3) and the large dispersion in [Sr/Fe] (Figure 1). Three possible explanations are:

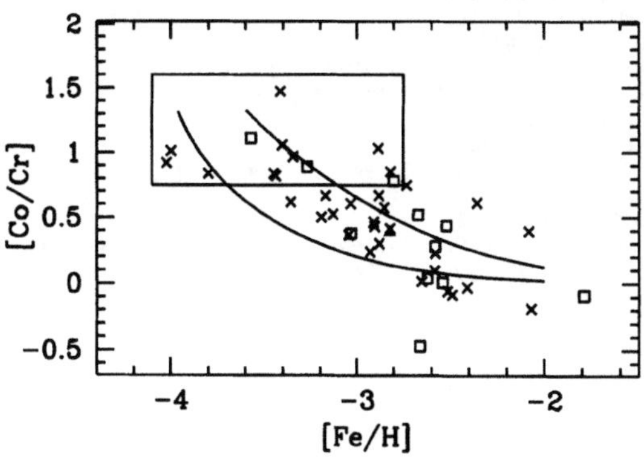

Figure 3. [Co/Cr] versus [Fe/H] from MPSS95. The loci indicate the composition path of material with high [Co/Cr] when diluted with normal iron-peak composition. The box indicates the approximate location of a primordial composition resulting from the first generation of SN.

1. Metallicity-dependent iron-peak yields (MPSS95). As the metallicity increased from [Fe/H]=−4 to −2.5 the Co, Cr, and Mn SNII yields slowly changed. This model requires many generations of SN events, and may be inconsistent with only 20 SNII events required by [Fe/H]= −2.5.
2. SN iron-peak yields are tightly correlated with explosion energy (Ryan et al. 1996). The most energetic SN events, with high [Co/Cr] yield ratios, mix their ejecta with more pure hydrogen than low-energy (low [Co/Cr]) SN events. This naturally leads to a trend of [Co/Cr] with [Fe/H], but suffers from the difficulty imposed by the small observed dispersion in [Fe/H] at a given [Co/Cr]. Figure 3 suggests that the dispersion in mixing of SN ejecta with pure hydrogen is approximately 0.3 dex.
3. The clouds evolved in isolation, with no exchange or addition of new material, and the ejecta of the first generation SN (formed from zero metallicity gas) was characterized by a high [Co/Cr] ratio (Searle & McWilliam 1997). In this model the second generation stars form over a large [Fe/H] range, but with high [Co/Cr] ratios (indicated by a box in Figure 3). Subsequent SN ejecta were characterized by normal [Co/Cr] ratios, and "diluted" the first generation ejecta, as indicated by the loci in Figure 3. The dilution loci form a bottle neck, with a fairly small dispersion in [Fe/H] at [Co/Cr]∼0.5 dex.

Note that #3 is consistent with the stochastic model, and small numbers of SN events, used to understand the [Sr/Fe] dispersion. This mechanism is astonishingly similar to the results of Ostriker & Gnedin (1997) who predicted population III in primordial gas clouds with star formation from cooling by H_2 gas, followed by re-ionization and a subsequent cooling for population II. The mass of the star-forming clouds in their model was 10^6–10^7 $M_\odot$, and reached a metallicity of [Fe/H]∼−3.7.

3. Summary

- Supernovae element yield ratios show a large dispersion.
- Stars near [Fe/H]=−3.5 may reflect the composition of individual SN events, born in clouds of mass 10^5–10^6 $M_\odot$. If not, then stars with [Fe/H]≤−5 should exist, with a larger [Sr/Fe] dispersion than seen in the most metal-poor stars presently known.
- The chemical composition of stars below [Fe/H]=−2.5 reflects either an early population II or population III, consistent with the population III predictions of Ostriker & Gnedin (1997).
- Improved abundance measurements from superior spectra and an increased sample of extremely metal-poor stars would be useful for testing these ideas.

References

Audouze, J., & Silk, J. 1995, *ApJ*, **451**, L49
Baraffe, I., & Takahashi, K. 1993, *A&A*, **280**, 476
Beers, T.C., Preston, G.W., & Shectman S.A. 1992, *AJ*, **103**, 1987
Gilroy, K.K., Sneden, C., Pilachowski, CA, & Cowan, J.J. 1988, *ApJ*, **327**, 298
Gratton, R.G., & Sneden, C. 1990, *A&A*, **234**, 366
Gratton, R.G., & Sneden, C. 1991, *A&A*, **241**, 501
Gratton, R.G., & Sneden, C. 1994, *A&A*, **287**, 927
Griffin, R., Griffin, R., Gustafsson, B., & Vieira, T. 1982, MNRAS, *198*, **637**
Käppeler, F., Beer, H., & Wisshak, K. 1989. *Rep. Prog. Phys.*, **52**, 945
Lambert, D.L. 1987, *Astron. Astrophys.*, **8**, 103
Mathews, G.J. & Cowan, J.J. 1990, *Nature*, **345**, 491
McWilliam, A., Preston, G.W., Sneden, C., & Searle, L. 1995, *AJ*, **109**, 2736 (MPSS95)
McWilliam, A. 1997, *AJ*, in press
Ostriker, J.P. & Gnedin, N.Y. 1997, *ApJ*, **486**, 581
Pagel, B.E.J. 1968. In *Origin and Distribution of the Elements*, ed. L.H. Ahrens, pp 195–204
Ryan, S.G., Norris, J.E., & Beers, T.C. 1996, *ApJ*, **471**, 254
Ryan, S.G., Norris, J.E., & Bessell, M.S. 1991, *AJ*, **102**, 303
Searle, L. & McWilliam, A. 1997, in progress
Spite, M. & Spite, F. 1978, *A&A*, **67**, 23
Tinsley, B.M. 1979, *ApJ*, **229**, 1046
Truran, J.W. 1981, *A&A*, **97**, 391

ABUNDANCE RATIOS IN METAL-POOR GLOBULAR CLUSTERS: DEEP MIXING AND ITS EFFECT ON STELLAR POPULATIONS OF THE GALACTIC HALO

ROBERT P. KRAFT
UCO/Lick Observatory
University of California, Santa Cruz, CA 95064

1. Primordial vs Evolutionary Scenarios

Only a bit more than 25 years ago, it seemed possible to assume that all Galactic globular clusters were chemically homogeneous. There were indications that star-to-star Fe abundance variations existed in ω Cen, but this massive cluster appeared to be unique. Following Osborn's (1971) initial discovery, Zinn's (1973) observation that M92 asymptotic giant branch (AGB) stars had weaker G-bands than subgiants with equivalent temperatures provided the first extensive evidence that there might be variations in the abundances of the light elements in an otherwise "normal" cluster. Since then star-to-star variations in the abundances of C, N, O, Na, Mg and Al have been observed in all cases in which sample sizes have exceeded 5-10 stars, e.g., in clusters such as M92, M15, M13, M3, ω Cen, M10 and M5. Among giants in these clusters one finds large surface O abundance differences, and these are intimately related to differences of other light element abundances, not only of C and N, but also of Na, Mg and Al (cf. reviews by Suntzeff 1993, Briley et al 1994, and Kraft 1994). The abundances of Na and O, as well as Al and Mg, are anticorrelated. Prime examples are found among giants in M15 (Sneden et al 1997), M13 (Pilachowski et al 1996; Shetrone 1996a,b; and Kraft et al 1997) and ω Cen (Norris & Da Costa 1995a,b).

These observed anticorrelations almost certainly result from proton- capture chains that convert C to N, O to N, Ne to Na and Mg to Al in or near the hydrogen fusion layers of evolved cluster stars. But which stars? An appealing idea is that during the giant branch lifetimes of the low-mass stars that we now observe, substantial portions of the stellar envelopes have been cycled through regions near the H-burning shell where proton-capture nucleosynthesis can occur. This so-called "evolutionary" scenario involving deep envelope mixing in first ascent red giant branch (RGB) stars has been studied by Denissenkov & Denissenkova (1990), Langer & Hoffman (1995), Cavallo et al (1996, 1997) and Langer et al (1997). The mixing mechanism that brings proton-capture products to the surface is poorly understood (Denissenkov & Weiss 1996, Denissenkov et al 1997, Langer et al 1997), but deep mixing driven by angular momentum has been suggested (Sweigart & Mengel 1979, Kraft 1994, Langer & Hoffman 1995, Sweigart 1997). Arguments in favor of the evolutionary picture include the constancy of C+N+O in M13 (Brown et al 1991, Smith et al 1996) and M15 (Sneden et al 1997), the constancy of Mg+Al in M13 (Shetrone 1996a), the steady average decline of C with evolutionary state in M92 (Carbon et al 1982, Langer et al 1986), M15 (Trefzger et al 1983), and NGC 6397 (Briley et al 1990), and the shift toward higher Na abundances with advancing evolutionary state in M13 (Pilachowski et al 1996, Kraft et al 1997).

On the other hand, the so-called "primordial" scenario attributes these required abundance redistributions to a preceding generation of higher mass cluster stars, whose interiors can more easily attain the very high temperatures needed to sustain advanced proton-capture syntheses. Several lines of evidence suggest the need for a primordial scenario. For example, M5 giants that have high O abundances (Sneden et al 1992) in some instances also have high N abundances (Briley & Smith 1993, Smith et al 1997), which suggests that there are primordial variations of N. Even more strikingly, there are strong indications that the main sequence stars in 47 Tuc exhibit a positive correlation between N and Na (Briley et al 1996). It is hard to understand how deep mixing could provide a mechanism for altering surface composition among main sequence stars. Finally, M15 giants exhibit a bi-modal distribution of Eu (Sneden et al 1997), an element that surely was

J. Andersen (ed.), Highlights of Astronomy, Volume 11A, 53–57.

synthesized in a high temperature Type II supernova environment, and which could not have been altered by deep mixing in low-mass red giants. (Interestingly, these Eu variations are completely unrelated to the variations of O, N, Na, Mg and Al also seen in this cluster).

The Sweigart & Mengel (1979) mixing theory predicts that the efficiency of deep mixing should increase with decreasing metallicity. Observations of carbon isotope ratios (Briley et al 1994) and of Na and O abundances of red giants in relatively metal-rich clusters such as M71 (Sneden et al 1994) tend to support this idea (cf Bell et al 1979, Dickens et al 1979, Bell & Dickens 1980). Considered together, the observations currently suggest that all clusters begin with primordial variations in at least some proton capture elements, and that these variations are modified or enhanced by deep mixing in the giant star stage of evolution, and that the degree of modification in general increases with decreasing metallicity. Thus the "debate" between the validity of primordial and evolutionary scenarios may be no debate at all. Both scenarios are needed to explain what we see in the spectra of globular cluster stars.

2. Deep Mixing and the Second Parameter Problem

The "second parameter problem" arises from the observation that the morphology of the horizontal branch (HB) in a cluster color-magnitude diagram can vary among clusters having the same Fe-peak metallicity (Sandage & Wildey 1967, van den Bergh 1967). The triplet of clusters M3, M13 and NGC7006 provides one of the best examples: they have Fe-peak metallicities ([Fe/H] $\sim$ -1.6) differing by less than 0.1 dex (Kraft et al 1993, Smith et al 1997), and although their RGB sequences are quite similar, their HB morphologies are distinctly different. The HB of M3 is populated more-or-less uniformly from (B-V) $\sim$ -0.2 all the way to the RGB, and contains nearly 200 RR Lyraes. The HB of M13, on the other hand has no stars redward of the RR Lyrae domain and almost no RR Lyraes; its HB is confined to stars lying blueward of the instability strip. The HB of NGC7006 is opposite to that of M13, in that a large fraction of its HB population lies redward of the RR Lyraes. Another factor besides metallicity must determine the HB morphology, and this unknown factor has come to be called the "second parameter".

Following several years of discussion and debate, cluster age has emerged as the most likely factor controlling the second parameter (Searle & Zinn 1978; Preston et al 1991; Lee et al 1988,1994; Zinn 1993). If age differences between globular clusters provide the correct explanation, important consequences follow for the history of the Galaxy, since the clusters having "excessively" red HBs for their metallicity are preferentially found in the outer Galactic halo. This in turn implies that the age of the halo increases with decreasing Galactocentric distance.

Whether age is the unique driver of the second parameter effect has not been determined. The outer halo clusters Pal 12, Rup 106, Arp 2 and Ter 7 are almost certainly 3-4 Gyr younger than typical Galactic globular clusters (Gratton & Ortolani 1988, Stetson et al 1989, Buonanno et al 1990, Da Costa et al 1992). However, these four anomalously young clusters have kinematics associated with the Magellanic Stream, and may therefore be objects in the process of being accreted to the Galaxy. As for the other halo clusters, Richer et al (1996) have considered 35 for which there are well-established deep CCD-based color-magnitude arrays, and have concluded that there is no compelling evidence for an age gradient or for age being the unique driver of the second parameter. Similar conclusions have been reached also by Stetson et al (1996).

If age is not entirely responsible for the second parameter effect, what else is plausible as an explanation? High resolution spectroscopic studies have revealed an interesting distinction between the giants in the "blue HB" clusters M10 and M13 (Kraft et al 1992, 1993, 1995), the "normal HB" cluster M3 (Kraft et al 1992), and the "red HB" cluster NGC7006 (Smith et al 1997). First-ascent RGB stars in M10 and especially in M13, in contrast with M3 and especially NGC7006 giants in the same evolutionary stage, show far more evidence for deep mixing of their envelopes through the interior CNO-cycle hydrogen-burning shell. Of importance in the present context is that the processing that converts Ne into Na and Mg into Al may also increase He at the expense of H, so that if mixing progresses as far as the Al (Shetrone 1996a, Smith et al 1997) and Na observed enhancements indicate, the He mass fraction in the envelopes of some HB progenitor giants in M13 and M10, in contrast to most giants in M3 and NGC7006, might be raised sufficiently to affect the HB morphology (Langer & Hoffman 1995, Sweigart 1997).

These results suggest that blue HBs will be generated if the precursor red giants dredge up large amounts of CNO-processed material in which the He mass fraction has been increased. Since classical theories of stellar evolution do not predict the existence of such deep mixing, some non-standard mechanism must be sought, and the most promising candidate seems to be some kind of rotationally induced instability. Indirect support for this mechanism is found in the observation (Peterson 1983, Peterson et al 1995) that some HB stars in M13 exhibit rotational velocities up to V sin i $\sim$ 40 km/s, whereas HB stars in M3 show little evidence for the existence of rotation. Further indirect evidence suggesting that He might be enhanced as a result of deep mixing comes from the study of M15 giants and HB stars. Some M15 giants exhibit striking deficiencies of O and Mg with corresponding enhancements of Na and Al (Sneden et al 1997). At the same time, masses derived for very blue HB stars in M15 (Moehler et al 1995) are difficult to account for unless the He/H ratio is increased to a value considered unacceptable by these authors, but which would be feasible if the atmospheres of those stars had been enriched in He by amounts consistent with the Al and Na enhancements discussed by Sneden et al (Langer et al 1997). Furthermore, if some giants mix more than others, the resulting increase in the He/H ratio would produce a progessive "blueing" of stars in their locations along the HB. HB morphology is extremely sensitive to He abundance (Rood 1973, Dorman 1992, Sweigart 1997); variations in He mass fraction may account for the widths of HBs in clusters such as M13 (Sweigart 1997).

3. Deep Mixing in Globular Cluster Giants vs Halo Field Giants

A further stellar populations "anomaly" emerges when one compares the [el/Fe]-ratios of light elements in metal-poor giants of the halo field with those of globular clusters. Some degree of deep mixing surely takes place in the field giants, since the $^{12}C/^{13}C$-ratio approaches its equilibrium value with advancing evolutionary state (Sneden et al 1986). But when field and cluster giants are compared at comparable [Fe/H] and evolutionary state, one finds no evidence either for O and Mg depletions or Na and Al enhancements such as one finds in globular clusters (Kraft 1994, Pilachowski et al 1996, Shetrone 1996a, Kraft et al 1997). Does this mean that field halo giants have, on the average, less angular momentum than cluster giants? And if some clusters have HB stars with enhanced He as a result of deep mixing, whereas field HB stars have no such enhancement, should not one exercise caution when comparing HB stars in clusters with HB stars in the halo field? Of special interest is the morphology of the halo field HB. It is known that the mean metallicity of the halo field is similar to that of M3, M13 and M10 (e.g. Laird et at 1993), but is the HB morphology of the field similar to one of these clusters or some combination of all three? Unfortunately, numerous selection effects (cf Kinman et al 1994) thwart attempts to establish the field morphology with accuracy sufficient to address this question.

4. Concerns and Caveats

4.1. THE MG VS AL ANTICORRELATION AND DEEP MIXING

Shetrone (1996b) demonstrated that among M13 giants in which Al abundances were very high, Mg was depleted at the expense not of ^{25}Mg or ^{26}Mg, but rather of the ^{24}Mg isotope. If present nuclear reaction rates and giant star models are correct (Cavallo et al 1996, Langer et al 1997), production of ^{27}Al from ^{24}Mg requires a hydrogen-burning shell temperature near 70MK, whereas the models achieve a maximum temperature of only 55MK. Langer et al (1997) suggested that hydrogen shell burning might after all be unstable (Bolton & Eggleton 1973, von Rudloff et al 1988), thus providing episodes of elevated shell temperature; Langer et al and Dennisenkov et al (1997) also questioned the reliability of some of the nuclear reaction rates (Zaidins & Langer 1997). Until these issues have been explored further, the need for the destruction of ^{24}Mg, rather than $^{25,26}Mg$, remains an obstacle to full acceptance of deep mixing as a driver of the second parameter effect. Current attempts to explain the ^{24}Mg vs ^{27}Al anticorrelation as a combination of primordial and evolutionary effects appear promising (Dennisenkov et al 1997).

4.2. "POLLUTION" ALONE IS PROBABLY NOT THE ANSWER

On a purely primordial scenario, one might imagine that the over-abundances of N, Na and Al seen in many globular cluster giants result from the pollution of the pre-existing cluster medium by the outpourings of massive stars. Presumably that "pre-existing" medium would ab initio reflect the abundance ratios associated with the ejecta of Type II supernovae, e.g., overabundances of O and alpha elements such as Mg, Ca, Si and Ti. But the globular cluster giants we now see are also seriously deficient in O, C and Mg. On a purely primordial picture, one would have to find some way to create and maintain subregions within the gas having these deficiencies until significant numbers of low mass stars could be formed from them. This requirement does not seem to have been addressed in the literature.

4.3. WHENCE COMETH THE "EXCESS" ANGULAR MOMENTUM OF CLUSTER STARS?

If angular momentum were conserved in shells as evolution proceeds, then the progenitors of rapidly rotating cluster HB stars would have to have had values of V sin i of nearly 100 km/s when they were objects on the main sequence. Nothing is known of the present rotational velocities of cluster main sequence stars; field subdwarfs are known to be slow rotators (Smith 1996). Could cluster stars be spun up by stellar encounters, or perhaps by "swallowing" stellar or planetary companions as their radii increase owing to stellar evolution? Such a process must take into account an essential constraint: virtually all giants in M13 mix as they approach the red giant tip, so any mechanism advanced has to be universal in character.

4.4. IF NOT MERIDIONAL CIRCULATION, WHAT THEN?

Sweigart and Mengel (1979) suggested that deep mixing was driven by meridional circulation which could not take place on the first giant branch until the hydrogen-burning shell had advanced through the abundance discontinuity left behind at the close of main sequence core hydrogen burning. Although dependent on metallicity, the luminosity at which this event takes place is near the level of the HB. However, in M92 (Langer et al 1986) and NGC 6397 (Briley et al 1990) there is strong evidence that C is being depleted already in stars at the base of the giant branch, well before meridional circulation can be invoked as a mixing mechanism.

References

Bell, R. A., Dickens, R.J., Gustafsson, B. 1979, *ApJ*, **229**, 604.
Bell, R. A. & Dickens, R. 1980, *ApJ*, **242**, 657.
Bolton, A. J. & Eggleton, P. P. 1973, *A&A*, **24**, 429.
Briley, M. M., Bell, R.A., Hoban, S., Dickens, R.J. 1990, *ApJ*, **359**, 307.
Briley, M. M. & Smith, G. H. 1993, *PASP*, **105**, 1260.
Briley, M. M., Bell, R.A., Hesser, J.E., Smith, G.H. 1994, *Can. J. Phys.*, **72**, 772.
Briley, M. M., Smith, V.V, Suntzeff, N.B., Lambert, D.L., Bell, R.A., Hesser, J.E. 1996, *Nature*, **383**, 604.
Brown, J. A., Wallerstein, G., Oke, J.B. 1991, *AJ*, **101**, 1693.
Buonanno, R., Buscema, G., Fusi Pecci, F., Richer, H.B., Fahlman, G.G. 1990, *AJ*, **100**, 1811.
Carbon, D. F., Langer, G.E., Butler, D., Kraft, R.P., Trefzer, Ch., Suntzeff, N.B. 1982, *ApJS*, **49**, 207.
Cavallo, R. M., Sweigart, A.V., Bell, R.A. 1996 *ApJ*, **464**, L79.
Cavallo, R. M., Pilachowski, C.A., Rebolo, R. 1997 *PASP*, **109**, 226.
Da Costa, G., Armandroff, T.E., Norris, J.E. 1992, *AJ*, **104**, 154.
Denissenkov, P. A. & Denissenkova, S. N. 1990, *Sov. Astr. Lett.*, **16**, 275.
Denissenkov, P. A. & Weiss, A. 1996, *A&A*, **308**, 773.
Denissenkov, P. A., Weiss, A., Wagenhuber, J. 1997, *A&A*, **320**, 115.
Dickens, R., Bell, R.A., Gustafsson, B. 1979, *ApJ*, **232**, 428.
Dorman, B. 1992, *ApJS*, **80**, 701.
Gratton, R. G. & Ortolani, S. 1988, *A&AS*, **73**, 137.
Kinman, T. D., Suntzeff, N.B., Kraft, R.P. 1994, *AJ*, **108**, 1722.
Kraft, R. P. 1994, *PASP*, **106**, 553.
Kraft, R. P., Sneden, C., Langer, G.E., Prosser, C.F. 1992, *AJ*, **104**, 645.
Kraft, R. P., Sneden, C., Langer, G.E., Shetrone, M.D. 1993, *AJ*, **106**, 1490.
Kraft, R. P., Sneden, C., Langer, G.E., Shetrone, M.D., Bolte, M. 1995, *AJ*, **109**, 2586.
Kraft, R. P., Sneden, C., Smith, G.H., Shetrone, M.D., Langer, G.E., Pilachowski, C.A. 1997, *AJ*, **113**, 279.
Laird, J. B. et al 1993, *ASP Conf. Ser. 48*, 95.
Langer, G. E., Kraft, R.P., Carbon, D.F., Friel, E., Oke, J.B. 1986, *PASP*, **98**, 473.
Langer, G. E. & Hoffman, R. 1995, *PASP*, **107**, 1177.
Langer, G. E., Hoffman, R.E., Zaidins, C.S. 1997, *PASP*, **109**, 244.
Lee, Y.-W., Demarque, P., Zinn, R. 1988, *Calibration of Stellar Ages*, ed. A.G.D. Philip (Schenectady, Davis), 149.
Lee, Y.-W., Demarque, P., Zinn, R. 1994, *ApJ*, **423**, 248.
Moehler, S., Heber, U., de Boer, K.S. 1995, *A&A*, **294**, 65.
Norris, J. E. & Da Costa, G. S. 1995a, *ApJ*, **441**, L81.
Norris, J. E. & Da Costa, G. S. 1995b, *ApJ*, **447**, 680.

Osborn, W. 1971, *Observatory*, **91**, 223.
Peterson, R. C. 1983, *ApJ*, **275**, 737.
Peterson, R. C., Rood, R.T., Crocker, D.A. 1995, *ApJ*, **453**, 214.
Pilachowski, C., Sneden, C., Kraft, R.P., Langer, G.E. 1996, *AJ*, **112**, 545.
Preston, G. W., Schectman, S.A., Beers, T.C. 1991, *ApJ*, **375**, 121.
Richer, H. B. et al 1996, *ApJ*, **463**, 602.
Rood, R. 1973, *ApJ*, **184**, 815.
Sandage, A. R. & Wildey, R. 1967, *ApJ*, **150**, 469.
Searle, L. & Zinn, R. 1978, *ApJ*, **225**, 357.
Shetrone, M. D. 1996a, *AJ*, **112**, 1517.
Shetrone, M. D. 1996b, *AJ*, **112**, 2639.
Smith, G. H., Shetrone, M.D., Bell, R.A., Churchill, C.W., Briley, M.M. 1996, *AJ*, **112**, 1511.
Smith, G. H., Shetrone, M.D., Briley, M.M., Churchill, C.W., Bell, R.A. 1997, *PASP*, **109**, 236.
Smith, G. H., Kraft, R.P., Sneden, C., Shetrone, M.D., Fullbright, J. 1997, *AJ*, (submitted).
Smith, V. V. 1996, private comm.
Sneden, C., Pilachowski, C.A., Vandenberg, D.A. 1986, *ApJ*, **311**, 826.
Sneden, C., Kraft, R.P., Prosser, C.F., Langer, G.E. 1992, *AJ*, **104**, 2121.
Sneden, C., Kraft, R.P., Langer, G.E., Prosser, C.F., Shetrone, M.D. 1994, *AJ*, **107**, 1773.
Sneden, C., Kraft, R.P., Shetrone, M.D., Smith, G.H., Laner, G.E., Prosser, C. 1997, *AJ*, (in press).
Stetson, P. B., Hesser, J.E., Smith, G.H., Vandenberg, D.A., Bolte, M. 1989, *AJ*, **97**, 1360.
Stetson, P. B., Vandenberg, D.A., Bolte, M. 1996, *PASP*, **108**, 560.
Suntzeff, N. B. 1993, *ASP Conf. Ser. 48*, 167.
Sweigart, A. V. 1997, *ApJ*, **474**, L23.
Sweigart, A. V. & Mengel, J. G. 1979, *ApJ*, **229**, 624.
Trefzger, C. F., Langer, G.E., Carbon, D.F., Suntzeff, N.B., Kraft, R.P. 1983, *ApJ*, **266**, 144.
van den Bergh, S. 1967, *AJ*, **72**, 70.
von Rudloff, I. R., Vandenberg, D.A., Hartwick, F.D.A. 1988, *ApJ*, **324**, 840.
Zaidins, C. & Langer, G. E. 1997, *PASP*, **109**, 252.
Zinn, R. 1973, *ApJ*, **182**, 183.
Zinn, R. 1993, *ASP Conf. Ser. 48*, 38.

SURVEYS FOR METAL-POOR STARS IN THE GALAXY: A NEW WINDOW ON THE LOW-ABUNDANCE UNIVERSE

T.C. BEERS
Department of Physics & Astronomy
Michigan State University
E. Lansing, MI 48824 USA

1. Introduction

Measurement of the abundances of the light and heavy elements in stars of the Milky Way galaxy is the cornerstone for the study of numerous aspects of chemical evolution in galaxies and the Universe. We stand poised to enter an era of rapid understanding, as new-generation telescopes with apertures in the 8m-10m class enable astronomers to obtain high-resolution, high-signal-to-noise near-UV, optical, and IR spectra of the stars which have locked up the chemical history of our Galaxy in their outer atmospheres. It is thus appropriate to review present surveys for the low-metallicity stars of our Galaxy, as the stars we uncover today will be studied so intensively in the coming decades.

2. The Complex Environment of the Milky Way

Astronomers' view of the formation history of the Milky Way has changed dramatically in recent years, as we have recognized the importance of what appears to be an ongoing process of heirarchical assemblage within the Local Group. The terminology in modern use reveals this complexity. The visible components of our Galaxy, once summarized as the Bulge, Disk, and Halo, are now subdivided into the Bulge, the Bar, the Thin Disk, the Thick Disk, the Metal-Weak Thick Disk, the Low Halo, and the High Halo. Stars (*e.g.*, the BMPs, see Preston *et al.* 1994), as well as entire globular clusters (Layden and Sarajedini, 1997), have been discovered in the halo of the Milky Way which are likely to have been donated in the past by accreted dwarfs like Sagittarius. Thus the halo itself, once thought to be the home of exclusively "old" stars, probably contains stars of a wide range of ages (at least several billion years), even among the most metal-deficient stars (see Preston 1994 for a detailed discussion of one such star).

To a great degree, the realization of complexity has been driven by moderate- to high-resolution spectroscopic observations of stars with metal abundances $[Fe/H] < -2.5$, *lower* than the most metal-deficient globular clusters. As recently as ten years ago, such stars were not thought to exist in any great number. However, long-term survey efforts by a number of groups have smashed through this "chemical barrier," and forged toward ever-lower-abundance stars, presently as metal-deficient as $[Fe/H] \sim -4.0$. Below we summarize the known samples of these stars, comment briefly on their characteristics, and close with a discussion of future followup studies of these extremely metal-poor stars.

3. Present Surveys for Metal-Poor Stars

Beers *et al.* (1992) describe an objective-prism/interference-filter technique for the identification of large numbers of candidate low-metallicity stars in the Galactic halo. The HK survey, as it has come to be known, is based on some 300 prism plates covering roughly 1/6th of the sky in the northern and southern hemisphere. Visual inspection of the survey plates (with a 10 X microscope) has resulted in the selection of some 10,000 low-metallicity candidates in the magnitude range $10.5 \leq B \leq 15.5$. For the past 15 years, follow-up medium-resolution (1–2 Å) spectroscopy and

J. Andersen (ed.), Highlights of Astronomy, Volume 11A, 58–61.

TABLE 1. Summary of the Metal-Poor Star Sample

Sample	N ≤ -1.5	N ≤ -2.0	N ≤ -2.5	N ≤ -3.0	N ≤ -3.5	N ≤ -4.0
LCL/RN	578	256	79	11 (29)	4	1 (1)
HK (BPSII)	726	540	197	29 (73)	5	1 (1)
HK (NEW)*	1100	700	300	50 (150)	10	4 (8)
TOTAL*	2500	1500	600	100 (250)	20	5 (10)

* Approximate numbers

broadband *UBV* photometry of these stars has been obtained with 1m- to 2m-class telescopes from Las Campanas, Kitt Peak National Observatory, the European Southern Observatory, Siding Springs Observatory, and the Canary Islands. The observational efforts from Siding Springs, the Canary Islands, and Kitt Peak are essentially finished; observations from ESO will be completed by spring 1998. Spectra for roughly half of the low-metallicity candidates have been obtained to date; results based on some 3000 spectra will soon be submitted for publication. Broadband photometry has been obtained for some 3500 of these stars (Preston *et al.* 1991; Norris *et al.* 1997); narrowband photometry is available for a smaller subset, in particular the most metal-deficient stars (Schuster *et al.*, 1996) .

The HK survey is not the only source for stars of extremely low metal abundance. Other large survey efforts have contributed stars with [Fe/H] < -1.5, primarily based on proper-motion selection criteria. In Table 1 we summarize the available stars from three programs. The first set are the kinematically-selected samples of Ryan and Norris (1991) and Carney *et al.* (1994). The second is the published list of stars from the HK survey (Beers *et al.*, 1992), labelled here as BPS II (note that abundances for these stars have been estimated anew using a more recent calibration; see Beers *et al.* 1997). The third subsample is the combined set of HK candidates from recent (as yet unpublished) followup efforts.

Note that for the stars in Table 1 with [Fe/H] ≤ -3.0 and ≤ -4.0 we have listed parenthetically the numbers of stars with [Fe/H] ≤ -2.8 and ≤ -3.8, respectively, in order to emphasize the effect which small changes in metallicity estimates might have on the cumulative numbers of stars below a given abundance. The factor of two to three discrepancy seen in the numbers of stars with [Fe/H] ≤ -2.8 and [Fe/H] ≤ -3.0 underscores the need for a uniform set of high-resolution determinations of the abundances of ALL stars in our sample with [Fe/H] ≤ -2.5.

4. Interpretations

Several general conclusions have already been reached from analysis of the nature of the low metallicity tail of the Metallicity Distribution Function (MDF), based on the above surveys. First, the so-called "Simple Model" of galactic chemical evolution (Hartwick, 1976) fails in the sense that is predicts too many metal-poor stars with abundances less than [Fe/H] $= -3.0$. The apparent limit of metal deficiency among presently-observable stars in the Galaxy is [Fe/H] ~ -4.0, as foreshadowed from the analysis of Audouze and Silk (1995). If the Galaxy was able to produce significant numbers of stars with abundances below -4.0, they are presumably fainter than $B = 15.0$, and located well outside the solar neighborhood, beyond the reach of present-day surveys.

Tantalizing departures from expectations of a "smooth" MDF have been noticed among stars with [Fe/H] < -2.0 – an "overdensity" of stars in the metallicity range $-2.8 <$ [Fe/H] < -2.6 may be the signpost of the typical metallicity level obtained after the first generations of massive supernovae contributed their processed material into the interstellar medium (see Figure 1). Alternative explanations of this feature are possible. For example, it might be associated with the metallicity level reached by typical dwarf galaxies which have been subsequently disrupted and dispersed into the halo of the Milky Way.

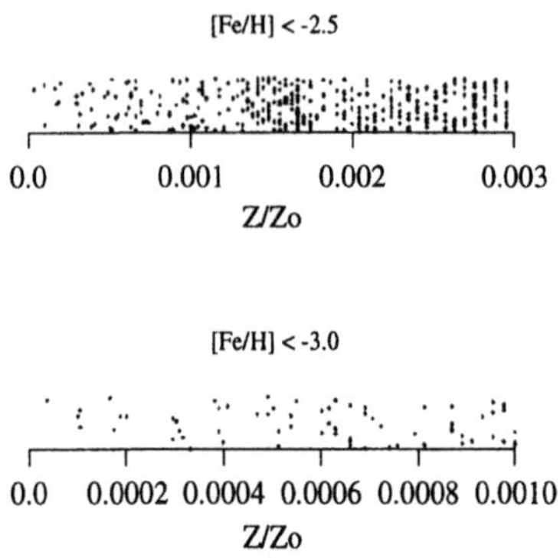

Figure 1. Density plots of observed abundances for stars in the combined sample. The position of each star has been randomized along the vertical direction.

5. Future Surveys

As the sample of low-metallicity stars enlarges, greater complexities have been revealed. We can expect this trend to continue in the near future. Thus, we should endeavor to devise new approaches for increasing the samples of stars identified as having the lowest possible elemental abundances. Several efforts are already underway to accomplish this task, and others will no doubt be launched. In the rush toward ever lower stellar metallicities, it is well to remember that stars of modest metal deficiency are also of great importance. For example, the transition from stars with disk-like to halo-like kinematics lies in the metallicity regime $-2.0 \leq$ [Fe/H] ≤ -1.0 (see Beers and Sommer-Larsen 1995). This is the same region where the dominant contributors to the chemistry of our Galaxy appears to change from primarily the ejecta from Type II supernovae to that of Type I supernovae.

The original HK survey plates are presently being scanned with the APM machine in Cambridge, in collaboration with Mike Irwin. Each plate typically yields 5000 classifiable spectra, hence we expect the final sample of stars available for automatic inspection to exceed one million, down to a magnitude of $B \sim 15.0$. A neural network is presently being implemented, and trained with the extensive set of metal-poor stars available from the visual survey. This approach is greatly aided by the existence of moderately-accurate color information obtained by recently-calibrated photographic surveys of the northern and southern sky. We anticipate the identification of additional metal-weak stars which were missed in the visual classification pass, in particular among the cooler stars where metal-deficiency is more difficult to recognize. As the peak of the halo MDF is at [Fe/H]~ -1.5, we also expect to identify large numbers of stars centered around this abundance.

A new HK survey is planned with the UK Schmidt telescope, hopefully beginning in 1998. This new survey will consist of approximately 100 (unwidened) spectral plates, covering 3600 square degrees of the southern sky, to be taken with an interference filter which with a wider passband than the original survey. Test exposures indicate that we might expect to obtain roughly 50,000 classifiable spectra per plate down to a magnitude of $B \sim 18.0$. This deep sample of metal-poor stars will enable an *in-situ* study of the distant halo of our Galaxy, and should be ideal for investigations of possible changes in the MDF of stars far from the solar neighborhood. Other interesting objects, such as field horizontal-branch stars up to 20-30 kpc from the Sun, should be readily identifiable as well.

Another large survey, known as the ESO/Hamburg Survey (Reimers and Wisotzki 1997; Christlieb *et al.* 1997; Wisotzki *et al.* 1997), is presently yielding substantial numbers of metal-poor star candidates based on automated scans of some 425 plates covering 9000 square degrees of sky down to $B \sim 17$. Of particular importance, the ESO/Hamburg survey plates include coverage of portions of the southern sky which were not sampled by the HK survey. Thus, this survey should provide a large number of newly-discovered metal-poor stars with apparent magnitudes which are suitable for rapid follow-up high-resolution spectroscopic analysis (*i.e.*, in the magnitude range $13.5 \leq B \leq 14.5$).

6. Future Followup

Long lists of candidate metal-poor stars are only the first step. Plans are presently underway for conducting followup spectroscopic and photometric observations of as many of these newly-identified metal-poor stars as possible. Telescopes of aperture 1.5-2m are ideal for moderate-resolution (1-2 Å) spectroscopy of single stars, in the "one-by-one" mode which has proven so successful for followup of the HK candidates. New collaborative efforts with observatories having suitable equipment is a vital step in order for this long-term work to succeed. Multiple-star moderate-resolution spectroscopy (50-100 stars at a time) is now becoming possible using wide-field fibre spectrographs such as Watson's FLAIR (Parker, 1997). In order to fully exploit this approach, plans are underway for the design and construction of automatic fiber positioning equipment for FLAIR.

Intermediate-resolution (0.2-0.3 Å) spectroscopy in the "single-star" mode of thousands of metal-poor stars is presently being planned for the Hobby-Eberly telescope. These "quick-look" spectra should be ideal for the rapid identification of the most interesting candidates for futher detailed study at high resolution.

High-resolution (0.03-0.1 Å), high signal-to-noise ($S/N > 100/1$) spectroscopy in the single-star mode is the capstone project of the analysis of the most metal-deficient stars (see, *e.g.*, Ryan *et al.* 1996). Essentially all of the 8m-10m class telescopes will have suitable equipment for such work. What is required, however, is the commitment of the community in order to set aside the several MONTHS worth of large telescope time required to assemble this valuable data set. The ultimate reward will be high-quality data for samples of 500-1000 stars of the Milky Way with metallicities $[Fe/H] < -2.5$, from which vital clues to the evolution of elemental abundance patterns in the Universe can be extracted.

References

Audouze, J., and Silk, J.: 1995, 'The First Generation of Stars: First Steps Toward Chemical Evolution of Galaxies', *ApJ*, **451**, L49–L52.

Beers, T.C., and Sommer-Larsen, J.: 1995, 'Kinematics of Metal-Poor Stars in the Galaxy', *ApJS*, **96**, 175–221.

Beers, T.C., Preston, G.W., and Shectman, S.A.: 1992, 'A Search for Stars of Very Low Metal Abundance. II. ', *AJ*, **103**, 1987–2034.

Beers, T.C., Rossi, S., Norris, J.E., Ryan, S.G., and Shefler, T.: 1997, 'Estimation of Stellar Metal Abundance. II. Refinements of the CaII K Technique and the Auto-Correlation Technique', in preparation.

Carney, B.W., Latham, D.W., Laird, J.B., and Aguilar, L.A.: 1994, 'A Survey of Proper Motion Stars. XII. An Expanded Sample', *AJ*, **107**, 2240–2289.

Christlieb, N., Grasshoff, G., Nelke, A., and Wisotzki, L.: 1997, 'Automatic Classification of Digitized Objective-Prism Spectra', in Wide-Field Spectroscopy, ed. Kontizas *et al.*, (Kluwer: Dordrecht), 109–113.

Hartwick, F.D.A.: 1976, 'The Chemical Evolution of the Galactic Halo', *ApJ*, **209**, 418–423.

Layden, A.C., and Sarajedini, A.: 1997, 'The Globular Cluster M54 and the Star Formation History of the Sagittarius Dwarf Galaxy', *ApJL*, **486**, L107–L110.

Norris, J.E., Ryan, S.G., and Beers, T.C.: 1997, 'A Search for Stars of Very Low Metal Abundance. III. UBV Photometry of Metal-Weak Candidates', in preparation.

Parker, Q.A.: 1997, 'FLAIR on the UK Schmidt Telescope', in Wide-Field Spectroscopy, ed. Kontizas *et al.*, (Kluwer: Dordrecht), 25–31.

Preston, G.W.: 1994, 'CS 22873-139: A Very Metal-Poor Main-Sequence Spectroscopic Binary with Colors Indicative of Intermediate Age', *AJ*, **108**, 2267–2272.

Preston, G.W., Beers, T.C., and Shectman, S.A.: 1994, 'The Space Density and Kinematics of Metal-Poor Blue Main Sequence Stars Near the Solar Circle', *AJ*, **108**, 538–554.

Preston, G.W., Shectman, S.A., and Beers, T.C.: 1991, 'Photoelectric UBV Photometry of Stars Selected in the HK Objective-Prism Survey', *ApJS*, **76**, 1001–1031.

Reimers, D., and Wisotzki, L.: 1997, 'The ESO/Hamburg Survey', *The Messenger*, **88**, 14–19.

Ryan, S.G., and Norris, J.E.: 1991, 'Subdwarf Studies. II. Abundances and Kinematics from Medium Resolution Spectra', *AJ*, **101**, 1835–1864.

Ryan, S.G., Norris, J.E., and Beers, T.C.: 1996, 'Extremely Metal-Poor Stars. II. Elemental Abundances and the Early Chemical Enrichment of the Galaxy', *ApJ*, **471**, 254–278.

Schuster, W.J., Nissen, P.E., Parrao, L., Beers, T.C., and Overgaard, L.P.: 1996, '$uvby - \beta$ Photometry of High-Velocity and Metal-Poor Stars VIII. Stars of Very Low Metal Abundance', *A&A*, **117**, 317–334.

Wisotzki, L., Bade, N., Engels, D., Groote, D., Hagn, H.-J., Kohler, T., and Reimers, D.: 1997, 'The Hamburg Objective-Prism Surveys for Bright Quasars', in Wide-Field Spectroscopy, ed. Kontizas *et al.*, (Kluwer: Dordrecht), 373–379.

ABUNDANCES OF METAL-POOR STARS AND THE FORMATION OF THE HALO

R. CAYREL
Observatoire de Paris et C.N.R.S. (URA 335)
61, avenue de l'Observatoire
75014 Paris, France

Abstract. The origin of the elements in the early Galaxy has been thought to be fairly well understood as primordial matter polluted by ejectae of type II supernovae. Some recent observational results have led to reconsider in more detail this statement. As a result, it is confirmed in some respects, but curiously, when the same metallicity can be found both in the halo and in the thick disk, the stars contaminated by SN Ia ejectae are more the halo stars than the thick disk stars.

1. Introduction

The greatest breakthrough in the field covered by this joint discussion, which has occurred in the last years, is the detailed analysis of a bunch of halo stars in the metallicity range $-4 <$ [Fe/H] < -2.5 (McWilliam *et al.* 1995, Sneden *et al.* 1996, Ryan *et al.* 1996), made possible by the discovery of over 50 extremely metal-poor stars by the H&K Ca II lines survey of Beers, Preston and Shectman (BPS survey in brief), (Beers *et al.*, 1992).

Because the subject has been covered by the principal author of this breakthrough in the first talk, this morning, I shall speak of other things in my talk.

I briefly, mention in section 2, a new controversy on the nature of the process which has produced the bulk of the neutron-capture elements in the extreme population II. In section 3, I discuss a fairly new discovery, concerning the α-elements over iron peak elements ratio in metal-poor stars. Section 4 deals with the question of the how to properly formulate the problem of the chemical evolution of the Galaxy. Section 4 concludes the talk.

2. *r*-process versus *s*-process: reopening the case?

It has been known from many years (Truran, 1981), that there was strong indications that neutron-capture elements in very metal-poor stars were chiefly produced by the *r*-process only, whereas it is well established that both processes *r* and *s* have contributed to the solar system chemical composition. The case has been reopened by a paper by Magain (Magain, 1995), who claimed that the very metal-poor star HD 140283 had most of its barium generated by the *s*-process. The argument is based on the analysis of the profile of the 4554 Å line. The width of the line depends upon the proportion of *r*-process and *s*-process barium, the hyperfine structure of the odd isotopes producing an extra-broadening of the line. Contrary to expectation, the width of the line was favoring the *s*-process rather than the *r*-process. Another piece of evidence, in the opposite direction comes from the analysis of the even more metal-poor star CS 22892-052(Sneden *et al.*, 1996), which is very enhanced in neutron-capture elements with respect to iron, and which wonderfully fits the proportions of solar *r*-elements only. It was worth studying more carefully if the conclusions of Magain, at the limit of the S/N ratio of the data, were inescapable. This was done by P. François and W. Gacquer, with new observations taken at CFHT. The results will be presented next week at Symposium 187, where you are invited to hear about the issue.

J. Andersen (ed.), Highlights of Astronomy, Volume 11A, 62–65.

3. The new [α/Fe] mess

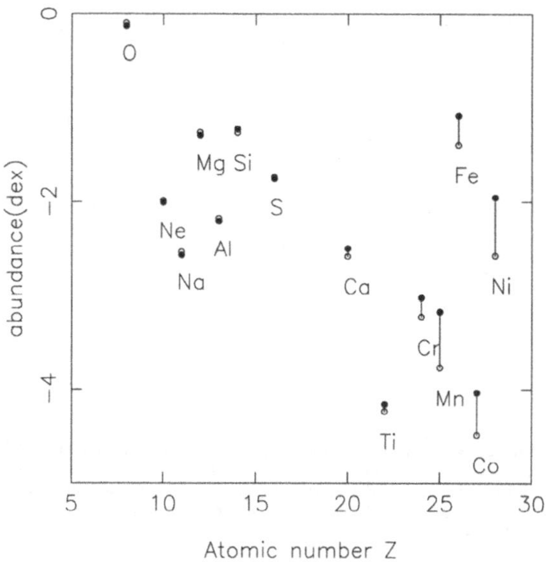

Figure 1. According to Tsujimoto *et al.* (1995) the proportion of several elements by mass, with respect to their sum, is shown (i) for the ejectae of SNe II, averaged over an initial mass function of Salpeter, between 10 and 50 $M_\odot$ (open circles); (ii) for the same ejectae mixed with the ejectae of SNe Ia in the proportion of 0.15 SN Ia per SN II. This is about the ratio producing the best fit to the solar composition. Note that the elements up to Ca are little affected by the contribution or no-contribution of SNe Ia, whereas the iron-peak elements are strongly affected.

One of the pillar of stellar nucleosynthesis is the fact that the α-elements are less deficient than iron by about 0.4 dex in metal-poor stars, with respect to the solar composition. This fact has recently been shown to be true also for extremely metal-deficient stars (McWilliam *et al.*, 1995). The current interpretation from 1980 (Tinsley, 1980), is that pop. I has been enriched both by the yields of SNe II and SNe Ia, whereas pop. II was enriched by SNe II only, because of the time-lag for the production of SNe Ia. The comparison of the yields predicted for the two types of supernovae , is the base for this interpretation, which can be even made more refined, according to the α-element under consideration, as they have each a different production ratio. Figure 1 shows this effect. The yields are from Tsujimoto *et al.* 1995. The circles represent the fractional abundances for pure SNe II ejectae, averaged over a Salpeter initial mass function. The fractional abundance is by mass of the element with respect to all the elements given, in decimal exponent.The dots are the same data for a mixture of SNe II and SNe Ia, with 0.15 SN Ia per SN II (approximately what reproduces the solar composition). It is clear that the proportions are not affected from O to Ti, whereas the iron peak elements are enhanced by the SN Ia contribution. This behaviour is statistically verified, but not in all details. For example, [Cr/Fe] and [Ni/Fe] should be different in metal-poor stars (with the usual notation [X]= $log(X(star)/X(Sun))$) and this is not the case. Recently attention has been drawn onto apparent exceptions to this rule. (King, 1997) found that the common proper motion pair HD 134439 and 134440, which have [Fe/H] = -1.5 and extreme kinematical properties (U=310,V=-467, W=44, in km/s with respect to the LSR), have practically [α-el/Fe]= 0. They have interpreted the fact as evidence for an accreted star, with a different chemical history. Carney *et al.* 1997 have discovered another α-element singular halo star BD + 80° 245, which is even α-element-poor instead of being α-element-rich. Its kinematics is (U=-173,V=-333, W=+255), also fairly extreme, and has [Fe/H] = -1.8. The authors also suggest that the star had a chemical history different from ordinary galactic halo stars. The star is also overdeficient in barium. The story could end there if Nissen&Schuster (1997) had not found, in studying halo and thick disk stars of overlapping metallicity, that [α-el/Fe] was just the contrary of what one may have naïvely

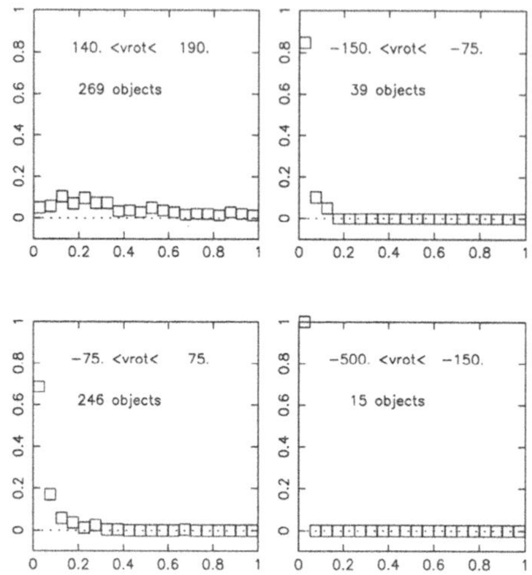

Figure 2. **Metallicity histogram as a function of angular momentum. The sample is that of Carney *et al.* 1994. The abcissa is the metallicity, in fraction of the solar metallicity. The ordinate is the fraction of objets in each bin. The size of the bins is $0.05Z_\odot$. Note the striking difference between the histogram of stars with a thick disk kinematics (140. $< V_{rot} <$ 190.) and the 3 histograms of the halo stars. Note also the continuous increase of the first bin fraction when the angular momentum decreases.**

expected. In kinematically selected halo stars two thirds of the objects have [α-el/Fe]≈ 0 , whereas it has the high value 0.3 to 0.4 (usually found in the halo) in the thick-disk stars! We are of course here also at metallicities unusally large for halo stars. This time, it becomes difficult to slip the problem under the carpet in invoking an "accretion" event. The stars have not a singular kinematics, and as suggested by the authors, the meaning of this observation might be that halo regions, with a particularily low rate of astration, may well have formed stars with a substantial contamination by SNe Ia, even at low metallicy. Actually the fit of the abundances is very consistent with the SNe Ia yields for Mg, Si and Ca. A few globular clusters, as Palomar 12 and Ruprecht 106 share this property, and are reputed 2 or 3 Gyr younger than the bulk of globular clusters. Therefore, the lesson is that low-metallicity does not necessarily means an age large enough to have avoided contamination by SN Ia, and that a period has existed in which the halo was still forming stars, after the disk had started star formation. The author of this presentation does not believe that the prograde or retrograde motion of the stars is a key element in this α element issue.

We present in fig.2, the evolution of the metallicity histogram of the stars of the (Carney *et al.*, 1994) sample, as a function of the angular momentum of the objects. The sample with velocity rotations typical of the thick disk shows a fairly flat histogram, with a mean metallicity half solar, but a shallow maximum at one fifth solar. The typical halo, taken by the selection $-75 < V_{rot} < +75$, in order to get rid of thick disk contamination, has a completely different histogram, with a sharp peak in the first three bins, and practically no star beyond metallicity 0.2 solar. The bulk of the retrograde halo has a similar distribution, more sharply peaked (85 per cent in the first bin). Finally the extreme retrograde halo, $V_{rot} < -150$, has all the objects in the first bin (note however that the total number of objects becomes small and prone to statistical fluctuation). The conclusion is that the retrograde halo is, on average, very metal-poor, and is the most primitive component of our Galaxy, even if it has produced later, a few rare α-el. rich objects.

4. A global view of galaxy evolution

The concept of accreted system is more and more heard in the context of galactic evolution. Actually there is not a single proved example of accretion, in a strict acception of this term, which is the encounter with a body with a positive energy, ending up by dynamical friction, as gravitationaly bound to the Galaxy. The example of the Sagittarius spheroidal galaxy (Ibata *et al.*, 1994), often uncorrectly described as an accretion event, is not an accretion, but the *merging* of a substructure which has always been there. The mere fact that the kinematics of halo stars shows that the gas from which they have originated was moving both in prograde and retrograde motions, proves that this gas had to be clumped in smaller substructures, as expected from cosmology. Therefore, the question is not if a given object has been accreted or not but *how long it has been trapped in a substructure before dissolving into the general galactic field.* It is of course beyond the scope of this talk to further elaborate on this subject, but the point deserved to be mentioned.

5. Conclusion

• The great event of these last years is the access to a stellar population substantially more metal-poor than the most metal-poor globular clusters , in the range $-4 <$[Fe/H]< -2.5.
• The case for a formation of the neutron-capture elements only by the r-process in very metal-poor stars has been strengthened by the discovery of a very metal-poor star with a superb enhanced spectrum of r-process rare earths (CS 22892-052). On the other side, the analysis of the profile of a barium line in HD 140283 has opened a controversy on which you will hear during the symposium 187 (W. Gacquer), next week.
• The statement that [α-el/Fe] is a function of [Fe/H] in the Galaxy, has been recently shaken by counter-examples. It must be admitted that a high [α-el/Fe] is the signature of an early bird in the Galaxy, whatever is its metallicity, which, at any time, has a strong scatter in the Galaxy.
• The concept of "accretion" must be used more cautiously. It is improperly used for example for the Sagittarius Ibata&Gilmore spheroidal galaxy: because the dynamical study of the Milky Way - Sagitarius system shows that it is a bound system, the term of "merging" of a substructure would be more appropriate. For the chemical history of the Galaxy the important question is then to know how long the substructures have evolved as independant systems, before dissolving into the general field.

References

Beers T.C., Preston G.W., Shectman S.A, 1992, *AJ*, 103,1987
Carney, B.W., Latham, D.W., Laird, J.B., Aguilar, L.A., 1994, *AJ*, **107**, 2240
Carney, B.W., Wright, J.S., Sneden, C., Laird, J.B., Aguilar, L.A., Latham, D.W., 1997, *AJ*, **114**, 363
Ibata, R.A., Gilmore, G., Irwin, M.J., 1994 ,*Nature*, 370, 194
King, J.R., 1997, *AJ*, 113, 2302
McWilliam A., Preston G.W., Sneden C., Shectman S.A., 1995, *AJ*, **109**, 2757
Magain, P.,1995, *A&A*, 297,686
Ryan S.G., Norris J.E., Beers T.C., 1996, *ApJ*, **471**, 254
Sneden C., McWilliam A., Preston G.W., Cowan J.J., Burris D.L., Armosky B.J., 1996, *ApJ*, **467**, 819
Tinsley, B., 1980, *Fundam.Cosmic Phys.*, 5, 287
Truran, J.W., 1981, *A&A*, **97**, 391
Tsujimoto, T., Nomoto, K., Yoshii, Y., Hashimoto, M., Yanagida, S., Thielemann, F.-K., 1995, *MNRAS*, **277**, 945

ELEMENT ABUNDANCE RATIOS IN GALACTIC BULGE STARS

R. MICHAEL RICH
Department of Astronomy, Columbia University
Mail Code 5242, New York, NY, 10027

I review the current status of measurements of element ratios in metal rich stars. These include giants in the Galactic bulge, metal rich Galactic globular clusters, and high velocity dwarfs in the Solar vicinity. There is evidence for enhancement of Mg and Ti in the bulge. While s-process elements have Solar abundance in bulge giants, they are deficient in the metal rich, high velocity dwarfs. There are no secure measurements of oxygen abundance in the bulge, but present data indicate no evidence either for extremely low or high [O/Fe] in bulge giants. The current data are limited by resolution and S/N, and I outline goals for observations with the next generation of ground-based telescopes.

1. Introduction

If the central bulge of the Milky Way formed rapidly, one would expect that enrichment could occur only through massive star SNe, with little or no contribution from core deflagration SNe. In this scenario, one expects alpha-capture elements to be enhanced at a given [Fe/H] compared to the Solar vicinity: element abundance ratios may record something of the history of star formation. The idea is developed in Wheeler, Sneden & Truran (1989) and its application to bulges and elliptical galaxies is developed in Matteucci & Brocato (1990). As Kraft reviews in this volume, the stellar evolution effects that modify surface abundances in globular cluster stars do not appear to be a problem for field giants.

The Galaxy actually has a number of metal rich stellar populations, and their evolutionary histories vary greatly and properties of these stars may be quite different. There are the well known giants in the central bulge of the Galaxy, particularly well observed in Baade's Window ($l = -4^o, b = -8^o$). There are also a substantial number of globular clusters at high metallicity in the Galactic Bulge and disk (Ortolani et al. 1995) and metal rich open clusters in the disk (Phelps & Janes 1996). Grenon (1989) has identified a group of high velocity, metal rich dwarfs in the disk, and he suggests that they may have originated in the bulge. Needless to say, there is much effort needed on a systematic study of element ratios in all of these populations. Feltzing (also in this volume) reviews the growing body of data on metal rich stars in the disk. Further details are found in the review of McWilliam (1997).

2. The Galactic Bulge

2.1. [FE/H]

Because of the distance and reddening, it has been difficult to establish the abundances of the bulge giants. The first low resolution efforts by Whitford & Rich (1983) and Rich (1988) found that the bulge giants were in general metal rich, with a wide range in metal abundance. The Rich (1988) scale found a median [Fe/H] of twice Solar, but the abundances were based on the sum of the Fe and Mg_2 Faber indices. Only with higher resolution, higher S/N data, it has been possible to approach the true [Fe/H] scale, which has a median of -0.15 dex (McWilliam & Rich, 1994; MR94). However, there remain some concerns: Geisler & Friel (1992) confirmed the Rich (1988) scale with Washington photometry. Further, given that light elements in the disk are known to have an abundance gradient, it is hard to understand how the bulge could be more metal poor than the Solar neighborhood. There is reason to be concerned that at the $R = 17,000$ of the MR94 study, the derived continuum may have been erroneously low due to unresolved iron lines, and the

J. Andersen (ed.), Highlights of Astronomy, Volume 11A, 66–69.

iron abundance may actually be much higher, in agreement with the low resolution analysis (where total equivalent widths of the strong lines might actually be easier to measure). There are 2 recent studies that strengthen confidence in the Fe abundance scale. Castro et al. (1996) analyzed Keck Hires spectra (R=38,000; S/N=30) of the metal rich disk giant μLeo and the bulge giant BW IV-167. Due to incomplete wavelength coverage and low S/N, it was possible to only measure the iron abundance. However, the bulge giant and μLeo are indistinguishable, with [Fe/H]=+0.47 ±0.03 dex. Their results are in excellent agreement with MR94's analysis of the same star and one may therefore conclude that doubling the resolution did not change the result. Peterson & Green (1997) have analyzed a hot horizontal branch star member of the metal rich old open cluster NGC 6791. Hot enough so that difficult sources of blanketing (such as CN) are not a problem, the analysis finds [Fe/H] = +0.3 – 0.4 dex, with $[\alpha/\mathrm{Fe}]$ enhanced by +0.2 dex. They also find that the cool giants in the cluster are indistinguishable from μLeo.

At the present time, then, the iron abundance scale looks reasonably secure. The possibility that the continuum is blanketed by weak Fe I lines can be largely ruled by these 2 studies. MR94 also found that CN lines could have the pernicious effect of causing a spuriously high Fe abundance, as many of the vibration/rotation lines coincide with Fe I lines at low resolution. Again, these two studies serve to confirm that the MR94 analysis (and their list of clean lines) does not suffer from this problem. It would be useful to use 8 and 10 m class telescopes to obtain echelle spectra of hot HB stars in metal rich Galactic Center globular clusters such as NGC 6553 and 6528 and to compare giants in these clusters with those of the bulge.

2.2. ALPHA-CAPTURE ELEMENTS

One would most like to know if oxygen is enhanced in the bulge, as it is the most abundant alpha-capture element produced in the hydrostatic shells of massive pre-supernova stars. Unfortunately, the single usable metastable line at 6300.3Å suffers from blending with an Sc II line at 6300.7 and with numerous weak telluric features in the regions. For stars of low radial velocity, the night sky emission line cannot be accurately subtracted, and the feature is lost. MR94 found no evidence for any dramatic oxygen enhancement, a result confirmed in the more careful effort of McWilliam, Rich, & Tomaney (1998).

As discussed in MR94, magnesium abundances are especially hard to measure; few gf values are available, and some of the Mg lines give systematically discrepant abundances. The MR94 abundances were typically based on spectrum synthesis of 3 lines per star. Approximately 4 to 6 lines of Ca and Si, respectively, were usable by MR94, although some of these lines are uncomfortably strong in metal rich stars. With its good representation of many medium strength lines, Ti I was well measured in the MR94 stars.

MR94 find a puzzling enrichment pattern for the alpha-capture elements, in which Mg and Ti approximately follow the expected enhancement, but Si and Ca follow the Solar neighborhood disk enrichment line (Fig. 1). MR94 compared the measured line strengths at constant T_{eff} for Ti lines in the bulge giants and in a large sample of 671 K giants (McWilliam 1990) and the enhancement of Ti in the bulge giants is clear. Calculated yields from Type II SN models covering a wide range of parameter space do not produce ejecta in which Mg and Ti are selectively enhanced relative to Ca and Si (Nomoto et al. 1997).

Low resolution studies by Terndrup et al. (1995) and Sadler et al. (1996) argue that Mg must be enhanced in the majority of bulge giants, and the Mg enhancement was introduced by MR94 to explain the discrepant R88 abundance scale. There is always the risk with these low resolution studies (as is the case for the giant ellipticals) that they confirm the expected result. The element ratios will be considered secure in the bulge when they have been determined at high resolution and high S/N for a larger sample.

2.3. HEAVY ELEMENTS

Because s-process elements are produced in the envelopes of AGB stars, their presence indicates that the population had an enrichment timescale of at least 10^8 yr; in the extreme metal poor halo, s-process is very deficient; MR94 find Solar abundances for La, Y, and Ba in the bulge giants. The r-process is associated with massive star SNe, and the abundance analysis of the peculiar extreme metal poor star CS22892-052 (Sneden et al. 1996) supports this association. Eu was measured in most of the bulge giants by MR94, and found to be mildly enhanced.

2.4. LIGHT ELEMENTS

MR94 find one super Li-rich giant, a curiosity since the canonical theories explaining such an enhancement (e.g. Scalo, Despain, & Ulrich 1975) require evolved luminous stars on the AGB, presumably the progeny of stars more massive than the Galactic bulge is known to have (Ortolani et al. 1995). High Na and Al abundances in the bulge giants are most likely related to the abundance anomalies seen in globular clusters giants (see Kraft in this volume). If present, such anomalies may raise a red flag concerning the alpha-capture elements as well although this is still not considered to be a problem for giants in field populations.

3. Relating the Bulge to Other Populations

It would be interesting to know if any other populations share the chemical evolution history of the bulge, or if there are any bulge stars that migrate through the Solar neighbhorhood. Given the similarity in colors between bulges and inner disks established from photometic surveys (Balcells & Peletier 199X), could the inner disk and bulge share a similar detailed chemical history?

The most obvious population to start with are the globular clusters in the bulge whose spatial and abundance distributions are now known to be very similar to that of the bulge field stars (Bica, Barbuy, & Ortolani, 1997). The only published high resolution study (Barbuy et al. 1992) finds [M/H]=−0.2; Barbuy's continuing analysis of this cluster suggests a slightly lower metal abundance and modest +0.3 dex enhancements in Mg and Ti.

Grenon (1989) has identified metal rich stars in the Solar Neighbhorhood with high eccentricity orbits, and he suggests they are local interlopers from the bulge. This idea has gained some momentum with the discovery of the bar-like nature of the bulge, which might allow some stars on chaotic orbits to be ejected. Barbuy & Grenon (1990) found enhanced oxygen in these high velocity stars, appearing to strengthen their connection to the bulge. Two studies of these metal rich disk stars have recently appeared; the results now make a bulge connection less likely. Castro et al. (1997) analyzes a small sample of high metallicity, high velocity dwarfs from Grenon's sample. She finds that [O/Fe] declines with [Fe/H], falling below the Solar value in metal rich stars. Mg and Si follow iron, while Ca and Ti decline like oxygen. In contrast to the bulge, the s-process elements are subsolar (−0.5 dex). These results are largely confirmed (but the alpha-capture elements follow iron) in the larger sample of Feltzing & Gustafsson (1998), which is not a kinematically selected sample. In contrast to Edvardsson et al. 1993, Feltzin & Gustafsson find no evidence for correlations between abundances and kinematics.

As far as Mg and Ti are concerned, one cannot rule out common chemical evolution for the Castro et al. disk stars and the bulge giants. The star G161-29 also has a strong Eu line. However, the 0.5-1.0 dex difference in s-process between the bulge and disk stars remains as a considerable obstacle to their association.

As a footnote, I mention the work of Sellgren et al. (1997) on Galactic center supergiants (she gives a full report in the proceedings of IAU Symposium 187). She finds [Fe/H] $\approx$ Solar for 11 stars, but the stars in question have formed only in the last 10^7 yr, and probably more reflect bar-driven gas inflow rather than the chemical evolution of the bulge.

4. Discussion

Although the bulge may have formed early and quickly, such a history is not reflected in dramatic enhancements of the alpha-capture elements thought to be produced in massive star SNe. It would be interesting to isolate the most metal poor stars in the bulge and to compare their detailed compositions with extreme stars of the outer halo. If the bulge formed along with the very first stellar generation, the s-process elements should be deficient in the most metal poor stars. If the material that formed the bulge was processed in some prior generation of stars (e.g. in the halo) then one might expect the s-process elements to tend toward Solar. In fact, the low s-process in the metal rich disk stars is puzzling, since their tendancy to have lower oxygen at higher [Fe/H] follows the Solar vicinity enrichment line and would be consistent with a slower enrichment (hence less ancient composition).

In further considering the history of the bulge, the work of Ortolani et al. (1995) leaves little doubt that the field population of the bulge 500 pc distant from the nucleus (Baade's Window) is as old as the metal rich globular clusters. There is also strong (and ever accumulating) evidence for the bar-shaped morphology of the bulge field population. However, theoretical simulations of bars suggest that they do not survive for a Hubble Time. The chemistry does not provide a strong constraint, except to rule out the most extreme starburst scenarios. There is clearly great need for more spectroscopy of bulge giants at a range of galactocentric distance. The Galactic center has a more complex star formation history, and detailed abundance analysis of stars representing the range of population ages will be of great interest in the coming years. Finally, the abundances of the old disk stars and old open clusters such as NGC 6791 are of great interest in that they illustrate that stellar populations can attain high metallicity in the disk, at lower density and larger galactocentric distance than one usually associates with metal rich populations. The starburst events that lead to high metallicity may depend more on the local environment than on the global environment, or the initial mass function might vary with time. Both NGC 6791 and the Grenon sample are $\approx 10^{10}$ yr old, and it is not clear why the ISM could reach such high metallicities early on given that the ISM is not so metal rich today. Such issues lay clear the importance of having a dedicated campaign to measure abundance ratios using the next generation of large telescopes.

References

Barbuy, B., Castro, S., Ortolani, S., & Bica, E. 1992, *A&A*, **259**, 607
Barbuy, B., Ortolani, S., Bica, E., & Renzini, A., Guarnieri, M.D. 1997,*IAU Symp. 187*, 203
Barbuy, B., Bica, E., & Ortolani, S. 1997. A&A, *in press*
Castro, S., Rich, R.M., McWilliam, A., Ho, L.C., Spinrad, H., Filippenko, A.V., & Bell, R.A. 1996, *AJ*, **96**, 1319
Castro, S., Rich, R.M., Grenon, M., Barbuy, B., & McCarthy, J.K. 1997, *AJ*, **114**, 376
Geisler, D., & Friel, E.D. 1992, *AJ*, **104**, 128
Grenon, M. 1989, *Astrophys. Space Sci.* , **156**, 29
Janes, K.A., & Phelps, R.J., *AJ*, **111**, 1604
Matteucci, F., & Brocato, E. 1990, *ApJ*, **365**, 539
McWilliam, A., 1990, *ApJS*, **74**, 1075
McWilliam, A. 1997, *ARAA*, **35**, 000
McWilliam, A., Rich, R.M. 1994, *ApJS*, **91**, 749 (MR94)
McWilliam, A., Tomaney, A., & Rich, R.M. 1998, *in preparation*
Nomoto, K., et al. 1997b, *NuPhA*. **616**, 79
Ortolani, S., et al. 1995, *Nat*, **377**, 701
Peletier, R.F., & Balcells, M. 1996, *AJ*, **111**, 2238
Peterson, R., & Green, E.M. 1997, *in preparation*
Rich, R.M. 1988, *AJ*, **95**, 828
Sadler, E.M., Rich, R.M., & Terndrup, D.M. 1996, *AJ*, **112**, 171
Scalo, J.M., Despain, K.H., & Ulrich, R.K. 1975, *ApJ*, **196**, 805
Sellgren, K. et al. 1997, *in preparation*
Sneden, C., McWilliam, A., et al. 1996, *ApJ*, **467**, 819
Terndrup, D.M., Sadler, E.M., & Rich, R.M. 1995, *JA*, **110**, 1774
Wheeler, J.C., Sneden, C., & Truran, J.W., Jr. 1989, *ARAA*, **27**, 279
Whitford, A.E., & Rich, R.M. 1983, *ApJ*, **274**, 723

THE MOST METAL-RICH DWARF STARS IN THE GALACTIC DISK

SOFIA FELTZING
Royal Greenwich Observatory & Insitute of Astronomy, Cambridge, U.K.

The metal-rich dwarf stars in the solar neighbourhood have historically not attracted as much attention as the metal-poor (halo) stars which tell us about the early phases of the chemical evolution of our galaxy. The properties of metal-rich stars are highly important when we try to interpret integrated spectra from metal-rich stellar populations, such as the Bulge and elliptical galaxies. A small group of so called super metal rich stars (SMR) have played a significant role in shaping the conceptions of such populations. Famous examples are the dwarf HR1614 and the giant μ Leonis. In the review by Taylor (1996) SMR stars are discussed in great detail, in particular the reality of extremly high [Fe/H]. Taylor (1996) find that no giant stars fulfil the criteria fulr SMR-ness that he sets and only a handful of dwarf stars do, and most of them are possible candidates rather than firm detections.

A few recent studies have studied known SMR candidates and stars with high [M/H] (as derived from photometry), Feltzing & Gustafsson (1997), Castro et al. (1997), Gonzalez (1997), McWilliam & Rich (1994), Castro et al. (1996). A first conclusion is that there are indeed quite a few stars with [Fe/H] > 0.2 dex (the definition of SMR, see Taylor 1996). We will here review these studies in some detail.

The combination of abundance ratios with kinematical data can detail the picture even further. For example we can study stars on highly eccentric orbits which trace the evolution in the galactic disk closer to the galctic centre. Not much is known about these stars but we did have some very intriguing observations; Barbuy & Grenon (1990) (hereafter BG90) found that dwarf stars on very eccentric orbits contained much more oxygen than what was expected from standard models of galactic chemical evolution of the disk, Edvardsson et al.(1993) (hereafter E93) found large spreads and "upturns" for certain elements, Na, Si, Ti, Al, for stars with 0.0dex < [Fe/H] < 0.2 dex.

1. Feltzing & Gustafsson (1997)

To further explore the questions sketched above we studied 47 G and K dwarf stars with photometric metallicities > 0.1 dex. Magnitudes and metallicities were combined with accurate velocity data (UVW) in the selection of stellar samples. One sample trace the chemical evolution in the disk interior to the Sun, $V_{\rm LSR} < -50$ km/s and/or $Q_{\rm LSR} > 60$ km/s (where $Q_{\rm LSR} = \sqrt{U^2_{\rm LSR} + V^2_{\rm LSR} + W^2_{\rm LSR}}$). $V_{\rm LSR} < -50$ km/s singles out stars with mean perigalactica less than 7 kpc. Another set of stars represent the solar orbit, $Q_{\rm LSR} < 30$ km/s. To explore the oxygen abundances in metal-rich disk dwarfs on highly eccentric orbits we added 3 stars from the sample from BG90. High resolution, R~80000, and high S/N, ~200, spectra of the stars were obtained with the 2d-coudé echelle spectrograph at the 2.7m McDonald telescope. For full details of selection and abundance analysis (standard LTE) see FG97. The abundance ratios are presented in Fig. 1.

We find that the relative abundances of O, Na, Si, and Ti relative to Fe do not vary with the star's present mean distance to the galactic centre (see e.g. Fig. ??). These results imply that there is no significant difference in the chemical evolution of the different stellar populations for stars with [Fe/H] > 0 dex. We find that [O/Fe] continues to decline with increasing [Fe/H], Fig. 2a, while [Si/Fe] and [Ca/Fe], however, seem to stay constant. A real ("cosmic") scatter in [Ti/Fe] at a given [Fe/H] is suggested as well as a decreasing abundance of the s-elements relative to Fe for the most metal-rich dwarfs. The "upturn" in [Na/Fe] vs [Fe/H] found by E93 is confirmed and reinforced, but its origin is still uncertain (metallicity dependent supernova yields may play an important role). We find no such trends for Al or Mg.

J. Andersen (ed.), Highlights of Astronomy, Volume 11A, 70–73.

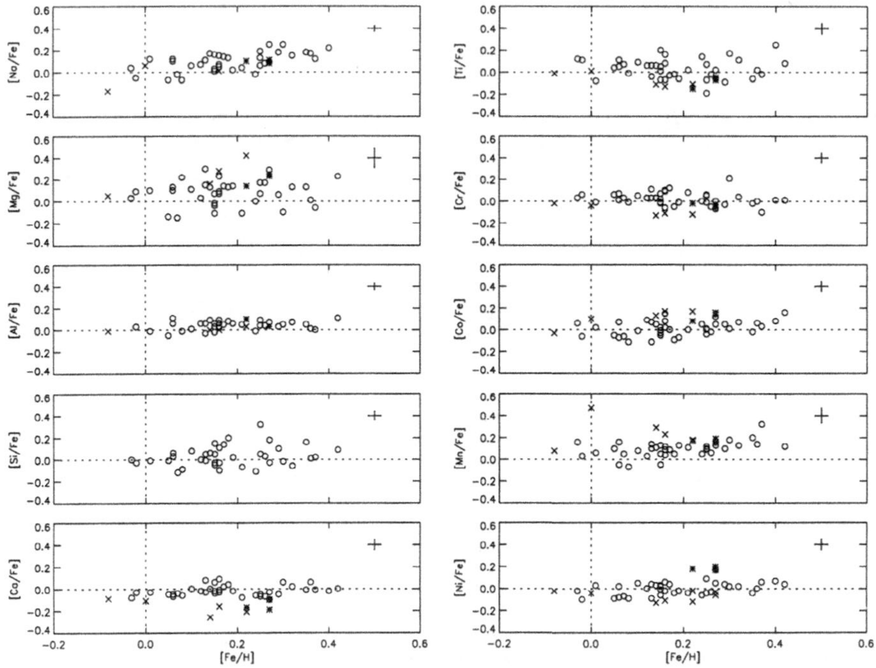

Figure 1. Abundances from Feltzing & Gustafsson (1997) relative to Fe. × symbols denote the five K dwarf stars while * symbols denote the stars in common with Barbuy & Grenon (1990).

An interesting question is now what difference in [Na/Fe] one would have reason to expect for stars formed at different $R_{\rm m}$. Using the data of Edvardsson et al. (1993) we estimate that the minimum value of d[Na/Fe]/d$R_{\rm m}$ is about 0.05 dex/kpc. The results obtained by FG97 supports this and indicate that the difference in [Na/Fe] is, in fact, at the most 0.05 dex.

In this connection, one should also note that orbital diffusion may well mask the possible differences between stars formed at 6 kpc and 8 kpc. E.g., stars of solar age in an orbit with $R_{\rm m} = 8.5$ kpc may have migrated from an orbit with $R_{\rm m} = 6 - 7$ kpc (cf. Wielen et al. 1996). The mixture of stars with different original orbits, in combination with a radial galactic gradient of [Fe/H], was proposed by Wielen et al. (1996) to explain the unexpectedly high scatter in [Fe/H] of 0.2 dex for solar type stars, with similar age and similar present $R_{\rm m}$, found by Edvardsson et al. (1993). From the work of Wielen et al. (1996) we estimate that two samples of stars with $\langle R_{\rm m} \rangle = 8.5$ and 6.5 kpc, respectively, would then be mixed by orbital diffusion so much that the population effects in abundances only would show up to about half the expected size as compared with the situation if orbital diffusion is not present. Although the reason for the great inhomogeneities in the gravitational potential, needed to account for the orbital diffusion of this magnitude, is not known, we conclude that the effects looked for by dividing the total sample of stars according to the velocity criteria used here, might be diminished considerably by this phenomenon.

2. Other studies

Barbuy & Grenon (1990) studied oxygen abundances for 11 dwarf stars on highly eccentric orbits. They find the stars to be enriched in oxygen as compared to expectations from models of galactic chemical evolution for stars of this metallicity, Fig. 2a.

Edvardsson et al. (1993): in their total sample of 189 dwarf stars, 53 have [Fe/H] > 0.0 dex (out of which 4 are above 0.2 dex). They found interesting features, for these stars, such as the upturns in Na, Mg and Al. They also found large spreads for certain elements, Si, Mg, Ti. The

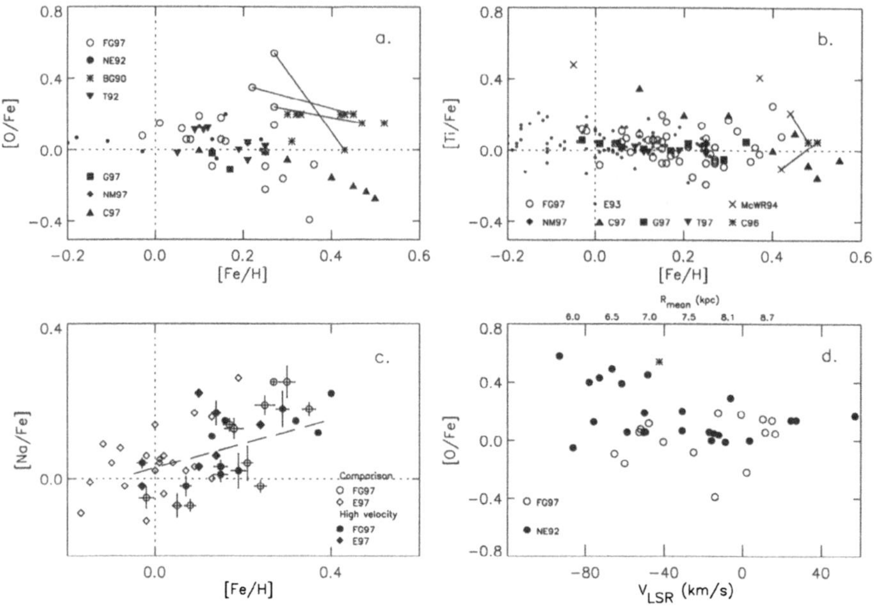

Figure 2. **a.** **Oxygen abundances from several studies. Results for the three stars studied by both FG97 and BG90 are connected with full lines. b. Titanium abundances from several studies. Giant stars studied by both McWR94 and C96 are connected with full lines. μ Leo is the star with highest [Fe/H] in the study by C96. c. The sodium data from FG97 devided into subsamples and compared with samples from E93 selected in the same way as for FG97. The dashed line indicates a least-square fit to all the stars in FG97. Errorbars are showed for the stars from FG97. d. Oxygen abundances from FG97 and NE92 as a function of V_{LSR}. V_{LSR} can be translated to mean distance to galactic centre. Distances are indicated on the top of the diagram. Acronyms used in the figures; BG90 = Barbuy & Grenon (1990), C96 = Castro et al. (1996), C97 = Castro et al. (1997), E93 = Edvardsson et al. (1993), FG97 = Feltzing & Gustafsson (1997), G97 = Gonzalez (1997abc), McWR94 = McWilliam & Rich (1994), NE = Nissen & Edvardsson (1992), NM97 = Neuforge & Magain (1997), T97 = Tomkin et al. (1997).**

question was raised weather or not this was a cosmic spread. E93 found a group of stars enriched in Fe and in Na, Mg and Al, dubbed the NaMgAl stars. **Tomkin et al. (1997)** re-investigated, with new spectra, nine of the stars studied in E93, six of which had been identified as NaMgAl stars. They find no evidence for these stars to be enriched in Na, Mg or Al. This is, for Na, contradictory to the findings of FG97. The main conclusion, however, is that there does not exist a group of stars enriched in Na, Mg and Al.

Further results on metal-rich dwarfs have been published by **Gonzalez (1997abc)**, **Neuforge-Verheecke & Magain (1997)** (α Centauri System) and **Castro et al. (1997)**. The results from these studies nicely confirm the findings by FG97, Fig. 2a and b.

Gonzalez has explored the link between high metallicity in planetary system candidates (e.g. $\rho^1$55 Cnc, 51 Peg). Tentative results from his investigations point to a correlation between metal-richness and the presence of planets.

Castro et al. (1997) investigated 9 stars selected in the same way as for BG90, i.e. very metal-rich stars on highly excentric orbits. From Geneva photometry their metallicities are between +0.3 and +0.6 dex; this is confirmed by the abundance analysis. A mixture of oscillator strengths are used. In general their results confirm the findings by FG97. One would expect these stars to show a similar abundance pattern for O as the BG90 stars. This is not the case, see Fig. 2a. While the BG90 stars clearly are overabundant in oxygen these additional 9 stars seem to be underabundant and follow the same trend as FG97. [Na/Fe] is $\sim$ 0.5 dex higher than for the stars studied by E93 and FG97 in the same [Fe/H] range. The reason for this is unclear. It is also not possible to directly address the question of abundances as a function of velocity, since velocity data for these stars have not yet been published.

Here we will only discuss two studies of giant stars. McWilliam & Rich (1994) derived abundances for 12 giant bulge stars. This work clearly showed that high dispersion is necessary for derivation of reliable abundances in giants. The McWilliam & Rich (1994) study suggests that the metallicity of the stars in the Bulge is, in the mean, slightly less than solar and with a large spread and that Si and Ti may be overabundant as compared to the local disk stars. On the other hand the Castro et al. (1996) study (at still higher resolution) of two giants, one from McWilliam & Rich (1994) and μ Leo, show the stars to be very similar to disk dwarfs of the same [Fe/H], Fig. 2b.

3. Concluding remarks

To conclude we are most probably seeing, in the most metal-rich disk dwarfs, the natural end of "normal" galactic chemical evolution in the disk. The abundance trends of [Element/Fe] vs [Fe/H] up to [Fe/H] = 0.2 dex are now well established for a large number of elements (O, Na, Mg, Al, Ca, Ni etc). These trends fit nicely into the overall picture of chemical evolution in the galactic disk as it is currently understood, apart from, possibly, O, Na, Si and Ti.

The full power of combining abundance data and kinematical data in studies of stellar populations has not yet been fully exploited, nor have the particular problems that may hamper such investigations, e.g. migration and its origin, been put on a sure footing. This needs to be done before data for stars on highly eccentric orbits can be reliably interpreted.

We cannot, so far, infer much about the bulge population(s) from locally studied stars. This may not even be possible, but nevertheless studies of stars on highly eccentric orbits are of great interest since they may tell us about the evolution of the inner disk. This in combination with future direct studies of abundances in the bulge stars will help to discriminate between different formation scenarios for bulges, i.e. are they separate populations related to the disks (suggested by colours of disks and bulges Balcells and Peletier 1996) or halos (suggested by kinematics e.g. Ibata and Gilmore 1995). For the time being we can only conclude that the bulge giants (McWilliam & Rich 1994) do not look any different in terms of abundance ratios, at a given [Fe/H], than do the local dwarfs (taking the errors into account) and that the abundance trends for stars from the inner disk are highly uncertain and have to be confirmed before one can make any firm conclusions on the relation between disks and bulges in spiral galaxies.

Spectra of such metal-rich stars as discussed here are very crowded and there is a clear need for further studies at high resolution and high signal to noise so that continua can be reliably defined and blends excluded. This is now possible for large samples of local dwarf stars, as demonstrated by E93 and FG97. For giant stars in the Bulge the new spectrographs on 8m telescopes (e.g. UVES on VLT) will make it possible to observe larger samples of stars in reasonable integration times. In particular we should study the metal-rich dwarf stars on highly eccentric orbits and the Bulge giants to gain clues to the formation scenario of the Bulge and bulges in other galaxies. We also need to build up statistics for local dwarfs with [Fe/H] between 0.2 and 0.4 dex and to confirm or disconfirm the very highest iron abundances derived.

References

Balcells & Peletier, 1996, *AJ*, **111**, 2238
Barbuy B., Grenon M., 1990, in *Bulges of Galaxies, ESO/CTIO workshop*, ed. Jarvis, B.J., Terndrup, D.M., 83
Castro S., Rich R.R., Grenon M., et al., 1997, *AJ*, **114**, 376
Castro S. Rich R.R., McWilliam A., et al. 1996, *AJ*, **111**, 2439
Edvardsson B., Andersen J., Gustafsson B., et al., 1993, *A&A*, **275**, 101
Feltzing S., Gustafsson B., 1997, *A&AS*, accepted
Gonzalez G., 1997a, *PASP*, in press
Gonzalez G., 1997b, preprint, to appear in proceedings of "Brown Dwarfs and Extrasolar Planets Workshop"
Gonzalez G., 1997c, *MNRAS*, **285**, 403
Ibata R., Gilmore G., 1995, *MNRAS*, **275**, 605
Mcwilliam A., Rich M., 1994, *ApJS*, **91**, 749
Neuforge-Verheecke C., Magain, P., 1997, *A&A*, accepted
Taylor B.J., 1996, *ApJS*, **102**, 105
Tomkin J., Woolf V.M., Lambert D.L., Lemke M., 1995, *AJ*, **109**, 2204
Wielen R., Fuchs B., Dettbarn C., 1996, *A&A*, **314**, 438

CHEMICAL EVOLUTION OF ELLIPTICAL GALAXIES

Mg_2 Gradients and G-Dwarf Problem

N.ARIMOTO
Institute of Astronomy, University of Tokyo
Mitaka, Tokyo 181, Japan

AND

C.KOBAYASHI
Department of Astronomy, School of Science, University of Tokyo
Yayoi, Bunkyo-ku, Tokyo 113, Japan

1. Introduction

Stellar populations of elliptical galaxies were studied extensively in the recent past (eg., Arimoto & Yoshii 1987; Buzzoni et al. 1992; Worthey 1994; Kodama & Arimoto 1997). However, colours and line-strengths analysed by these approaches came mainly from central regions of galaxies and possible variations of stellar population structure within a galaxy were entirely ignored. Kodama & Arimoto (1997) recently show that a colour-magnitude (CM) relation of ellipticals in Coma cluster (Bower, Lucey, & Ellis 1992), and the CM relation of cluster ellipticals in general, should originate from a difference in mean stellar metallicities, as suggested by Arimoto & Yoshii (1987), and cannot be due to an increase of mean stellar age towards luminous galaxies as claimed by Worthey (1994). It is quite often stated that any theory of galaxy formation should explain the origin of CM relation, or in other words, the mass (luminosity) - metallicity relation of elliptical galaxies. However, one should not forget the fact that the CM relation is defined only for stars in the central regions of galaxies; it is possible that the CM relation itself is an observational artifact, since ellipticals show in general radial gradients of metallic-line strengthes, decreasing from a centre towards outer regions (eg., Faber 1977; Davies, Sadler, & Peletier 1993; Gonzalez 1993; Carollo & Danziger 1994). Indeed, Gorgas & Gonzalez (1996) found that ellipticals with larger central Mg_2 indices tend to have steeper Mg_2 gradient, which suggests that global metallicities of ellipticals are perphaps universal and are independent of galaxy luminosities.

In this article we briefly present our latest results derived from a study of Mg_2 gradients of 114 ellipticals available in the literature and address ourselves to a reality of global CM relation of ellipticals. Most galaxies are in the field and details of data, including the references, will be given elsewhere.

2. Global CM Relation

Arimoto et al. (1997) have shown that the mean stellar metallicity of an elliptical galaxy is approximately given by the projected metallicity measured at one effective radius (r_e), if the luminosity profile is given by de Vaucouleurs law and the Mg_2 gradient is well fitted by an exponential law. We have calculated the mean stellar metallicity of each galaxy by actually integrating the observed Mg_2 profile and have confirmed that Mg_2 at $r = r_e$ is indeed a good measure of the mean stellar metallicity. Although Mg_2 index is not usually measured beyond $r = r_e$, we assume that it decreases exponentially outwards. However, we believe that this would not intorduce any significant uncertainty in our results, since it is very unlikely that Mg_2 decreases abruptly beyond $r = r_e$ and even in such a case the resulting metallicity is overestimated no more than twice. Figure 1 shows Mg_2 at the galaxy centre (top) and at an effective radius (bottom). Mg_2 indices are transfered into the metallicity (*but not iron abundance*) [Fe/H] by using the formula given by Buzzoni et al. (1992), where we assume that all galaxies are 15 Gyrs old. In the top panel, a larger sample is taken from Davies et al. (1987), and a small one is ours. There is no systematic difference between

J. Andersen (ed.), Highlights of Astronomy, Volume 11A, 74–77.

the two sampled galaxies; we thus believe that our sample represent a fair population of ellipticals in the field. The bottom panel gives the mean stellar metallicities of 114 ellipticals which we analysed. Obviously, the global metallicities are much lower than the central values; for example, the histogram peaks at $Mg_{2,e} \simeq 0.24$ ($[Fe/H]_e \simeq -0.3$) which is about 1/3 of the peak location of the central values at $Mg_{2,c} \simeq 0.31$ ($[Fe/H]_c \simeq +0.2$). We note that the mean metallicity of elipticals is not universal, instead it varies from $[Fe/H] \sim -1.0$ to $\sim +0.2$. This indicates that the metallicity gradients of luminous ellipticals are not steep enough to reduce the global metallicity as low as that of faint dwarf ellipticals.

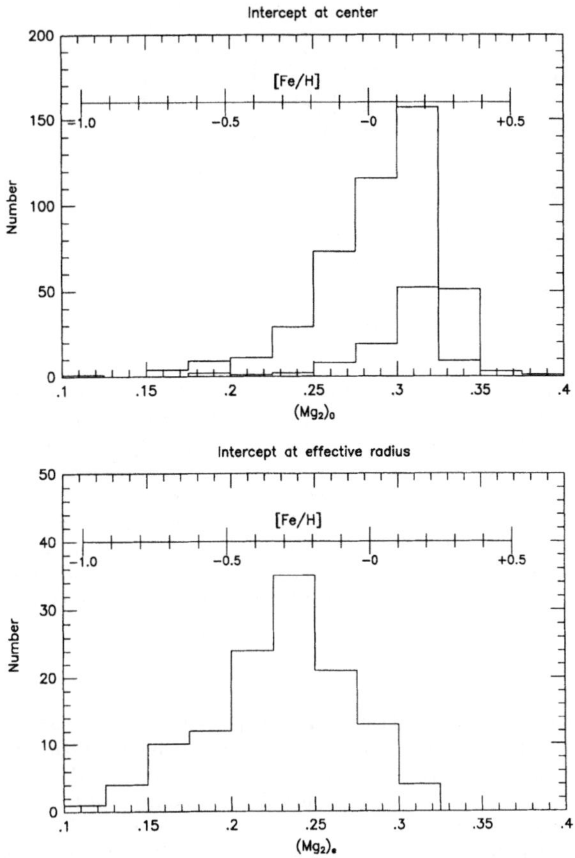

Figure 1. Central Mg_2 indices of 114 ellipticals (top). Larger sample is taken from Davies et al.'s (1987). Mg_2 indices of 114 ellipticals at one effective radius (bottom).

Figure 2 shows a plot of 114 ellipticals in the $Mg_{2,e}$ versus $\log \sigma_0$ (central velocity dispersion) diagram. Although the relationship is not so tight as that of the central Mg_2 index, the global metallicity increases significantly from $[Fe/H]_e \sim -0.6$ at $\log \sigma_0 = 1.9$ to $[Fe/H]_e \sim -0.1$ at $\log \sigma_0 = 2.6$. This could be compared to the equivalent of $Mg_{2,c}$ versus $\log \sigma_0$ relation; $[Fe/H]_c \sim -0.4$ at $\log \sigma_0 = 1.9$ and $[Fe/H]_c \sim +0.3$ at $\log \sigma_0 = 2.6$. Therefore, we conclude that elliptical galaxies do have the global CM relation, but the slope is less steep than the central CM relatioin.

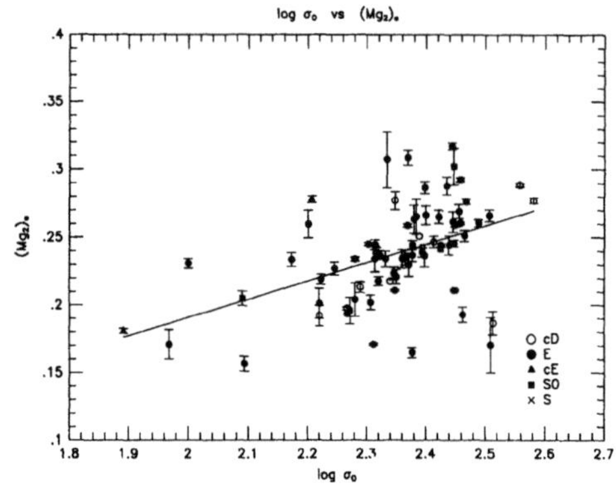

Figure 2. $Mg_{2,e}$ versus $\log \sigma_0$ relation of 114 ellipticals.

In Fig.2 one may immediately notice that the scatter around the $Mg_{2,e} - \log \sigma_0$ relation is unexpectedly large. Possible causes for the dispersion are as follows: 1) variation due to different observation facilities, 2) different correction for velocity dispersion gradient, 3) deviation from de Vaucouleurs luminosity profile, 4) abrupt drop of Mg_2 index beyond $r = r_e$, 5) deviation from an exponential fit for Mg_2 at the central region, 6) errors in effective radius, and 7) gradient of M/L ratio. We have checked each of them in detail and found none of them could cause the dispersion appearing in Fig.2. This leads us to conclude tentatively that the dispersion around the relationship between the mean stellar metallicity and the central velocity dispersion, which should be an indicator of galactic mass, could be real and come directly from the different history of galaxy formation.

3. G-dwarf Problem

The observed luminosity profile and the Mg_2 gradient are the 2D projected ones. By using Abell integral, we have deprojected these 2D distributions and derive the 3D luminosity profile and the 3D metallicity distribution for NGC4472, a giant elliptical galaxy in our sample. If we assume that all stars within a spherical shell $(r, r + dr)$ have an identical metallicity given by the 3D-[Fe/H] distribution thus derived, and if the variation of M/L ratio along the radius is negligible, it is possible to calculate the number fraction of stars with a specfied [Fe/H] value and to derive the stellar metallicity distribution which is found to be quite useful to assign chemical evolution history of the solar neighbourhood disc of the Milky Way Galaxy (eg., Köppen & Arimoto 1990). **A similar kind of chemical evolution analysis is now possible for elliptical galaxies.** We have found that the metallicity distribution of NGC4472 is in the range $-0.5 \leq [Fe/H] \leq +0.9$, but the metallicity of bulk of stars is within $-0.2 \leq [Fe/H] \leq +0.3$ and takes a sharp peak at $[Fe/H] \simeq 0.0$. The metallicity distribution of NGC4472 is somewhat similar to the G-dwarf metallicity distribution in the solar neighbourhood disc, but is shifted systematically by $\Delta[Fe/H] \sim +0.3$. *Unexpectedly, NGC4472 is lacking metal-poor stars of $[Fe/H] \leq -0.2$, a clear sign of the G-dwarf problem.* The G-dwarf problem is usually solved by invoking significant amount of gas infall with a rate similar to that of star formation. The same may apply for elliptical galaxies.

The location of the peak of metallicity distribution gives directly the yield of heavy elements synthesis (Köppen & Arimoto 1990). If a galaxy lost a part of heavy elements to the intragalactic space, the peak metallicity should be read as an effective yield, but it is unlikely that a giant elliptical galaxy NGC4472 had lost a significant amount of gas and metals since its gravitational potential is quite large. Therefore, the peak metallicity should give the true yield, and if so, the yield of NGC4472 is about solar and is almost twice larger than the yield in the solar neighbourhood disc. Since the latter is well reproduced by the Salpeter initial mass function (IMF), our result clearly indicates that the IMF of NGC4472, and perphaps all ellipticals, is less steeper than the Salpeter IMF, thus confirming the claim first made by Arimoto & Yoshii (1987).

4. Conclusions

Elliptical galaxies with larger central Mg_2 index tend to have steeper Mg_2 gradient, but this does not smear out the global CM relation. However, the scatter around the global CM relation is surprisingly large and we tentatively conclude that this dispersion may be associated with the formation mechanism of elliptical galaxies.

The 3D metallicity distribution of NGC4472 shows a symptom of typical G-dwarf problem, ie., a lack of metal-deficient stars. This suggests that elliptical galaxies formed from infalling gas clumps towards the central region of galactic gravitational potential, and that bulk of stars formed intensively with a flat IMF within 1-2 Gyrs, perphaps, within less than 0.5 Gyrs.

References

Arimoto, N., Matsushita, K., Ishimaru, Y., Ohashi, T., & Renzini, A., 1997, *ApJ*, **477**, 128
Arimoto, N. & Yoshii, Y., 1987, *A&A*, **173**, 23
Bower, R.G., Lucey, J.R., & Ellis, R.S., 1992, *MNRAS*, **254**, 601
Buzzoni, A., Gariboldi, G., & Mantegazza, L., 1992, *AJ*, **103**, 1814
Carollo, C.M., & Danziger, I.J., 1994, *MNRAS*, **270**, 523
Davies, R.L., Sadler, E.M., & Peletier, R.F., 1993, *MNRAS*, **262**, 650
Davies, R.L., Burstein, D., Dressler, A., Faber, S.M., Lynden-Bell, D., Terlevich, R.J., & Wegner, G., 1987, *ApJS*, **64**, 581
Faber, S.M., 1977, in The Evolution of Galaxies and Stellar Populations, eds., .T.Tinsley & R.B.Larson (New Heaven: Yale Univ. Press) p.157
Gonzalez, J.J., 1993, PhD thesis, Univ. of Califolnia
Gorgas, J., & Gonzalez, J.J., 1996, *ASP Conference Series*, **86**, 225
Kodama, T., & Arimoto, N., 1997, *A&A*, **320**, 41
Köppen, J., & Arimoto, N., 1990, *A&A*, **240**, 22
Worthey, G., 1994, *ApJS*, **95**, 107

ABUNDANCE RATIOS IN COMPOSITE STELLAR POPULATIONS WITH SPECIAL EMPHASIS ON ELLIPTICAL GALAXIES

UTA FRITZE - VON ALVENSLEBEN
Universitätssternwarte Göttingen
Geismarlandstr. 11, 37083 Göttingen, Germany

1. Introduction

Abundance Ratios are observed in a wide variety of instances addressed in this volume: in **Gas** and **Stars**, Simple Stellar Populations (**SSPs**) like star clusters, and, finally, in Composite Stellar Populations (**CSPs**) like E/S0 galaxies (Bulges, spheroidal galaxies, Merger Remnants, etc.) on which the focus of this contribution will be. I'll briefly review the observational situation and the current interpretation and then cast some doubt on the latter from our own chemically consistent spectrophotometric modelling results.

2. Observations and Standard Interpretation

As defined in the Lick system, absorption line indices like Fe5270, Fe5335, Mg_2, Mgb, and many others have been measured by various groups on large numbers of early type galaxies and bulges. For the nuclei of giant field ellipticals Fe5270 and Fe5335 are seen to correlate with Mg_2, but the slopes of Fe5270 and Fe5335 versus Mg_2 are found to be rather shallow with considerable scatter, they do not follow the relation for globular clusters, and they are flatter for nuclei of different galaxies than within one galaxy (see e.g. Faber *et al.* 1992, Worthey *et al.* 1992, Davies *et al.* 1993). **Gradients** in Mg_2 tend to be stronger for larger Es, gradients in Fe correlate with gradients in Mg_2. Mg_2 correlates with velocity dispersion σ while $\langle Fe \rangle := \frac{1}{2}(Fe5270 + Fe5335)$ does not (or only weakly). Ellipticals with fine structure or peculiar core kinematics tend to have lower Mg_2 than predicted for their luminosity by the luminosity – metallicity relation (see e.g. Schweizer *et al.* 1990, Bender *et al.* 1993, Carollo & Danziger 1994).

For a large sample of E/S0s in nearby **clusters** Jørgensen 1997 confirms the trends for field ellipticals and shows that Mg_2 and $\langle Fe \rangle$ both correlate with projected cluster surface density and that Mg_2 strongly correlates with M/L while $\langle Fe \rangle$ does so only weakly.

Comparing the relative strengths of Fe5270, Fe5335, and Mg indices measured on CSPs with those observed in Galactic reference stars generally leads to the conclusion that **magnesium seems to be enhanced with respect to iron** in several cases. Typical values quoted for the nuclei of massive boxy **Ellipticals** are $[Mg/Fe] \sim 0.4 - 0.6$, while $[Mg/Fe] \sim 0$ for the lower luminosity disky ellipticals. [Mg/Fe] seems to increase with increasing velocity dispersion, luminosity, metallicity, and – for cluster ellipticals – with increasing cluster surface density, while inside galaxies [Mg/Fe] seems to be constant (see Bender 1997 for an extensive review). For **Bulges** the situation is less clear. While Sadler *et al.* 1996 find $[Mg/Fe] \sim 0$ for $[Fe/H] \geq 0$ and $[Mg/Fe] \sim +0.2$ for $[Fe/H] < 0$, Jablonka *et al.* 1996 report a range of [Mg/Fe] values from 0 to +0.5 with no difference between low and high luminosity bulges. **Spheroidal Galaxies** have $[Mg/Fe] \sim 0$ (Gorgas *et al.* 1997), the centers of luminous **S0s** resemble ellipticals with $[Mg/Fe] > 0$ (Fisher *et al.* 1996).

The **Standard Interpretation** of these results relies on the assumptions that

1. relative strengths of absorption indices reflect element abundance ratios in the stellar population and
2. that these stellar element ratios reflect abundance ratios in the ISM out of which the stars were formed.

J. Andersen (ed.), Highlights of Astronomy, Volume 11A, 78–81.

If, and only if, both assumptions were correct (I'll show below that this is not the case for CSPs), an observed stellar $[Mg/Fe]_* > 0$ would reflect an ISM abundance ratio $[Mg/Fe]_{ISM} > 0$ at the birth of (the bulk of) the stellar population.
Mg, like other α-elements, is a typical SNII product that is returned to the ISM with the death of the first massive stars on a timescale of 10^6 yr. Fe has significant ($\gtrsim 50\%$) contributions from SNIa, it is restored to the ISM on a timescale of several 10^8 yr (see e.g. Matteucci 1994).
$[Mg/Fe]_{ISM} > 0$ would have either (or several) of the following implications for the **central regions** of galaxies:

- an IMF favoring High Mass Star Formation (**SF**), at least in early evolutionary stages or for a stellar Pop3
- a characteristic timescale for SF $t_* < t_{SNI} \sim 3 \cdot 10^8$ yr
- a lower fraction of SNI progenitor binary stars
- mass segregation and incomplete mixing (since only nuclear regions are observed)
- low metallicity stellar yields (favour Mg over Fe for SNII)

Whatever effect is invoked, it must be stronger in massive Es, S0s (and possibly bulges) than in low mass systems (see e.g. Faber *et al.* 1992, Worthey *et al.* 1992, Matteucci 1994, Bender *et al.* 1993, Thomas *et al.* 1997 *in prep*). The SF timescale $t_* \sim 3 \cdot 10^8$ yr is much shorter than the collapse time of order ~ 1 Gyr in the classical initial collapse SF scenario for ellipticals (Larson 1974), it reminds of the duration of starbursts in spiral-spiral mergers or of the lifetime of the Infrared Ultraluminous phase. Intriguingly, ASCA measurements seem to imply $[Mg/Fe] > 0$ for the hot ICM in galaxy clusters despite the fact that the timescale for ICM enrichment is of the order of a Hubble time.

Stellar abundance ratios in CSPs depend on the SF history, the IMF, any eventual pre-enrichment (e.g. Pop3, halo $\rightarrow$ disk) or dilution (metal-poor infall) of the ISM **before** the birth of the stars dominating the light at the index wavelengths today **and** they also depend on possible modifications inside the stars through nucleosynthesis and mixing. Clearly, abundance ratios of the sun or solar neighborhood reference stars reflect the local SF history, IMF, and all possible pre-enrichment and dilution effects of the local ISM. A serious complication comes from the fact that e.g. the iron index $\langle Fe \rangle$ is as sensitive to iron as it is to global metallicity (which is dominated by the SNII product oxygen), as shown by Tripicco & Bell 1995.

3. Chemically Consistent Modelling of Galaxy Evolution

Starting from a gas cloud with primordial abundances and specifying a SF rate (as a function of time or current gas content) and an IMF – the two fundamental parameters of our model – successive generations of stars are formed with metallicities increasing from $Z = 0$ to some present value. Their distribution over the HRD is calculated from stellar evolutionary tracks for various metallicities. Colour and absorption index calibrations in terms of (T_{eff}, log g, [Fe/H]) and a library of model atmosphere spectra for various metallicities are used to describe the photometric and spectral evolution in a chemically consistent way (Einsel *et al.* 1995, Möller *et al.* 1997). Stellar yields are taken from nucleosynthesis models for a variety of metallicities. Solving a modified form of Tinsley's equations including SNI contributions from carbon deflagration white dwarf binaries in the way prescribed by Matteucci 1994 we obtain the time evolution of ISM abundances and abundance ratios for a series of elements. We properly account for the finite lifetime of each star, i.e. we do not use an Instantaneous Recycling Approximation. The SF history appropriate for the type of galaxy together with the IMF determine the evolution of ISM abundances and abundance ratios, the age and metallicity distributions, and the colour and index evolution of a composite stellar population.

While by now the first reasonably complete sets of stellar evolutionary tracks, yields and model atmospheres for various metallicities are becoming available, the influence of non-solar abundance ratios is still far from being explored. A few recent efforts include Barbuy 1994 and Barbuy *et al.* 1995 who take the relative proportions of stars in various evolutionary stages from observed CMDs of globular clusters with various [Mg/Fe] and combine with Kurucz's model atmospheres (only available for $[Mg/Fe] = 0$). Borges *et al.* 1995 give calibrations for Mg_2 and NaD in terms of (T_{eff}, log g, [Fe/H], **and [Mg/Fe]**), and Weiss *et al.* 1995 show that for a 0.9 $M_\odot$ star the transition from an initial composition with $[Mg/Fe] = 0$ to $[Mg/Fe] = +0.4$ increases its Mg_2 index at fixed metallicity by as little as 0.03.

No complete grid in stellar mass, [Mg/Fe], or [α/Fe] is available, neither for stellar evolutionary tracks, nor for yields, colour or index calibrations, not to mention model atmospheres.

A problem with stellar yields for non-solar metallicities is that they depend on several poorly known quantities, as e.g. $\Delta Y/\Delta Z$, explosion energies, mass loss rates, remnant masses, etc., and so, might not be very reliable yet. Beyond the discrepancies between various authors there is agreement, however, that for type II SNe the ratio of the Mg and the Fe yields, integrated over any plausible IMF, increases considerably with decreasing metallicity. Chemically consistent modelling of the ISM abundance evolution thus results in higher $[Mg/Fe]_{ISM}$ ratios in early evolutionary phases as compared to models using solar metallicity yields only. If the bulk of stars in an elliptical were formed on a short timescale we would expect them to have an enhanced Mg-to-Fe ratio frozen in. If this were then reflected in the stellar absorption line ratio of its CSP is an open question (see *caveats* below and also Tantalo *et al.* 1997). In a spiral-spiral merger scenario for the formation of an E/S0 galaxy including a strong burst of SF the low metallicity stars of the gas-rich late-type spiral progenitors are all expected to have incorporated $[Mg/Fe]_{ISM} > 0$ and the stars forming in the burst on a timescale of $\sim 10^8$ yr will have built in the rapidly increasing $[Mg/Fe]_{ISM}$ during the burst (Fritze - v. A. & Gerhard 1994a). This effect is seen in the spectra of young star clusters (SSPs) in the merger remnant NGC 7252 (Schweizer & Seitzer 1993, Fritze - v. A. & Burkert 1995).

With appropriate SF histories our models for the evolution of ISM abundances and abundance ratios were used with solar metallicity yields to interprete the redshift evolution of MgII- and CIV-QSO absorption line systems that originate in galaxy haloes (Fritze - v. A. *et al.* 1989, 1991) and of Damped Lyα Absorbers (= DLAs) presumed to be (proto-)galactic disks (Fritze - v. A. & Fricke 1995). First results from our chemically consistent models in comparison to DLA abundances are presented by Lindner *et al.* 1998.

4. Results for Absorption Indices

For SSPs of various metallicities we obtain the time evolution of colours, luminosities and indices. Using colours together with absorption indices allows to largely disentangle the age-metallicity degeneracy, since – as first shown by Worthey 94 – each colour and each index has its specific sensitivities to age and metallicity (Fritze -v. A. & Burkert 1995). Model results include theoretical calibrations of indices, as e.g. Mg_2, in terms of [Fe/H] (Kurth *et al.* 1997, *in prep.*).

For the CSP of an elliptical galaxy, our model e.g. for a "classical L* – Elliptical" (1 zone, no dynamics with a characteristic timescale for SF $t_* = 1$ Gyr as indicated for an initial collapse scenario) gives a broad metallicity distribution with a luminosity-weighted mean metallicity in the V-band $\langle Z\rangle_V \sim \frac{1}{2}\cdot Z_\odot$. While the metallicity distribution of our global galaxy model falls off sharply towards higher than solar metallicities it has a broad wing towards metallicities lower than $\frac{1}{2}\cdot Z_\odot$ (Möller *et al.* 1997). Similar results from independent modelling approaches are reported by Greggio 1997 and Arimoto 1997 (*this Joint Discussion*). The stellar metallicity you see depends on the wavelength of your observation. For an elliptical, it is higher by $\sim 40\%$ in K than it is in the V-band. Needless to say, luminosity weighted mean stellar metallicities in general differ from the ISM metallicity. Taking a consistent metallicity distribution into account, our model reaches the observed range of $[MgFe]_* := \mathrm{Log}\,\sqrt{\mathrm{Mgb}\cdot\langle\mathrm{Fe}\rangle} \sim 0.4 - 0.6$ (see e.g. Gonzalez 1993) **without** [Mg/Fe] > 0, neither for stellar tracks, nor for index calibrations (Möller *et al.* 1997). With stellar evolutionary tracks from the Padova group, our chemically consistent "classical elliptical galaxy" model reaches $Mg_2 \geq 0.30$ and Fe5335 ≥ 3.1 at ages ≥ 15 Gyr. Typical values observed in the nuclei of E/S0 galaxies are in the range $0.18 \leq Mg_2 \leq 0.37$ and $2.0 \leq$ Fe5335 ≤ 3.5 (Worthey *et al.* 1992, Gonzalez 1993). In comparison with observational data it should be kept in mind that those always refer to central regions of galaxies ($R < (\frac{1}{2} - 1)\cdot R_e$) and are thus expected to give upper limits only to our global galaxy models.

Increasing the timescale for SF we find that all models with $t_* \leq 5$ Gyr and an average metallicity $\langle Z\rangle_V > \frac{1}{5}\cdot Z_\odot$ reach the observed range of $[MgFe]_* \sim 0.4$ - 0.6. Models with $t_* = 5$ Gyr and $\langle Z\rangle_V \geq \frac{1}{2}\cdot Z_\odot$ reach $Mg_2 \geq 0.30$ and Fe5335 ≥ 2.7 at ages ≥ 15 Gyr.

Revisiting our spiral-spiral merger models with interaction-triggered starbursts of various strengths (Fritze -v. A. & Gerhard 1995a,b) it turns out that in case of a strong burst (like the one in NGC 7252) merger remnants reach the observed $[MgFe]_* \sim 0.4$ - 0.6 about 2 Gyr after the burst for $\langle Z\rangle_V = Z_\odot$ and about 4 Gyr after the burst for $\langle Z\rangle_V = \frac{1}{2}\cdot Z_\odot$. Merger remnants keep lower Mg_2

and Mgb than ellipticals at fixed M_B for about 3 Gyr in agreement with observations (Schweizer *et al.* 1990, Bender *et al.* 1993) but can reach even the highest Mg_2 and Fe index values observed in ellipticals about 4 – 5 Gyr after the burst.

Spirals with SF truncation (via tidal stripping, sweeping or harassment) but without a burst – which dynamically might transform into spheroidal galaxies – reach the observed $[MgFe]_* \sim 0.4$ - 0.6, rapidly for $\langle Z \rangle_V = Z_\odot$, i.e. early type spirals, but hardly for $\langle Z \rangle_V = \frac{1}{2} \cdot Z_\odot$ (Sbc or later). They feature average elliptical galaxy values for Mg_2 and Fe5335 in case of early type spirals but values close to the lower limit of ellipticals (as appropriate for their low luminosity) in case of late type spirals (Fritze - v. A. 1997, *in prep.*).

All this is readily understood looking at the Lick calibrations (Gorgas *et al.* 1993). They directly show that $[MgFe]_* \gtrsim 0.4$ for giants later than K0 ($\frac{1}{2} \cdot Z_\odot$) or later than G6 ($Z_\odot$), so that whenever these stars dominate the light $[MgFe]_*$ - values like those observed in giant Ellipticals are to be expected.

5. Conclusions

In addition to giving a very brief overview on observations and current interpretations of abundances and abundance ratios in composite stellar populations I stressed that elliptical galaxies are really composite in terms of stellar metallicities (and ages) and require a chemically consistent description that takes into account the range of initial stellar abundances. I showed that while for stars or SSPs, like star clusters, absorption indices do reflect ISM abundance ratios at birth, this is not the case for CSPs. Conclusions concerning the timescale for SF or the IMF based on Mg_2 and Fe index data for elliptical galaxies or bulges may well be true, but need not.

Acknowledgements

I greatfully acknowledge a travel grant (Fr 916/4-1) from the Deutsche Forschungsgemeinschaft.

References

Barbuy, B., 1994, *ApJ*, **430**, 218
Barbuy, B., de Freitas Pacheo, J. A., Idiart, T., 1995, IAU Symp. **171**, p. 340
Bender, R., 1997, in *The Nature of Elliptical Galaxies*, 2nd Mt. Stromly Symp. *in press*
Bender, R., Burstein, D., Faber, S. M., 1993, *ApJ*, **411**, 153
Borges, A. C., Idiart, T. P., de Freitas Pacheco, J. A., Thevenin, F., 1995, *AJ*, **110**, 2408
Carollo, C. M., Danziger, I. J., 1994, *MNRAS*, **270**, 523 + 743
Davies, R. L., Sadler, E. M., Peletier, R. F., 1993, *MNRAS*, **262**, 650
Einsel, C., Fritze - v. Alvensleben, U., Krüger, H., Fricke, K. J., 1995, *A&A*, **296**, 347
Faber, S. M., Worthey, G., Gonzalez, J. J., 1992, IAU Symp. **149**, 255
Fisher, D., Franx, M., Illingworth, G., 1996, *ApJ*, **459**, 110
Fritze - v. Alvensleben, U., Burkert, A., 1995, *A&A*, **300**, 58
Fritze - v. Alvensleben, U., Fricke, K. J., 1995, IAU Symp. **164**, p. 457
Fritze - v. Alvensleben, U., Gerhard, O. E., 1994, *A&A*, **285**, 751 + 775
Fritze - v. Alvensleben, Krüger, H., Fricke, K. J., Loose, H.-H., 1989, *A&A*, **224**, L1
Fritze - v. Alvensleben, Krüger, H., Fricke, K. J., 1991, *A&A*, **246**, L59
Gonzalez, J. J., 1993, PhD Thesis, Univ. of California, Santa Cruz
Gorgas, J., Faber, S. M., Burstein, D., Gonzalez, J. J., Courteau, S., Prosser, C., 1993, *ApJS*, **86**, 153
Gorgas, J., Pedraz, S., Guzman, R., Cardiel, N., Gonzalez, J. J., 1997, *ApJ*, **481**, L19
Greggio, L., 1997, *MNRAS*, **285**, 151
Jablonka, P., Martin, P., Arimoto, N., 1996, *AJ*, **112**, 1415
Jørgensen, I., 1997, *MNRAS*, **288**, 161
Larson, R. B., 1974, *MNRAS*, **166**, 585
Lindner, U., Fritze - v. Alvensleben, Fricke, K. J., 1998, in 13th IAP Colloquium *Structure and Evolution of the IGM from QSO Absorption Line Systems*, eds. P. Petitjean, S. Charlot, *in press*
Matteucci, F., 1994, in *Panchromatic View of Galaxies*, eds. Hensler *et al.* , Editions Frontières, p. 277
Möller, C. S., Fritze - v. Alvensleben, U., Fricke, K. J., 1997, *A&A*, **317**, 676
Sadler, E. M., Rich, R. M., Terndrup, D. M., 1996, *AJ*, **112**, 171
Schweizer, F., Seitzer, P., 1993, *ApJ*, **417**, L29
Schweizer, F., Seitzer, P., Faber, S.M., Burstein, D., Dalle Ore, C.M., Gonzalez, J.J., 1990, *ApJ*, **364**, L33,
Tantalo, R., Bressan, A., Chiosi, C., 1997, *astro-ph/9705060 + 9710101*
Tripicco, M. J., Bell, R. A., 1995, *AJ*, **110**, 3035
Weiss, A., Peletier, R. F., Matteucci, F., 1995, *A&A*, **296**, 73
Worthey, G., 1994, *ApJS*, **95**, 107
Worthey, G., Faber, S. M., Gonzalez, J. J., 1992, *ApJ*, **398**, 69

METALLICITY DISTRIBUTIONS IN EXTRAGALACTIC GLOBULAR CLUSTER SYSTEMS: CONSTRAINTS ON GLOBULAR CLUSTER AND GALAXY FORMATION

JEAN P. BRODIE
Lick Observatory
University of California, Santa Cruz, CA 95060, U.S.A.

1. Introduction

The merger model for elliptical galaxy formation has received increasing attention since it was first suggested by Toomre & Toomre (1972). Van den Bergh (1984) pointed out a problem with the idea that elliptical galaxies were formed by simply combining two, or more, spiral galaxies. He noted that the specific frequency (S_N, number of globular clusters per unit galaxy light) is systematically lower for spirals than for ellipticals. Schweizer (1987) suggested that globular clusters (GCs) might be expected to form in the merger process, thereby alleviating or possibly eliminating the S_N problem. Ashman & Zepf (1992) developed this idea into a merger model for GC formation with testable predictions.

We recently examined this model in the light of new HST and ground-based imaging data on the blue and red sub-populations of GCs in elliptical galaxies (Forbes, Brodie & Grillmair 1997). We concluded that the merger model for GC formation has serious problems, particularly in explaining the characteristics of GCs in giant elliptical galaxies with high S_N. A multi-phase collapse scenario was suggested as more consistent with the available evidence.

Below we describe two "case studies" which illustrate how optical spectroscopy with the new generation of 10m-class telescopes can help constrain both GC and galaxy formation models. Some of the issues that can be addressed with spectroscopic data include the following:

- Do GCs in other galaxies show the same abundance ratios as those in the Local Group?
- Are GCs in elliptical galaxies more metal rich than in spirals, as would seem to be indicated by their colors?
- What is the upper limit on [Fe/H]?
- Can abundances, abundance ratios, ages and kinematics identify GCs that formed during a merger?

Broad-band colors suffer from a well-known age-metallicity degeneracy and so have been of limited value in addressing these issues.

The first case study is an examination of the GCs in NGC 1399. The second is a spectroscopic study of the proto-globular cluster candidates in NGC 1275.

2. The NGC 1399 Globular Cluster System

NGC 1399 is the cD galaxy at the center of the Fornax cluster at a distance of 19.3 Mpc. Using the Keck telescope, we recently obtained 5 Åresolution spectra of 18 GCs in NGC 1399 (Kissler-Patig etal. 1998).

In general, we found no anomalies in the strengths of absorption line indices sensitive to metals (e.g. Mg, Fe), nor in the more age-sensitive Balmer-line indices (Figure 1). However, our sample contained two clusters for which: 1) Mg, Fe and Na were stronger than in any Milky Way or M31 GC. 2) There was tentative evidence for a Mg/Fe overabundance, similar to that seen in giant elliptical galaxies. Worthey et al. (1992) discussed possible explanations for this effect in galaxies, perhaps the most plausible of which was a variable (flatter) IMF. 3) Hβ and Hγ (effectively the only two Balmer lines in the spectral range) were abnormally strong. A similar effect was also seen

J. Andersen (ed.), Highlights of Astronomy, Volume 11A, 82–85.

in a few M31 GCs (Brodie & Huchra 1991), although at a lower metallicity. The Balmer lines are too strong to be reproduced by the spectral synthesis models (Worthey 1994, Fritze v. Alvensleben & Burkert 1995) at any age or metallicity. A component of blue horizontal branch stars might offer at least a partial explanation. Such stars are not represented in the models because they were not thought to be present in metal-rich clusters. However, Rich et al. (1997) recently discovered blue horizontal branches in two metal-rich GCs in the Milky Way, suggesting that a revision of the models may be in order.

Since the two "anomalous" NGC 1399 clusters are also very metal rich (see below), they must have formed from solar metallicity (i.e. enriched) gas, which may indicate that they are also younger than the rest. Judging by the photometric color distribution (these peculiar clusters are in the extreme red tail of the distribution), these clusters constitute less than 5% of the total GC population.

We calculated mean metallicity, [Fe/H], for all 18 clusters using the method described in Brodie & Huchra (1990). The Mg line-strengths were first corrected for a non-linear dependence of Mg index strength on metallicity for metallicities near and above solar, according to the models of Worthey (1994). With the exception of the two peculiar clusters, the range in metallicity seen in the NGC 1399 GCs is very similar to the range seen in the MW and M31: The metal-poor clusters have metallicities similar to those of the MW halo and the metal-rich clusters have metallicities similar to MW bulge/disk GCs. This suggests that the vast majority of NGC 1399 GCs formed from processes similar to those which formed the MW and M31 GC systems, and at roughly the same epoch (i.e. they are old).

V-I was found to correlate well with [Fe/H] and, although individual metallicities cannot be determined to better than 0.5 dex, a mean metallicity for a GC ensemble can be determined to about 0.3 dex from broad-band colors. However, the slope of the V-I vs. metallicity relation for the combination of Milky Way and NGC 1399 GCs was found to be twice as flat as the relation for Milky Way GCs alone. The consequence of this result is a much more homogeneous picture for extragalactic GC systems. Metallicity estimates for very red GCs are now roughly solar, rather than very super-solar as was previously thought, and metallicity differences between galaxies' GC systems are much less significant.

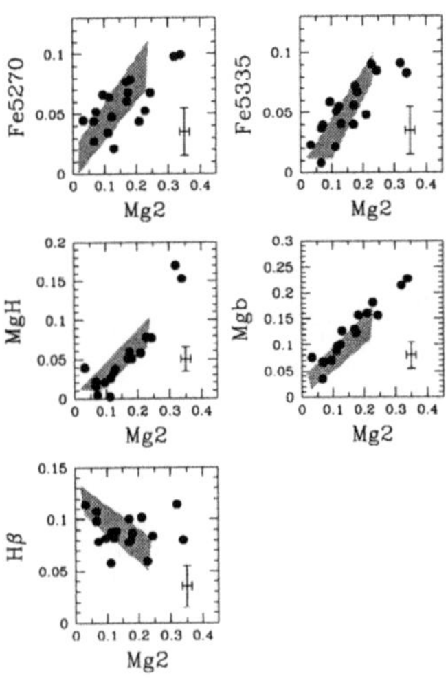

Figure 1. Mg2 versus various other indices, with the range spanned by Milky Way and M31 globular clusters shown as shaded region (taken from Brodie & Huchra 1990 and Burstein et al. 1984). Apart from two very metal-rich objects, all globular clusters are consistent with abundances found in the Milky Way and M31 globular clusters.

In summary, it appears that NGC 1399 and its GC system formed at early times. The same formation processes that were responsible for forming the MW GCs probably formed the vast majority of NGC 1399 GCs. Only a small percentage need have formed later from solar metallicity gas, e.g. in a merger.

3. Proto–globular Cluster Candidates in NGC 1275

The discovery, with HST imaging, of proto–globular cluster candidates in NGC 1275 (Holtzman et al. 1992) was regarded by many as a major success of the merger model which predicted that newly–formed clusters should be observable in currently or recently merging systems. The NGC 1275 clusters constitute an important test of GC formation models. NGC 1275 is the peculiar cD galaxy at the center of the Perseus cluster. It shows evidence for a merger history and may indeed be undergoing a merger at present. It also has one of the largest know cooling flows. GC formation in cooling flows (Fabian et al. 1984) received some support from the photometric study of the NGC 1275 system by Richer et al. (1993).

Keck spectroscopy of 5 proto–globular clusters in NGC 1275 (Brodie et al. 1997) revealed that the candidates are not HII regions, are clearly dominated by early A–type stars, and are not similar to young or intermediate age Magellanic Cloud or MW open clusters.

The Balmer absorption lines were found to be too strong to be consistent with any of the standard IMF (Salpeter or Scalo), solar metallicity, Bruzual & Charlot stellar evolutionary models at any age (Figure 2). The preliminary Bruzual & Charlot (1997) models indicate that no appreciable increase in equivalent width can be achieved by changing the metallicity. However, a 2–3 $M_\odot$ IMF, adopted to simulate a flatter IMF, reproduces the observed equivalent widths and colors and indicates an age of $\sim$ 500 Myr for these objects.

The Fritze v. Alvensleben and Kurth (1997) models are able to reproduce the strong Balmer equivalent widths, with best agreement at 350 Myr. At this age the model colors are about 0.2 mag too blue to be consistent with the observed colors.

The proto–globular cluster candidates are extremely centrally concentrated. The entire sample (some 60 objects) is within 8 kpc of the nucleus and the brightest clusters are all within 2 kpc. We find a low velocity dispersion, $\sim$ 200 km/s, for our sample of 5 of the brighter candidates. This is lower than estimates of the stellar central velocity dispersion, $\sim$ 300 km/s (Lazareff et al. 1989).

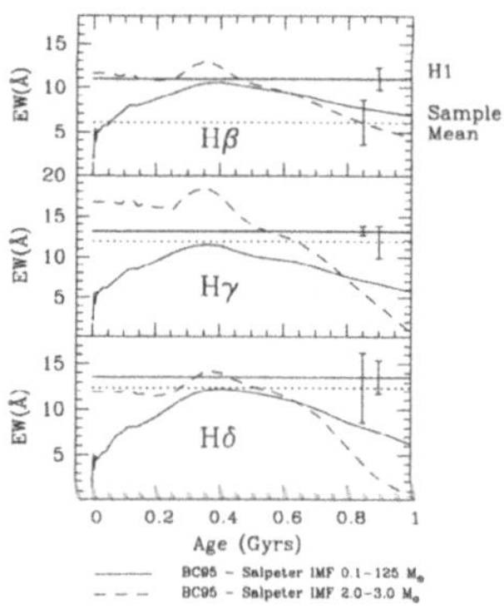

Figure 2. The time evolution of Balmer line equivalent widths for populations with different IMFs, produced by the Bruzual and Charlot models circulated in 1995. Also shown are the Balmer line equivalent widths measured for the brightest cluster, H1, (solid horizontal line) and the mean equivalent widths for the other clusters in the sample (dotted horizontal line). The value for Hβ is depressed, particularly in the lower signal–to–noise spectra, by under-subtracted Hβ emission in the galaxy background.

3.1. CONSTRAINTS ON CLUSTER ORIGINS

The spectroscopic information allows us to set some interesting constraints on the origin of these objects. We can clearly rule out formation in a continuous cooling flow. The spatial scale of the clusters is very much less than the cooling flow radius (by a factor of 60 for our spectroscopic sample and a factor of 20 for the entire candidate sample) and their age is very much greater than the cooling time ($\sim$ 10 Myr). Moreover, a star formation rate in excess of 400 $M_\odot$/yr is deduced from the cooling flow (Allen & Fabian 1997) which implies a steep IMF to explain the absence of high mass stars. We note that, if anything, the cluster's IMF may be biased *against* low mass stars.

It is equally clear that these clusters did not form in widespread shocks from merging galaxies. They are centrally concentrated and, if they formed far from the center and later fell in, a high rather than a low velocity dispersion would be expected.

It appears, then, that the clusters formed in a discrete event some 500 Myr ago. This may have been induced by a merger which provided the fuel for a short–lived gas inflow episode. It is not, however, clear to what extent processes such as these may have affected the global characteristics of GC systems. The resultant clusters are not distributed like old GCs in central cD galaxies which are significantly more diffuse than the galaxy light. If they do indeed have an IMF which is biased against low mass stars, they may fade very rapidly. A pure A–star population would fade away in only $\sim 10^9$ yr.

4. Summary

- The observational evidence fails to match merger model expectations in several respects and a multi–phase collapse scenario may be more consistent with the observed properties of extragalactic GC systems.
- The vast majority of GCs in NGC 1399 probably formed by the same processes and at the same epoch as those responsible for forming the MW GC system. Only a small subset (<5%) need have formed more recently (e.g. in a merger).
- The newly–defined V-I versus [Fe/H] relation suggests that metallicity differences between galaxies' GC systems are much less significant than previously thought and that there are no very super–solar GCs.
- The proto–globular cluster candidates in NGC 1275 probably formed in a merger–induced gas infall episode some 500 Myr ago. However, their properties are such that they may not represent formation processes that had a significant effect on the global properties of GC systems.

Acknowledgments

This research was funded by HST grants GO.05990.01-94A and GO.05920.01-94A and by faculty research funds from the University of California, Santa Cruz.

References

Ashman, K.M., & Zepf, S.E. 1992, ApJ, 384, 50
Allen, S.W., & Fabian, A.C. 1997, MNRAS, 286, 583
Brodie, J.P., & Huchra, J.P. 1990, ApJ, 362, 503
Brodie, J.P., & Huchra, J.P. 1991, ApJ, 379, 157
Bruzual, G., & Charlot, S. 1997, in preparation
Burstein, D., Faber, S.M., Gaskell, C.M., Krumm, N. 1984, ApJ, 287, 586
Fabian, A.C., Nulsen, P.E.J., & Canizares, C.R. 1984, Nature, 310, 733
Forbes, D., Brodie, J.P., & Grillmair, C. 1997, AJ, 113, 1652
Fritze-v. Alvensleben, U. & Burkert, A. 1995, A&A, 300, 58
Fritze-v. Alvensleben, U. & Kurth, O. 1997, in preparation
Harris, W.E. 1996, AJ, 112, 1487
Holtzman, J.A., et al. 1992, AJ, 103, 691
Kissler-Patig, M., Kohle, S., Hilker, M., Richtler, T., Infante, L., Quintana, H. 1997, A&A, 319, 470
Kissler-Patig, M., Brodie, J.P., Schroder, L.L.,Forbes, D.A., Grillmair, C.G., & Huchra, J.P. 1998, AJ in press, January issue
Brodie, J.P., Schroder, L.L., Huchra, J.P., Phillips, A.C., Kissler-Patig, M., & Forbes, D.A. 1997, *AJ*. submitted
Lazareff, B., Castets, A., Kim, D.W., & Jura, M. 1989, ApJ, 336, L13
Rich, R.M., Sosin, G., Djorgovski, S.G., et al. 1997, ApJ, 484, L25
Richer, H.B., Crabtree, D.R., Fabian, A.C., & Lin, D.N.C. 1993, AJ, 105, 877
Toomre A., Toomre J., 1972, ApJ 178, 623
Worthey, G., Faber, S.M., & Gonzales, J.J. 1992, ApJ 398, 69
Worthey, G. 1994, ApJS, 95, 107

ABUNDANCES IN THE GASEOUS GALACTIC HALO

ULYSSES J. SOFIA
Villanova University
Department of Astronomy and Astrophysics, Villanova, PA 19085 USA

Abstract. The well measured gas-phase abundances in the low halo suggest that this region of the Galaxy has total (gas plus dust) metal abundances which are close to those in the solar neighborhood. The gas-phase abundances in the halo are generally higher than those seen in the disk, however, this affect is likely due to the destruction of dust in the halo clouds. Observations of high velocity clouds (HVCs) in the halo suggest that these clouds have metal abundances which are substantially lower than those measured for the local interstellar medium. These determinations, however, are often of lower quality than those for the low halo because of uncertainties in the hydrogen abundances along the sightlines, in the incorporation of elements into dust, and in the partial ionization of the clouds.

1. Introduction

The abundance of elements in the gas and dust phases of the Galactic disk have been well studied. Data obtained with the *Copernicus* satellite in the 1970's gave us much insight into interstellar medium (ISM) (Jenkins 1987). The Goddard High Resolution Spectrograph (GHRS) aboard the Hubble Space Telescope (HST) has refined these studies through its enhanced resolution over *Copernicus*. This better allows *individual* absorbing clouds to be studied rather than average line of sight properties (Spitzer & Fitzpatrick 1993, Sofia, Cardelli & Savage 1994).

Halo gas studies naturally lagged behind exploration of the Galactic disk. The distance to the sources are greater and the velocity structure of the absorption is needed to distinguish the halo from the disk gas. Studies of halo abundances are important for exploring a region of the Galaxy which appears to differ greatly from the disk. The stellar population of the halo is quite different from the disk suggesting that the beginning and end products of stellar lifecycles (the ISM) are different (Jenkins & Wallerstein 1996). Also, differences in the kinematics (Danly et al. 1992) and ionization characteristics (Savage & Massa 1987) lead one to conclude that the ISM in the halo may differ from that in the disk.

Studies of elemental abundances in halo gas have greatly progressed since the launch of HST in 1990. The GHRS aboard the HST allowed ultraviolet absorption features toward halo stars to be observed at a resolution of 3.5 km s^{-1}. Since halo gas along a sightline can be separated from the local gas through differences in radial velocities, the high-resolution GHRS spectra allow gas from the halo to be well isolated. High resolution spectra also have the advantage of determining if, and possibly how much, saturation is affecting an absorption feature (Savage & Sembach 1991, Jenkins 1996). The advantage of HST's ultraviolet spectral coverage is that this region contains a large number of resonance transitions of species dominant in the neutral ISM.

2. The Low Halo

Sembach & Savage (1996) have studied the abundance of 8 elements in the low halo ISM in 10 absorbing regions toward 6 Galactic and 1 extragalactic sources. The data which they use are original as well as from Sofia, Savage & Cardelli (1993), Spitzer & Fitzpatrick (1993, 1995), Savage *et al.* (1992, 1993), Savage & Sembach (1995), Cardelli *et al.* (1991, 1994, 1995) and Sofia *et al.* (1994). The elements studied varied in their dust incorporation characteristics from extremely low

J. Andersen (ed.), Highlights of Astronomy, Volume 11A, 86–89.

depletions (sulfur) to moderate (Mg, Si and Mn) and highly depleted (Cr, Fe, Ni and Ti) species. They find that the low-halo gas abundances are generally higher than the disk abundances when compared to zinc. Zinc is a low depletion species (Roth & Blades 1995, Sembach et al. 1995) and can be used as a proxy for hydrogen. The apparently greater metal abundances in the halo could either reflect a higher overall abundance of the elements, or less depletion onto interstellar grains in the halo.

Figure 1 shows the logarithmic abundances in the 7 components of 5 sightlines which Sembach & Savage believe contain only halo gas (not all elements are available for each absorption component). The quantity plotted is the logarithmic abundance ratio in the cloud compared to the cosmic abundance ratio or

$$\left[\frac{X}{H}\right] \sim \left[\frac{X}{Zn}\right] = log\left[\frac{N(X)}{N(ZnII)}\right] - log\left(\frac{X}{Zn}\right)_{cosmic}. \quad (1)$$

The measurement errors are also plotted for each data point. The adopted cosmic abundances in the Figure are from solar ratios, although it is not clear that this is appropriate (Sofia et al 1994, Snow & Witt 1996, Fitzpatrick 1996). It is obvious from the Figure that the abundances of the elements relative to Zn (or hydrogen) are quite uniform for the low halo. The solid squares represent the weighted average for each element in the halo sightlines. It is unlikely that the abundance ratios would be so uniform if the intrinsic interstellar abundances (i.e. gas + dust) were not the same for all of the sightlines.

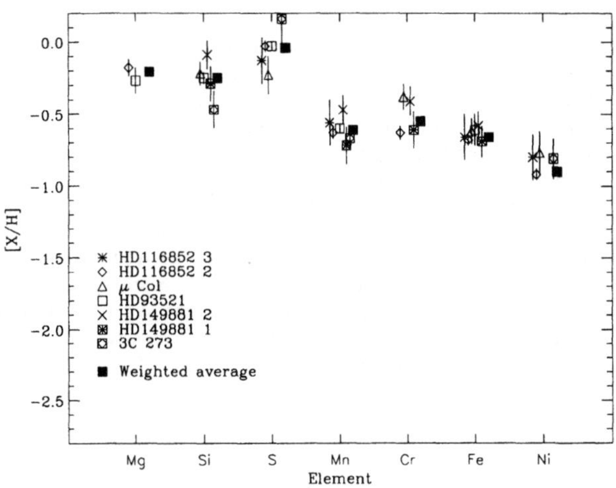

Figure 1. Logarithmic abundance ratios for low halo sightlines as compared to the solar abundance ratios.

The mean values plotted for the relative logarithmic halo abundances are listed in Table 1 for cosmic abundances represented by the sun and by the ISM values suggested by Snow & Witt (1996). It is clear from the Table that either of the cosmic abundance scenarios produces depletions which can be reasonably explained by dust incorporation of the missing atoms. Sembach & Savage (1996) in fact argue that the depletions in these clouds likely represent the composition of resilient dust grain cores. This is supported by the smooth transition of depletion patterns from disk to halo sightlines (see Figure 6 in Sembach & Savage 1996).

What is likely occurring in the low halo is that the composition of the ISM is very much like that in the Galactic disk. This may not be surprising since neutral H observations show that the turbulent motions of the ISM may well mix the disk and low halo (Lockman & Gehman 1991).

TABLE 1. Measured Relative Halo Abundances

Element	$[X/H]_{sun}$	$[X/H]_{ISM}$
Mg	-0.21	-0.02
Si	-0.25	+0.03
S	-0.04	+0.14
Mn	-0.61	?
Cr	-0.55	-0.38
Fe	-0.66	-0.38
Ni	-0.90	-0.70

3. High Velocity Clouds

High velocity clouds (HVCs) in the halo may represent a population of ISM gas which has been isolated from the low halo and disk material, and therefore may not contain the same elemental abundances. There are problems with HVC abundance measurements, however. As with the low halo, in order to derive accurate ISM abundances one must be aware of phenomena which can mimic apparent changes in metallicity.

Observations of HVCs with abundances which are apparently as low as 10% of the solar values are extremely intriguing and suggestive of unenriched material. The magnitude of the pitfalls in the metallicity determinations, however, is often difficult to quantify. For instance Bowen & Blades (1993) suggest that the low metal abundance they see in the HVC toward Mrk 205 may be partially or completely an ionization effect. Schwartz, Wakker & van Woerden (1995) discuss the possibility that depletion of material onto dust grains may significantly affect the interpretation of abundances. Lu, Savage & Sembach (1994) have abundance measurements in an HVC toward NGC 3783. They minimize depletion effects by observing sulfur (Spitzer & Fitzpatrick 1993) and of ionization effects by comparing S II to H I (Savage et al 1993). They find a sulfur-to-hydrogen ratio that is approximately 0.15 ± 0.05 that in the Sun, but question the reliability of the neutral H abundance used for the comparison. Their H I abundance was derived from 21 cm emission spectra. Radio beams are large relative to the stellar point source used for absorption studies, therefore they will measure H I outside of the line of sight. Since large variations of H I column densities occur over small (arcminute) regions within an HVC (Wakker & Schwartz 1991), using 21 cm column densities may lead to inaccurate conclusions.

4. Conclusions

The abundances that we can measure well, those in the low halo, suggest that composition in this region is much like that of the local ISM. HVCs, which we know are unusual by virtue of their radial velocities, often appear to have metallicities which are a small fraction of the local ISM or the Sun. These clouds may indeed be extremely subsolar in abundances, however they are intrinsically difficult to measure. One must be aware of of the limitations of these studies and interpret the results accordingly.

References

Bowen, D.V. and Blades, J.C. 1993, *ApJ*, 403, L55

Cardelli, J.A., Savage, B.D., Bruhweiler, F.C., Smith, A.M., Ebbets, D.C., Sembach, K.R. and Sofia, U.J. 1991, *ApJ*, 377, L57

Cardelli, J.A., Savage, B.D., Sofia, U.J., Keenan, F.P. and Dufton, P.L. 1994, *ApJ*, 420, L29

Cardelli, J.A., Sembach, K.R. and Savage, B.D. 1995, *ApJ*, 440, 221

Danly, L., Lockman, F.J., Meade, M. and Savage, B. D. 1992, *ApJS*, 81, 125

Fitzpatrick, E.L. 1996, *ApJ*, 473, L55

Jenkins, E.B. 1987, in *Interstellar Processes*, eds. D.J. Hollenbach and H. A. Thronson, Jr., Klewer Academic Publishers, Dordrecht, 533

Jenkins, E.B. 1996, *ApJ*, 471, 292
Jenkins, E.B. and Wallerstein, G. 1996, *ApJ*, 462, 758
Lockman, F.J. and Gehman, C.S. 1991, *ApJ*, 382, 182
Lu, L., Savage, B.D. and Sembach, K.R. 1994, *ApJ*, 426, 563
Roth, K.C. and Blades, J.C. 1995, *ApJ*, 445, L95
Savage, B.D., Cardelli, J.A. and Sofia, U.J. 1992, *ApJ*, 401, 706
Savage, B.D., Lu, L., Weymann, R.J., Morris, S.L. and Gilliland, R. 1993, *ApJ* 404, 124
Savage, B.D. and Massa, D. 1987, *ApJ*, 314, 380
Savage, B.D. and Sembach, K.R. 1991, *ApJ*, 379, 245
—— 1995, *ApJ*, 434, 145
Schwartz, U.J., Wakker, B.P. and van Woerden, H. 1995, *A&A*, 302, 364
Sembach, K.R. and Savage, B.D. 1996, *ApJ*. 457, 211
Sembach, K.R., Steidel, C.C., Macke, R.J. and Meyer, D.M. 1995, *ApJ*, 445, L27
Snow, T.P. and Witt, A.N. 1996, *ApJ*, 468, L65
Sofia, U.J., Cardelli, J.A. and Savage, B.D. 1994, *ApJ*, 430, 650
Sofia, U.J., Savage, B.D. and Cardelli, J.A. 1993, *ApJ*, 413, 251
Spitzer, L. and Fitzpatrick, E.L. 1993, *ApJ*, 409, 299
——, 1995, *ApJ*, 445, 196
Wakker, B.P. and Schwartz, U.J. 1991, *A&A*, 250, 484

ELEMENTAL ABUNDANCES IN QUASAR ABSORPTION LINE SYSTEMS

LIMIN LU AND WALLACE W.L. SARGENT
Caltech, 105-24, Pasadena, CA 91125, USA

AND

THOMAS A. BARLOW
Center for Astrophysics and Space Sciences, University of California, San Diego, C-0424, La Jolla, CA 92093, USA

Abstract. We discuss elemental abundances in quasar absorption line systems and their implications for galaxy formation and chemical evolution.

1. Introduction

Quasar absorption lines provide a means to probe the physical and chemical conditions in the gaseous component of the universe since the redshift of $z = 5$. The relevance of quasar absorption lines to the topic of this Joint Discussion, "Abundance Ratios in the Oldest Stars", is made clear by the following considerations: at $z = 2 - 5$, the lookback time is about 12-14 Gyrs (assuming 15 Gyrs for the present universe), which roughly corresponds to the time when the Galactic halo was formed. Consequently, studies of quasar absorption line systems at such redshifts may shed new light on the formation and early evolution of the Milky Way and of galaxies in general.

In this short article we will highlight some results concerning elemental abundances in high-redshift ($z > 2$) quasar absorption line systems and discuss their implications for galactic chemical evolution. The focus of the discussion will be on the Lyα forest absorption lines, which trace clouds in the intergalactic medium (IGM), and on the damped Lyα (DLA) absorption systems, which are widely accepted to be the high-redshift counterparts of present-day galaxies. Recent reviews of the general properties of Lyα forest clouds can be found in Weymann (1992), while that of DLA systems are given by Wolfe (1988, 1995).

2. Abundances in Lyα Forest Clouds: Evidence for Pop III Enrichment?

The IGM clouds were thought initially to have primordial compositions since early observations failed to detect any metal absorption lines at the redshifts of Lyα forest clouds. High resolution (FWHM$\sim$ 8 km/s), high S/N ($\sim$ 100) spectra taken recently with the 10m Keck telescopes (Cowie et al 1995; Tytler et al 1995; Songaila & Cowie 1996) clearly revealed the presence of weak C IV absorption at the redshifts of $\sim$ 70% of the Lyα forest clouds with neutral hydrogen column density N(H I)$\geq 3 \times 10^{14}$ cm^{-2} at $2.5 < z < 3.7$. The IGM clouds are known to be highly ionized (neutral fraction H I/H$\sim 10^{-4}$), probably through photoionization by the meta-galactic UV background. From modeling the ionization state of the clouds, Cowie et al (1995), Tytler et al (1995), and Songaila & Cowie (1996) deduced a mean metallicity of [C/H]~ -2 to -2.5, although the results are probably uncertain by as large as a factor of 10 owing to the large ionization corrections. There is also evidence for intrinsic variations in [C/H] from cloud to cloud (Rauch, Haehnelt, & Steinmetz 1997). A factor of 2-3 enhancement in the Si/C ratio seems to be indicated based on the Si IV/C IV ratios observed in some of the clouds (Songaila & Cowie 1996). Similar enhancement has been seen in Galactic halo stars (cf, Wheeler, Sneden, & Truran 1989), suggesting that the metals in the IGM clouds were probably produced by massive stars.

IGM clouds with N(H I)$\geq 3 \times 10^{14}$ cm^{-2} contain nearly half of the total baryon at $z = 2 - 4$ according to cosmological simulations (Miralda-Escude et al 1996; Bi & Davidsen 1997). Hence much

J. Andersen (ed.), Highlights of Astronomy, Volume 11A, 90–93.

of the universe was already enriched by $z > 4$. IGM clouds with lower N(H I) may also contain metals, but the corresponding C IV lines would be too weak to be detectable in existing observations. The metals could have been made locally (i.e., expelled from nearby star-forming regions at the same redshift), or by an earlier generation of stars (Population III). Theoretical considerations (eg, Ostriker & Snedin 1996) suggest that Pop III star formation is expected before "normal" galaxies were formed. Hence the Galaxy may have formed out of metal-enriched gas rather than primordial material. Recent observations of extremely metal poor halo stars (McWilliam et al 1995; Ryan, Norris, & Bessell 1996) revealed some very interesting abundance anomaly at [Fe/H]< -2.5, which may reflect the inhomogeneity resulting from the enrichment by only a few supernovae (Audouze & Silk 1995). These stars could be the survivors from the Pop III generation.

3. Abundances in DLA Galaxies: Implications for Galactic Chemical Evolution

Abundance determination for DLA systems began in the late 70's (Smith, Jura, & Margon 1979), but it was only recently that abundance estimates for a large number of systems became available. A summary of the DLA abundance measurements and discussions of their implications can be found in Lauroesch et al (1996). Results from two new surveys have since appeared in press, one by Lu et al (1996) who studied abundances for a large set of elements in a sample of DLA systems using the 10m Keck I telescope, and one by Pettini et al (1997a,b; 1994) who studied the Zn and Cr abundances in a sample of DLA systems. The new surveys represent a significant improvement over previous measurements in terms of both the quantity and quality of the measurements. We will therefore draw material primarily from these new surveys in the following discussion. It is important to point out that the exact nature of high-redshift DLA galaxies remains unclear at present.

3.1. THE DISTRIBUTION OF METALLICITY AS A FUNCTION OF REDSHIFT

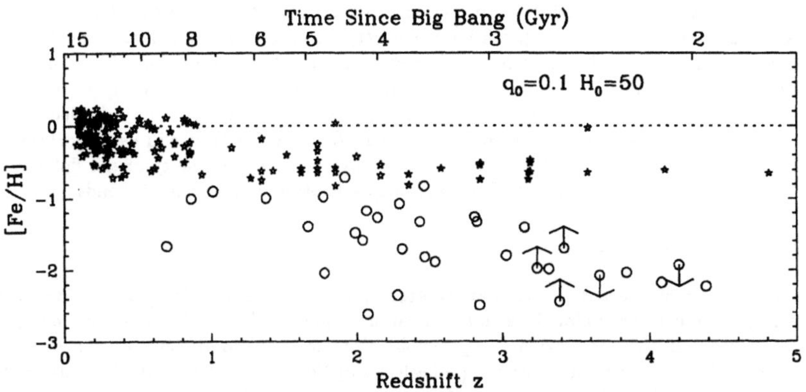

Figure 1. [Fe/H] vs z for damped Lyα systems (open circles) and for Galactic disk stars (star symbols; Edvardsson et al 1993), where [Fe/H]$\equiv$log(Fe/H)$_{DLA}-$log(Fe/H)$_\odot$. The stellar ages have been converted into redshifts for a $q_0 = 0.1$ and $H_0 = 50$ cosmology.

Firgure 1 shows the distribution of [Fe/H] vs z for DLA galaxies (open circles). The typical measurement uncertainty on [Fe/H] is ~ 0.1 dex assuming negligible ionization corrections. The sample comprises of the 23 DLA systems from Lu et al (1996), which includes some measurements from the literature, plus 14 new measurements from unpublished Keck observations, including 5 from Wolfe and Prochaska. Some of the new measurements are still preliminary and may be subject to change in the final analysis. However, the changes (if any) are expected to be small and should have no effect on any of the conclusions. The "$\star$" symbols in figure 1 map out the similar relation for a sample of Galactic disk stars in the solar neighborhood (Edvardsson et al 1993).

In general, the DLA galaxies have $-2.5 <$[Fe/H]< -1, corresponding to 1/300 to 1/10 solar metallicity. The N(H I)-weighted mean metallicity is $<$[Fe/H]$>\simeq -1.5$ at $\langle z\rangle = 2.5$. The low metallicities are consistent with them being young galaxies in the early stages of chemical enrichment. The N(H I)-weighted mean metallicity in term of Zn is about a factor of 2-3 (0.4 dex) higher (Pettini et al 1997b), possibility suggesting that some of the Fe atoms are locked up in dust grains in the DLA galaxies (Pettini et al 1997a; also section 3.2).

DLA galaxies appear significantly less metal-enriched than the Galactic disk in its past. Taken at face value, this argues against the suggestion that DLA galaxies are high-redshift (proto-) galactic disks (Wolfe 1988). Rather, the metallicities of the DLA systems are very similar to those of globular clusters and nearby dwarf galaxies. This result may be potentially in conflict with the evidence that DLA systems appear to show kinematics characteristic of fast-rotating ($v_{rot} \sim 250$ km/s) disks (Prochaska & Wolfe 1997). However, the rotating disk interpretation of the DLA kinematics may not be the only possible explanation (Haehnelt, Steinmetz, & Rauch 1997). In addition, the significance of the metallicity discrepancy between the DLA galaxies and the Milky Way disk may still be questioned for several reasons. (1) The ages of the stars in figure 1 at $z > 2$ are uncertain by 2-3 Gyrs from measurement errors alone, and there may also be systematic effects (Edvardsson et al 1993). (2) The true metallicities of DLAs may be higher than that indicated by [Fe/H] if significant dust depletion of Fe has occurred, although this effect alone cannot explain the discrepancy entirely because the inferred depletion of Fe is only ~ 0.4 dex (section 3.2). (3) DLA systems may preferentially probe regions of disk galaxies (assuming they exist at the relevant redshifts) beyond the equivalent of the solar circle owing to their larger absorption cross sections. The metallicity gradient known to exist in local spiral disks (Vila-Costas & Edmunds 1992) may explain the discrepancy (Ferrini, Molla, & Diaz 1997).

The mean metallicity of DLA systems increases gradually from $z > 4$ to $z \sim 2$ as expected. However, there is a factor of ~30 scatter in [Fe/H] at $2 < z < 3$, presumably reflecting differences in the formation epoch/star formation history of the galaxies and/or a mixture of morphological types. The metallicity distribution appears to reach a "plateau" value of [Fe/H]~ -2 to -2.5 at $z > 4$. Coincidentally, this "plateau" metallicity is identical (within the measurement uncertainties) to that found for the IGM clouds at similar redshifts (section 2). This coincidence suggests that the metals in DLA galaxies with [Fe/H]~ -2 to -2.5 may simply reflect those in the IGM, possibly produced by Pop III stars. If this interpretation is correct, then significant star formation did not start in DLA galaxies until $z \sim 3-4$. Such an inference is consistent with the decline in the neutral gas content of DLA systems at $z > 3$ (Storrie-Lombardi, McMahon, Irwin 1996), presumably because DLA galaxies are still being formed at such high redshifts; and with the rapid decline in the space density of quasars at $z > 3$ (Schmidt, Schneider, & Gunn 1995). It will be important to study more DLA systems at the highest redshift possible to confirm the reality of the "plateau" metallicity, and to improve the accuracy of the metallicity determination for IGM clouds.

3.2. RELATIVE ABUNDANCES

The abundance ratios of selected elements are plotted against the [Fe/H] of the DLA systems in figure 2, which represents an updated version of a similar figure in Lu et al (1996) but incorporating our new unpublished observations. In principle, abundance ratios of elements can be used to diagnose the nature of the nucleosynthesis (Wheeler et al 1989): were the heavy elements made by Type II supernovae, Type Ia supernovae, AGB stars, or some combination of these processes? However, abundances derived from interstellar absorption lines may not reflect the total abundance of the elements since many elements (e.g., Si, Fe, Cr, Mn) are prone to condensation onto dust grains (Jenkins 1987). Keeping this in mind, the relative abundance patterns shown in figure 2 are very similar to those seen in metal-poor halo stars (cf, Wheeler et al 1989), showing a low N/Fe ratio, an overabundance of Si and S relative to Fe (i.e., enhancement of α-process elements over iron-peak elements), and an underabundance of Mn relative to Fe (i.e., the odd-even effect). The implication is that metal productions in most of the DLA systems under study were probably dominated by massive stars via Type II supernova. However, the observed Zn/Fe ratios in DLA systems, $<$[Zn/Fe]$>\simeq 0.4$ dex, are inconsistent with the abundance pattern seen in halo stars, where [Zn/Fe]$\simeq 0$ for all stars with [Fe/H]> -3. Since Zn is relatively unaffected by dust, while 80-99% of the Fe atoms are usually removed from the gas phase by condensation onto dust grains (Jenkins 1987), the super-solar Zn/Fe ratios found in DLAs suggest that the [Fe/H] values may underestimate the true metallicities of the DLA galaxies by ~ 0.4 dex owing to dust depletion.

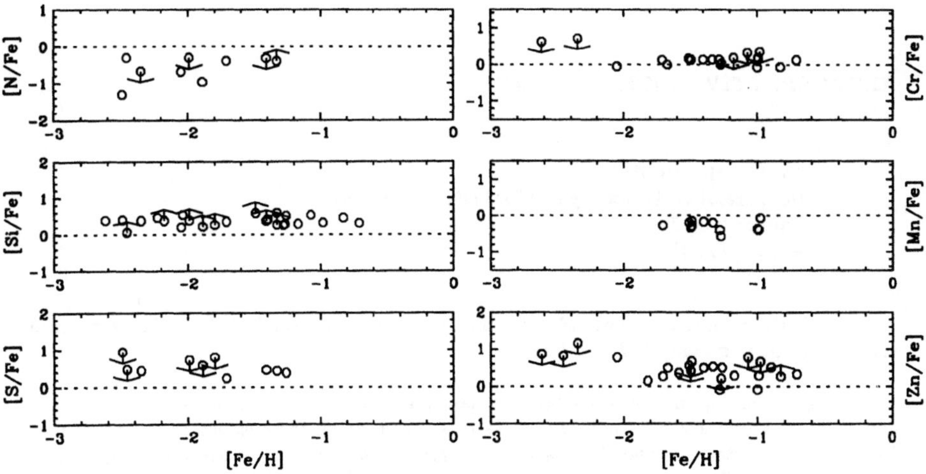

Figure 2. Relative abundance patterns of damped Lyα systems.

Consequently, a full understanding of the DLA abundance patterns must take into account the effect of dust depletion. This issue is discussed in detail elsewhere (cf, Lu et al 1997).

Acknowledgements

We thank Art Wolfe and Jason X. Prochaska for providing their abundance measurements in advance of publication.

References

Audouze, J., & Silk, J. 1995, ApJ, 451, L49
Bi, H., & Davidsen, A.F. 1997, ApJ, 479, 523
Cowie, L.L., Songaila, A., Kim, T.S., & Hu, E.M. 1995, AJ, 109, 1522
Edvardsson, B., Andersen, J., Gustafsson, B., Lambert, D.L., Nissen, P.E., & Tomkin, J. 1993, A&A, 275, 101
Ferrini, F., Molla, M., & Diaz, A.I. 1997, ApJ, 487, L29
Haehnelt, M. G., Steinmetz, M., & Rauch, M. 1997, ApJ, in press
Jenkins, E. B. 1987, in Interstellar Processes, ed. D.J. Hollenbach and H.A.Thronson, Jr. (Dordrecht. Reidel), p533
Lauroesch, J.T., Truran, J.W., Welty, D.E., & York, D.G. 1996, PASP, 108, 641
Lu, L., Sargent, W.L.W., & Barlow, T.A. 1997, in Cosmic Chemical Evolution (IAU Symp. No. 187), in press
Lu, L., Sargent, W.L.W, Barlow, T.A., Churchill, C.W., & Vogt, S. 1996, ApJS, 107, 475
McWilliam, A., Preston, G.W., Sneden, C, Searle, L. 1995, AJ, 109, 2757
Miralda-Escude, J., Cen, R., Ostriker, J.P., & Rauch, M. 1996, ApJ, 471, 582
Ostriker, J.P., & Snedin, N.Y. 1996, ApJ, 472, L63
Pettini, M., Smith, L.J., Hunstead, R.W., & King, D.L. 1994, ApJ, 426, 79
Pettini, M., King, D.L., Smith, L.J.,& Hunstead, R.W. 1997, ApJ, 478, 536
Pettini, M., Smith, L.J., King, D.L., & Hunstead, R.W. 1997, ApJ, in press
Prochaska, J.X., & Wolfe, A.M. 1997, ApJ, in press
Rauch, M., Haehnelt, M.G., Steinmetz, M. 1997, ApJ, 481, 601
Ryan, S.G., Norris, J.E., & Beers, J.C. 1996, ApJ, 471, 254
Schmidt, M., Schneider, D.P., & Gunn, J.E. 1995, AJ, 100, 68
Songaila, A, & Cowie, L.L. 1996, AJ, 112, 335
Smith, H.E., Jura, M., Margon, B. 1979, ApJ, 228, 369
Storrie-Lombardi, L.J., McMahon, R.G., & Irwin, M.J. 1996 MNRAS, 283, L79
Tytler, D., Fan, X.-M., Burles, S., Cottrell, L., Davis, C., Kirkman, D., & Zuo, L. 1995, in *QSO Absorption Lines*, ed. G.Meylan (Springer-Verlag), p289
Vila-Costas, M.B., & Edmunds, M.G. 1992, MNRAS, 259, 121
Weymann, R.J. 1992, in The Environment and Evolution of Galaxies, eds. J.M. Shull & H.A. Thronson, Jr. (Kluwer), p213
Wheeler, J.C., Sneden, C., & Truran, J.W. 1989, ARA&A, 27, 279
Wolfe, A.M. 1988, in QSO Absorption Lines: Probing the Universe, eds. Blades, Turnshek, and Norman (Cambridge Univ Press), p297
Wolfe, A.M. 1995, in QSO Absorption Lines, ed. G.Meylan (Springer-Verlag), p13

A RETROSPECTIVE INTRODUCTION

DAVID L. LAMBERT
Department of Astronomy & McDonald Observatory
University of Texas
Austin, TX, USA

"It takes little talent to see clearly what lies under one's nose, a good deal of it to know in which direction to point that organ." W. H. Auden

Listening to the sequence of fine talks that have made up this Joint Discussion, I have been impressed by the veritable flood of new observations that bear on the question of the abundance ratios in the oldest stars. The flood is marked by depth and diversity, as 'noses' are being pointed in highly profitable directions. As an example of depth, I would note the extensive investigations reviewed by Andy McWilliam of the chemical compositions of *very* metal-poor stars where [Fe/H] $\simeq$ -3 marks the upper end of the range defining the class. To characterize diversity, I draw attention to the analyses of quasar absorption line systems so ably discussed by Limin Lu who showed that the abundance ratios for the gas in such high z systems resemble ratios found from Galactic metal-poor stars. Of course, to many at this Joint Discussion the wealth and diversity did not come as a surprise. Indeed, I suspect that many of us pursue the topic of the composition of the first generations of stars because of the wealth and diversity of observational constraints that may now spawn a unified interpretation. Hopefully, the marvellous talks given here will pull in a few new minds - young and old - to ponder nucleosynthesis in very young galaxies and in our Galaxy in particular.

1. Insights into Evolution

"Truth is Catholic, but the search for it is Protestant." W. H. Auden

Comprehensive measurements of abundance ratios are undertaken to provide novel insights into the evolution of stellar systems and stars. I think it fair to say that this is the primary reason why we strive to measure abundance ratios. Several speakers have alluded to applications of abundance ratios. A few examples must suffice to illustrate this well known point.

- **Red giants.** High-resolution spectroscopy, as reviewed by Bob Kraft, of red giants in globular clusters provides evidence of compositional changes attributed to deep mixing beyond the modest levels predicted by the first dredge-up of standard stellar models. Luminosity dependent evolution of the carbon and nitrogen abundance including the $^{12}C/^{13}C$ ratio point to mixing. Changes in sodium and aluminium abundances point to contact between the convective envelope and deep hot regions in which the Ne-Na and Mg-Al H-burning cycles operate.
- **Mass transfer and binaries.** Among very metal-poor stars, there appears to be a rather high frequency, as Tim Beers discussed, of very carbon-rich stars. It is tempting to identify these as members of a binary system in which the initially more massive star evolved to become a carbon AGB star before losing mass at a copious rate. Transfer of mass to the companion produced the now C-rich dwarf star. Is this the complete story for the very metal-poor C-rich stars?
- **Chemical Evolution of Stellar Systems.** Stars by their nucleosynthesis and mass loss exert the controlling influence on the evolution of the chemical composition of a system. If we understand stellar nucleosynthesis theoretically or empirically, we may learn a lot about evolution of a stellar system from a study of abundance ratios. In this context, stellar system ranges from open clusters, globular clusters, to galaxies of all kinds including intra-galactic gas. In the case of our Galaxy and other nearby galaxies, we may use abundance ratios to

J. Andersen (ed.), Highlights of Astronomy, Volume 11A, 94–98.

probe evolution of sub-systems in the galaxy: for example, the halo, disk, and bulge. Roger Cayrel discussed the evolution of the halo. Michael Rich reviewed the compositions of stars in our bulge. At the very lowest metallicities, [Fe/H] ≤ -3, evidence of a Population III may turn up – a population of objects that were the first sources of nucleosynthesis after the Big Bang and which seeded the primordial gas before galaxies were formed.

2. Occam's Razor

"It is vain to do with more what can be done with less." William of Occam

This admonition has certainly been accepted by many who have tried to relate abundance ratios to a model of the Galaxy's evolution. In turn, the acceptance has been encouraged by the seemingly simple evolution of the abundance ratios.

For the Galaxy, this evolution is historically portrayed as the evolution of a ratio [X/Fe] versus [Fe/H] where X is an element of interest. When halo and disk stars are combined in such a plot, the resulting relation is generally smoothly varying and exhibits little scatter in [X/H] at a given [Fe/H] over the metallicity range readily explored, say [Fe/H] ≥ -2. This assertion is borne out by, for example, the reviews by Wheeler, Sneden & Truran (1989) and Lambert (1989). Iron is a product of both Type II and Type Ia supernovae. Then, from the evolutionary point of view, a key element to contrast with iron is one produced primarily synthesised by Type II supernovae but not Type Ia supernovae. Oxygen is the ideal example but unfortunately despite a variety of spectroscopic indicators of the oxygen abundance (permitted and forbidden O I lines, and molecular OH ultraviolet and infrared lines) uncertainties continue to becloud the run of [O/Fe] in metal-poor stars. As a substitute, the so-called α-elements - Mg, Si, Ca, and suprisingly Ti - are taken.

The run of [α/Fe] with [Fe/H] is considered constant at about + 0.35 for [Fe/H] ≤ -1 and to decline smoothly for higher metallicities reaching 0.0 at about the solar metallicity (i.e., [Fe/H] = 0). There is, it has been said, little to no 'cosmic' scatter about the mean relation. For example, McWilliam (1997) reviews measurements of [α/Fe] of halo stars and concludes "the full range of measured [α/Fe] ratios is well represented by $+0.37\pm0.08$." This seems to overlook McWilliam et al.'s (1995a,b) results for stars with [Fe/H] $\leq$ -3.5 showing a large scatter in [α/Fe]. Ryan, Norris & Beers (1996) report increased scatter starting at a slightly higher metallicity.

The standard simple interpretation of the run of [α/Fe] with [Fe/H] would delight William of Occam: the ratio in halo stars reflects the composition of processed material from Type II supernovae, and the decline about [Fe/H] = -1 is due to the additional contribution of ejecta from Type Ia supernovae with a much lower α/Fe ratio than the Type II SN. The halo [α/Fe] ratios are in fair agreement with those expected of Type II supernovae originating from metal-poor massive stars (Timmes, Woosley & Weaver 1995).

Exploration of *very* metal-poor stars adds a scatter to the smooth run of [α/Fe] vs [Fe/H] established for [Fe/H] $\geq$ -2 (McWilliam et al. 1995a,b; Ryan et al. 1996). In part, the scatter likely arises because the very metal-poor stars formed out of gas contaminated by ejecta from just one to a few Type II supernovae. In particular, the yield of iron from Type II supernovae is very dependent on the 'mass-cut' between the neutron star and the ejecta. This mass cut may be dependent on individual properties of supernovae while the yield of α-elements is likely less dependent on local properties and certainly on the location of the mass-cut.

A goal in studying the most metal-poor stars is to try and extrapolate stellar compositions to the birth of the Galaxy and so obtain observational constraints on the proto-Galaxy's composition. Was the proto-galactic gas mixed with ejecta from a pre-galactic population of stars, the so-called Population III?

3. A Few Impressions

As an observer of this sequence of talks, I was left with a few striking impressions and questions. A distillation follows:

- **Accuracy of Abundance Data.** Very little was said on this topic! My immediate impression was that this was regrettable but certainly understandable given the pressure of time. One may readily list potential sources of error, especially those pernicious systematic errors- e.g., model atmospheres, non-LTE effects, basic atomic data, stellar parameters. Several sources of error and uncertainty may be reduced with a well designed differential analysis, a point perhaps not adequately made today.

  The key seems to be to convey to the audience a fair impression of the accuracy of abundance data without suppressing mention of potential sources of error. I am reminded at this point of Evan Skillman's (1997) impression that "the confidence in abundances derived from nebular spectra is vastly different between people that work in the field and those that do not, in the sense that people that work in the field hold nebular abundance measurements to be quite reliable while outsiders consider them suspect." "For those of us in the field, there is nothing more interesting than a small bit we don't understand". Skillman avers that outsiders, however, may view the nebular abundances with suspicion because undue emphasis is given by "people in the field" to errors and uncertainities. Amen!

- **Surveys.** This Joint Discussion has beautifully highlighted the need for surveys for candidates to be observed and analysed for their chemical composition. This applies to QSOs as well as to stars. A preeminent example is Tim Beers' search for ultra-metal poor stars which has been responsible - almost solely - for the continuing flow of data on compositions of stars with [Fe/H] $\leq$ -3. Perhaps out of this flow will come evidence for Population III, a pre-galactic or protogalactic generation of 'objects' be they stars of normal masses or super-massive objects. If this generation included stars of normal mass, we should find low mass main sequence objects of pure hydrogen and helium. Of course, super-massive objects will be detected only through their nucleosynthetic consequences for the protogalactic gas and the first generation of normal stars.

  Surveys of larger samples of less extremely metal-poor Galactic stars are likely to bear fruit, especially if the abundance analyses are undertaken with errors controlled and minimized by observations of control samples. Here, the recent work by Nissen & Schuster (1997) comes to mind. In their sample of stars with [Fe/H] between about -0.8 and -1.2, they found a group of stars deficient in [α/Fe] by about 0.1 dex with other anomalies (e.g., [Ni/Fe] about 0.1 dex below the normal value of 0.). These stars appear to share kinematic properties. This example may be an illustration of what awaits us when large samples are analysed to a precision of 0.05 dex or better. If the halo was built up by agglomeration of dwarf galaxies, globular clusters, and even smaller systems, abundance anomalies with respect to the traditional run of [X/Fe] with [Fe/H] are likely. Certainly, variations of [X/H] with age are expected but age estimates are essentially unobtainable for the surviving low mass stars.

- **Globular Clusters.** Chemical compositions of stars in Galactic globular clusters were ably reviewed by Bob Kraft with illustrations aplenty drawn from analyses based on high-resolution spectra acquired with the Keck telescope, as well as 3-4 meter telescopes. As a usually remote observer of the globular cluster scene, I am struck by two questions.
  - First, why do the stars of a given cluster with the glaring exception of ω Cen and the likely exception of M22 have the same metallicity? In well-observed clusters, the upper limit on the range in [Fe/H] is less than 0.1 dex while the cluster to cluster difference is about 1.5 dex at its extreme. Theoretical proposals to account for the mono-metallicity of globular clusters exist - see Brown, Burkert & Truran (1991,

1995) and Murray & Lin (1992). In these scenarios, the present stars are a second generation formed from gass polluted by ejecta from a first generation now lost from the cluster.

* Second, are the abundance anomalies for light elements (C to Al, say) among stars of a given cluster entirely the result of internal nucleosynthesis and mixing within individual stars or were they in part imprinted at the birth of the stars? Bob Kraft discussed this question which is often succintly phrased as 'evolutionary versus primordial' abundance anomalies. When a clear variation of an anomaly (say, the $^{12}C/^{13}C$ ratio) with the luminosity of the red giant is observed, it is plausible to attribute the anomaly to deep mixing by a giant's convective envelope. On the other hand, when abundance anomalies are seen on the subgiant branch or close to the main sequence (see, for example, Briley et al. 1996) , the more plausible explanation is surely that the gas from which stars formed was not chemically homogeneous. (And we should not forget the potential role of binary stars and mass transfer between the stars.)

In thinking about these questions, I am struck by the almost complete lack of field halo stars with anomalies like those seen among cluster stars. This to me is a pointer to a primordial ('environmental') contribution to these anomalies. Surely, gas enriched to the observed metallicity of a cluster by a now extinct population of stars may contain chemical inhomogeneities. The observed anomalies are indicative of H-burning by the CNO, Ne-Na, and Mg-Al cycles. It is an interesting question if the ejecta from stars formed of truly pristine gas, pure H and He with a dab of Li, can contain these anomalies when the star had no C,N, O etc. If not and if there is an enviromental component to the abundance anomalies, it would seem that either the present stars are third generation not second generation, as the above-referenced theories on cluster formation imply, or the gas was not pristine but was contaminated with C, N O etc. Alternatively, the abundance anomalies of main sequence stars may result from accretion of the ejecta of more massive second generation stars.

- **Mass Cut between Massive and Intermediate Mass Stars.** Massive stars die as Type II supernovae. Intermediate (and low) mass stars evolve to become asymptotic giant branch stars and die by shedding their outer envelope to leave the stelar core as a white dwarf. Elements beyond the iron-peak are synthesised by these stars: r-process nuclides in Type II supernovae (probably) and s-process nuclides in AGB stars. The mass cut is presently put at 8-10$M_\odot$ based on observations of white dwarfs in open clusters. To my knowledge there is no quantitative understanding of the mass cut.

  Mass loss is likely a determining factor. Since mass loss rates driven by radiation pressure on dust grains or gas are metallicity dependent, one may wonder if the mass cut is metallicity dependent too. If it is, the production ratio of r/s nuclides will vary with metallicity. Is this an observable effect by careful characterization of appropriate [X/Fe] vs [Fe/H] relations? Such an exercise must recognise that the character of the s-process itself evolves with [Fe/H]: the total exposure to neutrons increases with decreasing metallicity. In this connection, it would be useful to test the oft made assumption that the r-process provides a solar distribution of r-process nuclides irrespective of the metallicity of the supernova.

- **Unification!** 'Unification' in astronomical parlance is usually applied to Seyfert galaxies, AGNs and assorted other odd things. Here, I appropriate the word to recall the resemblance between the abundance ratios of the oldest Galactic stars and of the absorption line systems revealed by the high-z QSOs. To first-order the resemblance is no surprise; nucleosynthesis by Type II supernovae is a universal phenomenon! By using red-shift and a cosmological model, it is possible now to plot abundances against a clock: [X/H] versus time is surely more informative than [X/Fe] versus [Fe/H]. Of course, the gaseous disks (presumably) that are sampled by QSO lines of sight belong to a family of galaxies unlike the stars in our Galaxy that *now* at least belong to a single galaxy. The sampling of a galaxy population is likely to compromise the simplicity of abundance versus clock plots. Particularly valuable information will be provided if the galaxy responsible for an absorption line system can be identified by deep imaging.

4. An Introductory Retrospection

The name of the English poet and man of letters, W. H. Auden, has appeared above. I was introduced to his works by an imaginative master at the grammar school who was faced with an unhappy group of senior schoolboys specialising in science but required by the edict of the headmaster to take 1 hour of English a week. Oh how we grumbled at the time! But that hour a week has led to many happy hours of reading in the subsequent 40 years.

A memory of the school classroom came back recently when I read an account of a visit by W. H. Auden to Dartmouth College. The then elderly Auden was entertained before delivering an address. "When he arrived at the auditorium where he was to deliver a paper on Romanticism, he tripped over the top step while mounting to the stage and his typescript spilled across the platform. Putting the pages back in correct order was of course beyond him. So, with his Oxford aplomb, he pulled them together out of order and read whatever page came to the surface. The audience thought the essay had many fine insights but was a bit lacking in structure, though charming." (Hart, 1997)

There is a sense in which we are following in Auden's faltering footsteps. Abundance ratios of the oldest stars provide us with a chemical history of the Galaxy. The pages of the history book are not ordered by metallicity. Nonetheless, we have provided 'fine insights' into the Galaxy's early evolution. In the case of the QSO absorption line systems, we may feel that the pages are numbered by z but we may doubt that they have come from a single book rather than a library of volumes.

Present activities as highlighted in this Joint Discussion augur well for the future. Contemporary ideas about the early evolution of galaxies may be vindicated as the abundance analyses of stars and gas, near and far, are extended in depth and breadth. Perhaps, as history shows, present ideas will prove to be off the mark. But let's leave the last word to W. H. Auden:

His boneless worm-like ancestors would be amazed
At the upright position, the breasts, the four-chambered heart,
The clandestine evolution in the mother's shadow.

References

Briley M., Smith V.V., Suntzeff N.B., Lambert D.L., Bell R.A., Hesser, J.A., 1996, *Nature*, **383**, 604
Brown, J.H., Burkert, A., Truran J.W., 1991, *ApJ*, **376**, 115
Brown, J.H., Burkert, A., Truran J.W., 1995, *ApJ*, **440**, 666
Hart J., 1997, *The New Criterion*, February issue
Lambert D.L., 1989, in *Cosmic Abundances of Matter*, ed. C. J. Waddington, AIP Conf. Proc., **183**, 168
McWilliam A., 1997, *ARAA*, **35**, 503
McWilliam A., Preston G.W, Sneden C., Searle L., 1995a, *AJ*, **109**, 2757
McWilliam A., Preston G.W, Sneden C., Schectman S., 1995b, *AJ*,**109**, 2736
Murray S.D., Lin D.N.C., 1992, *ApJ*, **400**,265
Nissen P.E., Schuster W.J., 1997, *A&A*, **326**, 751
Ryan S.G., Norris J.E., Beers T.C., 1996, *ApJ*, **471**. 254
Skilmann E. 1997, in *Stellar Astrophysics for the Local Group*, ed A. Aparico et al., Cambridge Univ. Press, in press
Timmes F.X., Woosley S.E., Weaver T.A., 1995, *ApJS*, **98**, 617
Wheeler J.C., Sneden C., Truran J.W., 1989, *ARAA*, **27**, 279

JD2

Dwarf Galaxies: Probes for Galaxy Formation and Evolution

Chairpersons and Editors: **E. Brinks & T. X. Thuan**

MULTI–SPECTRAL STUDIES OF THE NEARBY DWARF GALAXIES UGCA 86 AND LMC/SMC

G. M. RICHTER
Astrophysikalisches Institut Potsdam
An der Sternwarte 16, D–14482 Potsdam, Germany
gmrichter@aip.de

M. BRAUN
ESA–VILSPA ISOPHOT IDT
Villafranca del Castillo, Satellite Tracking Station
Apartado 50727, E–28080 Madrid, Spain
mbraun@iso.vilspa.esa.es

AND

R. ASSENDORP
Astrophysikalisches Institut Potsdam
An der Sternwarte 16, D–14482 Potsdam, Germany
rassendorp@aip.de

1. Background

UGCA 86 is an irregular dwarf galaxy in the IC 342 / Maffei I group, just next to the Local Group. It was first mentioned by Zwicky (1968) as VII Zw 009, but not contained in his "Catalogue of Selected Compact Galaxies and of Post–eruptive Galaxies" (1971). It was independently rediscovered by Nilson (1974) and Rots (1979) as UGCA 86 and A 0355 resp. Rots found it by HI observations, and from peculiarities in the HI morphology and kinematics he suspected that it was interacting with IC 342. Thus, the tentatively interesting items: a starforming, low surface brightness dwarf galaxy in an interacting system (one of the nearest), triggered us to engage in more detailed studies.

In a first step, we made detailed surface photometry in U, B and V (Richter et al. 1991). UGCA 86 proved to be one galaxy (which was not trivial; Saha & Hoessel 1991 discussed if it could be a chance superposition or a collision of two independent galaxies, due to the very different appearence of the southern and the central starburst; Miller & Hodge 1992 and the distance measurements of Karachentsev & Tikhonov 1993 support our result) with the typical exponential brightness profile of a spheroidal dwarf galaxy, and contains at least two starburst regions of very different color: a central red one and a blue one in the southern outskirts. There is an infrared source in the IRAS Point Source Catalogue coincident with UGCA 86. The amount of dust indicated by this source is in very good quantitative agreement with what would be required to redden the central starburst by the observed color difference compared to the southern burst. Nevertheless, the straightforward hypothesis, that the color of the central burst is due to dust extinction, is contradicted by the improved, higher resolution data.

2. Improved data

In order to improve the resolution, we deconvolved the IRAS data by the Pyramid Maximum Entropy method (Bontekoe et al. 1994). The results show at least two main components: a source near the centre, but definitely not coincident with the red central starburst; and a source coincident with the southern blue starburst with a long extension. Compared with an Hα image[1], the central

[1] by courtesy of A. Aparicio and C. Alart, who obtained it with the 2.2 m telescope at Calar Alto

J. Andersen (ed.), Highlights of Astronomy, Volume 11A, 101–102.

IR–source coincides with a clump of HII–regions and diffuse Hα radiation not prominent in the broadband images. Thus, the red color of the central starburst is not due to dust absorption, but obviously to its higher age compared to the southern one. The extension of the southern IR source coincides with a chain of Hα–filaments, which are in good agreement in direction and bending with the instreaming gas suggested by the HI observations of Rots (1979). Thus, it seems very likely that the southern burst is shock–induced by gas infall.

X–ray observations by ROSAT yielded a point source near the centre which could not yet be definitely identified. The luminosity ($2 \ldots 7 \times 10^{38}$ergs^{-1}; spectrum badly defined due to high foreground absorption) is at the upper end of the LF of the known source classes with exception of the AGNs. Since a compact star cluster (extent and luminosity similar to a globular) lies within the error circle, it is likely a LMXB.

3. High resolution FIR data for SMC/LMC

Because the SMC is similar in size and luminosity to UGCA 86, but is nearer (and thus allows higher linear resolution) by a factor of 30, we began a MaxEnt deconvolution of all IRAS observations (survey and pointed) of the SMC/LMC system[2]. It is not yet completed due to the extreme computational requirements. A first, preliminary result is that some of the shell–like structures in the LMC may be connected to the Mach–cones of high velocity gas clouds which splashed through the disk (Braun 1996).

References

Bontekoe, T. R., Koper, E., Kester, D. (1994) *A&A*, **284**, 1037.
Braun, M. (1996) *Astron. Nachr.*, **317**, 369.
Karachentsev, I. D., Tikhonov, N. A. (1993) *A&A Suppl.*, **100**, 227.
Miller, B. W., Hodge, P. (1992) in D. Hollenbach, H. Thronson, J. Shull (eds), 90. NASA Conf. Publ. 31901.
Nilson, P. (1974) *Catalogue of Selected non–UGC Galaxies.* Report No. 5, Uppsala Astronomical Observatory.
Richter, G. M., Schmidt, K.–H., Thänert, W., Stavrev, C., Panov, P. (1991) *Astron. Nachr.*, **312**, 309.
Rots, A. H. (1979) *A&A*, **80**, 255.
Saha, A., Hoessel, J. G. (1991) *A.J.*, **101**, 465.
Zwicky, F. (1968) *List No. VII.* CalTech.
Zwicky, F. (1971) *Catalogue of Selected Compact Galaxies and Post–eruptive Galaxies.* Guemligen, Switzerland.

[2]in cooperation with Sterrenkundig Instituut "Anton Pannekoek", Amsterdam and SRON, Groningen

RELATION BETWEEN BCDS AND BLUE LSBGS

NILS BERGVALL, GÖRAN ÖSTLIN, ARNAUD PHARASYN AND JARI RÖNNBACK
Astronomical Observatory, Box 515, S–75120 Uppsala, Sweden

AND

JOSEFA MASEGOSA
Instituto de Astrofísica de Andalucía, P.O. Box 3004, 18080 Granada, Spain

1. Introduction

Two important questions are connected to the evolutionary history of blue compact dwarf galaxies (BCDs): the triggering mechanism of the starburst and the properties of the precursor(s) and successor. The halos of many BCDs show red colours indicating that the host galaxy is in fact of early type. But where did the gas come from? Was it there from the start in a quiescent mode or was it fed by a merger with an intergalactic gas cloud or a gas–rich dwarf? Is there a general answer to this question or are there different families of BCDs?

The BCDs generating most of the interest are those with low chemical abundances, typically 10% solar. The only plausible precursors in this case are metal–poor dIs, LSBGs or pure HI clouds (e.g. Thuan & Seitzer 1979, ApJ 231, 680). In a comparison between LSBGs and BCDs Taylor et al. (e.g. 1997, ApJ 480, 524; 1996, ApJS 107, 143) found an excess of HI rich companions in the environments of BCDs which they interpret as a support of tidal triggering. We have studied a few massive ($\sim 10^9 \mathcal{M}_\odot$) metal–poor blue compact galaxies (BCGs). These are too bright to be classified as dwarfs but define an extension of the BCD family. For simplicity we will in the following use the common abbrevation BCD. In a search for possible precursors we have studied a sample of blue LSBGs with metallicities similar to BCDs (5–15%).

2. HI masses and chemical abundances

Fig. 1 shows the $\mathcal{M}_{\rm HI}/{\rm M}_B$ relation for BCDs (Thuan and Martin 1981, ApJ 247,823) applying the restriction B-V$\leq$0.6; LSBGs (from McGaugh et al. 1997, ApJ 481, 689) with the restriction $\mu_{B,0} \geq$22.0 and B-V$\leq$0.6; and blue LSBGs (Bergvall et al., in prep.). Least–square fits to the data show that the regression lines for the two types of galaxies run nearly parallel. If we induce starformation in an LSBG to produce a BCD, the shift indicates an increase in luminosity of 1.5 mag. Fig 1b shows the metallicity–luminosity diagram for nearby dwarf galaxies (Skillman et al. 1989, ApJ 347, 875), blue LSBGs (Rönnback and Bergvall 1994 A&AS 108, 193; 1995, A&A 302, 353) and four luminous BCDs with low metallicities (Bergvall et al., in prep.). The positions of the two most popular young galaxy candidates, I Zw 18 and SBS 0335–052 are also indicated. These have $\mathcal{M}_{\rm HI}/{\rm L}_B$ similar to LSBGs whereas the more massive ones follow the BCD relation in Fig 1a. For the dIs we have used the M–Z relationship from Skillman et al. (1989). A similar fit to the LSBGs (dotted line), assuming that its slope is the same as for the dwarfs, gives a significant shift towards higher luminosities and/or lower Z. If we now, on the basis of the HI data assume that we, by turning on a burst in the LSBG, increase its luminosity by 1.5 mag and at the same time *demand that the metallicity does not increase* we hit the next line running through the BCDs (dashed line). Thus, the combined HI data and metallicities support a relationship between BCDs and blue LSBGs. If we want to use the dIs in the diagram to fabricate the low metallicity BCDs we have to increase the luminosity by 3 mags. This makes it much more difficult since: 1) the star formation efficiency has to be nearly 100% and 2) the resulting $\mathcal{M}_{\rm HI}/{\rm L}_B$ will be much higher than the observed value for BCDs (the $\mathcal{M}_{\rm HI}/{\rm L}_B$ for dIs is lower than for LSBGs).

J. Andersen (ed.), Highlights of Astronomy, Volume 11A, 103–104.

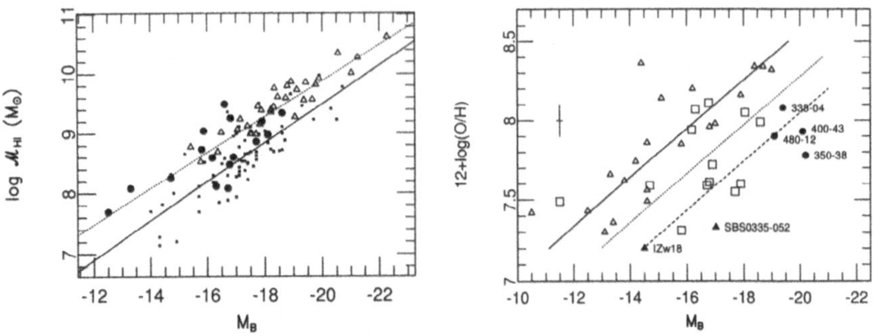

Figure 1. **a) The $\mathcal{M}_{HI}/M_B$ relationship for blue LSBGs (subsample of McGaugh et al., triangles) blue LSBGs (circles) and BCDs (Thuan & Martin; dots). The regression lines for the samples are indicated. b) The metallicity–luminosity relationship for dIs (Skillman et al., triangles), blue LSBGs (Rönnback & Bergvall, squares), the BCDs ESO 338–04, 350–38, 400–43 and 480–12 (Bergvall et al., filled circles), I Zw 18 and SBS 0335–052 (filled triangles)**

3. Photometric properties

Scale lengths and central surface brightnesses can often put useful constraints on the evolution of different types of galaxies. Disk type LSBGs are particularly simple systems in this respect. The central surface brightness of massive BCDs is $\approx$ 4 magnitudes brighter than a LSBG while the required increase in total luminosity according to the above is 1.5 mag. We thus have to focus the starburst towards the centre and steepen the luminosity profile. At the same HI mass the BCDs thus should have a shorter scale length, in accordance with observations. In such galaxies one would hope to detect the precursor disk at large radii and faint levels. Exponential profiles have been detected in several cases (e.g. Papaderos et al. 1996, A&AS 120,207) but the derived central surface brightness of this component is significanly brighter than the mean value for LSBGs ($\mu_{B,0} \approx$ 23–23.5). Normally the measured profiles stop at $\mu_B \approx$ 26–27. From a study of 'our' four luminous BCDs, however, we have found that at an even fainter level, $\mu_B \approx$ 28–29, the profile will continue to flatten out. When combining near–IR colours with optical we find that in two cases, ESO 400–G43 and ESO 350–IG38, the outer profile has colours of an early type galaxy while in the other two cases it may very well be an LSBG disk. To make further constraints, we have made comparisons between their observed broadband colours and those predicted by our spectral evolutionary models (SEMs). We find that the ages of the 'dE' (=dSph) component in the first two cases is of the order of 10 Gyr. In the other two cases the colours are similar to those of blue LSBGs, indicating a continuous star formation history.

Strong support for the presence of a dE host comes from recent HST observations of ESO 338–IG04 (Östlin et al., in prep.), in which we have detected a system of globular clusters (GCs). Our photometry is sufficiently good so that we can use the SEMs to derive age estimates. The results point at three distinct age groups of the GCs, one old, one of an age of $\sim$ 2 Gyr and one young. From the luminosity function of the old GC population, corrected for incompleteness, we can derive the total number of old GCs. Comparing this with the luminosity of the underlying galaxy, we find it to be consistent only with a very GC rich dE, with $M_B \sim -18$.

4. Central kinematics

We (Östlin et al., in prep.) recently used Fabry–Perot observations to study the Hα velocity field in a sample of BCDs and their companions, including the ones discussed above. The general result is that the velocity field in the companions is quite ordered. The main components, however, show a higher complexity. The most likely interpretation is that mergers are frequent and probably responsible for the starburst activity. From these results we draw the conclusion that an attractive scenario for the ignition of a BCD is that a merger is taking place — involving an LSBG and sometimes a dE (possibly containing a gas disk).

VIRGO AND OTHER LATE-TYPE DWARFS

N. BROSCH, A. HELLER AND E. ALMOZNINO
Wise Observatory & Dept. of Astronomy and Astrophysics
Tel Aviv University, Tel Aviv 69978, Israel

We attempt to solve the question of star formation triggers and star formation laws by studying samples of simple objects and defining carefully the possible external effects. Among the star formation (SF) triggers there are some that can operate only in large disk galaxies. These are shear instabilities and density waves, and we can eliminate them if we restrict the sample to diskless objects of low mass. Such galaxies, which do show star formation, are late-type dwarf galaxies (DGs).

Other SF triggers are related to the neighborhood a galaxy finds itself in. Such triggers are galaxy-galaxy collisions and galaxy-intracluster matter interactions (stripping, shocks, etc.). These also can be eliminated by properly choosing the sample to study; one selects galaxies from neighborhoods of widely different densities (of galaxies) and compares their SF parameters.

We selected our first samples in a region of relatively high galaxy density, where a complete morphological classification of objects was available, and suitable depth in brightness sampling could readily be achieved. This is the Virgo cluster, where Binggeli, Sandage and Tammann (1985, BST) provide a sample of more than 2000 DGs. Among those, some 25% are of late type and are classified by BST as blue compact dwarfs (BCDs), as Magellanic irregulars (Im) of five possible sub-classes, or as combinations of those two classifications.

We study four samples of DGs in the Virgo cluster selected on the basis of average blue surface magnitude as an indicator of recent star formation activity. All galaxies are brighter than m_B=17.21. The selection of Virgo late-type DGs carries an additional advantage, as all the objects have total HI measurements from Arecibo by Hoffman *et al.* (1987, 1989). This allows us to consider also the raw material reservoir size's influence on the SF.

Our study relies primarily on CCD surface photometry in BVRI and Hα done at the Wise Observatory (WiseObs). Here we present results of a study of BCDs (N=17) and low surface brightness (LSB) DGs (N=29) in Virgo, and explain our plans to study a complementary sample in the extreme field. The BCD sample is the 1996 PhD thesis of EA and the LSB sample is AH's PhD subject.

The integrated photometry of BCDs showed, through a comparison with predictions of evolutionary synthesis programs, that it is possible to understand these systems as resulting from burst evolution of DGs. In general, we found evidence for (at least) one SF burst and an older population, perhaps older than a few 10^8 yrs, which also fits an SF burst model. One object could be the result of a single, recent SF burst with no older population required. We found no significant differences among the high-surface-brightness (HSB) sample galaxies as a function of total HI or other observables. Preliminary results for the low-surface- brightness (LSB) sample indicate a similar lack of correlation with total HI.

The observational data for LSB DGs have been completed in Hα, and we present here preliminary results regarding the statistics of HII regions and a comparison with similar parameters for BCDs. The HII regions were detected from "net Hα" images of the galaxies, totalling about 90 minutes of exposure in the line filter.

J. Andersen (ed.), Highlights of Astronomy, Volume 11A, 105–106.

We find that the LSB sample shows present star formation activity, in the form of HII regions, in more than 50% of the sample objects. The sizes of HII regions are similar to those found for the HSB sample. The difference among HSB and LSB galaxies appears to be in the intensity of the starbursting activity and in the relative coverage of the LSB galaxies by HII regions, which is much smaller than for the HSB sample.

We find that whereas for "pure" BCDs ~63% of their (broad-band delimited) area is covered by HII regions, this fraction drops to only 20% for the LSBs (it is 30% among the 'mixed morphology' galaxies). Only a weak positive correlation is revealed between the presence of HII regions among LSBs and the total amount of HI, but this requires further investigation. Finally, we found no dependence of the presence of HII regions in late-type dwarf galaxies of our joined (BCD+LSB) sample with Virgocentric distance. This argues *against* galaxy interactions being significant triggers of SF.

Acknowledgements

We acknowledge support by a Center of Excellence grant from the Israel Science Foundation.

References

Binggeli, B., Sandage, A., and Tammann, G.A. (1985) *Astron. J.*, **90**, 1681
Hoffman, G.L., Glosson, J., Helou, G., Salpeter, E.E., and Sandage, A. (1987) *Astrophys. J. Suppl.*, **63**, 247
Hoffman, G.L., Williams, H.L., Salpeter, E.E., Sandage, A., and Binggeli, B. (1989) *Astrophys. J. Suppl.*, **71**, 701

DWARF ELLIPTICAL GALAXIES IN THE CEN A AND SCULPTOR GROUPS

H. JERJEN AND K.C. FREEMAN
Mount Stromlo and Siding Spring Observatories
Private Bag, Weston Creek PO, ACT 2611, Canberra, Australia

AND

B. BINGGELI
Astronomical Institute of the University of Basel
Venusstrasse 7, 4102 Binningen, Switzerland

1. The "missing" dEs

Côté (1995) surveyed the nearby Centaurus A (Cen A) and Sculptor (Scl) group regions in the search for gas-rich, dwarf irregular galaxies (Irr) which are associated with these galaxy aggregates. 36 Irrs have been found and confirmed as new group members via 21cm/H_α redshifts. In finding this large number of Irrs, the question rises whether systems from the other dwarf galaxy family, the gas-poor dwarf ellipticals (dE), also exist in the two groups. So far nothing is known about dE membership but taking into account the morphology-density relation for dwarfs (Binggeli et al. 1990) a number ratio Irr/dE$\approx 1-2$ would be expected. Prompted by these "missing" dEs we started a large program to identify and study dEs in Cen A and Scl. The main scientific goals are (1) to find dE candidates and to establish group membership via velocity or distance and (2) to estimate the luminosity function (LF) of dwarf galaxies to the faintest possible luminosities. For the two groups, being respectively at a mean distance of 4.0 Mpc and 2.5 Mpc, we will be able to reach dwarf galaxies 4–5 magnitudes fainter than in the previous studies in the Virgo and Fornax clusters and the five groups Leo I, Dorado, NGC1400, NGC5044, and Antlia (Sandage et al. 1985, Ferguson & Sandage 1990). We briefly outline here first results from the Cen A group. All details about the complete survey including Scl will be presented in Jerjen et al. (1997)

2. How to find them

A list of 13 dE candidates in the Cen A group was compiled based on a visual inspection of 45 SERC IIIa-J films covering the entire group area. The galaxies were selected based on the characteristic dE morphology, namely a low surface brightness light distribution without gas or star formation features. We did CCD imaging of all objects in the B and R-band at the 2.3m telescope at Siding Spring. A total integration time per galaxy and per filter of 3600s was required to obtain photometric and structure parameters. Côté used radial velocities as distance indicators for her gas-rich dwarfs. This is not applicable to our gas-poor dEs with $SB_{B,eff} \sim 24\,\mathrm{mag\,arcsec}^{-1}$. Alternative distance determinations for dEs in general necessitate resolving the system into stars. We decided to explore the surface brightness fluctuation (SBF) method developed by Tonry & Schneider (1988) which has been proven to be very efficient and reliable in obtaining distances to giant elliptical galaxies. However, our application differs from previous ones, in that we are working on low surface brightness objects and are photon-limited by the sky rather than by the galaxy itself. Therefore we worked in the R-band instead of the generally prefered I-band. The sky is significantly redder in R-I than the red giants causing the fluctuation and this choice of filter avoids the fringing that occurs with thinned CCDs in the I-band. Five Cen A candidates were selected for the analysis. They have been observed under good seeing conditions ($\sim 1.5''$) and cover a range in mean effective surface brightness ($23^m < SB_{B,eff} < 25^m$) and effective radius ($15'' < r_{B,eff} < 35''$). The SBF analysis followed in main parts the procedure described in Tonry & Schneider. After preparing the image

J. Andersen (ed.), Highlights of Astronomy, Volume 11A, 107–108.

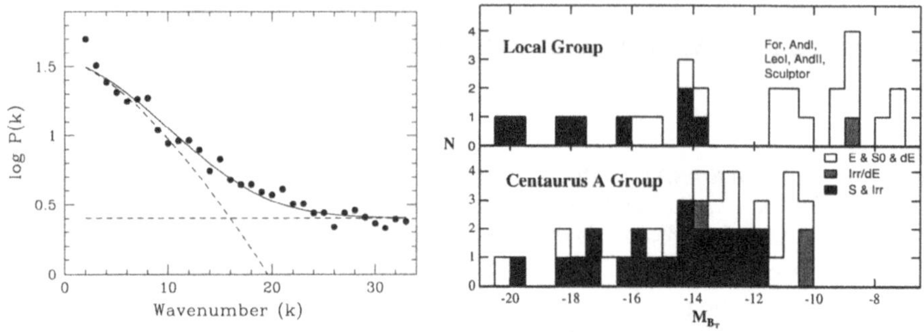

Figure 1. (a) Observed power spectrum of the dE candidate ESO384-016. The solid line is the best fit to the data, which is the sum of a PSF convolved and a scale free noise component (dashed lines). (b) LFs of the Cen A group and the LG. Some faint dEs of the LG are indicated to illustrate the luminosity limit of our survey.

of each galaxy its spatial power spectrum (PS) was derived (see Fig. 1a). The PS consists of a PSF convolved component (fluctuation power) and a scale free white noise component (photon and RON noise). The amplitude of the fluctuation power gives the magnitude of a "typical star" in the galaxy, $\bar{m}_R$. All five galaxies yielded strong signals corresponding to $26.6 < \bar{m}_R < 27.1$. Most of the published SBF studies are carried out in the I-band (e.g. Tonry et al. 1997) and no empirical calibration of $\bar{M}_R$ exists yet. We estimate this quantity with the theoretical isochrones from Worthey's models (1994). For a typical [Fe/H]=-1.4 and allowing for different star formation histories (Da Costa 1997) we get $\bar{M}_R = -1.15(\pm 0.03)$. Four galaxies turned out to be Cen A group members covering the distance range 3.6 <D(Mpc)< 4.4. The other dE is slightly behind the group at D=7.3 Mpc. An independent proof of group membership came from redshifts we measure for the two brightest sample galaxies. For instance, the dE candidate ESO384-016 has $v_\odot = 561\,\mathrm{km\,s^{-1}}$ and $D_{SBF} = 4.1$ Mpc which is in perfect agreement with data for well-known Cen A group galaxies: NGC5253 ($v_\odot = 404\,\mathrm{km\,s^{-1}}$; D= 4.1 Mpc, Saha et al. 1995) and NGC5128 ($v_\odot = 562\,\mathrm{km\,s^{-1}}$; D= 3.7 Mpc, Soria et al. 1996). Even though five dE distances only are available now, there is strong qualitative evidence from the morphological similarity between confirmed and unconfirmed dEs that the eight remaining candidates are likely to be Cen A group members too. Fig. 1b shows the LF of all known Cen A group galaxies, assuming all dEs but one (for which we have the distance) are associated with the group. If true, we have found dwarf systems as faint as the Fornax or Leo I dwarfs in the Local Group.

Acknowledgements

HJ and BB are grateful to the Swiss National Science Foundation for the financial support.

References

Binggeli, B., Tarenghi, M., Sandage, A., 1990, A&A, 228, 42
Côté, S., 1995, PhD, Australian National University
Da Costa G.S., 1997, in *The Second Stromlo Symposium: The Nature of Elliptical Galaxies*, eds. M. Arnaboldi, G.S. Da Costa, P. Saha, ASP Conf. Series, vol. 116, p. 270
Jerjen, H., Freeman, K.C., Binggeli, B., 1997, in preparation
Sandage, A., Binggeli, B., Tammann, G.A., 1985, AJ, 90, 1759
Saha A., Sandage A., Labhardt L., Schwengeler H., Tammann G.A., Panagia N., Macchetto F.D., 1995, ApJ, 438, 8
Soria R., et al., 1996, ApJ, 465, 79
Tonry J.L., Schneider D.P., 1988, AJ, 96, 807
Tonry, J.L., Blakeslee, J.P., Ajhar, E.A., Dressler, A., 1997, ApJ, 475, 399
Worthey, G., 1994, ApJS, 95, 107

AUTOMATED SEARCH FOR LSB GALAXIES IN THE NEARBY UNIVERSE

Z. MORSHIDI, J.I. DAVIES AND R.M. SMITH
Dept. of Physics and Astronomy, University of Wales, Cardiff
P.O.Box 913, Cardiff CF2 3YB, U.K.

1. Introduction

We have carried out a large area survey for nearby LSB galaxies to try and answer the following questions:

1. How are LSB galaxies distributed relative to the *normal* HSB galaxies? Do they have the same clustering properties suggesting that they trace the same mass distribution?
2. What are the implications of the clustering properties on galaxy formation scenarios?
3. Can they account for faint blue galaxies at higher redshifts?
4. Is there any difference between the field and cluster LSB galaxies?

2. The Survey

We have devised an automated technique to pick out local LSB galaxies using 65 UKSTU photographic plates corresponding to 2187 sq.°. We have imposed stringent selection criteria in order to minimise the background contamination. Our galaxies have $\mu_0 \geq 22.5$ B mag arcsec^{-2}, $\alpha \geq 3.0$ arcseconds and $B_T \leq 20.0$. Our survey areas are confined to the Fornax Cluster, the NGC1400, Sculptor and Dorado Groups and the fields between them. The survey, which was carried out in three stages over a period of 2 years, has produced detections of 2435 LSB galaxies (32 galaxies per field). We have estimated the background contamination in three ways: first, by a limited redshift sample, secondly, by Ferguson's (1989) classification on the Fornax plate and finally, by a numerical simulation of background galaxies. The first method produced an estimate of 14 galaxies per field, while the second and the third methods produced the same estimate of 19 galaxies.

3. Results

Figure 1 shows the spatial distributions of the dwarf galaxies. 19 galaxies representing the background contaminants have been randomly removed from each field. The data for the bright galaxies were taken from the RC3 (de Vaucouleur 1991). The results of the dwarf simulations indicate that, if the local Luminosity Function of Driver (1994) is correct, we should have detected 54 dwarf galaxies in each field if the intrinsic surface brightness follows the $\mu_0 - M_B$ relation and 68 galaxies if they have a flat intrinsic surface brightness distribution. Yet our data indicate an average detection rate of 13 nearby galaxies per field. On average, we have detected less galaxies than that predicted by the Driver model. Either Driver's Luminosity Function is incorrect or the dwarf galaxies in the field have faded below our detection threshold *and* the field galaxies have different surface brightness distributions than those in the clusters. Babul & Rees (1992) suggest that the dwarf ellipticals in the clusters can maintain their luminosities while those in the field are more susceptible to gas loss thus ceasing their star formation activities and, consequently, fading more quickly than the cluster dwarfs.

J. Andersen (ed.), Highlights of Astronomy, Volume 11A, 109–110.

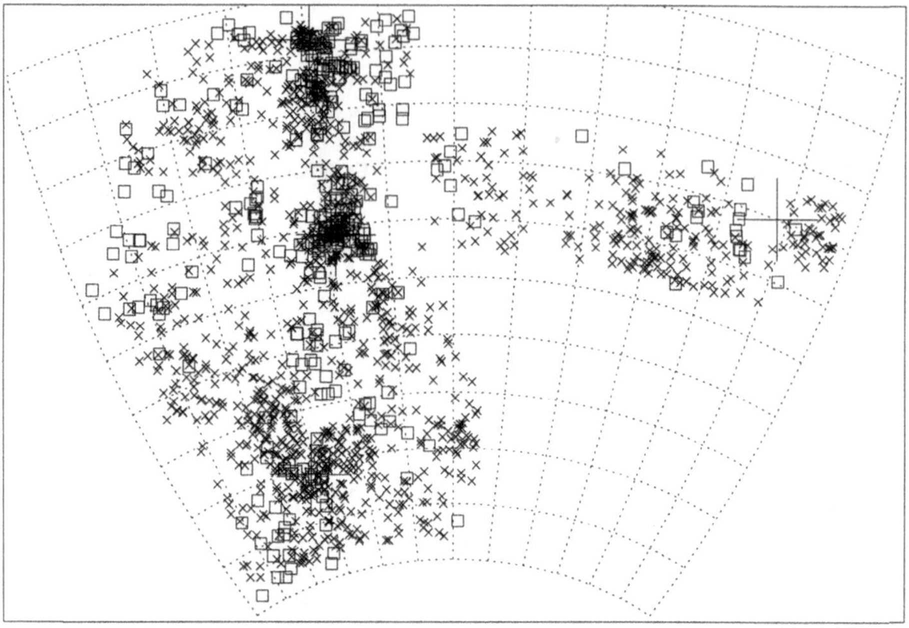

Figure 1. The distribution of nearby LSB galaxies (×) and the bright galaxies (□) in an equal-area plot. The large + are the centres of the cluster and groups.

References

Babul, A. and Rees, M.J. (1992) *MNRAS*, **255**, 346
de Vaucouleur, G.H. (1991) *Third Ref. Cat. of Bright Galaxies*, Springer-Verlag, New York
Driver, S.P. (1994) PhD Thesis, University of Wales Cardiff
Ferguson, H.C. (1989) *AJ*, **98**, 367
Freeman, K.C. (1970) *ApJ*, **160**, 811
Loveday, J., Peterson, B.A.,Efstathiou, G. and Maddox, S.J. (1992) *ApJ*, **390**, 338

THE DISTRIBUTION OF BCDGS IN VOIDS

ULRICH LINDNER AND KLAUS J. FRICKE
Universitätssternwarte, Geismarlandstr. 11, D-37083 Göttingen, Germany

AND

JAAN EINASTO AND MARET EINASTO
Tartu Astrophysical Observatory, EE-2444 Tõravere, Estonia

1. Introduction

Generally the density of the environment is important for the formation and evolution of galaxies: bright elliptical galaxies reside in the centers of galaxy clusters, whereas late type galaxies preferentially populate small groups, filaments or the outskirts of rich clusters of galaxies. Furthermore, fainter galaxies define smaller voids than bright galaxies. Our investigation of void catalogues in the Northern Local Void region shows that large voids are interlaced by systems of faint galaxies dividing them into smaller subvoids (Lindner *et al.*, 1995). This **void hierarchy** is an important property of the Large Scale Structure in the nearby Universe and can help to solve problems concerning the distribution of dwarf galaxies and BCDGs.

The questions whether dwarf galaxies are homogeneously distributed in voids or if their distribution reveals any structure is crucial for some theories of galaxy formation, e.g. the biased galaxy formation scenario (Dekel & Silk, 1986). Moreover, an investigation of the environment of BCDGs may clarify the cause of starbursts and address the question whether BCDGs may constitute a special population of galaxies in voids.

2. Results

We investigated the radial distribution of dwarf galaxies and BCDGs inside spherical voids from our void catalogues and performed nearest neighbor tests to study possible differences in the distribution of BCDGs and "normal" dwarf galaxies with respect to bright galaxies.

2.1. THE RADIAL DISTRIBUTION OF GALAXIES IN VOIDS

Figure 1 reveals that most faint galaxies are concentrated near void walls. This result has also been found using other methods (Kuhn *et al.*, 1997). The radial distributions are almost identical for all three void catalogues suggesting that the void hierarchy continues to fainter than $M = -18.8$ galaxies (Lindner *et al.*, 1996). The radial distribution of real galaxies in voids differs significantly (99 % confidence level) from random samples which rules out biased galaxy formation scenarios (Dekel & Silk, 1986).

2.2. NEAREST NEIGHBOR ANALYSIS

In Figure 2 we studied unbinned cumulative distributions of nearest neighbor distances for different kinds of galaxies (BCDGs, all galaxies, galaxies brighter than −18.8 and random samples). KS tests confirm that BCDGs are nearer to faint galaxies than to bright galaxies. The nearest neighbor distance distributions for BCDGs and all galaxies are similar (KS: $\approx 76\,\%$) and differ significantly (99.9% confidence level) from random samples. The median nearest neighbor distance for BCDGs is about $0.7\,h^{-1}$ Mpc, which corresponds to a characteristic scale for loose groups or filaments of galaxies.

J. Andersen (ed.), Highlights of Astronomy, Volume 11A, 111–112.

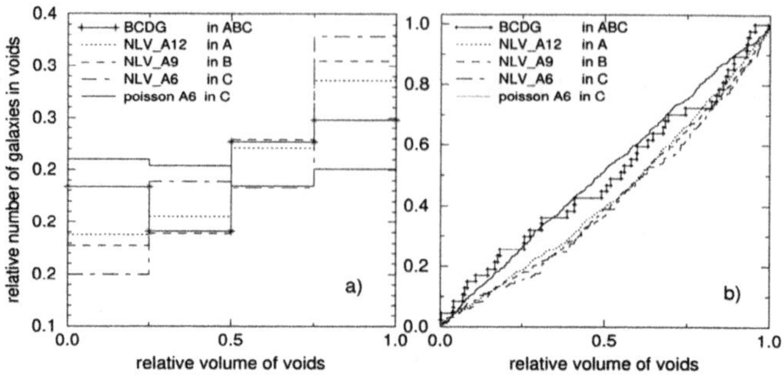

Figure 1. **a) differential and b) unbinned cumulative radial distribution of galaxies in voids**

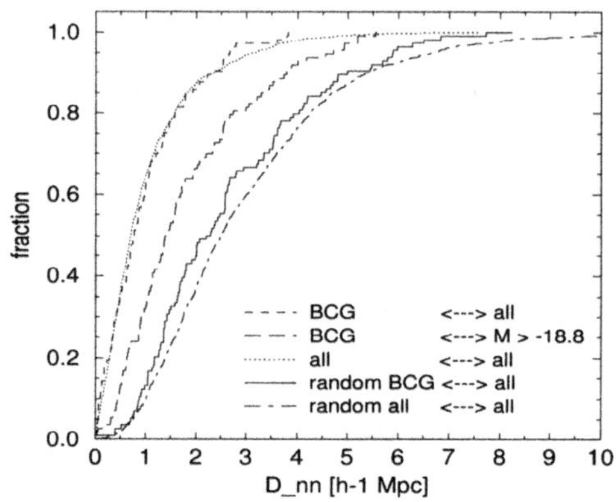

Figure 2. **unbinned cumulative distribution of nearest neighbor distances D_nn**

3. Conclusion

It is generally assumed that BCDGs are LSB dwarf galaxies undergoing a phase of enhanced star formation (Krüger *et al.*, 1991). We found that BCDGs reside preferentially in intermediate density regions, i.e. they avoid high density regions and are not very isolated. This suggests that star bursts may be triggered by neighboring galaxies: high density environment implies high SFR and thus all gas is already consumed. No BCDGs are found in low density environment, suggesting that close neighbors may be needed for BCDG activity. The similarity of the distributions indicates that BCDGs are not an exceptional population of isolated galaxies in voids as it has been suggested previously (Pustil'nik *et al.*, 1995).

References

Dekel, A., & Silk, J., 1986, ApJ, 303, 39
Krüger, H., Fritze-von Alvensleben, U., Loose, H.-H., & Fricke, K.J., 1991, A&A 242, 343
Kuhn, B., Hopp, U., & Elsässer, H., 1997, A&A 318, 405
Lindner, U., Einasto, J., Einasto, M., Freudling, W., Fricke, K.J., & Tago, E., 1995a, A&A 301, 329
Lindner, U., Einasto, M., Einasto, J., Freudling, W., Lipovetsky, V.A., & Richter, G., 1995b, A&A 314, 1
Pustil'nik, S.A., Ugryomov A.V., Lipovetsky, V.A., Thuan, T.X., & Guseva, N.G., 1995, ApJ, 443, 449

MID-IR EMISSION OF LOW-METALLICITY GALAXIES

S. PLANTE
Dept. de Physique, Université Laval
Cité Universitaire, Ste-Foy, PQ, Canada G1K 7P4

M. SAUVAGE
CEA/DSM/DAPNIA/Service d'Astrophysique
C.E. Saclay, 91191 Gif sur Yvette CEDEX, France

AND

D. KUNTH
Institut d'Astrophysique
98bis Bd Arago, 75014 Paris, France

1. The influence of star formation in NGC 595

NGC 595 is a giant HII region located in the western part of the spiral galaxy M 33. It is the second in importance in this galaxy, after NGC 604. At 0.84 Mpc, *HST* is able to resolve its stellar content. Malumuth *et al.* (1996) obtained *HST* UV, U, B and V images of this region and derived an ionizing luminosity of 5×10^{50} phot s^{-1}, and an average reddening $E_{B-V} = 0.36 \pm 0.28$ mag. The stars are mostly concentrated in the central part of the region, where little emission of gas is seen (the ionized gas lies more in a shell around the stars, figure 1a). Wilson & Scoville (1993) showed the molecular gas to be situated in the south-east part of the region, just outside of the bright knot of stars. Viallefond *et al.* (1986) found a reddening gradient in the north-east/south-west direction by observing the HI gas, which was confirmed by Malumuth *et al.* (1996) with stellar photometry.

We obtained *ISO* images for NGC 595 in the 5.0 to 8.5 μm range. The emission in this spectral range is dominated by the so-called *PAH* bands. Current interpretation of these has them originating from stochastically heated molecules. Two of these bands are located in the range observed, at 6.2 μm and 7.7 μm. Stochastic heating implies that the in-band flux is directly proportional to the number of photons absorbed by the molecules. For typical HII regions, Cohen *et al.* (1989) found 0.58 for the $I_{6.2}/I_{7.7}$ in-band ratio. However many processes, ionization, dehydrogenation, can modify this ratio. Furthermore, an underlying continuum is present though its exact origin is unknown.

2. Data Reduction

Three different filters have been used to cover this wavelength range: LW4 (5.5-6.5 μm), LW5 (6.5-7.0 μm) and LW6 (7.0-8.5 μm). The first one essentially contains the 6.2 μm band, the second a pseudo-continuum in-between and the third, the 7.7 μm band. We used CIA[1] to reduce the data. Cosmic rays where removed with a multi-resolution based algorithm. We used an iterative technique to build a flat-field from the LW6 image, which had the best signal-to-noise ratio. We used this flat-field to correct our three images, as the wavelength range is small. Transient correction has been done with an inversion technique which takes into account the fact that we had different integration times for each of the filters. For details on ISOCAM data reduction, see Starck *et al.* (1998).

[1]CIA, CAM Interactive Analysis, is a joint development by the ESA Astrophysics Division and the ISOCAM consortium led by the ISOCAM P.I., C. Cesarsky, Direction des Sciences de la Matière, C.E.A., France.

J. Andersen (ed.), Highlights of Astronomy, Volume 11A, 113–114.

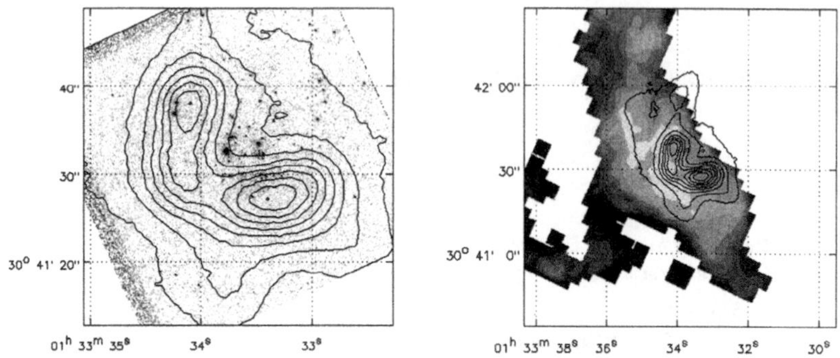

Figure 1. For both figures, North is on top and East is to the left, and the equinox is J2000.0. **(a) - left:** Hα image on WFPC2 B-band image from Malumuth *et al.* (1996). Contours and grey levels are linearly spaced. **(b)- right:** In grey levels we represent the LW4/LW6 color ratio (logarithmic). For display reasons, the ISO maps have been resampled to 0.3″ pixels but the real pixel size is 3″. Overlaid on this map is the Hα image of **(a)**. Although the two distributions are related, the ISO emission is more extended and displaced in the south-east direction.

3. Analysis

Metcalfe *et al.* (1996) obtained an *ISO* spectrum of Haro 3, a Wolf-Rayet dwarf galaxy. Their spectrum shows clearly the two PAH bands. The spectrum of NGC 595 should in principle be similar, as both regions harbor the same energy sources (i.e. massive stars). The images we obtained show that the flux density in the LW5 filter (the pseudo continuum) is higher than that in the LW4 filter (the 6.2 μm band) and almost as high as that in the LW6 filter (the 7.7 μm band). We tried to reconstruct for each pixel the 5.0-8.5 μm spectrum of NGC 595 by fixing the FWHM of the two PAH bands (0.4 μm for the 6.2 μm band and 0.9 μm for the 7.7 μm), the shape of the continuum and the integrated flux for each spectral band as taken from our images. The restrictions for the FWHM and the continuum are backed-up by a number of *ISO* observations in different environnements. We were not able to create a spectrum showing two emission bands on top of a continuum. This could be caused by an oversimplification of the problem, by a contamination of the LW5 filter by the 7.7 μm band or by the real absence of the 6.2 μm emission band. The second alternative is ruled out by the study of the relative distribution of the flux in the LW5 and LW6 filter: if the 7.7 μm band was contaminating the LW5 filter, we should see a spatial correlation between the LW6 and the LW5 flux, which is not the case. The first alternative is probably ruled out also as we varied all the parameters of the fit without being able to reproduce a spectrum with two emission bands. Follow-up ISO observations are planned to put more constraints on the in-band intensities.

Whatever produces the emission in the three ISO filters, the intensity ratios we measure are not constant over the whole region (see fig 1b where we show the LW4/LW6 ratio). This implies that different excitation mechanisms contribute emission throughout the region, or that the dust properties vary spatially.

Finally, although the emission in the ISO filters resembles that in Hα, the emission maximum in the Hα band is not cospatial with the maximum in the ISO filters. The maximum emission in the ISO filters is in fact at the location of the peak CO emission, therefore this indicates that either the Hα emission suffers substantial extinction, or the stars have mostly cleared out the HII region of dust.

References

Cohen, M., Bregman, J., Witteborn, F.C., Wooden, D.H., Rand, D.M., de Muizon, M.J., Allamandola, L. J., & Tielens, A.G.G.M. 1989, *Infrared Spectroscopy in Astronomy*, pp. 149-154
Malumuth, E.M, Waller, W.H., & Parker, J.W.M. 1996, AJ, 111, 1128
Metcalfe, L. *et al.* 1996, A&A, 315, 105
Starck, J.L. *et al.*, in preparation
Viallefond, F., Goss, W.M., van der Hulst, J.M., & Crane, P.C. 1986, A&AS, 64, 237
Wilson, C.D., & Scoville, N. 1992, ApJ, 385, 512

THE NEARBY YOUNG DWARF GALAXY SBS 0335–052: PRIMORDIAL GAS AND LYMAN-ALPHA EMISSION

TRINH XUAN THUAN AND YURI I. IZOTOV
Astronomy Department, University of Virginia
Charlottesville, VA 22903, USA
AND
Main Astronomical Observatory, Kiev 252650, Ukraine

Ever since their discovery, blue compact dwarf (BCD) galaxies have been thought of as excellent candidates for being young galaxies, i.e. systems presently undergoing one of their very first bursts of star formation. This is mainly because BCDs are very metal-deficient, the metallicities of their ionized gas ranging between $Z_\odot/50$ and $Z_\odot/3$, which makes them the least chemically evolved galaxies in the Universe. Other evidence, such as the very high fractional neutral hydrogen gas content and the lack of an evident underlying old stellar population on optical images, also point to the relative youth of some BCDs.

Thuan, Izotov & Lipovetsky (1997) have argued SBS 0335–052 to be a young galaxy on the basis of the following evidence: 1) *HST* imaging of the BCD shows its underlying extended low surface brightness component to have an irregular and filamentary structure, suggesting that a significant part of the emission (~1/3) comes not from an underlying stellar population, but from ionized gas. Any underlying stellar population must be younger than ~10^8yr. Propagating star formation occurs in a chain of 6 super-star clusters with ages ranging between 4 and 30 Myr. 2) The underlying component shows unusually blue colors consistent with gaseous emission, in contrast to most BCDs which possess an underlying red component. 3) VLA 21 cm observations show the BCD to be embedded in an extraordinarily large HI cloud with dimensions some 64 × 24 kpc (the typical size of HI envelopes around BCDs is more like a few kpc in each dimension).

HST UV spectrophotometry of SBS 0335–052 by Thuan & Izotov (1997) shows a broad damped Lyα absorption giving a HI column density of N(HI) = (7.0±0.5) × 10^{21} cm^{-2}, one of the largest HI column densities known in a galaxy. The UV spectrum also shows many absorption lines of heavy elements, such as O I λ1302, Si II λ1260, λ1304, and S II λ1251, λ1254, λ1259, at the redshift of the BCD. Comparison with high resolution quasar spectra suggests that the O I λ1302, Si II λ1304 and S II λ1251, λ1254 absorption lines are probably not saturated and can be used to derive column densities. If these lines originate from the neutral gas envelope of SBS 0335–052, then the abundances of oxygen, silicon and sulfur are respectively 37000, 4000 and 116 times lower than the solar values. These highly discrepant deficiency factors suggest that the heavy element absorption lines are produced not in the H I, but in the H II region. If so, oxygen and sulfur abundances can be derived from the absorption lines which are in agreement, within a factor of 2, with those derived from the optical emission lines, giving a metallicity ~ $Z_\odot/40$ for the HII region. If the heavy element absorption lines all come from the H II region, then the HI cloud in SBS 0335–052 is primordial, composed of pristine gas unenriched in heavy elements by previous star formation.

How do the properties of SBS 0335–052 compare with those of young, high-redshift galaxies? The star formation rate (SFR) in the BCD is moderate, equal only to ~ 0.5 $M_\odot$yr^{-1}, about one order of magnitude lower than the SFRs derived by Steidel et al. (1996) for their high-redshift star-forming galaxies. SBS 0335–052 contains an equivalent number of ~ 4500 O7 stars, so that its ultraviolet luminosity is ~ $10^9 L_\odot$, in the low range of ultraviolet luminosities (10^9 – $10^{10} L_\odot$) found by Lanzetta et al. (1996) in their study of high-redshift ($2.5 \leq z \leq 6$) galaxies in the Hubble Deep Field. The sizes of the star-forming regions in these galaxies are ~ 1 kpc, which compare well with that in SBS 0335–052. These luminosities and sizes are modest in comparison to luminous non-dwarf galaxies because star formation in the high-redshift objects occurs in small, concentrated regions

J. Andersen (ed.), Highlights of Astronomy, Volume 11A, 115–116.

rather than in galaxy-sized ones. This is in contrast to the situation in a BCD like SBS 0335–052 where star formation occurs over the whole galaxy. As for the metallicities, the ones deduced for high-redshift systems are generally in the range spanned by BCDs, but with the vast majority of the known systems being significantly more chemically evolved than SBS 0335–052. For example, Pettini et al. (1997) found a mean metallicity of $\sim Z_\odot/13$ for 10 damped Lyα systems in the redshift range z = 0.7 to 3.4. Thus, except for their higher star formation rates, the star-forming regions in high-redshift galaxies have properties which are very similar to those in young nearby dwarf galaxies like SBS 0335–052. It is still not yet clear why SBS 0335–052 did not participate in star formation and metal-enrichment at the same epoch as the majority of the galaxies, given that the collapse time of its H I envelope is only $\sim 5\times10^8$yr, much shorter than the Hubble time.

A most interesting result is the detection in the spectrum of SBS 0335–052 of strong Si IV λ1394, λ1403 lines with P Cygni profiles. A minimum amount of metals is needed to provide the necessary opacity to drive the stellar winds originating from massive O stars which are responsible for the P Cygni profiles. The problem is especially acute in the case of SBS 0335–052 where, as suggested above, the OB stars seen now in the starburst probably formed out of pristine primordial gas, unenriched in heavy elements by previous star formation. Since the present star formation has proceeded in 6 separate bursts of duration less than a few Myr, we can imagine a scenario where the stars born in the later bursts and responsible for the stellar winds, formed from gas already enriched in heavy elements by supernovae resulting from the previous bursts, although it is not clear whether the metals produced in supernovae have had enough time to mix with the pristine gas. Another possibility is to postulate that somehow the evolution of massive stars leads to an increase of their surface metallicity and hence to the onset of a stellar wind.

What are the implications of the non-detection of Lyα emission in SBS 0335–052 concerning the escape of Lyα photons from a galaxy? It has been suggested that there should be an increase in the Lyα/Hβ flux ratio as the metallicity of the galaxy decreases, since presumably low metallicity objects contain less dust, and hence suffer less Lyα photon destruction by dust. The observational evidence for such a correlation is mixed. Terlevich et al. (1993) did find one, while Giavalisco et al. (1996) found a very weak one. The fact that there is no Lyα emission in the two most metal-deficient BCDs known, I Zw 18 (Kunth et al. 1994) and SBS 0335–052 argues strongly against the existence of such correlation. Thus metallicity, and hence dust, cannot be the only factor determining the escape of Lyα photons in SBS 0335–052, although HST optical pictures show clear evidence of dust in the BCD (Thuan et al. 1997). Lequeux et al. (1995) have argued that Lyα photons can escape when the neutral material where the absorption occurs is outflowing with respect to the star-forming region. While there is evidence for fast gas motions in SBS 0335–052 with velocities up to $\sim$1500 km s^{-1}, the H I gaseous envelope appears to be static with respect to the H II region, as the 21 cm and emission-line velocities are in good agreement. Thus, with its extremely large H I column density, the redistribution of Lyα photons in SBS 0335–052 by multiple scattering over the large volume of the H I cloud probably plays also an important role in diminishing the intensity of the Lyα line. The orientation of the flattened H I cloud may also play a role. Because the HI envelope is seen edge-on, Lyα photons escape more easily along directions perpendicular to the line of sight than along it.

References

Giavalisco, M., Koratkar, A. and Calzetti, D. 1996, ApJ, 461, 831
Kunth, D., Lequeux, J., Sargent, W.L.W. and Viallefond, F. 1994, AA, 282, 709
Lanzetta, K.M., Yahil, A. and Fernandez-Soto, A. 1996, Nature, 381, 759
Lequeux, J., Kunth, D., Mas-Hesse, J.M. and Sargent, W.L.W. 1995, AA, 301, 18
Pettini, M., King, D.L., Smith, L.J. and Hunstead, R.W. 1997, ApJ, 478, 536
Steidel,C.C., Giavalisco, M., Pettini, M., Dickinson, M. and Adeberger, K.L. 1996, ApJ, 462, L17
Terlevich, E., Diaz, A.I., Terlevivh, R. and Garcia Vargas, M.L. 1993, MNRAS, 260, 3
Thuan, T.X. and Izotov, Y.I. 1997, ApJ, November 10
Thuan, T.X., Izotov, Y.I., and Lipovetsky, V.A. 1997, ApJ, 477, 661.

EXTENDED HI STRUCTURES IN THE IRREGULAR GALAXY NGC 4449

HUGO VAN WOERDEN
Kapteyn Institute, Postbus 800, 9700 AV Groningen, The Netherlands
DEIDRE A. HUNTER
Lowell Observatory, 1400 W Mars Hill Road, Flagstaff AZ 86001, USA
AND
ERIC M. WILCOTS AND JAY S. GALLAGHER
Washburn Observatory, 475 N. Charter Street, Madison WI 53706, USA

NGC 4449 is a bright (B = 9.5), nearby (4 Mpc) irregular galaxy with a high surface brightness and strong star-formation activity; for photos see Hunter and Gallagher (1989). Using the Effelsberg 100-meter dish, van Woerden et al. (1975) – and later also Bajaja et al. (1994) – found its HI distribution to extend, at the level 2×10^{19} atoms cm^{-2}, over a full degree, or ten times the D_{25} diameter. Van Woerden et al. also mentioned that the HI velocity gradient in this outer disk was opposite to that measured at Westerbork over the innermost 4 arcminutes.

With the VLA D-array, we have used a mosaic of 3×3 pointings to map the HI at resolutions of $1.0'$ and 10 km s^{-1} over a field (FWHM) of 1° diameter (Hunter et al. 1998). The HI distribution (Fig. 1) shows the following features: 1) a bright inner body, centred on the optical galaxy and dominated by an annulus of 5 kpc diameter; 2) a symmetric elliptical – probably highly inclined – disk of 35×15 kpc diameter, with strong condensations at its NE and SW ends; 3) a set of narrow streamers, consisting of straight segments making abrupt angles, and containing several bright clouds. The HI mass in the inner body and elliptical disk together is 12×10^{8} $M_{\odot}$; the streamers (about 7×10^{8} $M_{\odot}$ of HI) are surrounded by a diffuse HI distribution, missed by the VLA interferometers but detected by the Effelsberg single-dish.

While the kinematics inside 4 kpc radius is complex, the elliptical disk outside 8 kpc radius appears to rotate at about 75 km s^{-1}, in a period of order 1 Gyr; its velocity gradient (increasing from NE to SW) is confirmed to be opposite to that in the inner body. The velocity field of the streamers is highly ordered; the velocity difference of 160 km s^{-1} (South - North), if interpreted as rotation, would imply a total mass of 10^{11} $M_{\odot}$ in the system; however, it is not clear that the streamers actually rotate around NGC 4449. With 100 km s^{-1} velocity. the travel time along the filament from the SW corner to the NE end would be about 1 Gyr. Velocity dispersions are about 10 km s^{-1} in the brighter parts of disk and streamers; the Northern joint and the SW crossing have broader or complex profiles. The time scale for expansion of the streamers is of order 200 Myr – unless self-gravity keeps them together.

What caused these peculiar structures? Even if the extended distribution found at Effelsberg represents delayed accumulation of primordial gas, we think the complex set of streamers must be due to external, tidal perturbations. The HI distribution is strikingly similar to that in the M81-M82-NGC3077 triplet (van der Hulst 1979, Yun et al. 1994), and three bodies may indeed be required to explain the sharp corners. The analogon of NGC 3077, at a closely corresponding position, may be provided by the gas cloud (a failed galaxy?) at the SW corner, where two filaments cross. But where is the third body? We consider several possibilities.

The nearby dwarf-irregular DDO 125, just $37'$ SSW of NGC 4449, appears undisturbed and has fairly normal global properties, including 1×10^{8} $M_{\odot}$ of HI and $M(\text{HI})/L_{\text{B}} = 0.4$. Could it once have had 8 times more gas and twice as many stars, and lost most of its gas and half of its stars over a period of 10^{9} years, without looking disturbed? If spread over an area of 100×5 kpc^{2}, the stars would produce an almost-undetectable surface brightness of only 28 mag arcsec^{-2}. Other, major possible perturbers are at distances exceeding $4^{\circ} = 300$ kpc, implying timescales of order 3 Gyr, possibly too long compared to that of the streamers.

J. Andersen (ed.), Highlights of Astronomy, Volume 11A, 117–118.

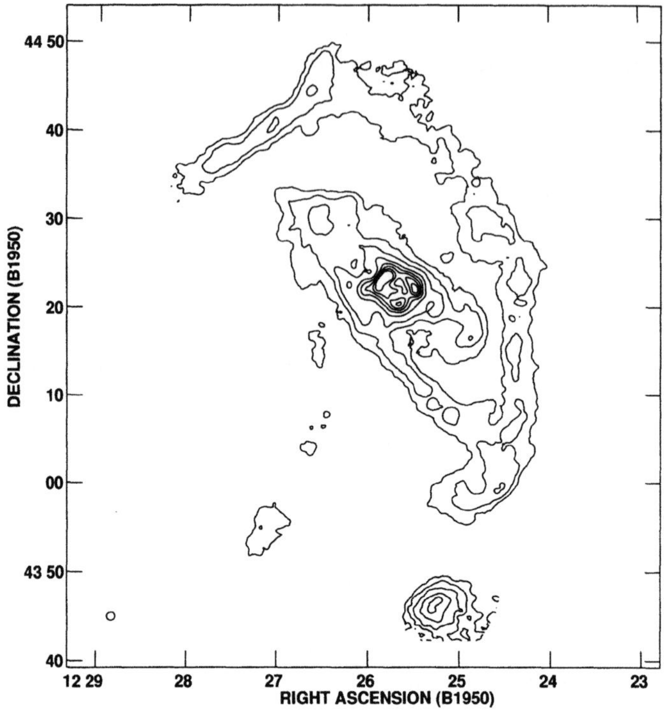

Figure 1. Distribution of HI column density; contour levels are 1, 2.5, 5, 10, 15, 20, 25 and 30 times 0.68×10^{20} atoms cm^{-2}. The ellipse at lower left is the half-maximum contour of the synthesized beam.

Or has the third body been swallowed by NGC 4449? Such a merger might have caused its current strong star formation, the counter-rotation of the inner HI, and its irregular distribution. The streamer running from the inner disk to the SW corner might then be a Toomre & Toomre (1972)-type bridge. Possible clues may come from measurement of the rotation of the inner stellar galaxy, from age analysis of its star clusters, and from a detailed study of the inner HI body and its possible interaction with the other subsystems.

Acknowledgements

The VLA is part of the U.S. National Radio Astronomy Observatory, operated by Associated Universities, Inc., under contract with the National Science Foundation. HvW thanks the Leids Kerkhoven-Bosscha Fonds and Washburn Observatory for support of his visit to Madison and to the IAU Assembly. DAH and EMW acknowledge support by NSF under Grants No. AST-9616940 and 9616907.

References

Bajaja, E., Huchtmeier, W.K., & Klein, U. 1994, A&A, 285, 388
Hunter, D.A., & Gallagher, J.S. 1989, Science, 243, 1557
Hunter, D.A., Wilcots, E.M., van Woerden, H., Gallagher, J.S., & Kohle, S. 1998, ApJ, in preparation
Toomre, A., & Toomre, J. 1972, ApJ, 178, 623
van der Hulst, J.M. 1979, A&A, 75, 97
van Woerden, H., Mebold, U., & Bosma, A. 1975, in La Dynamique des Galaxies Spirales, ed. Weliachew, L., Paris:CNRS, p. 483
Yun, M.S., Ho, P.T.P., & Lo, K.Y. 1994, Nature, 372, 530

THE ISM IN NEARBY DWARF GALAXIES

L. M. YOUNG
Astronomy Department, University of Illinois
1002 W. Green St., Urbana, IL 61801, USA

AND

K. Y. LO
Institute of Astronomy and Astrophysics, Academica Sinica
P. O. Box 1-87, Nankang, Taipei, Taiwan 11529, ROC

1. Introduction

In recent years, color-magnitude diagrams have revealed that nearby dwarf galaxies have a fascinating variety of star formation histories. The fact that the dwarfs have periods of star formation activity and inactivity raises several questions. How do the properties of the interstellar medium (ISM) in a galaxy affect its star formation rate? Are the interstellar media of dwarf galaxies similar to those in spirals? And what can we learn about dwarf galaxies' evolution from their gas contents?

We investigate these issues through observations of the neutral interstellar medium in a small sample of nearby dwarf galaxies. These include two dwarf irregulars (Leo A and Sag DIG), two dwarf ellipticals (NGC 185 and NGC 205), and two intermediate irregular/spheroidal galaxies (LGS 3 and the Phoenix dwarf), whose distances are $\approx$ 1 Mpc or less. HI synthesis images of these galaxies probe their atomic media at resolutions of 60–160 pc. Observations of CO emission in the dwarf ellipticals clarify the relationship between atomic and molecular gas in those galaxies.

2. Properties of the ISM

HI in our own spiral Galaxy is made up of two phases, a cold phase and a warm phase. Four dwarf galaxies with current or recent star formation also show evidence of a cold atomic gas component. For example, the HI spectra in the irregulars Leo A and Sag DIG can be decomposed into high-dispersion (σ = 8–10 km/s) and low-dispersion (σ = 3–5 km/s) components. We argue (Young & Lo 1996a, 1997b) that these components are probably analogous to Galactic cold and warm neutral phases. Theoretical models of the HI medium can explain the presence of the two HI phases in the dwarf galaxies, even though the gas-phase metallicity and interstellar UV field are much lower in the dwarfs than in our own Galaxy. Twenty to thirty percent of the HI in those irregulars is in the cold component, with the remaining HI in the warm component; star formation seems to be associated with the cold HI component.

The dwarf elliptical galaxies NGC 185 and NGC 205 also show low-dispersion HI clumps associated with molecular clouds and dust patches (Young & Lo 1996a, 1996b). LGS 3, the sample galaxy with the least amount of current star formation, shows only a high-dispersion warm HI phase. These results suggest that a cold atomic component is necessary for, or at least commonly associated with, star formation. The presence of a *cold* HI component should be a better diagnostic of conditions appropriate for star formation than measures of the *total* HI content.

J. Andersen (ed.), Highlights of Astronomy, Volume 11A, 119–120.

3. Evolution of Elliptical and Spheroidal Galaxies

The neutral gas contents of early-type galaxies are vital clues to understanding their evolution. For example, some giant elliptical galaxies do have significant HI, and the kinematics of this gas generally suggest that the gas was acquired in mergers or interactions (van Gorkom 1992). In contrast to the giant ellipticals, HI in the dwarf ellipticals NGC 185 and NGC 205 has a compact distribution and relatively low angular momentum (Young & Lo 1997a). Thus, an internal origin for the gas in the dwarfs (stellar mass loss) is possible; mergers or interactions are not necessarily implicated.

The HI contents of the Local Group dwarf spheroidal galaxies also offer clues about their evolution. Current observations have shown very little HI in these galaxies (e.g. Knapp et al. 1978; Mould et al. 1990). It has been proposed that a strong burst of star formation activity might have expelled all of the ISM out of the galaxy (e.g. Dekel & Silk 1986). The presence of gas in the intermediate irregular/spheroidal LGS 3 conflicts with this idea (Young & Lo 1997b). Either LGS 3 has not had a burst of star formation activity, or the ISM is more difficult to expel than current models predict. We also detect two clouds of HI near the Phoenix dwarf spheroidal (Young & Lo 1997b). However, the velocity of the galaxy is not known and we cannot be sure whether the HI clouds are associated with Phoenix or perhaps with the Magellanic Stream.

References

Dekel, A., & Silk, J. 1986, ApJ, 303, 39
Knapp, G. R., Kerr, F. J., & Bowers, P. F. 1978, AJ, 83, 360
Mould, J. R., Bothun, G. D., Hall, P. J., Staveley-Smith, L., & Wright, A. E. 1990, ApJ, 362, L55
van Gorkom, J. 1992, in *Morphological and Physical Classification of Galaxies*, ed. G. Longo et al. (Dordrecht: Kluwer), p. 233
Young, L. M., & Lo, K. Y. 1996a, ApJ, 462, 203
Young, L. M., & Lo, K. Y. 1996b, ApJ, 464, L59
Young, L. M., & Lo, K. Y. 1997a, ApJ, 476, 127
Young, L. M., & Lo, K. Y. 1997b, ApJ, in press

HI DISTRIBUTION IN EXTREME DWARF IRREGULARS AND DWARF SPHEROIDALS

CLAUDE CARIGNAN
Département de Physique and
Observatoire Astronomique du Mont Mégantic,
Université de Montréal, C.P. 6128, Succ "centre-ville",
Montréal, Québec, Canada H3C 3J7

1. Introduction

Recent studies (Puche & Westpfahl 1994, Young & Lo 1996) have shown that the distribution of HI in some extreme low luminosity dwarf irregular galaxies (e.g. M81dwA, Holmberg I, Leo A) tends to have a ring–like (or shell–like) distribution which suggests that a single burst of star formation could expell most of the remaining ISM (or at least a large fraction of it) from the system. In view of this, Puche & Westpfahl (1994) suggested that in dwarf spheroidal galaxies, the HI should be found at large radii since no young stellar population is observed in most of them.

2. HI in Dwarf Spheroidals

Recently, HI was observed close to the dwarf spheroidals Phoenix (Carignan, Demers & Côté 1991) and Tucana (Oosterloo, Da Costa and Staveley–Smith 1996). While the situation for Phoenix is ambiguous due to the proximity along the line–of–sight of the Magellanic Stream, the large HI mass derived ($\sim 10^6$ $M_\odot$) for Tucana suggests that this gas is more likely to be a foreground Galactic cloud.

In Sculptor, some quantity of HI ($\sim 5 \times 10^3$ $M_\odot$) was first detected at the same systemic velocity than the optical ($\sim$ 110 km s^{-1}), using single dish Parkes (HPBW $\simeq 15'$) observations. However, much more HI gas was uncovered after mapping the dSph with the Australia Telescope Compact Array. Both global profiles are shown in Fig. 1. From the synthesis observations, a total HI mass of $\simeq 2 \times 10^4$ $M_\odot$ is derived, for an $M_{HI}/L_B = 0.02$. As can be seen in Fig. 2, most of the HI is mainly in two clouds having a ring–like structure and lies outside the optical, as predicted by Puche & Westpfahl (1994).

The morphology of the HI distribution and the fact that the gas is located further out in dSph's than in the low–mass dI's suggests that there may exists an evolutionary link between those two types of dwarfs.

References

Carignan, C., Demers, S. and Côté, S. (1991) *ApJ*, **381**, L13
Oosterloo, T., Da Costa, G.S. and Staveley–Smith, L. (1996) *AJ*, **112**, 1969
Puche, D. and Westpfahl, D. (1994) in *ESO/OHP Workshop on Dwarf Galaxies*, eds. G. Meylan & P. Prugniel (Garching: ESO), p. 273
Young, L.M. and Lo, K.Y. (1996) *ApJ*, **462**, 203

J. Andersen (ed.), Highlights of Astronomy, Volume 11A, 121–122.

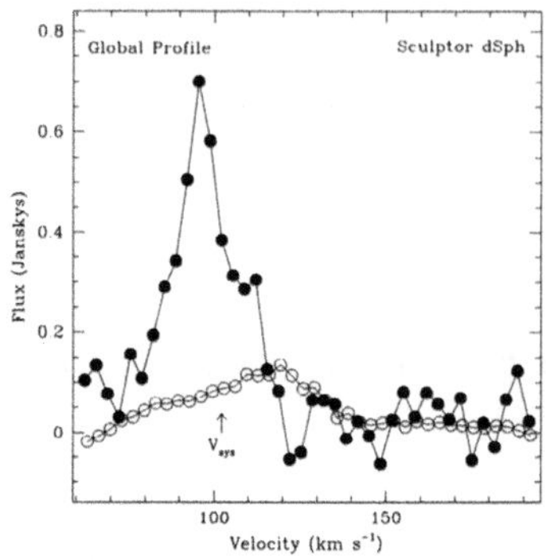

Figure 1. HI profiles of Sculptor from the Parkes (○) and the ATCA (•) observations.

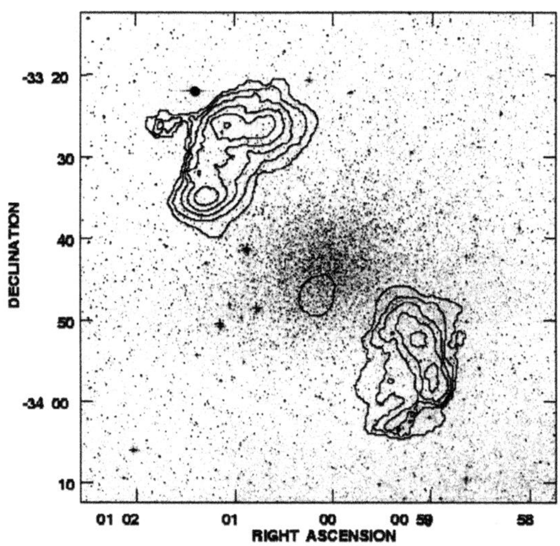

Figure 2. Sculptor HI surface density distribution superposed on an optical image.

FAR-INFRARED [C II] PROPERTIES OF LOW-METALLICITY GALAXIES

K. MOCHIZUKI
Department of Astronomy, University of Texas
Austin, TX 78712, USA

T. ONAKA
Department of Astronomy, University of Tokyo
Bunkyou, Tokyo 113, Japan

AND

T. NAKAGAWA
The Institute of Space and Astronautical Science
3-1-1 Yoshinodai, Kanagawa 229, Japan

1. Introduction

The line flux ratio of the [C II] 158 μm fine-structure ($^2P_{3/2} \rightarrow ^2P_{3/2}$) emission to CO rotational emission is a probe of the intensity of the interstellar UV radiation in galaxies. The largest values for the [C II]/CO (J = 1–0) line ratios are found in starburst galaxies, none of which, however, show values larger than 6×10^3 (Stacey et al. 1991). The exception ([C II]/CO $\simeq 2 \times 10^4$) are galaxies of Magellanic type which have low metallicity, an intense interstellar UV field, and irregular morphology (Table 1).

Mochizuki et al. (1994) showed that the two conditions, a low metallicity and an intense UV field are expected to result in the large [C II]/CO ratios in the Magellanic types. These conditions are likely to be common in more massive galaxies as well, but at an early epoch in their evolution. This suggests that bright [C II] emission can be a good indicator of young galaxies. However, the metal-poor galaxies previously observed in the [C II] line were exclusively Magellanic types. Observations of different types of dwarf galaxies are necessary to help us understand the physical conditions in the interstellar medium of low-metallicity systems.

2. Observations and Results

We observed the dwarf galaxies IC 5152 and I Zw 36 with LWS (the Long Wave Spectrometer) on ISO (the Infrared Space Observatory). The aperture of LWS was approximately 80 arcseconds; the spectral resolution was $\lambda/\Delta\lambda = 300$ in the grating mode we used. Compared to the Magellanic types, IC 5152 lacks intense star-forming activity ($M_B = -14.8$ mag, which is 20 times fainter than

TABLE 1. Metallicity (oxygen abundance) and [C II]/CO (J = 1–0) line flux ratios of galaxies. For our Galaxy and the LMC, the average values are presented.

Object	12 + log[O/H]	[C II]/CO (J = 1–0)	reference for [C II] data
Galaxy	8.91	1.3×10^3	Nakagawa et al. 1995
starbursts	–	$\simeq 6 \times 10^3$	Stacey et al. 1991
LMC	8.34	2.3×10^4	Mochizuki et al. 1994
Magellanic types	$\simeq 8.3$	$\simeq 2 \times 10^4$	Lord et al. 1995
I Zw 36	7.93	$\lesssim 3 \times 10^3$	this work

J. Andersen (ed.), Highlights of Astronomy, Volume 11A, 123–124.

expected from the relation between M_B and metallicity; Skillman, Kennicutt, and Hodge 1989); I Zw 36 is a blue compact.

Our observations show that IC 5152 has a [C II] flux of 7×10^{-16} W m^{-2}; the [C II]/B band flux ratio is 1/3 of that in the LMC, although IC 5152 is faint in the B band. This is an example of a metal-poor galaxy, but without an intense interstellar radiation field, which shows a lack of bright [C II] emission.

For I Zw 36, we obtained an upper limit of 1×10^{-16} W m^{-2} (3σ) in the [C II] line. This, combined with the millimeter observation of Tacconi and Young (1987), implies a line flux ratio of [C II]/CO (J = 1–0) $\leq 3 \times 10^3$. This upper limit is comparable to the ratio typically observed in our Galactic plane (Table 1), although I Zw 36 has both a very low metallicity and a high UV surface brightness.

3. Discussion

In a metal-poor galaxy, its lower dust abundance allows CO-dissociating UV photons to penetrate deeper into each molecular cloud; this makes for a large [C II]/CO line ratio (Maloney and Black 1988). However, our observations show that the [C II]/CO flux ratio also depends significantly on conditions other than metallicity.

A molecular cloud with a high gas density can produce a small [C II]/CO line ratio even with a low metallicity and an intense UV field, because the high density enhances the $C^+ \rightarrow CO$ conversion rate. The PDR (Photon-Dominated Region) model of Mochizuki (1996) reproduces a line ratio of [C II]/CO (J = 1–0) $\leq 3 \times 10^3$ at a gas density of $n_H \geq 10^4$ cm^{-3} (Table 2), when incident UV intensity is 10 times that of the solar neighborhood value, metallicity is at the LMC value, and each molecular cloud in a galaxy is 10^3 solar masses. Thus we suggest that the smaller line ratio in I Zw 36 results from a high gas density, which is compatible with its compact morphology.

High-J CO rotational lines observed toward distant galaxies (c.f. Ohta et al. 1996) indicate that these galaxies have a large amount of high-density molecular gas. This suggests that young galaxies may have an interstellar medium more similar to that in I Zw 36 than to that in Magellanic types.

TABLE 2. The derived line luminosity ratio of [C II]/CO (J = 1–0) based on a PDR model, with the assumed gas densities.

parameter	case 1	case 2	case 3
gas density [cm^{-3}]	10^3	$10^{3.5}$	10^4
[CII]/CO(J = 1–0)	2.3×10^4	7.5×10^3	2.4×10^3

References

Lord, S. D., et al. 1995, in Airborne Astronomy Symposium on the Galactic Ecosystem: From Gas to Stars to Dust, ed. Haas, M. R., et al. (ASP Conf. Ser., 73), 151
Maloney, P., and Black, J. H. 1988, ApJ, 325, 389
Mochizuki, K., et al. 1994, ApJ, 430, L37
Mochizuki, K. 1996, Ph. D. Thesis, University of Tokyo
Nakagawa, T., et al. 1995, ApJ, 455, L35
Ohta et al. 1996, Nature, 382, 426
Skillman, E. D., Kennicutt, R. C., and Hodge, P. W. 1989, ApJ, 347, 875
Stacey, G. J., et al. 1991, ApJ, 373, 423
Tacconi, L. J., and Young, J. S. 1987, ApJ, 322, 681

STAR FORMATION HISTORIES OF LOCAL GROUP DWARF GALAXIES

E.K. GREBEL
Astronomisches Institut Universität Würzburg
Am Hubland, D-97074 Würzburg, Germany

1. Introduction

The Local Group contains 37 currently known or probable member galaxies within a radius of 1.8 Mpc (Grebel 1997a). 31 of these galaxies are dwarf galaxies with $M_B > -17$ mag. The most frequent galaxy types are low-mass, gas-poor dwarf spheroidal (dSph) galaxies, which are dominated by old and intermediate-age populations, and more massive, gas-rich dwarf irregular (dIrr) galaxies with recent star formation. DSph galaxies have been found mostly in close vicinity to the large spiral galaxies in the Local Group, while dIrr galaxies tend to be more distant and isolated.

Knowledge of the star formation histories of these nearby galaxies will help us to evaluate the influence of galaxy mass, metallicity, and location with respect to other galaxies on galaxy evolution, to explore different star formation processes, and to understand distant, unresolved galaxies.

2. Methods to study star formation histories

Due to their proximity Local Group galaxies can be resolved into individual stars, which allows us to study their star formation histories in great detail. A variety of mostly photometric methods is employed ranging from studies of the stellar content to synthetic colour-magnitude diagrams. Poorly known and variable reddening, distance uncertainties, crowding, and lack of main-sequence photometry for older populations often make the derivation of star formation histories difficult. Spectroscopy and radio studies provide additional information on spectral types, abundances, and kinematics.

Proper motions are essential for determining orbits and constraining past and future interactions, but are extremely difficult to obtain due to the required large epoch differences.

3. Spatial variations in star formation

Stellar population studies in nearby galaxies often concentrate on one single field within a galaxy, which is then taken to be representative of the entire galaxy. However, even the lowest-mass galaxies known, the dSphs, turn out to show spatial variations of star formation rates and times.

For example, in the center of the predominantly intermediate-age dSph Fornax star formation lasted at least 5 Gyr with considerable enrichment (Beauchamp et al. 1995, Grebel 1997b), while only 570 pc north of the center star formation proceeded for only about 2 Gyr with hardly any change in metallicity (Grebel 1995, 1997a). A comprehensive picture of the star formation history requires large-area coverage with deep photometry.

4. Compiling star formation histories of all Local Group dwarves

Using own and all available literature data of stellar population studies in Local Group galaxies, I derived their star formation histories and visualized them through population boxes (Grebel 1997a). A population box (Hodge 1989) is a 3-D representation of the star formation history of a galaxy showing the evolution of star formation rate and metallicity with time. Care has to be taken since input data are quite heterogeneous and differ in data quality, methods, and models used.

J. Andersen (ed.), Highlights of Astronomy, Volume 11A, 125–126.

In summary, I find that all Local Group dwarf galaxies

- contain old (> 10 Gyr) and intermediate-age (1–10 Gyr) populations irrespective of morphological type,
- vary widely in star formation histories, metallicities, fractions, and ages of their subpopulations even within the same morphological type, and
- show complex star formation histories even within their older populations, and spatial variations in age and metallicity.

Objectives for future studies include accurate distances, accurate main-sequence turnoff ages, large-area coverage, and proper motions. The results will help us to understand star formation mechanisms, the role of interactions, and the formation history of the Local Group as a whole, and can then be extrapolated to distant galaxies.

References

Beauchamp, D., Hardy, E., Suntzeff, N.B., Zinn, R. (1995), *AJ*, **109**, pp. 1628–1633
Grebel, E.K. (1995), PhD Thesis, Bonn University
Grebel, E.K. (1997a), Ludwig–Biermann–Award Lecture, *Reviews in Modern Astronomy*, **Vol. 10**, pp. 29–60
Grebel, E.K. (1997b), Invited review at the AAS topical session on "Old Stellar Populations Beyond the Milky Way", *BAAS*, **Vol. 29**, p. 825
Hodge, P.W. (1989), *ARA&A*, **Vol. 27**, pp. 139–159

COSMOLOGY WITH NEARBY DWARF GALAXIES: THE VIEW FROM HST

ELINE TOLSTOY
Space Telescope — European Coordinating Facility
Karl-Schwarzschild str. 2
D-85748 Garching bei München, Germany

1. Colour-Magnitude Diagrams: What they tell us.

Stellar Evolution theory is based upon well understood physics and provides clear predictions as to how a Colour-Magnitude Diagram (CMD) will change due to effects of age and metallicity. The theory has been tested by looking at nearby coeval star clusters. The power of applying CMD analysis to galaxies has been demonstrated in studies of the Carina dSph (Smecker-Hane et al. 1996). In Carina the observation of separate, distinct Main Sequence (MS) Turnoffs has forced us to believe that this small, nearby companion of our Galaxy has had a very complex star formation history. No similar *direct* evidence for "bursting" behaviour on a global scale has been seen indisputably in larger systems. Partly this is due to the greater distances of larger systems, but also to the complications in distinguishing old star formation events (>1 Gyr old) in systems which are *currently forming stars.* Where we lack MS turnoffs we have to resort to statistical modeling of the CMD. This has been applied using a a number of different approaches, but all generally based on Monte-Carlo techniques (e.g. Tosi et al. 1992; Bertelli et al. 1992; Tolstoy & Saha 1996). However, even using these more sophisticated analysis techniques, it is difficult to find unique solutions. This is mostly due to the age-metallicity degeneracy on the Red Giant Branch (RGB). The RGB is usually the most populated, easiest to observe phase of stellar evolution. The Carina CMD reveals the dangers of blindly interpreting the RGB, because from the RGB alone it is impossible to extract the information revealed by the MS Turnoffs.

2. Recent Results from the Hubble Space Telescope

The HST is an invaluable resource for understanding the evolution of the resolved stellar populations in nearby galaxies. Crowding problems plague ground-based studies of all but the closest, least densely populated systems. The spatial resolution of HST (0.1″) minimises image crowding and makes it possible to detect fainter and thus older stars, and look farther back into the past star formation history of a galaxy. It has been postulated that nearby dwarf irregular galaxies are the most likely candidates for the remanents of the high redshift *faint blue galaxy* population (Koo, Guzman, this volume). The only way to test this unambiguously is to make deep CMDs, detecting more than just the RGB.

Recent HST CMDs are starting to provide a more detailed understanding of galaxy evolution (e.g. Armandroff, Grebel, this volume). The more carefully we look at nearby star-forming galaxies, the more detail we find. A few recent examples of HST/WFPC2 results are:

Sextans A (Dohm-Palmer at al. 1997): This appears to be a typical low metallicity (Z=0.001) star forming galaxy. It's WFPC2 CMD shows a galaxy that has most recently (over the last Gyr) been actively forming stars with a rate that has fluctuated with time by a factor of 2-3 across the galaxy. This is one of the first CMDs to definitively show the gap between the MS and the Blue Loop (BL) Helium Burning stars enabling a more accurate disentangling of the recent star formation history. This is a testament to the superior resolving powers of HST.

Pegasus dI (Gallagher et al. 1997, submitted): From the WFPC2 CMD of this galaxy we see it has a similar metallicity to Sextans A, but a much lower current rate of star formation. From the well defined RGB, and AGB as well as Red Clump stars it is clear that this galaxy *has*

J. Andersen (ed.), Highlights of Astronomy, Volume 11A, 127–128.

been forming stars further back in time than the observed MS, but the age range and intensity of the star formation cannot be disentangled without a reliable model for the chemical evolution versus time of this galaxy. If a rapid early chemical enrichment is assumed, with very little further self-enrichment, then this can be only a 3–4 Gyr old galaxy. However if we assume a more gradual enrichment, the galaxy can contain much older stars.

Leo A (Tolstoy et al. 1997, in preparation): The WFPC2 CMD shows that this galaxy has a low current star formation rate, although not as low as Pegasus, and an extremely low metallicity (Z=0.0004). As in Sextans A, there is a well defined BL sequence, well separated from the Main Sequence. We also see what appears to be a Red Clump. What is remarkable about this CMD is the extremely blue, and narrow RGB. It is necessary to go to the lowest metallicity stellar evolution tracks available to model this, making it impossible to quantify the possible effects of chemical evolution. The red limit of the RGB, gives an age limit for the oldest stars of about a few Gyr. Another result from this new CMD, and the detection of a Red Clump, is that the Cepheid distance to Leo A (Hoessel et al. 1994) is clearly wrong, by about 2.5 magnitudes. The reasons for this are unclear.

VII Zw403 (Lynds et al. 1997, in preparation): This is the nearest example of a Blue Compact Dwarf (BCD) Galaxy. The comparison between the WFPC2 CMDs of VII Zw403 and Sextans A is very revealing. The young, high mass stellar populations look very similar, including the differentiation between MS and BL. However the large difference in ratio of young stars in the MS+BL and those in the RGB+AGB between these systems shows that these galaxies have had very different star formation histories in the recent past. In addition to a highly populated RGB VII Zw403 also has a strong AGB sequence. The uncertainties in modeling the AGB leads to a reluctance to rely on its interpretation, except to note that since the AGB is a very short-lived phase, to see so many is a strong indication that a "burst" of star formation has occured in the recent past (1–5Gyr ago). Modeling of the RGB supports (although not unambiguously) the idea that a burst of star formation occured within the last 2 Gyr (Hunter et al. 1996). Interestingly VII Zw403's CMD looks very similar to the CMD of the nearby dwarf irregular galaxy, NGC 6822 (Gallart et al. 1994), which supports the idea that BCDs and low surface brightness dIs are the same objects (Bergvall, this volume).

3. The Big Picture

To fit these results into a scenario which will provide an overall understanding of dwarf galaxy evolution Ferrara & Tolstoy (1997, in preparation) have created a model which uses supernovae and stellar winds to create significant turbulent pressure in the ISM of a galaxy which effectively regulates star formation efficiency. The result of this is that the *total* mass of a galaxy determines how many stars a galaxy will form over time. The more massive a galaxy (total mass, dark plus visible) the more easily HI gas can stay concentrated in the centre, and thus star formation is more efficient, and the metallicity of the galaxy is enhanced. Thus the scatter in metallicity between galaxies of the same HI gas mass can be accounted for by variations in the dark matter content. From this model the *observed* global properties of mass and metallicity of dwarf galaxies seen today can be accounted for (e.g. van Zee this volume). It was also found, and subsequent hydro-simulations confirm (Ferrara & MacLow 1997, in preparation), that it is very difficult to lose gas from a galaxy through supernovae explosions. Thus the total mass of a galaxy does not only regulate the mass loss from the galaxy, but also the overall star formation efficiency within the galaxy.

References

Bertelli G., Mateo M., Chiosi C., & Bressan A. (1992) ApJ, 388, 400
Dohm-Palmer R., Skillman E., Saha A., Tolstoy E., Mateo M., Gallagher J., Hoessel J., Chiosi C. (1997) AJ, *in press*
Gallart C., Aparicio A., Chiosi C., Bertelli G., Vilchez J. M. (1994) ApJ, 425, 9
Hoessel J. G., Saha A., Krist J, Danielson G. E. (1994) AJ, 108, 645
Hunter D. A., Lynds R., Tolstoy E., O'Neil E. (1996) BAAS, 189, 3802
Smecker-Hane T.A., *et al.* (1996) *From Stars to Galaxies*, eds. Leitherer et al., p.328
Tolstoy E. & Saha A. (1996) ApJ, 462, 672
Tosi M., Greggio L., Marconi G., Focardi P. (1991) AJ, 102, 951

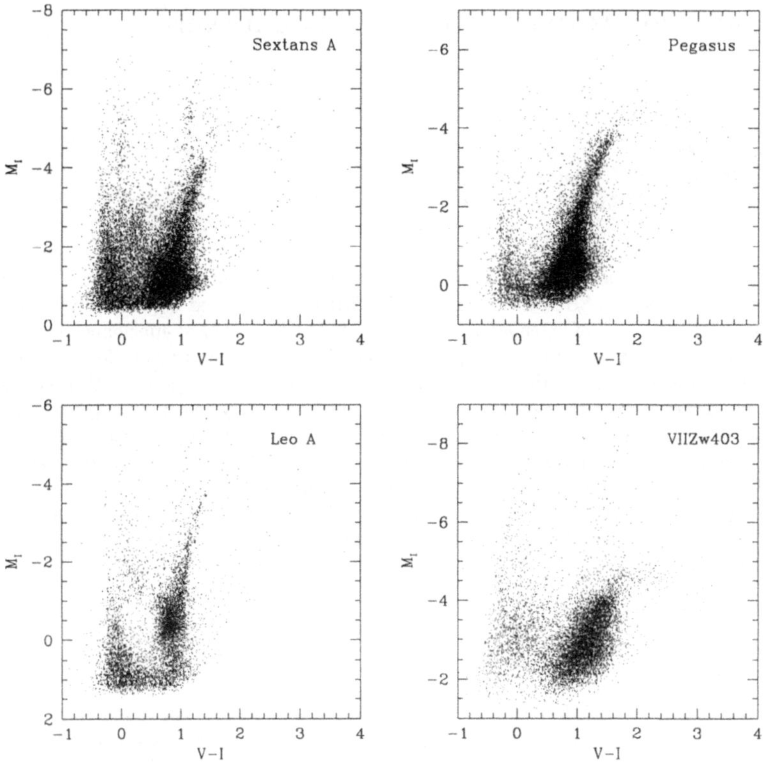

Fig. 1: HST-WFPC2 Colour-Magnitude Diagrams for four nearby star forming dwarf galaxies, as described in the text. The Red Clump is seen at $M_I \sim -0.5$, and the tip of the RGB is $M_I \sim -4$. The bifurcation of the "blue plume" into Main Sequence and the Blue Loop is most clearly seen in the Sextans A CMD.

STAR FORMATION ACTIVITY IN HIGH M_H/L_B GALAXIES

L. VAN ZEE
National Radio Astronomy Observatory
PO Box O, Socorro, NM 87801, USA

Abstract. The results of an investigation of the evolutionary status of fifteen gas–rich, low surface brightness dwarf galaxies (LSBDGs) are presented. LSBDGs are defined by unusually high values of M_H/L_B and the presence of extended HI envelopes. At both the present epoch and the past, the star formation process appears to be inefficient: the LSBDGs are underluminous for their HI mass and very little active star formation is currently observed. Analysis of global optical colors and elemental enrichments indicate that these objects are not "young" systems; rather, star formation has been occurring for several Gyr. One clue to the star formation history of these objects is that the global gas density is significantly lower than the Toomre instability criterion throughout the gas disk. Local peaks in the HI surface density, however, approach the instability criterion and are correlated with sites of active star formation. Thus, while star formation appears to be inhibited globally, the local gas distribution plays a crucial role in regulating the star formation activity in these low mass galaxies.

1. Introduction

As part of an investigation of the onset of star formation in low mass galaxies, we have undertaken a study of the evolutionary status of fifteen isolated, gas–rich, low surface brightness dwarf galaxies (LSBDGs) which appear to have inefficient star formation processes. The LSBDGs are distinguished by unusually high values of M_H/L_B and the presence of extended HI envelopes (van Zee *et al.* 1995). Both of these physical properties suggest that the star formation process is inefficient in the LSBDGs since they have not been able to convert their abundant supply of gas into stars. One possibility is that these galaxies are still in the process of collapse and thus have not yet processed the majority of their gas. Alternatively, the star formation process may be inefficient because the global gas density falls below the critical density for star formation.

To address these possibilities, we have examined the gas distribution and constrained the evolutionary status of the dominant stellar populations of a sample of LSBDGs. High spatial and spectral resolution HI synthesis observations were undertaken of 6 LSBDGs with the VLA in order to investigate the global and local gas surface densities (van Zee *et al.* 1997a). Deep optical images of 15 LSBDGs were obtained with the KPNO 0.9m; optical colors place constraints on the age of the dominant stellar population (Bruzual & Charlot 1993).

2. Results and Discussion

Based on UBVR colors, the LSBDGs have undergone quasi–continuous star formation for at least the last few Gyr, yet their current star formation rates are extremely low (van Zee *et al.* 1997b). These galaxies are not "young" systems. Rather, star formation has been underway for a significant period of time, but the process of converting gas into stars appears to be inefficient.

Analysis of the HI synthesis images reveals that the majority of the HI gas appears to be below the density threshold required for star formation (van Zee *et al.* 1997a). Thus, the star formation process appears to be inhibited in these objects because the local gas surface density rarely crosses the critical density threshold. Similar studies of a small sample of "normal" dwarf galaxies, however, suggest that the HI gas in these systems is also well below the critical threshold.

J. Andersen (ed.), Highlights of Astronomy, Volume 11A, 130–131.

In fact, it is impossible to distinguish between these two types of galaxies based on HI content alone.

These results suggest that the local gas density plays a critical role in regulating star formation activity. Star formation may be inhibited in dwarf galaxies because the local gas density rarely crosses the gravitational instability threshold. However, it is also possible that star formation is inefficient in the LSBDGs simply because massive star formation has never been induced via an interaction. In the future, it will be beneficial to investigate the star formation histories and gas distributions of a large sample of apparently isolated dwarf galaxies which do have efficient star formation processes in order to distinguish between these mechanisms.

References

Bruzual, G., & Charlot, S. 1993, ApJ, 96, 877
van Zee, L., Haynes, M.P., Giovanelli, R. 1995, AJ, 109, 990
van Zee, L., Haynes, M.P., Salzer, J.J., Broeils, A.H. 1997a, AJ, 113, 1618
van Zee, L., Haynes, M.P., Salzer, J.J. 1997b, AJ, in press

DWARF ELLIPTICAL GALAXIES IN THE M81 GROUP: THE STRUCTURE AND STELLAR POPULATIONS OF BK5N AND F8D1

TAFT E. ARMANDROFF
Kitt Peak National Observatory, National Optical Astronomy Observatories, P.O. Box 26732, Tucson, Arizona 85726, U.S.A.

NELSON CALDWELL
F. L. Whipple Observatory, Smithsonian Institution, P.O. Box 97, Amado, Arizona 85645, U.S.A.

G. S. DA COSTA
Mount Stromlo and Siding Spring Observatories, The Australian National University, Private Bag, Weston Post Office, ACT 2611, Australia

AND

PATRICK SEITZER
Department of Astronomy, University of Michigan, Ann Arbor, Michigan 48109, U.S.A.

1. Discussion

The M81 group is one of the nearest groups of galaxies, but its properties are quite different from those of the Local Group. It has therefore provided a different environment for the evolution of its member galaxies. We have carried out a CCD survey of the M81 group to search for analogs to Local Group dwarf elliptical (dE) galaxies. All the M81 dwarfs previously identified in photographic surveys were recovered and we also discovered several new systems whose surface brightnesses fall within the range found for Local Group dE's.

We have obtained HST WFPC2 images through the F555W and F814W filters of two M81 group dE's: BK5N and a new system, designated F8D1. The resulting color–magnitude diagrams show the upper two magnitudes of the red giant branch. The I magnitudes of the red giant branch tip in both galaxies yield distances that are consistent with membership in the M81 group. Surface brightness and total magnitude measurements indicate that BK5N and F8D1 have similar central surface brightness (24.5 and 25.4 mag arcsec^{-2} in V, respectively), but F8D1's larger length scale results in it being 3 magnitudes more luminous than BK5N. BK5N lies on the relation between central surface brightness and absolute magnitude defined by Local Group dwarf ellipticals, but F8D1 does not. F8D1 is more luminous for its central surface brightness than the relation predicts, similar to the large low surface brightness dwarf galaxies found in, for example, Virgo. The mean color of the giant branch is used to establish the mean abundance of each galaxy. F8D1, the more luminous galaxy, is significantly more metal rich ([Fe/H] $\approx$ –1.0) than BK5N ([Fe/H] $\approx$ –1.7). Both BK5N and F8D1 lie on the relation between absolute magnitude and metal abundance defined by Local Group dwarf ellipticals. However, as regards the relation between central surface brightness and metal abundance, BK5N again follows the Local Group dwarfs, while F8D1 deviates significantly from this relation. This suggests that the total amount of luminous matter is more fundamental in controlling metal enrichment than the surface density of luminous matter. We have also used the color width of the giant branch compared with the photometric errors to establish abundance ranges in both galaxies, the sizes of which are comparable to those in Local Group dE's.

From the numbers and luminosities of asymptotic giant branch stars more luminous than the red giant branch tip, we infer that, again like many of the Local Group dE's, both BK5N and F8D1 have had extended epochs of star formation. F8D1 contains stars as young as 3 – 4 Gyr, while in BK5N stars as young as ~8 Gyr are present. The fractions of intermediate-age population, at

J. Andersen (ed.), Highlights of Astronomy, Volume 11A, 132–133.

$\sim$30%, are similar in both galaxies. Further, one globular cluster has been found in F8D1, but none are present in BK5N. These numbers of clusters are consistent with our expectations from Local Group dwarfs.

Overall, we find that BK5N is similar in all respects to the Local Group dE's. Thus, in spite of the different environments of the M81 group and the Local Group, dwarf galaxies of very similar properties are present in each group. F8D1, on the other hand, has no Local Group counterpart and is indeed rare even in large clusters of galaxies. Our study of its stellar population is the first for a large low surface brightness dwarf.

The full text of this paper is scheduled for publication in the February 1998 issue of the Astronomical Journal.

NEW LIGHT ON THE ORIGIN OF NEBULAR HE II EMISSION IN YOUNG STARBURSTS

DANIEL SCHAERER
Space Telescope Science Institute, 3700 San Martin Drive
Baltimore, MD 21218, USA

1. Discussion

In the Local Group few H II regions exhibit nebular He II λ4686 emission, indicative of unusually high excitation. Approximately eight such nebulae are known, all located in low metallicity environments (IC 1613, SMC, LMC) except the Galactic ring nebula G2.4+1.4 (Esteban *et al.* 1992). The best studied case is the nebula S 3 surrounding the WO3 star DR 1 (Garnett *et al.* 1991, Kingsburgh & Barlow 1995). Except for two cases (N 44C, N159 F; Stasińska *et al.* 1986, Pakull & Angebault 1986) the nebulae are asociated with early WN (cf. Niemela *et al.* 1991) and WO stars.

So far approximately 50 to 60 extragalactic H II regions showing nebular He II λ4686 are known (Campbell *et al.* 1986, Izotov *et al.* 1994, 1997a). Most of them are found in H II galaxies, BCDs and related objects from low metallicity samples used for determinations of the primordial helium abundance. The most prominent case is the low metallicity (Z) record holder I Zw 18. He II emission is typically on the order of 1-5 % of Hβ.

Different ionization mechanisms (photoionization by hot stars, shock excitation, photoionization by X-rays) have been put forward to explain the required hard spectrum (Garnett *et al.* 1991).

The first quantitative approach to address this problem are the evolutionary synthesis models of Schaerer (1996) and Schaerer & Vacca (1997), which include well tested evolutionary tracks and recent spherically expanding non-LTE atmosphere models appropriate for WR and O stars. *These models predict nebular emission due to hot WN and WC/WO stars* (He II λ4686/Hβ up to 0.01-0.02). The classification of the ionizing sources agrees well with the majority of the nebulae discussed above. As shown by Schaerer (1996) the models are also in good agreement with observations of WR galaxies showing the characteristic broad WR emission (mostly He II λ4686 and/or C IV λ5808) and/or the presence of nebular He II.

If we assume instantaneous bursts, no contamination from an underlying population and ionization bounded regions, the Hβ equivalent width indicates the age of the population. Comparing the observations with our recent synthesis models (Schaerer 1996, Schaerer & Vacca 1997) we find that the vast majority of the extragalactic H II regions with nebular He II falls in the age range where WR stars are predicted to be present. *This suggests that nebular emission may be related to WR stars in all cases.*

The above suggestion is supported by the recent discovery of WC and WN stars in I Zw 18 (Izotov *et al.*, 1997b, Legrand *et al.* 1997). Furthermore the observed nebular He II emission can be fully explained if the ionizing flux of DR 1 (Kingsburgh & Barlow) is representative for the observed WC stars in I Zw 18.

To examine this case in more depth we have calculated new evolutionary tracks and synthesis models for Z=0.0004 of I Zw 18. We obtain a mass limit for the formation of WR stars of $M_{\rm WR} \gtrsim 90\,M_\odot$. Predictions for the WR bumps and nebular He II λ4686 for an instantaneous burst with a Salpeter IMF extending to 150 $M_\odot$ are shown in Figure 1. The line intensities (bottom) seem to indicate a difficulty for the models. However, the observed equivalent widths are in fair agreement, which shows that the above comparison is hampered by different spatial extensions of the gas and stars, as also recognized by Izotov et al. In particular the nebular 4686 can well be explained with a Salpeter IMF extending to a large, but not unreasonable, upper mass cut-off. *Although not being*

J. Andersen (ed.), Highlights of Astronomy, Volume 11A, 134–135.

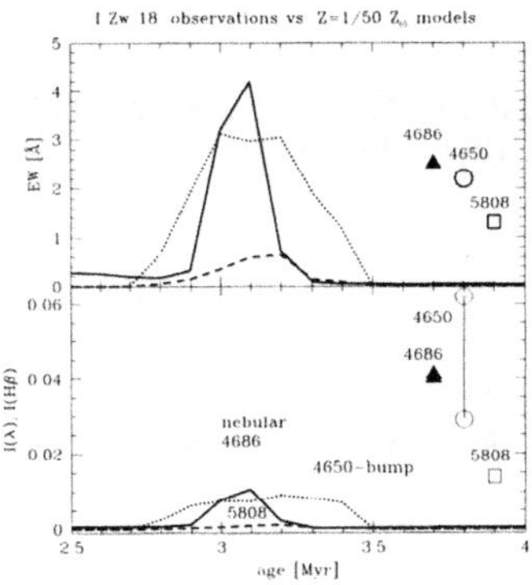

Figure 1. **Predicted nebular He II λ4686 (solid), 4650 bump (dotted), 5808 bump (dashed) for an instantaneous burst with a Salpeter IMF and $M_{up} = 150\,M_{\odot}$ (top: equivalent widths, bottom: line intensity $I/H\beta$). Observations from Izotov et al. (1997b) and Legrand et al. (1997) are shown on the right side. As discussed in the text $I(4650)/H\beta$ shows important differences between the two observations.**

a conclusive proof, the results obtained so far support the hypothesis that WR stars are responsible for nebular 4686 emission.

The observations of Izotov et al. (1997b) and Legrand et al. differ in the region of the 4650–4686 WR bump. While the former find broad emission centered at 4686 Å, the latter exclude any broad 4686 emission and detect a bump at 4645 Å, attributed to C III emission from WC stars. These differences could, e.g., be due to the different slit positions. As a consequence the exact content of WN and WC stars in I Zw 18 remains uncertain.

The observed strength of the total 4650–4686 WR bump and C IV λ5808 are well reproduced by the models. However, we predict broad O V λ5590 emission with an equivalent width of approximately 1 Å, which does not seem to be observed. Indeed at this low Z the single star evolutionary models predict WN, no WC, but instead oxygen rich stars (WO). This is due to the fact that at low Z the He-burning core (identified with WC or WO stars according to the C, O, and He composition) is revealed at a more chemically advanced stage given the low expected mass loss. Roche lobe overflow in binary systems may have lead to the formation of WC stars. Alternatively the presence of WC stars in I Zw 18 might indicate a deficiency in the evolutionary tracks, which have so far not been tested at metallicities below the SMC. Systematic studies of the WR (WN, WC, and possibly also WO !) and O star populations in very low metallicity H II galaxies are very promising in this respect. Future analysis of stellar populations in such environments should greatly improve our knowledge of massive star evolution in the early universe.

Acknowledgements

Last minute support from the IAU and a travel grant from the Swiss Society for Astronomy and Astrophysics are greatly acknowledged.

References

Campbell, A., Terlevich, R., Melnick, J. 1986, MNRAS , 223, 811
Esteban C., *et al.* 1992, A&A , 259, 629
Garnett D.R., Kennicutt R.C., Chu Y.-H., Skillman E.D. 1991, ApJ , 373, 458
Kingsburgh R.L., Barlow M.J. 1995, A&A , 295, 171
Izotov Y.I., *et al.* 1997b, ApJ, 487, L37
Izotov Y.I., Thuan T.X., Lipovetsky V.A. 1994, ApJ, 435, 647
Izotov Y.I., Thuan T.X., Lipovetsky V.A. 1997a, ApJS , 108, 1
Niemela V.S., *et al.* 1991, IAU Symp. 143, eds. K.A. van der Hucht, B. Hidayat, p. 425
Pakull M.W., Angebault, L.P. 1986, Nature, 322, 511
Legrand F. 1997, *et al.*, astro-ph/9707279
Schaerer D. 1996, ApJ , 467, L17
Schaerer D., Vacca W.D.W. 1997, ApJS , in press
Stasińska G., Testor G., Heydari-Malayeri M. 1986, A&A , 170, L4

THE LOW-MASS STELLAR LUMINOSITY FUNCTION OF THE 30 DOR STARBURST CLUSTER

HANS ZINNECKER
Astrophysikalisches Institut Potsdam
An der Sternwarte 16, D–14482 Potsdam, Germany

Abstract.

Diffraction limited near-infrared H-band (1.6 μm) NICMOS HST images are scheduled to be obtained in mid–October 1997 of the young cluster NGC 2070 (age 3.5 Myr) in the 30 Dor giant HII region in the LMC. The aim is to search for the low-mass ($M < 2\ M_{\odot}$) low-luminosity, red pre-Main Sequence stellar population and to establish the H-band infrared luminosity function. With the NICMOS we can now determine whether the IMF in this prototypical extragalactic starburst cluster is deficient in subsolar low-mass stars or not. The best ground-based data can sample only $M > 2\ M_{\odot}$. In principle, NICMOS in the H-band (F160W) is sensitive enough to reach a magnitude of ~ 23.5 in a relatively short integration time, which indeed corresponds to the fantastic possibility to detect young stellar objects with masses near the hydrogen burning limit ($M=0.1\ M_{\odot}$) according to pre-Main Sequence evolutionary models. Even if we could reach only $H = 22.5$ (i.e. $M=0.4\ M_{\odot}$), our observations will still go a long way in directly answering, by star counts, whether the IMF in starburst galaxies is low-mass deficient or not, with all the corresponding far-reaching implications.

The observations would also tell us whether the 30 Dor cluster can be regarded as a prototype young globular cluster. This possibility would be ruled out, if we found NGC 2070 to be low-mass deficient, because old globular clusters do have a rich population of low-mass stars.

J. Andersen (ed.), Highlights of Astronomy, Volume 11A, 136.

BLUE COMPACT GALAXIES AND THE PRIMORDIAL HELIUM ABUNDANCE DETERMINATION

Y.I. IZOTOV
Main Astronomical Observatory, Ukrainian National Academy of Sciences
Goloseevo, Kiev-22, 252650, Ukraine

AND

T.X. THUAN
Astronomy Department, University of Virginia
Charlottesville, VA 22903, USA

Abstract. We use a sample of 45 low-metallicity H II regions to determine the primordial helium abundance Y_p. We have carefully investigated the physical effects which may make the He I line intensities deviate from their recombination values such as collisional and fluorescent enhancements, underlying He I stellar absorption and absorption by Galactic interstellar Na I. By extrapolating the Y vs. O/H linear regression to O/H = 0, we obtain Y_p = 0.245±0.002. Our Y_p gives $\Omega_b h_{50}^2$ = 0.06±0.01.

1. Introduction

In the standard hot big bang model of nucleosynthesis (SBBN), four light isotopes, D, ^{3}He, ^{4}He and ^{7}Li, were produced by nuclear reactions a few seconds after the birth of the Universe. Given the number of relativistic neutrino species and the neutron lifetime, the abundances of these light elements depend on one cosmological parameter only, the baryon-to-photon ratio η, which in turn is directly related to the density of ordinary baryonic matter Ω_b. Thus precise abundance measurements of the four light elements can provide information about the mean density of ordinary matter in the Universe.

The primordial mass fraction Y_p of ^{4}He is usually derived by extrapolating the Y – O/H and Y – N/H correlations to O/H = N/H = 0 using various samples of dwarf irregular and blue compact galaxies (BCGs). These galaxies are the least chemically evolved galaxies known, so they contain very little helium manufactured by stars after the big bang. Because the Y_p – η relation has a very small slope, Y_p has to be determined with exquisite precision.

2. Contamination by stellar absorption

Several physical mechanisms discussed in Izotov *et al.* (1994, 1997) may make the He I line intensities deviate from their recombination values. While for the majority of BCGs, collisional and fluorescent enhancements of the He I line intensities are small, they can play an important role in some galaxies. The best way to take into account these effects is to use several of the brightest He I emission lines in the optical range (λ3889, λ4471, λ5876, λ6678 and λ7065) and to solve self-consistently for the electron number density in the He$^+$ zone and the optical depth in the He I λ3889 line to reproduce the theoretical He I line intensity recombination ratios.

Underlying He I stellar absorption can be important in some galaxies. Neglecting it can lead to misleadingly low helium mass fractions. A case in point is the NW component of I Zw 18 where the use of the He I λ6678 line gives Y = 0.233±0.008 and that of the He I λ4471 line $Y \leq$0.200, an unphysically low value. Izotov & Thuan (1997) have argued that the NW component of I Zw 18 should not be used for Y_p determination, but rather its SE component which is much less subject to

J. Andersen (ed.), Highlights of Astronomy, Volume 11A, 137–138.

underlying He I stellar absorption. The derived value for the SE component is considerably higher $Y = 0.243 \pm 0.009$.

3. Results and Discussion

The results of the fits for the Y – O/H and Y – N/H linear regressions for our sample are given in Table 1. Our Y_p gives $\Omega_b h_{50}^2 = 0.06 \pm 0.01$. In the framework of standard big bang nucleosynthesis theory, our derived Y_p is consistent with a low value of the D/H abundance as measured by Tytler *et al.* (1996) in absorption systems toward quasars, while it is inconsistent with the high D/H abundance reported by Songaila *et al.* (1994).

TABLE 1. Maximum Likelihood Linear Regressions

Method	Oxygen	Nitrogen
All He I lines	0.2451±0.0024 + 43±27(O/H)	0.2459±0.0017 + 882±429(N/H)
He I λ6678	0.2433±0.0044 + 65±54(O/H)	0.2451±0.0031 + 1197±954(N/H)

References

Izotov, Y. I., Thuan, T. X., & Lipovetsky, V. A. 1994, *ApJ*, **435**, p. 647
Izotov, Y. I., Thuan, T. X., & Lipovetsky, V. A. 1997, *ApJS*, **108**, p. 1
Izotov, Y. I., & Thuan, T. X. 1997, *ApJ*, in press
Songaila, A., Cowie, L. L., Hogan, C. J., & Rugers, M. 1994, *Nature*, **368**, p. 599
Tytler, D., Fan, X.-M., & Burles, S. 1996, *Nature*, **381**, p. 207

2D CHEMODYNAMICAL SIMULATIONS OF LOW-MASS GALAXIES

G. HENSLER AND A. RIESCHICK
Institut für Theor. Physik and Astrophysik
Universität Kiel, D-24098 Kiel, Germany

Because of their low gravitational energy, low-mass galaxies are seriously affected by energetical processes in their interstellar medium, such as supernova explosions, or by gravitational perturbations, e.g., by neighbouring galaxies. This can reasonably explain their variety of morphological types. If the evolutionary timescales of galaxies are predominantly determined by internal processes, the multi-phase character as well as star-gas interactions and phase transitions have to be taken into account. For this purpose we have developed a numerical treatment of the dynamical behaviour of gas and stars, which also accounts for the metal dependence of some processes and which can trace the chemical evolution for different elements. This so-called chemodynamical treatment is described in detail in Theis et al. (1992) and Samland et al. (1997). It considers three stellar components and devides the gas into clouds (CM, with a mass spectrum) and a hot intercloud medium (ICM). Since the element enhancement of the interstellar medium is produced by different processes with different lifetimes of their progenitors, O, Fe, and N are used as tracer elements to represent supernovae type II (SNeII), type Ia (SNeIa), and planetary nebulae (PNe) contributions. While supernovae form the ICM, PNe only attribute to the CM so that only mixing effects of both gas phases can alter abundance ratios.

Due to limited computer capacities the first chemodynamical simulations of dwarf galaxies could be performed only one-dimensionally so far (see e.g., Hensler et al. 1993, 1998). The recently developed two-dimensional chemodynamical code CoDEx (Samland 1994) was first applied to massive disk galaxies and produced models of which a particular one could represent various chemical and structural observations of the Milky Way with striking agreement (Samland et al. 1997). Here we present the first 2D chemodynamical results for one dwarf galaxy model with $10^9 M_\odot$ and discuss the implications briefly with respect to the importance of abundance mixture processes. The simulations start with a Plummer-Kuzmin density distribution (Satoh 1980). A stationary dark matter halo with a density distribution according to Burkert (1995) is added. The DM central density amounts to $0.062 M_\odot pc^{-3}$. The initial spin parameter λ is 0.045.

Besides their morphological differences, dwarf galaxies are characterized by a low metal content and low N/O abundance ratios of almost $log(N/O) = -1.4$ over a large range of O (Matteucci 1985) which is representative for primary N production. Figs. 1-3 show the densities of N and O in the CM and the ICM, respectively. Because of N being mixed into the CM by PNe, while O is blown into the ICM by SNeII, the density distribution in the CM is different for both elements in the inner parts of the galaxy. While an abundance difference is also expected for the ICM, figs.2 reveal an outflowing ICM structure which is loaded by O as well as by N. The N enrichment is due to the evaporation process acting on clouds which are embedded in low-density and high-temperature gas like in the center of the model. Fig.3 shows this evaporation rate and the $log(N/O)$ from this simulation of -1.3 is already close to the observed mean value of dwarf galaxies and within the scatter. Condensation of the ICM into clouds closes the material cycle in galaxies, so that the same N/O ratio is appearing in stars as well as in HII regions. Already this first qualitative result provides valuable insight into the importance of self-consistent chemodynamical models which can account for the abundance trends seen in dwarf galaxies.

J. Andersen (ed.), Highlights of Astronomy, Volume 11A, 139–140.

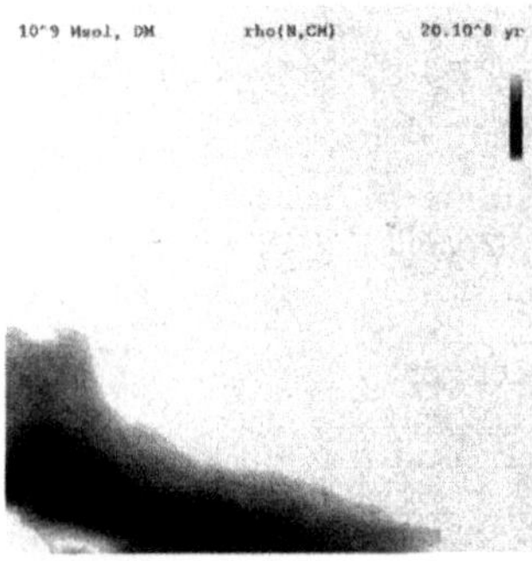

Fig. 1a.

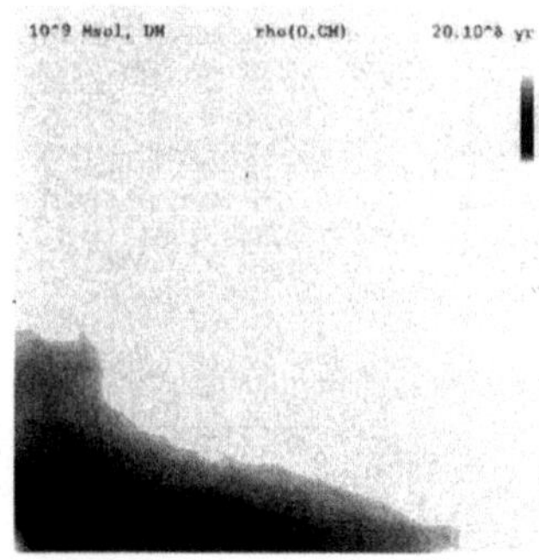

Fig. 1b.

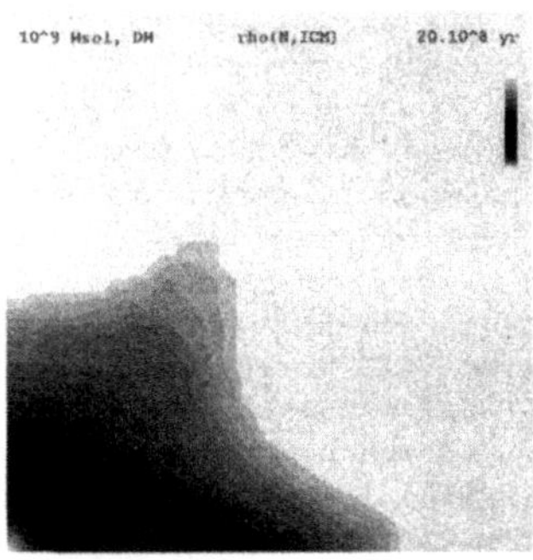

Fig. 2a.

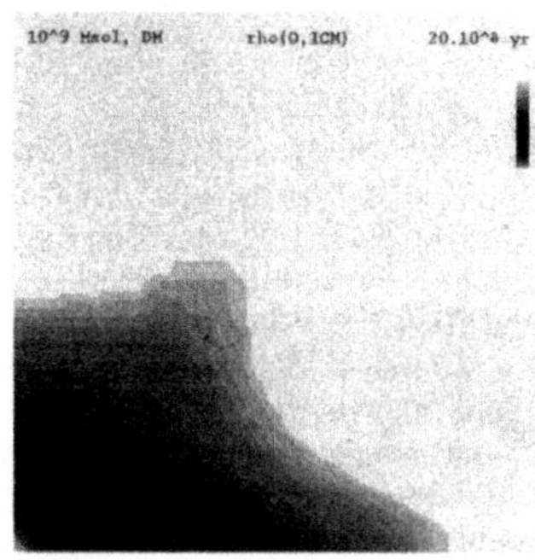

Fig. 2b.

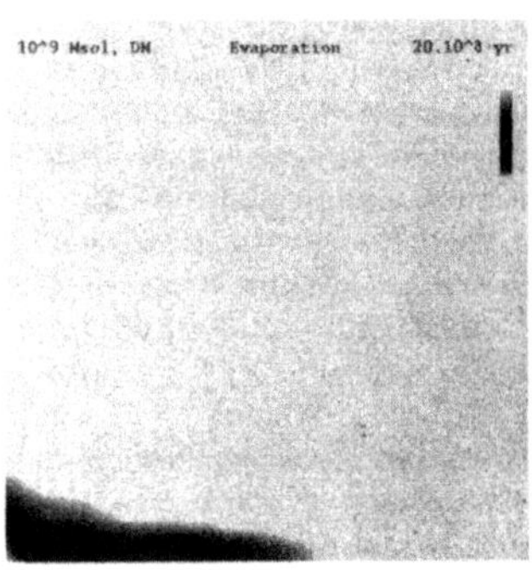

Fig. 3.

References

Burkert, A. 1995, ApJ, **447**, L25

Hensler, G., Theis, C., Burkert, A. 1993, in Proc. 3d DAEC Meeting, ed. S. Stasińska & D. Alloin (Paris: l'Obs. de Paris), p. 229

Hensler, G., Theis, C., Gallagher, J.S. 1998, ApJ, in prep.

Matteucci, F., Tosi, M. 1985, MNRAS, **217**, 391

Samland, M. 1994, Ph.D. thesis, Univ. of Kiel, Germany

Samland, M., Hensler, G., Theis, C. 1997, ApJ, **476**, 544

Satoh, C., 1980, PASJ, **32**, 41

Theis, C., Burkert, A., Hensler, G. 1992, A&A, **265**, 465

Fig. 1-3. One quadrant of 6.8 kpc × 6.8 kpc size between rotation axis (left boundary) and equatorial plane (lower boundary) of a $10^9 M_\odot$ dwarf galaxy with a DM halo (see text) as seen after 2 Gyrs, showing the mass densities of N (figs. a) and O (figs. b) in the CM (figs. 1) and the ICM (figs. 2) as well as the evaporation rate (fig. 3). The grey intensities are chosen arbitrarily reaching the maximum values at the center and the minimum (close to 0.) in the outermost region (light gray).

TIDAL DWARF GALAXIES

P.-A. DUC
ESO, Karl-Schwarzschild Straße 2
D-85748 Garching bei München, Germany

I.F. MIRABEL
CEA, SAp, C.E. Saclay
91191 Gif-sur-Yvette Cedex, France

AND

E. BRINKS
Departamento de Astronomía, Univ. de Guanajuato
Apdo. Postal 144, Guanajuato, Gto. 36000, Mexico

1. Introduction

The life and evolution of galaxies are dramatically affected by environmental effects. Interactions with the intergalactic medium and collisions with companions cause major perturbations in the morphology and contents of galaxies: in particular stars and gas clouds may be gravitationally pulled out from their parent galaxies during tidal encounters, forming rings, tails and bridges. This debris of collisions lies at the origin of a new generation of small galaxies, the so-called "tidal dwarf galaxies" (hereafter TDGs). Such an exotic way of forming galaxies was put forward by Schweizer (1978) and by Mirabel et al. (1992), who clearly observed the genesis of a star-forming object, out of material tidally expelled from the interacting system NGC 4038/39 ("The Antennae"). Recent studies, based on optical and HI observations, have shown that TDGs actually form a class of "recycled" objects with some properties similar to the more classical dwarf irregulars (dIrr) and blue compact dwarf galaxies (BCDs).

2. Case studies

Fig. 1 presents different examples of interactions: spiral-spiral collisions (NGC 4038/39), merger between spirals (NGC 7252) and encounters involving early-type galaxies (Arp 105, "The guitar galaxy"; NGC 5291, "the seashell galaxy"). Long tidal tails are clearly seen emanating from the parent galaxies. At their tip, at distances up to 100 kpc from the nuclei, small irregular objects are found with absolute magnitudes typical of dwarf galaxies. They host blue compact clumps that we identified with star-forming regions. Their spectra show emission lines typical of HII regions, ionized by massive OB stars younger than 10 Myrs. Given the time scale for the formation of clumps in tidal tails — typically 1 Gyr —, the stars at the tip of the antennae must have been born in situ. The contours of the HI column density are superimposed on the CCD images. It is striking on these images that the central regions of the parent galaxies contain little atomic gas, whereas the optical tails, and especially the tidal dwarfs at their tip, are associated with HI clouds as massive as 5×10^9 $M_\odot$(see also Hibbard & van Gorkom, 1996).

3. Properties of tidal dwarf galaxies

We have carried out multi-wavelength observations to study the properties of about 20 TDGs. Optical imaging and spectroscopy were obtained with the CFHT at Mauna Kea and the NTT at la Silla; near-infrared imaging was performed at the ESO/MPI 2.2m and HI observations were obtained at the VLA.

J. Andersen (ed.), Highlights of Astronomy, Volume 11A, 141–144.

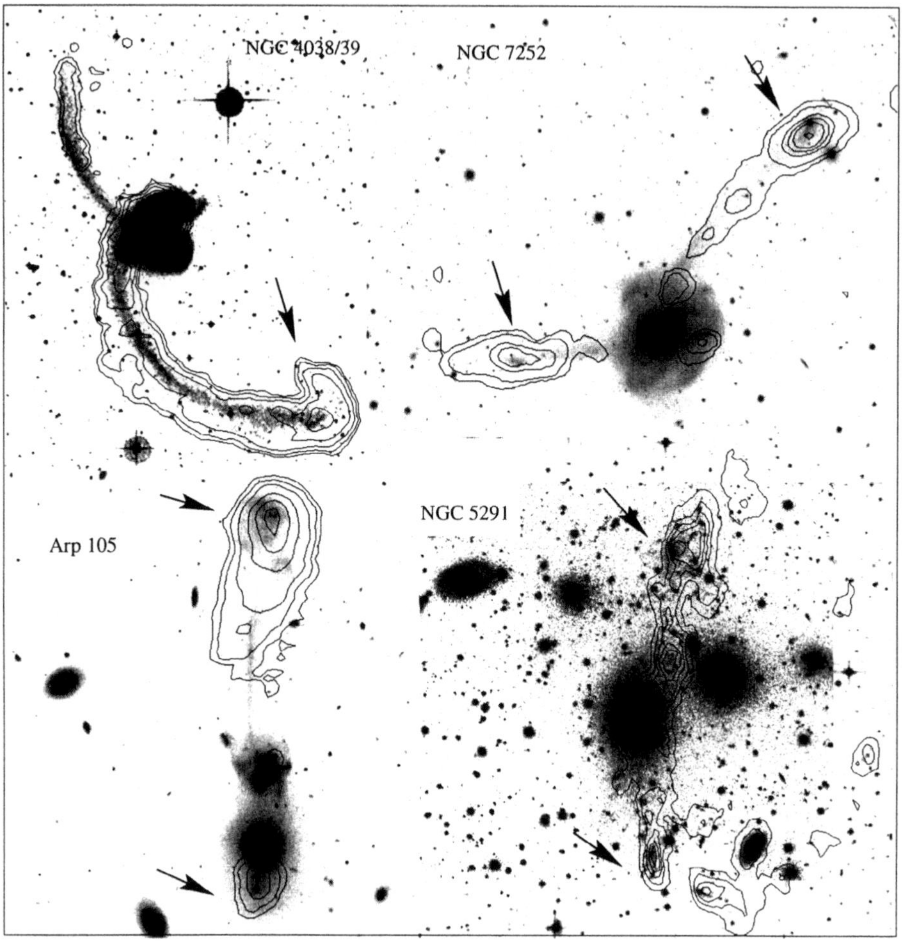

Figure 1. **Tidal dwarf galaxies in four interacting systems. They are seen at the end of optical tidal tails or HI plumes and are indicated by arrows. The CCD images are from the Digital Sky Survey, Duc (1995), and Duc & Mirabel (1997). The HI maps (black contours) are from Hibbard et al. (1997), Hibbard et al. (1994), Duc et al. (1997a), and Malphrus et al. (1997)**

3.1. STELLAR POPULATIONS

Tidal dwarf galaxies are made of two stellar components: young stars, formed from the recent collapse of expelled HI clouds, and an older star population, also tidally pulled out from the disks of their interacting parent galaxies. Using our aperture photometry and spectroscopic data (Duc et al., 1997b), we could estimate the relative proportion of both populations and conclude that TDGs actually split into two categories 1) extremely young objects, forming their first generation of stars (e.g., the dwarfs around NGC 5291, Duc & Mirabel, 1997), with high star formation rates equivalent to that observed in Blue Compact Galaxies, and 2) galaxies dominated by an old stellar population formed in the disk of their progenitors, which look like dwarf irregulars (e.g., Arp 105). In both classes of objects, the current star–forming episode, fuelled by an important HI reservoir, is younger than 10 Myrs.

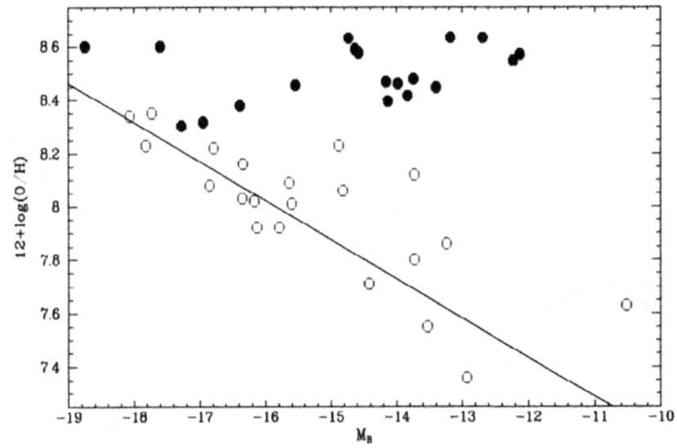

Figure 2. Oxygen abundance vs. absolute blue magnitude for our sample of tidal dwarfs (black points) and a sample of isolated dwarf galaxies (open points; from Richer & McCall, 1995).

3.2. METALLICITY

Fig. 2 shows the oxygen abundance vs. absolute magnitude of a sample of TDGs and nearby dIrr. Clearly TDGs are more metal rich than classical dwarfs of the same luminosity: $Z_{\odot}/3$ on average, a value typical of the outer regions of spirals. They actually do not follow the correlation found for field dwarf and giant galaxies between luminosity (hence mass) and metallicity. Being "recycled" objects, formed from pre-enriched material, tidal dwarfs inherited from their parents their relatively enriched interstellar medium.

3.3. DYNAMICS AND ORIGIN

Little is still known about the internal dynamics of tidal dwarfs. First indications are that the most massive TDGs may be gravitationally bound (Hibbard et al., 1997, Malphrus et al., 1997). Hints for rotation were found in the HI cloud associated with a TDG in Arp 105 (see Fig. 3b). Strong velocity gradients of the ionized gas were also measured (Fig. 3a). This suggests that some objects in tidal tails may already be dynamically independent. Further kinematical studies would be necessary to verify this assertion and estimate their dark matter content, predicted to be low in numerical simulations (Barnes & Hernquist, 1992).

Models for the formation of TDGs put forward two mechanisms: a local dynamical instability in the old stellar populations of tidal tails, followed by accretion of gas (Barnes & Hernquist, 1992) or the collapse of a supermassive cloud triggering precipitous star formation activity (Elmegreen et al., 1993). The fact that the HI component seems to dominate the mass of TDGs (Duc & Mirabel, 1997) favors the second hypothesis. The HI potential well is large enough to trap the old star population observed in some TDGs.

4. The fate of tidal dwarf galaxies

Do tidal dwarf galaxies contribute significantly to the overall population of dwarf galaxies? The answer to this fundamental question depends on the frequency of tidal interactions between galaxies and on the survival time of tidal dwarfs. The latter is limited by the hostile environment of TDGs, in the vicinity of giant parent galaxies. They may fall back on their progenitors on time scales of 1 Gyr, as pointed out by Hibbard & Mihos (1995), or be tidally disrupted. It is therefore expected that only the most massive TDGs and those that are far away enough from their progenitors will survive. This limits the number of galaxies produced to one or two per colliding system. It is therefore striking that the luminous star–forming TDGs that are found are located at the end of the optical tails, at distances from the nucleus larger than 50 kpc.

From an observational point of view, the census of TDGs is not an easy task. TDGs should obviously be searched for in the environment of interacting galaxies. Hunsberger et al. (1996) claim from the analysis of photometric data that half of the dwarf galaxies in the Hickson compact

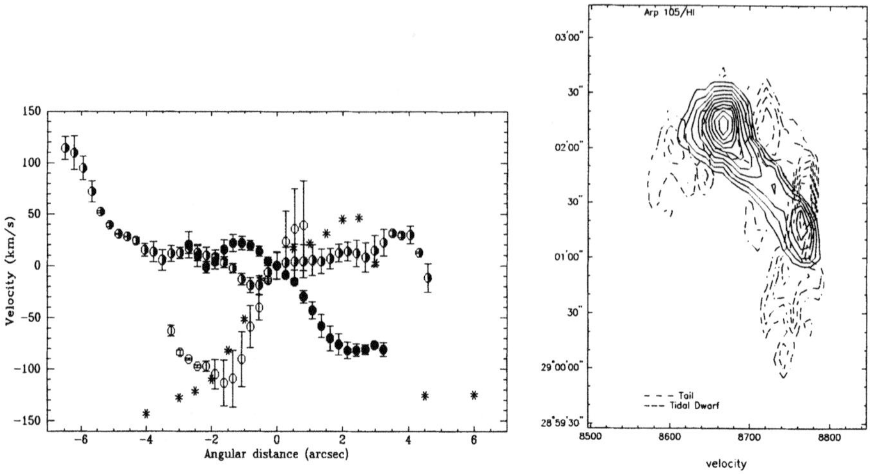

Figure 3. **a) Velocity profiles of the ionized gas for several tidal dwarfs; adapted from Duc & Mirabel (1997) b) Position-Velocity diagram of the HI northern tidal tail in Arp 105. Two components were identified: the expanding HI tidal tail (dashed contours) and a kinematically decoupled, possibly rotating, component (plain contours) associated with the forming tidal dwarf observed at the tip of the tail (see Fig. 1); adapted from Duc et al. (1997a)**

groups could be of tidal origin. However, one should note that once the stellar/gaseous bridge between the parent and child galaxies has dissipated, it is difficult to re-establish a link between the two. Our study has shown that a good genetic fingerprint of TDGs is their higher metallicity. In this respect, several studies have put forward trends for dwarf galaxies in groups or clusters to be more metal rich than field dwarfs (Bothun et al. 1985; Vilchez 1995). Since the collision rate is enhanced in denser environments, it is tempting to argue that a significant fraction of dwarfs in clusters could be recycled objects. A bimodal star formation history is also a strong signature for tidal dwarfs. Evolutionary Synthesis Models simulating a burst of star formation on top of the underlying component of old galaxies reproduce well the TDG star formation history and will give constraints for their future evolution (Fritze - v. Alvensleben & Duc, 1997).

References

Bothun, G. D., Mould, J. R., Wirth, A., and Caldwell, N. 1985, *AJ,* **90**, 697
Duc, P.-A. 1995, *Ph.D. thesis*, Université Paris VI
Duc, P.-A., Brinks, E., Wink, J. E., and Mirabel, I. F. 1997a, *A&A,* **326**, 537
Duc, P.-A., Fritze-v. Alvensleben, U., and Mirabel, I. F. 1997b, in *Extragalactic Astronomy in the Infrared*, eds. G. Mamon, T. Thuan, and T. Tran, Editions Frontières
Duc, P.-A. and Mirabel, I. F. 1997, *A&A, preprint*
Elmegreen, B. G., Kaufman, M., and Thomasson, M. 1993, *ApJ,* **412**, 90
Fritze-v.Alvensleben, U. and Duc, P.-A. 1997, in *Abstract Book XXIIIrd IAU General Assembly*, JD2-050P
Hibbard, J., van der Hulst, J., and Barnes, J. 1997, *in preparation*
Hibbard, J. E., Guhathakurta, P., van Gorkom, J. H., and Schweizer, F. 1994, *AJ,* **107**, 67
Hibbard, J. E. and Mihos, J. C. 1995, *AJ,* **110**, 140
Hibbard, J. E. and van Gorkom, J. H. 1996, *AJ,* **111**, 655
Hunsberger, S. D., Charlton, J. C., and Zaritsky, D. 1996, *ApJ,* **462**, 50
Malphrus, B., Simpson, C., Gottesman, S., and Hawarden, T. G. 1997, *AJ, preprint*
Mirabel, I. F., Dottori, H., and Lutz, D. 1992, *A&A,* **256**, L19
Richer, M. G. and McCall, M. L. 1995, *ApJ,* **445**, 642
Schweizer, F. 1978, in *Structure and Properties of Nearby Galaxies*, eds. E. Berkhuijsen and R. Wielebinski (Reidel, Dordrecht), p. 279
Vilchez, J. M. 1995, *AJ,* **110**, 1090

DISTANT COMPACT NARROW EMISSION LINE GALAXIES AS PROGENITORS OF TODAY'S SPHEROIDAL GALAXIES

DAVID C. KOO AND RAFAEL GUZMÁN
UCO/Lick Observatory, University of California,
Santa Cruz, CA 95064, USA

1. Introduction

Dwarf galaxies at high redshifts are important to study for a variety of reasons. By dwarf, we mean galaxies with low-mass, though galaxies with low-luminosities or small sizes are also commonly referred to as dwarfs. Several groups have suggested that such galaxies may be major contributors to faint blue galaxies, whose nature remains unclear. Dwarfs are viable candidates for faint blue galaxies if many undergo strong bursts of star-formation at redshifts $z \sim 1$ (Babul and Ferguson 1996) or even lower redshifts $z \sim 0.3$ (Cowie, Songaila, and Hu 1991; Broadhurst et al. 1988) and fade or disappear by today; if they have a much steeper luminosity function (Driver et al. 1994) than generally adopted in faint galaxy models; or if they represent small pre-merger fragments of larger galaxies today (Guiderdoni and Rocca-Volmerange 1990; Broadhurst, Ellis, and Glazebrook 1992), as might be expected in standard cold dark matter models with hierarchical galaxy formation.

2. Observations

As one approach to find distant dwarfs, we have been studying a sample of distant, very compact galaxies. They were originally discovered serendipitously as part of a survey for QSO's, with candidates selected on the basis of being stellar-like in apparent size (i.e., unresolved on ground-based images) and of having colors atypical of Galactic stars (Koo and Kron 1988). These candidates had very blue colors, and from low-resolution spectra gathered at Kitt Peak National Observatory, they were found to have moderate redshifts ($0.1 < z < 0.7$) and thus luminosities close to that of typical (L^*) galaxies. Since they had very strong, narrow emission lines more indicative of active star formation than of AGN's, we have dubbed these objects as compact narrow emission line galaxies (CNELG).

More recently, we have exploited the Wide Field Cameras (I and II) on the Hubble Space Telescope to measure their tiny sizes and to examine the morphology of CNELG's (Koo et al 1994, Guzmán et al. 1996, Guzmán et al. 1997b). With a spectral resolution of 8 $\mathrm{km\,s^{-1}}$ and using the Keck 10-m Telescope, the HIRES echelle spectrograph (Vogt et al. 1994) provides the most powerful diagnostic, namely high resolution kinematics. The vast majority of CNELG spectra reveal very narrow velocity widths of 30-70 $\mathrm{km\,s^{-1}}$ in the strong emission lines (Koo et al. 1995). Although winds, dust obscuration, and poor representation of the gravitational potential by the luminous star formation regions may all invalidate the use of the emission line widths for probing the gravitational potential, the available data to date shows a fairly tight ratio of 0.7 +/- 0.1 between such optical line-width measures and radio measures of HI motions (Telles and Terlevich 1993). The HI kinematics presumably sample most of the true potential. Making the assumption that our line widths, after an upward correction of 40%, are meaningful measures of the true potential, and when combined with HST size data, we are able to obtain a direct measure of mass to determine whether they are low-mass dwarfs or more massive but very compact galaxies.

J. Andersen (ed.), Highlights of Astronomy, Volume 11A, 145–146.

3. Results

We now highlight the key findings of our CNELG project. Despite their relatively high luminosities ($M_B \sim -21$), CNELG's are found to be genuine dwarfs, i.e., with low masses of about 10^9 $M_\odot$ or 1% that of equally luminous galaxies. Moreover, HST data indicate that their half-light radii of 0.3 to 0.4 arcsecs correspond to 1-3 kpc, again qualifying CNELGs as dwarfs by their sizes. Although CNELGs are nominally some of the best candidates for the bursting dwarfs proposed by Babul and Ferguson (1996), most CNELGs are brighter than the most luminous of the predicted bursting dwarfs by nearly a factor 10.

So what are CNELGs? To decipher their nature and relationship to various types of galaxy classes, their basic galaxy properties of luminosity, size, surface brightness, velocity dispersions, mass, M/L, etc., have been compared to that found for known local populations of galaxies. Evidence from these scaling relations are then combined with line ratios and strengths and very blue colors that support a strong burst of star formation; radial profiles that resemble exponentials; redshifts that imply a wide range of lookback times of several to many Gyr; and rough volume densities of about 1% of galaxies. In other words, a consistent picture emerges. We find that CNELGs are not blue compact dwarfs, clumpy irregulars, or starbursts, but resemble instead the more luminous and compact subset of the class known as H II galaxies. As discussed by Guzmán (in these proceedings), the CNELGs also appear to be a subset of the much larger pool of faint, blue, high surface brightness galaxies selected from the Hubble Deep Field Flanking Fields and studied by Phillips et al. (1997) and Guzmán et al. (1997a). The CNELGs may also be the lower-luminosity counterparts to the compact galaxies found at very high redshifts $z \sim 3$ by Steidel et al. (1996) and Lowenthal et al. (1997). Assuming that the past integrated star formation does not overwhelm the new stars, CNELGs are expected to fade by several magnitudes. This fading places the CNELGs squarely on top of the locations of the spheroidals in the parameters of luminosity vs. size or surface brightness vs. velocity width (Koo et al. 1995; Guzmán et al. 1996)

4. Summary

Kormendy and Bender (1994) have highlighted the importance and mystery of the origin of spheroidal galaxies like NGC 205, while Skillman and Bender (1995) have emphasized a "crisis" in understanding why some low mass systems become or remain irregulars while others become spheroidals. Our HST and Keck work on CNELGs, which have luminosities comparable to that of Andromeda but tiny sizes and *masses* like that of its companion NGC 205, strongly support the possibility that some, if not all, of these distant, strongly-starbursting dwarfs will eject their remaining gas and fade to become today's spheroidals like NGC 205. The underlying physical cause for the intense starburst remains, however, a puzzle.

References

Babul, A., and Ferguson, H. C. 1996, *ApJ*, **458**, 100
Broadhurst, T. J., Ellis, R. S., and Glazebrook, K. 1992, *Nature*, **355**, 55
Broadhurst, T. J., Ellis, R. S., and Shanks, T. 1988, *MNRAS*, **235**, 827
Cowie, L. L., Songaila, A., and Hu, E. M. 1991, *Nature*, **354**, 460
Driver, S. P., et al. 1994, *MNRAS*, **266**, 155
Guiderdoni, B., and Rocca-Volmerange, B. 1990, *A&A*, **227**, 362
Guzmán, R. et al. 1996, *ApJ*, **460**, L5
Guzmán, R. et al. 1997a, *ApJ*, **489**, 559
Guzmán, R. et al. 1997b, *ApJ*, submitted
Koo, D. C. et al. 1994, *ApJ*, **427**, L9
Koo, D. C. et al. 1995, *ApJ*, **440**, L49
Koo, D. C., and Kron, R. G. 1988, *ApJ*, **325**, 92
Kormendy, J., and Bender, R. 1994, ESO/OHP Workshop on *Dwarf Galaxies*, ed. G. Meylan and P. Prugniel (Garching:ESO) 161
Lowenthal, J. D., et al. 1997, *ApJ*, **481**, 673
Phillips, A. C. et al. 1997, *ApJ*, **489**, 543
Skillman, E. D., and Bender, R. 1995, *5th Texas-Mexico Conf. on Astrophysics*
Steidel et al. 1996, *AJ*, **112**, 352
Telles, E. and Terlevich, R. 1993, *Ap&SS*, **205**, 49
Vogt, S. et al. 1994, Proc. *SPIE*, **2198**, 362

COMPACT GALAXIES AT Z = 0.2–1.3: IMPLICATIONS FOR GALAXY EVOLUTION AND THE STAR FORMATION HISTORY OF THE UNIVERSE

R. GUZMAN, A. C. PHILLIPS, J. GALLEGO AND D. C. KOO
UCO/Lick Observatory, Board of Astronomy and Astrophysics
University of California, Santa Cruz, USA

Abstract. We study the global properties of 51 compact field galaxies with redshifts $z \sim 0.2 - 1.3$ and apparent magnitudes $I_{814} < 23.74$ in the flanking fields of the Hubble Deep Field. All these galaxies have angular half-light radii $r_e < 0.5$ arcsec. Keck spectra covering ~4000–9000 Å, combined with HST I_{814} images and Keck V-band images, were used to derive redshifts, $V_{606} - I_{814}$ colors, absolute blue magnitudes (M_B), linear half-light radii (R_e), blue average surface brightnesses within R_e (SB_e), velocity widths (σ), virial masses (M), mass-to-light ratios (M/L), excitations ($O[III]/H\beta$), and star formation rates (SFR). The results of this study can be summarized as follows:

(i): Only 12% of the 51 compact galaxies have absorption-line dominated spectra, while 88% show strong, narrow emission lines, similar to the so-called CNELGs (e.g., Koo, this volume).

(ii): Despite being very luminous (i.e., $L_B \sim L^{\star}$; see figure 1a), compact emission-line galaxies are low-mass stellar sytems (i.e., $M \leq 10^{10}$ $M_{\odot}$, typically; see figure 1b).

(iii): Roughly 60% of the compact emission-line galaxies have colors, sizes, surface brightnesses, luminosities, velocity widths, excitations, star formation rates, and mass-to-light ratios characteristic of young, star-forming HII galaxies (see figures 1 and 2). The remaining 40% form a more heterogeneous class of evolved starbursts, similar to local disk starburst galaxies.

(iv): Without additional star formation, galaxy evolution models predict that HII-like distant compacts will fade to resemble today's spheroidal galaxies such as NGC 205 (Koo, this volume).

(v): Our sample implies a *lower limit* for the global comoving SFR density of ~0.004 $M_{\odot}$ yr^{-1} Mpc^{-3} at $z = 0.55$, and ~0.008 $M_{\odot}$ yr^{-1} Mpc^{-3} at $z = 0.85$. These values, when compared to a *similar* sample of local galaxies, support a history of the universe in which the SFR density declines by a factor ~10 from $z = 1$ to today (see figure 3). From the comparison with the SFR densities derived from previous data sets, we conclude that compact emission-line galaxies, though only ~20% of the general field population, may contribute as much as ~45% to the global SFR of the universe at $0.4 < z < 1$.

A full description of these results can be found in Phillips et al. (1997) and Guzmán et al. (1997).

Acknowledgements. This project is a collaborative effort of the DEEP team at UC Santa Cruz (http://www.ucolick.org/~deep/home.html). R. Guzmán would like to thank the scientific committee for their kind invitation to participate in this Joint Discussion. Funding for this project is credited to NASA grants AR-06337.08-94A, AR-06337.21-94A, GO-05994.01-94A, AR-5801.01-94A, and AR-6402.01-95A from the Space Telescope Institute, and NSF grants AST 91-20005 and AST 95-29098.

References

Guzmán R., Gallego J., Koo D.C., et al. 1997, ApJ, to appear in November
Phillips A.C., Guzmán R., Gallego J., et al. 1997, ApJ, to appear in November

J. Andersen (ed.), Highlights of Astronomy, Volume 11A, 147–148.

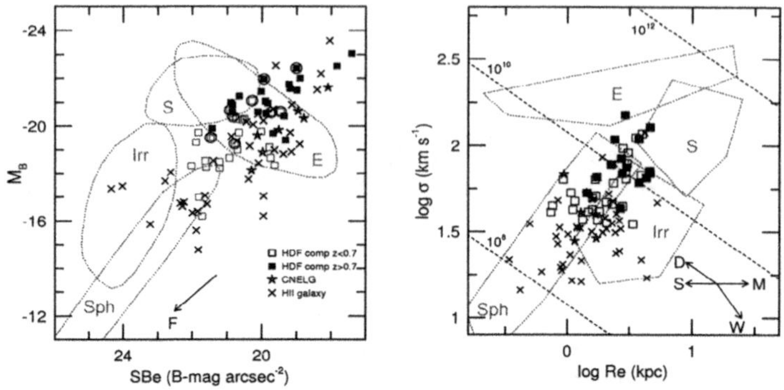

Figure 1. (a): SB_e vs. M_B. Open squares: compacts at $z < 0.7$; filled squares: compacts at $z > 0.7$; circles: absorption-line compacts; crosses: nearby HII galaxies; dotted lines indicate the locus occupied by various other types of local galaxies; the arrow (F) represents the direction of fading. (b): R_e vs. σ. Symbols as before; dashed lines represent constant mass-lines in $M_\odot$; the arrows represent the effects of dissipation (D), mergers (M), stripping (S) and winds (W) on R_e and σ. We adopt $H_0 = 50$ km s^{-1} Mpc^{-1} and $q_0 = 0.5$.

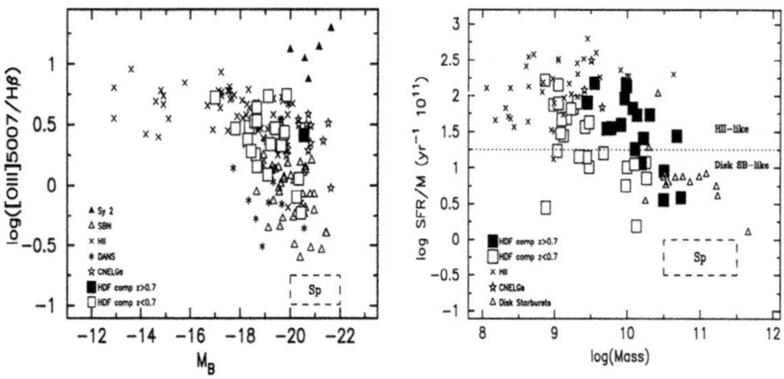

Figure 2. (a): M_B vs. $[OIII]/H\beta$. Local galaxies: DANS=Dwarf Amorphous Nuclear Starbursts; SBN=Starburst Nuclei; Sy2:=Seyfert 2 galaxies; HII=HII galaxies. Dashed lines represent the approximate location of spiral galaxies. (b): M vs. SFR/M. Symbols as before; the dotted line represents the division between HII-like and disk starburst-like galaxies adopted in our classification.

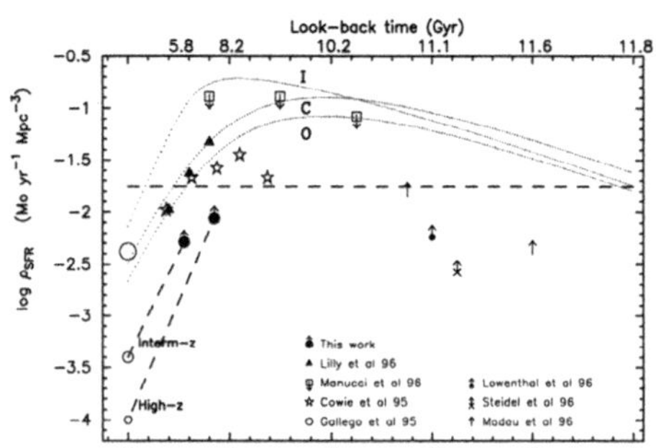

Figure 3. SFR density vs. redshift. The two large filled circles are the estimates for compact galaxies at $z < 0.7$ and $z > 0.7$, respectively. These values should be compared to the open circles labelled "Interm-z" and "High-z, which represent the values for *similar* samples of nearby compact galaxies. Dotted lines represent Pei & Fall's models (1995). The dashed line represents the fiducial value. We adopt $H_0 = 50$ km s^{-1} Mpc^{-1} and $q_0 = 0.5$.

JD3

Precession-Nutation and Astronomical Constants in the Dawn of the 21st Century

Chairpersons and Editors: **V. Dehant & T. Fukushima**

INTRODUCTION OF *JD* 3 ON 'PRECESSION, NUTATION AND ASTRONOMICAL CONSTANTS IN THE DAWN OF THE 21ST CENTURY'.

V. DEHANT
Observatoire Royal de Belgique
3, avenue Circulaire
B-1180 Brussels, Belgium
email: V.Dehant@oma.be

AND

T. FUKUSHIMA
National Astronomical Observatory of Japan
2-21-1, Ohsawa, Mitaka, Tokyo, 181 Japan
email: toshio@nao.ac.jp

Precession/nutation part

Due to the adoption of the new International Celestial Reference Frame (ICRF), the Earth's Orientation Parameters (EOP) will be revised and their definitions will need to be re-examined and clarified. This implies that precession/nutation formulation will be also revised in the future.

The precession/nutation theories for a non-rigid Earth suffer from a lack of dissipation in the core and from a mismodeling of the ocean and of the atmospheric effects. The scientific community is examining these questions. The IAU community is consequently not yet ready to adopt a new precession/nutation geophysical model but the users may use the International Earth Rotation Service (IERS) empirical series.

In order to review those questions and prepare the future research, the Scientific Organizing Committee (SOC: P. Bretagnon, V.A. Brumberg, N. Capitaine, V. Dehant (Chair), T. Fukushima, E. Groten, H. Kinoshita, B. Kolaczek, D.D. McCarthy, P.K. Seidelmann and P.T. Wallace) has proposed invited talks on the current situation concerning :
(1) the formulation of precession/nutation (N. Capitaine, see paper 1),
(2) the planetary theories and their relation to precession/nutation (P. Bretagnon, see paper 2),
(3) the precession/nutation for a rigid Earth (J. Souchay and H. Kinoshita, see paper 3),
(4) the DExxx JPL ephemerides precision and accuracy (E.M. Standish, see paper 4),
(5) the observations of the Celestial Ephemeris Pole (CEP) and in particular the pole offset from which precession/nutation corrections can be derived (M. Feissel and A.M. Gontier, see paper 5),
(6) the relativistic effects on precession/nutation (S.A. Klioner and M. Soffel, see paper 6),
(7) the best available nutation series from the theoretical and observational point of view (P.M. Mathews and T.A. Herring, see paper 7),

The SOC wanted also that one invited speaker (J. Kovalevsky, President of Division 1) presents the consequences for the astronomers of a change in the precession/nutation model that will follow the adoption of the new ICRF (see paper 8 of J. Kovalevsky and D.D. McCarthy).

Astronomical constants part

The present situation, the progress and plans concerning (1) the best estimation of astronomical constants, (2) the standards of fundamental astronomy, and (3) the relativity in celestial mechanics and astrometry were further topics of the Joint Discussion 3. The SOC has proposed three additional invited talks on these topics, respectively presented by (1) D. McCarthy (see paper 9), (2) P.T. Wallace (see paper 10), and (3) V.A. Brumberg and B. Guinot (see paper 11).

J. Andersen (ed.), Highlights of Astronomy, Volume 11A, 151–152.

discussion

The invited talks have prepared us to an interesting discussion.

Resolutions B4, B5, B6 and B7 are the consequences of what has been presented or discussed.

Indeed, Resolution B4 is concerned with the imperfections of the non-rigid Earth nutation theories that still need an important effort from the scientific community.

Resolution B5 asks for a new WG to be formed in order to examine and clarify the consequences for precession/nutation from the adoption of the new ICRF.

Resolution B6 concerns some clarifications in the frame of general relativity, consequences of the reference systems and time scales.

Resolution B7 is a resolution issued from two JD: JD 3 (this one) and JD 7 on the new ICRF; it recommends that high-precision observing programs be organized in order to maintain the ICRF at the highest possible accuracy and in order to obtain high-accuracy precession/nutation. This resolution explicitly mentioned Very Long Baseline Interferometry (VLBI) and Lunar Laser Ranging (LLR), while other techniques as the Global Positioning System (GPS) may also play an important role in the future.

list of invited talks

1. Capitaine N., "Overview, Formulation and Current Situation for Precession/Nutation."
2. Bretagnon P., "The Planetary Theories and the Precession of the Ecliptic."
3. Souchay J. and Kinoshita H., "The theory of the Nutation for a Rigid Earth Model: Current State of the Situation."
4. Standish E.M., "Fundamental Arguments of the Current Nutation Theory; Dynamical Reference Frame."
5. Feissel M. and Gontier A.-M., "Observations of the Celestial Motion of the Earth's Pole."
6. Klioner S.A. and Soffel M., "Relativistic Considerations for Precession and Nutation."
7. Mathews P.M. and Herring T.A., "Observational and Theoretical Modeling of Nutation."
8. Kovalevsky J. and McCarthy D.D., "Astronomical Effects of Current Changes in Fundamental Astrometric References."
9. McCarthy D.D., "Latest Best Estimates of Astronomical Constants."
10. Wallace P.T., "SOFA (Standards Of Fundamental Astronomy); Progress and Plans."
11. Brumberg V.A., Guinot B., and RCMA SWG members, "General Relativity and the IAU Resolutions; Report of the IAU WGAS Sub-Working Group on Relativity in Celestial Mechanics and Astrometry (RCMA SWG)."

OVERVIEW, FORMULATION AND CURRENT SITUATION FOR PRECESSION-NUTATION

N. CAPITAINE
Observatoire de Paris, URA1125,
61, avenue de l'Observatoire, 75014 Paris, France
E-mail:capitaine@obspm.fr

Abstract. This paper describes the conventional models for precession and nutation, the current procedure for taking them into account in the reduction of highly accurate observations, the recent IAU recommendations on this topic, as well as the current situation for theory and observations. This emphasizes the imperfections in the conventional models and in the current procedure which are not in consistency either with the adoption of the new International Conventional Celestial Reference System (ICRS) or with the high accuracy and resolution of the current observations of Earth rotation. This addresses the question of the adoption of a new formulation and of an improved model combining precession and nutation of the equator with respect to the ICRS as well as the adoption of a revised definition of the Celestial Ephemeris Pole.

1. Current situation for precession and nutation

1.1. CURRENT CONVENTIONAL MODELS

The conventional models for precession/nutation since 1984 are based upon :

(1) the 1976 IAU System of Astronomical Constants adopted by the IAU XVI GA of the IAU, including values, at J2000.0, for the constant of precession and the mean obliquity,
(2) numerical expressions for the precession quantities w.r.t. epoch J2000.0 (Lieske *et al.*, 1977),
(3) the 1980 IAU theory of nutation (Seidelmann, 1982) providing the nutation in longitude and obliquity (Kinoshita, 1977; Wahr, 1981) for the Celestial Ephemeris Pole, CEP.

These conventional models are associated with the use of :

- the FK5 as the Conventional Celestial Reference System (i.e. the Reference Frame at J2000.0 as realized by the FK5 catalogue + the use of (1), (2), (3)),
- of the relationship between Greenwich Mean Sidereal Time (GMST) and UT1 as given by Aoki *et al.* (1982).

These conventional models have been shown, for more than ten years, to present deficiencies w.r.t. VLBI and LLR observations and to be therefore inadequate for highly accurate observations. Consequently, IERS monitors the so-called "celestial pole offsets" as the observed estimates of the corrections to the 1976 IAU precession and 1980 IAU nutation.

1.2. RECENT IAU RECOMMENDATIONS

At the 1994 XXII GA of the IAU, following the recommendations of the Joint Discussion 14 untitled "Towards the establishment of the astronomical standards" and of the Joint Discussion 19 untitled "Nutation", two resolutions (C1 and C6) have been adopted and an IAU/IUGG WG on "Nutation for a non-rigid Earth Nutation theory" has been established.
The IAU resolution C1 :

- urged that observations of the offset of the celestial pole w.r.t. the pole defined by the 1980 theory of nutation be made with the most precise techniques available,

J. Andersen (ed.), Highlights of Astronomy, Volume 11A, 153–157.

- asked the IERS to provide an empirical model for corrections to the 1980 theory of nutation to be used for *a priori* estimates of the CEP offsets.

The IAU resolution C6 recommended that :

- the present 1976 IAU System of Astronomical Constants be retained,
- an IAU File of Current Best Estimates of Astronomical Constants be established.

1.3. CURRENT SITUATION

The 1994 IAU resolutions C1 and C6 have been followed.

The celestial pole offsets are monitored on a regular basis by the IERS with an improving accuracy (0.2 milliarcsecond (*mas*) in 1996 for daily values) and an interim precession-nutation model, the "IERS 1996 Theory of Precession/Nutation", has been provided in the IERS Conventions 1996 (McCarthy, 1996) that match the observations with uncertainties of ± 1 *mas*.

The 1996 IERS Conventions have provided a list of IERS Numerical Standards for Current Best Estimates of a few constants (including the precession constant, p_1, and the mean obliquity, ϵ_0, at J2000.0), the 1976 IAU System of Astronomical Constants being retained for all the other constants. However, consistency has to be ensured between the IERS Numerical Standards for p_1 and ϵ_0 and the numerical values for precession of the IERS 1996 Precession/Nutation theory.

The IAU/IUGG WG on "Non-rigid Earth Nutation theory" has given rise to a large number of studies for improving the theory of nutation for a Rigid Earth Model as well as for the computation of the geophysical nutations and relativistic effects. A large number of estimations of the rate of precession, obliquity rate and amplitudes of nutation have also been performed since 1994, using the most accurate data.

Much progress have therefore been done since the previous IAU General Assembly in the accuracy of the available models and observations for precession and nutation and it is clear that the conventional models have now to be re-considered in order to be in consistency with :

- the new available planetary theories,
- the new theories of Earth Rotation (rigid body, geophysical nutations and relativistic effects),
- the new International Conventional Celestial Reference System (ICRS) adopted by this IAU General Assembly,
- the present observations (accuracy and method of observations).

2. Current procedures for taking precession and nutation into account

2.1. FORMULATION FOR PRECESSION

The current formulation for precession is based on the general precessional motions as defined by Woolard and Clemence (1966) and on the accumulated precessional angles as defined by Lieske *et al.* (1977). The numerical expressions for the precession quantities have been given as functions of two time parameters in Julian centuries, t from the basic epoch (J2000.0) to the date and T, the difference between an arbitrary fixed epoch and the basic epoch. Simplified expressions, when the arbitrary epoch is chosen to be J2000.0 (i.e. $T = 0$) are given in the IERS Conventions.

A few comments can be made on the current formulation for precession :

- it combines the precession of the equator and the precession of the ecliptic (i.e. planetary precession) into the IAU "Constant of precession", p_1, (the speed of general precession in longitude at J2000.0) and into the quantities of precession, the basic quantity, $p_A = \Lambda_A - \Pi_A$, the general precession in longitude, being the difference between two precession angles, one along the ecliptic of date and one along the ecliptic of epoch,
- the motion of the pole of the ecliptic, used in the quantities of precession, has been derived from Newcomb's theory of the Sun (1894) with only improved values for planetary masses, whereas improved planetary theories are available (Simon *et al.*, 1994).

The usual precession quantities depend both upon the precession of the Earth's equator with respect to a fixed celestial frame, due to perturbing torques produced by the Moon, the Sun and the planets on the oblate Earth and upon the motion of the ecliptic of date with respect to a fixed celestial frame, due to planetary perturbations. A better consistency between the models and the real motions would require to clearly separate, in the precession quantities, the part which is due to the motion of the equator with respect to the ecliptic of epoch and the part which is due to the motion of the ecliptic of date, and in the same way to express the angles of nutation, from the ecliptic of epoch.

2.2. FORMULATION FOR NUTATION

The basic variables used in the theory of the rotation of the Earth around its centre of mass are either Euler angles between the Terrestrial Reference System (TRS) and the Celestial Reference System (CRS) (Woolard, 1953), or Andoyer canonical variables plus Oppolzer terms (Kinoshita, 1977). They provide the precession and nutation of the equator in longitude ψ and obliquity ϵ respectively, as well as the angle of Earth's rotation in space, ϕ, as : $\psi_A + \Delta\psi$, $\epsilon_A + \Delta\epsilon$, $\phi_A + \Delta\phi$, the quantities being separated into a secular term (e.g. ψ_A for precession), and a sum of periodic terms represented here by Δ (e.g. $\Delta\psi$ for nutation in longitude).

The current solution for the motion of the equator is referred to the ecliptic of date and represent therefore the motion of the equator with respect to a moving plane and not to the ICRS.

2.3. CURRENT PROCEDURE

The coordinate transformation to be used from the TRS to the CRS at the date t can be written as : $[CRS] = PN(t)R(t)W(t)[TRS]$, where $PN(t)$, $R(t)$ and $W(t)$ are the transformation matrices arising from the motion of the CEP in the CRS, from the rotation of the Earth and from the motion of the CEP in the TRS respectively.

The current procedure, which appears as option 1 in the IERS Conventions (McCarthy 1996), separates the celestial motion of the CEP into two transformation matrices P and N for precession and nutation respectively and makes use of the equinox for realizing the intermediate reference frame of date. It uses, the precession quantities z_A, ζ_A, θ_A given by Lieske *et al.* (1977) in the transformation matrix P, the nutation quantities in longitude, $\Delta\psi$, and obliquity, $\Delta\epsilon$, referred to the mean ecliptic and equinox of date in the transformation matrix N, and apparent Greenwich Sidereal Time, GST, as a function of UT1 in the transformation matrix R.

The numerical values for the precession quantities at date t are those of Lieske *et al.* (1977) with improved values for the precession rate of the equator in longitude and obliquity associated with the IERS 1996 series of nutation. The numerical values for the nutation angles at date t are those of the IERS 1996 series of nutation, but the IERS "celestial pole offsets" are the observed differences with respect to the conventional celestial pole position defined by the IAU model for precession and nutation. The numerical expression of GST is obtained from the relationship between GMST and UT1 (Aoki *et al.*, 1982) and the expression of the "equation of equinoxes", with two complementary terms since 1 January 1997.

The following comments can be made on the current procedure :

(1) it artificially separates the motion of the equator into precession and nutation; such a separation, due to technical constraints belonging to past, is no more necessary with the modern observations and computations,
(2) it refers to the "true equinox of date" and therefore GST includes both Earth's rotation and accumulated precession-nutation of the equinox along the moving equator; this is for historical reasons and is related to the use of the FK5, which has to provide the pole and the equinox at each date, which is no more the case of the ICRS,
(3) classical solutions and quantities for precession and nutation refer to the ecliptic of date instead of a fixed plane to be related to the ICRS,
(4) the conventional precession and nutation model is clearly inadequate and an improved model is necessary.

The "1996 Theory of precession/nutation" is an improved model as compared to the IAU conventional models; however, it is not meant to replace the IAU model, but rather to be used for prediction purposes, as well as when accurate *a priori* estimates of nutation are necessary. Improved theoretical series could be used, but the advantages of the use of a conventional model which may include few thousands of terms has to be compared with that of the IERS 1996 series.

An other point to be stressed is that the arguments of nutation have to be in consistency with the planetary theories used in the solution for nutations in order not to introduce spurious differences in the nutation quantities (Chapront-Touzé and Chapront, 1994).

2.4. ALTERNATIVE PROCEDURE

An other procedure can be used for taking into account precession and nutation of the equator. The celestial pole coordinates (X, Y) (Capitaine, 1990), which include both precession and nutation of the equator and moreover refer to the ICRS, would be variables more consistent to the real motions.

The relationship between these quantities and the classical quantities for precession and nutation is such that (Capitaine, 1990) :

$$\begin{aligned} X &= \chi_o + \sin(\omega_A + \Delta\epsilon_1)\sin(\psi_A + \Delta\psi_1) \\ Y &= \eta_o + \sin\epsilon_o\cos(\omega_A + \Delta\epsilon_1) + \cos\epsilon_o\sin(\omega_A + \Delta\epsilon_1)\cos(\psi_A + \Delta\psi_1) \end{aligned}$$

χ_o, η_o being the celestial offsets between the computed pole for J2000.0 and the pole of the ICRS, ω_A the inclination of the mean equator on the fixed ecliptic and index 1 being for nutation quantities referred to the fixed ecliptic. The development as function of time of these quantities can also be directly provided by the theory if it uses such kind of variables.

Concerning Earth rotation, the stellar angle $\theta = \theta_o + \int_{t_o}^{t} \omega_3\, dt$, kinematically defined by Guinot (1979) such that its time derivative is equal to the Earth's angular velocity, ω_3, around the axis of rotation can advantageously replace GST; it makes an explicit use of the "non-rotating origin" (NRO), which has to be implicitly used to express any accumulated angle along the moving equator (such as the accumulated precession and nutation in the complete expression of GST).

This alternative procedure is described in option 2 of the IERS Conventions.

3. The Celestial Ephemeris Pole

The intermediate pole defines an intermediate reference direction in the transformation between the CRS and the TRS. In the IAU-1964 series of nutation, the considered pole was the pole of instantaneous rotation which has been shown to be not observable by the classical astrometric observations. In the IAU-1980 series of nutation, the CEP has been adopted as being a pole closer to the actually "observed pole" at that time. A conceptual definition has been given for that point so that it has no diurnal motion w.r.t. TRS or CRS. The corresponding conventional definition is the realization of this pole by the precession-nutation model. This leads to several practical realizations of the CEP :

(1) model IAU-1976 precession + IAU-1980 nutation,
(2) model (1) + estimated celestial offsets.

The realization (1) shows the imperfections due to the models IAU-1976 and IAU-1980 as diurnal terms in the TRS, with amplitudes of the order of 10 *mas* (total). The realization (2) has the imperfections due to the adjustment procedure and for example, every diurnal prograde motion is absorbed in the residuals if they are not *a priori* estimated.

The conceptual definition given for the CEP is no more consistent with the theory of nutation which now includes diurnal terms of the CEP in the CRS (Bretagnon *et al.*, 1997), as well as with the model for pole motion which includes diurnal motion of the CEP in the TRS (Gross, 1993; Herring and Dong, 1994).

It is no more consistent with the sub-daily determinations of the Earth Orientation Parameters (EOP), with a level of accuracy of the order of less than 1 *mas*, as given through intensive campaigns, which make less clear the distinction between high-frequency polar motion, dominated by diurnal and semi-diurnal periodic signals and nutation dominated by diurnal retrograde polar motion.

The use of the CEP has therefore to be re-discussed and a new definition of the "actually observed pole" is necessary to be consistent with the new theory including diurnal terms both in space and within the Earth and according to new observing procedures.

4. Conclusion : Questions to the Joint Discussion

A new ICRS has been adopted by the IAU which is no more linked to the pole and the equinox. Present techniques of observations, which are sensitive to the orientation of the equator with respect to the ICRS, are improving in accuracy and resolution. New theoretical developments are available for precession and nutation.

The questions which have to be addressed are therefore the following :

(1) Is it necessary to keep a formulation combining precession of the equator and of the ecliptic ? (i.e. keep the constant of precession as a primary constant of the IAU System of Astronomical Constants ?)
(2) Is it necessary to adopt an improved model for the ecliptic pole motion ?
(3) Is it time to adopt an improved, consistent and combined formulation and model for the precession and nutation of the equator referred to the ICRS ?
(4) Is it time to abandon the equinox as the point of reference on the equator ?
(5) How can the definition of the CEP be revised to be consistent with the present theory of nutation and pole motion as well as to the present observing procedures ?

References

Aoki, S., Guinot, B., Kaplan, G.H., Kinoshita, H., McCarthy, D.D. and Seidelmann, P.K. (1982) *Astron. Astrophys.*, **105**, pp. 359.
Bretagnon, P., Rocher, P. and Simon, J.-L. (1997) *Astron. Astrophys.*, **319**, pp. 305.
Capitaine, N. (1990) *Celest. Mech. Dyn. Astr.*, **48**, pp. 127.
Chapront-Touzé, M. and Chapront, J. (1994) in *Journées Systèmes de Référence spatio-temporels*, ed. N. Capitaine, Observatoire de Paris.
Guinot, B. (1979) in *Time and the Earth's Rotation*, eds. D.D. McCarthy and J.D. Pilkington, D. Reidel Publishing Company, pp. 7.
Herring, T.A. and Dong, D. (1994) *J. Geophys. Res.*, **99**, B9, pp. 18 051.
Kinoshita, H. (1977) *Celest. Mech.*, **15**, pp. 277.
Lieske, J.H., Lederle, T., Fricke, W. and Morando, B. (1977) *Astron. Astrophys.*, **58**, pp. 1–16.
McCarthy, D.D. (ed.) (1996) *IERS Conventions*, IERS Technical Note, **21**, Observatoire de Paris.
Seidelmann, P.K. (1982) *Celest. Mech.*, **27**, pp. 79.
Simon, J.L., Bretagnon, P., Chapront, J., Chapront-Touzé, M., Francou, G. and Laskar, J. (1994) *Astron. Astrophys.*, **282**, pp. 663.
Wahr, J.M. (1981) *Geophys. J. R. Astron. Soc.*, **64**, pp. 651.
Woolard, E.W. (1953) *Astr.Pap.Amer. Eph. Naut. Almanach*, **XV**, I, pp. 1–165.
Woolard, E.W. and Clemence, G.M. (1966) *Spherical Astronomy*, Academic Press, New-York.

THE PLANETARY THEORIES AND THE PRECESSION OF THE ECLIPTIC

P. BRETAGNON
Bureau des longitudes, URA 707
77, avenue Denfert-Rochereau
F-75014 Paris, France

1. Introduction

In this paper, I give the present state of the analytical planetary theories by describing the general theories and the secular variation theories, the variations of the ecliptic with respect to the ecliptic J2000, the utilization of the analytical planetary theories in the calculation of the precession-nutation of the equator and in the calculation of the expressions of transformation between the barycentric and geocentric reference systems. At last, I describe the construction of new planetary theories undertaken at the Bureau des longitudes.

The analytical planetary theories arise in two forms: the general theories give, with a low accuracy, the variations of the elements of the planets over several million years; the secular variation theories reach a high accuracy over time spans of a few thousands of years.

In all these solutions, the motion of the planets is represented with 6 elements: a, the semi major axis, λ, the mean longitude and the variables $k = e\cos\varpi$, $h = e\sin\varpi$, $q = \sin\frac{i}{2}\cos\Omega$, $p = \sin\frac{i}{2}\sin\Omega$ where e represents the eccentricity of the orbit, ϖ the longitude of the perihelion, i the inclination of the orbit about the ecliptic J2000 and Ω the longitude of the node. It is not possible to represent the variations of the planetary orbits with the elements e, ϖ, i and Ω because, for most of the planets, e and i can reduce to zero and therefore ϖ and Ω are not continuous functions.

2. General planetary theories

In the general theories, the planetary elements are represented as Fourier series with an additive constant for the semi major axis and a linear function in time for the mean longitude.

$$\begin{aligned}
a_i &= a_{i0} + \sum_{\Phi^*} A_{i,\Phi^*}\cos\Phi^* + \sum_{\Phi} A_{i,\Phi}\cos\Phi \\
\lambda_i &= \lambda_{i0} + \lambda_{i1}t + \sum_{\Phi^*} B_{i,\Phi^*}\sin\Phi^* + \sum_{\Phi} B_{i,\Phi}\sin\Phi \\
k_i &= \sum_{k=1}^{8} \lambda_{ik}M_k\cos\psi_k + \sum_{k=1}^{8} M_{i,\psi_k}\cos\psi_k + \sum_{\Phi^*} M_{i,\Phi^*}\cos\Phi^* + \sum_{\Phi} M_{i,\Phi}\cos\Phi \\
h_i &= \sum_{k=1}^{8} \lambda_{ik}M_k\sin\psi_k + \sum_{k=1}^{8} M_{i,\psi_k}\sin\psi_k + \sum_{\Phi^*} M_{i,\Phi^*}\sin\Phi^* + \sum_{\Phi} M_{i,\Phi}\sin\Phi \\
q_i &= \sum_{k=1}^{8} \mu_{ik}N_k\cos\theta_k + \sum_{k=1}^{8} N_{i,\theta_k}\cos\theta_k + \sum_{\Phi^*} N_{i,\Phi^*}\cos\Phi^* + \sum_{\Phi} N_{i,\Phi}\cos\Phi \\
p_i &= \sum_{k=1}^{8} \mu_{ik}N_k\sin\theta_k + \sum_{k=1}^{8} N_{i,\theta_k}\sin\theta_k + \sum_{\Phi^*} N_{i,\Phi^*}\sin\Phi^* + \sum_{\Phi} N_{i,\Phi}\sin\Phi
\end{aligned}$$

The Φ arguments are linear combinations of the angles $\bar{\lambda}_j$ (coefficient r_j), ψ_j, and θ_j. $\bar{\lambda}_j$ represents the argument of the planetary longitude $\lambda_0 + \lambda_1\,t$, ψ_j the argument of the Lagrange solution in eccentricity, θ_j the argument of the Lagrange solution in inclination connected to the planet j. The periods of the angles λ_j are the periods of the planets included between 88 days and 165

J. Andersen (ed.), Highlights of Astronomy, Volume 11A, 158–162.

years if we consider 8 planets from Mercury to Neptune. The periods of the angles ψ and θ are included between 46 000 years and 2 million years. The matrices (λ_{ik}) and (μ_{ik}) are the matrices of the eigenvectors of the Lagrange solution and the part of the solutions which contains (λ_{ik}) and (μ_{ik}) is the Lagrange-Laplace solution of degree 1 in eccentricity or inclination; the other terms are of higher degree. In these expressions, the Φ^* arguments correspond to the case in which all the coefficients of the $\bar{\lambda}_j$ are zeros, that is to say to the long period terms. We have hereunder, for example, the long period parts of the variable k of the Earth with the periods in years; h is represented by a similar expression with sines instead of cosines. The time t is counted in thousands of Julian years from J2000.

$$
\begin{array}{rlr}
k = & 0.004\,260 \cos(1.561\,714 + 0.027\,2157\,t) & 230\,866 \\
& + 0.016\,135 \cos(3.471\,568 + 0.036\,1472\,t) & 173\,822 \\
& + 0.010\,246 \cos(5.610\,357 + 0.083\,8008\,t) & 74\,978 \\
& - 0.013\,399 \cos(5.414\,083 + 0.086\,7938\,t) & 72\,392 \\
& + 0.018\,173 \cos(0.540\,233 + 0.020\,8860\,t) & 300\,832 \\
& + 0.001\,439 \cos(2.286\,358 + 0.136\,4670\,t) & 46\,042 \\
& + 0.000\,598 \cos(2.066\,687 + 0.015\,2880\,t) & 410\,988 \\
& + 0.000\,013 \cos(1.302\,756 + 0.003\,2650\,t) & 1\,924\,406 \\
& + 0.000\,182 \cos(2.704\,153 + 0.084\,9704\,t) & 73\,946 \\
& + 0.000\,276 \cos(5.926\,229 + 0.091\,6930\,t) & 68\,524 \\
& - 0.003\,367 \cos(4.901\,937 + 0.081\,8946\,t) & 76\,723 \\
& + 0.000\,378 \cos(5.217\,809 + 0.089\,7868\,t) & 69\,979 \\
& - 0.002\,354 \cos(6.122\,503 + 0.088\,7000\,t) & 70\,836 \\
& + 0.000\,857 \cos(5.098\,211 + 0.078\,9016\,t) & 79\,633 \\
& + 0.000\,599 \cos(5.806\,631 + 0.080\,8078\,t) & 77\,755 \\
& - 0.000\,174 \cos(2.310\,136 + 0.030\,5229\,t) & 205\,852 \\
& + 0.001\,007 \cos(2.723\,146 + 0.032\,8400\,t) & 191\,327 \\
& - 0.000\,124 \cos(3.725\,634 + 0.091\,3001\,t) & 68\,819 \\
& - 0.000\,337 \cos(2.583\,195 + 0.033\,5454\,t) & 187\,304 \\
& + \ldots & \\
& + \text{periodic terms} &
\end{array}
$$

These expressions keep an accuracy of about 10^{-4} for k and h and therefore for the eccentricity over several million years.

The long periods of this solution being in between 46 000 years and 2 million years, it is possible to develop the sines and the cosines with respect to time and to represent the solution as polynomials in time over 1000 years. We get the following polynomials here limited to the fifth degree

$$
\begin{aligned}
k &= -.003\,742 - .000\,828\,7\,t + .000\,027\,0\,t^2 + 11\,588 \times 10^{-10}\,t^3 - 231 \times 10^{-10}\,t^4 - 7 \times 10^{-10}\,t^5 \\
h &= -.016\,286 - .000\,608\,7\,t - .000\,034\,6\,t^2 + 8\,596 \times 10^{-10}\,t^3 + 312 \times 10^{-10}\,t^4 - 5 \times 10^{-10}\,t^5 \qquad (1)
\end{aligned}
$$

In the same way, we have the long period parts of the variable q of the Earth with the period in years and a similar expression for p with sines. The accuracy of the inclination of the ecliptic about the ecliptic J2000 given by such expressions is about $10''$ over several million years.

$$
\begin{array}{rlr}
q = & 0.006\,037 \cos(0.247\,373 - 0.027\,2026\,t) & 230\,977 \\
& + 0.002\,582 \cos(5.369\,126 - 0.032\,8269\,t) & 191\,404 \\
& + 0.010\,016 \cos(4.366\,638 - 0.091\,2870\,t) & 68\,829 \\
& + 0.003\,802 \cos(4.878\,784 - 0.086\,3878\,t) & 72\,732 \\
& + 0.013\,775 \cos(1.877\,845) & \\
& + 0.001\,405 \cos(2.210\,841 - 0.127\,0650\,t) & 49\,449 \\
& - 0.000\,866 \cos(5.554\,460 - 0.014\,5436\,t) & 432\,024 \\
& - 0.000\,650 \cos(3.548\,901 - 0.003\,3522\,t) & 1\,874\,347 \\
& - 0.000\,378 \cos(3.854\,492 - 0.096\,1862\,t) & 65\,323
\end{array}
$$

$+\ 0.000\,499 \cos(4.170\,364 - 0.088\,2940\,t)$	71 162
$+\ 0.001\,741 \cos(5.075\,058 - 0.089\,3808\,t)$	70 297
$-\ 0.001\,192 \cos(4.562\,912 - 0.094\,2800\,t)$	66 644
$-\ 0.000\,124 \cos(0.995\,795 - 0.023\,8954\,t)$	262 945
$+\ 0.000\,903 \cos(4.620\,704 - 0.036\,1341\,t)$	173 885
$-\ 0.000\,401 \cos(5.509\,077 - 0.033\,5323\,t)$	187 377
$+ \ldots$	
+ periodic terms	

The development with respect to time of the long period terms of q and p gives the polynomials

$$
\begin{aligned}
q &= -0.001\,132\,5\,t + 0.000\,012\,28\,t^2 + 12\,661 \times 10^{-10}\,t^3 - 147 \times 10^{-10}\,t^4 - 3 \times 10^{-10}\,t^5 \\
p &= -0.000\,096\,0\,t + 0.000\,046\,89\,t^2 - 5\,124 \times 10^{-10}\,t^3 - 245 \times 10^{-10}\,t^4 + 3 \times 10^{-10}\,t^5 \qquad (2)
\end{aligned}
$$

The precision of the general theories is limited because of the very large number of arguments, linear combinations of 24 components and these solutions cannot reach the level of accuracy of the present observations. On the contrary, the secular variation theories represent the motion of the planets with a very high precision.

3. Secular variation theories

The secular variation theories build, in a direct way, the polynomial part as well as the Fourier and Poisson series S_j, with sines and cosines, which only depend on the 8 planetary mean longitudes. In the expression

$$x = x_0 + x_1 t + \ldots + x_{j+1} + S_0 + tS_1 + \ldots + t^j S_j$$

with

$$S_j = \sum_i (a_{j,i} \sin(\sum_{k=1}^{8} r_k \bar{\lambda}_k) + b_{j,i} \cos(\sum_{k=1}^{8} r_k \bar{\lambda}_k))$$

x_0 is an integration constant, the other quantities are calculated perturbations.

For the Earth, we obtain the polynomial parts of the variables k, h, q and p

$$
\begin{aligned}
k = &-0.003\,740\,8165 - 0.000\,822\,6742\,t + 0.000\,027\,6246\,t^2 + 11\,696 \times 10^{-10}\,t^3 \\
&-270 \times 10^{-10}\,t^4 - 7 \times 10^{-10}\,t^5 \\
h = &-0.016\,284\,4766 - 0.000\,620\,2965\,t - 0.000\,033\,8263\,t^2 + 8\,510 \times 10^{-10}\,t^3 \\
&+277 \times 10^{-10}\,t^4 - 5 \times 10^{-10}\,t^5 \\
q = &-0.001\,134\,6887\,t + 0.000\,012\,3731\,t^2 + 12\,654 \times 10^{-10}\,t^3 - 137 \times 10^{-10}\,t^4 - 3 \times 10^{-10}\,t^5 \\
p = &-0.000\,101\,8038\,t + 0.000\,047\,0200\,t^2 - 5\,417 \times 10^{-10}\,t^3 - 251 \times 10^{-10}\,t^4 + 5 \times 10^{-10}\,t^5 \qquad (3)
\end{aligned}
$$

3.1. COMPARISON OF THE POLYNOMIAL PARTS OF THE EARTH'S ELEMENTS WITH THE GENERAL THEORIES

It is interesting to compare the polynomials (3) with the long period terms of the general theories obtained from the developments (1) and (2) with respect to time.

For the Earth over 1000 years, the differences for the variables k and h and therefore for the eccentricity are about 10^{-5}; for q and p, the differences are about 5×10^{-6} which corresponds to 2 arcseconds for the inclination.

In this comparison, we can see that it is easy to obtain a low accuracy secular variation solution from a general theory but, of course, the contrary is not possible. Over 1000 years, to the variables k and h correspond the polynomial parts of the eccentricity and of the longitude of the perihelion

$$
\begin{aligned}
e &= 0.016\,708\,6342 - 0.000\,420\,3654\,t - 0.000\,012\,6734\,t^2 + \ldots \\
\varpi &= 102^\circ\!.937\,348\,08 + 11\,612''\!.352\,90\,t + 53''\!.275\,77\,t^2 + \ldots \qquad (4)
\end{aligned}
$$

Let us recall that, over 100 000 years, the longitude of the perihelion is not a continuous function and that the secular variation solutions represent the planetary motion over only 1000 years. When the eccentricity becomes zero, we have a discontinuity of the perihelion of 180 degrees.

Therefore the secular term of $11''.61235290$ per year in the formula (4) is not the frequency of a periodic term with a 111 605 year period reckoned from the equinox J2000 or with a 20 937 year period reckoned from the equinox of date.

There are not such periods in the solar system and there is no meaning to introduce such periods in the heliocentric motion of the Earth or in the precession-nutation theory.

3.2. ACCURACY OF THE SECULAR VARIATION THEORY OF THE EARTH

The integration constants of the VSOP82 (Bretagnon, 1982) and VSOP87 (Bretagnon and Francou, 1988) solutions have been determined by fitting to the numerical integration DE200 (Standish, 1982) of the JPL. The comparison, between 1900 and 2000, of VSOP with DE200 shows differences of a few mas, 5 mas for the longitude of the Earth and about 10^{-8} for the other variables. The inclination of the true ecliptic about the ecliptic J2000 is defined with an accuracy of 3 mas but the main reason of the uncertainty of the analytical and numerical planetary solutions comes from the quality of the observations used to determine the integration constants. Thus, between DE200 and DE403, we have differences which reach 30 mas for the Earth over 1900-2000.

The accuracy of VSOP82 is therefore consistent with the quality of the observations used in DE200.

4. Utilization of the analytical theories in the calculation of the precession-nutation

This accuracy of the VSOP solutions allows us to build the contributions of the Sun and the planets to the analytical precession-nutation series at the level of 0.1 μas. The perturbations of the terrestrial equator due to the Moon are calculated with an uncertainty of about 1 μas by using the analytical solution ELP. The comparison for the three Euler angles ψ, ω and φ between the analytical precession-nutation solution SMART97 (Bretagnon *et al.*, 1997a, 1997b) and a numerical solution using the numerical integration DE403/LE403 gives residuals are 2 μas for ψ, 0.6 μas for ω, and 2 μas for φ over 55 years between 1968 and 2023.

Note that, as in the analytical solution of the Sun, there is no meaning to introduce the 20 937 year period term in the nutation series. One find also, in the classical nutation series, similar period arguments which are in phase every 20 937 years or 10 468 years such as the arguments of period 18.6 years, 6 months, one year. There is no meaning to keep 2 arguments with similar periods that it is impossible to discriminate over the 20 year time span of the high precision observations.

5. Utilization of the analytical theories in the calculation of the expressions of transformation between the barycentric and geocentric reference systems

The analytical solutions of the planetary motions are used to calculate the expressions of transformation between the barycentric and geocentric reference systems. Several solutions have been built at the first order in c^{-2} where c is the light velocity.

The analytical solution for $TCB - TCG$ (Fairhead and Bretagnon, 1990) has been compared with a numerical integration of the JPL. The differences are smaller than 3.4 nanoseconds over 1900-2000.

Fukushima has completed the analytical solution by taking into account the post-Newtonian effects and the perturbations due to the asteroids. The comparison to numerical integrations he has performed gives an accuracy of the analytical solution of 1.8 nanosecond over the time span 1980-2000.

6. New planetary theories

We have seen that the VSOP82 and VSOP87 planetary solutions allow us to determine the variations of the rigid Earth equator with an accuracy of 1 μas and the relation between TCB and TCG with an accuracy of 1.8 nanosecond.

Nevertheless, it is useful to have today better analytical planetary theories, specially to provide very precise relations between the barycentric and geocentric coordinates, relations which are essential to the reduction of the observations VLBI, SLR, LLR.

Moisson (Bureau des longitudes) has undertaken the construction of a new planetary theory by using recent values of the planetary masses. The first step of the work gives solutions three times more precise than VSOP82. The improvement is particularly important for the variables q and p which define the inclination and the longitude of the node of the ecliptic. In a comparison with DE403, the residuals in inclination do not exceed 0.1 mas over 1900-2000 which is 30 times more precise than the VSOP82 solution. His new solution shall be expressed in two forms: as a function in Barycentric dynamical time (TDB) and as a function in Barycentric coordinate time (TCB). This is necessary because, till now, in the expression of $TCB - TCG$, the time is TCB but the potential due to all the bodies of the solar system except the Earth and the velocity of the Earth are functions in TDB which involves an error depending on c^{-4} not negligible when we consider the post-Newtonian effects.

References

Bretagnon, P. (1982) *Astron. Astrophys.*, **114**, pp. 278.
Bretagnon, P. and Francou, G. (1988) *Astron. Astrophys.*, **202**, pp. 309.
Bretagnon, P., Rocher, P. and Simon, J.-L. (1997a) *Astron. Astrophys.*, **319**, pp. 305.
Bretagnon, P., Francou, G., Rocher, P. and Simon, J.-L. (1997b) *Astron. Astrophys.*, in press.
Fairhead, L. and Bretagnon, P. (1990) *Astron. Astrophys.*, **229**, pp. 240.
Standish, E.M. (1982) *Astron. Astrophys.*, **114**, pp. 297.

THE THEORY OF THE NUTATION FOR A RIGID EARTH MODEL: CURRENT STATE OF THE SITUATION

J. SOUCHAY
Observatoire de Paris/DANOF, URA 1125 du CNRS, 61 Avenue de l'Observatoire, F-75014, Paris; E-mail:"Souchay@obspm.fr"

AND

H. KINOSHITA
Tokyo National Astronomical Observatory, Ohsawa 2-21-1, Mitaka-Shi Tokyo 181, Japan

1. Introduction

Whereas no particular attention was paid to the theory of the nutation for a rigid Earth model, for more than a decade after the adoption by the International Astronomical Union (IAU) of coefficients as calculated and listed by Kinoshita (1977), an increasing number of studies have been done in the recent years aiming to improve this theory. The improvement became necessary mainly because of the big parallel improvement of the VLBI observations itself, which leads to present determinations of some coefficients of nutation at the level of a few $10\mu as$. Therefore the amelioration of the theory of the nutation for a rigid Earth model can be divided in two aspects: one is to consider a smaller level of truncation of the coefficients of nutation; the other is to evaluate in the best way the coefficients already taken into account, particularly the leading coefficients which are typically those subject to the largest absolute differences.

As considering the first aspect, it seems that presently a level of truncation of $0.1\mu as$ for both $\Delta\psi \sin\varepsilon$ and $\Delta\varepsilon$ seems quite reasonable, for the global nutation obtained by summing the coefficients under this level does not exceed $\pm 3\mu as$, which is far below the level of the noise of the residuals between the most accurate VLBI observations and the theory for a non-rigid Earth model, whose the absolute amplitude is typically of the order of $0.1mas$, after fitting a few coefficients whose the modeling is not yet satisfactory, because of complex geophysical effects (see for instance Herring, 1991, Souchay *et al.*, 1995).

As considering the second aspect, an agreement between the global nutations or the amplitudes of the coefficients as calculated by different authors, if not constituting an unquestionable proof of the validity of these coefficients, is a big step in order to insure this validity. Moreover, an additioning agreement between the analytical determination of the nutation and the determination obtained independently by the way of numerical integration would prove both that the analytical computations have been carried out properly, and that no contribution has been forgotten, at the level of truncation considered.

In Table 1, we summarize main of the various studies which have been done concerning the nutation for a rigid Earth model since 1985, with the possibility that the list be not complete. We mention when these studies resort to analytical computations, numerical integration, or both of them. We can notice the increasing number of works after a reconstruction of the theory by Kinoshita and Souchay (1990).

2. Recent improvements

Before the reconstruction of the theory of the nutation for a rigid Earth model by Kinoshita and Souchay (1990) the difference between the nutations given by numerical integration and the analytical tables (Kinoshita, 1977) exceeded $1mas$ as was shown by Kubo and Fukushima (1988)

J. Andersen (ed.), Highlights of Astronomy, Volume 11A, 163–167.

TABLE 1. List of various studies dealing with rigid Earth nutation, from 1985. the symbol "A" (third column) is for "analytical", the symbol "N" for "numerical".

AUTHOR(S)	YEAR	A	N	REMARKS AND CONTENTS
Kubo and Fukushima	1987		X	Comparison with tables Kinoshita (1977)
Schastok *et al.*	1988		X	Comparison with tables Kinoshita (1977)
Zhu and Groten	1989	X		J_2, J_3, second order, tables, fig. axis truncation: $0.01mas$
Schastok *et al.*	1989		X	Comparison with tables Kinoshita and Souchay (1990)
Kinoshita and Souchay	1990	X		Complete theory,J_2, J_3, second-order, planetary effects, triaxiality, precession, H_d, tables 3 axes, truncation $5\mu as$
Souchay and Kinoshita	1991		X	Comparison with tables Kinoshita and Souchay (1990)
Hartmann and Soffel	1994	X		Direct planetary effect from tidal potential
Williams	1994	X		Precession, H_d, tilt-effect, obliquity rate
Hartmann *et al.*	1995	X		J_3 effect from tidal potential
Williams	1995	X		Direct planetary effects, tables, truncation: $0.1\mu as$
Souchay and Kinoshita	1996	X		Corrections and new developments (I), lunisolar J_2, indirect planetary effects, corrections, truncation: $0.1\mu as$
Souchay and Kinoshita	1997	X		Corr. and new developments (II), obliquity rate, J_3, J_4, direct planetary effects, truncation $0.1\mu as$
Hartmann and Soffel	1997			Complete theory from tidal potential (except diurnal and sub-diurnal), J_2, J_3,J_4 planetary effects, triaxiality, precession, H_d tables 3 axes, truncation $0.5\mu as$
Roosbeek and Dehant	1997	X	X	Complete theory (except diurnal and sub-diurnal), J_2, J_3, second order effects, planetary effects, triaxiality, precession, H_d tables RDAN97, 3 axes, truncation $0.1\mu as$
Bretagnon *et al.*	1997	X	X	Complete theory, J_2, J_3,J_4, triaxiality, $C_{3,i}$, $C_{4,i}$, $C_{5,i}$, planetary effects, precession, H_d tables, 3 axes, truncation $< 0.01\mu as$
Folgueira *et al.*	1997	X		Nutations from $C_{3,i}, S_{3,i}$
Folgueira *et al.*	1997	X		Nutations from $C_{4,i}, S_{4,i}$
Souchay *et al.*	1997	X		Corr. and new developments (III), complete theory J_2, J_3,J_4, $C_{3,i}$, $C_{4,i}$, second-order, planetary effects, triaxiality, precession, H_d, tables "REN-2000", 3 axes, truncation $0.1\mu as$
Souchay	1997		X	Numerical integration: comparisons with tables REN-2000, comparisons inter-tables : REN-2000 (Souchay *et al.*, 1997), RDAN97 (Roosbeek and Dehant, 1997) SMART97 (Bretagnon *et al.*, 1997)

as well as by Schastok *et al.* (1987). Clearly some big discrepancies occurred for the terms with argument Ω (18.6 yr) and 2Ω (9.3 yr) both in longitude and in obliquity.

The reason is that two kinds of effects were not taken into account in the analytical tables. One is the spin-orbit coupling effect, firstly pointed out by Kubo (1982), which consists in an interaction between the rotation of the Earth and the orbital motion of the Moon, and the other is the crossed-nutation effect. They were both calculated simultaneously by Zhu and Groten (1989) as well as by

Kinoshita and Souchay (1990), in the frame of the Hamiltonian theory for which the Earth-Moon system is considered as a whole (the results between these authors were quite in accordance). They are at the origin of the considerable reduction of the differences between the nutations as calculated by numerical integration and those calculated starting from the analytical tables (Schastok *et al.*, 1989; Souchay and Kinoshita, 1991). Nevertheless some small differences remained for the leading components with period 18.6 *yr*, 9.3 *yr* semi-annual and fortnightly, which suggested that some corrections remained to be done in the analytical calculations.

A significant one was pointed out by Williams (1994), which consisted in including out-of-phase components for the nutations with arguments Ω and 2Ω both for $\Delta\psi$ and $\Delta\varepsilon$, due to the planetary tilt-over effect (the Moon's orbit is precessing with respect to an axis which is slightly shifted with respect to the axis of the ecliptic). Recently a significant additioning out-of-phase component in $\Delta\psi$ for the Ω term has been noticed by Bretagnon *et al.* (1997). The related ratio ρ for this out-of-phase component to the in-phase one has been evaluated by Souchay *et al.* (1997) analytically with the following formula: $\rho = -\frac{2\dot{\varepsilon}}{\dot{\Omega}} \tan 2\varepsilon_0$ for $\Delta\psi$, which gives a $0.248mas$ amount, and matches perfectly the difference of $0.250mas$ found by Bretagnon *et al.* (1997). Roosbeek and Dehant (1997) made also a detailed analysis of the contributions to the out-of-phase components for the Ω term, which are quite in agreement with Bretagnon *et al.* (1997) and Souchay *et al.* (1997). Moreover Souchay *et al.* (1997) recalculated accurately the second-order tilt-effect and crossed-nutation effects at the level of $0.1\mu as$ instead of $5\mu as$ (Kinoshita and Souchay, 1990) and thus showed that they are at the origin of significant amplitudes around the semi-annual and fortnightly frequencies.

Concerning the nutations due to second-order harmonics of the geopotential, Kinoshita (1977) showed that the triaxiality of the Earth, characterized by the $C_{2,2}$ and $S_{2,2}$ tesseral harmonics, give birth to several coefficients at the level of $0.01mas$, with nearly semi-diurnal period. They were calculated up to $5\mu as$ by Kinoshita and Souchay (1990) who calculated also the coefficients due the influence of the geopotential J_3 which could no more be neglected at this level of truncation, with a leading coefficient at the level of $0.1mas$. Souchay and Kinoshita (1997) recalculated these two kinds of nutations up to $0.1\mu as$, with quite a good agreement concerning the second effect with the results of Hartmann *et al.* (1996). Bretagnon *et al.* (1997) showed that the $C_{3,i}$, $S_{3,i}$, $C_{4,i}$, $S_{4,i}$ harmonics produce also a significant number of coefficients up to $0.1\mu as$ with a leading one closed to $40\mu as$. Their period is close to diurnal, semi-diurnal or 1/3 diurnal. Folgueira *et al.* (1997a, 1997b) found results in total agreement with Bretagnon *et al.* (1997) for these coefficients, but by the way of a theory based on the Hamiltonian canonical equations, thus completing the series related to this theory at the level of $0.1\mu as$ (Souchay and Kinoshita, 1996; Souchay and Kinoshita, 1997; Souchay *et al.*, 1997). They showed that the Oppolzer terms, which make the difference between the angular momentum axis and the figure axis, are much larger than the respective coefficients for the angular momentum axis, which is a specific property of these terms.

A study of the direct action of the planets on the nutation have been done firstly by Vondrak (1983), then by Kinoshita and Souchay (1990), Hartmann and Soffel (1994), Williams (1995), Souchay and Kinoshita (1997). The agreement between these three last studies when comparing the coefficients up to $0.1\mu as$ is remarkable, with only a few differences exceeding this level of truncation (Souchay and Kinoshita, 1997).

3. Comparison between global theories and numerical integrations

We can notice recently the accomplishment of four works aiming to the construction of complete or quasi-complete series of the nutation for a rigid-Earth model. They have been done and described precisely by Hartmann and Soffel (1997), Bretagnon *et al.* (1997), Roosbeek and Dehant (1997) and Souchay *et al.* (1997). The series are quoted respectively as HS97, SMART97, RDAN97 and REN-2000. In Table 2 we summarize some of the main particularities of the series, for which we can remark fundamental differences in the way they have been constructed. For instance the series REN-2000 (Souchay *et al.* , 1997) are computed starting from Hamiltonian canonical equations (Kinoshita, 1977; Kinoshita and Souchay, 1990) in opposite to the other series for which the Dynamical Euler's equations are used. Moreover Bretagnon *et al.* (1997) for the establishment of their series SMART97 do not separate the different contributions to the nutation, which is not the case of the other authors who were concerned about making a detailed analysis of each of the effects involved.

TABLE 2. Comparison between 4 recent rigid Earth nutation theories.

Item	Hartmann and Soffel (1997)	Roosbeek and Dehant (1997)	Bretagnon et *et al.* (1997)	Souchay *et al.* (1997)
Series name	HS97	RDAN97	SMART97	REN-2000
Basic equations	Tidal potential	Euler's. Eq.	Euler's Eq.	Hamiltonian
Basic ref. syst.	Fixed	Fixed	Fixed	Moving
Way of study	each contribution	each contribution	global	each contribution
J_2	-	-	-	-
J_3	-	-	-	-
J_4	-	-	-	-
$C_{2,2}$, $S_{2,2}$	-	-	-	-
$C_{3,1}$, $S_{3,1}$	no	no	-	-
$C_{4,1}$, $S_{4,1}$	no	no	-	-
J_2 Planetary	-	-	-	-
Truncation limit	$0.5\mu as$	$0.1\mu as$	$< 0.01\mu as$	$0.1\mu as$
Number of terms		≈ 1500	10000	≈ 1500
Tables	3 axes	3 axes	3 axes	3 axes
Comparison with Numerical. Int.	no	same paper a few μas r.m.s.	same paper a few μas r.m.s.	this paper a few μas r.m.s.
Dyn. ellip. $H_d \times 10^3$	3.273792489	3.2737670	3.2737671	3.2737548

We can also remark that only two series, SMART97 (Bretagnon *et al.*, 1997) and REN-2000 (Souchay *et al.*, 1997) can be considered as really complete up to $0.1\mu as$, for they are the only ones to include the effects due to the second-order harmonics of the geopotential $C_{3,i}, S_{3,i} (i = 1, 2, 3)$ and $C_{4,i}, S_{4,i} (i = 1, 2, 3, 4)$, which are characterized by quasi-diurnal or sub-diurnal coefficients with about thirty of them having an absolute amplitude exceeding $1\mu as$ (Bretagnon *et al.*, 1997; Folgueira *et al.*, 1997) for $\Delta\psi$. The work of Hartmann and Soffel seems to be the more accurate and complete done so far in the attempt to make the relationships between the coefficients of the series of the tidal potential (Hartmann and Wenzel, 1995), and the coefficients of the nutation. Unfortunately, their method is not well suitable for the determination of low-frequency nutations because the uncertainty in the determination of the coefficients grows with their period, as was already noticed by Souchay (1993) when comparing analytical nutations.

The comparison with a numerical integration has been done at the same time as the analytical computations of the coefficients by Roosbeek and Dehant (1997) for the series RDAN97 and by Bretagnon *et al.* (1997), for the series SMART97. It is satisfactory at the level of a few microarcseconds in the two cases, but the residuals for a short-time span are particularly small in the second study, which is a proof that the analytical computations have been done with a very good internal accuracy. Moreover a specific study of the comparison of the series REN-2000 both with numerical integration and with the series RDAN97 and SMART97 has been done independently by Souchay (1997). The results found in this last study show that the residuals between analytical nutation from the series REN-2000 and numerical integration have a $3.2\mu as$ r.m.s and a $5.5\mu as$ r.m.s. respectively for $\Delta\psi \sin\varepsilon$ and $\Delta\varepsilon$ for a 134 years time span. The residuals are much noisy at high frequency than those obtained by Bretagnon *et al.* (1997), but this might be only due to the truncation of the series. REN-2000 truncated at $0.1\mu as$ contain no more than 1537 terms and 1192 terms respectively for $\Delta\psi$ and $\Delta\varepsilon$, whereas the series SMART97 contain several thousands of coefficients.

Anyway the agreement between the three series can be considered as very satisfactory. For instance Souchay (1997) showed that the residuals are about $5\mu as$ for $\Delta\psi \sin\varepsilon$ and $\Delta\varepsilon$ still for a 134 years time span, when comparing the global nutations of the figure axis from J2000.0, as calculated from REN-2000 and SMART97, after fitting only the 18.6 yr leading component. Moreover he showed that the absolute difference concerning this term, essentially due to different values adopted to the dynamical ellipticity of the Earth H_d, is less than $10\mu as$ when adopting the same value of H_d. The residuals between the nutations as computed from RDAN97 and REN-2000, although a little less convincing, are still at the level of a few microarcseconds r.m.s.

Therefore we can conclude that a big step has been done recently in the frame of rigid Earth nutation theory, and that the coherence between the various studies is established, thus leading to the possibility of a definitive adoption of series at the level of truncation of $0.1\mu as$.

References

Bretagnon, P., Rocher, P. and Simon, J.L. (1997) *Astron. Astrophys.*, **319**, pp. 305.

Folgueira, M., Souchay, J. and Kinoshita, H. (1997) *Celest. Mechanics*, submitted.

Folgueira, M., Souchay, J. and Kinoshita, H. (1997) *Celest. Mechanics*, to be submitted.

Hartmann, T. and Soffel, M. (1994) *Astron. J.*, **108**, pp. 1115.

Hartmann, T., Williams, J.G. and Soffel, M.. (1996) *Astr. J.*, **111**, pp. 1400.

Hartmann, T. and Wenzel, H.G. (1995) *Geophys. Res. Letters*, **22**, pp. 3553.

Hartmann, T. and Soffel, M. (1997) *Celest. Mech.*, submitted.

Herring, T. (1991) in: *Proceedings of the 127th. Colloquium of the International Astronomical Union, Reference Systems*, ed. J.A. Hughes, C.A. Smith, G.H. Kaplan, U.S. Naval Observatory, Washington.

Kinoshita, H. (1977) *Celest. Mech.*, **15**, pp. 277.

Kinoshita, H. and Souchay, J. (1990) *Celest. Mech.*, **48**, pp. 187.

Kubo, T. (1982) *Celest. Mech.*, **26**, pp. 96.

Kubo, Y. and Fukushima, T. (1988) in: *The Earth's Rotation and Reference Frames for Geodesy and geodynamics*, ed. A.K. Babcock and G.A. Wilkins, pp. 331.

Roosbeek, F. and Dehant, V. (1997) *Celest. Mechanics*, submitted.

Schastok, J., Soffel, M. and Ruder, H. (1987) in: *Proc. IUGG symp. U4, Geophysical Monograph*, 59, pp. 17–20.

Schastok, J., Soffel, M. and Ruder, H. (1989) *Celest. Mech.*, **47**, pp. 219.

Souchay, J. and Kinoshita, H. (1991) *Celest. Mech.*, **52**, 45

Souchay, J. (1993) *Astron. Astrophys.*, **276**, pp. 266.

Souchay, J., Feissel, M., Bizouard, C., Capitaine, N. and Bougeard, M. (1995) *Astron. Astrophys.*, **299**, pp. 277.

Souchay, J. and Kinoshita, H. (1996) *Astron. Astrophys.*, **312**, pp. 1017.

Souchay, J. and Kinoshita, H. (1997) *Astron. Astrophys.*, **318**, pp. 639.

Souchay, J., Loysel, B., Kinoshita, H. and Folgueira, M. (1997) *Astron. Astrophys.*, submitted.

Souchay, J. (1997) *Astron. Jour.*, submitted.

Vondrak, J. (1983) *Bull. Astr. Inst. Czechosl.*, **33**, pp. 26.

Williams, J.G. (1994) *Astron. J.*, **108**, pp. 711.

Williams, J.G. (1995) *Astron. J.*, **110**, pp. 1420.

Zhu, S.Y. and Groten, E. (1989) *Astron. J.*, **98**, pp. 1104.

FUNDAMENTAL ARGUMENTS OF THE CURRENT NUTATION THEORY

DYNAMICAL REFERENCE FRAME

E.M. STANDISH
Caltech/JPL
JPL 301-150; Pasadena, CA 91109 USA
ems@smyles.jpl.nasa.gov

1. The Ephemerides

The latest JPL planetary and lunar ephemerides, DE405, are now referenced to the ICRF with an accuracy of about 1 *mas*. This has been accomplished mainly by fitting the ephemerides to 18 VLBI observations of the Magellan Spacecraft in orbit around Venus, 1990–1994, and to 2 VLBI observations of the Phobos Spacecraft in its approach to Mars, 1989. The orientation of DE405 is discussed in more detail elsewhere in this volume (Standish, 1997). Since all of the parameters of the inner solar system are now determined to 1 *mas* or better, one should be able to extract numerically the fundamental arguments of the nutation theories to the level of 1 *mas*.

There are two ways of extracting the ecliptic, for example, from a numerical ephemeris: 1) one computes the node and obliquity of the instantaneous ecliptic at multiple points in time and then fits these with analytic functions, or 2) one fits an analytical planetary theory to the ephemerides and then computes the node and obliquity from the theory's parameters. This paper relates a short example using method #1 and concludes that method #2 is probably more preferable.

2. Numerical determination of $\overline{\Omega}$ and $\overline{\epsilon}$

Using the "inertial" definition (Standish, 1981) of the instantaneous ecliptic, $\Omega_I \equiv \tan^{-1}(\hat{h}_x/-\hat{h}_y)$ and $\epsilon_I \equiv \sin^{-1}(\hat{h}_z)$, different fits were made to values of the node and obliquity, computed over various time intervals. One can judge the validity of the resultant values by the appearance of the post-fit residuals and by how much the resultant values themselves vary from one to the other. As in Standish (1982), there were 56-58 periodic terms, 7 mixed-secular terms, and 3-8 time-polynomial terms; the time intervals used were 1500 AD – 2500 AD, 1000 AD – 3000 AD, and 3000 BC – 3000 AD. The best results seem to be $\overline{\Omega} = 0''.052 \pm 0''.010 + 10''.556(\pm 0''.001)T + \ldots$ and $\overline{\epsilon} = 23^o\ 26'\ 21''.409(\pm 0''.010) - 46''.811(\pm 0''.001)T + \ldots$

3. Discussion

Relatively large uncertainties were assigned to the mean values because it seems that a long interval (5000 years or more), centered at J2000, would be necessary to produce trustworthy results at the 0''.001 level. However, it is possible that the analytical theories, containing thousands of terms, developed at the Bureau des Longitudes, could be fit to the present existing intervals to give milliarcsecond results. Otherwise, it will be necessary to extend DE405 another 4000-5000 years into the future.

References

Standish, E.M. (1981) Two Differing Definitions of the Dynamical Equinox and the Mean Obliquity. *Astron. Astrophys.*, **101**, L17-18.
Standish, E.M. (1982) Orientation of the JPL Ephemerides, DE200/LE200, to the Dynamical Equinox of J2000. *Astron. Astrophys.*, **114**, pp. 297–302.
Standish, E.M. (1997))Linking the dynamical reference frame to the ICRF. elsewhere in this volume.

J. Andersen (ed.), Highlights of Astronomy, Volume 11A, 168.

OBSERVATIONS OF THE CELESTIAL MOTION OF THE EARTH'S POLE

M. FEISSEL AND A.-M. GONTIER
Central Bureau of IERS, Paris Observatory, CNRS/URA1125
61, avenue de l'Observatoire
75014 Paris, France

Abstract. The accurate observation of the celestial motion of the Earth's pole requires careful connection of the VLBI observables with quantities and reference frames related to the theoretical description of precession-nutation models. The analysis of the derived celestial pole offsets beyond the semi-empirical IERS(1996) precession-nutation model shows variability of some components, such as the seasonal term and a $\sim$ 430 day term.

1. Reference systems, conceptual definitions and VLBI observables

The celestial motion of the Earth's pole is observed in the International Celestial Reference System (ICRS, Arias *et al.*, 1995) recommended for use by the IAU starting with 1 January 1998. The realization of this system, the International Celestial Reference Frame (ICRF), is described in Ma *et al.* (1997). A description of the consequences in the current astronomical procedures induced by this change is given by Kovalevsky and McCarthy (1997).

The observations consist of time series of VLBI-derived coordinates of the Celestial Ephemeris Pole (CEP) in the ICRS. The CEP is by convention a pole that has no diurnal motion either in space or in the Earth. The IAU 1980 Theory of Nutation (Seidelmann, 1982) models the periodic celestial motion of the CEP relative to the mean pole at J2000.0 corresponding to the 1976 IAU precession model (Lieske *et al.*, 1976). In VLBI data analysis, the IAU 1976-1980 models are used as a prioris and the deviation of the actual celestial direction of the CEP is estimated from 24h observing sessions. The corresponding parameters are the so-called celestial pole offsets, $d\psi$ and $d\varepsilon$. As the VLBI data analysis is referred to the ICRS while the celestial pole offsets are referred to the mean pole at J2000.0, the time series of the celestial pole offsets include a constant term that is due to the fact that the mean pole at J2000.0 is not in coincidence with the ICRS pole. The shift, shown in Figure 1, is estimated to be 17.24 ± 0.01 mas in the direction 12h and 5.1 ± 0.01 mas in the direction 18h (IERS 1997). For reference, the FK5 pole position (see Kovalevsky and McCarthy, 1997) is also shown; the FK5 pole was meant to be a realization of the mean pole at J2000.0.

Very Long Baseline Interferometry (VLBI) is the only technique that can provide the tie to a celestial reference system, a terrestrial reference system and the Earth orientation parameters simultaneously. The modeling of VLBI observables, the *delay* and the *delay rate*, is made in the framework of General Relativity. It includes the transformation from terrestrial to celestial coordinate systems (McCarthy, 1996). Mass redistribution in the solid Earth and the oceans by global tides affect both universal time and polar motion and modify the Earth's orientation. Although these effects are ignored in the context of the definition of the CEP, they have to be considered. Diurnal and semi-diurnal variations in polar motion and UT1 are therefore taken into account (see Herring, 1991). The major sets of VLBI analysis software have been intensively tested and intercompared. They are consistently implemented within 1 ps for the delay and 1 fs/s for the delay rate, one order of magnitude better than the observations (Jacobs *et al.*, 1997).

J. Andersen (ed.), Highlights of Astronomy, Volume 11A, 169–172.

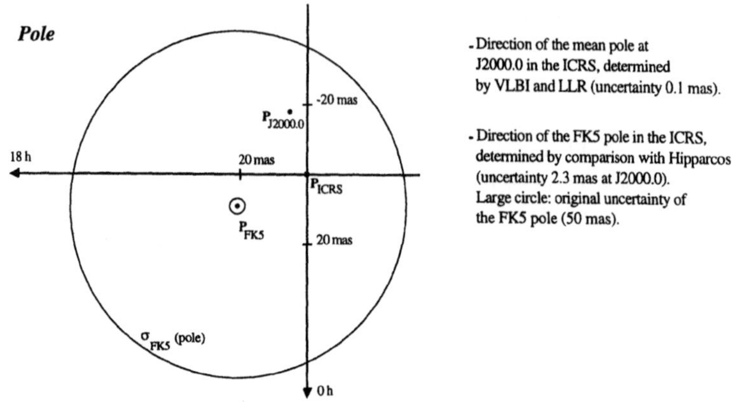

Figure 1. Physical and conventional directions in the International Celestial Reference System, ICRS.

2. Analysis of celestial pole offsets time series, 1984-1996

The series analyzed here, EOP(GSFC)97 R 01, was obtained by Ma *et al.* (1997), over 1979-1996 from all existing VLBI observations. The time series of celestial pole offsets derived from VLBI observations show evidence of errors in the IAU (1980) Theory of Nutation and in the IAU (1976) Precession that are to a large extent corrected in the IERS Conventions (1996) semi-empirical model, as described by Mathews and Herring (1997). The latter model accounts for 99.4% of the variance of the observed differences with the conventional IAU models. However, the signal remaining after this correction is still one order of magnitude larger than the noise of a single observation (±0.1 mas in 1996). The remaining variance is mainly localized in high frequencies as illustrated in Table 1.

TABLE 1. Precision and noise level of VLBI celestial pole results, 1984-1996 EOP(GSFC) 97 R 01 (Ma *et al.*, 1997) - IERS Conventions (1996) Unit: 0.001"

Years	uncertainty of a single value		weighted root mean square residual							
			all periods		with 433d term estimated		for periods longer than 30d		for periods shorter than 30d	
	ψ	ε	ψ	ε	ψ	ε	ψ	ε	ψ	ε
1984-85	1.04	0.42	1.40	1.17	1.11	1.14	0.67	0.25	0.89	1.11
1985-86	0.76	0.30	1.38	1.59	1.13	1.31	0.54	0.22	0.99	1.29
1986-87	0.71	0.25	1.46	1.46	1.11	1.34	0.48	0.20	1.00	1.32
1987-88	0.67	0.26	1.49	1.59	1.24	1.49	0.43	0.21	1.17	1.48
1988-89	0.54	0.23	1.69	1.68	1.45	1.42	0.46	0.16	1.38	1.41
1989-90	0.41	0.17	1.43	1.57	1.36	1.33	0.38	0.14	1.31	1.32
1990-91	0.34	0.14	1.47	1.47	1.34	1.26	0.35	0.13	1.29	1.25
1991-92	0.30	0.13	1.42	1.30	1.31	1.15	0.33	0.13	1.27	1.14
1992-93	0.26	0.11	1.34	1.29	1.18	1.15	0.31	0.13	1.14	1.14
1993-94	0.19	0.08	0.98	1.19	0.82	1.01	0.25	0.12	0.78	1.00
1994-95	0.15	0.06	0.76	0.74	0.63	0.66	0.21	0.09	0.59	0.65
1995-96	0.16	0.06	0.80	0.84	0.69	0.79	0.23	0.11	0.65	0.78

The residual motion not accounted for by the IERS Conventions (1996) model is plotted on

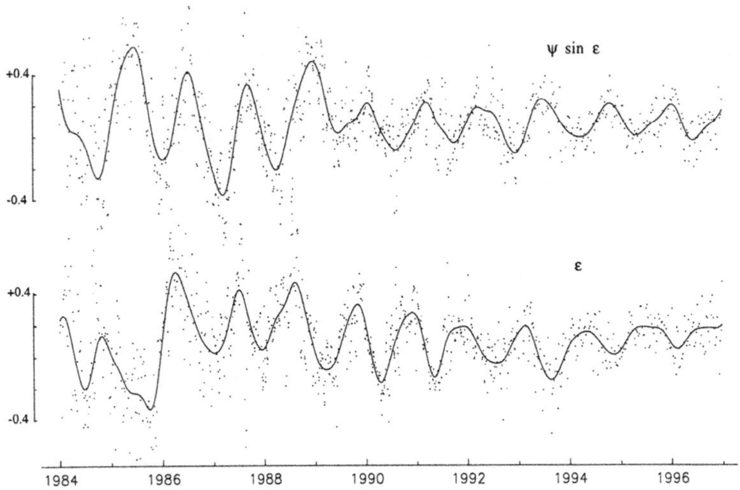

Figure 2. Variations in the celestial pole offsets not accounted for by the IERS Conventions (1996) precession-nutation model. The dots are the individual determinations. The solid line is a Vondrak smoothing that keeps periods longer than 300 days. Unit: 0.001"

Figure 2. The oscillation shown by the solid line is generally considered to be the effect of the Free Core Nutation (FCN). The theoretical predictions of the FCN period span a few days around 433 days, and active research is made by different groups to explain that its amplitude seems to vary with time, or with the time interval considered (e.g. Dehant *et al.*, 1997, and other papers in this JD proceedings). If one considers a unique FCN period (433 days) over 1984-96, the mean amplitudes (in $\psi \sin \varepsilon$ or ε) determined over the two-year intervals of Table 1 range from 0.32 to 0.06 mas, with formal uncertainties of 0.07 mas at the beginning and 0.03 mas at the end of the period covered.

The IERS Conventions (1996) precession-nutation model gives a constant amplitude for the 365.22 day term. While the rigid-Earth term has only in-phase components, the observed annual nutation has statistically significant out-of-phase components. In addition, both the in-phase and out-of-phase components have significant variations from year to year, as shown on Figure 3. These variations are ascribed to atmospheric effects.

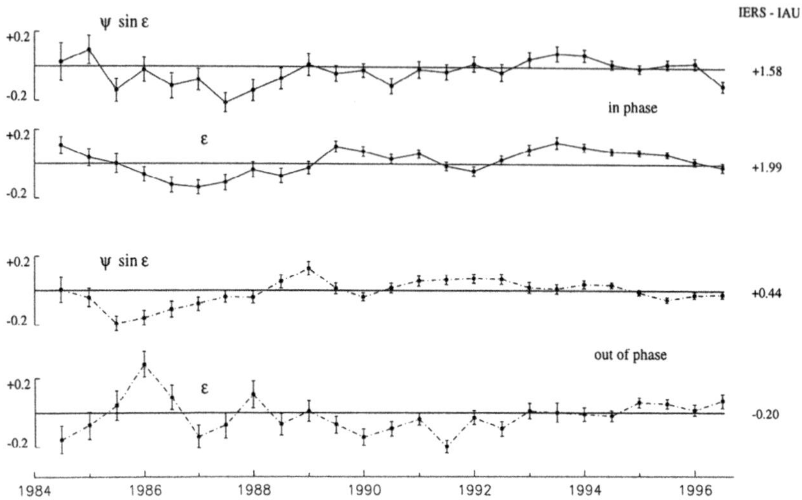

Figure 3. Year to year variations in the components of the 365.22 day nutation term with their standard deviations, relative to the mean IERS corrections, given on right of the figure. The plotted values are determined over one-year intervals, shifted by 0.5 year. Unit: 0.001"

References

Arias, E.F., Charlot, P., Feissel, M. and Lestrade, J.-F. (1995) The extragalactic reference system of the International Earth Rotation Service, ICRS, *Astron. Astrophys.*, **303**, pp. 604–608.

Dehant, V., Feissel, M., Defraigne, P., Roosbeek, F. and Souchay, J. (1997) Could the energy near the FCN and the FICN be explained by luni-solar or atmospheric forcing?, *Geophys. J. Int.*, **130**, pp. 535–546.

Herring, T.A. (1991) The ZMOA-1990 nutation series, in "Reference Systems", Proceedings of the IAU Colloquium 127, eds A. Hugues, C. Smith, G. Kaplan, U.S. Naval Observatory, Washington, pp. 175–166.

IERS (1997) 1996 IERS Annual Report, Observatoire de Paris, Paris, pp. II-38.

Jacobs, C.S., Sovers, O.J., Gordon, D., Ma, C. and Gontier, A.-M. (1997) The accuracy of the ICRF: an intercomparison of VLBI analysis software, JD 7 this volume.

Kovalevsky, J. and McCarthy, D.D. (1997) Astronomical effects of current changes in fundamental astrometric references, JD 3 this volume.

Lieske, J.H., Lederle, T., Fricke, W. and Morando, B. (1977) Expressions for the Precession Quantities Based upon the IAU (1976) System of Astronomical Constants, *Astron. Astrophys.*, **58**, pp. 1–16.

Ma, C., Arias, E.F., Eubanks., T.M., Fey, A.L., Gontier, A.-M., Jacobs, C.S. and Sovers, O.J. (1997) Formation of the ICRF, JD 7 this volume.

Ma, C. and Baver, K.D. (1997) Site Positions and Velocities, and Earth Orientation Parameters from the NASA Space Geodesy Program-GSFC: Solutions GLB1069 and GLB1070,*1996 IERS Annual Report*, description available under anonymous ftp: hpiers.obspm.fr file iers/annreports/1996AR/gsfc.r

Mathews, P.M. and Herring, T.A. (1997) Observational and Theoretical Modeling of Nutation, JD 3, this volume.

McCarthy, D.D. (ed) (1996), *IERS Conventions*, IERS technical Note, **21**, Observatoire de Paris, Paris, pp. 20–39.

Seidelmann, P.K. (1982) 1980 IAU Theory of Nutation: The Final Report of the IAU Working Group on Nutation, *Celestial Mechanics*, **27**, pp. 79–106.

RELATIVISTIC CONSIDERATIONS FOR PRECESSION AND NUTATION

S.A. KLIONER AND M. SOFFEL
Lohrmann Observatorium, Technische Universität Dresden, Mommsenstraße 13, D-01062 Dresden, Germany

Abstract. The whole scope of problems related with the rotational motion of celestial bodies is briefly discussed. Relativistic modeling of the Earth rotation is considered from a conceptual point of view. Relativistic effects in rotational equations of motion of an extended body in general relativity are discussed. Numerical values of the effects are given.

1. What do we call Earth rotation and how do we model it?

Definition of rotational motion of an extended body being fairly simple in Newtonian mechanics is not obvious at all in the framework of general relativity. Theoretically speaking the concept of a precessing extended *rigid* body in general relativity encounters fundamental difficulties and cannot be introduced even in the first post-Newtonian approximation. From a practical point of view, however, the rotational motion of the Earth even at the Newtonian level is defined operationally through the time-dependence of geocentric quasi-inertial coordinates of observing sites. An analogous operational definition can be applied in general relativity. To this end, we need a physically adequate geocentric reference system to model local physics of the Earth. Then time-dependence of coordinates of observing sites relative to that geocentric reference system gives relativistic operational definition of the rotation of the Earth.

2. Relativistic reference systems

Nowadays there are two well developed formalisms for the construction of relativistic astronomical reference systems: the Brumberg-Kopeikin formalism (see, e.g., Brumberg, 1991) and the DSX formalism (Damour, Soffel, Xu, 1991, 1992, 1993). Although the two approaches look quite different at first glance, the main serious difference between them is the concept of mass multipole moments of some gravitating body. Here, the DSX formalism makes use of the so-called Blanchet-Damour moments that have direct physical significance for an isolated body. It is mainly because of this fact that the DSX scheme is superior to the Brumberg-Kopeikin formalism.

The two principal reference systems needed to model the Earth rotation are the barycentric reference system of the solar system and the local geocentric reference system.

The barycentric reference system is natural to model translational motion of the center of mass of the Sun and planets as well as light propagation from the distant sources (planets, stars or quasars) to the observers in the solar system. This reference system is used, for example, to construct the JPL planetary ephemerides and to define the International Celestial Reference System (ICRS).

In principle, any reference system covering a region of space-time under consideration can be used to model any kind of physical phenomena in that region. However, in some reference systems the mathematical description of physical processes is simpler than in others. Moreover, relativistic models involve a set of parameters (say, masses of planets, coordinates of observing stations, coordinates of planets or satellites at some initial moment of time) which are usually defined only with respect to that reference system. Nevertheless, it is usually preferable if these parameters could be reasonably interpreted from a physical point of view. A local geocentric reference system is physically adequate to model physical phenomena occurring in the vicinity of the Earth. Influence of the external masses (the Sun, planets, etc.) reduces in this reference systems to tidal effects. This

J. Andersen (ed.), Highlights of Astronomy, Volume 11A, 173–176.

makes a mathematical description of local physical phenomena in this reference system especially simple. On the other hand, coordinate parameters defined in that local reference system could then be physically interpreted at some reasonable level of accuracy in terms of local physical phenomena and without the necessity to remember always their relativistic coordinate nature (e.g., time dependence of the geocentric coordinates of observing sites can be interpreted as a result of the rotational motion of the Earth and some geophysical drifts). The local geocentric reference system is used to describe translational motions of Earth's satellites, to model Earth rotation (including precession-nutation angles), and therefore, to define the International Terrestrial Reference System.

Both reference systems are defined by the structure of their metric tensors. Coordinate transformations between the reference systems are indispensable to relate coordinates of objects, bodies and observers defined in those two reference systems and are given in detail in the formalisms.

3. Precession-nutation angles

Newtonian theory of Earth rotation heavily uses various angles between various planes and lines, and their time-dependence to describe the Earth's rotation. The concept of a plane, a line and an angle between them can be meaningfully introduced only in Newtonian physics based upon Euclidean space. Curved pseudo-Riemannian manifolds that we are dealing with in general relativity do not allow us to consider planes, lines and angles in a physically meaningful way. However, forgetting about physical meaning we can use three space coordinates of some reference systems to introduce "coordinate planes", "coordinates lines" and "coordinates angles" formally in the same way as we do in Euclidean geometry. Those concepts being defined only in one reference system can still be useful from a formal mathematical point of view.

By its construction the local geocentric reference system is physically adequate for modeling of any physical processes occurring in the vicinity of the Earth, and in particular, for modeling the rotation of the Earth itself. In this geocentric reference system (T, X, Y, Z) for the sake of continuity the coordinate plane ($T = T_0$, $Z = 0$) may , e.g., be related with some equator at some epoch T_0. The vectors of the post-Newtonian spin, angular velocity or figure axes (see below) then define the precession-nutation angles in that system. The coordinate transformations between the geocentric and barycentric reference systems enable one to define the orientation of the Earth with respect to distant celestial objects.

On the other hand, the concept of a coordinate plane has fundamental restrictions as compared to its Newtonian (Euclidean) counterpart. Thus, a coordinate plane is defined in one specific reference system. If we transform that plane into some other reference system it would be no longer a coordinate plane in that other reference system. Some agreements could be adopted to work with coordinate planes defined in different reference systems (see, e.g., Brumberg, Bretagnon, Guinot, 1996). However, for example, the concept of time-dependent plane (say, moving ecliptic) *cannot* be defined with the post-Newtonian accuracy in a reasonable way.

4. Rotational equations of motion

In the local geocentric reference system one can derive rotational equations of motion of the Earth, which take the same functional form as in Newtonian physics

$$\frac{d}{dt} S^i = L^i + \mathcal{O}(c^{-4}), \tag{1}$$

where S^i is the post-Newtonian spin, and L^i is the post-Newtonian tidal torque (see, e.g., Damour, Soffel, Xu, 1993 and Klioner, 1996 for details). These equations are valid without any assumption on shape and composition of the Earth. Both spin and torque coincide with their Newtonian counterparts in the Newtonian limit. The torque L^i vanishes for case of an isolated body. Both S^i and L^i contain many relativistic contributions. However, in practice we determine the value of S^i directly from observations rather than compute it from the distribution of density, pressure, etc. within the Earth. Therefore, only the contributions in the torque are important for practical applications. One can show (see, Damour, Soffel, Xu, 1993) that L^i can be expanded as

$$L^i = \sum_{l=0}^{\infty} \frac{1}{l!} \left(\epsilon_{iab} M_{aL} G_{bL} + \frac{1}{c^2} \frac{l+1}{l+2} \epsilon_{iab} S_{aL} H_{bL} \right), \tag{2}$$

where $L = i_1 i_2 \dots i_l$ is the multiindex, M_L are the Blanchet-Damour mass multipole moments, S_L are spin moments needed to Newtonian accuracy only, and G_L and H_L are the electric- and magnetic type tidal moments of external gravitational field. Both M_L and S_L are parameters of the gravitational field of the Earth and should be determined from observations. The moments G_L and H_L can be computed from planetary and lunar ephemerides. For practical purposes the expansion (2) can be truncated as usual considering magnitudes of the multipole and tidal moments.

One should note, that the post-Newtonian spin and the rotational equations of motion (1) are not unique (see, Damour, Soffel, Xu, 1993 for a discussion). Therefore, in practice one has to adopt a definition of spin by some agreement. Say, the form of expansion (2) of the torque L^i makes the definition of S^i unique.

5. Angular velocity and tensor of inertia

Eq. (1) is sufficient to discuss precession and nutation of the spin. For many reasons it is also important to consider precession and nutation of the angular velocity and figure axis. Therefore, we need a relativistic definition of the tensor of inertia and angular velocity.

A variety of theoretical approaches can be followed to get the same post-Newtonian definition of the tensor of inertia C^{ij} (from restricted rigid body models (Soffel, 1994) to relativistic Tisserand-like axes of a deformable Earth (Klioner, 1996)). This allows one to split the spin S^i into a product of the tensor of inertia and corresponding angular velocity

$$S^i = C^{ij}\, \omega^j. \tag{3}$$

The post-Newtonian tensor of inertia C^{ij} is defined as integral over the volume of the body. Again, although the definition of C^{ij} contains a number of explicit relativistic terms usually we do not compute C^{ij} from the distribution of density, pressure, etc. within the Earth, but determine the values of C^{ij} from observations. Therefore, for practical purposes we can simply use the fact that the spin S^i can be represented in the form given in (3).

The definition of angular velocity of a deformable body is not unique already in Newtonian mechanics (one can consider Tisserand axes and principal axes of inertia which give two different definitions of angular velocity). This ambiguity is aggravated in general relativity by the ambiguity in the definition of the relativistic tensor of inertia C^{ij} for the non-isolated Earth. Therefore, one more agreement is needed.

6. Rigid-multipole model

A rigid Earth plays a very important role as a first order approximation in the Newtonian theory of Earth's rotation. This rigid model (1) crucially simplifies the mathematical description of the rotational motion and (2) is not too far from reality, so that the effect of non-rigidity can be than added to the model by means of perturbation theory. The reason why the rigid body model substantially simplifies the rotational equations of motion is that both the mass multipole moments $M_L^{\rm Newt}$ and the tensor of inertia $C^{ij}_{\rm Newt}$ rotate rigidly. In other word there exist a rigidly rotating reference system $Y^i = P^{ij}(T)\, X^j$, where $P^{ij}(t)$ is some time-dependent orthogonal matrix, where both $M_L^{\rm Newt}$ and $C^{ij}_{\rm Newt}$ are constant.

This nice feature can obviously be saved in the post-Newtonian framework. One might define "rigid-multipole" models by means of a series of *assumptions*

$$C^{ij}(t) = P^{ia}\, P^{jb}\, \overline{C}^{ab}, \qquad \overline{C}^{ij} = \mathrm{diag}(\mathcal{A}, \mathcal{B}, \mathcal{C}), \quad \mathcal{A}, \mathcal{B}, \mathcal{C} = \mathrm{const}, \tag{4}$$

$$M_{i_1 i_2 \dots i_l} = P^{i_1 a_1}\, P^{i_2 a_2} \dots P^{i_l a_l}\, \overline{M}_{a_1 a_2 \dots a_l}, \qquad \overline{M}_{a_1 a_2 \dots a_l} = \mathrm{const}, \quad l \geq 2, \tag{5}$$

$$S_{i_1 i_2 \dots i_l} = C^{k i_1 i_2 \dots i_l}\, \omega^k, \quad l \geq 2, \tag{6}$$

$$C^{k i_1 i_2 \dots i_l} = P^{kj}\, P^{i_1 a_1}\, P^{i_2 a_2} \dots P^{i_l a_l}\, \overline{C}^{j a_1 a_2 \dots a_l}, \qquad \overline{C}_{a_1 a_2 \dots a_l} = \mathrm{const}, \quad l \geq 2, \tag{7}$$

and the orthogonal matrix $P^{ij}(t)$ is related to the angular velocity ω^i from (3)

$$\omega^i(t) = \frac{1}{2}\,\varepsilon_{ijk}\,P^{mj}(t)\,\dot{P}^{mk}(t). \tag{8}$$

Relations (6)–(7) for the higher spin moments $S_L, l \geq 2$ and for C_{iL} are only necessary to Newtonian accuracy since they appear only in relativistic terms of (2).

Note, that in Newtonian theory we derive or prove (4)–(8) from the fundamental property of rigidity of the body. Here, on the contrary, in general relativity we define the model by *assuming* the properties of C^{ij}, M_L and S_L not considering some more fundamental property of the body.

The experience of the Newtonian models of Earth's rotation shows that the phenomenological model (4)–(8) can be used as a first-order approximation for a description of the *global* rotational motion of the Earth. As in Newtonian theory, the model is a basis for considering the effects of non-rigidity in the rotation of the Earth *as a whole*. However, the model does not constrain at all *internal* structure and dynamics of the Earth and completely insufficient for, e.g., a post-Newtonian theory of elasticity, which still waits for a rigorous treatment.

7. Observational consequences

There are three main observational consequences of general relativity in the rotation of the Earth:

1. In practice we prefer to use so-called kinematically nonrotating geocentric reference system which rotates relative to a locally inertial reference system. This results in Coriolis and centrifugal forces in the right-hand side of (1). These forces gives additional effects in the orientation of the spin of the Earth which are called geodetic precession ($\sim 1.9''$ per century) and nutation (with an amplitude of 0.15 mas and a period of one year). (see, e.g., Bizouard, *et al.*, 1992 for details).

2. Explicit relativistic terms in L^i results in additional periodic relativistic effects amounting to ~ 1 microarcsecond (see, Bizouard, *et al.*, 1992). It is to note, however, that the analogous effect in the rotation of the Moon is 2-3 order of magnitude larger and may amount to ~ 1mas (see, Bois, Vockrouhlicky, 1995).

3. The key relation of Newtonian mechanics

$$\left(M^{ij}\right)_{\text{Newton}} = \left(\frac{1}{3}\,\delta^{ij}\,C^{ss} - C^{ij}\right)_{\text{Newton}} \tag{9}$$

relating the quadrupole moment of the gravitational field M^{ij} and the tensor of inertia C^{ij} can be shown to be violated in general relativity. This means that, for example, J_2, related to M^{ij}, cannot be easily related to the dynamical ellipticity H, which is derived from C^{ij}, the effect being of order of $\delta J_2/J_2 \sim 10^{-9}$ (see, Klioner, 1997).

References

Bizouard C., Schastok, J., Soffel M.H. and Souchay J. (1992) Etude de la rotation de la Terre dans le cadre de la relativité générale: premiere approche. In: *Proc. Journées Systèmes de Référence Spatio-temporels 1992*, ed. N. Capitaine, Observatoire de Paris, pp. 76.

Bois, E. and Vockrouhlicky, D. (1995) Relativistic spin effects in the Earth-Moon system. *Astron. Astrophys.*, **300**, pp. 559.

Brumberg V.A. (1991) *Essential Relativistic Celestial Mechanics.* Adam Hilger, Bristol.

Brumberg V.A., Bretagnon P. and Guinot, B. (1996) Astronomical Units and Constants in the General Relativity Framework. *Celest. Mech. Dyn. Astron.*, **64**, pp. 231.

Damour T., Soffel M. and Xu Ch. (1991) General-Relativistic Celestial Mechanics. *Phys. Rev. D.*, **43**, pp. 3273.

Damour T., Soffel M. and Xu Ch. (1992) General-Relativistic Celestial Mechanics. *Phys. Rev. D.*, **45**, pp. 1017.

Damour T., Soffel M. and Xu Ch. (1993) General-Relativistic Celestial Mechanics. *Phys. Rev. D.*, **47**, pp. 3124.

Klioner, S.A. (1996) Angular Velocity of Rotation of Extended Bodies in General Relativity. In: *Dynamics, ephemerides and astrometry in the solar system*, eds. S. Ferraz-Mello, B. Morando and J.E. Arlot, Kluwer, Dordrecht, pp. 309.

Klioner, S.A. (1997) On the problem of post-Newtonian Rotational Motion. In: *Dynamics and Astrometry of Natural and Artificial Celestial Bodies*, eds. I.M. Wytrzyszczak, J.H. Lieske and R.A. Feldman, Kluwer, Dordrecht, pp. 383.

Soffel, M. (1994) The problem of rotational motion and rigid bodies in the post-Newtonian framework. *unpublished notes*.

OBSERVATIONAL AND THEORETICAL MODELING OF NUTATION

P. M. MATHEWS
Department of Physics, University of Madras, India

AND

T. A. HERRING
Department of Earth, Atmospheric and Planetary Sciences, MIT, USA

1. Observational Nutation Results: IERS 1996 Conventions series

The IERS nutation series was developed as a nutation series which would match observations and could be used with geodetic systems which have sensitivity to nutations but are not able to determine accurately corrections to the IAU-1980 nutation series. After experimentation with different methods for parameterizing the nutation estimation problem, the best fit to the VLBI observed nutation angle data was found by estimating selected coefficients in the Mathews *et al.* (1991a) nutation response function (see equation (1) below), along with other parameters to be discussed below. The estimation of the response function coefficients was also found to account for mantle anelasticity and the loading part of the ocean tide correction and therefore no corrections were applied for these effects. Their contributions should be absorbed into the estimated response coefficients. These initial analyses also showed: setting the inner-core response to zero yielded better fits to the nutation angle data than the Mathews *et al.* (1991b) theoretical response and resonance frequency; the amplitude of the RFCN free mode appeared to be time dependent; and the prograde annual nutation was significantly affected by the S1 thermal tide which is not proportional to the S_1 gravitational potential.

The data set used to estimate the IERS conventions series was the GSFC nutation angle data set with 2040 pairs of adjustments to the IAU 1980 nutations in obliquity and longitude spanning the interval between 1979 and 1995. Here we also compare the IERS conventions series with a US Naval Observatory analysis of VLBI data between 1979 and August 1997. This data set consists of 2152 pairs of nutation angle estimates and extends by 1.8 years the interval covered by the GSFC data set. It also contains less pre-1984 data than the GSFC data set which accounts for the smaller increase in total number of observations than would be expected for the additional time interval covered.

The parameters estimated to model the GSFC data were R, R' and the retrograde Free Core Nutation (FCN) resonance strength R_{FCN} and resonance frequency σ_{FCN}. All these parameters were estimated as complex quantities. In addition, we estimated linear rates of change of the longitude and obliquity, the in- and out-of-phase components of prograde annual nutation, and free terms with the frequency of the RFCN at two year intervals after 1984. One free term was estimated for all the data before 1984 because the density and quality of the data prior to 1984 was not sufficient to support the 2-year estimates. The fundamental arguments of Simon *et al.* (1994) were used with the rigid-Earth nutation series, along with planetary nutations, of Souchay and Kinoshita (1996) (electronic version of October 23, 1995). An additional correction was applied due to annual modulation of the geodetic precession (Fukushima, 1991).

Table 1 gives the differences for the largest nutations between the IERS series and the IAU-1980 series and the analysis of the USNO data set. In general, the agreement between the USNO analysis and the IERS series is consistent with the uncertainties of the former. One exception to this agreement is the 13.66 day prograde nutation where the USNO analysis value differs from the GSFC value by 32 μas. This difference appears to be due to different treatment of the ocean-

J. Andersen (ed.), Highlights of Astronomy, Volume 11A, 177–181.

driven diurnal and semidiurnal rotation changes by the two groups. Recent GSFC analyses are more consistent with the USNO analysis suggesting that a common model in now in use.

2. Theoretical Models

Predictions of the amplitudes of the circular components of nutation are based on theoretical treatments of the nutations of an idealized (oceanless, elastic) ellipsoidal Earth model with axial symmetry under excitation by the spectral components of the tide generating potential (TGP), and of the contributions to the amplitudes from ocean tides, mantle anelasticity, and other mechanisms (e.g., electromagnetic and topographic couplings of the fluid outer core (FOC) to the mantle and to the solid inner core (SIC)).

A solid accomplishment from recent work is the attainment of microarcsecond (μas) level accuracy in the rigid Earth nutation series, needed for the understanding of nutation observations at current levels of precision. (The nutation of a rigid Earth is determined by the TGP and the Earth's dynamical ellipticity $e \equiv (C - A)/A$). Much effort has also gone into the modeling of deviations from hydrostatic equilibrium structure for the ellipsoidal Earth and into the computation of their effects on nutation.

Progress during the last several years in respect of other aspects of the theory has not, however, been significant. In regard to mantle anelasticity, the most commonly used models are based on a constant value of mantle Q from the 1 second reference period of PREM to some longer period T_s (say, 200 sec) and a $T^{-\alpha}$ dependence of Q for the range of periods from T_s to tidal. The values to be assigned to T_s and α are not well determined yet, and recent computations of anelasticity contributions to nutation amplitudes, using various plausible sets of values of T_s and α, follow the approach of Wahr and Bergen (1986). Estimation of the contributions from ocean tidal loading has not benefited much from the substantial improvements stemming from the use of TOPEX/POSEIDON data in the modeling of ocean tides, because no significant improvement has resulted therefrom in the precision of estimates of the amplitude (relevant to nutation) of the degree 2 tesseral part of the ocean tide function. The contributions from core-mantle boundary couplings will be considered further below.

2.1. RIGID EARTH NUTATION THEORY:

The amplitudes of spectral components of the rigid Earth's nutation, computed to 0.1 μas accuracy by independent methods by Bretagnon *et al.* (1997), Souchay and Kinoshita (1997), and Roosbeek and Dehant (1997), are in almost complete agreement. (Some of these have out-of-phase parts–a recent discovery, see Williams (1994).) The process of evaluation of the rigid Earth amplitudes involved consideration of a number of direct and indirect planetary effects, the effects of nutation on the TGP, the "J_2 tilt" and triaxiality effects, etc. These play an essential role in the accurate determination of the Earth's dynamical ellipticity e from the observed rate of precession of the equator. The values obtained for e by the different groups (and reflected in the respective nutation series) are in close agreement, and is just about 0.006% higher than the value used in the IAU 1980 theory of nutation. The new rigid Earth series have been used by Mathews and Buffett (unpublished, 1997) and by Dehant and Defraigne (1997) in their very recent computations for the deformable Earth with nonhydrostatic structure (see below).

2.2. ELLIPSOIDAL EARTH MODEL:

The e determined as indicated above is over 1% larger than the value for the ellipsoidal Earth model constructed (from PREM, for instance) under the assumption of a hydrostatic equilibrium structure; and the dynamical ellipticity e_f of the fluid core as inferred from the free core nutation (FCN) resonance period of about 430 days needed to explain the observed retrograde annual nutation amplitude (Herring *et al.*, 1986, Gwinn *et al.*, 1986) is nearly 5% larger than the hydrostatic equilibrium value. Such nonhydrostatic structure was taken account of in the analytic nutation theory of Mathews *et al.* (1991a,b) by simply using the values thus inferred for these ellipticity parameters. In recent work following this approach (Mathews and Buffett, unpublished, 1997), the ellipticity contributions to deformability parameters that also enter the analytical theory have been taken into account, along with the effects of the irrotational part of the core fluid flow on these parameters (Buffett *et al.*, 1993). In the treatment of nutation theory along the lines of Wahr (1981), deviations from hydrostatic equilibrium have to be introduced by modifying the sets of surfaces of equal density/rheology/potential throughout the mantle as well as the boundary surfaces (CMB, 670 km discontinuity). Defraigne (1996) has used the viscous forces associated with mantle convection flows to produce the nonhydrostatic values e and e_f (or rather, an equivalent flattening of the CMB) referred to above, as well as agreement with certain other types of observational data; nutation amplitudes based on this model have been computed by Dehant and Defraigne (1997). A nonhydrostatic structure having the desired e_f but not derived from a dynamical model has been employed by Schastok (1997) in computing nutation amplitudes. The amplitudes computed by these authors for a few of the most important nutations are compared in Table 1 with recent observational estimates.

TABLE 1. Theoretical predictions and observational estimates of nutation amplitudes: Differences from IAU 1980 in mas

Period		Theoretical		predictions	IERS1996	Observed
(Days)		DD	S	MB	H	H
13.66	*ip*	−0.18	−0.14	−0.060	−0.145	−0.113 ± 0.006
	op	−0.14	−0.17	−0.023	−0.130	−0.139 ± 0.006
− 13.66	*ip*	−0.02	−0.02	−0.023	−0.019	−0.014 ± 0.006
	op	0.01	0.02	0.007	0.015	0.032 ± 0.006
182.62	*ip*	0.67	0.67	0.644	0.606	0.593 ± 0.006
	op	0.45	0.49	0.554	0.510	0.514 ± 0.006
− 182.62	*ip*	−0.02	−0.02	−0.040	−0.065	−0.050 ± 0.006
	op	0.05	0.07	0.034	0.047	0.048 ± 0.006
365.26	*ip*	−0.02	0.07	0.036	−0.012	−0.017 ± 0.006
	op	−0.03	−0.02	−0.022	−0.124	−0.138 ± 0.006
− 365.26	*ip*	−1.98	−1.94	−1.984	−1.978	−1.976 ± 0.013
	op	0.12	0.11	−0.298	−0.322	−0.344 ± 0.006
6798.38	*ip*	−0.03	0.07	0.056	−0.099	−0.038 ± 0.027
	op	0.01	0.10	0.073	0.054	0.062 ± 0.006
−6798.38	*ip*	−2.34	−2.62	−2.740	−2.749	−2.782 ± 0.027
	op	−0.86	−1.01	−1.324	−1.490	−1.524 ± 0.006

DD: Dehant and Defraigne (1997), S: Schastok (1997), MB: Mathews and Buffett (unpublished, 1997), H: Estimates by Herring discussed above from VLBI data; *ip*: in-phase, *op*: out-of-phase. All the predictions include the respective authors' estimates of contributions from anelasticity and ocean tides. Ocean tidal *current* effects are ignored by MB, who include, however, Buffett's (1992) estimates of CMB electromagnetic coupling contributions. The sigmas are computed such that χ^2/f of the fit to the nutation angle data is unity where f is the degrees of freedom.

The near agreement of the predictions for the retrograde annual in-phase amplitude with the observed value in all three cases is the result of adjusting e_f or the equivalent CMB flattening to achieve such agreement. For a number of other frequencies, differences among the three predictions are rather large (tens to hundreds of μas), presumably due to differences in various aspects of the modeling. It is worth noting, however, that inclusion of the CMB electromagnetic coupling effects (Buffett, 1992) in the computations of Mathews and Buffett (unpublished, 1997) has helped not only to nearly eliminate the discrepancy of ≈ 0.45 mas found between the other authors' results for the retrograde annual out-of-phase amplitude and the observational estimate, but also to reduce the discrepancies in the 18.6 year amplitudes (both in-phase and out-of-phase). Nevertheless, overall agreement with the observational estimates cannot be considered satisfactory for any of the theoretical treatments presented.

2.3. INVERSE MODELING:

Least squares fitting of a formula derived from general theory to nutation amplitudes estimated from observations yields a "best fit" formula that can be used for prediction of celestial pole offsets, and the estimates obtained for parameters in the formula by the fitting procedure may provide useful constraints on Earth models. Resonance formulae for the transfer function $T(\sigma)$ from the nutation amplitude $\tilde{\eta}_{rigid}(\sigma)$ for a rigid Earth to $\tilde{\eta}(\sigma)$ for a deformable Earth are most convenient for such inverse modeling.

The resonance formula due to Mathews *et al.* (1991a), good for an oceanless Earth, is

$$T(\sigma) \equiv \frac{\tilde{\eta}(\sigma)}{\tilde{\eta}_{rigid}(\sigma)} = R + R'(1+\sigma) + \sum_\alpha \frac{R_\alpha}{\sigma - \sigma_\alpha}, \tag{1}$$

wherein σ is the tidal excitation frequency, and the sum is over the relevant normal modes: the Chandler wobble (CW), the retrograde free core nutation (FCN), and the inner-core-induced prograde free core nutation (PFCN). (The frequency dependence of the ocean tide response to the TGP complicates the frequency dependence of $T(\sigma)$ for an Earth with oceans.) Sum rules following from the general theory connect R and R' to the resonance frequencies σ_α and coefficients R_α. (All frequencies are in cycles per sidereal day (cpsd)). One may obtain the following reduced resonance formula by using these sum rules:

$$T(\sigma, e\,|\,e_0) \equiv \frac{\tilde{\eta}(\sigma, e)}{\tilde{\eta}_{rigid}(\sigma, e_0)} = (e_0 - \sigma)\left(\frac{e}{e_0}\right)\left\{\frac{1}{1+e} + (1+\sigma)\sum_\alpha \frac{N_\alpha}{\sigma - \sigma_\alpha}\right\}, \tag{2}$$

where e_0 is the value used for e in the rigid Earth nutation series employed, and the possibility is left open that the actual e for the Earth might be slightly different. Estimates obtained for the resonance parameters N_α and σ_α may be used, in turn, to estimate or constrain ellipticity, deformability, and other basic parameters of the Earth and its core regions, since analytical expressions in terms of such basic parameters which give close approximations to the N_α and σ_α are available.

A least squares fit (carried out by Mathews) of the reduced resonance formula to VLBI estimates of the in- and out-of-phase amplitudes of 16 prograde-retrograde pairs of nutations and of the precession rate (obtained by Herring), has yielded estimates for the first time for a number of parameters besides the FCN period. The results given below were obtained from a fit done after removal of estimated anelasticity, ocean loading, and CMB electromagnetic coupling contributions from the nutation amplitude data before fitting. They may therefore be compared directly to values computed for standard Earth models. (The CMB coupling correction significantly improved the fit.) The estimates obtained were: $e = (32845520 \pm 16) \times 10^{-10}$; the coefficient of the PFCN resonance $N_{PFCN} = (191 \pm 37) \times 10^{-6}$, indicating a firm detection of this inner-core-induced mode with $\sigma_{PFCN} = -0.99882 \pm 0.00095$ corresponding to a best estimate of 844 days (with a rather large uncertainty) for the PFCN period in space; $N_{FCN} = (5155 \pm 14) \times 10^{-5}$; FCN period $= -433.0 \pm 0.3$ days, close to the period inferred by Herring et al. (1986) from the retrograde annual nutation alone; Chandler wobble period $= 390.9 \pm 0.9$ days. Inferences about the Earth's properties may be drawn from the estimates with the aid of approximate analytical expressions for the resonance parameters: $\sigma_{CW} = -eN_{CW} = (A/A_m)(e - \kappa)$; $\sigma_{FCN} = -1 - (A/A_m)(e_f - \beta)$; $N_{FCN} = (A_f/A_m)(1 - \gamma/e)$; $\sigma_{PFCN} = e_s\{1 - (P + Q\rho_f^{ICB} e_s)(1 - R\rho_f^{ICB})\}$, where P, Q, R are certain parameters of a standard Earth, A, A_m, A_f are moments of inertia, and e, e_f, e_s dynamical ellipticities, of the Earth and of its mantle, fluid outer core, and solid inner core regions (indicated by subscripts m, f, s, respectively), and κ, β, γ are deformability parameters (see Mathews *et al.*, 1991a). In particular, the 844 day period found for the PFCN, which contrasts with the computed value (476 days) based on PREM under hydrostatic equilibrium, would suggest a lower ellipticity for the inner core or a larger density contrast across the inner core boundary (ICB) than in PREM, apart from a possible role for magnetic coupling across the ICB. The estimates for the other resonance parameters confirm the already known excess ellipticity of the fluid core as well as of the whole Earth; other small deviations from PREM (hydrostatic) of κ and γ are indicated, but it is not clear that they are statistically significant.

The postfit residuals when only the in-phase parts of the amplitudes are fitted are small, with an rms of 12 μas. When the out-of-phase parts are also sought to be fitted by allowing the resonance parameters to be complex, the rms of residuals doubles, essentially because of a few large out-of-phase residuals: about 120 μas in the prograde annual which is almost certainly due to the S_1 thermal atmospheric tide, and a similar residual in the 18.6 year prograde and about 40 μas in the 9.3 year prograde which need to be explained. The unreduced resonance formula which involves two extra adjustable parameters provides a slightly better fit (rms of residuals 21 μas) and may be used where the best fit is the only objective; however, the estimates obtained for the parameters in this formula which does not respect the sum rules are not suitable for geophysical interpretation.

3. Concluding remarks

Vast improvement has taken place over the past few years in the precision of nutation data obtained from VLBI observations, especially in respect of the long period (18.6 years and 9.3 years) components; and the demands placed thereby on the accuracy of the rigid Earth nutation series for the purpose of geophysical modeling have been essentially met. Parameter estimates from inverse modeling of the observed nutations have begun to yield more information about the Earth's deep interior than was available heretofore, and suggest, in particular, a closer look at the electromagnetic couplings at the CMB and the ICB. The efforts to improve forward modeling may well be helped by such insights, and are also likely to help constrain the considerable freedom which remains in the modeling of mantle anelasticity. Theoretical evaluation of atmospheric effects on nutations remains an unfinished task, as does the reduction of uncertainties in estimates of ocean tide effects including, in particular, the effect of ocean tidal currents. Finally, points of concern regarding the amplitudes of long period nutations, especially their out-of-phase parts, need to be resolved.

References

Bretagnon, P., Rocher, P. and Simon, J.-L. (1997) Theory of rotation of the rigid Earth. *Astron. Astrophys.*, **319**, pp. 305–317.

Buffett, B.A. (1992) Constraints on magnetic energy and mantle conductivity from the forced nutations of the Earth. *J. Geophys. Res.*, **97**, pp. 19,581–19,597.

Buffett, B.A., Mathews, P.M., Herring, T.A. and Shapiro, I.I. (1993) Forced nutations of the Earth: Contributions from the effects of ellipticity and rotation on the elastic deformations. *J. Geophys. Res.*, **98**, pp. 21,659–21,676.

Defraigne, P. (1997) Geophysical model of the dynamical flattening of the Earth in agreement with the precession constant. *Geophys. J. Int.*, **130**, pp. 47–56.

Dehant, V. and Defraigne, P. (1997) New transfer functions for nutations of a non-rigid Earth. *J. Geophys. Res.*, accepted.

Fukushima, T. (1991) Effects of geodesic precessions annual effect. *Astron. Astrophys.*, **244**, pp. L11.

Gwinn, C.R., Herring, T.A. and Shapiro, I.I. (1986) Geodesy by radio interferometry: Studies of the forced nutations of the Earth, 2. Interpretation. *J. Geophys. Res.*, **91**, pp. 4755–4765.

Herring, T.A., Buffett, B.A., Mathews, P.M. and Shapiro, I.I. (1991) Forced nutations of the Earth: Influence of inner core dynamics, III. Data analysis. *J. Geophys. Res.*, **96**, pp. 8259–8273.

Herring, T.A., Gwinn, C.R. and Shapiro, I.I. (1986) Geodesy by radio interferometry: Studies of forced nutations of the Earth, 1. Data analysis, *J. Geophys. Res.*, **91**, pp. 4745–4754.

Mathews, P.M., Buffett, B.A., Herring, T.A. and Shapiro, I.I. (1991a) Forced nutations of the Earth: Influence of inner core dynamics, I. Theory. *J. Geophys. Res.*, **96**, pp. 8219–8242.

Mathews, P.M., Buffett, B.A., Herring, T.A. and Shapiro, I.I. (1991b) Forced nutations of the Earth: Influence of inner core dynamics, II. Numerical results and comparisons. *J. Geophys. Res.*, **96**, pp. 8243–8257.

Roosbeek, F. and Dehant, V. (1997) RDAN97: An analytical development of the rigid Earth nutation series using the torque approach. *Celest. Mech.*, submitted.

Schastok, J. (1997) A new nutation series for a more realistic model Earth. *Geophys. J. Int.*, **130**,pp. 137–150.

Simon, J.-L., Bretagnon, P., Chapront, J., Chapront-Touzé M., Francou, G. and Laskar, J. (1994) Numerical Expressions for Precession Formulae and Mean Elements for the Moon and Planets. *Astron. Astrophys.*, **282**, pp. 663–683.

Souchay, J. and Kinoshita, H. (1996) Corrections and new developments in rigid-Earth nutation theory, I. Lunisolar influence including indirect planetary effects. *Astron. Astrophys.*, **312**, pp. 1017–1030.

Souchay, J. and Kinoshita, H. (1997) Corrections and new developments in rigid-Earth nutation theory, II. Influence of second-order geopotential and direct planetary effect. *Astron. Astrophys.*, **318**, pp. 639–652.

Wahr, J.M. (1981) The forced nutations of an elliptical, rotating, elastic, and oceanless Earth. *Geophys. J. R. Astron. Soc.*, **64**, pp. 705–727.

Wahr, J.M. and Bergen, Z. (1986) The effects of mantle anelasticity on nutations, Earth tides and tidal variations in rotation rate, *Geophys. J. R. Astron. Soc.*, **87**, pp. 633–668.

Williams, J.G. (1994) Contributions to the Earth's obliquity rate, precession, and nutation. *Astron. J.*, **108**, pp. 711–724.

ASTRONOMICAL EFFECTS OF CURRENT CHANGES IN FUNDAMENTAL ASTROMETRIC REFERENCES

J. KOVALEVSKY

OCA/CERGA - UMR/CNRS 6527
Av. Copernic, F 06130–Grasse, France

AND

D. MCCARTHY

U.S. Naval Observatory
3450 Massachusetts Ave, Washington DC 20390, USA

Abstract. We present a first description of the changes in the current procedures that will have to be performed in order to take into account the adoption of the ICRS as the new reference system and new nutation-precession series. A number of choices will have to be made: they are presented in this paper.

1. Introduction

Although this Joint Discussion concerns the specific problem of a new nutation model associated with an improved constant of precession, it is not possible to separate these changes from the simultaneous changes occurring with the adoption of the extragalactic reference system ICRS and its realizations in radio as well as in visual wavelengths (ICRF and Hipparcos catalogue). At this time, there is a fundamental change in the concepts, the consequences of which have not yet been fully analyzed, and this paper is a contribution to an intellectual effort that should be undertaken in the next few years to establish new procedures that astronomers will have to adopt to reduce astrometric observations.

Starting from the beginning of the 21st century, the situation will be the following:

• The reference system to which all positions and motions should be referred will be the International Celestial Reference System (ICRS) (Ma and Feissel, 1997) which is based upon the assumption that positions of distant extragalactic objects have no global rotation. The corresponding coordinate axes will be fixed in space and close to the FK5 J2000 equator and equinox within the intrinsic errors of the latter.

• The new nutation series will be given in the new reference system.

• The precession formulae will consist of the additional terms to be added to the nutation in order to describe the motion of the celestial pole in the ICRS.

• The complicated situation which arises from the different realizations of the equinox (dynamical or FK5 catalog equinox) will disappear, and the equinox will be replaced by the single fixed origin of the International Celestial Reference Frame (ICRF).

Let us comment upon the practical consequences of these changes.

J. Andersen (ed.), Highlights of Astronomy, Volume 11A, 182–186.

2. Change of fundamental catalog

The transformation from the FK5 catalog to Hipparcos cannot be reduced to a couple of rotations, one at epoch and another proportional to time. At the level of accuracy that is reached, the regional errors of FK5 are significant and, depending upon how they are modeled, the result of the rotations are different. The transformation formulae must also include some developments as functions of right ascension and declination to be applied to the coordinates of FK5 in order to obtain them in the ICRS. However, such a work remains to be finalized. The difficulty encountered is that the results depend significantly with the choice of the stars entering into the comparison. In particular, double stars and stars presenting a non-uniform proper motion should be excluded. Some stars may even not have been identified as such, but the proper motion in FK5, obtained from observations over several decades, may be different from the quasi-instantaneous proper motion obtained by Hipparcos, which may include some effect due to an orbital motion smoothed out in the FK5.

In any case, even if, despite considerable regional effects, one is looking for only the global orientation and spin differences between Hipparcos and FK5 catalogs, one must introduce in their determination, some representation of regional errors. This is what was done by Mignard and Froeschlé and the result is published in Volume 3 of the catalogue (ESA 1997, Section 19.2), with non-linear effects computed but not considered as sufficiently significant yet to be published.

The results are the following :

1 - Orientation of the FK5 in the Hipparcos system referred to ICRS

$$\varepsilon_x = -18.8 \pm 2.3 \text{ mas}$$

$$\varepsilon_y = -12.3 \pm 2.3 \text{ mas}$$

$$\varepsilon_z = +16.8 \pm 2.3 \text{ mas}$$

The transformation matrix from FK5 system to Hipparcos is

$$R_\varepsilon = \begin{pmatrix} 1 & \varepsilon_x & -\varepsilon_y \\ -\varepsilon_x & 1 & \varepsilon_z \\ \varepsilon_y & -\varepsilon_z & 1 \end{pmatrix}$$

2 - Spin of the FK5 system in the Hipparcos system referred to ICRS

$$\omega_x = -0.10 \pm 0.10 \text{ mas/year}$$

$$\omega_y = +0.43 \pm 0.10 \text{ mas/year}$$

$$\omega_z = +0.88 \pm 0.10 \text{ mas/year}$$

The rotation matrix R_ω is similarly at time t :

$$R_\omega = \begin{pmatrix} 0 & \omega_x & -\omega_y \\ -\omega_x & 0 & \omega_z \\ \omega_y & -\omega_z & 0 \end{pmatrix} (t - 1991.25)$$

It is useful, at this point, to remind that the Hipparcos system is the ICRS with an uncertainty of $\sigma_\varepsilon = \pm 0.60$ mas at epoch (1991.25) and $\sigma_\omega = \pm 0.25$ mas/year in spin.

It is to be noted, that the preceding values concern only the systems of coordinates. There are a large regional differences, which are in some cases larger than 100 mas in position and 4 mas/year in proper motions, so that these formulae should not be used to correct the positions of individual stars.

Similar formulae can be derived for the epoch J2000 although they will be less precise, due to the time degradation of the Hipparcos catalog and of the FK5. Similar techniques should be applied to other catalogs based upon the FK5, taking into account, when necessary, the formulae describing the difference between regional errors of the FK5 and the other catalog. For instance, for the PPM Catalogue, the results are the following (ESA 1997, Section 19.3) :

$$\varepsilon_x = -22.5 \pm 2.4 \text{ mas}$$

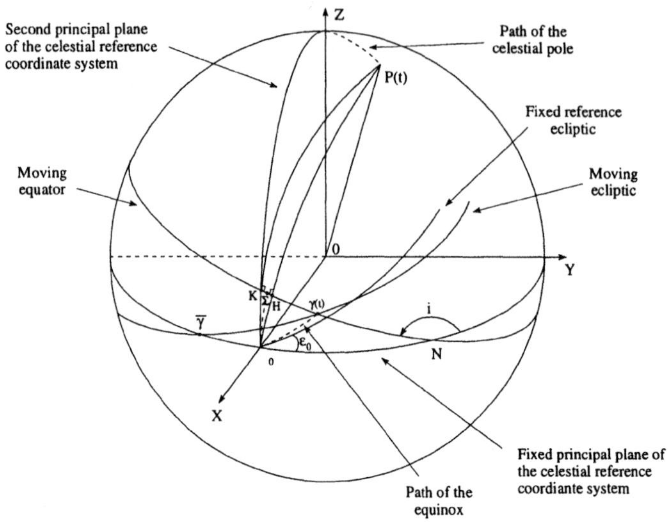

Figure 1. Possible origins of the equatorial frame. For the clarity of the figure, $t < 0$.

$$\varepsilon_y = -7.0 \pm 2.4 \text{ mas}$$
$$\varepsilon_z = +16.8 \pm 2.4 \text{ mas}$$
$$\omega_x = -0.66 \pm 0.07 \text{ mas/year}$$
$$\omega_y = +0.84 \pm 0.07 \text{ mas/year}$$
$$\omega_z = +0.14 \pm 0.07 \text{ mas/year}$$

3. The equator and the ecliptic

Due to precession, the celestial equator, defined as the plane perpendicular to the rotation axis of the Earth, will move freely in the new coordinate system and in the future, will tend to deviate increasingly from the fixed XOY reference plane (see Figure 1).

The ecliptic plays no role in the fundamental system. It may be useful still to give expressions for the position of the node $\bar{\gamma}$ of the ecliptic on the fixed XOY plane. It is, important to note, that it cannot be considered as an origin on the ecliptic. However, it is useful to have a fixed reference coordinate system close to the present ecliptic system for use in describing the motions in the solar system. The best choice might be that it is derived from the $OXYZ$ axes by a rotation around OX of an angle ε_0 equal to the mean obliquity of the ecliptic at J2000. This would mean that γ_0 is also the origin of the fixed solar system reference frame.

4. The true equatorial reference frame

For observations from the ground, one must use a *true equatorial reference frame* (TERF) defined by the orientation of the Earth in space. The system is defined by the true equator at the time of observation, perpendicular to the direction of the instantaneous axis of rotation of the Earth (the celestial pole). Another solution is to take the Celestial Ephemeris Pole. What follows does not depend upon the choice between these two poles. The position of the true equator may be defined by two angles:

- the angle between the origin γ_0 of the ICRF and the ascending node of the equator, N,
- the inclination i of the equator on the reference XOY plane.

A third angle will define the position of the origin of angles reckoned along the equator. Several definitions could be adopted for the position of this point.

Let us list five possibilities which may be proposed.

(a) The origin $\gamma(t)$ is the intersection of the moving ecliptic on the moving equator. This is the true equinox, which is not adequate because it is not defined with a precision comparable with the definition of the equator.

(b) The intersection K of the moving equator with the fixed XZ plane. It moves on the sky on this plane with an amplitude equal to ε, the obliquity of the ecliptic, with a period of 26000 years.

(c) The intersection H of the moving equator with the celestial meridian defined by the celestial pole $P(t)$ at time t and γ_0. With respect to ICRS, H moves in 26000 years on a curve whose amplitude is of the order of $2\pi(1 - \cos\varepsilon)$.

(d) A point Σ on true equator such that $\Sigma N = \gamma_0 N$.

(e) The point σ on the moving equator such that its instantaneous motion has no rotation component around the moving polar axis. This is the departure point (also presented as the non-rotating origin) which is such that no correction due to the system of coordinates is to be added to $d(UT)/dt$ in order to represent the absolute rate of rotation of the Earth. It is defined by

$$Q = \sigma N - \gamma_0 N = \frac{1}{2}\int_{t_0}^{t}\left(\frac{dx}{dt}y - x\frac{dy}{dt}\right)dt$$

where x, y and z are, in the ICRS, the direction cosines of the vector $OP(t)$ (Capitaine *et al.*, 1986). The quantity Q can be directly computed from the expressions of precession and nutation in the fixed reference system. All other choices of the origin but the last one, introduces spurious terms in UT1.

It is to be noted that for all solutions except the first, the ecliptic plays no role in the definition of the reference frame nor in the origin of the moving celestial coordinates attached to the true celestial pole.

5. Precession and nutation

These quantities referred to the ICRS are now representative of a single physical phenomenon: the motion of the Earth's pole on the sky. In particular, precession is no more corrupted by the motion of the ecliptic or the rotation of the Galaxy. As a consequence, there is no longer any reason to separate them and one should produce a single set of formulae with the short (nutation), long and secular (precession) terms of the motion of $P(t)$. If σ is chosen as the origin on the moving equator, the reference expression for Q should be provided together with the nutation-precession series. These quantities describe the relation R between the ICRF and the ITRF as a function of time. It would be also useful to provide consistent expressions of $\gamma_0 N, N\sigma$ and i.

Let us call $\psi = \gamma_0 N$ and $\phi = N\sigma$. Let S be a star with coordinates α, δ in the fixed system and α', δ' in the moving equatorial system with its origin in σ, the transformation from α, δ to α', δ' is

$$\begin{pmatrix} \cos\delta'\cos\alpha' \\ \cos\delta'\sin\alpha' \\ \sin\delta' \end{pmatrix} = R \begin{pmatrix} \cos\delta\cos\alpha \\ \cos\delta\sin\alpha \\ \sin\delta \end{pmatrix}$$

where R is given by

$$\begin{pmatrix} \cos\phi\cos\psi - \cos i\sin\phi\sin\psi & \cos\phi\sin\psi + \cos i\sin\phi\cos\psi & \sin i\sin\phi \\ -\sin\phi\cos\psi - \cos i\cos\phi\sin\psi & -\sin\phi\sin\psi + \cos i\cos\phi\cos\psi & \sin i\cos\phi \\ \sin i\sin\psi & -\sin i\cos\psi & \cos i \end{pmatrix}$$

Note : if the origin is not σ, but one of the points defined in (b), (c) or (d) in Section 4, this formula does not change, provided that ϕ is differently defined.

6. Sidereal time

The definition of sidereal time will depend upon the choice that is made for the origin on the moving equator. It must be consistent with the adopted definition of UT1. If the adopted origin of UT1 is the departure point σ, the sidereal time is strictly proportional to UT1 with no more secular terms introduced into the expression of the ratio between the length of the day and 24 hours of sidereal time, and in order to avoid misinterpretation, the expression *stellar angle* might be preferably used. Its definition would be the hour angle of the origin σ reckoned from the origin of terrestrial longitudes. Its derivative with respect to time is the angular velocity of the Earth around its polar axis.

7. Stellar positions

The positions of stars on a date t reckoned from J2000 in the ICRS have to be corrected only for proper motion and parallaxes as well as, if deemed useful, the light deflection and the perspective acceleration. The observation of a star from the Earth will provide coordinates in the ITRF. Applying the ITRF-ICRS transformation matrix, one will obtain directly its position in the fixed ICRS for the epoch of observation. Any change of epoch will be made applying only the above mentioned corrections.

Remark - One of the changes with respect to the present situation is that no intermediary *mean equinox of data* or *mean sidereal time* need to be introduced. With the advent of computers, the conventional relation R and its inverse will be available for the time t of observation and their application straightforward to go from the true equatorial system to the International Celestial Reference System and vice-versa.

8. Conclusions

The adoption of the ICRS will impose some requirements on users for consistency in the implementation of precise reference frames. It will also provide an opportunity to make changes that will simplify future use of reference systems.

It will be necessary, now, for users to allow for the corrections to the offsets of the celestial pole with respect to the current IAU adopted theories of precession and nutation. The procedure to transform between a terrestrial frame and the ICRS is not consistent with the current IAU theories at the level of a milliarcsecond. Adoption of the ICRS will also require that users of frames other than Hipparcos account for significant systematic errors in the positions and motions of objects in the non-ICRS frames.

Adoption of the ICRS will also allow us to dispose of the poorly defined dynamical equinox in the definition of astronomical reference systems. A more straightforward approach will be possible in the determination of the orientation of the astronomical reference systems with respect to terrestrial systems. A much more accurate and consistent theory of precession/nutation can also be adopted as well as an improved definition of sidereal time.

References

Capitaine, N., Guinot, B. and Souchay, J. (1986) A non-rotating origin on the instantaneous equator: definition, properties ans use. *Celest. Mech.*, **39**, pp. 283–307.

ESA (1997) *The Hipparcos and Tycho Catalogues*. ESA Publ. SP-1200, Vol. 3.

Ma, C. and Feissel, M. (eds) (1997) *IERS Technical Note*, **23**, Observatoire de Paris.

LATEST BEST ESTIMATES OF ASTRONOMICAL CONSTANTS

D.D. MCCARTHY
U.S. Naval Observatory
3450 Massachusetts Ave. NW
Washington, DC 20392-5420, USA

1. Abstract

Systematic errors caused by the use of inconsistent models and constants currently limit the accuracy of the most precise astro-geodetic observations. It is important, therefore, to provide conventional estimates of astronomical constants for general use in astronomy and geodesy. Establishment of such a set raises issues involving the accuracy of the constants, their internal consistency, and the reference system in which they are expressed. Future sets of constants will be updated at regular intervals to provide the most accurate values at that instant. Careful consideration must be given to their internal consistency. It is likely that they will be provided only in electronic format with references to their determination and to procedures for their use.

2. Introduction

Establishment of a conventional set of astronomical constants involves consideration of several issues.

Accuracy - Generally researchers require the most accurate estimates available. However, in some instances it may be more desirable to make use of estimates which may not be the most accurate but may be "conventional."

Consistency - Internal consistency of a set of constants is a desirable, if not critical, quality of a set of constants. Some researchers may require constants which are consistent with themselves but may not be the most accurate.

Frequency of updates - The question of how often the conventional set of constants should be updated is particularly important for researchers when it is now possible to make "best" estimates available instantaneously by electronic means. In some cases it is also possible to re-analyze large amounts of data rather quickly using revised constants and models. On the other hand, some may require constants which will be "stable" over an extended period of time.

Availability - Particularly for current best estimates, users need to have easy and fast access. It may also be useful to have access to explanatory information regarding the constants available at the same location.

Minimal Set - Rather than update a large set of constants, it may be more logical to provide a minimum set from which a more extensive set could be derived. If this course is followed it would be useful to provide proper procedures to be used in these derivations.

3. Requirement for constants

Modern observations making use of the most precise techniques such as very long baseline interferometry, laser ranging to satellites or the Moon, and GPS provide valuable information for astronomical reference systems, geodesy, and geophysics. Unfortunately, the analyses of these ob-

J. Andersen (ed.), Highlights of Astronomy, Volume 11A, 187–190.

Figure 1. Systematic Errors in x.

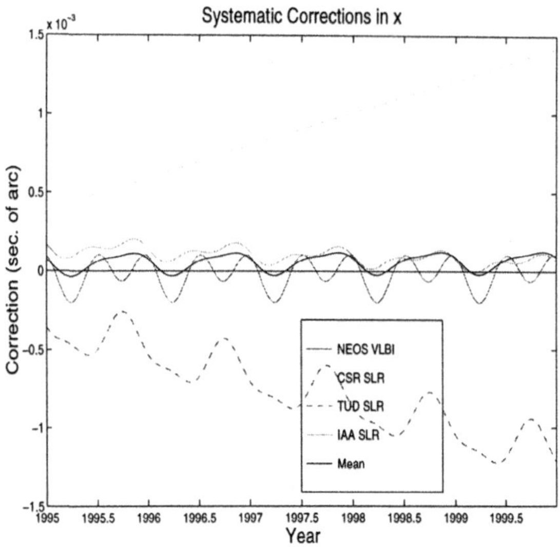

servations are limited by discrepancies in the models and constants used. One example of this fact is shown in Figure 1 which shows the large systematic differences among the estimates of the polar coordinate x by different analysis centers.

4. Relevant Constants

Version 1.3 of the 1986 CODATA Recommended Values of Physical Constants (Cohen and Taylor, 1986) is currently in widespread use. It is available electronically at

http://physics.nist.gov/funcon.html.

A new general adjustment of these values is currently in progress. This 1997 Adjustment is anticipated to be available after September, 1998.

The International Association of Geodesy (IAG) has established a Special Commission on Fundamental Constants (Burša, 1995). Its mission is to provide a set of current best estimates of geodetic parameters for the 1999 General Assembly of the International Union of Geodesy and Geophysics (IUGG). The IUGG has had a long tradition of quadrennial updates to the current best estimates of geodetic parameters.

The International Earth Rotation Service (IERS) has continued its publication of constants in its IERS Conventions (McCarthy, 1996). MERIT standards (Melbourne *et al.*, 1983) were the first in a series which has been continued with the publication of the IERS Standards in 1987, (McCarthy, 1989), 1992 (McCarthy, 1992) and the IERS Conventions in 1996. They are available electronically at

http://maia.usno.navy.mil/conventions.html.

5. IERS Conventions

The IERS Conventions are designed to be used in transforming between celestial and terrestrial reference frames. Accuracy is the foremost consideration for the models and constants which are included. As a result there may be small inconsistencies between the chapters of this publication. The values of the constants contained in Chapter 4 are listed in Table 1. It is organized into 5 columns: item, value, uncertainty, reference, comment. All values are given in terms of SI units (Le Système International d'Unités (SI), 1991). The SI second, the basic unit of the TT time scale, is specifically assumed. If the TDB time scale is used, new units of time, t_{TDB}, and length, ℓ_{TDB}, are implicitly defined by the expressions (Seidelmann and Fukushima, 1992)

$t_{TDB} = t/(1\text{-}L_B)$, and

$\ell_{TDB} = \ell/(1\text{-}L_B)$,

where t and ℓ are the SI units and LB is a derived constant given in Table 1. The 1976 IAU System of Astronomical Constants (Astronomical Almanac for the Year 1984) is adopted for all astronomical constants which do not appear in Table 1.

Some geodetic parameters are affected by tidal variations. The values given above are in the zero-tide system so that they correspond to a realistic time-averaged crust. This is done to be consistent with XVIII IAG General Assembly Resolution 16.

TABLE 1. IERS Numerical Standards.

ITEM	VALUE	UNCERTAINTY	REF.	COMMENTS
c	299 792 458 ms^{-1}	Defining	[2]	Speed of light
L_B	$1.550\ 519\ 748 \times 10^{-8}$	4×10^{-17}	[4]	Average value of d(TCB)/d(TT)-1
L_C	$1.480\ 826\ 845\ 7 \times 10^{-8}$	1×10^{-17}	[4]	Average value of d(TCB)/d(TCG)-1
L_G	$6.969\ 290\ 3 \times 10^{-10}$	1×10^{-17}	[4]	Average value of d(TCG)/d(TT)-1
W_0	626 368 56.85m^2s^{-2}	1.0 m^2s^{-2}	[1]	Potential of the geoid
τ_A	499.004 783 8061s	0.000 000 02s	[7]	Astronomical unit in seconds
$c\tau_A$	149 597 870 691m	30m	[7]	Astronomical unit in meters
P	5 029$''\!.$0965/century	0$''\!.$3/century	[5]	General precession in longitude at J2000.0
ϵ_0	84 381$''\!.$412	0$''\!.$005	[7]	Obliquity of the ecliptic at J2000.0
a_E	6 378 136.49m	0.10 m	[1]	Equatorial radius of the Earth
1/f	298.256 42	0.00001	[1]	Flattening factor of the Earth
$J_{2\oplus}$	$1.082\ 635\ 9 \times 10^{-3}$	1.0×10^{-10}	[1]	Dynamical form-factor of the Earth
G	$6.672\ 59 \times 10^{-11} m^3kg^{-1}s^{-2}$	$8.54 \times 10^{-15} m^3kg^{-1}s^{-2}$	[3]	Constant of gravitation
$GM_\oplus$	$3.986\ 004\ 418 \times 10^{14} m^3s^{-2}$	$8 \times 10^5 m^3s^{-2}$	[1]	Geocentric gravitational constant
μ	0.0123 000 345	5×10^{-10}	[7]	Moon-Earth mass ratio
ω	$7.292\ 115 \times 10^{-5}$rad s^{-1}	variable	[1]	Nominal mean angular velocity of the Earth
$GM_\odot$	$1.327\ 124 \times 10^{20} m^3s^{-2}$		[6]	Heliocentric gravitational constant
g_e	9.78 0327 ms^{-2}	$1 \times 10^{-6} ms^{-2}$	[1]	Mean equatorial gravity
$R_0 = GM_\oplus/W_0$	6 363 672.461 m	0.1 m	[1]	Geopotential scale factor

6. Future Maintenance

A three to four year cycle in the updating of a conventional set of astronomical constants appears to be appropriate. The IERS plans to update its Conventions using one editor, an editorial board, and subject matter experts. Constants will be included in this document and these estimates should be carefully coordinated with the International Astronomical Union. Efforts should be made to make the constants internally consistent but not at the expense of accuracy. Ever improving observational precision demands the most accurate constants and models. An issue remaining to be resolved is the reference frame in which to express these parameters. This largely involves issues of relativity related to geocentric or barycentric reference frames. A logical solution is to provide useful transformations between systems along with the adopted values. Future sets of constants may also include relationships between the constants, perhaps in the form of a user manual. In this way users may make use of a minimal set to derive more extensive sets.

Future issues of the latest best estimates of the astronomical constants will most likely be provided as a world wide document for easy access by users. Hopefully this could then provide links to references involving the constants, software to be employed in the use of the constants and the relationships to other constants.

References

Astronomical Almanac for the Year 1984 (1984) U.S. Government Printing Office, Washington, DC.

[1] Burša, M. (1995) Report of the I.A.G. Special Commission SC3, Fundamental Constants, XXI, I.A.G. General Assembly.
[2] Cohen, E.R. and Taylor, B.N. (1973) *J. Phys. Chem. Ref. Data*, **2**, p. 663.
[3] Cohen, E.R. and Taylor, B.N. (1986) The 1986 Adjustment of the Fundamental Physical Constants. *CODATA Bulletin*, Bureau International des Poids et Mesures.
[4] Fukushima, T. (1994) Time Ephemeris. *Astron. Astrophys.*, **294**, pp. 895–906.
[5] Herring, T. (1996) KSV_1996_3 Precession/Nutation. personal communication.
Le Système International d'Unités (SI) (1991) Bureau International des Poids et Mesures, Sèvres, France.
McCarthy, D.D. (ed.) (1989) IERS Standards, *IERS Technical Note*, **3**, Observatoire de Paris, Paris.
McCarthy, D.D. (ed.) (1992) IERS Standards, *IERS Technical Note*, **13**, Observatoire de Paris, Paris.
McCarthy, D.D. (ed.) 1996, IERS Conventions, *IERS Technical Note*, **21**, Observatoire de Paris, Paris.
Melbourne, W., Anderle, R., Feissel, M., King, R., McCarthy, D., Smith, D., Tapley, B. and Vicente, R. (1983) Project MERIT Standards, U.S. Naval Observatory Circular 167.
Seidelmann, P.K. and Fukushima, T. (1992) Why New Time Scales? *Astron. Astrophys.*, **265**, pp. 833–838.
[6] Standish, E.M. (1994) Report of the IAU WGAS Sub-group on Numerical Standards.
[7] Standish, E.M., Newhall, X.X., Williams, J.G. and Folkner, W.F. (1995) JPL Planetary and Lunar Ephemerides, DE403/LE403. JPL IOM 314.10-127, *Astron. Astrophys.*, to be submitted.

SOFA: STANDARDS OF FUNDAMENTAL ASTRONOMY

PROGRESS AND PLANS

P.T. WALLACE
Starlink Project
Rutherford Appleton Laboratory
Chilton, Didcot, Oxon OX11 0QX
United Kingdom

1. Introduction

At its 1994 General Assembly, the International Astronomical Union resolved to introduce new arrangements to establish and maintain an accessible and authoritative set of constants, algorithms and procedures that implement standard models used in fundamental astronomy. The initiative was the responsibility of the Working Group on Astronomical Standards (WGAS), part of Division 1 of the IAU, and the set of procedures and constants was named SOFA, for "Standards Of Fundamental Astronomy". The background to the IAU Resolution is given by Fukushima (1995) and Wallace (1996).

The proposed scheme includes the following elements:

- The **Maintenance Committee**, whose initial tasks are to:
  - establish a revision mechanism;
  - begin the continuous revision of the "best estimates";
  - discuss the possible revisions of the astronomical constants (in particular the precession constant);
  - assess the possibilities for coordination with other fundamental-astronomy and geodesy activities, for example IERS and IAG (International Association of Geodesy).
- The **Relativity Subgroup**, which reviews the definitions of certain units and astronomical constants to ensure consistency within the framework of General Relativity.
- The **SOFA Reviewing Board**, whose initial tasks are to:
  - invite bids from institutions willing to operate the IAU/SOFA distribution center;
  - assess the bids and nominate a center;
  - start reviewing algorithms, procedures and software.
- The **SOFA Center**, located at a suitable institution. The Center organizes, maintains, and distributes the SOFA algorithms and procedures.

The present status of the SOFA scheme is that the Reviewing Board has been set up, the site for the Distribution Center has been identified, and preliminary software work is underway.

2. The SOFA Review Board

In 1995 February, the author was invited by the WGAS to chair the SOFA reviewing board (which has since become known as the "SOFA Review Board"). A process of consultation then took place, clarifying various aspects of the SOFA initiative itself and the role of the Board. This led to an announcement inviting the nomination of candidates for Board membership, in *IAU/WGAS Circular 104*, 1995 November. By 1996 February, the Board's membership had been agreed, and ratified by the WGAS and its Advisors. The current Board membership is as follows:

Wim Brouw	Australia Telescope National Facility
Anne-Marie Gontier	Paris Observatory

J. Andersen (ed.), Highlights of Astronomy, Volume 11A, 191–193.

Catherine Hohenkerk	Royal Greenwich Observatory
Wen-Jing Jin	Shanghai Observatory
George Kaplan	United States Naval Observatory
Zinovy Malkin	Inst. Appl. Astron., St Petersburg
Dennis McCarthy	United States Naval Observatory
Jeffrey Percival	University of Wisconsin
Patrick Wallace	Rutherford Appleton Laboratory

As a by-product of the trawl for Board members, a number of individuals were identified who were prepared to act as "reviewers" of new SOFA products.

With no independent funding, the SOFA Review Board relies on the host institutions of its members to provide the necessary resources. SOFA's needs in terms of computer facilities have, at least up to now, been modest, and easily met by existing equipment. Effort is, and is likely to remain, the most serious bottleneck. A further constraint is funding for travel. The bulk of the Board's work so far has been carried out by electronic mail, and this will continue. However, two meetings have also taken place, and most Board members were able to attend at least one of them. The first meeting was at United States Naval Observatory (USNO) in 1996 September, followed a month later by one held at the Paris Observatory.

In the months leading up to the Washington and Paris meetings, discussion via electronic mail had produced a measure of agreement on various matters of detail, and a draft Fortran coding standard had been prepared. The meetings allowed this work to be consolidated, and significant further progress was made. Among the decisions reached were:

- The SOFA Collection will, to start with at least, consist only of callable subprograms, as opposed to freestanding applications. All such items are to be called, simply, "routines", ignoring distinctions such as function *versus* subroutine.
- No existing package (SLALIB (Wallace 1997) and NOVAS (Kaplan 1996) were among those considered) fits the requirements of SOFA, though such packages are important sources of code and test examples.
- Before work on the fundamental-astronomy routines themselves could begin, a **toolkit** of vector/matrix routines would have to be provided. Once they were in place, the first tranche of **canonical** SOFA routines could be developed. These would comprise the established IAU standards, supplemented by *de facto* standards where necessary to fill in gaps. Later, alternative algorithms and packagings would be developed, to suit the needs of general users. These would include **manager** routines, which would string together SOFA calls to provide higher-level capabilities, giving non-specialist users easy access to correct procedures.
- On programming languages, it was recognized that SOFA must address all users and cannot, for example, afford to provide only Fortran code or C code. This diversity would be managed by regarding SOFA algorithms expressed in different languages as independent products, with packaging and interfacing optimized for the language concerned.
- With several levels of SOFA routine and multiple programming languages, it was recognized that it would be necessary to proceed in stages. The toolkit would be developed first, initially in Fortran 77, followed soon after by ANSI C. This would then allow canonical SOFA products (precession and nutation models, timescale conversion *etc.*) to be developed in Fortran and C, in parallel. The high-level manager routines could then be developed, probably only in C. Throughout the process, the object-oriented principle would be kept in mind, with the possibility in the long run that C++, Java or some other OO language would take over completely.

3. The SOFA Center

The Announcement of Opportunity for institutions to bid to become the SOFA Center was published at the beginning of 1997, in *IAU Bulletin 79* and elsewhere. By the 1997 April 30 deadline, two bids had been received, from Rutherford Appleton Laboratory (RAL) and the United States Naval Observatory (USNO) respectively. In 1997 June (with representatives of the two bidding institutions excluded from the voting) the Board endorsed RAL as the SOFA Center.

The SOFA Center will be operated in parallel with the Starlink service and will employ the same techniques as Starlink's "Software Store". This is a self-service facility, enabling enquirers to obtain software and documentation directly through the World Wide Web. The Web is expected to remain the principal medium for disseminating SOFA products; however, supplementary distribution services using magnetic, optical or printed media may also be introduced if there is a demand.

It is intended that the service provided by the SOFA Center will be monitored by a management board. This board will present reports and assessments at each triennial General Assembly of the IAU. The management board will initially be the WGAS. Because IAU Working Groups exist only for a limited period, it will be necessary in due course to establish a permanent SOFA Management Board, perhaps in the form of a new IAU Commission. Steps to bring about this change will be taken prior to the IAU General Assembly in 2000.

4. The First SOFA Algorithms

Apart from the "toolkit" routines, the following set of algorithms will be the first SOFA products to appear:

- IAU 1976 precession.
- IAU 1980 nutation.
- UTC/TAI transformation.
- TCB/TDB/TT transformation.
- UT1/GST transformation (mean and apparent).
- Closed-formula planetary ephemerides.
- Star-catalog-entry to space-motion transformation.
- Catalog transformation from B1950 to J2000.
- Solar-system barycenter coordinates of an Earth-based observer.
- Apparent/topocentric/observed star places for an Earth-based observer.

The great care with which the SOFA routines will have to be designed and evaluated means that rapid progress cannot be expected, and it is likely to take several years to provide even this basic set. However, successive refinements of existing models, such as new precession and nutation series, will be relatively easy to introduce quickly. Selection of new models will, of course, involve full consultation with the specialist community, and great care will be taken to ensure consistency with other standards and conventions, such as those of the International Earth Rotation Service.

An important feature of the SOFA software will be the accompanying documentation. This will include test examples, allowing other implementors of the same algorithms to check for consistency and accuracy, something which is very difficult to do at present.

References

Fukushima, T. (1995) Report of the IAU WGAS Subgroup on Standard Procedures. In: *Highlights of Astronomy*, **Vol. 10**, ed. I. Appenzeller, pp. 185–189.
Wallace, P.T. (1996) In: *Astronomical Data Analysis Software and Systems V*, A.S.P. Conf. Ser., **Vol. 101**, eds. George H. Jacoby and Jeanette Barnes (San Francisco: ASP), pp. 207.
Wallace, P.T. (1997) *Starlink User Note 67.36*.
Kaplan, G.H. (1996) http//aa.usno.navy.mil/AA/software/novas/novas_info.html.

GENERAL RELATIVITY AND THE IAU RESOLUTIONS

Report of the IAU WGAS Sub-Working Group
on Relativity in Celestial Mechanics and Astrometry (RCMA SWG)

V.A. BRUMBERG
Chaiman of the RCMA SWG
Institute of Applied Astronomy, 8, Zhdanovskaya st., 197042 St. Petersburg, Russia

AND

P. BRETAGNON, N. CAPITAINE, T. DAMOUR, T.M. EUBANKS, T. FUKUSHIMA
B. GUINOT, S.A. KLIONER, S.M. KOPEIKIN, A.V. KRIVOV, P.K. SEIDELMANN
AND M.H. SOFFEL
Members of the RCMA SWG

1. Introduction

RCMA SWG was appointed by the IAU WGAS (Working Group on Astronomical Standards) in accordance with IAU Resolution C6 (1994) with the aim 'to provide definitions of the astronomical units, of the quantities linking these astronomical units to the units of the International System (SI), and of other astronomical quantities, compatible with the theory of General Relativity'. It is evident that the relativistic aspects of units of measurement cannot be isolated from the more general problem of astronomical constants and fundamental astronomy concepts in the relativistic framework. Therefore, along with the problem of units the main topics of discussion of RCMA SWG concerned also the IAU (1991) Resolutions on References Systems (RSs) and Time Scales (TSs) and their interpretation in IERS Standards (1992) and IERS Conventions (1996). In what follows we tried to summarize the results of these discussions.

Nowadays it is common practice to determine from observations the Parameterized Post-Newtonian (PPN) parameters γ and β, which reflect possible violations of Einstein's General Relativity Theory. The IAU (1991) Resolutions, in fact, recommend to use the value $\gamma = 1$ of General Relativity to model the observations. This is justified by the fact that no observational data have shown a statistically significant violation of General Relativity. Therefore this theory, which is the simplest theory of gravitation satisfying all phenomena with the current accuracy of observations, can be recommended for applications. However, the RCMA SWG wishes to stress that it would be against basic scientific principles to stop improving the determination of the PPN parameters or to test General Relativity by other means.

The present status of fundamental astronomy, several years later after the adoption of new IAU (1991) resolutions on reference systems and time scales, leads to the following remarks:

1) the currently employed terminology should be changed to be consistent in the relativistic framework;

2) presently, because of achievable accuracy, relativity has to be taken into account in modeling virtually all kinds of modern astronomical and geodynamical observations. Here, the qualitative difference between the space-time of General Relativity and the Newtonian absolute space and time should never be forgotten. The quasi-Newtonian standpoint of speaking about (Newtonian-theory) + (small correction terms) often leads to misunderstandings or even wrong or ambiguous results. On the other hand, it is much more advantageous to follow basic principles of relativistic modeling which are extremely clear and easy to understand and to use.

J. Andersen (ed.), Highlights of Astronomy, Volume 11A, 194–199.

As an illustration of the present status, we shall consider in Section 3 the IERS Standards (1992) and IERS Conventions (1996).

2. IAU (1991) Resolutions

2.1. NEED OF HIGHER ORDER TERMS

Recommendations I, II and III of IAU (1991) Resolution A4 define the barycentric reference system (BRS) and the geocentric reference system (GRS) in the framework of General Relativity by specifying the corresponding four-dimensional metric forms ds^2. The coordinate time-scales of these forms are TCB and TCG, respectively. Only lowest-order terms of order c^{-2} (c being the light velocity in vacuum) are indicated in the time and spatial components of the metric tensors related to BRS and GRS. These terms are sufficient for present astrometric applications but, in a decade or so, the metric forms of BRS and GRS should be given more accurate by

(1) adding into the time and spatial (c^{-2})-components of the GRS metric a linear (with respect to the geocentric coordinates) term describing the acceleration of the geocenter in the barycentric reference system relative to a fictitious test-particle located in the geocenter and moving under the influence of the gravitational field of the solar system with except for the Earth (such term is taken into account in the high-precision analysis of Earth's satellite motion, e.g., Huang *et al.*, 1990);

(2) specifying a post-Newtonian definition of the geocenter, e.g., by setting the Barycentric Distance (BD) mass-dipole of the Earth to null;

(3) adding the (c^{-3})-order terms for the mixed components of the BRS and GRS metric tensors (this enables one to specify the choice of the space-time coordinates and to distinguish, particularly, between harmonic and PPN standard coordinates; in case of the GRS this permits also to distinguish between dynamically and kinematically nonrotating versions of the GRS; the two versions differ due to relativistic precession including a secular part (precession in astronomical sense) and periodic one (which could be called nutation); the relativistic precession is due to the gravitational field of all the bodies of the solar system, the influence of the Sun being dominating;

(4) indicating the (c^{-4})-order terms in the time components of the BRS and GRS metric tensors (they are necessary to take into account the (c^{-4})-order terms in the time-scale relationships (see, e.g., Fukushima, 1995), i.e. to define TCB and TCG more rigorously than it is done by means of the first approximation expressions in the notes for IAU (1991) recommendations).

Let us note that the terms of order c^{-4} in the time component of the BRS metric tensor and the terms of order c^{-3} in the mixed components are actually taken into account nowadays to derive the post-Newtonian equations of motion of the solar system bodies used to produce modern ephemerides. These terms give herewith effects of the same order of magnitude as the terms explicitly specified by the IAU. The term of order of c^{-4} in the GRS metric is also used to get the equations of motion of Earth's satellites recommended by IERS Standards (1992) and Conventions (1996).

2.2. NEED OF CLARIFICATION IN USING TERRESTRIAL TIME, TT

Returning to the present status of the time-scales it should be noted that the recommendation IV of the IAU (1991) Resolution A4 declares TT as a time-like argument to be used for geocentric ephemerides. This resulted in some confusion about the roles of TCG and TT. IAU (1994) resolution C7 did not remove such a confusion. On the one hand, this resolution defines the epoch J2000.0 and the Julian century in terms of TT. On the other hand, it recommends to develop new ephemerides in terms of the time-like arguments TCB and TCG. Moreover, presently TT is often misunderstood as a time-scale of GRS. This usually leads to altering the spatial coordinates of the GRS (see the following section) in explicit contradiction with the IUGG Resolution 2 (1991) based just on the IAU (1991) Resolution A4.

Hence, even if it is possible to postpone a more accurate exposition of the BRS and GRS metrics one should avoid the confusion in using TCG and TT as soon as possible.

3. IERS Conventions (1996)

IERS maintains two basic reference frames, the International Celestial Reference Frame (ICRF) and the International Terrestrial Reference Frame (ITRF). According to the language used in both IERS Standards (1992) and IERS Conventions (1996), ICRF represents a materialization of the BRS constructed by means of VLBI observations in conformity with IAU (1991) resolutions. However, instead of the BRS coordinate time TCB, spatial coordinates $\vec{x}$ and mass-factors GM, IERS Standards (1992), IERS Conventions (1996), as well as JPL DE/LE, use

$$\mathrm{TDB} = (1 - L_B)\,\mathrm{TCB}, \;\; (\vec{x})_{\mathrm{TDB}} = (1 - L_B)\,\vec{x}, \;\; (GM)_{\mathrm{TDB}} = (1 - L_B)\,GM \tag{1}$$

with $L_B \approx 1.5505197 \times 10^{-8}$ given in the IAU (1991) resolutions for the relationship between TCB and TDB of JPL DE/LE ephemerides. Actually each version of JPL DE/LE, being a bit different materialization of the BRS, provides slightly different values of L_B. This does not influence the values of $(GM)_{\mathrm{TDB}}$ at a significant level since currently typical precision of the mass factors are $10^{-10} - 10^{-11}$ and only a few leading digits of L_B play a role. However, it is anyway advantageous for many reasons to publish the transformation of the TDB as defined by each version of the DE/LE ephemerides and the TT together with the ephemeris itself. This transformation (which defines in particular L_B) can be considered as a "time part" of a relativistic space-time ephemeris.

Scaling of the spatial coordinates and the mass factors in (1) is performed to keep the value of the light velocity as well as the equations of motion of the solar system bodies invariant under transformation from TCB to TDB.

From the theoretical point of view, the ITRF maintained by IERS is supposed to be a realization of the GRS$^+$, a geocentric system rotating with the Earth and resulting from application of Euclidean rotation of the spatial axes of the GRS envisaged by the IAU/IUGG (1991) resolutions. But instead of the coordinate time TCG, spatial coordinates $\vec{w}$ and mass factors GM related to the GRS, IERS Standards (1992) use actually

$$\mathrm{TT} = (1 - L_G)\,\mathrm{TCG}, \;\; (\vec{w})_{\mathrm{TT}} = (1 - L_G)\,\vec{w}, \;\; (GM)_{\mathrm{TT}} = (1 - L_G)\,GM \tag{2}$$

with $L_G \approx 6.9629 \times 10^{-10}$ as given by IAU (1991) resolutions. The value of L_G is slightly different for various geoid models. Again, the scaling of the spatial coordinates and the mass factors is aimed to keep the value of the light velocity and the GRS equations of motion (e.g., those of the Earth's artificial satellites) invariant under the transformation from TCG to TT.

Transformations (1) are often considered as resulted from the change of the units of time from the SI unit to the 'TDB unit'. Just similarly, transformations (2) are meant in IERS Standards (1992) as resulted from the change of the units of time from the SI unit to the 'TT unit'. In this report transformations (1) and (2) are regarded as determining new quantities with the same units of measurement. This simplifies the notations and satisfies Recommendation III of the IAU (1991) Resolution A4 demanding the consistency of the units of measurement of the coordinate times with the SI unit of time. The detailed comparison of these two approaches to interpret transformations like (1) and (2) may be found in Guinot (1997).

In such a way, the time scale TT and the quantities related with TT (e.g., $(GM)_{\mathrm{TT}}$ and $(\vec{w})_{\mathrm{TT}}$) should be considered below as other physical quantities related with TCG, GM and $\vec{w}$ by (2).

An important particular question concerns the value of the geocentric constant GM_E. IERS Standards (1992) give two values for the geocentric constant

$$GM_E \text{ in TT units} = 3.986004418 \times 10^{14}\; m^3 s^{-2} \tag{3}$$

and

$$GM_E \text{ in TCG units} = 3.986004415 \times 10^{14}\; m^3 s^{-2}\,. \tag{4}$$

The meaning of these values remains unclear considering that the scaling relation (2) implies $(GM)_{\mathrm{TT}} < (GM)$. IERS Conventions (1996) reproduce the value GM_E coinciding with the IERS Standards (1992) value in 'TT units'. In doing this, IERS Conventions (1996) gives no definite information about the units used herewith. RCMA SWG wishes to stress that the current situation may lead to significant discrepancies between the versions of the ITRF constructed by using different observational techniques. This question should be clarified and an unambiguous value of GM_E should be published.

Hence, the GRS^+ being adequate to the IAU/IUGG (1991) resolutions differs from ITRF as maintained by IERS Standards (1992) by introducing L_G factor. In spite of the fact that IERS Conventions (1996) claim to achieve a better compatibility of ITRF with the IAU (1991) resolutions, the actual realization of ITRF in IERS Conventions (1996) represents a significant step backward as compared with IERS Standards (1992). The problem is to clarify what kind of spatial coordinates are wanted for ITRF. There are two main options:

(1) spatial coordinates of the GRS corresponding to its coordinate time TCG;

(2) locally measurable distances in the infinitesimal vicinity of an observer's site in analogy to the observer's proper time.

The first option corresponds to the metric adopted by the IAU/IUGG (1991) resolutions. The second option is actually realized by the VLBI formula recommended by IERS Conventions (1996). Implicitly, it makes use of the fact that any metric in General Relativity may be reduced in the infinitesimal vicinity of any point to the Minkowski form where all four coordinates may be regarded as physically measurable time and distances (proper time and proper distances). Coordinate time of such a reference system coincides with the proper time of an observer staying at its origin, while the coordinates correspond to physically measurable distances as far as infinitely small (reasonably small, in practice) space regions are considered. Such a local observer's reference system should be constructed for each observer on the Earth separately. While such an approach is quite suitable for constructing local geodetic networks, it is absolutely wrong for the whole Earth, leading to inconsistencies when sufficiently long distances are to be considered. Let us, however, note that the current level of accuracy allows one to speculate that we could introduce such 'locally measurable' spatial coordinates for the whole Earth. At a level of accuracy the transformations of spatial coordinates can be represented by a scaling $\vec{w}_{\text{locally measurable}} = (1 + L_G)\vec{w}$ (note the factor $1 + L_G$, not $1 - L_G$). Many other terms which could not be represented as a scaling are an order of magnitude smaller and neglected here. An improvement in accuracy by one order of magnitude, which is very likely in the nearest future, will reveal the inconsistency and will make it very clear that such an approach is erroneous not only from the theoretical point of view, but also from the very practical one. Moreover, the procedures of IERS Conventions (1996) for LLR and SLR observations, which still use the scaled GRS coordinates $(\vec{w})_{\text{TT}}$ do not agree with the procedure for VLBI observations (the same is most likely true for the GPS analysis). The discrepancy between the models amounts to 5 mm, at least. Although the present errors of these techniques exceed the 5 mm level, this internal incompatibility of various techniques used in ITRF is not admissible.

For all these reasons RCMA SWG cannot consider IERS Conventions (1996) procedures as satisfying the IAU (1991) resolutions.

4. Astronomical Units

Along with the SI units of time (s), mass (kg) and length (m), defined as proper quantities, astronomical units (AS) may be used when appropriate. The astronomical unit of time $d = 86400\,s$ is defined directly by its relationship to the SI unit of time. As for AS units of mass and length are concerned, one should distinguish between their definitions and their relationships to the corresponding SI units. The astronomical unit of mass is defined as the mass of the Sun M_S. In order to find its relationship to the SI unit of mass, one may use the laboratory determined value of G ($= 6.672 \times 10^{-11}\ m^3\ kg^{-1}\ s^{-2}$ in the IAU (1976) system of constants) and the value of GM_S ($= 1.32712438 \times 10^{20}\ m^3\ s^{-2}$ in the IAU (1976) system of constants) determined from the relativistic equations of motion of planets and planetary radar observations. In such a way one gets M_S ($= 1.9891 \times 10^{30}\ kg$ from the values above) in terms of the SI units of mass.

As far as the astronomical unit of length A is concerned, it is defined now by the condition $k = 0.01720209895$ provided that the units of measurement are d, M_S and A. Its relationship to the SI unit of length resulting from this definition is

$$A = (GM_S/k^2/d^2)^{1/3}, \tag{5}$$

GM_S being a value in SI units discussed above. In such manner one gets the numerical value of A ($= 1.49597870 \times 10^{11}$ m for the values above). Although Eq. (5) originally related with the third Kepler law is 'Newtonian' by its internal philosophy, it can be considered also in the relativistic framework as simply a numerical definition to be retained for the reasons of continuity. Note also, that the accuracy of M_S expressed in the SI units is much lower than that of A and d.

Hence, General Relativity affects not the definitions of the astronomical units of mass and length, but only the numerical values of M_S and A in the SI units. These numerical values are not fixed. On the contrary, they are in permanent improvement with the increasing observation precision and by addition of new terms (Newtonian and relativistic contributions) into the model of the planetary motion. Thus, the values of GM_S and A have been significantly improved since 1976 and, for example, the values corresponding to each version of the JPL DE/LE ephemerides can be calculated using the lists of constants delivered together with each ephemeris.

The only principal question here is to choose some explicit relativistic definition of the mass of the Sun. First, there exist many reasonable definitions of the mass of a body in General Relativity (which do not differ quantitatively for the Sun at current level of accuracy). Second, the mass of a body cannot, generally speaking, be considered as constant, while the present definitions of the astronomical units of mass and length assume M_S to be constant. Let us note that the latter problem is by no mean specific to General Relativity: the Solar mass is variable also due to, say, solar wind and emission, and these effects (giving a change of about 10^{-13} per year) are several orders of magnitude larger than possible relativistic variability. However, within the present accuracy one may forget about this problem.

5. Recommendation

The foregoing analysis enables us to formulate the following draft of the recommendation to be considered at the XXIIIrd IAU General Assembly.

The XXIIIrd General Assembly of the IAU

considering that

- a relativistic solar system barycentric four-dimensional coordinate system with its coordinate time scale TCB was defined by IAU Resolution A4 (1991),
- a relativistic geocentric four-dimensional coordinate system with its coordinate time scale TCG was defined by IAU Resolution A4 (1991) and IUGG Resolution 2 (1991),
- the basic physical units of space-time in all coordinate systems were recommended by IAU Resolution A4 (1991) to be the SI second for proper time and the SI meter for proper length,

noting that

- practical realization of barycentric and geocentric coordinate systems in many groups (see IERS Standards, 1992) is based on time scales TDB and TT instead of TCB and TCG, respectively, and involves the scaling factors $1 - L_B$ and $1 - L_G$ for the spatial coordinates and mass factors GM in barycentric and geocentric systems, respectively, L_B and L_G being given in IAU Resolution A4 (1991),
- even more complicated scaling factors are introduced in the VLBI model of IERS Conventions (1996),
- currently employed definitions of fundamental astronomy concepts and astronomical constants are based on Newtonian mechanics with its absolute space and absolute time leading to ambiguities in dealing with relativistic effects,

recommends that

- the spatial coordinates of the Barycentric and Geocentric Reference Systems as defined the IAU (1991) resolutions be used for celestial and terrestrial reference frames, respectively, without any scaling factors,
- the final practical realizations of the coordinate systems for use in astronomy and geodesy should be consistent with the systems defined by IAU/IUGG (1991) resolutions,

- the use of TT for convenience of observational data analysis should not be accompanied by scaling of the spatial geocentric coordinates,
- algorithms for determination of astronomical constants and definitions of fundamental astronomy concepts should be explicitly given within the basic reference systems envisaged by IAU/IUGG (1991) resolutions,
- IAU WGAS continue the consideration of relativistic aspects of the concepts, algorithms and the constants of fundamental astronomy.

References

Brumberg, V.A., Bretagnon, P. and Guinot, B. (1996) Astronomical Units and Constants in the General Relativity Framework. *Celes. Mech. Dyn. Astron.*, **64**, pp. 231–242.

Brumberg, V.A. (1995) Reference Systems and Astronomical Constants in General Relativity. In: *Earth Rotation, Reference Systems in Geodynamics and Solar System*, Journées Systèmes de Référence Spatio-temporels 1995, eds. N. Capitaine, B. Kolaczek and S. Debarbat, Warsaw, pp. 29–36.

Brumberg, V.A. (1997) Ephemeris Astronomy Definitions and Constants in General Relativity. IAU Colloquium No. 165, Poznan, In: *Dynamics and Astrometry of Natural and Artificial Celestial Bodies*, eds. I.M. Wytrzyszczak, J.H. Lieske and R.A. Feldman, Kluwer, Dordrecht, pp. 439.

Capitaine, N. and Guinot, B. (1995) Astronomical Units and Constants in a Relativistic Framework. *Highlights of Astronomy*, **10**, pp. 201.

Fukushima, T. (1995) Time Ephemeris. *Astron. Astrophys.*, **294**, pp. 895–906.

Guinot, B. (1997) Application of General Relativity to Metrology (International report), *Metrologia*, **34**, pp. 261–290.

Huang, C., Ries, J.C., Tapley, B.D. and Watkins, M.M. (1990) Relativistic Effects for Near-Earth Satellite Orbit Determination. *Celes. Mech. Dyn. Astron.*, **48**, pp. 167–185.

Klioner, S.A. (1996) Angular Velocity of Extended Bodies in General Relativity. In: *Dynamics, Ephemerides and Astrometry in the Solar System*, eds S. Ferraz-Mello, B. Morando and J.E. Arlot, Kluwer, Dordrecht, pp. 309–320.

Klioner, S.A. (1995) Relativistic Effects in Orientation of Astronomical Reference Frames. In: *Earth Rotation, Reference Systems in Geodynamics and Solar System* Journées Systèmes de Référence Spatio-temporels 1995, eds. N. Capitaine, B. Kolaczek and S. Debarbat, Warsaw, pp. 175–178.

Klioner, S.A. (1997) On the Problem of Post-Newtonian Rotational Motion. IAU Colloquium No. 165, Poznan, In: *Dynamics and Astrometry of Natural and Artificial Celestial Bodies*, eds. I.M. Wytrzyszczak, J.H. Lieske and R.A. Feldman, Kluwer, Dordrecht, pp. 383.

This list of references in no way claims to be complete. It enables one to get a more detailed information about the topics considered above and to find further references on general relativity applications in fundamental astronomy.

JD4

Challenges in Atomic Physics for Cosmic X-Ray Spectroscopy

CANCELLED

JD5

Preserving the Astronomical Windows

Chairperson and Editor: **S. Isobe**

JD5 "PRESERVING THE ASTRONOMICAL WINDOWS"

Supporting Division: X(Radio astronomy)
Supporting Commissions: 40(Radio astronomy), 50(Protection of existing and potential observatory sites)
Co-supporting Commissions: 8(Positional astronomy), 21(Light at the night sky), 46(Teaching of astronomy), 51(Bioastronomy:search for extraterrestrial sites)
SOC members: N. Brouillet(France), B. Hidayat(Indonesia), Jingyao Hu(China), S. Isobe (Chairperson, Japan), Ch. Leinert(Germany), K. Mattila(Finland), D. McNally(UK), WJ. Percy
(Canada), W. Wamsteker(Spain), and J. Whiteoak(Australia)

Held on August 21 and 22, 1997, during the XIIIrd General Assembly in Kyoto.

In these decades astronomy has been developing very much because observations in optical, radio, and the other wavelength domains as well as neutrino, cosmic ray, and in-situ observations have been extensively carried out collaborating with theoretical works. However, it is also true that astronomical observational conditions have become worse and worse during these same period. Astronomical observers are suffering light pollution, radio interference, and space debris. IAU had realized these problems and set up a commission to deal these matter in 1978 : that is the Commission 50 "Protection and identification of existing and potential observational site. The Commission organized the IAU Colloquium No. 112" (D. Crowford 1991) in Washington D. C, USA, from 13 to 16 August 1988, and extensively discussed the matters. Some number of astronomers worked so hard to reduce the problems by contacting and collaborating with the other organizations such as CIE, COSPAR, URSI, ITU, IAD, and UNESCO, UN. One of these activities was a joint meeting of UNESCO, ICSU, IAU, and CSR held in Paris from June 30 to July 2 1992, that is "Adverse Environmental Impact on Astronomy" (D. McNally 1994). However, since most of the observational astronomers have been busy to work out their own observations, the problems have still become worse.

The JD5 is a meeting inside the IAU. The main purpose of this JD is to know present situations of different interferences, to find ways to reduce the problems, and most importantly to make many IAU members realize the problems. Fortunately, we could have a talk by L. Woltjer, the President of the IAU, relating investments of astronomy and those loss by the problems dealed in this JD5. This is an important point which we, astronomers, should always think about Astronomy uses some billion US dollars per year and our loss becomes some fractions of this value.

We had 23 talks including the President, 9 paster papers, and nearly 200 audiences who attended at least some session(s). Therefore, the JD5 was successful with many discussions.

To approach public people, different organizations, different governments, and UN, on the problems, there are many ways, but it is inevitably necessary for us, astronomers, to do much efforts. Astronomers actively working on the problems gave reports during the JD5, and the followings are short summaries of their talks. Those full papers describing in much detail will be published in the other book (Isobe and Hirayama 1998). It is highly expected that all the IAU members start to consider the problems which may kill astronomical observations.

Syuzo Isobe (Chairman of JD5)

1. ADVERSE ENVIRONMENTAL IMPACT ON ASTRONOMY

D. McNally (University of London Observatory)

J. Andersen (ed.), Highlights of Astronomy, Volume 11A, 205–218.

This paper sets out to urge the quantification of adverse environmental impact on astronomy (AEI). Up to the present AEI has been treated anecdotally. This is not a criticism - the demonstrations have been compelling, particularly so in the case of light pollution. However, we must now move from demonstration to effective action. That effective action will impact the legitimate activities of others. Therefore astronomy will have to present evidence capable of convincing not just those sympathetic to the astronomical community, but the community in general that astronomy has a legitimate grievance and that the redress being sought will be effective in improving astronomical observing condition.

In order to be convincing we must quantify. We must be able to state, for example, what background levels of light pollution are, what the normal range of the level is and how the level is changing from year to year. All claimed adverse impact should have some measure to express the real nature of the impact. Only when such measures are in place can we argue for legal protection.

The problems to be faced in quantification must not be minimized. The problems are considerable and each adverse impact has its own unique set. Some of the problems are outlined in the paper. Unfortunately time is not on our side. Every year sees a rebirth of the concept of advertising from space in one form or another. Sooner or later, unless we are supremely vigilant, a proposal will be accepted and the consequences will be disastrous for our science. Hence the thrust of the EC Resolution before this General Assembly.

There is now great investment in astronomy as a science. That investment has produced some of the most exciting astronomy in all history. One only has to think of the spectacular discoveries of the HST, the advance in basic astrometric data represented by HIPPARCOS, and the unparalleled spectroscopic data base from IUE. We are currently looking forward to the fruits of the 10 m generation optical telescopes and sustained adventures in radio astronomy. Yet that tantalisingly brilliant future is at risk.

Unfortunately the interests of astronomy are now in conflict with other human activities. Astronomy has retreated to its mountains, deserts and into space. It can no longer retreat-there is nowhere left to go. The other competing interests are well heeled profitable and will claim enormous benefits for society. Some of the benefits are real but not all. We as astronomers will have to come to terms with the rest of society- but by the same token, the rest of society will have to come to terms with us - if astronomy is to survive as a major science. It would be a great disaster for the whole of science were the science of astronomy to be curtailed by the pollution produced terrestrially. Loss of investment would be disaster, loss of cosmic perspective would be a catastrophe.

2. IMPACT ON RADIO ASTRONOMY

M. Morimoto (Space Science Course, Kagoshima University)

The radio sky has continuously supplied "exotic objects" to astronomy, such as quasars, pulsars etc. After a while these objects proved to be important constituents of the Universe and provided a new look of the Universe.

Cosmic Background Radiation is the remnant of the blackbody radiation of Universe when it was cooled to about 3000k. Density of quasars is higher in the distant part of the Universe and proves the Universe was more crowded in the early stage.

Spectral line emission from molecules formed in the interstellar gas clouds shows processes of star formation in the Galaxy and galaxies.

Technology in radio telescope in also continuously in a rapid progress. Recent success of HALUCA satellite and VSOP has opened a new era of high resolution radio imaging through space VLBI. Accuracy of source position measurements is approaching 1 microarcsec as level and trigonometry will soon cover whole Galaxy and beyond.

These are only a small portion of achievements of radio astronomy, but prove that it continues to contribute important parts of the human knowledge of the Universe.

On the other hand, this exciting field of science is critically endangered by the more and more intense usage of radio spectrum for communication and other purposes.

Frequencies below ~ 300 MHz (m-waves and longer)

Low frequency radio emission represents low density plasmas in Universe, which is difficult to observe by other means. Examples are outer corona of the sun, envelope of the Galaxy and galaxies etc. Developments in array technology, image processing etc has enabled more sophisticated observations especially with higher angular resolution.

However, the situation may NOT allow the human being to enjoy this improvement fully. Protected bands are very narrow here, because this is the most crowded frequency band and allocation is given in terms of very small widths.

High resolution observations will become possible by use of arrays of space baselines. A different set of protection criteria will become necessary.

300–30000 MHz (dm to cm waves)

This is the band where most of the important discoveries were done and developments in new technology are in progress, but on the other hand, the most intensively used band by communications and other purposes. New demands such as satellite systems for personal pocket phones etc are rapidly growing but there is no resources of unused bands.

Situation will become more serious as more and more new demands will require space to earth transmission with relatively high power and to cover essentially whole surface to the earth. If some substantial measure NOT taken, human eyes in this important spectrum will be black out.

Above 30 GHz (mm and sub-mm waves)

Molecules in the interstellar clouds display a very large number of spectral lines mainly through rotational transition. It provides an unique possibility of direct observations of cool gas phase in Universe. Interference situation is not YET serious.

The band is not used very actively because of the cost and technical demands will be pointed to this area very soon.

Frequency as the treasure of Nature

Originally, communications lived in a small area, perhaps below 50MHz. After the 2nd World War, the "big march" to the higher frequencies started and conquered the whole microwave region, which was, although sparsely, already inhabited by radio astronomy. Conquers allowed the original inhabitant to survive within narrow protected areas.

Frequency is like land. It is finite and easy to use up. It belongs to Nature. It is NOT owned by anyone.

Developments in the technology is at present encourage demands to USE the new frequency bands rather than protecting it, or in other words, Nature is protected only be the technical difficulties to use the band rather than the human wisdom.

Conclusion

We are about to lose the radio sky, which has continuously provided fascinations to the human being. Just like in other frontiers of the conservation of Nature, a use of the human wisdom to protect Nature rather than destroying is only the solution out or it.

3. LIGHT POLLUTION: ITS DAMAGE TO EDUCATION AND CULTURE

J. R. Percy (Erindale Campus, University of Toronto)

One could argue that light pollution has no impact on education or culture, because so few people care about real science. Perhaps people prefer bright lights, and "space art". The word *education*, however, comes from the Latin word *to lead*, and *culture* obviously refers to cultivation. It is our responsibility to educate the public in astronomy, and in the issues related to light pollution and related topics, and to cultive their appreciation of the night sky, and the universe.

Astronomy is important to education and culture because of its deep historical roots, its practical and philosophical applications, its aesthetic and emotional appeal (the beauty of the universe, and the sense of shared exploration and discovery), and its message about our place in time and space, and our cosmic roots and environment. In the classroom, it demonstrates the observational approach to the scientific method, and is a wonderful tool for teaching concepts of light, and gravity. It attracts young people to science and technology, and promotes public interest in science. Like all science, however, it is best taught through inquiry—simple, "hands-on" (or "eyes-on") activities. This is difficult in light-polluted skies.

Ironically, light pollution provides interesting opportunities for student projects, because it has scientific, technological, and societal aspects. Several groups have developed such projects, including co-operative, Internet-based projects in which students measure light pollution with simple devices, and then send their measurements to a central database. Students can also study light sources in their neighbourhood, using simple transmission diffraction gratings—a good cloudy-night activity!

Light pollution also hinders the recruitment, training, and work of amateur (volunteer) astronomers. Amateurs make important contributions to astronomy research and education. Amateurs in many parts of the world—including Japan—have carried out important studies of light pollution, thanks to their skill and motivation. They are our allies—"grass-roots" voices for the preservation of the astronomical environment.

The issue of space debris is somewhat more difficult, since we have no control in the matter; most people, however, can relate to the problem, because of widespread media reporting of space activities. The issue of the pollution of the electromagnetic spectrum is even more difficult, because people have no direct experience with it, and they somehow feel that scientists can overcome any technological problem of this kind. Public education is an essential part of any solution.

What can we do about light pollution? We cannot just replace the real sky by the planetarium sky, or the computer screen—useful as these may be. We cannot expect people to travel to dark sites, even though the rich and motivated can do so. We cannot do all astronomy from space, even though this is a popular misconception. We must promote astronomy education and culture, and encourage our colleagues and students to do the same. We must educate ourselves about the preservation of the astronomical environment, and "spread the word". We must work with partners: teachers, amateur astronomers, lighting engineers, and the media, to achieve these goals. We must support organizations such as IAU Commission 50, and the International Dark-Sky Association. We must promote an appreciation of the night sky by making skygazing a part of our courses, and other activities. Few people who have seen a dark sky are unmoved.

4. NATURAL OPTICAL SKY BACKGROUND

Ch. Leinert[a] and **K. Mattila**[b] ([a] Max-Planck-Institut für Astronomie, Heidelberg, Germany; [b] University of Helsinki Observatory, Finland)

4.1. INTRODUCTION

Knowledge of the natural dark night sky brightness is required if one wants to set meaningful limits to the disturbance by artificial illumination. Knowledge of the behaviour of the natural dark night sky brightness is also required if one wants to monitor the sky for existing or changing light pollution. Therefore, before coming to quantitative results of night sky brightness measurements, we give an overview on the different components of night sky brightness and their different behaviour. From this, recommendations will result for the planning of future sky brightness monitoring programmes, as they are requested in the first of the proposed IAU resolutions to result from this joint discussion.

4.2. BRIGHTNESS OF THE COMPONENTS OF THE NIGHT SKY

The light of the night sky comes from several "layers" at vastly different distances. Neglecting the small contribution from extragalactic background light and the diffuse galactic light, which approximately can be taken as a 20% addition to the light of the stars, the main contributors to the night sky brightness are—the integrated starlight due to our own galaxy (15–200 S_{10} [1 S_{10} unit = one 10th magnitude star per square degree = $2.17 \cdot 10^{-8}$ W/m^2 sr μm at B.]), —the zodiacal light, due to scattering of solar radiation on interplanetary dust (30–100 S_{10}), airglow emission from the high atmospheric layers (50–180 S_{10}), and scattering of these components in the troposphere. The light pollution, of course, results from troposheric backscattering. Tropospheric scattering contains two components: Rayleigh scattering, with almost isotropic distribution of scattered brightness, and Mie scattering by aerosols, preferentially scattering into the forward direction ± 30 °. These two components are present in the artificial sky illumination as well, i.e. the observable sky brightening towards a city is not the whole story. There is an almost isotropic Rayleigh component as well!

4.3. TEMPORAL VARIATIONS

Most of the variations are due to Airglow, sometimes by a gradual decrease during the night, but fluctuations and continuous brightness increases occur as well. Differences from night to night may be substantial (20–50%), and this is a most significant effect to be considered when measuring changes in light pollution!. The variations with solar cycle are of similar size (≈50%) and correlated to the solar 10.7 cm flux.

There are indirect temporal variations caused by the passage of Milky Way and the zodiacal light, which may introduce spurious variations in measurements taken at different hours or days. Some change in scattered light intensity will also occur by the diurnal motion of these extended light sources.

There is a slight annual variation of zodiacal light, by about ± 10% at high ecliptic latitudes, due to the Earth's orbital motion within the interplanetary dust cloud.

4.4. BRIGHTNESS OF THE NIGHT SKY

Measurements of dark night sky brightness at different observatories lead to best values of typically 22 mag/square arcsec in V, 23 mag/square arcsec (≈ 100 S_{10}) in B. The spectrum of the night sky is dominated by strong airglow lines e.g. at 557.7 nm [OI], 589 nm [NaI], 630 nm [OI] and OH emission at longer wavelengths. Comparatively dark spectral bands are from 380 nm to 400 nm, 450 nm to 520 nm, 640 nm to 710 nm, and around 820 nm. If a choice has to be made, try to protect these bands!

4.5. RECOMMENDATIONS FOR LIGHT POLLUTION MEASUREMENTS

Therefore, if one wants to avoid mixing up natural changes with man-made increases in monitoring of light pollution, the following measures are recommended:

- Perform measurements at preselected, known positions on the sky, document all relevant coordinates, avoid stars or correct for them.
- Measurements should cover a sufficiently long interval (several hours per night/ several nights) to be free of "random" airglow variations within a night and from night to night.
- Measurements should cover a substantial fraction of the solar 11-year cycle or, at least, take the effect of solar activity into account when comparing measurements from different years.

5. FIELD SURVEY OF OUTDOOR LIGHTING IN JAPAN

K. Narisada, K. Kawakami

The Illuminating Engineering Institute of Japan conducted a field survey concerning the outdoor lighting in 1995, in various areas of the country. The aim of the survey was to find which types of luminaires in existing lighting installations are wasting the greatest amount of upward flux into the sky. The survey was conducted in six of the seven zones, proposed in an earlier draft of the guide of the CIE, i.e. E1, E2a, E3a, E3b, E4a and E4b. On the basis of the results of the survey, the proportion of the upward flux per 100 hectares was compared separately for eight types of luminaires, such as the luminaires for traffic roads, residential areas, decorative (shopping) streets, projectors, luminous signs, etc. As a result, it was found that the major sources of the upward flux were the luminaires for decorative streets and luminous signs and to be improved their photometric characteristics.

6. BILATERAL AGREEMENTS ON LIMITS TO OUTDOOR LIGHTING; THE NEW CIE RECOMMENDATIONS, THEIR ORIGIN AND IMPLICATIONS

D. A. Schreuder

Summary

Sky glow presents itself as a background luminance over the sky, against which the astronomical objects are to be observed. The interference of astronomical observations is caused by the resulting reduction in luminance contrast. The glow is caused by non-directional scatter of light by particles in space and in the atmosphere. Part of the light, and part of the particles are natural, and part is man-made. The '(natural) background radiation' is defined as the radiation (luminance) resulting from the scatter of natural light by natural particles. For earth-bound observatories, the background luminance is the limit for sky glow. It is customary to express the sky glow (from man-made sources) as a percentage of the natural background luminance.

The main sources of stray light that may interfere with astronomical observations are:
. lighting of industrial sites, airports and building sites;
. road and street lighting;
. advertising signs
. floodlighting of buildings, discos and monuments
. lighting of billboards
. lighting of greenhouses
. lighting of sports facilities
. area lighting of sales areas, parking lots, farm yards, railroad yards etc.

The Commission Internationale de l'Eclairage CIE set up 'Guidelines for minimizing sky glow; A CIE Technical Report'—presently in print. Most recommendations are expressed in values of the Upward Light Output Ratio—installed (ULORinst). The recommendations are based on three principles:

* the requirements for upward light emission are related to the activities in the 'zone' under consideration ('zoning');

* the requirements for upward light emission are most stringent in the 'night'; they may be relaxed in the 'evening' ('curfew');

* the lighting requirements in nearby zones must be taken into account (distance relations).

The future plans for the CIE Technical Committee CIE TC 4-21—in close collaboration with IAU Commission 50—include:
. preparing Draft ISO-Standards for measuring methods and for limiting values of sky glow
. preparing a draft manual on the theory and practice of road lighting for astronomers and for highway authorities.
. collect routine sky brightness measurements from various observatories
. contribute to the education of astronomy and the role of lighting there-in.

7. U.S PERSPECTIVE FOR INTERFERENCE TO RADIO ASTRONOMY

T. E. Gergely (Division of Astronomical Sciences, National Science Foundation)

During the last decade a number of new radio telescopes were built, and several are currently in various construction, design or planning stages. These instruments represent a substantial scientific investment, and radio astronomy continues to contribute many new ideas and results to astronomy. At the same time, the regulatory environment related to the radio spectrum has undergone considerable liberalization in the USA, as well as internationally. In the USA responsibility for spectrum utilization is shared by the National Telecommunication and Information Administration (NTIA), that oversees Government uses, and the Federal Communication Commission (FCC), that looks out for private sector spectrum needs. US spectrum management principles, as stated by the FCC, include the fostering of competition, flexibility of spectrum use, licensing and spectrum fee policies geared towards supporting the value of the spectrum and looking out for both the global market context and the public interest. Regulatory support of radio astronomy under the last item on this list is certainly not very high among FCC priorities. Similar situation prevails at the International Telecommunication Union (ITU), which regulates the uses of the radio spectrum internationally.

The last decade also has seen an expansion of satellite based radio services. Satellite technologies with substantial spectrum demands include the mobile satellite services, digital audio broadcasting, direct-to-home broadcasting and advanced fixed and fixed satellite services. This last name is a euphemism for satellite systems currently planned for distributing Internet and related material and other high data rate applications. The large expansion of satellite services, simultaneously with the relaxation of regulations, threatens radio astronomy. Considerable spectrum has been reallocated to satellite downlinks, and unavoidably (and sometimes avoidably) these are found next or near enough to radio astronomy bands so that spurious and/or out-of-band emissions spill over, and make astronomical observations difficult or impossible. The increasing spectrum congestion at cm and dm wavelengths, the wide bandwidths needed for Internet applications, and the availability of mm-wave technology are beginning to be felt in the mm and sub-mm regions of the spectrum, which astronomers considered their exclusive domain until recently, regardless of the allocation table.

Solutions to these problems will not be easy. Astronomers will need to educate the general public, as well as the regulatory authorities. In doing so, they must keep in mind that many new and satellite based services are extremely popular. The astronomy community will also have to get more involved in the various regulatory fora, and establish good contacts with industry. To do so will require that more human as well as financial resources be dedicated to spectrum management activities.

8. JAPANESE PERSPECTIVE FOR INTERFERENCE TO RADIO ASTRONOMY

K. Kawaguchi (NRAO, National Astronomical Observatory)

Interference to radio astronomical observations becomes serious, especially in relatively low frequency regions. Since general discussions on interference are given in other papers of this conference, the present paper describes some examples of interference detected in recent astronomical observations and future perspective.

I. Spurious in fringe detection between Kashima and VSOP in 1600 MHz region Recently, ISAS (Institute of Space and Aeronautics) launched a spacecraft HALCA with 8-m radio telescope for very long baseline interferometer. The first fringe test was carried out on 1997 May 7 for VSOP (VLBI Space Observatory Program) using the Usuda 64-m and Kashima 34-m ground radio telescopes and HALCA space telescope. In the test, the Kashima telescope received extended spurious noises at 1642.5 MHz, which made it difficult to detect fringes between HALCA and Kashima. Then the observation frequency was shifted to 1670–1686 MHz to avoid spurious in Kashima, and fringes were successfully detected between Kashima and HALCA. The band is protected as radio-astronomical band, but it is noted that such strong spurious are present for interference the astronomical observations.

II. Spurious in Taurus Molecular Cloud-1 observations with Nobeyama 45-m telescope The telescope site Nobeyama was chosen by considering to be free from radio interference. The site is surrounded with mountains and sparsely populated. However, it is found that recent development of various communication systems give large interference to radio observations below 30 GHz region. Especially when we observe quiet, cold dark astronomical source like TMC-1, where spectral line width are narrow as 0.6 km sec^{-1} (60 kHz at 30 GHz), it becomes very difficult to discriminate real lines from artificial spurious emission, even if a position switching method is employed for astronomical objects.

III. 95-GHz cloud profiling radar CRL (Communication Research Laboratory, Japan) and NASDA (National Space Development Agency of Japan) have project of a cloud profiling radar system from satellite for investigation of global warming which can be emotive issues. The radar system will use 3 mm radiation by considering scattering effect and established technology levels.

The frequency is also very important for radio astronomical observations. The following interference will occur,

(1) Main lobe—main lobe coupling
In the case of the Nobeyama 45-m telescope, the received power from the radar is estimated to be about 1 W by considering the sizes of radar beam and telescope. Since the destruction power for SIS (Superconductor-Insulator-Superconductor) junction is estimated about 10 mW, the radar power is two order of magnitude larger than the destruction power.

(2) Main lobe (telescope)—side lobe (radar) coupling
This is the case that a telescope sees a radar side lobe. Usually an SIS mixer receiver is working with a local power of 10^{-9} W, and higher local power gives a change in bias point and results in increase of noise temperature.

(3) Side lobe—side lobe coupling
This is the case that a telescope receives a power less than receiver saturation level, but the radar radiation is observed as a spurious line at the exact radar frequency. The telescope receives more power than harmful interference level (Pr = −203 dBW at 90 GHz region), for a shorter distance than 1080 km. If the satellite radar system wants to use the 95 GHz band, the band must be allocated to active sensing from satellite. From astronomical points of view, we are worry about such thing that many active services use the 95 GHz region after the allocation. The 95 GHz region is also not allocated for radioastronomical bands, but the effects (1) and (2) from radar make it difficult to observe in radio astronomical bands of 3 mm region. We must make efforts to reallocate the 95 GHz band for astronomical bands, because of protection of the best observing bands. Alternative way is to reserve the millimeter-wave telescope sites as protection zones.

9. SHARING THE RADIO SPECTRUM

R. J. Cohen (NRAL Jodrell Bank)

The radio spectrum is a common resource like air or water which can be used in many ways and for many purposes. Similarly it must be shared equitably. International use of radio is regulated by the International Telecommunications Union (ITU). The scientific use of radio for passive measurements such as radio astronomy or remote sensing (passive) was introduced into the ITU regulations at a late stage, and does not fit naturally into the framework. The needs of radio astronomy are inherently different from those of most other radio users.

The power levels reaching Earth from cosmic sources are generally very low and they are beyond the astronomer's control. Large antennas and sensitive receivers are needed to detect and measure them. This makes radio astronomy particularly vulnerable to interference from man-made emissions. The cosmic emissions are often of unknown character, such is the nature of research, and this can make it difficult to distinguish the wanted signal from interference. Radio astronomy needs quiet frequency bands in which to conduct measurements of the highest sensitivity. This is best achieved by world-wide allocations of passive frequency bands. Yet many other frequency bands are also of scientific interest. Only 2% of the radio spectrum below 50 GHz is allocated to radio astronomy, and only 30allocations are passive.

As the worldwide use of radio expands, radio astronomy is increasingly going to be confined to the officially allocated bands. Concern is now mounting at the state of those bands. Real transmitters cannot confine their emissions to sharply-defined frequency bands like those allocated by the ITU. Inevitably there are unwanted emissions which spill into other frequency bands, creating a type of radio pollution. The ITU has been slow to address this growing problem of unwanted emissions. Emissions from satellites are a particular concern for radio astronomy. Just one satellite can affect radio observatories all around the world, however isolated geographically and however well-shielded from terrestrial transmitters. The ITU has no regulations on the unwanted emissions from satellite transmitters.

Examples of interference to radio astronomy from satellites are growing steadily. In the 1980s the global satellite navigation systems GPS and GLONASS became well known to radio astronomers through their unwanted emissions into the frequency bands around 1.6 GHz used to observe spectral lines of OH. The solution to such problems takes many years to negotiate and to implement, once the system is already flying. This year the Iridium communications satellite system looks set to cause interference in the very same OH band, despite the experiences of GLONASS and despite efforts to avoid a similar problem with Iridium. At higher frequencies we have the Astra satellite over Europe interfering at 10.7 GHz with observations in a passive band.

My great concern is for the future. Altogether 80% of radio astronomy bands with a worldwide primary allocation are immediately adjacent to bands allocated for satellite downlinks! GLONASS and Astra may be only the beginning. The low Earth orbitting satellites (LEOs) of the future will fly in large constellations like Iridium, so that many will be visible simultaneously. The challenge will need to be addressed at many levels, from technical to regulatory. We are not opposed to satellite systems, but we do not want their downlinks right next to our spectral lines. We cannot stop the telecommunications revolution, but we must find ways to reduce its impact on radio astronomy.

10. KEEPING THE RADIO WINDOWS OPEN

W. A. Baan (Arecibo Observatory, IUCAF Secretariat Barrio Esperanza)

Passive scientific use of the radio spectrum has become more difficult due to harmful interference from active users in other spectral bands. The time has come for national administrations and spectrum managers to re-consider the protection afforded the non-profit passive services from market-oriented commercial users. The following urgent issues need to be addressed in order to assure adequate protection for the passive users operating in the Radio Astronomy and Earth Exploration services:

1. The Radio Regulations of the International Telecommunication Union —The Radio Regulations (RR) afford protection of all spectrum users on an equal basis and do not allow preferential treatment of one particular service. However, the passive Radio Astronomy and Earth Exploration users of the spectrum only observe natural emissions that are significantly weaker than the vast majority of man-made signals. Therefore, this "equality for all" rule from the RR cannot reasonably be interpreted as equal interference levels for all. It is essential that the ITU modernize and improve its protection standards for the passive services as it introduces and regulates existing and new space-based telecommunications industries.

2. Limits of Unwanted Emissions—Unwanted emissions from other spectrum users in adjacent or far-removed bands have become a limiting factor for the use of many passive bands. Passive users are intend on lowering the limits for out-of-band and spurious emissions. Although the protection of the passive services was one of its principal objectives, ITU-R Task Group 1-3 on "Spurious Emissions" has recommended standards for WRC-97 that do little to protect the passive services. Continued efforts within a new Task Group 1-5 are required to bring the standards for unwanted emissions in line with current radio technology.

3. Space Services—The explosive growth of the use of satellites by various services offers great benefits to mankind yet the current growth in number poses an ominous threat for scientific spectrum users. Downlink emissions can be particularly damaging for ground-based astronomical telescopes and already the emissions of GLONASS and ASTRA-1D have severely curtailed observations. The new IRIDIUM system poses a similar threat. In building space vehicles the highest engineering standards should not be sacrificed in favor of economic arguments. In particular, electro-magnetic environmental impact statements and extensive testing must be required before launching space-borne transmitter systems.

4. Spectrum Bands Above 30 GHz—The increasing commercial use of frequency bands above 30 GHz poses new problems for the scientific spectrum users. In particular, the protection of spectral neighbors will require higher engineering standards and vastly improved filter technology. Regulations for high frequency bands up to and beyond 1000 GHz should take into account the lessons learned at lower frequencies to prevent similar shortcomings.

5. Spectrum Splintering—The ITU Allocation Table contains spectrum designations for many "unlike" services with different operational standards in adjacent bands. Many of these bands are rather narrow and extensive coordination is required for these users to operate together and not interfere. Even today about seventy percent of all radio astronomy bands are threatened by satellite links and still new satellite downlink allocations are sought next to passive bands. Administrations need to strive to allocate 'like" services to broader band sections and avoid further splintering of the spectrum in increasingly narrow bands.

6. Sharing Considerations—Creative time sharing and geographical sharing schemes may provide limited access to certain bands, which are allocated to other services but have scientific potential. Geographical sharing schemes will require the creation of coordination zones or quiet zones by national administrations to provide local protection from ground-based services for sensitive observatories.

Peaceful coexistence in the spectrum for passive and active services is possible and depends on mutual respect for each other's activities. As spectrum occupation increases, the need to protect all passive and active spectrum users becomes more urgent. Only by mutual understanding and cooperation can we keep the radio windows to the Universe open and clean and in reserve for passive use by future generations.

11. THE SPACE DEBRIS ENVIRONMENT OF THE EARTH, AMOUNTS AND GROWTH

W. Flury (ESA/ESOC)

Since 1957, more than 3800 launches have led to more than 8000 trackable objects currently in orbit around the Earth. Surveillance, that is detecting and tracking space objects with radar, optical and infra-red sensors, is carried out by the space surveillance networks of the United States and Russia. The trackable objects have typically a minimum size of about 10 cm at low altitude and about 1 m at the geostationary orbit. Of the large number of catalogued objects fewer than about 500 are operational satellites. The remainder are abandoned satellites (21%), upper stages (16%), fragments of satellites and upper stages (45%), and mission-related objects (12%) such as lens covers, separation bolts or clamp-bands. A much larger number of smaller objects—unobservable by routine space surveillance—maybe 100,000 to 150,000 larger than 1 cm and thus capable of damaging a spacecraft, is in orbit. With special campaigns of powerful radar facilities, e.g. Haystack, FGAN, or the combined use of FGAN with the 100 m Effelsberg radio telescope, statistical information on 1 cm size objects in Low-Earth orbit (LEO) can be gained.

With an uncontrolled growth orbital debris could become a major hazard for all human activities in space. Already now orbital debris is a potential risk in two important regions, namely at low Earth altitude, and in the geostationary ring. An indication of the degrading quality of the space environment are the first confirmed collision of an operational spacecraft (Cerise) with a rocket fragment, avoidance manoeuvres of the US Space Shuttle and of unmanned spacecraft in LEO, and the more than 60 replacements of windows of the US Space Shuttle damaged by impacts of small-size debris and meteoroids. Space debris is also degrading astronomical observations as large scale exposures of the sky may contain satellite trails.

This presentation will provide an overview of the terrestrial space debris environment. Sources for the catalogued objects and small-size space debris will be reviewed. Models for the spatial distribution will be presented. Explosions and collisions as the main debris sources will be addressed. Finally, the long-term evolution and debris reduction measures will be reviewed.

12. ENVIRONMENTAL DISTURBANCES OF ASTRONOMICAL OBSERVATIONS

J. Kovalevsky (CERGA, Observatoire de la Côte d'Azur)

This presentation is aimed at describing several effects, generally of geophysical origin, which contribute, in addition to sky glow or electromagnetic wave interferences, to degrade astronomical observations from the ground.

The efficiency of an astronomical instrument must be considered together with all the atmospheric layers crossed by the incoming light. This optically active element affects the shape of the image and the apparent direction of the observed celestial body. Astrometric measurements may be significantly biased by un-modelled spurious refraction effects. But the most important source of disturbances is the local atmospheric turbulence which is the major factor of astronomical seeing. The height of the boundary layer is function of the existence of heat sources. In day-time, it is governed by the Sun, but in night-time, the presence of buildings, roads, vegetation of various types, has a definite influence on the quality of the atmospheric images, increasing their unstability and introducing sometimes an inclination of the atmospheric layers, causing abnormal refraction.

The chemical composition of the atmosphere has evidently a direct effect on spectroscopy. This is particularly important in millimeter and submillimeter band astronomy, in the research of various interstellar or circumstellar molecules. The atmospheric emissions and opacity in this wavelength domain are major perturbations in recognizing the spectral lines of these extraterrestrial emissions. Increase in opacity may also occur after major geophysical events such as aerosols or dust from volcanos.

Optical interferometres are very sensitive to any perturbation of the optical path lengths. This can be produced by deformations of the wavefront due to turbulence, and particularly long period deformations in the higher atmospheric layers. Another cause is the variation of the baseline, consequence of seismic activity. Very significant perturbations are produced by oceanic waves which and are important even several hundreds of kilometers away. Storms on the ocean may be sources of important deterioration of interferometric observations.

Some of the atmospheric dynamical effects may be compensated using active or adaptative optics. It has however to be noted that these techniques correct the wavefront only within a few arc seconds. This is certainly the solution when a single star is being observed, but does not apply for larger field studies. It is also conceivable that is the future, interferometric baselines could be corrected in real time from seismic simultaneous readings. Millimetric and submillimetric observations can get rid of a large part of spectroscopic pollution in hight altitude sites. But most of the ground based astronomical observations will continue to be highly sensitive to these environmental perturbations.

13. THE AVOIDANCE OF MAN-MADE POLLUTION IN INTERPLANETARY SPACE

C. S. L. Keay (University of Newcastle)

At the 20th General Assembly of the International Astronomical Union in Baltimore, U.S.A., in 1988 members of Commission 22 and several Commissions in what is now the Division of Planetary Sciences expressed deep concern that no work was being undertaken to identify and avoid pollution problems in interplanetary space. Commission 22 set up small a working group with I P Williams as convenor. It identified several problems requiring further study and recommended further investigation.

At the 22nd General Assembly in The Hague in 1994 a formal Working Group on The Prevention of Interplanetary Pollution was set up with members from each of the Commissions in the Division of Planetary Sciences. Care was taken to include members from each of the international powers with a presence in space operations and research. Astronomers from Australia, China P R, the Czech Republic, France, Japan, the Russian Confederation and the United States of America. Largely through e-mail contact the convenor, C S L Keay, drew on the expertise among the nominated members to expand the earlier report.

The principal areas of concern dealt with by the Group are:

* Environmentally harmful propellant residues;
* Unconfined debris from impacting objects;
* Pollution from explosives, particularly nuclear;
* Radionuclide pollution from nuclear power generators;
* Undesirable transfers of surface materials;
* Biocontamination prevention and quarantine measures.

It was stessed that the challenge is to devise cost-effective techniques and procedures acceptable to mission planners which will prevent or at the very least minimise pollution outcomes. To achieve success a program of careful preliminary research will be necessary, paying attention to the following points in regard to all aspects of space missions:

* Identify likely sources of undesirable pollution;
* Measure present levels and predict future levels;
* Assess future impact of pollution on research goals;
* Investigate likely severity and irreversibility;
* Formulate preventative or minimisation procedures;
* Set guidelines for assessing preservation values;
* Develop means to protect sensitive environments;
* Seek ways to increase awareness of the issues.

The working Group recommended the establishment of an international scientific committee to involve all relevant agencies and draw up suitable procedures and protocols for achieving success. It should be very much less costly to be proactive in this endeavour than to allow permanent harm to sensitive regions of the interplanetary environment.

14. THE PROCESS OF FREQUENCY MANAGEMENT, INTERNATIONAL TREATIES AND THE RESPONSIBILITY OF ASTRONOMERS

J. Tarter (SETI Institute)

Working with national and international regulatory bodies to allocate the radio spectrum, and monitoring regional applications for spectrum usage are increasingly becoming the cost of doing radio astronomy. Economic incentives are enormous, with new services constantly seeking additional allocations within the radio spectrum. Radio astronomers cannot depend on the goodwill of for-profit service providers to keep portions of the spectrum free for the use of the passive services (radio astronomy and remote sensing). At present, a small number of senior scientists and managers are struggling against this tide. They should not only be praised, and thanked, but joined. It is imperative that current and future generations of radio astronomers learn the business of frequency management and participate in the distribution of this scarce resource.

15. EDUCATING THE PUBLIC ABOUT PRESERVATION OF THE ASTRONOMICAL WINDOWS

W. T. Sullivan, III (Univ. of Washington)

If astronomers are going to succeed in their efforts to keep the skies free of light pollution and radio interference, then they must deal not only with lighting and radio engineers, business concerns, and politicians, but also the general public. If the public becomes persuaded that clear, pristine skies are worth having for the same sorts of reasons that we value drinkable water and breathable air, then we have gained the most important constituency. It is, after all, the public who pay the taxes, who buy the products, and who vote in elections. Just as with other environmental issues, if the public is won over, then the engineers and policy makers (eventually) must follow along.

In this paper I highlight efforts around the world that have been particularly noteworthy in the cause of reducing light pollution by educating the public. Radio interference is also of great harm to astronomers, but it will be covered only briefly in this paper because very little has been done to educate the public about the radio problem. This is because radio pollution is a more abstract and technical concept for the average person, as well as because many of the problems are addressable at the international (World Telecommunications Union) and national levels. With light pollution, however, the battles are more often on a local stage.

The premier organization fighting light pollution and educating the public about it is the International Dark-Sky Association (www.darksky.org/ ida/index.html), based in Tucson, Arizona, USA, and with over 2000 members in 69 countries. Over the past ten years IDA, largely through the person of David Crawford, has been able to not only preserve dark skies for observatories, but more significantly convince a great number of lighting engineers and public officials that it's in their own best interest to use lighting systems that are also beneficial to astronomers. Exterior lighting for streets and buildings that does not shine upwards saves a great deal of money, does its job much more efficiently and safely, and is aesthetically more pleasing. An alliance has been forged between those on all sides who are "enlightened" and it is steadily growing in its size and influence. For the cause IDA has produced 7 slide sets, 120 information sheets, and 2 videotapes.

The best way to check other major efforts around the world is to visit the Web sites linked to IDA's site. For example, in the United Kingdom the Campaign for Dark Skies (CDS) has also been effective in its educational and political efforts. Their booklet Starry, Starry Night, produced in conjunction with the Council for the Protection of Rural England, is first-rate in promoting dark skies and good lighting practices. CDS has also promoted "Star Watch UK", a project for schools and other youth groups to count the stars visible in a well-defined area of the sky. This type of program has also been done in Canada and the USA, and especially in Japan, where it was funded by the national Environmental Agency. Based on these amateur data, maps were constructed for sky brightness all around Japan. Even more than the maps, the real value of such projects is the increased awareness they bring to the public of deteriorating sky conditions. Many other examples, from countries such as Switzerland, Holland, Italy, and Spain, are discussed in the full paper.

16. PUBLIC EDUCATION TO PRESERVE DARK SKIES AND ASTRONOMICAL WINDOWS WITH EAVESDROPPING AND ROBOTIC TELESCOPES

J. E. F. Baruch (Robotic Observatories)

Abstract

It is suggested that the only guarantee for the preservation of dark skies is the guarantee that exists in the hearts of the people around the world. This paper notes how rising living standards and increasing GNP are associated with light pollution destroying the heritage of the night sky. It argues that it is necessary to decouple light pollution from rising living standards. A contribution to this process is to provide indices that are easily understood and useable for measuring the quality of the night sky.

It is shown how robotic telescopes can provide access to the night sky for people in the developed world through programmes which provide education, involvement in astronomy research and access to the leading edge of astronomy through the new technique of eavesdropping. These education programmes will help to bring an understanding of the value of our heritage that belongs to our grandchildren: The Night Sky.

17. THE ISSUES OF SPACE DEBRIS AND NEAR-EARTH OBJECTS AT THE UNITED NATIONS

H. J. Haubold (United Nations Office for Outer Space Affairs)

The focal point of the United Nations activities in the field of peaceful uses of outer space is the Committee on the Peaceful Uses of Outer Space (COPUOS), set up in 1959 by the General Assembly. Since its establishment, the Committee and its two subcommittees (the Scientific and

Technical Subcommittee and the Legal Subcommittee) have resulted in five legal instruments, all of which have entered into force, as well as four sets of principles adopted by the United Nations General Assembly on the conduct of space activities, including astronomical satellite missions (http://www.seas.columbia.edu/ ah297/un.html):

1. Treaties

The 1966 Treaty on Principles Governing the Activities of States in the Exploration and Use of Outer Space, including the Moon and Other Celestial Bodies ("Outer Space Treaty") provides that space exploration shall be carried out for the benefit of all countries, irrespective of their degree of development. It also seeks to maintain outer space as the province of all mankind, free for exploration and use by all States and not subject to national appropriation. The 1967 Agreement on the Rescue of Astronauts, the Return of Astronauts and the Return of Objects Launched into Outer Space ("Rescue Agreement") provides for aiding the crews of spacecraft in the vent of accident or emergency landing, and establishes a procedure for returning to a launching authority a space object found beyond the territorial limits of that authority. The 1971 Convention on International Liability for Damage Caused by Space Objects ("liability Convention") provides that the launching State is liable for damage caused by its space objects on the Earth's surface or to aircraft in flight and also to space objects of another State or person or property on board such objects. The 1974 Convention on Registration of Objects Launched into Outer Space ("registration Convention") provides that launching States shall maintain registries of space objects and furnish specified information on each space object launched, for inclusion in a central United Nations Register (maintained by the United Nations Office for Outer Space Affairs). The 1979 Agreement Governing the Activities of States on the Moon and Other Celestial Bodies ("Moon Agreement") elaborates in more specific terms the principles relating to the Moon and other celestial bodies set out in the 1966 Treaty and sets up the basis for the future regulation of exploration and exploitation of natural resources found on such bodies.

2. Principles

The Declaration of Legal Principles Governing the Activities of States in the Exploration and Use of Outer Space (1963). The Principles Governing the Use by States of Artificial Earth Satellites for International Direct Television Broadcasting (1982). The Principles Relating to Remote Sensing of the Earth from Outer Space (1986). The Principles Relevant to the Use of Nuclear Power Sources in Outer Space (1992).

3. Space Debris

Space debris are all man-made objects in Earth orbit or reentering the dense layers of the atmosphere that are non-functional with no reasonable expectation of their being able to assume or resume their intended functions or any other functions for which they are or can be authorized, including their fragments and parts. The item on space debris was included into the agenda of the Scientific and Technical Subcommittee at its 1994 session; since 1995 the Subcommittee continued its consideration of this item on a priority basis. Member States and relevant international organizations are currently providing information on practices in minimizing the creation of space debris and the effects of this environment on space systems, including those used for astronomical purposes (`http://www.un.or.at/OOSA_Kiosk/`).

4. Near-Earth Objects

Pursuing an understanding of Earth's interactions with near-Earth objects and adverse environmental impacts on astronomy have become issues of global research. In an effort to provide a scientific basis for future cooperative international research and space exploration, The Explorers Club and the United Nations Office for Outer Space Affairs organized an international conference on near-Earth objects (A. Carusi, T. Gehrels, and S. Isobe in Near-Earth Objects: The United Nations International Conference, Ed. J. Remo, Annals of the New York Academy of Sciences 822 (1997) 632 pp.). The forthcoming United Nations Conference on the Exploration and Peaceful Uses of Outer Space in 1999 at Vienna, Austria, may provide a forum to review such issues in depth.

The views, interpretations, and opinions presented in this paper do not necessarily reflect the position of the United Nations.

18. BILATERAL AGREEMENTS, ZONING, INTERNATIONAL PROTOCOL

S. Isobe (National Astronomical Observatory)

It is clear that conditions of astronomical observations become worse and worse : those are light pollution, radio interference, and space debris. However, causes of these worse astronomical conditions are created to make human life convenient. Therefore, our problems can not be solved straightforward. An way which astronomers have managed to keep good observational conditions is to escape from residential areas to country areas and to mountain areas, and may be in future to back side of the moon. An expansion of cities has produced bright sky glow because of unproper usage of lighting instruments and an increase of necessary radio bands for daily life has produced the higher level of radio noise for astronomical observations. Launching of useful satellites has produced large number of space debris.

In these situations it is hard to keep conditions of astronomical observations in a good shape only by efforts of astronomers. We have to work together with other organizations to develop a channel reaching to the majority of people. For a space debris problem, there are the IAD (Inter-Agency Space Debris coordination) composed of space agencies from many countries such as NASA, ESA, NASDA, ISAS, Russian one, and Chinese one, specialists for space debris at each organization are worry about debris collisions to working satellite near future, but decision makers do not consider it at a severe situation. Astronomers should work with those specialists. For a problem of radio interference, there are COSPAR, ITU, and URSI with which astronomers should work together.

For a problem of light pollution, there is CIE(Commission Internationale de l'Eclairage) with which IAU Commission 50 now jointly sets up a Technical Committee 4-21 under the CIE. To compromise requests by observational astronomers and lighting engineers, an idea of zoning was proposed by Paul Murdin in 1992 and is now included a guide line of CIE TC4-21 "Interference by Light of Astronomical Observations".

To make people realize difficult conditions of astronomical observations, astronomers themselves should produce scientific evidences to show the difficulties. Unfortunately, some certain fractions of astronomers do not show any interest in the problems. We should concentrate our efforts and set up international protocol. For the light pollution, the CIE Division 4 passed a resolution to minimize air glow in October, 1996, the IAU GA passed a resolution to support the CIE one in August, 1997. Then, it is expected that the CIE GA will to pass it in 1999, which will be a starting point to bring it as the ISO standard. When this international protocol will be once set up, all the governments will take different actions to reduce sky glow under the idea of zoning.

Astronomers can directly be against people producing useless light, but it is usually hard work since many people enjoy bright lighting. It is much short way to minimize light pollution that international protocol such as the ISO standard is set up. We can find out a similar way for problems of radio interference and space debris.

JD6

Interactions between Planets and Small Bodies

Chairpersons and Editors: **M. Ya. Marov & H. Rickman**

INTERACTIONS BETWEEN PLANETS AND SMALL BODIES: INTRODUCTION

M. MAROV
Keldysh Institute
Russian Academy of Sciences, Moscow 125047, Russia

AND

H. RICKMAN
Astronomiska observatoriet
Box 515, S–75120 Uppsala, Sweden

The exploration of our Solar System is rapidly growing in importance as a scientific discipline. During the last decades, great progress has been achieved as the result of space missions to planets and small bodies and improved remote-sensing methods, as well as due to refined techniques of laboratory measurements and a rapid progress in theoretical studies, involving the development of various astrophysical and geophysical evolutionary models, based in particular on the approach of comparative planetology. In the crossroads of astronomy and geophysics, recent years have seen a growing understanding of the importance of impact phenomena throughout the history of the Solar System and, therefore, the necessity to get more insight into the problem of interactions of planets and small bodies. This importance is clearly manifested by the observed cratering records of planetary surfaces and such dramatic events as the explosions of the comet P/Shoemaker-Levy 9 fragments in Jupiter's atmosphere in 1994, that of the Tunguska object over Siberia in 1908, and the Chicxulub event dating back to the end of the Cretaceous.

The significance of impact processes in planetary evolution is demonstrated not only by the obvious fact that small bodies scar the planetary surfaces, but also by the progressively collected evidence that they shape the evolution of the atmosphere and hydrosphere of the Earth and atmospheres of other planets. Indeed, deposition of cometary-like volatiles through impact processes is considered to be mainly responsible for the origin of secondary atmospheres on the inner planets. These impacts, at later times, probably have repeatedly interfered with the Earth's biological evolution and may once have been essential for starting this evolution. Currently part of the NEO ('Near-Earth Objects') population, closely approaching the Earth's orbit, represents a potential threat to our civilization. At the same time, the asteroidal and cometary populations have been continuously shaped by planetary gravity, resulting in a preferential flux of meteorites originating in resonant zones of the main belt, and, further out in the Solar System, the formation of dynamical structure in the recently discovered Kuiper belt. In addition, comets are known to be continually tossed around by close encounters with the planets, and an understanding of their dynamics is necessary in order to assess the cometary inventory of the Solar System and its slow "evaporation" into interstellar space. Indeed, comets provide the link of our Solar System with the Galaxy in several respects, e.g., by probing the material of the presolar cloud. Thus the various processes accompanying their interactions with the planets offer an opportunity to understand better both the nature of these primitive small bodies and their role in planetary cosmogony and Galactic evolution.

The explosive growth of knowledge in the field of Solar System exploration and the interdisciplinary character of the subject of impacts, bringing together planetary and galactic astronomers as well as specialists in geophysics, dynamics, and cosmochemistry, made the subject of this Joint Discussion very timely. The topic and program of the meeting brought interest of astronomers

J. Andersen (ed.), Highlights of Astronomy, Volume 11A, 221–222.

dealing with both observations and theory and allowed to give an outlook to the formation and evolution of planetary systems in general, being also relevant to the problem of Galactic evolution. Specific attention was given to near-Earth objects (NEO) with the involvement of historical retrospectives, NEO survey simulation and orbital element distribution, as well as an assesments of potential hazards for the Earth and improved facilities (either ground-based or space-born) to detect these objects most reliably and cost-efficiently.

Nineteen papers, including twelve invited and seven contributed, were presented at the three sessions, entitled "Impacts and Solar System Evolution", "Origin and Dynamics of Planet-Crossing Bodies", and "Observations of Impacts and Impactors". Forty-one posters covering a wider range of topics relevant to the title of JD 6 were selected and displayed during the poster session. The discussions benefitted from the broad participation of scientists from Japan as the hosting country where active research in most aspects of the topic are carried since a long time. For instance, five years earlier prof. S. Yabushita of Kyoto University organized an international workshop on "Dynamics and Evolution of Minor Bodies with Galactic and Geological Implications", which was very successful and had a clear impact on the development of international contacts and collaborations. The IAU XXIIIrd GA offered an invaluable opportunity to gather an even wider community for free and open-minded discussions, since the Pacific region including Australia houses an important part of the relevant research programmes.

Unfortunately, because of the strict time limit some authors were reluctant to prepare a paper for the IAU Proceedings. In addition, a limited page budget allowed to publish only short versions of the original presentations. But in order to cover all main topics discussed at JD 6 and hence to make the proceedings as comprehensive as possible, we decided as editors to complement the original papers and/or extended abstracts submitted for publication with an overview of other presented papers and posters. This overview comes after the papers and includes several sections where topically relevant papers are reviewed. We in no way attempted to modify the authors' original concepts or to use this review to promote our own ideas. Essentially the overview is based on the abstracts already published in the IAU GA abstract book, applying only minor editing.

A BRIEF SUMMARY OF KUIPER BELT RESEARCH

R. MALHOTRA
Lunar and Planetary Institute
3600 Bay Area Boulevard
Houston, TX 77058. USA

1. Historical background

Ideas about the contents of the Solar System beyond Neptune and Pluto can be traced back to at least Edgeworth (1943, 1949) and Kuiper (1951), who speculated on the existence of pre-planetary small bodies in the outer Solar System beyond the orbit of Neptune – remnants of the accretion process in the primordial Solar Nebula. The basis for the speculation was primarily the argument that the Solar Nebula was unlikely to have been abruptly truncated at the orbit of Neptune, and that in the trans-Neptunian accretion timescales were too long for bodies larger than about ~ 1000 km in radius to have formed in the 4.5 billion year age of the Solar System. Another important theoretical argument relevant to this region of the Solar System is related to the origin of short period comets. Fernández (1980) suggested that the short period comets may have an origin in a disk of small bodies beyond Neptune, rather than being "captured" from the population of long period comets originating in the Oort Cloud, the latter scenario having considerable difficulty reconciling the observed flux of short period comets with the exceedingly low efficiency of transfer of long period comet orbits to short period ones by means of the gravitational perturbations of the giant planets. The new scenario received further strength in the numerical work of Duncan et al. (1988) and Quinn et al. (1990) which showed that the relatively small orbital inclinations of the Jupiter-family short period comets were not consistent with a source in the isotropic Oort Cloud of comets but could be reproduced with a source in a low-inclination reservoir beyond Neptune's orbit. Duncan et al. named this hypothetical source the *Kuiper Belt*, and the name has come into common use in the last decade (although other names are also in use, e.g. Edgeworth–Kuiper Belt, and trans-Neptunian objects). A recent theoretical milestone was the work by Holman and Wisdom (1993) and Levison and Duncan (1993) on the long term stability of test particle orbits in the trans-Neptunian Solar System. This work showed that low-eccentricity, low-inclination orbits with semimajor axes in excess of about 43 AU are stable on billion year timescales, but that in the region between 35 AU and 43 AU orbital stability times range from 10^7 yr to more than 10^9 yr [see, for example, figure 1 in Holman (1995)]. Orbital instability in this intermediate region typically leads to a close encounter with Neptune which causes dramatic orbital changes, with the potential for subsequent transfer to the inner Solar System. Thus, this region could in principle serve as the reservoir of short period comets at the present epoch. However, the idea of a kinematically cold — i.e. low-eccentricity, low-inclination — population in this region is at odds with recent observations, and the question of the origin of short period comets remains unsettled at the present time.

On the observational side, published sky surveys relevant to the search for outer Solar System small bodies are listed in Table 1. The earliest is that of Clyde Tombaugh who carried out a search of the ecliptic sky in the 1930's from Lowell Observatory, a search that resulted in the discovery of Pluto. The most recent searches of David Jewitt and his students have used the new technology of CCDs to go to fainter magnitudes (but at the expense of covering only very small areas of the sky). The first Kuiper Belt object was discovered in 1992 by Jane Luu and David Jewitt, and given the designation *1992 QB*$_1$. Several dozen detections have been made since then in ground-based observations. In addition, Cochran et al. (1995) reported on a detection of small, very faint members of the Kuiper Belt using the Hubble Space Telescope (HST). However, this claim is questionable as the noise characteristics of the data were improperly accounted for (Brown et al., 1997). The HST

J. Andersen (ed.), Highlights of Astronomy, Volume 11A, 223–228.

experiment is scheduled to be repeated in the autumn of 1997, using the improved instrumentation aboard the Space Telescope.

TABLE 1. Observational surveys of the outer Solar system

Author(s)	limiting magnitude, m_R	Sky area deg^2	number detected	notes
Tombaugh (1961)	16.8	1530	1	Pluto
Kowal (1989)	19.5	6400	1	2060 Chiron
Luu & Jewitt (1988)	20.0	297	0	
Levison & Duncan (1990)	22.0	4.9	0	
Jewitt & Luu (1995)	24.8	1.2	7	1992 QB1,...
Williams et al. (1995)	20.5–23.5	0.5	2	
Irwin et al. (1995)	23.5	0.7	2	
Cochran et al. (1995)	28.1	0.001	29(?)	
Jewitt et al. (1996)	24.2,23.2	3.9,4.4	17	
Gladman & Kavelaars (1997)	24.8	0.05	0	

2. Summary of observations

The current inventory (as of August 1997) of KBOs consists of 53 objects detected at heliocentric distances of 28 AU to 43 AU. The physical properties of KBOs remain poorly known, as summarized below.

Color: Colors appear to be reddish, but quite diverse (Luu and Jewitt, 1997; Tegler and Romanishin, 1997).

Albedo: No albedo information is available, although albedos are thought to be similar to those of short period comets, typically 0.04.

Size: Physical diameters of the observed KBOs are in the range 100–500 km (as inferred from the observed brightnesses and the assumption of a geometric albedo=0.04). Owing the rather narrow range of sizes observed so far, the size distribution is not highly constrained; it is consistent with an inverse power law with index near 3 (cf. Jewitt et al., 1996).

Population: Taking account of the characteristics of the small field CCD surveys that these objects were detected in, it is estimated that there are $\sim 10^5$ KBOs brighter than magnitude 24 (or diameter greater than 100 km, in the 30–to–100 AU heliocentric distance range). This may be compared with the asteroid population which consists of only 230 bodies larger than 100 km in diameter. The Kuiper Belt population dwarfs that of the asteroid belt by more than 2 orders of magnitude.

Total mass: Using plausible power-law size distributions, the total mass of the Kuiper Belt in the 30–100 AU region is estimated to be 0.2–0.6 Earth-masses (Jewitt et al., 1996), consistent with the upper limit estimated from the orbit of Halley's comet (Hamid et al., 1968). We note, however, that the larger number is quite insecure as it is based upon the orbital characteristics of only one object, namely the brightest observed KBO, 1996 TL_{66} (Luu et al., 1997). The estimated Kuiper Belt mass is 1-to-2 orders of magnitude smaller than the primordial mass of solids in this region inferred from theoretical models of the Solar Nebula (cf. Tremaine 1990, Stern 1996).

Orbits: The orbital periods of KBOs are well in excess of 200 years. Therefore, observations over less than $\sim$ 5 years allow only very limited accuracy for the fitted orbits, and only about half of the detected KBOs have been fitted to orbits of reasonable quality (Marsden 1997). Nevertheless, some patterns are already evident in the distribution of their orbital elements (Fig. 1): it is clear that the orbital distribution is rather peculiar, highly non-uniform in semimajor axis, with nearly 50% of all objects librating in orbital resonances with Neptune, primarily the 3:2 resonance. There is a noticeable paucity of objects at semimajor axes less than 39 AU and also in the range 40 AU to 42 AU. The orbital eccentricities of the 3:2 resonant objects are surprisingly large, with the largest eccentricity objects being on deeply Neptune-crossing orbits (but protected from catastrophic close

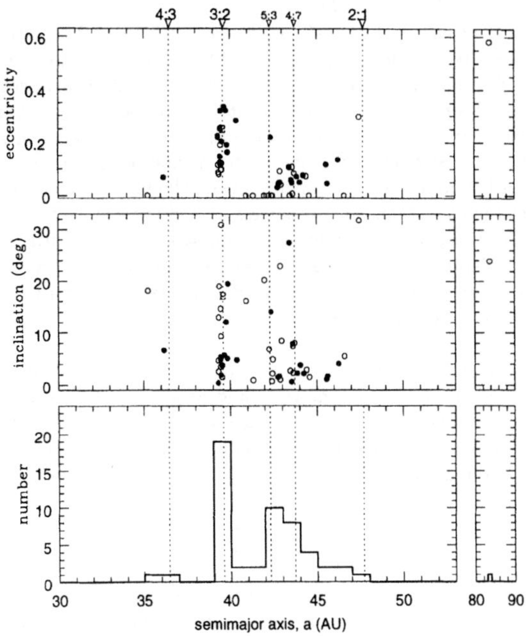

Figure 1. Orbital element distribution of detected KBOs (as of August 15, 1997). The open circles represent objects observed at only one opposition; filled circles those observed at multiple oppositions. The vertical dotted lines indicate the locations of Neptune's mean motion resonances.

encounters with that planet due to the resonance libration). Another surprise is the inclination distribution: inclinations range up to 31 degrees! Considering that ecliptic surveys are biased against high inclinations (cf. Jewitt et al. 1996), it appears that the Kuiper Belt is rather thick. It should be stated that there is possibly significant observational incompleteness in the distribution beyond $a \approx 42$ AU. Worthy of special note is the detection of one object, 1996 TL_{66}, on the most highly eccentric orbit which is also highly inclined and whose estimated semimajor axis is ~ 83 AU, the largest known so far (Luu et al. 1997). This object is possibly a member of a "scattered disk" population, a remnant of those Uranus-Neptune zone and/or Kuiper Belt planetesimals that were placed on chaotic but dynamically long-lived orbits after a close encounter with Neptune in the early history of the Solar System (Duncan and Levison 1997).

3. Orbital dynamics

A very large fraction of KBOs – roughly 40% – have orbits similar to that of Pluto, namely, locked in the 3:2 orbital resonance with Neptune, and have been informally named "Plutinos". The 3:2 resonance lock provides a dynamical protection mechanism against close encounters with Neptune by causing a libration of the longitude of perihelion about a center 90° removed from Neptune's longitude. (The reader is referred to Malhotra and Williams (1997) for a detailed discussion of Pluto's resonance librations.) A two-dimensional view of the global structure of the phase space in the neighborhood of Neptune's mean motion resonances has been obtained by means of surfaces-of-section within the planar restricted three-body model (Malhotra 1996). In particular, these views show that the stable zone in Pluto's neighborhood is very narrow, and is surrounded by a large "chaotic sea". Generally, each low order exterior orbital resonance of Neptune has a similar phase space structure: a narrow zone where eccentric orbits can librate stably, surrounded by a large chaotic zone. In addition to the mean motion resonances, important dynamical effects arise from secular resonances as well. Numerical and semi-numerical analyses of these are given in Duncan et al. (1995), Morbidelli et al. (1995) and Morbidelli (1997).

The overabundance of KBOs in the 3:2 resonance is a remarkable feature in light of the very small volume of phase space represented by the libration zone relative to the neighboring non-resonant (generally chaotic) phase space. Such an overabundance was actually predicted by the "resonance sweeping" theory for the origin of Pluto's Neptune-crossing resonant orbit (Malhotra 1993, 1995). In this theory, it is proposed that during the late stages of accretion of the giant planets, as part of the process of gravitational scattering and removal of residual planetesimals (the same process which is believed to have formed the Oort Cloud), an exchange of energy and angular momentum amongst the planets and planetesimals caused the giant planets to migrate radially. In particular, Neptune's outward orbital migration caused the trans-Neptunian population of small bodies to be swept by Neptune's orbital resonances. As a consequence, there was great mass erosion in the region up to about 50 AUs; the majority of the surviving small bodies were captured into Neptune's mean motion resonances and had their orbital eccentricities pumped up to modestly large values.

The process of resonance capture is a very delicate one and difficult to analyze mathematically. However, under certain simplifying assumptions and idealizations, first order perturbation theory for mean motion resonance sweeping and resonance capture into a $j : j+k$ mean motion resonance provides a relationship between the eccentricity pumping and the magnitude of outward migration of Neptune's orbit (Malhotra 1995):

$$\Delta e^2 \simeq \frac{j}{j+k} \ln \frac{a_N}{a_{N,i}} \qquad (1)$$

where a_N is Neptune's current orbital semimajor axis and $a_{N,i}$ is its value in the past at the time of resonance capture. Thus, from Pluto's current orbital eccentricity of 0.25, we can estimate that Neptune's orbit expanded by at least 5 AU. Furthermore, the observed KBOs in the 3:2 Neptune resonance have eccentricities up to 0.32; this increases the above estimate for Neptune's orbital expansion to 7 AU.

Numerical simulations of giant planet orbital migration and resonance sweeping of the trans-Neptunian region have corroborated the above picture (Malhotra 1995, 1997a; Holman 1995). These simulations also show that KBO orbital inclinations may also be excited to large values during resonance sweeping (Malhotra 1997b). However, we note that these simulations assumed an adiabatic, smooth process for planetary migration on timescales of a few million years, and treated KBOs as massless test particles, thereby possibly overestimating the resonance capture efficiency as well as neglecting the gravity and mutual scatterings or collisions of the KBOs and other bodies scattered by the giant planets.

Several additional suggestions have been made in the last year which are also relevant to understanding the origin of the dynamical structure of the Kuiper Belt. Levison et al. (1997) have pointed out the importance of the sweeping of secular resonances, particularly in the mass-loss phase of the primordial massive Kuiper Belt; and Morbidelli and Valsecchi (1997) have discussed the possible role of "large Neptune-scattered planetesimals" in causing a gravitational "stirring" of KBOs into unstable orbits, thereby depleting a large fraction of the primordial mass, and permitting only those objects that happened to fall into stable resonant niches to survive over the age of the Solar System.

4. Concluding remarks

Kuiper Belt research is in its infancy. This fact alone explains much of the excitement surrounding this field. But I think that the excitement is justified more substantively by the likelihood that the peculiar Kuiper Belt dynamical structure holds memory of the long term dynamical history of the outer Solar System. Indeed, it is possible to trace a detailed, quantitative trail from the present Kuiper Belt dynamical structure to dynamical processes in the early stages of planet formation, an era hitherto considered largely outside of observational constraint.

Observations to date have detected only the brightest and, hence, presumably the closest and largest KBOs. It is essential to obtain a more complete census of the KBO population, including its orbital and size distribution in order to make further progress in dynamical studies which will allow an elucidation of the relationship between these objects and Neptune, Triton, Pluto-Charon, the Centaur population, and the Oort Cloud, as well as the Jupiter-family short period comets, and the asteroids. It is also important to seek opportunities for physical studies of these objects, many of which are perhaps the least perturbed remnants of the planetesimal accretion process and may provide new constraints on conditions that existed in the primitive Solar Nebula.

5. Acknowledgements

This research was supported in part by NASA's Origins of Solar systems Research Program under grant number NAGW-4474, and was done while the author was a Staff Scientist at the Lunar and Planetary Institute which is operated by the Universities Space Research Association under contract no. NASW-4574 with the National Aeronautics and Space Administration.

6. References

Brown, M.E., Kulkarni, S.R., & Liggett, T.J. (1997) An analysis of the statistics of the HST Kuiper Belt Object search, *Astrophys. J.* (in press)
Cochran, A., Levison, H., Stern, S.A., and Duncan, M. (1995) The discovery of Halley-sized Kuiper belt objects using the Hubble Space Telescope, *Astrophys. J.* **455** pp. 342–346
Duncan, M.J., Quinn, T., Tremaine, S. (1988) The origin of short period comets, *Astrophys. J.* **328** pp. L69–L73
Duncan, M. and H.F. Levison (1997) A disk of scattered icy objects and the origin of Jupiter-family comets, *Science* **276** pp. 1670–1672
Duncan, M., H.F. Levison and S.M. Budd (1995) The dynamical structure of the Kuiper Belt, *Astron. J.* **110** pp. 3073–3081
Edgeworth, K.E. (1943) The evolution of our planetary system, *J. of the British Astron. Assoc.* **53** pp. 181–188
Edgeworth, K.E. (1949) The origin and evolution of the Solar system, *MNRAS* **109** pp. 600–609
Fernandez, J.A. (1980) On the existence of a comet belt beyond Neptune, *MNRAS* **192** pp. 481–491
Gladman, B. and Kavelaars, J.J. (1997) Kuiper Belt searches from the Palomar 5 m telescope, *Astron. & Astrophys.* **317** pp. L35-38
Hamid, S.E., Marsden, B. and Whipple, F. (1968) Influence of a comet belt beyond Neptune on the motions of periodic comets, *Astron. J.* **73** pp. 727-731
Holman, M.J. (1995) The distribution of mass in the Kuiper Belt, in *Proc. of the 27th Symposium on Celestial Mechanics*, eds. H. Kinoshita and H. Nakai, Tokyo, Japan
Holman, M.J. and Wisdom, J. (1993) Dynamical stability in the outer Solar system and the delivery of short period comets, *Astron. J.* **105** pp. 1987–1999
Irwin, M., Tremaine, S., Zytkow, A.N. (1995) A search for slow moving objects and the luminosity function of the Kuiper belt, *Astron. J.* **110** pp. 3082–3089
Jewitt, D., and J. Luu (1995) The solar system beyond Neptune *Astron. J.* **109** pp. 1867–1876
Jewitt, D., J. Luu and J. Chen (1996) The Mauna Kea-Cerro Tololo (MKCT) Kuiper Belt and Centaur Survey, *Astron. J.* **112** pp. 1225-1238
Levison, H.F. and Duncan, M.J. (1993) The gravitational sculpting of the Kuiper Belt, *Astrophys. J.* **406** pp. L35–L38
Levison, H.F., Stern, S.A., and Duncan, M.J. (1997) The role of a massive primordial Kuiper Belt on its current dynamical structure, *Icarus*, submitted.
Kowal, C. (1989) A solar system survey, *Icarus* **77** pp. 118–123
Kuiper, G.P. (1951) Origin of the Solar system, in *Astrophysics-A topical symposium*, J.A. Hynek (ed.), pp. 357–424, McGraw-Hill Book Co., NY
Luu, J. and Jewitt, D. (1988) A two-part search for slow moving objects, *Astron. J.* **95** pp. 1256–1262
Luu, J. and Jewitt, D. (1997) Color diversity among the Centaurs and Kuiper Belt objects, *Astron. J.* (in press)
Luu, J. *et al.* (1997) A new dynamical class of object in the outer Solar system, *Nature* **387** pp. 573–575
Malhotra, R. (1993) The origin of Pluto's peculiar orbit, *Nature* **365** pp. 819–821
Malhotra, R. (1995) The origin of Pluto's orbit: Implications for the Solar System beyond Neptune, *Astron. J.* **110** pp. 420–429
Malhotra, R. (1996) The phase space structure near Neptune resonances in the Kuiper Belt, *Astron. J.* **111** pp. 504–516
Malhotra, R. (1997a) Implications of the Kuiper Belt for the Solar system, *Planetary and Space Science*, in press.
Malhotra, R. (1997b) Inclination excitation by resonance sweeping of the Kuiper Belt, *Bull. of the American Astronomical Society* **29**(3) p. 1020
Malhotra, R., and J.G. Williams (1997) Pluto's heliocentric orbit, in *Pluto and Charon*, eds. S.A. Stern and D. Tholen, Univ. of Arizona Press, Tucson, AZ, in press.
Marsden, B.G. (1997) Kuiper belt: securing adequate orbital data, *Planetary and Space Science*, in press.
Morbidelli, A. (1997) Chaotic diffusion and the origin of comets from the 2/3 resonance in the Kuiper Belt, *Icarus* **127** pp. 1–12
Morbidelli, A., F. Thomas and M. Moons (1995) The resonant structure of the Kuiper Belt and the dynamics of the first five trans-Neptunian objects, *Icarus* **118** pp. 322–340
Morbidelli, A. and Valsecchi, G.B. (1997) Neptune scattered planetesimals could have sculpted the primordial Edgeworth-Kuiper Belt, *Icarus* **128** pp. 464–468
Quinn, T.R., Tremaine, S., Duncan, M.J. (1990) Planetary perturbations and the origin of short period comets, *Astron. J.* **355** pp. 667–679

Stern, S.A. (1996) On the collisional environment, accretion timescales and architecture of the massive primordial Kuiper belt, *Astron. J.* **112** pp. 1203–1211

Tegler, S.C. and Romanishin, W. (1997) The extraordinary colors of trans-Neptunian objects 1994 TB and 1993 SC, *Icarus* **126** pp. 212–217.

Tombaugh, C. (1961) The trans-Neptunian planet search, in *Planets and Satellites*, G. Kuiper & B.Middlehurst (eds.), Univ.of Chicago Press, Chicago, pp. 12–30

Tremaine, S. (1990) Dark matter in the Solar system, in *Baryonic Dark Matter*, D. Lynden-Bell and G. Gilmore (eds.), Kluwer Academic Publishers, Netherlands, p. 37–65.

Williams, I.P., O'Ceallaigh, D.P., Fitzsimmons, A., Marsden, B.G. (1995) The slow moving objects 1993 SB and 1993 SC, *Icarus* **116** pp. 180–185

A MECHANISM FOR ASTEROIDS TO AVOID CLOSE APPROACH AND COLLISION TO PLANETS

YOSHIHIDE KOZAI
Gunma Astronomical Observatory
Ootomo Branch Office of Gunma Prefecture, 1-18-7 Ootomo, Maebashi 371, Japan
e-mail: kozai@yso.mtk.nao.ac.jp

Abstract.

In this paper it is explained how asteroids with large aphelion distances avoid very close approach to Jupiter. Namely, since for asteroids with resonant mean motions the critical arguments librate as it has been known for many years, and for those with non-resonant mean motions the eccentricities change with the argument of perihelion so the distance to Jupiter cannot be very small.

Also some behaviours of Kuiper-belt asteroids are studied by a similar way and it is found that the libration region for the argument of perihelion appears only when the eccentricity is very large. Therefore, even though the eccentricity is very large it is possible for any Kuiper-belt asteroid to avoid very close approach to any of the planets.

1. Asteroids with large aphelion distances

It is known that all asteroids with semi-major axes between 3.9 and 5.3 AU have mean motions of 1:1, 7:6, 4:3 and 3:2 resonances as far as the numbered asteroids are concerned. However, as the critical arguments are librating around $\pm\pi/3$ for Trojan asteroids and around 0 for the others except for those of Hilda group of small eccentricity, they do not approach Jupiter very closely. It is also true for 2:1 and other resonance cases.

There are a few asteroids, for which the aphelion distances are large. Namely, for some of them they are larger than 5 AU, even though their semi-major axes are less than 3.7 AU, that is, their mean motions are non-resonant. In Table 1 non-resonant asteroids with 5 largest aphelion distances are listed with necessary data.

In the Table the maximum and minimum values of the eccentricities and the inclinations are computed by adopting a simple model (Kozai, 1962, 1979). Namely, it is assumed that all the disturbing planets are moving along circular orbits on the same plane and the disturbing function is averaged with respect to the mean anomaly of the asteroid and the mean longitudes of the disturbing planets.

In Table 2 similar data for resonant asteroids with 10 largest aphelion distances are given. The mean motions of the asteroids in the Table are in 7:6(a=4.7AU), 4:3(4.2AU), 3:2(4.0AU) and

TABLE 1. Data for non-resonant asteroids with large aphelion distance

No.	Q_M	Q	a	r_{90}	r_1	r_2	q_M	Q_m	i_m	i_M	e_m	e_M
6130	4.82	4.59	2.99	1.87	1.39	4.55	1.40	4.58	10.	23.	0.53	0.61
5324	4.88	4.78	2.96	1.71	1.23	3.66	1.42	4.51	13.	30.	0.52	0.65
7604	4.91	4.90	3.12	2.09	1.97	2.24	2.07	4.16	19.	35.	0.34	0.57
1373	5.30	4.51	3.40	2.34	2.86	3.23	3.05	3.05	29.	40.	0.32	0.56
5164	5.54	5.49	3.66	2.69	2.31	3.36	2.75	4.57	17.	32.	0.25	0.51

J. Andersen (ed.), Highlights of Astronomy, Volume 11A, 229–232.

TABLE 2. Data for resonant asteroids with large aphelion distances

No.	Q_M	Q	a	r_{90}	r_1	r_2	q_M	Q_m	i_m	i_M	e_m	e_M
2959	5.02	5.03	3.94	3.64	3.48	3.80	2.93	4.95	4.	7.	0.26	0.28
1877	5.03	4.78	3.94	3.65	3.30	4.39	3.33	4.56	15.	20.	0.16	0.27
2483	5.11	5.06	3.97	3.65	2.89	5.06	2.89	5.06	2.	6.	0.27	0.29
4446	5.11	5.10	3.98	3.65	3.37	4.00	3.05	4.91	7.	12.	0.23	0.28
1941	5.11	5.10	3.99	3.68	3.03	4.70	2.92	5.07	2.	5.	0.27	0.28
5661	5.18	4.90	3.95	3.57	3.01	4.90	3.01	4.90	8.	14.	0.24	0.31
1922	5.41	4.80	3.25	1.82	1.78	4.28	2.08	4.43	18.	40.	0.36	0.66
5370	5.60	5.46	3.35	1.83	1.27	4.82	1.55	5.14	7.	30.	0.54	0.67
6144	6.40	6.40	4.72	4.11	4.05	4.17	3.01	6.42	5.	4.	0.36	0.36
3552	7.58	7.26	4.23	1.58	1.35	4.50	2.64	2.64	25.	46.	0.61	0.79

2:1(3.3AU) resonances. In fact there are many high eccentricity asteroids in mean motion resonance regions except for the Hilda group.

2. Secular Variations

When Delaunay variables (μ being the product of the constant of gravitation and the solar mass),

$$\begin{array}{llll} L = \sqrt{\mu a}, & G = L\sqrt{1-e^2}, & H = G\cos i, \\ l = \text{mean anomaly}, & g = \varpi - \Omega, & h = \Omega, \end{array}$$

are introduced, they satisfy the canonical equations of three degrees of freedom with the Hamiltonian, F,

$$F = \frac{\mu^2}{2L^2} + \sum m'/\Delta, \tag{1}$$

where Δ is the mutual distance with one of the disturbing planets and m' is its mass as the fraction of the solar mass.

Because of the assumption, the Hamiltonian is symmetric around the z-axis after the averaging with respect to the fast moving angular elements, the mean anomaly of the asteroid, l, and the mean longitudes of the disturbing planets, λ. Then, h, the longitude of the ascending node, does not appear in the Hamiltonian, and therefore, H, the z-component of the angular momentum, becomes constant. Since l has been eliminated, L is constant. Then the equations are reduced to a system of one degree of freedom with the integral F=constant.

Then with the two parameters, L and H, G can be expressed as periodic functions of g, the argument of perihelion, in fact of $2g$. In practice instead of L, H and G,

$$a, \quad \Theta = \sqrt{1-e^2}\cos i, \quad X = \sqrt{1-e^2}, \tag{2}$$

are used and the Hamiltonian is replaced by,

$$F^* = \frac{1}{(2\pi)^2}\sum\int_0^{2\pi}\int_0^{2\pi}\frac{m'}{\Delta}\,dl\,d\lambda'. \tag{3}$$

For resonant asteroids the averaging is made under the assumption that the critical argument is fixed at 0. That is in Table 2, it is assumed that the critical argument for any asteroid is fixed.

For given values of a and Θ the values of F^* are computed for various sets of X and $2g$, and curves of equal F^* values – along which X as well as e and i vary as functions of $2g$ – can be drawn. In fact, X can take values between 1 and Θ which correspond, respectively, to $e = 0$ and $i = \arccos\Theta$ and to $e = \sqrt{1-\Theta^2}$ and $i = 0$. When Θ is nearly equal to 1, any curve of equal F^* value is a nearly horizontal line, namely, parallel to the $2g$-axis. However, when Θ is well below 1, the curves are no longer horizontal lines, and they show that the eccentricity is a minimum at $2g$ =0 and 2π, and a maximum at $2g = \pi$ and vice versa for the inclination.

By drawing such diagrams the variations of the two elements are traced. It is known that if the value of Θ is less than 0.8 a libration region appears, in which the argument of perihelion librates around $2g = \pi$, and the eccentricity and the inclination take their maximum and minimum values at $2g = \pi$.

In the Tables the first column gives the assigned number, the second and the third columns are, respectively, the maximum and the present aphelion distances and the fourth column expresses the semi-major axis. The fifth column, r_{90}, is the heliocentric distance of the ascending and the descending nodes of the orbital plane with respect to the reference plane when the eccentricity takes the maximum value, that is, at $2g = \pi$, and this distance is computed as $a(1 - e^2)$.

The sixth and seventh columns, r_1 and r_2, are the heliocentric distances of the ascending and the descending nodes computed by the present values of the eccentricity and the argument of perihelion. Then q_M and Q_m are the perihelion and aphelion distances for the minimum value of the eccentricity which corresponds to the case of $2g$=0. And finally the maximum and minimum values of the inclinations and the eccentricities are given.

For the asteroids (1373) and (3552) the arguments of perihelion do not make complete revolutions but librate around $\pi/2$ or $3\pi/2$ and the maximum and the minimum of the two elements occur there. Therefore, under q_M and Q_m the values of $a(1 - e_m^2)$, the heliocentric distances of the nodes, where $2l = \pi$, are given.

Table 1 shows that although the maximum and the present aphelion distances are large for the 5 asteroids, the heliocentric distances of the asteroids when they cross the reference plane, the orbital plane of the disturbing planets, are not so large as to approach Jupiter very closely. Still the largest heliocentric distances on the reference plane are 4.6 AU for some of the asteroids.

On the other hand, in Table 2 the heliocentric distances when the asteroids cross the reference plane are not necessarily small. In fact for (6144) the aphelion distance is as large as 6.4 AU and the inclination is small. However, since the critical arguments are librating, they do not approach Jupiter very closely.

In fact it is clear that the mechanism to avoid very close approach works for large eccentricity asteroids. And one can say that more eccentric asteroids are more stable in the sense that they do not approach Jupiter closely both for non-resonant and resonant cases.

3. Kuiper-belt Asteroids

It is already known that for asteroids of the Kuiper belt the eccentricities and the inclinations change as functions of the arguments of perihelion (Morbidelli, Thomas and Moons, 1995). However, they do not change so much as those for main-belt asteroids with similar values of Θ except for regions corresponding to the eccentricities such as 0.25, 0.52, 0.76 and 0.87, for which the perihelion distances are nearly equal to the semi-major axes of Neptune, Uranus, Saturn and Jupiter, respectively. In fact the value of F^* takes its maximum near one of such eccentricities and $2g = 0$.

Generally the value of F^* increases gradually as the value of $X = \sqrt{1 - e^2}$ decreases and there is no libration region except for cases of small values of Θ. However, for a case of resonant mean motion like the Pluto-Neptune case the value of F^* is decreased as the eccentricity is increased if the critical argument is librating around π. Therefore, a libration region appears in the diagram, and it is known that the argument of perihelion for Pluto is librating.

If the value of Θ is very small, libration regions appear as it is shown for the case of $a = 40$ AU and $\Theta = 0.20$ in Figure 1. Under the horizontal axis the value of $2g$ is plotted; on the left-hand side of the vertical axis the value of $X = \sqrt{1 - e^2}$ and on the right-hand side the values of the eccentricity and the inclination are given.

If for any asteroid the eccentricity and the argument of perihelion are in one of the libration regions near $2g = \pi$, as its perihelion is very far from the reference plane it never approaches any planet even though the eccentricity is very large.

In any case the mechanism discussed here plays an important role for highly eccentric orbit asteroids to avoid very close approach to major planets.

Finally it is remarked that since the eccentricities and the mutual inclinations of the disturbing planets are neglected any secular resonance is not taken into account in this paper.

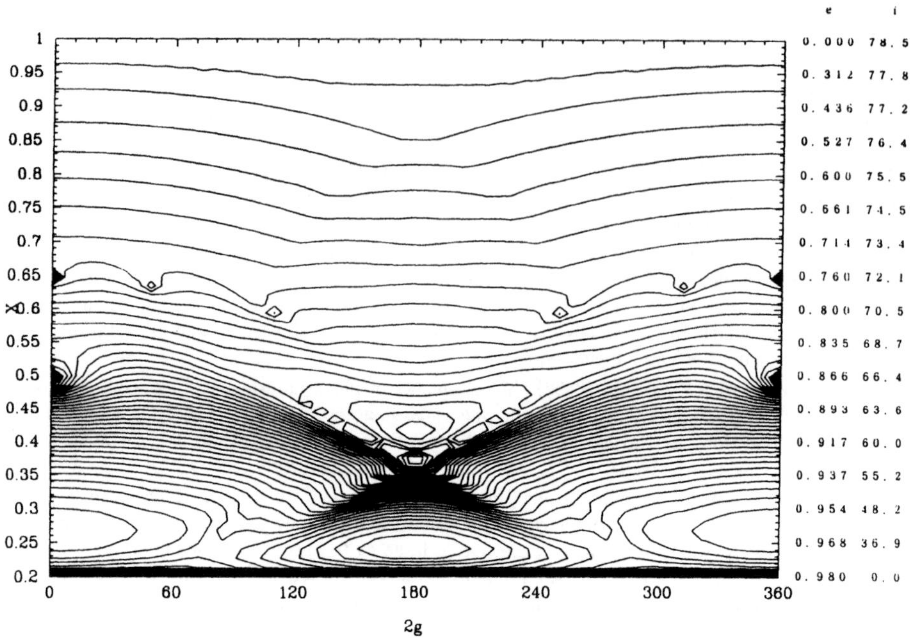

Figure 1. Diagrams for $a = 40$ AU and $\Theta = 0.20$.

References

Kozai, Y. 1962. *Astron. J.*, **67**, 591-198.

Kozai, Y. 1979. In *Dynamics of the Solar System*, ed. R.L. Duncombe, 231-237.

Morbidelli, A., Thomas, F. and Moons, M. 1995. *Icarus*, **118**, 322-340.

IMPACT-INDUCED ACTIVITY OF THE ASTEROID-COMET P/1996N2 ELST-PIZARRO: YES OR NO?

H. BOEHNHARDT
University Observatory
Scheinerstr. 1, D-81679 München, Germany

European Southern Observatory
Alonso de Cordova 3107, Santiago, Chile

Z. SEKANINA
Jet Propulsion Laboratory
California Institute of Technology
4800 Oak Grove Drive, Pasadena CA 91109, USA

A. FIEDLER
University Observatory
Scheinerstr. 1, D-81679 München, Germany

H. RAUER
DLR, Institute for Planetary Exploration
Rudower Chaussee 5, D-12484 Berlin, Germany

R. SCHULZ
European Space Agency
Keplerlaan 1, NL-2200 AG Noordwijk, The Netherlands

AND

G. TOZZI
Osservatorio Astrofisico di Arcetri
Largo E. Fermi 5, I-50125 Firenze, Italy

Abstract. P/1996 N2 Elst-Pizarro was discovered with a tail of dust in July 1996 and was thus classified as periodic comet. From the orbit point of view it belongs to the outer asteroid belt and is very likely a member of the Themis family, a group of bodies originating from an asteroid collision.

The absence of a coma and the presence of the needle-like anti-tail in August 1996 did not exclude the possibility of a sudden emission event which could have been caused by the impact of another minor body on the object. However, with the development of a normal tail between September and November 1996 it became obvious that the dust emission episode continued at least for 2 months after perihelion, a scenario that is typical for gas-driven cometary activity. However, it is still possible that a collision triggered the event.

The paper summarizes the observational pro and cons for impact-induced activity in this object, possibly the first ever detected comet-asteroid of the solar system.

1. The Object Type: Asteroid or Comet?

P/1996N2 Elst-Pizarro was discovered by Eric Elst on 7 August 1996 on 1-meter Schmidt telescope exposures taken on 14 and 16 July 1996 by Guido Pizarro at the European Southern Observatory ESO at La Silla, Chile (Elst and Pizarro, 1996). At the time of discovery as well as during the subsequent 1 1/2 months the object exhibited a straight and narrow, several arcmin long tail pointing almost toward the Sun (Pravec, 1996). Although no discernible coma was reported from the follow-up observations made in August 1996 (Offut, 1996), the object was classified by the

J. Andersen (ed.), Highlights of Astronomy, Volume 11A, 233–236.

Central Bureau for Astronomical Telegrams as a comet, mainly because of the continuous presence of the tail (Marsden, 1996a).

TABLE 1. Comparison of Orbital Elements of Elst-Pizarro with average elements of the Themis Family

	Elst-Pizarro	Themis Family
semi-major axis (AU)	3.156	3.146
eccentricity	0.167	0.168
inclination (deg)	1.38	1.87
reference	Marsden, 1996b	Kresák, 1971

However, as already noted by Marsden (1996b) after the initial orbit determination, the orbit of P/1996N2 Elst-Pizarro is typical as a main belt asteroid rather than a comet. Moreover, astrometric observations from 1979 and 1985 show these to be earlier detections of P/1996N2 Elst-Pizarro reported to be asteroid (Marsden, 1996b,c; McNaught & Hawkins 1996). Hence, P/1996N2 Elst-Pizarro could be a cometary object in the main asteroid belt and thus represent the first comet-asteroid ever detected. However, one should exercise a caution since the early 1996 observations did not unambiguously support the comet classification, e.g. no coma was found and no spectrum was available which could be used for a search of emission features typical for comets. Furthermore, there exists a remarkable coincidence between the semi-major axis, eccentricity and inclination of P/1996N2 Elst-Pizarro and those of the Themis asteroid family (see Table 1) suggesting that P/1996N2 Elst-Pizarro is possibly to be a member of this family. This family contains many C-type and some B- and F-type asteroids of typically low geometric albedo (0.01–0.06). Their members are likely to consist of primitive and metamorphic grades of carbonoceous chondritic material originating from a uniform CM-like parent body (Gradie et al., 1979; Chapman et al., 1989).

2. Collision-Induced Origin for the Tail?

The enigmatic early reports of a comet in the asteroid belt showing a tail without a coma led some people to speculate about a possible impact-induced origin of the tail (Balam, 1996). Furthermore, an initial analysis of the narrow sunward tail seen in August 1996 (Sekanina, 1996) did not exclude the possibility of a brief emission episode of micron to sub-micron size dust which should have taken place between 40 to 80 days after perihelion (which occurred on 16 April 1996). Finson-Probstein calculations of the tail geometry predicted the possibility of a second, "normal" dust tail pointing essentially away from the Sun (Boehnhardt, 1996). The scenario applied to the period from September to December 1996 in case the emission episode lasted fairly long. the normal tail was indeed first observed around mid September 1996 (Boehnhardt et al., 1996).

2.1. OBSERVATIONAL FACTS

In the following we report and summarize the preliminary results of our observations of P/1996N2 Elst-Pizarro collected between 23 August and 5 November 1996 at ESO La Silla, Chile. These target-of-opportunity observations were performed in the context of a joint programme of European scientists in support of the ROSETTA and ISO missions of the European Space Agency (ESA). They involved broadband BVR imaging and low dispersion spectroscopy (spectral range about 350 – 700 nm, spectral resolution about 0.8 nm/pixel) in the visible wavelength range with the focal reducers DFOSC at the Danish 1.5 meter telescope and EFOSC2 at the MPG/ESO 2.2 meter telescope.

When corrected for contributions from the dust tails (see below), the object's head in the available set of images appears to be stellar. This is concluded from a comparison of cross-cut profiles through the seeing disc of the nucleus with point-spread functions obtained from stars in the same images. The R band brightness of the nucleus of 19.1 mag corresponds to an equivalent radius of 2 – 3.5 km depending on the geometric albedo assumed (0.1 – 0.04). In co-added frames with a total integration time of more than 60 min, no coma is detected. In late August 1996 the straight sunward dust tail had a width of less than 7 arcsec, but extended well beyond the edge of the field of views of our instruments, i.e. it was longer than 8 arcmin which is more than 550000 km in projection at the "comet". The "normal" tail was pointing in approximately anti-sunward direction from mid-September to early November 1996. It showed a much shorter extension of less than 25 arcsec and appeared to be somewhat broader than the anti-tail (although still less than about 10 arcsec in width). The co-added images and the seeing disc analysis revealed the existence of a very weak and diffuse bridge of light between the normal tail and the anti-tail, which was best seen – though very faint – in October and November 1996, when the plane of the tail was inclined most to the direction from Earth (about 0.6 deg). A preliminary assessment of the spectra obtained on 23 August (with DFOSC) as well as on 9 September 1996 (with EFOSC2) revealed essentially reflected sunlight without superposed emission band features typical for comets.

Our observational results show that during our monitoring period P/1996N2 Elst-Pizarro exhibited dust tail phenomena without a parallel detectable coma activity.

2.2. DUST TAIL MODELLING

During our dust tail analysis of P/1996N2 Elst-Pizarro Monte-Carlo simulations were run for various dates of our observations. The models were compared with the images complemented by the measurements of the tail extension and geometry, i.e. with the position angles, the length and the width of both tails. As already mentioned by Boehnhardt et al. (1996), the simultaneous presence of the normal tail and the anti-tail in September 1996 argues for a dust production which continued for almost two months after perihelion passage. To be more specific, from the measured position angles of both tails it can be concluded that the emission episode producing the dust tails lasted from about 25 May to about 15 July 1996, i.e. from roughly 5 weeks post-perihelion to the time of discovery. The normal tail contained younger dust grains, with a low solar radiation pressure acceleration of less than 0.1 the solar attraction, while the anti-tail corresponds to an early part of the emission episode, with dust grains affected by larger radiation pressure. From the narrow width of the anti-tail and the overall geometry of both tails an expansion velocity of the grains of less than 2.5 m/s is inferred.

3. Conclusions and Speculations

Based upon the observational evidence and the modelling we conclude that both tails are not likely direct products of an impact event involving the comet-asteroid P/1996N2 Elst-Pizarro shortly after perihelion. Also ruled out can be the interpretation of the dust tails as trails of large dust particles ejected long ago. On the other hand, an interpretation in terms of normal cometary activity suffers clearly from the nondetection of a coma around the object. Also, the inferred expansion velocity is barely above the estimated velocity of escape from the object and considerably lower than typical ejection velocities for the microscopic dust grains whose existence in the tails is implied by the derived radiation pressure accelerations. A plausible scenario for the dust-emission episode may need to invoke an impact as a triggering mechanism that disturbed the surface layer and brought about a temporally limited activity of "icy dirt", a mixture of dust and ice in which the mass of dust dominated by orders of magnitude. Such activity could conceivably explain the extremely low expansion velocity observed. Gentle lifting or disintegration processes for dust may also be considered as potential explanation scenarios, but require a thorough modelling and comparison with the observations. And, of course, further observational monitoring and the analysis of possibly existing data covering the emission period of the object from May to July 1996 will be very useful to solve the two existing major questions about P/1996N2 Elst-Pizarro: the object type and the nature of the activity in 1996. For both issues a comparitive study with another intermediate candidate between comets and asteroids, i.e. 107P/Wilson-Harrington (McFadden et al. 1994), may be of great interest.

References

Balam, D. (1996) Asteroid or Comet?, *Sky and Telescope* **92**, p. 10

Bell, J.F., Davis, D.R., Hartmann, W.K., Gaffey, M.J. (1989) Asteroids: The Big Picture, *Asteroids II* (eds. R.P. Binzel, T. Gehrels, M. Shapley Matthews), Univ. Arizona Press, Tucson, pp. 921–945

Boehnhardt, H. (1996) Comet P/1996 N2 (Elst-Pizarro), *IAU Circular* 6473

Boehnhardt, H., Schulz, R., Tozzi, G., Z. Sekanina (1996) Comet P/1996 N2 (Elst-Pizarro), *IAU Circular* 6495
Chapman, C.R., Paolicchi, P., Zappala, V., Binzel, R.P., Bell, J.F. (1989) Asteroid Families: Physical Properties and Evolution, *Asteroids II* (eds. R.P. Binzel, T. Gehrels, M. Shapley Matthews), Univ. Arizona Press, Tucson, pp. 386–415
Elst, E., Pizarro, G. (1996) Comet P/1996 N2 (Elst-Pizarro), *IAU Circular* 6456
Gradie, J.C., Chapman, C.R., Williams, J.G. (1979) Families of Minor Planets, *Asteroids* (ed. T. Gehrels), Univ. Arizona Press, Tucson, pp. 359–390
Kresák, L. (1971) Orbital Selection Effects in the Palomar-Leiden Asteroid Survey, *Physical Studies of Minor Planets* (ed. T. Gehrels), NASA SP-267, Washington DC, pp. 197–210
Marsden, B.G. (1996a) Comet P/1996 N2 (Elst-Pizarro), *IAU Circular* 6456
Marsden, B.G. (1996b) Comet P/1996 N2 (Elst-Pizarro), *IAU Circular* 6457
Marsden, B.G. (1996c) Comet P/1996 N2 (Elst-Pizarro), *Minor Planet Electronic Circular* 1996-R07
McFadden, L.A. (1994) The Comet-Asteroid Transition: Recent Telescopic Observations, *Asteroid, Comets, Meteors 1993* (eds. A. Milani, M. Di Martino, A. Cellino), Kluwer Academic Publishers, Dordrecht, pp. 95–110
McNaught, R.H., Hawkins, M.R.S. (1996) Comet P/1996 N2 (Elst-Pizarro), *IAU Circular* 6473
Offut, W. (1996) Comet P/1996 N2 (Elst-Pizarro), *IAU Circular* 6456
Pravec, P. (1996) Comet P/1996 N2 (Elst-Pizarro), *IAU Circular* 6459
Sekanina, Z. (1996) Comet P/1996 N2 (Elst-Pizarro), *IAU Circular* 6459

AN UPDATE ON THE PROBLEM OF SMALL COMETS

J.C. BRANDT
Laboratory for Atmospheric and Space Physics, University of Colorado, Boulder CO, USA

AND

M.F. A'HEARN
Department of Astronomy, University of Maryland, College Park MD, USA

The evidence for a major population of small comets (SCs) is summarized in two steps. First, we briefly summarize our previous work. Second, we describe recent work that continues to find SCs. When new capabilities become available that can detect SCs, we expect to find them and this is occurring. At present, their detection is a haphazard process and we advocate a dedicated, optimized search program.

We define SCs as those icy bodies (i.e., sublimating bodies) with radius less than 1000 meters. Often the existence of SCs is inferred from effects ascribed to them. These "effects" include: lunar cratering; cratering on Ganymede; excess interplanetary hydrogen; delivery of volatile materials to the terrestrial planets; and a source of fragile bodies entering the Earth's atmosphere. While some of these indirect methods support the existence of SCs, direct observations are clearly preferred.

For our previous work on this subject we refer to (Brandt *et al.*, 1996; Brandt *et al.*, 1997). We found that SCs are often detected when special circumstances produce intrinsic or apparent brightenings. An example of the former is the detection of 15 comets very near the Sun by coronagraphs on the SOLWIND and SMM (Solar Maximum Mission) spacecraft. These comets are too faint to be detected when passing the Earth's orbit, but brighten considerably at very small heliocentric distances (see Marsden, 1989). These comets are probably part of the Kreuz group. Also note that sungrazing is a normal end state for many comet orbits (Asher *et al.*, 1994).

Examples of the latter situation occur when comets come very close to Earth. Comet 7P/Pons-Winnecke in 1927 has radii of 310 or 210 meters for assumed albedos of 0.05 or 0.10, respectively (see Yeomans, 1991). Comet C/1983 J1 (Sugano-Saigusa-Fujikawa) has a radius of 370 meters as determined by infrared, visual, and radar studies (Hanner *et al.*, 1987). Also, Comet 41P/Tuttle-Giacobini-Kresák has a radius of 800 meters for an albedo of 0.02 as reported by Roemer (1966). Indeed, Roemer noted: "The nuclei of most periodic comets appear to be less than 10 km in radius, and *many* may have radii of less than 1 km" (emphasis added).

We have also noted that an optimum search strategy should be based on the unique properties of comets: an extended source, with most of the light coming from the coma rather than the nucleus; motion with respect to the background stars; and characteristic spectral signature. Most existing search programs are designed to detect moving point sources and hence, are not detecting SCs efficiently.

The recent work consists of two parts. First, another spacecraft, SOHO, with coronagraphs is in operation. The SOHO/LASCO observers have found more than thirty comets very near the Sun and reported them in IAU Circulars. Second, research using the Planetary Camera (PC) of the Hubble Space Telescope's WF/PC 2 has yielded two more SCs. For an assumed albedo of 0.04 and assumptions about the phase law Lamy *et al.* (1997a,b) find radii of 340 meters for comet 45P/Honda-Mrkos-Pajdušáková and 600 meters for comet 46P/Wirtanen. They conclude that comet 46P/Wirtanen "...belongs to a class of small, highly active nuclei...."

We conclude that the existence of SCs is not in doubt, but their number and properties are very uncertain. Reasonable estimates place the number of 250-meter radius SCs in a shell between 1 and 2 AU heliocentric distance in the range 16 to 800. The distribution function $f(r) \propto r^{-3}$ (differential) probably holds well below 1000 meters, perhaps down to $\sim$100 meters before turning

J. Andersen (ed.), Highlights of Astronomy, Volume 11A, 237–238.

over. The turnover almost certainly does not occur at 1000 meters, the size at which our traditional methods of detection begin to fail. What is the origin of the SCs? Were they formed at the same time as the rest of the comets and simply represent the small radius/low mass part of the original distribution function? Are they largely the result of fragmentation processes that seem to account for the SCs seen very near the Sun? Or, do they originate in a variety of processes?

Because SCs are intrinsically faint, the search program must be optimized for efficiency. If the goal is to characterize SCs down to (say) 250 meters radius, we may need to search down to 100 meters radius because generally the radius is not observed directly.

When a suitable search program for SCs is completed, we can assess their utility as indicators of processes in the formation of the solar system, use them as clues to comet splitting events, and evaluate their potential importance in the interaction between planets and small bodies.

Finally, we note that our SCs are not related to the house-sized masses of water proposed by Frank and Sigwarth (1997a,b,c,d) to be impacting the Earth's upper atmosphere in large numbers. Sometimes these objects are called comets, but they are not. They don't sublimate to form comas or tails. Inappropriate use of a well-known word can only further confuse a contentious debate.

References

Asher, D.J., Bailey, M.E., Hahn, G., and Steel, D.I., (1994) Asteroid 5335 Damocles and its Implications for Cometary Dynamics, *MNRAS*, **267**, 26-42

Brandt, J.C., A'Hearn, M.F.,Randall, C.E., Schleicher, D.G., Shoemaker, E.M., and Stewart, A.I.F. (1996) On the Existence of Small Comets and their Interactions with Planets, *Earth, Moon, and Planets*, **72**, 243-249

Brandt, J.C., A'Hearn, M.F., Randall, C.E., Schleicher, D.G., Shoemaker, E.M., and Stewart, A.I.F. (1997) Small Comets (SCs): An Unstudied Population in the Solar System Inventory, in *Completing the Inventory of the Solar System*, ASP Conference Series, **107**, T.W. Rettig and J.M. Hahn, eds., pp. 289-297

Frank, L.A., and Sigwarth, J.B. (1997a) Transient Decreases of Earth's Far-Ultraviolet Dayglow, *Geophys. Res. Letters*, **24**, 2423-2426

Frank, L.A., and Sigwarth, J.B. (1997b) Simultaneous Observations of Transient Decreases of Earth's Far-Ultraviolet Dayglow with Two Cameras, *Geophys. Res. Letters*, **24**, 2427-2430

Frank, L.A., and Sigwarth, J.B. (1997c) Detection of Atomic Oxygen Trails of Small Comets in the Vicinity of the Earth, *Geophys. Res. Letters*, **24**, 2431-2434

Frank, L.A., and Sigwarth, J.B. (1997d) Trails of OH Emissions from Small Comets Near Earth, *Geophys. Res. Letters*, **24**, 2435-2438

Hanner, M.S., Newburn, R.L., Spinrad, H., and Veeder, G.J. (1987) Comet Sugano-Saigusa-Fujikawa (1983 V) – A Small, Puzzling Comet, *Astron. J.*, **94**, 1081-1087

Lamy, P.I., Toth, I., Jorda, L., Weaver, H.A., and A'Hearn, M. (1997a) The Nucleus and Inner Coma of Comet 46P/Wirtanen, submitted to *Astron. Astrophys.* special Comet Wirtanen issue

Lamy, P.I., Toth, I., A'Hearn, M. and Weaver, H.A. (1997b) Hubble Space Telescope Observations of the Nucleus of Comet 45P/Honda-Mrkos-Pajdušáková and Its Inner Coma, submitted to *Icarus*

Marsden, B.G. (1989) The Sungrazing Comet Group. II *Astron. J.*, **98**, 2306-2321

Roemer, E. (1966) The Dimensions of Cometary Nuclei, in *Nature et Origine des Comètes, Mém. Soc. Roy. Sci. Liège*, **Tome XII**, 23-28

Yeomans, D.K. (1991) *Comets*, Wiley & Sons, New York–Chichester–Brisbane–Toronto–Singapore.

ASPECTS OF THE TERRESTRIAL INFLUX OF SMALL METEOROIDS

A.D. TAYLOR AND W.G. ELFORD
Department of Physics and Mathematical Physics
University of Adelaide, Adelaide, 5005, South Australia

Abstract. This paper briefly summarises the contribution to the understanding of meteoroids and their orbits from recent radar studies of meteors. New data from VHF and UHF observations indicates that some meteoroids are ablating at heights up to 160 km and modelling work suggests that such meteoric ionization can only be deposited by meteoroids that ablate at temperatures below 1000° K. Meteoroids with entry speeds well in excess of 100 km/s are clearly interstellar and some tentative sources have been identified. The distributions of the orbital elements of earlier data have been reanalysed in the light of a new understanding of the selection effects in those radar studies, and the presence of a significant set of retrograde orbits identified. The new technique of observing meteors 'coming down the radar beam' has led to very accurate measures of deceleration and detailed analysis of fragmentation events.

1. Introduction

The bodies that contribute to the meteor phenomenon in the atmosphere are 6 – 10 orders of magnitude smaller than most objects that produce impact features on the surface of the planets. Nevertheless, the study of meteors is relevant to any discussion of impacts of large bodies with the planets as the sources of meteoroids are comets and asteroids, and the meteoroids carry some degree of 'memory' of their progenitors. In the case of meteors of cometary origin there is a close association between the orbits of many meteor streams and the orbits of the parent bodies. Further, observations of meteor phenomena over time scales as short as one year lead to significant statistics regarding the distributions of the mass, velocity, ablation characteristics, and orbital elements of these small bodies. Radar and optical (video) observations in recent years have indicated that a significant fraction of meteoroids ablating in the atmosphere have low ablation temperatures, and a small fraction are interstellar. Evidence for these new families of meteoroids is presented below, together with some comments on the way in which radar observations are giving a new understanding of the structure of stony meteoroids.

2. Meteor heights

Particles of mass $> 10^{-10}$ kg with a speed of 20 – 60 km/s ablate in the atmosphere to produce narrow trails of ionization detectable by radar, while the trails formed by more massive particles can be observed with video systems. VHF radars observe meteors up to a height of about 115 km, while MF radars have detected a substantial flux of meteors at heights up to 140 km (Steel and Elford, 1991). These height limits are determined predominantly by the radio scattering process rather than the actual meteor phenomenon. UHF incoherent scatter observations (EISCAT, 930 MHz) have detected short term enhancements in the level of ionization in a 1 km cube of atmosphere at 160 km that have been attributed to meteors (Pellinen-Wannberg and Wannberg, 1996). These observations imply that conventional meteor radars operating in the frequency range 30 – 50 MHz are observing only a fraction of the meteors occurring in the atmosphere. This conclusion is supported by a comparison of meteoroid fluxes inferred from radar observations and those determined from impacts on spacecraft, which suggests that about 90% of meteors are not detected by radar.

J. Andersen (ed.), Highlights of Astronomy, Volume 11A, 239–245.

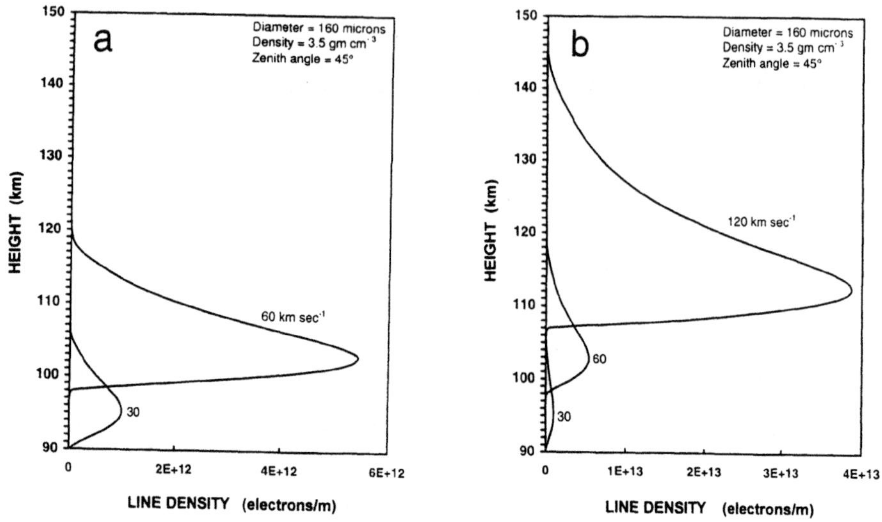

Figure 1. Effect of the incident velocity on the ionisation profile of a stony meteoroid.

3. Ablation Characteristics of Meteoroids

The heights at which stony meteoroids ablate can be estimated by applying some well established analysis procedures to small particles entering the atmosphere, being heated by collisions with air molecules and subsequently evaporating. The most comprehensive investigation is that of Love and Brownlee (1991) although the heating rate they determine is too fast due to an incorrect choice of specific heat for stony particles. Love and Brownlee used a value of 10^6 ergs/g·K which they attributed to Öpik (1958). The actual value given by Öpik (Appendix 2, p. 161) for 'meteoritic stone' is 9×10^6 ergs/g·K. Using a procedure similar to that of Love and Brownlee we have investigated the effects on the height and rate of production of ionization in a model meteor trail as we make significant changes in the mass, density and speed of the particle. The least significant effect is that due to a change in the density; a factor of two increase in particle density moves the whole ionization profile downward by about 2 km.

For a given particle a change in speed of entry has a very significant effect on the height and degree of ionisation of the meteor trail. In Figure 1a is shown the predicted electron line densities for a stony meteoroid entering the atmosphere at 30 and 60 km/s. Doubling the velocity increases the value of the peak line density by a factor of 5.3 and raises the height of the trail by about 8 km. Doubling the speed again to that of an interstellar meteoroid (say 120 km/s) has an even more dramatic effect on the height and intensity of the trail as is shown in Figure 1b. Thus interstellar meteors can be expected to produce very bright trails at heights above 105 km, and possibly occurring as high as 135 km.

Lebedinets (1991) has pointed out that meteoroids of low ablation temperature (say 500° – 1,000° K) may be very significant contributors to meteor trails occurring at heights above 110 km. To test this suggestion we have modelled the case of an organic polymer (CHON-type particle) of density ~ 1 gm/cc, that ablates at a temperature of about 500° K (cf. 2000° K for a stony particle). The result is shown in Figure 2, where it is evident that such a particle with an incoming speed of 60 km/s will produce ionization at heights between 120 and 170 km, although the peak ionization is about one third of that for a stony particle of the same mass. It is conjectured that the ionization detected by the EISCAT system was deposited by a particle of this type.

CHON-type particles probably contain only a small fraction of the metallic atoms Fe, Na, Mg, Ca, which are the main contributors to the optical emission from meteor trails formed by stony particles. Thus it is not surprising that optical studies have not revealed evidence of meteors formed by CHON-particles.

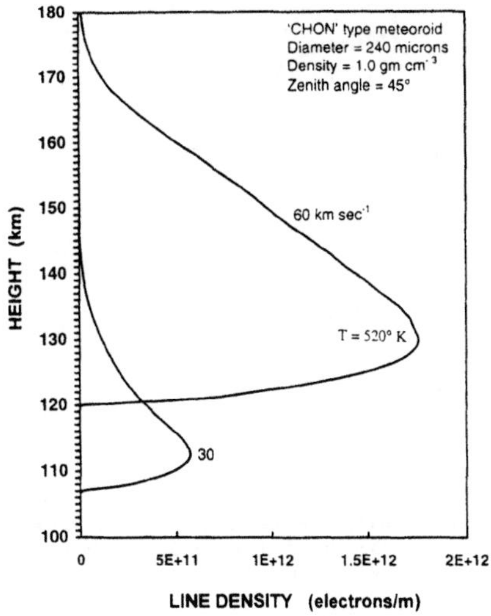

Figure 2. Effect of lowering the ablation temperature on the ionisation profile of a meteoroid.

4. Velocity distribution of meteoroids

The amount of ionisation produced by an ablating meteoroid is strongly dependent on the meteoroid speed as can be inferred from Figure 1. A survey by Bronshten (1983) of laboratory studies of the ionization coefficients of atoms that typify chondritic meteoroids indicates that if the mass and speed of a meteoroid are m and v, the maximum electron line density has a mass-velocity dependence of the form $m v^b$, where $b = 3.75$. The number of meteoroids, N, greater than a given mass $m_\circ$, is usually expressed by the relation

$$N(m > m_\circ) \propto m_\circ^{-\alpha}$$

where α is the cumulative mass index. Taylor (1995) has discussed the value of α and concludes that for the mass range of meteoroids associated with radar meteors, the most likely value for the mass index is 1.1. Whence, to normalise the velocity distribution to a constant mass we apply a weight of the form $(v_\circ/v)^{\alpha b}$, where $\alpha b = 4.13$. It is usual to choose $v_\circ$ as 30 km/s.

The most comprehensive set of observations of meteors formed by small meteoroids with a median mass of 10^{-4} gm observed under nearly uniform conditions during one year, was obtained during the 1968–69 Harvard Radio Meteor Project. The velocity distribution of 14,220 meteoroids corrected for the effects of velocity bias on the meteoric ionisation and the radar detectability of the meteor trails, was published by Southworth and Sekanina (1973), and has been used extensively by subsequent workers. Recently Taylor (1995) has shown that a typographical error in the computer code used by Southworth and Sekanina caused the flux of high speed meteoroids to be underestimated by a factor of about 100. Taylor's initial revision of the meteoroid velocity distribution is shown in Figure 3 as the full line. Subsequently, the effects on the velocity distribution of the initial radius of the meteor trail and the diffusion of the ionisation were included, based on the model ionisation profiles discussed above (Elford *et al.*, 1997). The new velocity distribution is shown in Figure 3 as filled circles.

5. Distribution of meteor radiants and orbits

The distribution of the radiants of meteors detected by the Harvard Radio Meteor Project is shown in Figure 4, where a weight has been applied to each observed radiant for the various selection effects described above. A correction for the flux enhancement due to the Earth's gravity has also been incorporated. Incomplete coverage of the southern-ecliptic hemisphere necessitates the assumption

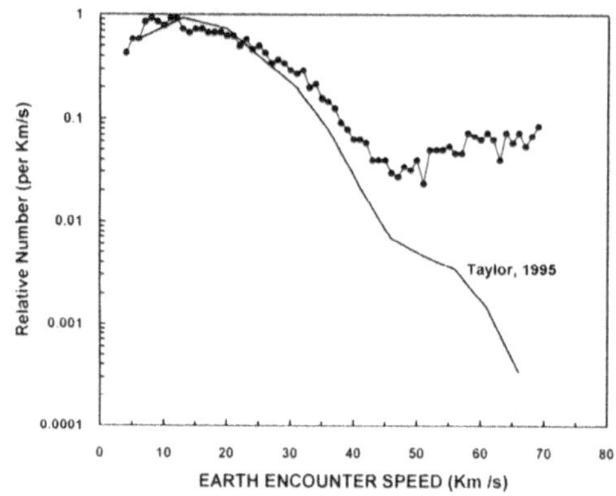

Figure 3. The speed distribution of meteoroids of median mass 10^{-4} gm encountering the Earth, derived from meteors observed during the Harvard Radio Meteor Project 1968–69. The continuous line is Taylor's improved correction for the effect of velocity on ionisation production during ablation. The filled circles include the effect of diffusion on the radar detectability of a meteor trail.

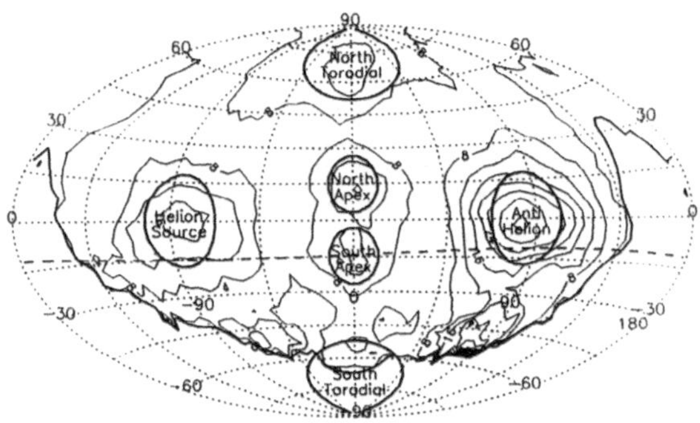

Figure 4. The radiant distribution of meteoroids with mass $> 10^{-4}$ gm encountering the Earth. The coordinate frame places the Earth at the centre, the solar (Helion) direction at a longitude of $-90°$, and the north ecliptic pole at $+90°$. The distribution is based on radar observations (see text) and has been corrected for selection effects associated with the radar detectability of meteors and the atmosphere. The heavy line is the limit of observation of the radar, and radiants south of the dashed line were observed for less than 10% of the time spent observing the region of the celestial north pole. The six radiant sources are overlaid with the one sigma radii appropriate to each source.

that the radiant distribution is symmetric about the ecliptic plane. Elford and Taylor (1997) have shown that linearly-polarised meteor echoes observed during the day may be significantly attenuated due to the effect of Faraday rotation. This is almost certainly the reason that the helion source in Figure 4 is of much lower strength than the anti-helion source, for one would expect to encounter as many meteoroids on highly elliptical low inclination orbits moving toward the vicinity of the Sun (anti-helion source) as there are moving away from the vicinity of the Sun (helion source). As we are uncertain of the value of the Faraday bias, we have not used the data described as the helion source but have assumed that the strength of this source is the same as that of the anti-helion source, and further that the distribution is symmetrical about the plane orthogonal to the plane of the ecliptic and passing through the apex and the poles.

Of the six major sources identified in Figure 4, the major ones are the helion and anti-helion contributed by meteoroids in prograde orbits, and the pair of apex sources contributed by meteoroids in retrograde orbits. The encounter speeds of 'helion/antihelion meteoroids' range between 10 and 40 km/s with a median value of 20 km/s, while the encounter speeds of the 'apex meteoroids' are $50 - 70$ km/s with a median value of about 60 km/s.

The distribution in space of the orbits that contribute to the data presented in Figure 4 is best

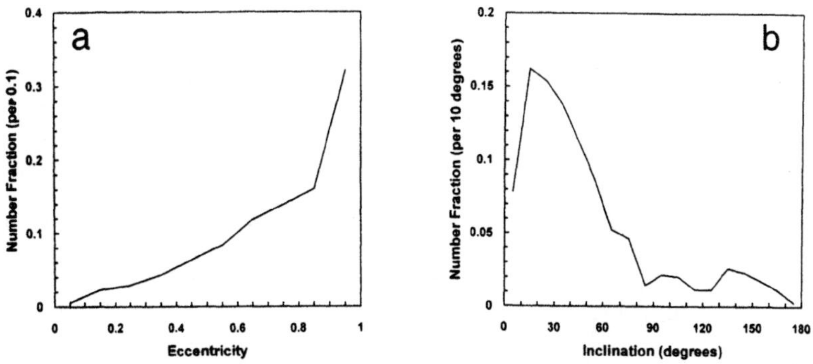

Figure 5. The distribution of the inclinations of the orbits of meteoroids contributing to the radiant distribution in Figure 4. Each orbit is weighted to account for the probability of a particle in the particular orbit colliding with the Earth. About 15% of the meteoroids move in retrograde orbits.

described in terms of the distributions of the orbital elements. In presenting such distributions it is necessary to take into account the probability of a particle in a particular orbit undergoing a collision with the Earth. This is achieved by weighting the data according to a procedure described by Kessler (1981).

The distribution of the eccentricities of the orbits is shown in Figure 5a. The majority of the orbits have eccentricities exceeding 0.4, and as the Poynting-Robertson effect on the motion of particles is to decrease the eccentricity of their orbits, there will be a tendency for points on this diagram to move from right to left over time.

The distribution of inclinations is shown in Figure 5b. The mean inclination is close to 40°. There is also a significant number of orbits with inclinations exceeding 90°; in fact the results imply that about 15% of meteoroids with masses $> 10^{-4}$ gm (diam > 200 microns) move in retrograde orbits. Both distributions given here are very different than those derived by Southworth and Sekanina (1973) who used an incorrect velocity bias, an underestimate of the effect of diffusion, and took no account of the effect of Faraday rotation.

6. Interstellar Meteors

A small but significant fraction of meteors detected by the AMOR radar in New Zealand have velocities in excess of 100 km/s, implying that they are of interstellar origin. From the set of 1508 such meteors collected in 1990 and 1991, with a mean speed of 164 km/s, Taylor *et al.* (1996) have used the variation in flux throughout the year to identify two discrete sources, as shown in Figure 6. One source (A) is associated with nearby A-type stars, and the other (B) with the Sun's

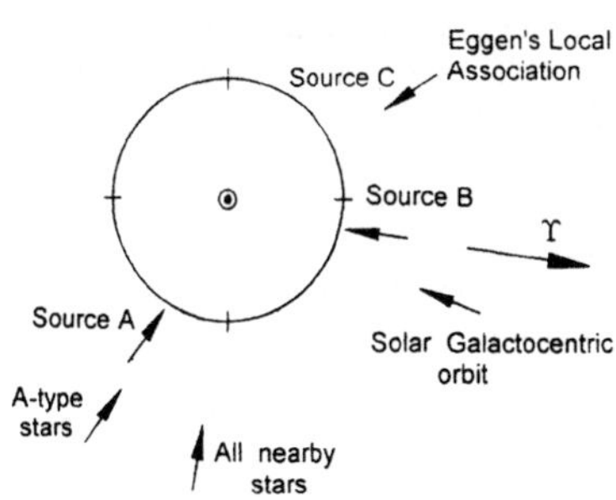

Figure 6. Possible sources of interplanetary meteoroids, projected on to the ecliptic. The symbol γ shows the direction of the vernal equinox. Source A is plausibly associated with the motion of the Solar System relative to the nearby A-type stars; source B is nearly aligned to the Sun's Galactocentric orbit; source C is in the direction of Eggen's local association of young star clusters and B-type stars.

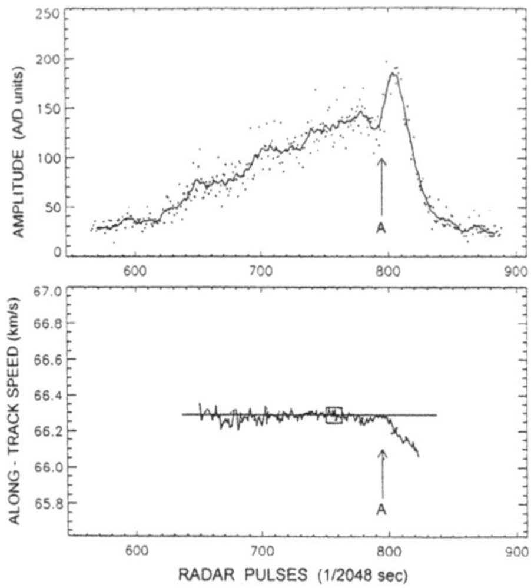

Figure 7. Amplitude of the meteor radar echo (top), and the speed of the ablating meteoroid determined from a 'head echo' observed using a narrow beam VHF radar. The meteoroid entered the radar beam at 21° to the beam axis and during the time interval 600-750 units moved at constant speed. The particle then underwent a catastrophic event at A and burnt out. The associated 'flare' in amplitude and the large deceleration suggests major fragmentation into many particles.

motion about the Galactic Centre. The data also suggest the presence of a third source (C), possibly associated with local B-type stars and young star clusters (Eggen's local association). The size of the interstellar particles producing these high velocity meteor trails is estimated to be in the range $15 - 40\,\mu$m.

7. Head echoes

Until recently it was generally accepted that radar detection of meteors was limited to those situations where the trail is orthogonal to the line of sight to the radar station. With this geometry, the strength of the echo is enhanced by the coherent scattering from the ionisation in a length of the trail about a Fresnel zone in extent. With the advent of narrow beam radars operating at VHF and UHF a new form of scattering has been observed from a small region near the head of a trail. A characteristic of this type of observation is the sequential occurrence of echoes in a number of range bins as the head of the trail advances down the radar beam, at angles up to 45° from the beam axis.

Zhou *et al.* (1995) have reported that this type of meteor head echo has been observed with the 430 MHz Arecibo incoherent scatter radar, and that a systematic study of such echoes was carried out on 'World Days' in 1991 and 1992. The echoes were observed over a height range of 80 to 120 km. The observations are consistent with general view that the meteors with the greatest heights are produced by meteoroids with the highest speeds.

Narrow-beam (2 – 3°) VHF radars have also revealed the presence of 'down the beam' head echoes (Taylor *et al.*, 1996). The time intervals between the appearance of the echoes in subsequent range bins give the line of sight velocity of the meteoroid. A better speed determination can be derived from the phase changes between the returned pulses. Although the meteoroid moves a distance equivalent to many wavelengths between radar pulses, the integer number of phase cycles can be determined by comparison with the line of sight velocity. The total phase-time record then gives the speed of the meteoroid along its trajectory, and also the angle between the trajectory and the axis of the radar beam. The observations give velocities of remarkable precision and make possible measurements of deceleration. Further, by sampling the amplitude of the echo in subsequent range bins an amplitude-time profile of the echo can be determined and related to the distribution of ionisation along the trail. An example of such an observation is shown in Figure 7, where the meteor shows a terminal flare, probably due to severe fragmentation (at A). The fragmentation is also associated with a rapid change in speed (deceleration) during the final stages of the echo. Prior to the fragmentation the speed of the meteoroid was 66.3 ± 0.05 km/s. The meteoroid crossed the radar beam at an angle of 21° to the beam axis.

In cases where a meteoroid undergoes gross fragmentation during ablation into two parts of relatively similar mass the resultant trails will give rise to two radio echoes that beat together. Such head echoes are not uncommon and further studies of the amplitude characteristics and the associated deceleration measurements of 'head echoes' can be expected to contribute to an improved understanding of the physical structure of meteoroids.

References

Bronshten, V.A. 1983. *Physics of meteoric phenomena*, Reidel, Dordrecht.

Elford, W.G., and Taylor, A.D. 1997. Measurements of Faraday rotation of radar meteor echoes for the modelling of electron densities in the lower ionosphere, *J. Atmos. Solar-Terr. Phys.*, **59**, 1021-1024.

Elford, W.G., Steel, D.I., and Taylor, A.D. 1997. Implications for meteoroid chemistry from the height distribution of radar meteors, *Adv. Space Sci.*, in press.

Kessler, D.J. 1981. Derivation of the collision probability between orbiting objects: the lifetimes of Jupiter's outer moons, *Icarus*, **48**, 39-48.

Lebedinets, V.N. 1991. Organic particles of cometary dust in space, *Adv. Space Res.*, **11(12)**, 149-153.

Öpik, E.J. 1958. *Physics of Meteor Flight in the Atmosphere*, Interscience Publishers, New York.

Love, S.G., and Brownlee, D.E. 1991. Heating and thermal transformation of micrometeoroids entering the Earth's atmosphere, *Icarus*, **89**, 26-43.

Pellinen-Wannberg, A., and Wannberg, G. 1996. Enhanced ion-acoustic echoes from meteor trails, *J. Atmos. Terr. Phys.*, **58**, 495-506.

Southworth, R.B., and Sekanina, Z. 1973. Physical and dynamical studies of meteors, *NASA Contractor Report CR-2316*, Washington DC, pp 106.

Steel, D.I., and Elford, W.G. 1991. The height distribution of radio meteors: comparison of observations at different frequencies on the basis of standard echo theory, *J. Atmos. Terr. Phys.*, **53**, 409-417.

Taylor, A.D. 1995. The Harvard Radio Meteor Project meteor velocity distribution reappraised, *Icarus*, **116**, 154-158.

Taylor, A.D., Cervera, M.A., Elford, W.G., and Steel, D.I. 1996. A new technique for radar meteor speed determination: inter-pulse phase changes from head echoes, in *Physics, Chemistry, and Dynamics of Interplanetary Dust*, B.Å.S. Gustafson and M.S. Hanner, eds., *ASP Conference Series*, **Vol 104**, 75-78.

Zhou, Q., Tepley, C.A., and Sulzer, M.P. 1995. Meteor observations by the Arecibo 430 MHz incoherent scatter radar – I. Results from time-integrated observations, *J. Atmos. Terr. Phys.*, **57**, 421-431.

MASS EXTINCTIONS, COMET IMPACTS, AND THE GALAXY

MICHAEL R. RAMPINO
NASA, Goddard Institute for Space Studies, New York, New York, 10025 USA, and Earth & Environmental Science Program, New York University, New York, New York, 10003 USA

AND

RICHARD B. STOTHERS
NASA, Goddard Institute for Space Studies, New York, New York, 10025 USA

Abstract. The hypothesis relating mass extinctions of life on Earth to impacts of comets whose flux is partly modulated by the dynamics of the Milky Way Galaxy contains a number of postulates that can be tested by geologic evidence and statistical analyses. In an increasing number of cases, geologic evidence for impact (widespread impact debris and/or large impact craters) is found at times of mass extinction events, and the record of dated impact craters has been found to show a significant correlation with mass extinctions. Statistical analyses suggest that mass extinction events exhibit a periodic component of about 26 to 30 Myr, and periodicities of 30 ± 0.5 Myr and 35 ± 2 Myr have been extracted from sets of well-dated impact craters. The evidence is consistent with periodic or quasi-periodic showers of impactors, probably Oort Cloud comets, with an approximately 30-Myr cycle. The best explanation for these proposed quasi-periodic comet showers involves the Sun's vertical oscillation through the galactic disk, which may have a similar cycle time between crossings of the galactic plane.

1. INTRODUCTION: IMPACTS AND MASS EXTINCTIONS

Numerous studies have now established that the mass extinction of life at the end of the Cretaceous Period (65 Myr ago), the so-called Cretaceous/Tertiary (K/T) boundary, was the result of the impact of an asteroid or comet ~ 10 km in diameter which created the ≥ 200 km diameter Chicxulub impact structure in the present Yucatán region of Mexico (Alvarez, 1986; Hildebrand *et al.*, 1995). Such an impact (releasing $> 10^{24}$ J or 10^8 Mt TNT equivalent) is calculated to cause a global catastrophe of enormous proportions, primarily related to dense clouds of fine ejecta, production of nitric oxides and acid rain, and smoke clouds from fires triggered by atmospheric re-entry of ejecta (Toon *et al.*, 1994). Other "target-sensitive" effects include enhanced greenhouse effect from atmospheric water vapor in an ocean impact, CO_2 released by impact into carbonate rocks, and cooling and acid rain from large amounts of sulfuric acid aerosols derived from calcium sulfate target rocks (Pope *et al.*, 1994; Toon *et al.*, 1994). If flood basalt eruptions are impact triggered (Rampino and Stothers, 1988), then volcanic perturbations of climate might also be expected.

Astronomical observations of Earth-crossing asteroids and comets and the cratering record of the inner planets allow estimates of the expected mean times between collisions of objects of various sizes with the planet (Shoemaker *et al.*, 1990). On the large-diameter end of the size spectrum, these data predict that in an ~ 100 Myr period, the Earth should be hit by several asteroids and comets larger than a few km in diameter (these release $\sim 10^{23}$ J or 10^7 Mt) and perhaps one ~ 10-km diameter ($\sim 10^{24}$ J, 10^8 Mt) object (most likely cometary) (Shoemaker *et al.*, 1990).

The likelihood that a 10^8 Mt event would have considerably greater global environmental impact than a 10^7 Mt event (Melosh *et al.*, 1990; Toon *et al.*, 1994), leads to the prediction of ~ 5 major mass extinctions, and about 25 ± 5 less severe pulses of extinction during the Phanerozoic Eon (the

J. Andersen (ed.), Highlights of Astronomy, Volume 11A, 246–251.

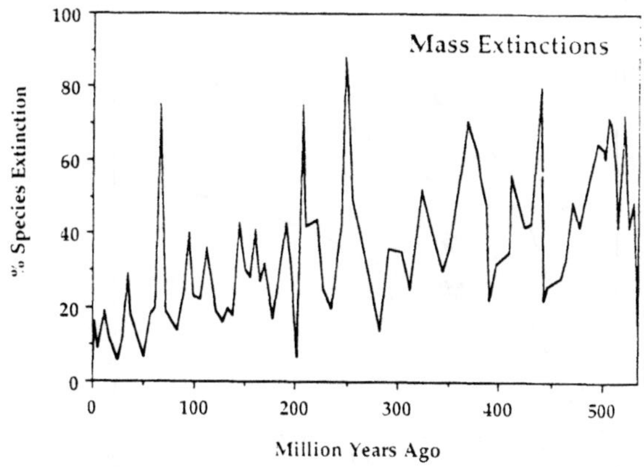

Figure 1. Percent extinction of marine species per geologic stage (or substage) during the Phanerozoic (data from Sepkoski, 1995).

TABLE 1. Large Dated Impact Craters (Grieve and Pesonen, 1996) with updates and Associated Extinctions (Sepkoski, 1995)

Name	Diameter (km)	Age (Myr)	Extinction	% Species
Popigai	100	35.7 ± 0.8	Late Eocene	30
Chesapeake	90	35.2 ± 0.3	Late Eocene	–
Chicxulub	200?	65.2 ± 0.4	K/T	76
Morokweng	100	145 ± 3	J/K	42
Manicouagan	100	214 ± 1	Late Triassic	76
Puchezh-Katunki	80	220 ± 10	Carnian	42

last 540 Myr). This agrees surprisingly well with the record of extinctions for that interval, which clearly shows 5 major mass extinctions and $\sim$ 20 less severe extinction pulses (Fig. 1).

Although the severity of impact-induced mass extinctions would probably depend on a number of variables (e.g. ambient climate conditions, susceptibility of fauna and flora, site of impact, etc.), the size and energy of the impactor is likely to be the most important factor. A theoretical "kill curve" relating mass extinctions and impacts (Fig. 2) has been proposed (Raup, 1992), and this curve can be compared with data representing specific large (> 80 km diameter) known impact craters with well-defined ages that overlap the ages of mass-extinction boundaries (when the full dating uncertainties in both are taken into consideration) (Table 1). We note that no other large, well-dated Phanerozoic craters are known. The observed points in Fig. 2 agree with the predicted curve within the envelope of error permitted by the geologic data, supporting at least a first-order relationship between large impacts and mass extinctions.

2. IS THERE A CORRELATION BETWEEN IMPACTS AND MASS EXTINCTION EVENTS?

The discovery of the K/T impact layer prompted the search for impact signatures at other geological boundaries. Among the materials considered diagnostic of impact are shocked minerals (including shocked quartz, stishovite, shocked zircons, etc.), impact glass (microtektites/tektites), microspherules with structures indicating high-temperature origin, and Ni-rich spinels. Shocked minerals and tektite glass are quite rare in the geologic record, and yet these materials have been reported in stratigraphic horizons close to the times of six recorded extinction pulses (Table 2) (Rampino *et al.*, 1997; Grieve, 1997). A recent preliminary report of shocked quartz from the Permian/Triassic boundary ($\sim$ 248 Myr) in Antarctica and Australia (Retallack *et al.*, 1996) is still

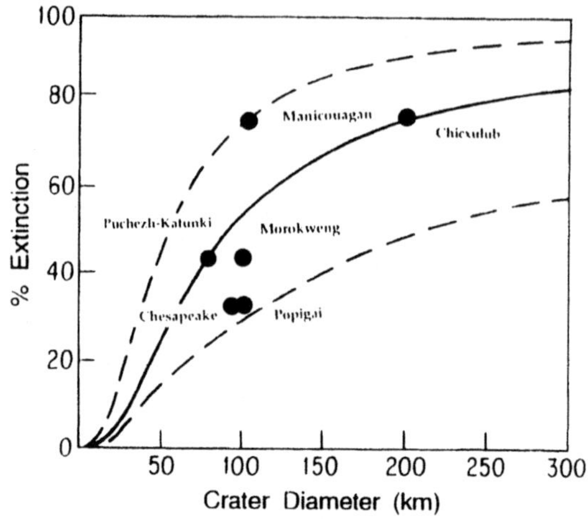

Figure 2. Kill curve for Phanerozoic marine species, with estimated error bars, plotted against estimated size of impact craters associated with extinctions of various magnitudes (assuming that the two are related) (Raup, 1992). Largest well-dated impact craters with ages overlapping mass extinction times are plotted for comparison (Table 1).

unconfirmed.

Seven of the ~ 25 extinction peaks in Fig. 1 [Pliocene (2.3 Myr), Late Eocene (36 Myr), Cretaceous/Tertiary (65 Myr), Jurassic/Cretaceous (144 Myr), Late Triassic (210 Myr), Late Devonian (365 Myr), and possibly the Proterozoic/Cambrian (540 Myr) boundary] are associated with large impact craters and/or some form of stratigraphic evidence of impacts (Tables 1 and 2). Several other extinction events are associated with "possible" evidence of impact consisting of iridium concentrations somewhat elevated with respect to background values; see tables in (Rampino *et al.*, 1997; Grieve, 1997).

The chances of accidentally matching a time series consisting of 24 extinction events (with the age error bars in this case taken as the difference in ages given by recent published time scales), with a time series consisting of the seven times of diagnostic stratigraphic evidence of impacts is $< 10^{-7}$. Studies thus far indicate that iridium anomalies above background and shocked quartz are rare in the geologic record, also suggesting that correlation of mass-extinction events and evidence of large impacts is very unlikely to be accidental.

Recently, Matsumoto and Kubotani (1996) showed a statistically significant correlation between the record of impact cratering of the last 250 Myr and the record of mass extinctions. Stothers (1993) had earlier found a significant correlation between impact crater ages and stratigraphic stage boundaries that are defined by faunal changes during the Cenozoic Era (the last 65 Myr). These studies support a general correlation between impacts and extinctions of various magnitudes from significant mass extinctions ($\geq 10^8$ to 10^7 Mt impacts) to the less severe faunal turnover events that mark stratigraphic stage boundaries ($\sim 10^6$ to 10^7 Mt impacts).

TABLE 2. Stratigraphic Evidence of Impacts at or near Extinction Boundaries (for references see text)

Age	Evidence
Pliocene (2.3 Myr)	Microkrystites, microtektites?
Late Eocene (36 Myr)	Microtektites (multiple), tektites, microspherules, shocked quartz
Cretaceous/Tertiary (65 Myr)	Microtektites, tektites, shocked minerals, stishovite, Ni-rich spinels, iridium
Jurassic/Cretaceous (142 Myr)	Shocked quartz
Late Triassic (210 Myr)	Shocked quartz (multiple?)
Late Devonian (368-365 Myr)	Microtektites (multiple)

3. ARE MASS EXTINCTIONS AND IMPACTS PERIODIC?

The mean occurrence rate of the 24 extinction pulses shown in Fig. 1 is one every $\sim$ 23 Myr. Raup and Sepkoski (1984, 1986) reported a statistically significant 26.4 Myr periodicity in extinction time series for the last 250 Myr, and a secondary periodicity of 30 Myr. Periods of $\sim$ 26 to 31 Myr have been derived using various subsets of extinction events (family and genus levels), different geologic time scales, and various methods of time-series analysis, and Rampino and Stothers (1986) reported detection of an $\sim$ 29 Myr component in the record of extinctions of non-marine vertebrates. The interpretation of these results has been a subject of considerable debate (e.g., Rampino and Haggerty, 1996). The detection of a periodic component in extinctions does not imply either that the record contains a strict periodicity, or that all extinction events follow a 26-Myr timescale; the extinction record might be a mixture of periodic and random events.

Fourier analysis of an extended record of 21 extinctions going back to about 515 Myr, more than doubling the length of the original time series, resulted in a spectrum with the highest peak at 27.3 Myr (Rampino and Haggerty, 1996). Fourier analyses on a series of truncated extinction time series starting from 0 to 540 Myr, and subtracting one extinction at a time back to 253 Myr produced a stable peak between 26.5 and 27.3 Myr which remained the dominant feature in the spectra (Rampino and Haggerty, 1996).

The impact-crater record has been interpreted as showing pulses at about 1, 35, 65, and 95 Myr ago (Shoemaker and Wolfe, 1986; Hut *et al.*, 1987). Time-series analyses of terrestrial impact craters have provided some evidence of a possible 28 to 32 million year periodicity in impacts (Rampino and Stothers, 1984a, 1984b, 1986; Alvarez and Muller, 1984; Shoemaker and Wolfe, 1986; Yabushita, 1991, 1992, 1996a, 1996b), and have been interpreted as suggesting that the impact record may be a mixture of periodic and random events (Trefil and Raup, 1987; Heisler and Tremaine, 1989; Fogg, 1989). Several studies have concluded that, considering the problems in dating, the observed differences in the formal periodicity detected in impact craters and mass extinctions are to be expected (Stothers, 1988, 1989; Fogg, 1989). Critics have argued, however, that the number of well-dated craters is still too small to extract a consistent statistically significant periodicity (Grieve and Shoemaker, 1994; Grieve and Pesonen, 1996).

4. A GALACTIC FORCING FOR PERIODIC IMPACTS AND EXTINCTIONS?

In the galactic models, the flux of long-period comets (comet showers) is modulated by the periodic passage of the solar system through the central plane of the Milky Way Galaxy (Rampino and Stothers, 1984a; Matese *et al.*, 1995). The original model (Rampino and Stothers, 1984a) proposed that the probability of encounters with molecular clouds that would perturb the Oort comet cloud causing comet showers would be modulated by the Sun's oscillation about the galactic disk. The rather flat distribution of clouds around the galactic plane suggests that an encounter would be more likely as the Sun passed through the plane region, and hence the encounters would be quasi-periodic, with a period equal to the time between plane crossings (Rampino and Stothers, 1986). Extensive numerical simulations have shown that this effect should be detectable in the terrestrial record of impact cratering with at least a 50% a priori probability (Stothers, 1985).

Matese *et al.* (1995) suggested that time modulation of the flux of new Jupiter-dominated Oort cloud comets could come from gravitational perturbations of the comet cloud by the adiabatically varying galactic tides during the in-and-out of plane oscillation; see also (Scalo and Smoluchowski, 1984). One remaining uncertainty is the most likely cycle time. The cycle time and modulation depend critically on the mass distribution in the galactic disk, particularly the distribution of dark matter in the disk. As Matese *et al.* (1995) have shown: (1) If there is no dark matter in the disk, then the mean plane crossing period is $\sim$ 44 Myr, and the peak-to-trough ratio in the comet flux is about 2.5 to 1. (2) If the current best estimate of local disk density is used, then the mean plane crossing period is $\sim$ 33 Myr, and the peak-to-trough flux ratio is about 4 to 1. (3) If the estimated errors of the various recent estimates in the disk matter density are considered, then the period could be as short as about 28 Myr, with a comet flux ratio of 4 to 1. The issue of dark disk matter is still unsettled (Gould *et al.*, 1996), and if there is substantial dark matter, the most likely remaining source may be very cold molecular clouds (Stothers, 1984; Lequeux *et al.*, 1993).

Using the current best estimate of disk mass density, Matese *et al.* (1995) found that the most recent times of peak comet flux were $\sim$ 1 Myr in the future, and also about 31, 65, and 98 Myr

TABLE 3. Results of Spectral Analyses of Impact Crater Ages

Smallest Diameter	N	Phase	Highest Peak (Myr)	Second highest Peak (Myr)
5 km	31	Free	30	35
		Fixed	30	35
35 km	11	Free	35	None
		Fixed	35	None
90 km	5	Free	36	29
		Fixed	36	30

ago, and that the cycle interval varied from 29.5 to 34.2 Myr over a 350 Myr run of their galactic model. Peak flux times lag the times of galactic plane crossing by about 1 Myr. Shoemaker *et al.* (1990) maintained that the largest impactors should be comets, and hence the largest craters should preferentially show the galactic modulation. Such pulses of increased comet flux would explain the stepped nature of some extinction events, clusters of similar-age craters, and multiple impact layers seen in the geologic record. Both galactic models must assume that shower comets come from the Oort cloud, in numbers moderately greater than background comets and asteroids (especially comets from the Kuiper belt).

In order to test the consistency of the latest cratering and extinction data with the galactic models, we performed a series of new linear spectral analyses on these data. Using the 11 mass extinctions over the last 250 Myr, we have performed the time-series analyses either allowing the phase to be a free parameter, or with fixed phasing, in which case we only used trial starting epochs in the range ± 1 Myr (the time of most recent galactic plane crossing). When the phase is allowed to vary, only one high spectral peak at 27 Myr is apparent. When the phase is fixed at the present time, the highest peak shifts to 28 Myr, but two somewhat smaller peaks at 32 and 35 Myr become considerably more prominent. These results suggest that although the average time interval between these mass extinctions is 25 Myr, the spacing is actually somewhat irregular, and so one need not accept 27 Myr as the "correct" period.

To test the crater record, we performed linear time-series analyses on sets of various size craters using the impact crater list of (Grieve and Pesonen, 1996) containing 32 craters, to which we added the Chesapeake crater (90 km, 35.2 ± 0.3 Myr), the Mjölnir crater (40 km, 142 ± 2.6 Myr), the Morokweng crater (> 100 km, 145 ± 3 Myr), and a correction for the ages of the Rochechouart (214 ± 8 Myr) and Popigai (35.7 ± 0.8 Myr) craters (Grieve, 1996, 1997; Rampino *et al.*, 1997). In all cases, the upper age limit was taken to be 250 Myr, and only those ages with errors of less than ± 10 Myr were used. Table 3 shows the results for the best-fitting periods obtained.

As Table 3 shows, we detected two peaks, a narrow peak at 30 ± 0.5 Myr and a broader peak at 35 ± 2 Myr. In a study designed to test the effects of the errors in the ages of the craters on the periods detected, Stothers (1988) showed that the errors alone were capable of shifting the dominant period between 30 and 35 Myr. We note, however, that the periods detected in the mass extinction and cratering records are statistically significant (at the 5% level) only when they are treated as periods known a priori.

Our studies show that the geologic data on mass extinctions and evidence of large impacts are consistent with a quasi-periodic modulation of the flux of Oort cloud comets with a mean period of about 30 to 36 Myr, as predicted by current galactic models. Further astronomical and geological studies should help to refine both the astronomical cycle time and the periodicity detectable in the geologic record.

References

Alvarez, W. 1986. Toward a Theory of Impact Crises. *Eos, Transactions of the American Geophysical Union* **67**, 649-658.

Alvarez, W. and R.A. Muller 1984. Evidence From Crater Ages for Periodic Impacts on the Earth. *Nature* **308**, 718-720.

Fogg, M.J. 1989. The Relevance of the Background Impact Flux to Cyclic Impact/Mass Extinction Hypotheses. *Icarus* **79**, 382-395.

Gould, A., Bahcall, J.N. and C. Flynn 1996. Disk M Dwarf Luminosity Function from Hubble Space Telescope Star Counts. *Astrophys. J.* **465**, 759-768.
Grieve, R.A.F. 1996. Chesapeake Bay and Other Terminal Eocene Impacts. *Meteoritics and Planetary Science* **31**, 166-167.
Grieve, R.A.F. 1997. Extraterrestrial Impact Events: The Record in the Rocks and the Stratigraphic Column. *Palaeoclimatology, Palaeogeography, Palaeoecology* **132**, 5-23.
Grieve, R.A.F. and L.J. Pesonen 1996. Terrestrial Impact Craters: Their Spatial and Temporal Distribution and Impacting Bodies. *Earth, Moon and Planets* **72**, 357-376.
Grieve, R.A.F. and E.M. Shoemaker 1994. The Record of Past Impacts on Earth. *Hazards Due to Comets & Asteroids* (T. Gehrels, Ed.), 417-462. Univ. of Arizona Press, Tucson.
Heisler, J. and S. Tremaine 1989. How Dating Uncertainties Affect the Detection of Periodicity in Extinctions and Craters. *Icarus* **77**, 213-219.
Hildebrand, A.R., Pilkington, M., Connors, M., Ortiz-Aleman, C. and R.E. Chavez 1995. Size and Structure of the Chicxulub Crater Revealed by Horizontal Gravity Gradients and Cenotes. *Nature* **376**, 415-417.
Hut, P., Alvarez, W., Elder, W.P., Hansen, T., Kauffman, E.G., Keller, G., Shoemaker, E.M. and P.R. Weissman 1987. Comet Showers as a Cause of Mass Extinctions. *Nature* **329**, 118-126.
Lequeux, J., Allen, R.J. and S. Guilloteau 1993. CO Absorption in the Outer Galaxy, Abundant Cold Molecular Gas. *Astronomy and Astrophysics* **280**, L23-L26.
Matese, J.J., Whitman, P.G., Innanen, K.A. and M.J. Valtonen 1995. Periodic Modulation of the Oort Cloud Comet Flux by the Adiabatically Changing Galactic Tide. *Icarus* **116**, 255-268.
Matsumoto, M. and H. Kubotani 1996. A Statistical Test for Correlation Between Crater Formation Rate and Mass Extinctions. *MNRAS* **282**, 1407-1412.
Melosh, H.J., Schneider, N.M., Zahnle, K.J. and D. Latham 1990. Ignition of Global Wildfires at the Cretaceous/Tertiary Boundary. *Nature* **343**, 251-254.
Pope, K. O., Baines, K.H., Ocampo, A.C. and B.A. Ivanov 1994. Impact Winter and the Cretaceous/Tertiary Extinctions: Results of a Chicxulub Asteroid Impact Model. *Earth and Planetary Science Letters* **128**, 719-725.
Rampino, M.R. and B.M. Haggerty 1996. The "Shiva Hypothesis": Impacts, Mass Extinctions, and the Galaxy. *Earth, Moon and Planets* **72**, 441-460.
Rampino, M.R., Haggerty, B.M. and T.C. Pagano 1997. A Unified Theory of Impact Crises and Mass Extinctions: Quantitative Tests. *Annals of the New York Academy of Sciences* **822**, 403-431.
Rampino, M.R. and R.B. Stothers 1984a. Terrestrial Mass Extinctions, Cometary Impacts and the Sun's Motion Perpendicular to the Galactic Plane. *Nature* **308**, 709-712.
Rampino, M.R. and R.B. Stothers 1984b. Geological Rhythms and Cometary Impacts. *Science* **226**, 1427-1431.
Rampino, M.R. and R.B. Stothers 1986. Geologic Periodicities and the Galaxy. *The Galaxy and the Solar System* (R. Smoluchowski, J. N. Bahcall and M. S. Matthews, Eds.), 241-259. Univ. of Arizona Press, Tucson.
Rampino, M.R. and R.B. Stothers 1988. Flood Basalt Volcanism During the Past 250 Million Years. *Science* **241**, 663-668.
Raup, D.M. 1992. Large-Body Impact and Extinction in the Phanerozoic. *Paleobiology* **18**, 80-88.
Raup, D.M. and J.J. Sepkoski, Jr. 1984. Periodicity of Extinctions in the Geologic Past. *Proceedings of the National Academy of Sciences USA* **81**, 801-805.
Raup, D.M. and J.J. Sepkoski, Jr. 1986. Periodic Extinctions of Families and Genera. *Science* **231**, 833-836.
Retallack, G.J., Seyedolali, A., Holser, W.T., Krinsley, D. and E.S. Krull 1996. Shocked Quartz at the Permian-Triassic Boundary in Australia and Antarctica. *Geological Society of America Abstracts with Programs* **28**, A-368.
Scalo, J.M. and R. Smoluchowski 1984. Galactic Gravitational Shock and the Extinction of Species. *Bulletin of the American Astronomical Society* **16**, 493-494.
Sepkoski, J.J., Jr. 1995. Patterns of Phanerozoic Extinction: A perspective From Global Data Bases. *Global Events and Event Stratigraphy in the Phanerozoic* (O.H. Walliser, Ed.), 35-51. Springer, Berlin.
Shoemaker, E.M. and R.F. Wolfe 1986. Mass Extinctions, Crater Ages, and Comet Showers. *The Galaxy and the Solar System* (R. Smoluchowski, J.N. Bahcall and M.S. Matthews, Eds.), 338-386. Univ. of Arizona Press, Tucson.
Shoemaker, E.M., Wolfe, R.F. and C.S. Shoemaker 1990. Asteroid and Comet Flux in the Neighborhood of Earth. *Geological Society of America Special Paper* **247**, 155-170.
Stothers, R.B. 1984. Mass Extinctions and Missing Matter. *Nature* **311**, 17.
Stothers, R.B. 1985. Terrestrial Record of the Solar System's Oscillation About the Galactic Plane. *Nature* **317**, 338-341.
Stothers, R.B. 1988. Structure of Oort's Comet Cloud Inferred From Terrestrial Impact Craters. *Observatory* **108**, 1-9.
Stothers, R.B. 1989. Structure and Dating Errors in the Geologic Time Scale and Periodicity in Mass Extinctions. *Geophysical Research Letters* **16**, 119-122.
Stothers, R.B. 1993. Impact Cratering at Geologic Stage Boundaries. *Geophysical Research Letters* **20**, 887-890.
Toon, O.B., Zahnle, K., Turco, R.P. and C. Covey 1994. Environmental Perturbations Caused by Asteroid Impacts. *Hazards due to Asteroids and Comets* (T. Gehrels, Ed.), 791-826. Univ. of Arizona Press, Tucson.
Trefil, J.S. and D.M. Raup 1987. Numerical Simulations and the Problem of Periodicity in the Cratering Record. *Earth and Planetary Science Letters* **82**, 159-164.
Yabushita, S. 1991. A Statistical Test for Periodicity Hypothesis in the Crater Formation Rate. *MNRAS* **250**, 481-485.
Yabushita, S. 1992. Periodicity in the Crater Formation Rate and Implications for Astronomical Modeling. *Dynamics and Evolution of Minor Bodies with Galactic and Geological Implications* (S.V.M. Clube, S. Yabushita and J. Henrard, Eds.), 161-178. Kluwer, Dordrecht.
Yabushita, S. 1996a. Are Cratering and Probably Related Geological Records Periodic? *Earth, Moon, and Planets* **72**, 343-356.
Yabushita, S. 1996b. Statistical Tests of a Periodicity Hypothesis For Crater Formation Rate-II. *MNRAS* **279**, 727-732.

VARIABILITY OF THE OORT CLOUD COMET FLUX: CAN IT BE MANIFEST IN THE CRATERING RECORD?

J.J. MATESE AND P.G. WHITMAN
Department of Physics, The University of Southwestern Louisiana
Lafayette, Louisiana, 70504-4210 USA

K.A. INNANEN
Department of Physics and Astronomy, York University, Ontario, Canada

AND

M.J. VALTONEN
Tuorla Observatory, Turku University, Finland

Abstract. We consider the subject of time dependence of the Oort cloud comet flux. Over long time scales the flux is likely to be dominated by the adiabatic galactic tide. This tide is substantially modulated as the Solar System moves in its galactic orbit. If Shoemaker was correct in his estimate that virtually all terrestrial craters of diameter >100 km are produced by long period comets, then the phase and plane crossing period of the Solar System about the galactic disk should be consistent with the ages of accurately dated large craters. A time series analysis of these ages in which the Solar oscillation phase is fixed to be consistent with observations indicates a maximal correlation for a period of 36 ± 2 Myr. This period is well within observational limits. If improvements in stellar velocity dispersion studies continue, it is possible that a sufficiently accurate determination of the Solar oscillation period can be found to unambiguously answer the following questions. Is the Solar oscillation cycle correlated with the time series of ages for large craters? If so, can we reject the hypothesis that the correlation is an artifact that could likely be reproduced by a random distribution of ages? We present evidence which suggests that if it is found that the data requires a plane crossing period in the range 36 ± 2 Myr, the answer to both of these questions will be affirmative.

1. Introduction

In the Rampino model (Rampino and Stothers, 1986; Rampino and Haggerty, 1996; Rampino *et al.*, 1997), as the Sun oscillates about the galactic disk, the Oort cloud of comets is more likely to receive an impulse from a molecular cloud as it passed the galactic midplane than when it is at maximum displacement from the midplane. Thus we can expect a quasi-periodicity in the modulation of the probability of these events. It has been argued that these events are too rare and the probability modulation too broad (Thaddeus, 1986) to produce a detectable signal. Subsequently it was shown (Matese *et al.*, 1995) that the continuous adiabatic tidal pertubation of the galaxy, which appears to dominate in making Oort cloud comets observable (Heisler, 1990), will undergo a similar modulation. But the objection that the modulation might be too broad remained.

Statistical arguments (Heisler and Tremaine, 1989; Grieve and Pesonen, 1996) indicated that for a small number of events (craters or extinctions), one could not expect a significant deduction about the reality of any putative period and phase if the convolution of dating uncertainties, period jitter and signal width was too large.

An entirely different problem consists of considering a mechanism (the production of long period comet impactors) that is known to be capable of producing these events and that is necessarily modulated quasi-periodically. The relevant question then is whether a known period and phase correlates with the time series of the events in a significant manner. The importance of answering

J. Andersen (ed.), Highlights of Astronomy, Volume 11A, 252–256.

this question is that it will shed light on the dynamical mechanism which dominates in producing these events. It must be emphasized that this problem has not been addressed statistically because the period has not been, as yet, adequately determined.

We now anticipate this problem and provide an analysis that is meant to condition the debate which will ensue when the galactic oscillation period is accurately bounded.

2. Analysis

In Matese *et al.* (1995), it was demonstrated that a substantial modulation in the adiabatically induced Oort cloud comet flux is necessary if there is to be consistency with galactic disk mass limits implied by observations of stellar velocity dispersions. If no dark disk matter exists the plane crossing period will be $\approx$ 45 Myr and the peak to trough flux variation will be $\approx$ 2.5 to 1. With a self consistent amount of compact dark disk mass, the period could be as low as 30 Myr with a variation in flux of 4 to 1. However periods that low or lower will imply that recent dark disk matter mass determinations are discrepant by more that 1 σ (Flynn and Fuchs, 1994). For discussion sake we shall allow that ultimately the period will be found to be somewhere in the range 25 Myr $\leq P \leq$ 45 Myr.

In determining a best correlated period we consider the set of craters satisfying the criteria that they are large (so that they will be dominated by long period comet impactors if Shoemaker *et al.* (1990) are correct), young (ages $\leq$ 250 Myr so that erosion is minimized) and accurately dated (uncertainties $\leq$ 3 Myr, so that the statistical dispersion is dominated by the width of the cycle flux). We are left with the following 9 craters.

TABLE 1. Large Accurately Dated Craters

Crater	D(km)	T(Myr)	ΔT(Myr)
Chicxulub	180	64.98	0.05
Morokweng	100	145	3
Manicouagan	100	214	1
Popagai	100	35.7	0.8
Chesapeake	90	35.2	0.3
Kara	65	73	3
Montagnais	45	50.5	0.8
Mjolnir	40	142	2.6
Manson	35	73.8	0.3

The analysis described makes no predictive claims as to the statistical significance of the best fit period obtained using this small number of craters. It will depend on the biases inherent in the interval choices for time, crater diameter and dating uncertainty. It will also depend on the cycle width of the putative dynamical mechanism. This does not mean that the results are without information content as we shall argue.

As a measure of goodness of fit we first determine the modulated flux produced by the galactic tide in a series of calculations in which the disk density is varied in such a way as to produce a sequence of flux versus time plots. The sequence gives the time dependent flux for 21 separate cases differing in mean plane crossing period by 1 Myr in a range 25 – 45 Myr.

In Figure 1 we show the flux for a single case with a mean period of 36 Myr. The peak flux times lag the galactic plane crossing times by $\approx$ 2 Myr and are not precisely periodic because of decreasing galactic density as the Sun recedes from the galactic core. The phase of the oscillations is restricted by observations which place the last previous plane crossing at $\approx$ 1 Myr in the past (and the next flux peak $\approx$ 1 Myr in the future). A perigalactic period of $\approx$ 180 Myr is observable in the data. The background shown is a combination of the adiabatic tidal effects of the large scale height old star population as well as an assumed contribution from stellar impulses of the outer Oort cloud affecting comets not accessible to the galactic tide. We do not show the larger but rarer contributions due to random stellar-induced showers from the inner Oort cloud which are estimated to add $\approx$ 20% to the total but dominate only 2% of the time (Heisler, 1990).

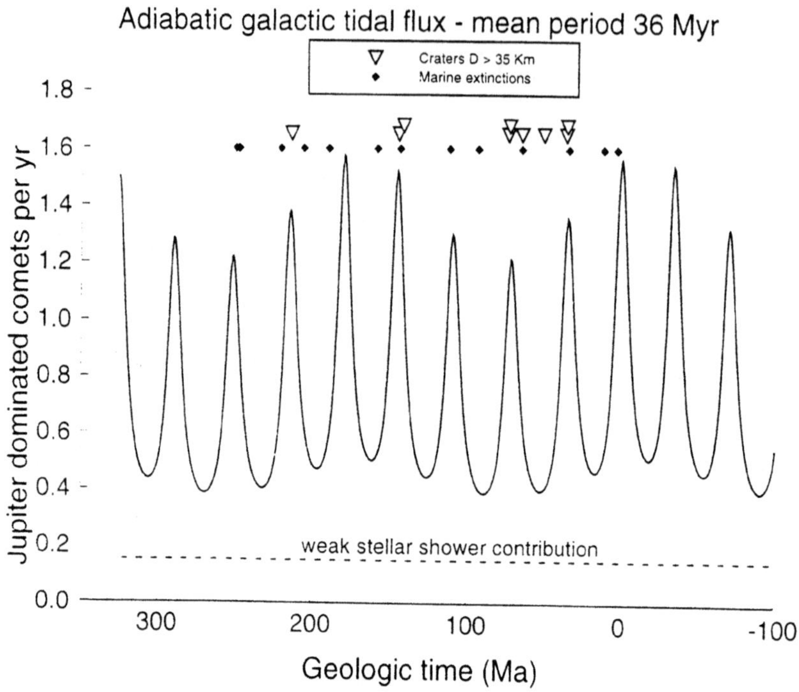

Figure 1. A model of the variable Oort cloud comet flux as modulated by the adiabatic galactic tide. Past time is positive here. Markers for 9 accurately dated large craters and 13 marine extinctions $\geq 20\%$ are shown.

The peaks above the background are attributable to the adiabatic tide of the compact component of the disk composed of molecular clouds, dust and young stars. Random molecular cloud impulses of the Oort cloud will have their probabilities modulated is a manner similar to that shown. The standard deviations of the peaks are $\approx 4 - 5$ Myr, but if the background is included the formal standard deviation of a complete cycle is closer to $7 - 8$ Myr. That is, roughly 2/3 of the model's flux occurs in a time interval of ≈ 15 Myr. Random strong stellar showers will broaden this interval while modulated strong molecular cloud showers will narrow it, but not below a value of ≈ 9 Myr.

We now locate the sequence of peak flux times τ_j for each mean period $\bar{P}$. The measure of correlation we use is

$$\chi^2(\bar{P}) = \frac{1}{N}\sum_{i=1}^{N}\left(\frac{T_i - \tau_j}{\bar{P}/2}\right)^2_{min}$$

This is constructed for each mean period using the crater agesin Table 1. These results are shown in Figure 2 for $N = 5$ and 9. For crater ages which perfectly anticorrelate with the peak flux times $\chi^2 = 1$, while $\chi^2 = 0$ for perfect correlation and $\overline{\chi^2} = 1/3$ for random crater ages. An alternative illustration is shown in Figure 3 in which we compare the crater ages and the peak flux times for each case.

One observes that the best correlation occurs for periods in the interval 36±2 Myr for both the largest 5 craters and all 9 craters. A secondary signal at $30 - 31$ Myr for the largest craters vanishes when smaller craters are included. The data are consistent with a random distribution of crater ages if the mean period is outside these intervals. These results can be compared with previous analyses in which the phase of the cycle was a free parameter and the cycle interval was uniform (Grieve and Pesonen, 1996). One can infer from Figure 3 that the best correlation for this crater data will occur within a period interval bounding 36 Myr using any goodness of fit measure if the phase is constrained to be near the present epoch. Using different data Matsumoto and Kubotani

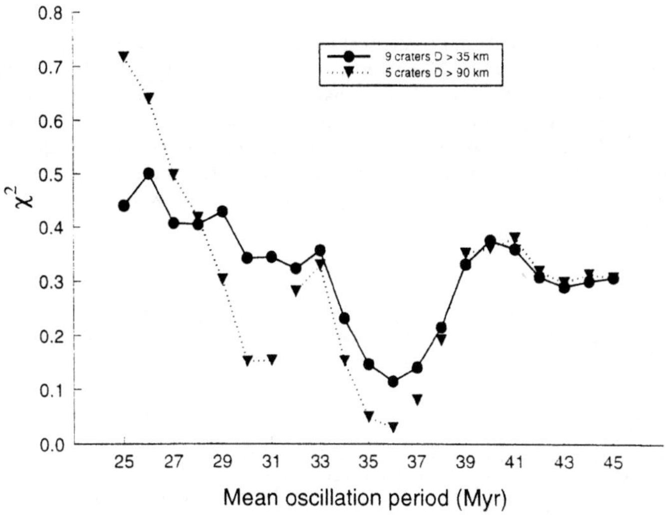

Figure 2. A correlation measure $\chi^2(\bar{P})$ for each of 21 mean periods in the range 25-45 Myr. Separately illustrated are the cases for the largest 5 and all 9 craters listed in Table 1.

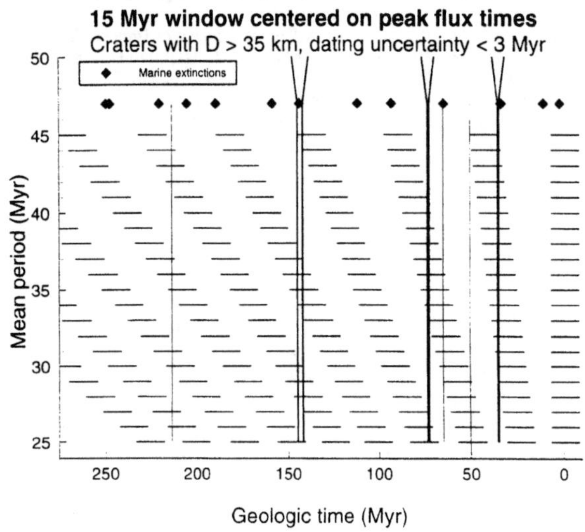

Figure 3. A comparison of 9 large accurately dated craters with a sequence of peak flux times τ for various mean periods $\bar{P}$ as predicted by the galactic oscillations model. A 15 Myr window is centered on each peak flux time as a reference.

(1996) have argued that there is a statistically significant correlation between peaks in cratering and extinction. This does not necessarily imply that they are quasi-periodic however.

3. Conclusions

Our interpretation of these results is that a galactic oscillations model in which the Solar cycle is manifest in the cratering record will be unsustainable as a working hypothesis if the cycle period is found to exclude the intervals mentioned above. Improvements in galactic disk mass determinations from stellar velocity dispersions can be obtained from Hipparcos data. Such studies are crucial in this regard.

What can we say if the period *is* found to be in these intervals? We have randomly selected a set of 9 hypothetical crater ages in a 250 Myr interval and calculated ${\chi^2}_{min}$ for $\bar{P}$= 35, 36 or 37 Myr. The process was repeated 10000 times. We find that 95% of the time the random set of crater ages had ${\chi^2}_{min} \geq 0.14$. The corresponding result for 5 randomly chosen ages is ${\chi^2}_{min} \geq 0.09$ for the 95% confidence level. In both cases these values exceed the correlation measure for the observed crater sets. This would not, in itself, indicate that the correlation between the oscillation cycle and the crater time series should be accepted as statistically significant. But it would indicate that such a correlation could not be readily rejected.

References

Flynn, C. and B. Fuchs 1994. Density of Matter in the Galactic Disc. *MNRAS* **270**, 471-479.

Grieve, R.A.F. and L.J. Pesonen 1996. Terrestrial Impact Craters: Their Spatial and Temporal Distribution and Impacting Bodies. *Worlds in Interaction: Small Bodies and Planets of the Solar System* (H. Rickman and M.J. Valtonen, Eds.), 357-376. Kluwer, Dordrecht.

Heisler, J. 1990. Monte Carlo Simulations of the Oort Comet Cloud. *Icarus* **88**, 104-121.

Heisler, J. and S. Tremaine 1989. How Dating Uncertainties Affect the Detection of Periodicity in Extinctions and Craters. *Icarus* **77**, 213-219.

Matese, J.J., P.G. Whitman, K.A. Innanen and M.J. Valtonen 1995. Periodic Modulation of the Oort Cloud Comet Flux by the Adiabatically Changing Galactic Tide. *Icarus* **116**, 255-268.

Matsumoto, M. and H. Kubotani. A Statistical Test for Correlation Between Crater Formation Rate and Mass Extinctions. *MNRAS* **282**, 1407-1412.

Rampino, M.R., and R.B. Stothers 1986. Geological Periodicities and the Galaxy. *The Galaxy and the Solar System* (R. Smoluchowski, J. N. Bahcall and M. S. Matthews, Eds.), 241-260. Univ. of Arizona Press, Tucson.

Rampino, M.R., and B.M. Haggerty 1996. The "Shiva Hypothesis": Impacts, Mass Extinctions and the Galaxy. *Worlds in Interaction: Small Bodies and Planets of the Solar System* (H. Rickman and M.J. Valtonen, Eds.), 441-460. Kluwer, Dordrecht.

Rampino, M.R., Haggerty, B.M. and T.C. Pagano 1997. A Unified Theory of Impact Crises and Mass Extinctions: Quantitative Tests. *Annals of the New York Academy of Sciences* (J.L. Remo Ed.) **822** 403-431.

Thaddeus, P. 1986. Molecular Clouds and Periodic Events in the Geologic Past. *The Galaxy and the Solar System* (R. Smoluchowski, J.N. Bahcall and M.S. Matthews, Eds.), 61-68. Univ. of Arizona Press, Tucson.

Shoemaker, E.M., R.F. Wolfe and C.S. Shoemaker. 1990. Asteroid and Comet Flux in the Neighborhood of the Earth. *Geological Society of America Special Paper 247*, 155-170.

SEARCHING FOR NEAS FROM EARTH OR SPACE

A.W. HARRIS
Jet Propulsion Laboratory
Caltech, Pasadena, CA 91109 USA

Abstract. [1] Different options for searching for near-Earth asteroids that may pose an impact hazard to the Earth in the future are discussed. The space option has not been studied in as much detail as the ground-based survey option, so a more detailed study of the space option might be appropriate. It is not evident, however, that a space-based system can be competitive or superior to the ground-based option, because past experience argues very strongly to the contrary.

1. Introduction

Impacts of asteroids or comets on the Earth pose a rare, but major threat to humanity (Chapman and Morrison, 1994). Based on this and other evaluations of the level of hazard posed by various sized events and their frequency, it appears feasible and cost effective to initiate a program to inventory all asteroids in Earth-crossing orbits, larger than $\sim$ 1 km in diameter, to determine if any one of the $\sim$ 1500 such objects thought to exist might be on a collision path with the Earth in the next century. Strategies for achieving this goal have been fairly thoroughly investigated (e.g. Morrison (ed.), 1992; Harris, 1997). These studies indicate that such a survey can be successfully completed from ground-based facilities in about 10 years, at a cost of about $100M.

In this paper, we consider the comparitive advantages of conducting such a search from space, versus from the ground. One can imagine several potential advantages of operations from space: (1) access to all wavelengths of the spectrum; (2) higher angular resolution; (3) continuous operation, that is, increased duty cycle; (4) lower sky background, i.e. improved signal-to-noise (S/N); and (5) access to more sky area. We shall discuss each of these individually below. Each of these potential advantages can be evaluated in terms of the difference in limiting magnitude of survey that could be achieved with a given system in space, compared to that on the ground. The "bottom line" is that it is practical and cost-effective to do the level of survey we have in mind at the present from the ground, and very unlikely to be faster or less expensive from space.

2. Comparison of Earth- vs. Space-Based Surveys

2.1. ACCESS TO ALL WAVELENGTHS OF THE SPECTRUM

For NEA surveys, the peak in the reflected light spectrum falls in mid-visible range, which is readily accessible from the ground. The best ground-based sites suffer less than 20% absorption at 2 air masses, and have a sky background level (mostly from air glow) only about a factor of 2 above that from space. These aspects will be addressed further in Sect. 2.4.

Thermal emission from asteroids at about 1 AU from the Sun peaks near 10 μm wavelength. This is less accessible from the ground, so space observations have a considerable advantage (cf. IRAS, ISO or SIRTF). Considering the difference in resolution of telescopes at optical vs. thermal IR, and the background sky levels at the two wavelengths, IR has a theoretical advantage of a factor of a few, but the key point is that the large element array detectors that would be needed to achieve this advantage do not exist. As will be shown below, what counts most is number of

[1]This paper is dedicated to the memory of Jurgen Rahe. This study was done at his request when he was Director of Solar System Exploration at NASA Headquarters.

J. Andersen (ed.), Highlights of Astronomy, Volume 11A, 257–261.

pixels collecting light simultaneously. Optical detectors with ~ 10 megapixels are readily available. IR detectors are presently limited to ~ 100 kilopixels (this estimate is thought to be generous), or two orders of magnitude behind optical. In addition, thermal IR detectors require liquid He cooling to achieve quantum-efficient performance. For all of these reasons, it is not believed detection by thermal emission is practical at this time, from space or from the ground.

2.2. HIGHER ANGULAR RESOLUTION

Detection of NEAs is noise-limited, that is, the lower the noise, the fainter the object that can be discerned. The noise level is proportional to the square-root of the background light level, which is proportional to the area of the resolution element (pixel or patch of pixels containing the image). Thus, if one can confine the image to a smaller angular area, one need only discriminate the signal above the background of a smaller patch of sky.

Let us compare a system in which the image is contained in a "box" of pixels 1″ on a side with a system of the same aperture, but containing the image within a box only 0.5″ across. With 4 times less background area, the background level (photon count) is only 1/4 as great, but the noise goes as square-root of photon count, so it is 1/2 the level, thus the 0.5″ system has twice the S/N level of the 1″ system. In order to recover the same S/N, the 1″ system would have to integrate 4 times longer, since S/N improves as the square-root of integration time. So the two systems could deliver the same threshold of detection if the 1″ system is allowed to use 4 times longer exposures.

In practical terms, we are limited by the number of detector elements we can have. So if these two systems each have the same number of pixels in their detector (CCD) arrays, then the 1″ system covers 4 times as much sky area in a single exposure as the 0.5″ system. Thus taking 4 times longer exposures, it covers sky area at exactly the same rate as the 0.5″ system, when operated to reach the same threshold detection level. This example illustrates a conclusion first reached by Greg Canavan (personal communication, 1995): as long as one avoids over-sampling the image, what counts most is the total number of pixels that can be placed in the focal plane. Applying this rule to the question of space vs. ground based surveying, the improved resolution available from space is only of value if we can place more resolution elements in the focal plane than is possible from the ground.

At present, we are limited by the size of CCD arrays, not by the number of resolution elements that can be delivered to the focal plane of a ground-based telescope. For example, presently contemplated ground-based systems are only utilizing ~ 2″ resolution, for the simple reason that higher resolution is not needed to completely fill the largest presently available CCD arrays. Thus the added resolution available from space is not necessary unless one were to consider a much more ambitious survey than we have dared to consider or try to justify.

2.3. CONTINUOUS OPERATION (INCREASED DUTY CYCLE)

System duty cycle is a very contentious question. From the ground, a telescope can only be used when the Sun is rather well below the horizon and when the Moon is below the horizon. Thus "dark" sky can only be had about 25% of the time, or ~ 180 hours a month. Typical observatory duty cycles, allowing for weather, equipment limitations, etc., are ~ 15%, or ~ 100 hours a month. In principle, one could operate 100% of the time from space, or ~ 720 hours a month, thus achieving a 7-fold improvement over typical ground-based operations, or even a 4-fold improvement over the theoretical limitation of "dark sky" from the ground.

Achieving this advantage, however, is fraught with difficulties. Perhaps the duty cycle of the HST is the best model we can take as a guide. That telescope spends only ~ 15% of its time collecting photons, the same fraction as a ground-based observatory. This is not for lack of desire for more HST time – it is because of practical limitations that had to be faced in designing the total system. It seems unlikely that a telescope attached to the International Space Station (ISS) can do any better, considering the manned space operations that will be going on continuously in the vicinity, pointing limitations, plus all the same other constraints as faced by HST.

In the absence of an engineering study, there is little more to be said about this issue, but here especially the burden of proof lies with ISS to demonstrate that anything even as good as a 15% duty cycle can be realized. Any improvement in duty cycle over a ground-base site can be translated to an improvement in limiting magnitude of otherwise equivalent systems. The absolute maximum corresponding to 100% duty cycle is equivalent to an improvement of $1^m.0$ of the threshold of detection. A factor of 4 improvement in duty cycle is equivalent to $0^m.75$ magnitude. Equal duty cycles, as suggested by HST experience, of course corresponds to no magnitude difference. It could (likely?) be even less, leading to a negative difference for a space-based system.

2.4. LOWER SKY BACKGROUND (IMPROVED S/N)

The lower sky background from space translates to a higher S/N for a given target, or to a fainter target brightness threshold for a given detection S/N. We can combine this with the modest improvement in signal due to lack of atmospheric absorption. The formulae earlier obtained (Harris, 1994, 1997) give the threshold magnitude level obtainable from a given system in terms of a threshold S/N that can be discriminated.

The numerical formulae have built into them background levels and atmospheric absorption appropriate for ground-based sites. If we assume that atmospheric absorption (at 2 air masses, as an average) amounts to 20%, and the sky background from space is a factor of 2 lower than from the ground, that implies that the signal from space is 1.2 times greater, and the noise is $\sqrt{2}$ times less, thus the S/N from space is improved by a factor of 1.7. We can thus "correct" for space operation by reducing the threshold for detection by a factor of 1/1.7, to obtain the appropriate limiting magnitude for a space based system. In the studies that have been done the limiting S/N for detection was normally taken to be 4.0. By using S/N = 4.0/1.7 = 2.4, the equivalent limiting magnitude for the same system from space can be calculated. The improvement is about $0^m.6$ magnitudes for otherwise equivalent systems.

2.5. ACCESS TO MORE SKY AREA

The last potential improvement from space operation is access to more sky area. It has been amply shown in several places (see, e.g., Bowell and Muinonen, 1994; Harris, 1997) that the optimum survey strategy is to cover as much sky area as possible. Ground-based observatories cannot reach down to low solar elongation, and a single observatory in either hemisphere necessarily misses a cone of sky area around the opposite hemisphere pole. In the various survey simulations that have been done, we have simulated reasonable sky access limitations for a ground-based telescope at $\sim 35°$ N latitude. More recently a simulation was done in which the horizon limitations were removed and instead a limit was set to the minimum solar elongation accessible.

A plot of the "completion curve" of such a system, using a minimum solar elongation of 45°, and comparing that to the one for a ground-based system at 35° N latitude is shown in Fig. 1. There is remarkably little difference between these curves. As described in (Harris, 1997) the relative completion curve can be scaled for systems of different capability, or to refer to objects of different size, by sliding them horizontally. As plotted, they refer to ground-based vs. space-based systems of equal magnitude threshold. By sliding the curves into rough coincidence, it is apparent that the advantage of extra sky area of the space based system is equivalent to only a difference of about 0.25 in limiting magnitude. That is, a space-based system reaching $20^m.0$ would be equivalent to a single-site ground-based system reaching $20^m.25$. This small difference is due almost entirely to the missing bit of sky in the south, thus siting a second ground-based telescope in the southern hemisphere should bring the two curves into almost total coincidence.

It should be noted that the simple addition of a second telescope, thereby doubling exposure times allowed, leads to an improvement of $\sim 0^m.4$ in threshold of coverage, even more important than covering the missing hemisphere. Why does the "southern hole" account for so little? The reason is that the main area of inaccessible sky is the band surrounding the galactic plane, which cannot be used because of image confusion. That band is about 40° wide (±20° galactic latitude), and is a much larger sky area than a 45° cone about the southern pole, or about the Sun, for that matter.

3. Discussion

Of all the above effects, (1) and (2) don't matter to first order, and (3), (4), and (5) lead to improvements which can be cast in terms of improvement in the threshold limiting magnitude for otherwise identical systems, space-based vs. ground-based, as it is summarized in Table 1.

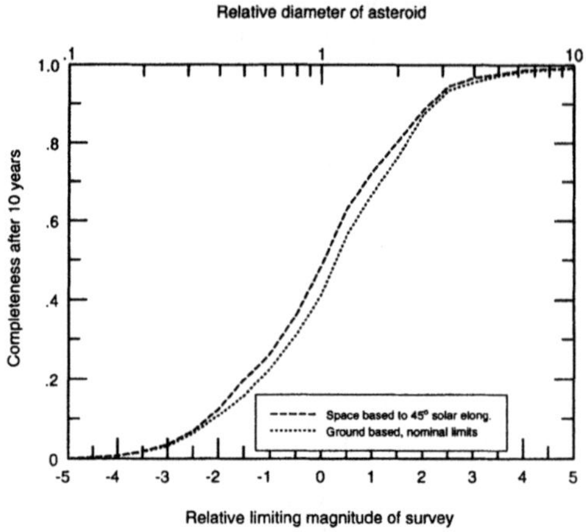

Figure 1. The relative survey completion curves for an Earth-based survey system compared to a space-based one with the same limiting magnitude capability (see text). The curves represent the fraction of NEAs detected in ten years vs. size of NEA (top scale), or equivalently, vs. system limiting magnitude for a given size object (bottom scale). As drawn, the curves could be taken to be completeness for 1 km objects, where the zero point of the system limiting magnitude scale is 18.5, or equivalently, the curves could represent the completeness vs. size in km (top scale) for a system of limiting magnitude 18.5. It is apparent from this plot that the advantage of space operation, with respect to additional sky accessible, is equivalent to only a difference of about 0.25 in limiting magnitude of the survey.

The above potential for improvement should be considered in comparison to simply augmenting ground-based systems. As it was already noted, siting a second telescope in the opposite hemisphere essentially eliminates the $0^m.25$ advantage due to greater sky area, so assuming that a second ground-based instrument would be placed in the opposite hemisphere, the range of potential advantage of a space-based system over a ground-based one is only $0^m.6 - 1^m.6$. How many ground-based telescopes, otherwise identical to the putative orbiting one, would be needed to achieve the same limiting magnitude? The answer is 3 to 20, corresponding to the range $0^m.6 - 1^m.6$.

One may reasonably doubt that any improvement over ground-based duty cycle is practically feasible, so we would place the "benefit" of space operation at best equivalent to 3 ground-based telescopes of the same kind. This must be balanced against the cost of development and operation of a space-based system. Building and operating a 1 m telescope in orbit for 10 years seems unlikely to be less expensive than the estimated cost of $100M estimated for a ground-based survey. A better estimate is unavailable, but a comparison with HST, which is only twice as large a telescope, is not encouraging. A comparison even with the HST operations budget, forgetting about the cost of construction and launch, is daunting.

Turning to ground-based systems, a set of about three $1-2$ m telescopes is about what it would take to do the Spaceguard Survey – that is, detect and catalog most NEAs larger than ~ 1 km in diameter, in a decade from the start of the survey. In a report to NASA, Shoemaker et al. (1995)

TABLE 1. Improvement in the threshold limiting magnitude for otherwise identical search systems, space-based vs. ground-based

Attribute	Improvement in limiting magnitude
Duty cycle	$< 0.0 - 1.0$
Darker sky, no atm. absorption	0.6
Greater sky area	0.25
Total improvement	$0.85? - 1.85$

estimated that such a system would cost $\sim$ \$100M to design, construct, and operate for a decade. This is about the cost of one Discovery mission, a very modest price indeed compared with other NASA programs, and minuscule compared to the ISS budget.

It is interesting to note that the cost of doing the survey is almost independent of time. At the present rate of expenditure, $\sim$ \$2M/year, it will take $\sim$ 50 years. At a budget of $\sim$ \$10M/year, the time can be shrinked to $\sim$ 10 years. One cannot shrink the time scale much under 10 years, because of the limits imposed by orbital dynamics: NEAs tend to be acceptably placed for discovery only about once in an orbit, of $4-5$ years period. A telescope in space could speed things up compared to the present 50-year time scale, but it could not improve much on the 10-year time scale of the \$100M ground-based system. The reason is simply the matter of time to develop a new system, plus the intrinsic minimum time scale of $\sim$ 5 years due to orbital mechanics. The earliest possible launch of an ISS telescope is 2001. Thus by the time an ISS based system could become operational, a ground-based survey could be nearing its midpoint, and "catching up" from space would be a near impossibility, regardless of how capable the system might be.

4. Conclusion

In the course of studying strategies for discovery surveys of near-Earth asteroids (NEAs), the possibility of doing a survey from space has been mentioned. We reported here a study of the relative effectiveness of operations from space vs. from the ground. Not surprisingly, the main factors determining effectiveness are (1) light collecting area (telescope aperture), (2) total number of resolution elements collecting light simultaneously (number of pixels for a well-scaled system), and (3) duty cycle of the system (fraction of time actually recording data). The theoretically greater sky area accessible from space makes little difference over a well-chosen ground site. Of these three factors, (1) favors ground operations over space, (2) is a technological limitation at the moment, until such time as large enough arrays of CCDs can be fabricated to capitalize on higher resolution available from space, and (3) has a theoretical advantage of a factor of a few for space operation, but has not been realized for any space-borne telescope yet (*cf.*, HST has an effective duty cycle of $\sim$ 15%, comparable to ground-based telescopes). The good news is, it appears possible to catalog most potentially hazardous NEAs down to $\sim$ 1 km in diameter with a modest ground-based survey, over a period of about a decade. Thus there is little reason to advocate a space survey for this purpose.

References

Bowell, E., and K. Muinonen 1994, Earth-crossing asteroids and comets: Groundbased search strategies, in *Hazards due to Comets and Asteroids*, T. Gehrels ed., U. Arizona press, Tucson, pp. 149-197.

Chapman, C.R., and D. Morrison 1994, Impacts on the Earth by asteroids and comets: Assessing the hazard, *Nature* **367**, 33-40.

Gehrels, T. (ed.) 1994 *Hazards due to Comets and Asteroids*, U. Arizona press, Tucson, 1300 pp.

Harris, A. 1994, CCD systems for searching for near-Earth asteroids, in *Seventy-Five Years of Hirayama Asteroid Families*, ASP Conference Series, **Vol. 63**, pp. 203-213.

Harris, A. 1997, Evaluation of ground-based optical surveys for near-Earth asteroids, *Planetary & Space Science*, in press.

Morrison, D. (ed.) 1995, *The Spaceguard Survey. Report of the NASA International Near-Earth-Object Detection Workshop*, Report prepared for NASA Office of Space Science and Applications.

Shoemaker, E.M. et al. 1995, *Report of the Near-Earth Objects Survey Science Working Group*, Report prepared for NASA Office of Space Science and Applications.

TIDAL SPLITTING OF COMETS IN EARTH'S VICINITY

H. RICKMAN
Astronomiska observatoriet
Box 515, S-75120 Uppsala, Sweden

AND

J.M. GREENBERG
Laboratory Astrophysics
University of Leiden, Postbus 9504, 2300 RA Leiden, The Netherlands

Abstract. We present a study of the influence of tidal splitting In order to clarify the role of comets as Earth impactors and bringers of organics to the young planet. Based on low values for the density and material strength, as indicated by Comet Shoemaker-Levy 9, we have modeled the evolution of a comet in an orbit leading to Earth impact. We find that cometary nuclei impacting the Earth with radii $\gtrsim 0.1 - 0.3$ km will split just before doing so. Nuclei of several km radius will split when passing within a distance $\lesssim 5 - 8$ Earth radii. Cometary nuclei with radii $\gtrsim 1$ km stand a much larger chance of being tidally split than of hitting the Earth. Consequently, streams of fragments and debris form in Earth-crossing orbits more often than large nuclei cause impacts. The mass distribution of impacting material is shifted toward sub-km objects, whereby the survivability of organics may be significantly increased. If the cometary impactors move in orbits where $\gtrsim 5$ passages within Earth's Roche zone are expected, we estimate that most of the material reaching Earth's surface will appear in the form of split fragments.

1. Introduction

The total mass of Earth impactors during the first ~ 700 Ma after the Moon-forming event has been estimated (Delsemme 1996; Chyba and Sagan 1996) as $\sim 10^{22}$ kg. The extended timescale of this bombardment points to comets coming from the outer regions of the planetary system. Comets contain a significant fraction of organic refractories in the shape of "CHON" material, consistent with Greenberg's model of the interstellar grains of star forming regions (Greenberg and Hage 1990). Mass spectroscopy of CHON material has revealed the presence of complex molecules of possibly prebiotic nature (Kissel *et al.* 1996).

Such organics may have contributed very importantly to the total inventory of prebiotic material, but a critical issue is the size-dependent survivability of cometary organics upon impact. For impactors from 100 m to 10 km diameter the fraction of surviving organics decreases with increasing size, and with this in mind we will now take a first look at the effects of tidal splitting on the size distribution of cometary impactors.

2. Critical Splitting Distance

Comet Shoemaker-Levy 9 (to be referred to as 'SL-9') split into a dozen major fragments as it passed through Jupiter's Roche sphere in July 1992. Further fragmentation later on (Sekanina *et al.* 1997) caused the number of observed subnuclei to reach more than 20. The material strength S can be estimated from:

$$\frac{\Delta_{SJ}}{R_J} = C\Big(\frac{\rho}{\rho_J}\Big)^{-1/3}\Big(1 + \frac{S}{P_c}\Big)^{-1/3} \tag{1}$$

J. Andersen (ed.), Highlights of Astronomy, Volume 11A, 262–265.

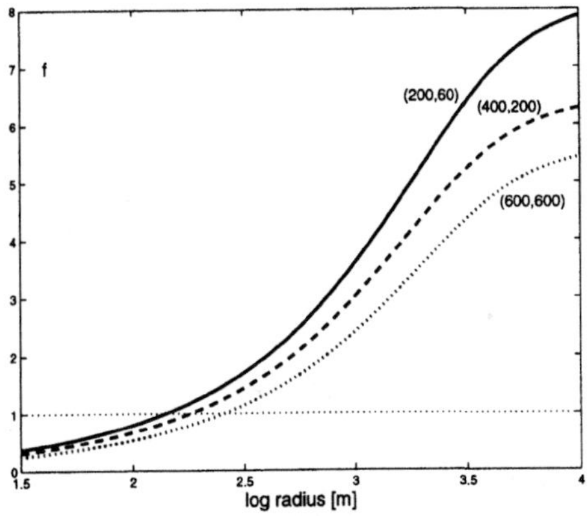

Figure 1. Splitting distance, in Earth radii, as a function of the radius of a cometary nucleus for three different combinations of (density, material strength) as indicated for each curve.

where $P_c = \frac{2}{3}\pi G\rho^2 R^2$ is the central gravitational pressure. Δ_{SJ}/R_J is the splitting distance expressed in jovian radii, and ρ and R are the cometary density and radius, respectively. For splitting near perijove ($\Delta_{SJ}/R_J \simeq 1.35$), and estimating $C = 2.7$ (see Harris 1996) and $R \sim 1$ km for the major fragments, one finds: $S \sim 100$ Pa for $\rho \sim 300$ kg/m^3 as estimated for 1P/Halley by Rickman (1989).

Greenberg *et al.* (1995) estimated the strength of a porous aggregate of submicron-sized, ice-mantled grains typical of the expected constituents of cometary material. The van der Waals bonding energy per unit volume was thus computed from the number of H_2O molecules in the regions of contact between neighbouring grains. The resulting relation between porosity and strength is consistent with the above values for ρ and S.

Eq. (1) may be used with the sole change from Jupiter's radius and density to those of the Earth in order to compute the splitting distance $f = \Delta_{SE}/R_E$ as a function of R for different values of ρ and S. Any comet with the assumed characteristics might split if its perigee distance were less than Δ_{SE}, based on the experience from SL-9. Some results are shown in Fig. 1. At large radii, the curves level off toward limiting values that represent the Roche limits for the respective densities. The plot also shows that $f = 1$ (splitting limit at Earth's surface) occurs for R between 140 and 260 m, based on the considered ranges of ρ and S. Some extra uncertainty comes from an unavoidable error bar on C due to the uncertainties of the splitting mechanism (see Sect. 3).

This indicates that cometary impactors with radii $\gtrsim 1$ km would split during the inbound passage of the Roche sphere. Taken at face value, these results even imply that every km-sized cometary impactor breaks up into pieces of $\lesssim 0.5$ km diameter before reaching the ground. As a consequence, comets would not cause any cratering on our planet!

3. Splitting Scenario

However, the outcome of course depends on the splitting scenario, which is as yet controversial. Sekanina *et al.* (1994) claim that effective breakup of SL-9 did not occur until 3^h after perijove. Such a delay would not leave time enough for breakup before impact on Earth (time of flight through Earth's Roche sphere $\lesssim 1^h$). More generally, some delay should always be expected between tidal stress and material yielding in models assuming a finite material strength, and it is indeed questionable if there would be time enough for full development of hierarchical splitting!

According to the strengthless rubble pile model of other authors (Asphaug and Benz 1996; Solem 1995), the spin state of the progenitor nucleus may have a decisive influence on the outcome. Some nuclei might hardly break up at all even if they were completely strengthless, while others might be shattered even more than SL-9. Bottke *et al.* (1997) have modelled the encounters of rubble-pile Earth-crossing asteroids with the Earth and found that for a multi-component splitting such as that of SL-9 to occur, a prograde spin and adequate spin axis orientation are required. Moreover,

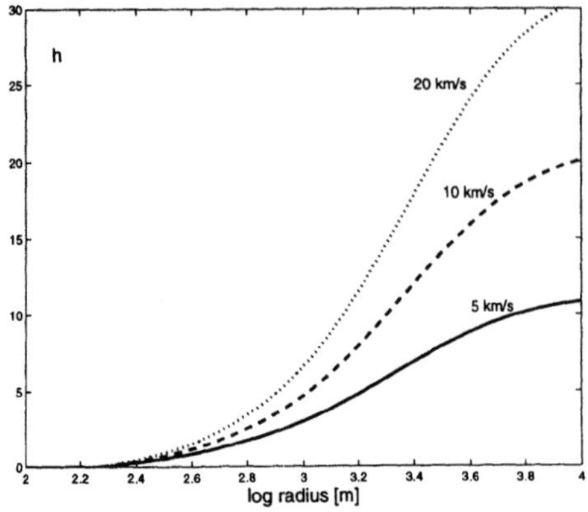

Figure 2. Ratio of cross-sections for splitting vs impact as a function of nuclear radius, assuming the intermediate values of density and strength of Fig. 1, for three different encounter velocities with respect to the Earth.

very slow encounters or very close approaches are anyway needed. Thus, in their scenario tidal splittings within Earth's Roche sphere are less important than we find under the assumption of instant splitting.

4. Splitting vs Impact Frequencies

In the range of radii where $f > 1$, comets may pass close enough to Earth to be tidally split without actually hitting it. Such encounters would be like the one of SL-9 with Jupiter in 1992, and breakup would be more likely to ensue even if delayed. Fig. 1 showed that, for large radii, the splitting cross-section is much larger than the collision cross-section.

To evaluate the frequency of splittings relative to impacts, we have to account for gravitational focussing. In terms of the above-defined value of f, the ratio of cross-sections for splitting and impact is given by:

$$h = \begin{cases} (f-1)\left\{1 + f/(1 + 2GM_E/R_E v_\infty^2)\right\} & ; \quad f \geq 1 \\ 0 & ; \quad f \leq 1 \end{cases} \tag{2}$$

Fig. 2 shows this ratio as a function of R for encounter velocities (v_∞) of 5, 10 and 20 km/s, spanning the range from low-eccentricity Near-Earth Object (NEO) orbits to high-eccentricity cometary orbits. It is evident that cometary nuclei with radii $\gtrsim 1$ km stand a much larger chance of being tidally split than of hitting the Earth. Judging from the SL-9 experience, streams of fragments and debris may then form in Earth-crossing orbits. This means, as time goes by, the deposition of cometary material by bodies of various sizes, typically much smaller than the progenitor nucleus.

5. Cometary Earth Impactors

How much of the deposited material will result from previous tidal splittings? This depends on *(a)* the a priori probability for a splitting event to occur, and *(b)* the fraction of the resulting fragments that later collide with the Earth.

Full computations remain to be done. We present only a first approach based on Poisson statistics for the number of passages through Earth's Roche sphere, given a certain expectance E. For cometary NEOs with a dynamical lifetime of $\sim 10^7$ orbits and inclinations $\sim 10°$, the expected number of passages is $E \sim 5$. Given an object that impacts the Earth, Fig. 3 shows the probability p that it has previously undergone tidal splitting as a function of R for three values of E. If the efficiency of fragment delivery to Earth is high, we find that most of the impactor mass originating in km-sized nuclei reaches the Earth in smaller units.

For active comets in their typical, unstable orbits, the situation is different. With a lifetime of $\sim 10^3$ orbits, the chance to have first a splitting, then an impact at a later encounter is practically

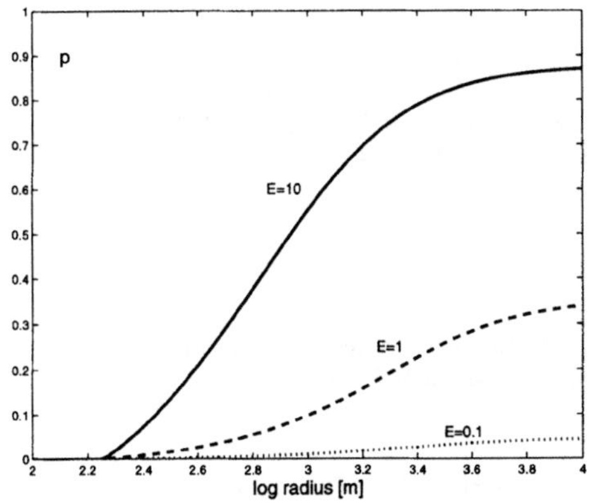

Figure 3. Probability of previous tidal splitting for cometary Earth impactors as a function of nuclear radius, based on simple Poisson statistics. The curves are for three different assumed values of the expectance of the number of passages through Earth's Roche sphere during the dynamical lifetime of the object.

nil. But in reality most such objects do not encounter the Earth at all, while those with a favourable geometry (node-crossers) stand a much higher chance. More detailed calculations are necessary to derive an accurate estimate of E for this case.

6. Conclusions

On the working hypothesis of instant splitting in response to the tidal stress, there should be a significant comminution of the cometary km-sized impactor population due to tidal splitting during close encounters preceding the one when impact occurs. Splitting of objects down to 0.1 km radius just before impact may also be of importance.

This means that some fraction, probably very significant, of the total cometary impactor mass is shifted from large objects into sub-km units. The increased abundance of air bursters (and micrometeorites) implies a large increase of the survivability of organic compounds with respect to previous estimates. These were based on an estimated mass distribution of observed comets, where large nuclei play a dominant role.

References

Asphaug, E. and Benz, W. (1996) Size, Density and Structure of Comet Shoemaker-Levy 9 Inferred by the Physics of Tidal Breakup, *Icarus*, **121**, 225–248

Bottke, W.F., Jr., Richardson, D.C. and Love, S.G. (1997) Production of Tunguska-Sized Bodies by Earth's Tidal Forces, *Planet. Space Sci.*, in press

Chyba, C.F. and Sagan, C. (1996) Comets as a Source of Prebiotic Organic Molecules for the Early Earth, in *Comets and the Origin and Evolution of Life*, eds. P.J. Thomas *et al.*, Springer-Verlag, pp. 147–173

Delsemme, A. (1996) The Origin of the Atmosphere and of the Oceans, in *Comets and the Origin and Evolution of Life*, eds. P.J. Thomas *et al.*, Springer-Verlag, pp. 29–67

Greenberg, J.M. and Hage, J.I. (1990) From Interstellar Dust to Comets: A Unification of Observational Constraints, *Astrophys. J.*, **361**, 260–274

Greenberg, J.M., Mizutani H. and Yamamoto, T. (1995) A New Derivation of the Tensile Strength of Cometary Nuclei: The Fit to Comet Shoemaker-Levy 9, *Astron. Astrophys.*, **295**, L35–L38

Harris, A.W. (1996) Effects of Shape and Spin on the Tidal Disruption of P/Shoemaker-Levy 9, in *Worlds in Interaction: Small Bodies and Planets of the Solar System*, eds. H. Rickman and M.J. Valtonen, Kluwer Acad. Publ., pp. 113–117

Kissel, J., Krueger, F.R. and Roessler, K. (1996) Organic Chemistry in Comets from Remote and In Situ Observations, in *Comets and the Origin and Evolution of Life*, eds. P.J. Thomas *et al.*, Springer-Verlag, pp. 69–109

Rickman, H. (1989) The Nucleus of Comet Halley: Surface Structure, Mean Density, Gas and Dust Production, *Adv. Space Res.*, **9**, (3)59–(3)71

Sekanina, Z., Chodas, P.W. and Yeomans, D.K. (1994) Tidal Disruption and the Appearance of Periodic Comet Shoemaker-Levy 9, *Astron. Astrophys.*, **289**, 607–636

Sekanina, Z., Chodas, P.W. and Yeomans, D.K. (1997) Secondary Fragmentation of Comet Shoemaker-Levy 9 and the Ramifications for the Progenitor's Breakup in July 1992, *Planet. Space Sci.*, submitted

Solem, J.C. (1995) Cometary Breakup Calculations Based on a Gravitationally-Bound Agglomeration Model: The Density and Size of Shoemaker-Levy 9, *Astron. Astrophys.*, **302**, 596–608

ENTRY FLASHES OF COMETARY FRAGMENTS IN JOVIAN UPPER ATMOSPHERE

S. TAKEUCHI
National Space Development Agency of Japan
Sengen, Tsukuba, Ibaraki 305, Japan

H. HASEGAWA
ASTEC, Inc.
Minami-cho, Shinjuku, Tokyo 162, Japan

AND

J. WATANABE
National Astronomical Observatory of Japan
Osawa, Mitaka, Tokyo 181, Japan

1. Introduction

Short lasting flashes, called as First Precursor (PC1), were observed by some ground-based near-infrared observations for the impacts of large-sized fragments of comet Shoemaker-Levy 9 (SL9) in July 1994. The impact detections by the spacecraft *Galileo* [2, 7] about 10 seconds after the PC1 detections by ground based telescopes, combined with the far-side impacts of SL9 fragments as viewed from Earth, suggest that the source of the PC1 should be located in the Jovian upper atmosphere above the limb, at which the atmospheric pressure is extremely low. Thus, an important problem on the PC1 is how does the falling cometary fragment, which is a huge meteorite, emit near-infrared in the extremely thin atmosphere. The ablation model, which is usually used for an impact bolide, can only estimate flux from the bolide in a dense atmosphere at visible wavelength. In this paper, we assume that the PC1s are thermal radiation from the fragments and attempt quantitative estimations of the PC1 fluxes using a simple entry flash model.

2. Observations of the PC1

A summary of the PC1 observations is given by Table 1. The precise position of impact site for each fragment is an important parameter and determines a minimum height of the fragment seen from Earth direction. We use the impact site estimations by Chodas and Yeomans [3], whose

TABLE 1. Summary of PC1 Observations and Geometry

Frag- -ment	Class [5]	PC1 Peak Flux ($Wm^{-2}\mu m^{-1}$/Jy)	Wavelength (μm) / Observatory	Meridian Angle(deg) [3]	Minimum Visible Height above 1bar(km)
A	2a	9.0×10^{-16}/0.0016	2.3 / Calar Alto [6]	65.4	556
G	1	1.8×10^{-14}/0.0048	0.89/ HST [5]	67.1	410
		$> 5.4 \times 10^{-14}$/0.010	2.34/ CASPIR [8]		
H	2a	7.9×10^{-14}/0.14	2.3 / Calar Alto [6]	67.5	367
K	1	3.0×10^{-13}/0.55	2.35/ Okayama [12]	68.3	304
L	1	2.0×10^{-13}/0.36	2.3 / Calar Alto [6]	68.9	250
Q1	2b	2.5×10^{-14}/0.045	2.3 / Calar Alto [6]	69.9	190
R	2b	2.4×10^{-13}/0.43	2.3 / Keck [4]	70.2	169
W	2c	6.0×10^{-13}/0.060	0.55/ HST [5]	71.2	117

J. Andersen (ed.), Highlights of Astronomy, Volume 11A, 266–269.

meridian angle from the mid-night longitude are shown at 5th column in Table 1, and calculate the minimum visible heights of the fragments. The accuracies of the meridian angles are about 0.1 degree (Chodas, personal communication), which corresponds to the uncertainty of the minimum visible height about some ten kilometers.

3. Entry Flash Model in the Upper Atmosphere

Because the PC1 emission was thermal continuum for AAT observation [9], we must know the temperature of PC1 radiative source in order to estimate the PC1 infrared flux. We assume simply that all the lost energy of the fragment due to the atmospheric drag is radiated from the surface of the fragment, so that we can estimate the radiative temperature using the following formulation.

The equations of motion and energy of the entry fragments are as follows;

$$\frac{du}{dt} = -\frac{C_d}{2m} A\rho V u\,, \tag{1}$$

$$\frac{dw}{dt} = -\frac{C_d}{2m} A\rho V w - g\,, \tag{2}$$

$$\frac{dE}{dt} = -\frac{C_d}{2} A\rho V^3\,, \tag{3}$$

where u and w are the horizontal and vertical velocity of fragment, V is the speed of fragment ($V^2\!=\!u^2\!+\!w^2$), A is the cross section of fragment ($A\!=\!\pi\, r^2$), r is the radius of fragment, m is the mass of fragment, E is the energy of fragment ($E\!=\!mV^2/2\!+\!mgz$), g is the acceleration due to the Jovian gravity (25 m/s^2), z is the altitude above the 1 bar level, ρ is the density of the Jovian atmosphere, C_d is the drag coefficient (1.2). It is assumed that all the lost energy of the fragment due to the atmospheric drag is radiated instantaneously as thermal emission;

$$-\frac{dE}{dt} = (4\pi r^2)\sigma T^4\,, \tag{4}$$

where σ is Stephan-Boltzmann constant, T is the radiative temperature or the surface temperature of the fragment. From the equations (3) and (4),

$$T = \left(\frac{C_d \rho V^3}{8\sigma}\right)^{1/4}. \tag{5}$$

The flux at wavelength λ, observed at Earth is

$$F_\lambda = B_\lambda(T)\, A\, \frac{1}{\Delta^2}\,, \tag{6}$$

where F_λ is the thermal flux of the fragment at wavelength λ, $B_\lambda(T)$ is the Planck function, Δ is the distance between Jupiter and Earth, which was 5.17AU (7.75×10^8km) at the SL9 impact days, July 1994.

Thus, given the density distribution of the Jovian atmosphere, the flux from a fragment with radius r can be calculated. V is about 60 km sec^{-1}.

4. Jovian atmosphere

We use two different models of the Jovian atmosphere. One is the distribution by Atreya et al. [1], which is called in general as the 'Orton' model. This model has a constant increment of temperature in the thermosphere (above 400 km). Another is the distribution sounded by the *Galileo* probe [10]. This model, called as the 'Probe' model in this paper, shows rapid increment of temperature around the mesopause (350–400 km) and constant, high temperature of 800 K in the thermosphere. But, below the height of 400 km, both atmospheric models have almost same distribution.

These two models are constructed for the Jovian equatorial latitude. Because the SL9 fragments impacted at the mid-latitude of Jupiter, we reconstruct the atmospheric models for the mid-latitude taking into account the difference between the surface gravities of the two latitudes. The vertical distributions of the temperature, the pressure and the density for both models are shown by Fig. 1.

Using the distributions of both models, the entry fluxes of fragments are calculated from equations (5) and (6). Fig. 2 shows the entry fluxes of 2.3μm for fragments with radii of 1 km, 500 m, and 200 m in the 'Orton' and 'Probe' models.

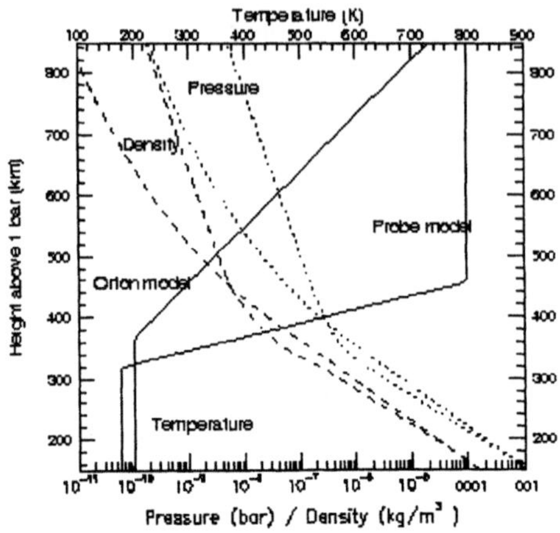

Figure 1. Vertical distributions of temperature (solid lines), pressure (dotted lines), and density (dashed lines) in the Jovian upper atmosphere. Higher values of the temperature, the pressure and the density show the distributions of the Probe model. Lower values show those of the Orton model.

5. Discussion and Summary

First, considering the uncertainty of the minimum visible height, Fig. 2 shows that the fragments H, K, and L, which are known to be the largest ones among the SL9 fragments, have radii of several hundred meters. Some estimations of the fragment sizes (e.g., Sekanina [11]) indicate that the diameters of the largest fragments are about 1 km or less and that the diameters of the medium sized fragments are a few hundred meters, which are consistent with the size estimations derived from the PC1 flux calculation in this paper. The visible (0.89μm) PC1 flux of the G impact can be also explained by the entry flash model and a fragment with a radius of several hundred meters, in spite of no figure of the result because of lack of space.

Next, we shall take into account three factors, refraction and methane absorption of the Jovian atmosphere and additional flux contributed from dust around the fragment nucleus, which are not incorporated in the entry flash model. The refraction of the atmosphere will decrease the minimum visible height some tens kilometers, but the methane absorption will take the "limb" of Jupiter up to higher altitude than the 1 bar level and will increase the minimum visible height by several tens of kilometers. Thus, these two factors have little effect for the PC1 flux calculation for now.

Also we don't think that the dust around the fragment nucleus has a large contribution to the PC1 flux, because of the weak PC1 flux of fragment Q_1. Q_1 was the brightest fragment among the 22 SL9 fragments [13], which means that there was a large amount of dust around the Q_1 fragment. However, the main event of the Q_1 impact was not largest and the size of the debris cloud after the Q_1 impact was a medium class (2b) [6, 5]. This suggests that the main event and the debris cloud depend on the fragment nucleus, and that the fragment Q_1 consists of a medium size nucleus and a lot of dust around the nucleus. Fig. 2 shows that the PC1 flux of the Q_1 is weak and that the size of the fragment Q_1 is 100 m or less. This is consistent with the suggestion by the main event and the debris, not with the pre-impact brightness, which means that the dust did not contribute to the PC1.

Last, we discuss the PC1 of the A impact, whose minimum visible height is so high that the difference between the two atmospheric models is important. For the Orton model, the estimated

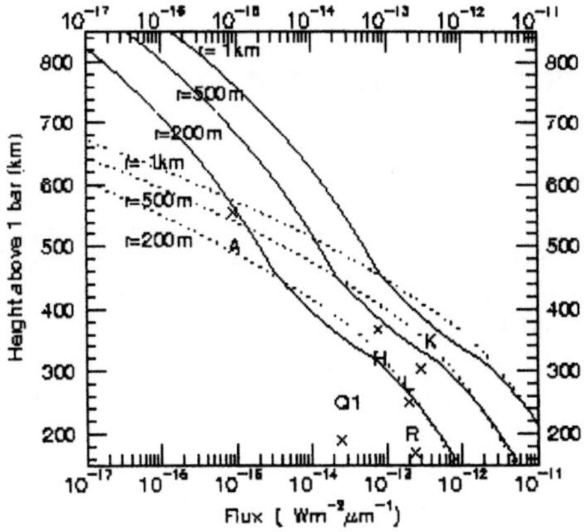

Figure 2. Entry fluxes at the near infrared of 2.3μm in the two atmospheric models. r is the radius of fragment. Solid lines show these fluxes for the Probe model, and dotted lines show those for the Orton model. Crosses are plotted for the observed PC1 fluxes and the minimum visible heights of the fragments.

radius of fragment A is larger than those of fragments H, K, and L. The probe model, on the other hand, indicates that fragment A is of medium class. Because fragment A appears to be of medium size, the probe model is relevant to the thermospheric model of Jupiter.

We conclude that the PC1 is the thermal radiation from the entry fragment due to the atmospheric drag. Also, the PC1 of fragment A, if the A fragment is a medium sized fragment with a radius of one or a few hundred meters, suggests that the Jovian thermosphere must be hot in the mid-latitude like in the equatorial latitude probed by the *Galileo* probe.

References

1. Atreya, S.K., et al. (1981), Jupiter: Structure and composition of the upper atmosphere, *ApJ*, **247**, L43–L47.
2. Chapman, C.R., et al. (1995), Preliminary results of Galileo direct imaging of SL-9 impacts, *GRL*, **22**, pp. 1561–1564.
3. Chodas, P.W., and Yeomans, D.K. (1996), The orbital motion and impact circumstances of Comet Shoemaker-Levy 9, in *The Collision of Comet Shoemaker-Levy 9 and Jupiter*, pp. 1–30, Cambridge University Press.
4. Graham, J.R., et al. (1995), The fragment R collision: W.M.Keck telescope observations of SL9, *Science*, **267**, pp. 1320–1323.
5. Hammel, H., et al. (1995), HST imaging of atmospheric phenomena created by the impact of comet Shoemaker-Levy 9, *Science*, **267**, pp. 1288–1296.
6. Herbst, T.M., et al. (1995), Near infrared imaging and spectroscopy of the SL9 impacts from Calar Alto, *GRL*, **22**, pp. 2413–2416.: Hamilton, D.P., et al. (1995), Calar Alto observations of the Shoemaker-Levy 9: Characteristics of the H and L impacts, *GRL*, **22**, pp. 2417-2420.
7. Martin, T.Z., et al. (1995), Observation of Shoemaker-Levy impacts by the Galileo Photopolarimeter Radiometer, *Science*, **268**, pp. 1875–18798.
8. McGregor, P.J., et al. (1996), CASPIR observations of the collision of comet Shoemaker-Levy 9 with Jupiter, *Icarus*, **121**, pp. 361–388.
9. Meadows, V., and Crisp, D. (1995), Impact plume composition from near-infrared spectroscopy, *Proceedings of European SL9/Jupiter Workshop*, pp. 239–244.
10. Seiff, A., et al. (1996), Structure of the atmosphere of Jupiter: Galileo probe measurements, *Science*, **272**, pp. 844–845.: Seiff, A., et al. (1997), Thermal structure of Jupiter's upper atmosphere derived from the Galileo probe, *Science*, **276**, pp. 102–104.
11. Sekanina, Z. (1996), Collision of comet Shoemaker-Levy 9 with Jupiter: impact study of two fragments from the timing of precursor events, *A&Ap*, **314**, pp. 315–327.
12. Takeuchi, S., et al. (1995), Near-IR imaging observations of the cometary impact into Jupiter, *GRL*, **22**, pp. 1581–1584.
13. Weaver, H.A., et al. (1995), The Hubble Space Telescope (HST) observing campaign on comet Shoemaker-Levy 9, *Science*, **267**, pp. 1282–1288.

OTHER PRESENTATIONS: AN OVERVIEW

This section overviews the papers which were presented at the oral and poster sessions but are not published individually in these Proceedings. The papers can be topically grouped in five blocks and they will be reviewed within these blocks, as follows: 1. Comets and Asteroids: Observations and Models; 2. Accretion and Formation Processes; 3. Dust Particles and Fluxes; 4. Near-Earth Asteroids: Observations and Dynamics; 5. Comet P/Shoemaker-Levy 9 and Related Phenomena. Abstracts of most of these papers were published in the IAU GA Abstract Book, Kyoto, August 17–30, 1997, pp. 24–31.

1. Comets and Asteroids: Observations and Models

P.R. Weissman and H.F. Levison discussed the origin and evolution of the unusual object 1996 PW which was discovered on August 9, 1996 by an automated search camera operating from Mt. Haleakala Observatory in Hawaii. Although asteroidal in appearance, it was soon determined that the object is in a near-parabolic orbit similar to that of a long-period comet. Speculation was focussed on whether the object is a dormant or extinct cometary nucleus, or an asteroid, and whether it is evolving into or out of the planetary system. The authors examined possible hypotheses and found that 1996 PW has most likely been a resident of the Oort cloud, probably ejected there early in the Solar System's history. However, they found it equally likely that 1996 PW is an extinct comet or an asteroid. Their calculations suggested that at least 1% of the population of the Oort cloud consists of asteroids from inside the orbit of Jupiter. Integration of the current orbit of 1996 PW forwards and backwards in time showed that it is likely to be ejected from the planetary system in ~200 returns, with a median lifetime of 7×10^5 years.

J.A. Fernández presented the results of a work, made in collaboration with G. Tancredi, J. Licandro and H. Rickman, on the population and size distribution of comets in the terrestrial planets zone. This study was aimed first of all to update the data set of absolute nuclear magnitudes of Jupiter family (JF) comets (periods $P < 20$ yrs). The results of recent CCD photometry of inactive or low-active comets have been compared with older photographic data, mainly taken by Elizabeth Roemer, in order to build a large data base of reasonably homogeneous quality. Constraints may then be placed on the nuclear sizes and fractions of active area by combining data on near-perihelion gas production rates with nuclear magnitudes, usually taken near-aphelion. These data may be used to study the size distribution and the distribution over perihelion distances (q) of the JF. Questions of special interest are if there exists a minimum size of an active JF comet and what fraction of such comets end up as extinct, asteroidal objects. It was shown that a large scatter exists in the absolute nuclear magnitudes of most JF comets and that those measured at the most remote locations are not always the faintest ones. Plausible reasons are both near-aphelion activity producing coma-contaminated magnitudes and, in some cases, errors of the coma subtraction in active comets near their perihelia. By means of a simple analytic model it was demonstrated that for heliocentric distances $r \lesssim 1.5$ AU the total magnitudes may be more than 5^m brighter than those of the bare nucleus for km-sized, not extremely inactive nuclei. Examples were given of how combined data on H_2O fluxes and nuclear magnitudes can place meaningful constraints on the nuclear radius (R_N), the active fraction and the visual albedo and also assess the reliability of the nuclear magnitudes in question. All the observed JF comets were shown to have $H_N < 19.5$ despite the recent discoveries of a large and ever-increasing number of much fainter near-Earth asteroids. It was thus concluded that comets with $R_N \lesssim 0.5$ km may not survive long enough to reach old dynamical states. Evidence was finally presented for a rapid growth of the JF comet population with increasing q. This was focussed around two points: *(a)* comets are on the average brighter for larger q (absolute total as well as nuclear magnitudes); *(b)* the discovery rate as a function of time shows a steep increase for $q \gtrsim 1.5$ AU.

The problem of early recovery of distant comets was discussed in a paper by O.R. Hainaut and

J. Andersen (ed.), Highlights of Astronomy, Volume 11A, 270–276.

K.J. Meech. As it is well known, observation of comets prior to the onset of the cometary activity gives direct measurements of some important physical parameters of the nuclei (size, rotation, shape). Moreover, the heliocentric distance, r, at which that onset takes place can directly be interpreted in terms of the physical nature of the volatile ice composing the nucleus. The authors have embarked upon a program to recover comets at large distances from the Sun. For this purpose, a very large CCD mosaic was built at the Institute for Astronomy ($8k^2$ pixels), providing a large enough field of view to encompass the uncertainty on the comet position. If the mosaic is installed on the UH 2.2m and CFHT 3.6m telescopes at Mauna Kea, it would enable to reach the required faint magnitudes ($R \sim 24 - 26$). Software has been developed to automatically search these huge images for moving objects. The observing methods and instrument, the search techniques and algorithms, and the first results of this program were presented.

Distant comet observations were also addressed by K.J. Meech and O.R. Hainaut. The authors reported on their long–term faint (to mag ≈ 30) distant comet observation program and the latest observations in this program, including recent data from Keck and HST. The goal was to obtain a uniform set of CCD images of short- and long-period comets as a function of heliocentric distance, r, comparison of activity levels and thus to get information about primordial and evolutionary differences between comets with different dynamical histories. Analysis of the extensive database of observations (≈ 50 comets over a range of r from 1 to ≈ 30 AU, on over 200 nights) showed a clear difference in activity level: Comets which have spent little time in the inner Solar System turned out intrinsically brighter than short-period comets, exhibiting dust comae and activity to large r where H_2O-ice sublimation is not possible. A comparison of these observations with laboratory experiments available on the condensation of low temperature ices, and new models of the solar nebula showed that the observations, lab experiments and theory could start to form a consistent picture of the early Solar System.

An application of a version of the method of apparent motion parameters, developed by Kiselev and Bykov at Pulkovo Observatory, to the analysis of CCD-observations of asteroids with the Spacewatch Telescope, was discussed by O.P. Bykov and P.P. Komarov. This method uses the basic Laplacian idea of preliminary orbit determination from positional observations. Its utilization showed that the method is best applied for near circular orbits within the range of $e < 0.2$, while elliptical or circular motion can be distinguished provided the angular acceleration of the body is available from CCD-observations. The authors suggested the use of this approach for calculations of circular orbits by observers, and to use the results for more detailed statistical investigations of the Main Belt Asteroid population. They also estimated that for the Kuiper Belt asteroids with small eccentricities the orbital parameters determined could remain valid within half a year after the last observation.

G. Tancredi, J.A. Fernández, J. Licandro and H. Rickman made an attempt to systematize the observational data on cometary activity. They compiled a new catalog of nuclear magnitudes to study physical properties of comets of the JF. As a source for the catalog they have used Kamél's (1991) compilation of data from many different sources, adding more recent data taken from the IAU Circulars, MPCs, ICQs and from their own observations. Among the few systematic observers of cometary nuclear magnitudes E. Roemer's long-term program (60's – 70's) is outstanding, having provided nuclear magnitudes for a large sample of JF comets from photographic plates. During the last decade the data set of nuclear magnitudes has been greatly enlarged and improved by using CCD detectors attached to medium- and large-size telescopes, and most of the recently published measurements of nuclear magnitudes came from J. Scotti who used a CCD camera attached to the Spacewatch Telescope. There is a tendency of the measured nuclear magnitudes to be fainter as JF comets are observed with CCD detectors. However, a few JF comets observed very far from the Sun (4–7 AU) showed brighter nuclear magnitudes than when they were observed closer in.

The coming of such an extraordinary comet as C/1995 O1 Hale-Bopp brought a very good opportunity of testing and improving cometary models. Indeed, due to the quantity of observations with which model output can be compared, the new data can be used to test the ability to do predictions and the dependency on initial parameters of the models. The results of one such model dealing with the problem of nucleus thermal evolution were discussed by M.T. Capria, A. Coradini and M.C. De Sanctis. In particular, it was attempted to reproduce phenomena such as the strong emission of CO and dust at large distance from the Sun.

In the paper presented by K. Muinonen and J.S.V. Lagerros the model of inversion of statistical shapes of small Solar System bodies was discussed. It has been hypothesized that the irregular shapes of such bodies can be modeled by using lognormal statistics (Gaussian random sphere). The Gaussian sphere is fully described by the mean radius and the covariance function of the logarithmic radius. The actual shape is thus fully described by the covariance function. Inverse methods were devised for deriving the covariance function from sample three-dimensional shapes. The inverse problem is challenging since the geometric centers (e.g., centers of mass for homogeneous bodies) do not necessarily coincide with the origins of generation. The amplitude-phase relationship was studied with the help of sample asteroid shapes generated by using the statistical parameters presently derived.

2. Accretion and Formation Processes

K. Yoshinaga, E. Kokubo and J. Makino dealt with the modeling of the long-term evolution of protoplanets. They admitted that although recent theoretical studies and N-body simulations of the system of planetesimals have shown that the planetesimals can grow to protoplanets of mass $\sim 1/10$ of the Earth-type planets in 10^6 years, the evolution of protoplanets after their formation is still largely unknown. Simulations of the evolution of protoplanets have been difficult because the timescale is very long ($\sim 10^7$ years) and the number of protoplanets is large (≥ 100). The authors reported the results of direct N-body simulation using a newly developed algorithm based on a time-symmetric integrator which can handle close encounters efficiently by means of individual timesteps. This approach allowed to achieve high accuracy for rather long timestep.

Knowledge on the evolution of molecular abundances in protoplanetary disks is indispensable for understanding the formation processes of comets and icy satellites from their molecular composition. The problem of evolution of molecular abundances in protoplanetary disks around young stars, in which planetary systems would be formed, was discussed in the paper by Y. Aikawa, T. Umebayashi, T. Nakano, and S.M. Miyama. It was assumed that in disk regions with surface density less than 10^2 g cm^{-2}, cosmic rays are barely attenuated even in the midplane of the disk and produce chemically active ions such as He^+ and H_3^+. Through reactions with these ions CO and N_2 are finally transformed into CO_2, NH_3, and HCN. According to the model, in the regions where the temperature is low enough for these products to freeze onto grain surfaces, large fractions of carbon and nitrogen are locked up in the ice mantles and is depleted from the gas phase on a time scale of several million years, whereas oxidized (CO_2) ice and reduced (NH_3 and hydrocarbon) ice coexist naturally in the disks. It was also shown that the molecular abundances both in the gas phase and in ice mantles varies with the distance from the central star.

A study of the depletion of the Hecuba gap in the asteroidal belt was presented by S. Ferraz-Mello, T.A. Michtchenko, D. Nesvorný, F. Roig, and A. Simula. Although gravitation alone was enough to form the gaps in the asteroid distribution at resonances, the Hecuba gap is the most difficult to explain. Indeed, simulations showed that some asteroids may remain in the resonant region for up to 1 billion years notwithstanding global stochasticity, and several well-known asteroids lie in the region. Some are just passing through but a few may remain there for the next billion years. The authors have investigated this problem using numerical integrations, Lyapunov exponent analysis, symplectic mappings and frequency map, taking as a framework for comparison the Hilda group remaining in the 3/2 resonance. The contrasting existence of a gap in one case and a group in the other was explained on the basis of different rates of chaotic transport. It was found that even in the most stable region of the Hecuba gap, diffusion is almost 2 orders of magnitude faster than in the region which accommodates the Hildas. The 5/2 near commensurability of Jupiter and Saturn was determinant in accelerating the diffusion rate in the Hecuba gap. Besides, if Jupiter and Saturn were secularly approaching the 5/2 resonance, the period of the great 5/2 inequality may have matched the asteroids' libration period, making the size of the stable regions still smaller and leading to an almost complete depletion of the gap. For instance, current stable asteroids would be late captures, when the Themis family formed 1 Gyr ago.

3. Dust Particles and Fluxes

K. Scherer and I. Mann investigated the influence of the interstellar medium on the dynamics of the dust particles in the outer heliosphere where the neutral gas density becomes larger than the solar wind density. Thus the drag force induced by the impact of the neutral gas on the dust particles is stronger than the electromagnetic or plasma Poynting-Robertson effect. It was also taken into account that, because of the axial symmetry of the inflowing neutrals, the induced drag force is asymmetric with respect to the orbital velocity. The first results of a study involving these effects were presented.

Beside the electromagnetic Poynting-Robertson effect, zodical dust particle motions are substantially influenced by plasma Poynting-Robertson drag forces induced by the solar wind passing over the dust particles. The dynamics of such particles in the region near the sonic surface of the solar wind was studied in a paper by M. Banaszkiewicz, H.J. Fahr, I. Mann, and K. Scherer. Calculations showed that the associated plasma drag coefficient strongly depends on whether or not the solar wind plasma is supersonic. Since this coefficient increases significantly with decreasing solar wind sonic Mach number, the study was focused on zodiacal dust dynamics in the region close to the sonic surface of the solar wind where the change from low to high Mach number flows occurs – most likely in the region near the ecliptic rather than at higher latitudes. On the basis of a parametrized 3-dimensional solar wind outflow model the zodical dust dynamics for particles was evaluated at different inclinations to demonstrate inclination-dependent radial migration periods. In the subsonic solar wind region it was taken into account that the plasma drag force has components normal to the orbital plane of the particles connected with the solar wind ion temperature anisotropies and inducing inclination drifts of the dust particles. Based on the results of calculations the authors concluded that observational studies of the zodiacal dust cloud close to the corona provide a diagnostic of the solar wind in its acceleration region.

A.V. Krivov and M. Banaszkiewicz investigated the sources of dust in the outer Saturnian system and delivery of dust material to Titan. They suggested neighboring Hyperion to act as the most effective dust supplier for Titan. Hypervelocity impacts of dust particles coming from the outermost moon Phoebe and bombardment by IDPs could produce impact ejecta from Hyperion's surface, an appreciable share of which would escape into planetocentric orbits initially close to that of Hyperion. The authors further assumed that, though such particles would initially be locked in a strong 4:3 mean motion resonance with Titan (so that encounters with this satellite would be prohibited) the solar radiation pressure and especially the plasma drag force would destroy the resonant locking. Once the resonance is broken, the orbits become unstable and experience multiple close approaches to Titan. Using numerical integrations, a statistical study of the grain trajectories was performed to find out the eventual fate of the debris and to construct a spatial distribution of dust in the Hyperion–Titan system. It turned out that most of the grains larger than $\sim 5\mu$m finally collide with Titan, whereas smaller particles either escape to interplanetary space or hit Saturn. A steady-state dust cloud in the Hyperion–Titan system is tilted off the equatorial plane of Saturn and has a structure that depends on the particle radii. Estimates of the dust influx to Titan showed that the upper limit of the income rate of the Hyperion particles may considerably exceed the direct influx of IDPs. Thus the influx of icy (H_2O) particles from Hyperion was suggested as a possible explanation of the observed abundance of CO_2 molecules in Titan's atmosphere.

New radiant areas of the Perseid meteor shower were studied by T. Yoshida, S. Suzuki, K. Suzuki, and T. Akebo. More than 100 Perseid meteors were recorded by using TV techniques at four stations on August 11/12 and 12/13, 1993. The lenses of MINOLTA f=135mm F2 (at two sites), CANNON f=85mm F1.2, and NIKON f=85mm F1.4 were used. A PC with a frame memory board was utilized in order to sample the video with a spatial resolution of 512×512 pixels and with 8 bit tone levels (i.e., 256 gray levels). The positional traceability was within $\pm$ 1 pixel at worst. Orbits of meteors were determined with an accuracy within $60''$ with respect to the reference stars. The results for new radiant areas of the Perseid shower observed in 1993 were summarized.

In the paper presented by K. Scherer, J.D. Anderson, J.L. Lau, D.C. Rosenbaum, and V.L. Teplitz the results of observation of meteoroids during the passage of Pioneer 10 through the Kuiper Belt (between 35 AU and 65 AU) were analysed based on discontinuities in the Doppler signal that were recorded. These jumps in the signal correspond to velocity changes of the spacecraft caused by impacts of Kuiper Belt meteoroids. The mass and size of four impacts were analyzed. It was found that the spatial distribution of the meteoroids seems to be located in discrete regions around mean motion resonances. The authors discussed the implications of the results for the Egdeworth–Kuiper Belt and planetary evolution.

4. Near Earth Asteroids: Observations and Dynamics

In his overview of asteroid dynamics A. Morbidelli first addressed several analytical and numerical studies of the last years (e.g., Farinella et al., 1994) which pointed out that the ν_6 secular resonance and the main mean motion resonances with Jupiter (3/1, 4/1, 5/2, 7/3) are very efficient in pumping the eccentricity of resonant asteroids up to Earth crossing – often Sun grazing – values. Gladman et al. (1997) have recently integrated numerically the evolution of a few thousand particles initially placed into such resonances and determined that their median lifetime is $\sim$ 2 Myr. It was shown that only particles which, under the effect of encounters with the terrestrial planets, happen to decrease their semimajor axis below 2 AU may live significantly longer (few times 10^7 years). The population of the known Near Earth Asteroids (NEAs), which is strongly biased towards objects on such long living orbits, has a halflife $\sim$ 10 Myr. Despite such short timescales, the population of NEAs up to $\sim$ 1 km in size should be in a sort of steady state, since $\sim$ 400 1-km asteroids should be injected per Myr into the ν_6 or 3/1 resonances (Menichella et al., 1996). On top of such a steady state the history of the Solar System should have been characterized by sporadic episodes of NEA overproduction associated to the major family-creating break up. According to Zappalà et al. (1997), some known families should have injected into resonance several 10^4 1-km asteroids, thus increasing temporarily the number of NEAs as well as the number of collisions with the Earth. However such "asteroid showers" should have lasted only $\sim$ 10 Myr, the rapid dynamical decay reducing the NEA population back to the steady state number on this timescale. Concerning multi kilometer NEAs, conversely, the injection rate into the main resonances seems to be too low to explain the existence of a few such bodies in the currently observed population. It may be conjectured that the origin of such asteroids is related to the existence of some slow channel, which may take asteroids out of the belt on timescales of $10^8 - 10^9$ yrs. Unfortunately, our knowledge of the dynamics of the asteroid belt over such long timescales is far from being exhaustive. Just a few results are beginning to form a very complicated puzzle. Milani and Farinella (1995) showed that asteroids close to the 5/2 resonance borders can be captured into resonance after several 10^7 yrs of slow chaotic diffusion, and the same seems to be true also for the 3/1 and ν_6 resonance up to a $\sim 10^8$ yrs timescale (Morbidelli and Gladman, 1997). Levison et al. (1997) have shown that the Trojan asteroids are not stable over the age of the Solar System, but some of them are presently leaking out of the 1/1 resonance, presumably encountering Jupiter in the close future. It is possible that the same is true for the asteroid belt, which could undergo a sort of "dynamical evaporation", thus supplying new – possibly large – objects to the NEA population. The existence of the Kirkwood gap associated with the 2/1 resonance can be fully understood only taking into account slow chaotic diffusion acting on $\sim 10^9$ yrs (Nesvorný and Ferraz-Mello, 1997; Morbidelli, 1996). This shows the importance of studying the dynamics over a timescale comparable to the age of the Solar System in order to understand the present structure of the asteroid belt and of the NEA population.

From the modeling of migration of small bodies from the Kuiper Belt it was revealed that under the gravitational influence of planets some bodies from the transneptunian region can reach the orbit of the Earth in several tens of million years. This problem was addressed in the paper of S.I. Ipatov. He found that the time interval during which the perihelion distance decreases from 10 to 1 AU can be less than 0.5 Myr. The limits of variations in orbital elements even for non-resonant Kuiper-belt objects highly depend on the initial orbital orientations of their orbits. Therefore, small variations in orbital elements due to collisions and mutual gravitational influence of the Kuiper Belt objects can cause large variations in orbits under the gravitational influence of the giant planets. Estimates of the mutual gravitational influence of the transneptunian objects show that variations of semimajor axes for some of them could exceed 1 AU during the age of the Solar System. The average intrinsic collision probability was obtained as about 10^{-21} km^{-2}yr^{-1} for $a = 40$ AU.

An intriguing problem of the possible influence of non-gravitational forces in decoupling cometary orbits from Jupiter was addressed by D. Asher. He emphasised that, although the orbits of most JF comets either overlap or at least closely approach the orbit of Jupiter, comets do exist within, and clearly decoupled from, the Jovian orbit. This generally permits a greatly increased dynamical timescale in the innner Solar System. The most notable example is 2P/Encke. As the non-gravitational force may not act in a consistent direction for long times, the question arises as to whether random small changes in orbital elements (particularly the semi-major axis) induced non-gravitationally can eventually produce the large orbital element changes necessary to decouple orbits from Jupiter. Numerical integration was used to quantify the likelihood of decoupled orbits resulting from Jupiter approaching orbits (and the timescale of this process), as a function of the size of small orbital element changes applied.

Near Earth Objects (NEOs) have different collision probabilities to the Earth depending on their sizes; to escape global or continental scale catastrophe, the activity should be at first targeted to detection of all NEOs. S. Isobe on behalf of the Japan NEO Team discussed a strategy to detect the NEOs using ground-based or lunar-based facilities. It was emphasised that all NEOs with diameters larger than 1 km can be detected during their passages through the asteroid belt by ground-based telescopes which are and will be in operation. All NEOs with diameters between 100 m and 1 km can be detected by space-borne or lunar-based telescopes, but it takes over 50 years after completion of these telescopes. Although the collision probability of a NEO with a diameter of 100 m with the Earth is once every several thousand years, its collision may occur within 50 years, during which it must be attempted to detect all these NEOs. In this regard the Team's close approach estimation of NEOs was referred to, which argues that 40% of NEOs approach the Earth from an area within 30° from the Sun. This means that collision will occur without any warning time. Therefore, space-borne or lunar-based continuous observations in the area are needed. In Japan, there is a discussion to build a lunar station in which the NEO detection telescope(s) are also considered. A strategy of the detection to minimize the possibility of being hit by a NEO was introduced and discussed.

The program of NEO observation in Beijing Astronomical Observatory was discussed by J. Zhu, R. Liu, J. Chen, X. Zhou, Z. Zheng, R. Chen, L. Deng, H. Yan, Z. Jiang, and Y. Li. This BAO Schmidt CCD Asteroid Program (SCAP) began in May 1995, and until Jan. 1997 over 1,000 asteroids were discovered. The SCAP recently found minor planet 1997 BR, an Apollo-type Earth crossing asteroid, whose orbit has an intersection with Earth's orbit within 0.0005 AU. It was reported that SCAP is going to pay more attention to NEO observations in 1997, and its relatively large CCD field of view ($58' \times 58'$) at the 60/90cm f/3.0 Schmidt telescope is suitable for follow-up observations to recover some lost NEOs.

5. Comet P/Shoemaker-Levy 9 and Related Phenomena

A.S. Hojaev, V.N. Dudinov, A.P. Zheleznyak, V.V. Shevchenko, and S.I. Ivanov presented the results of complex observations of the comet SL9 impact on Jupiter at High Altitude Maydanak Observatory, Uzbekistan. Multicolor photographic, polarimetric and speckle imaging observations of the Jovian clouds were carried out using the 1.5m reflector AZT-22 and the 1.0m Zeiss reflector to study the aftermaths of the cometary fragment impacts. Both integral and individual spectral band observations were performed, using the combination of ray and polarization filters. More than 150 high resolution photographic images and about 70 speckle series with more than 10 000 images of Jupiter were obtained with the 1.5m telescope. From the observations with the 1.0m telescope the parameters of a flash event above the limb produced by the S-impact fireball were estimated. The polarization map of Jupiter shows a North-South asymmetry and differences in the polarization indices for the impact sites. Analysis of halo dynamics and morphology of the dark spots formed by the impacts was performed, and their mean velocities were derived. Zonal wind velocities resulting in halo deformation during 10–12 rotational periods were determined to be $30 - 35$ m/s for jovicentric latitudes 42–44 deg. A phenomenological model for the impact events was proposed.

H. Hasegawa, S. Takeuchi, A. Mori, N. Yamamoto, and J. Watanabe analysed time variations of aerosols in the Jovian stratosphere produced by the SL9 impacts with the planet. The aerosol layer was observed as bright clouds in near-infrared methane absorption bands between 2.0 and 2.4 μm. Observations were made in May 1995 and July 1996 with near-infrared camera and spectrometer OASIS at the Okayama Astrophysical Observatory. A radiative transfer model was developed to reproduce the observed spectral features. The model includes multiple scattering by the stratospheric particles and tropospheric ammonia cloud layer, and molecular absorptions by hydrogen and methane. It allowed to determine the vertical distribution and optical thickness of the layer and its time variation, including the column density and aerosol distribution two years after the impact. The column density of the layer decreased by one order of magnitude between 1995 and

1996.

In the paper by D.C. Boice and J. Benkhoff the near-nucleus environment of the icy conglomerate fragments of comet SL9 was discussed. This environment was characterized during the two-year period between breakup and impact with Jupiter using a dusty coma model that included gas release from three separate sources (surface, subsurface, and distributed dust). It was assumed that the breakup exposed fresh surface layers undepleted in volatile ices, such as, H_2O, CO, CO_2, CH_3OH, and the time-dependent gas release from porous layers as they become progressively depleted in volatiles was investigated. The entrainment of dust by the calculated CO gas release was found to be consistent with observations of dust comae and tails of the fragments, such as, 1) the continuous release of dust into the coma prior to impact, 2) large grain sizes ranging from 10 to 500 μm in the dust comae, 3) estimates of the total dust release rates leading to dust-to-gas mass fractions on the order of two or less when compared with model CO release rates, and 4) upper limits imposed by observations of CO, OH and CO^+ column densities. The results of several comet simulations were presented to compare and contrast the effects of model parameters with observations.

References

Farinella., P., Froeschlé, C., Froeschlé, C., Gonczi, R., Hahn, G., Morbidelli, A. and Valsecchi, G.B. 1994, *Nature* **371**, 314-317

Gladman, B.J., Miglorini, F., Morbidelli, A., Zappalà, V., Michel, P., Cellino, A., Froeschlé, C., Levison, H.F., Bailey, M. and Duncan, M. 1997, *Science* **277**, 197-201

Kamél, L. 1991, *The Comet Light Curve Catalogue/Atlas*, Uppsala Univ.

Levison, H.F., Shoemaker, E.M. and Shoemaker, C.S. 1997, *Nature* **385**, 42-44

Menichella, M., Paolicchi, P. and Farinella, P. 1996, *Earth Moon & Planets* **72**, 133-149

Milani, A. and Farinella, P. 1995, *Icarus* **115**, 209-217

Morbidelli, A. 1996, *Astron. J.* **111**, 2453-2461

Morbidelli, A. and Gladman, B.J. 1997, *Meteoritics & Planet. Sci.*, submitted

Nesvorný, D. and Ferraz-Mello, S. 1997, *Icarus*, in press

Zappalà, V., Cellino, A., Gladman, B.J., Manley, S. and Miglorini, F. 1997, *Science*, submitted

JD7

The New International Celestial Reference Frame

Chairperson and Editor: **L. V. Morrison**

THE NEW INTERNATIONAL CELESTIAL REFERENCE FRAME

Joint Discussion 7, Friday 22 August 1997

ORAL PRESENTATIONS

The following papers were read at the Joint Discussion. The text of the papers is printed on the following pages in the order of their presentation.

Paper		
00	Progress of the reference frame	L.V. Morrison (UK)
01	Formation of ICRF	C. Ma (USA) & F. Arias (Ar)
03	Optical counterpart of ICRF : Hipparcos	L. Lindegren (S) & M.A.C. Perryman (Nl)
05	Tycho: extension of optical reference frame	E. Høg (Dk)
06	Referring Schmidt surveys to ICRF	M. Lattanzi (USA)
07	High density requirements	R. Cannon (Au)
08	Extension of optical reference frame: ground based	N. Zacharias (USA)
09	: space based	S. Röser (D)
10	Linking Hipparcos to ICRF	J. Kovalevsky (F)
11	Linking dynamical reference frame to ICRF	E.M. Standish (USA)
12	Linking FK5 to ICRF	F. Mignard (F)
13	Maintenance of ICRF - Radio	A.L. Fey (USA)
14		C.S. Jacobs/O.J. Sovers (USA)
15	- Optical	N. Zacharias (USA)
16	- Complex optical objects	C. Worley (USA)
17	- Statistical treatment	H. Schwan (D)

POSTER PAPERS

Abstracts of the poster papers are printed after the oral papers under the following sectional headings:

The International Celestial Reference System (ICRS)
Accuracy of the International Celestial Reference Frame (ICRF)
Linking Hipparcos to the ICRS
Linking the dynamical frame to the ICRS
Extending the Hipparcos frame to fainter magnitudes
Extending the Hipparcos frame to the infrared
Astrolabe catalogues.

J. Andersen (ed.), Highlights of Astronomy, Volume 11A, 279.

THE PROGRESS OF THE REFERENCE FRAME

L.V. MORRISON
Royal Greenwich Observatory, Cambridge, UK

F.W. Argelander proposed to a meeting of the Astronomischen Gessellschaft in 1867 that an accurate catalogue of all stars down to 9th magnitude should be compiled. The sky was divided into zones of declination and observations were undertaken with transit circles at a number of observatories in an international collaboration. The importance of fitting these zonal observations to one global reference frame was realized, and for this purpose A. Auwers produced a *Fundamental-Catalog* (FC) for the Northern and Southern hemispheres in 1879 and 1883, respectively. The accuracy of the positions was about half of an arcsecond. This was the first international celestial reference frame.

This reference frame was improved over the next hundred years by continued international cooperation which was later coordinated through the IAU. These efforts culminated in the FK5 which was issued in 1988. The progress in accuracy from the FC to the FK5 is shown in Figure 1. Meanwhile, radio astronomy through the technique of Very Long Baseline Interferometry (VLBI) began to produce positions of extragalactic sources with an accuracy of a milliarcsecond (mas), which is almost two orders of magnitude better than that of the optical positions of the stars in the FK5.

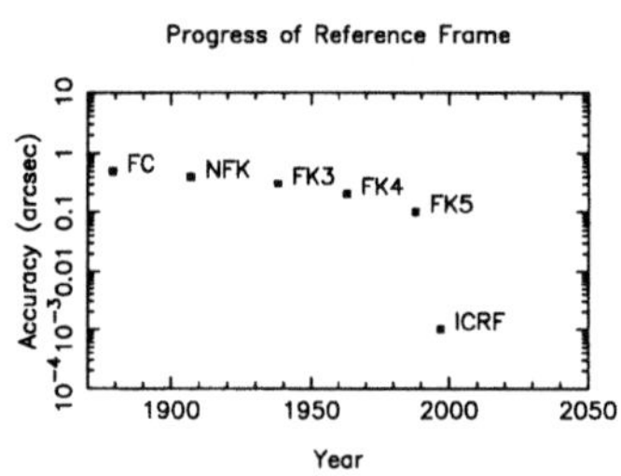

Figure 1. Improvement in accuracy of the reference frame from the first Fundamental-Catalog (FC) of Auwers in 1879 to the ICRF in 1997.

The inherent problem of a reference frame based on optical positions of nearby stars is that a complete model of the dynamics of these stars, as well as the motion of the Earth's equator (precession), are required in order to establish a non-rotating frame. Systematic errors in the observations compound the problem, and the elimination of these was one of the objectives of the Hipparcos space mission adopted by ESA in 1980.

By a remarkable coincidence of events, we are now in possession of radio and optical catalogues which both have accuracies of 1 mas or better. The radio catalogue of 608 positions which has been constructed in the past year and published as IERS Technical Note 23 will at this IAU General Assembly be proposed as the primary international celestial reference frame (ICRF). The Hipparcos Catalogue which was issued in June 1997 and is linked to the ICRF with an accuracy of 0.6 mas at epoch 1991.25 will be proposed as the realization of the ICRF at optical wavelengths.

At this Joint Discussion, we will hear how these catalogues were constructed and linked, how their accuracy is to be maintained, and how the reference frame is to be transferred to other catagories of objects.

J. Andersen (ed.), Highlights of Astronomy, Volume 11A, 280.

FORMATION OF THE INTERNATIONAL CELESTIAL REFERENCE FRAME

C. MA
NASA, Goddard Space Flight Center, Greenbelt, MD 20771, USA

E.F. ARIAS
Facultad de Ciencias Astronomicas y Geofisicas de La Plata; Observatorio Naval Buenos Aires; CONICET, Argentina

T.M. EUBANKS AND A.L. FEY
US Naval Observatory, Washington, USA

A.-M. GONTIER
Central Bureau of IERS, Observatoire de Paris

AND

C.S. JACOBS AND O.J. SOVERS
JPL, Pasadena, USA

1. Introduction

The goal of the work described here is to create the definitive catalogue for the new International Celestial Reference Frame (ICRF) using the best data and methods available at the time the work was done. This work is the joint cooperative effort of a subgroup of the IAU Working Group on Reference Frames which was formed expressly for this purpose in February 1995. The authors of this report constituted the subgroup. A fuller account of this report can be found in the introduction to the ICRF catalog (IERS 1997).

The ICRF of 608 sources presented here is based on essentially all the VLBI observations accumulated over about 15 years in several worldwide programs. Dual frequency Mark III data have both geodetic and astrometric applications. Most of the data (95% of nearly 2 million observations) were acquired primarily for geodetic purposes. The major geodetic programs include: NASA's Crustal Dynamics Project/Space Geodesy Program and USNO's NAVEX sessions for the terrestrial reference frame, as well as IRIS, NAVNET and NEOS sessions for monitoring Earth rotation. The geodetic programs have used the brightest radio sources, gradually concentrating on the most compact as sensitivity improved. These geodetic sources were also the foundation of astrometric work because of the large number of observations for the ~150 most commonly used. The astrometric programs which densify the sky include the Radio-Optical Reference Frame sessions done by US Naval Research Laboratory (NRL) and USNO and the space navigation efforts of Jet Propulsion Laboratory (JPL).

The subgroup has taken an empirical approach in the selection of data, analysis, estimation of errors, and categorization of the final results. The characterization of a radio source, i.e., its position, how it was treated in the analysis, and whether it was suitable for use as a defining object, was derived entirely from the VLBI data and analysis and not from any other information. This approach leads to a rigorous selection of defining objects and a reliable realization of the ICRF as a set of relative positions oriented to the axes of the International Celestial Reference System, ICRS (Arias et al., 1995).

It was considered essential that the realization of the ICRF be derived from a single analysis, even if imperfect, rather than from a combination catalogue made of several VLBI solutions.

J. Andersen (ed.), Highlights of Astronomy, Volume 11A, 281–286.

2. Preparation for analysis

As suggested above, the sources can be characterized along several lines. The most important are variations in position seen in the data and number of observations. The underlying conceptual basis of this type of realization of the celestial reference frame is that positions are invariant with time. Therefore, the first task was to ensure that this condition was not significantly violated.

A series of solutions was made. In each solution the positions of all sources except for a small test set were estimated as time-invariant parameters. The position of each test source was re-estimated for each day the source was observed. Each source was treated as a test source in some solution. The complete set of source positions as functions of time was analysed to determine which sources had statistically significant variations in their positions. In addition, in another solution apparent linear motion was estimated simultaneously for all sources with sufficient data, generally two or more observing sessions. Two modes of variation were recognized. For sources with sufficient data to derive statistically significant apparent linear motions, a source was considered problematical if the apparent motion exceeded 50 μas/yr and was larger than the formal error.

Other sources were rejected if the weighted root-mean-square of the source position variations exceeded 0.5 milliarcsecond (mas) in arc length and were also more than 4σ based on the formal position errors. Since these two classes of sources showed undesirably large position variation, they were treated differently from other sources in the final analysis. To accommodate their position variations without deforming the rigid geometry of the remaining sources, the positions of these sources were adjusted separately for each session in which they were observed.

The variations in position seen in the dataset were not comprehensively analyzed to determine the causes. The position variations for some sources show clear correlations with changes in mapped structure and to some extent the position changes can be derived from the structure, but there was no strong evidence of any regularly repeated behavior.

3. Configuration of the ICRF analysis

The configuration of the ICRF analysis was developed as a balance between competing goals: the most data and the least systematic error, the best models and available options, the largest number of useful estimated parameters and computer speed, and the like.

The most important choices were related to data and modeling/estimation. To extract the most information from the data set both delay and delay rates were used, and only observations below 6 degrees were excluded. The troposphere was modeled using the MTT mapping function (Herring, 1992). Time-variable troposphere gradients were estimated.

The primary geodetic parameters, the station positions, were estimated separately for each session. In this way any non-linear motion of the stations does not affect the integrity of the invariant source positions. The unadjusted a priori models for geophysical effects and precession/nutation generally followed the *IERS Standards 1992* (McCarthy, 1992). The VLBI theoretical model was the so-called consensus model given in the *IERS Conventions 1996* (McCarthy, 1996).

The parameters were estimated using arc-parameter elimination (Ma et al., 1990), which is an incremental least-squares method that can accommodate large numbers of parameters if they are associated only with particular data intervals. In the ICRF analysis several classes of parameters were adjusted. For each observing session, the adjusted arc parameters included:

1. positions of 'arc' sources with identified excessive apparent motion or random variation,
2. two celestial pole offsets to account for errors in the standard precession/nutation models,
3. positions of the stations,
4. the rate of UT1 relative to a good a priori time series,
5. 1-hour troposphere parameters described above,
6. troposphere gradients in the E-W and N-S directions, linear in time,
7. quadratic clock polynomials for the gross clock behavior,
8. 1-hour clock parameters similar to the 1-hr troposphere parameters, and
9. necessary nuisance parameters such as clock jumps and baseline clock offsets.

Certain parameters were adjusted as invariant quantities from the entire dataset. These global parameters included:

1. invariant source positions,

2. geometric axis offsets for all fixed antennas (small effect), and
3. 252 parameters for Earth rotation variations in the diurnal and semidiurnal bands caused by ocean tides (small effects).

After a series of test solutions to refine various aspects of the analysis, a final solution was run which included 1.6 million pairs of delays and rates from August 1979 through July 1995.

4. Determination of realistic error

Given the very large number of observations for some sources, the error contribution from their observation noise was very small and not a meaningful measure of uncertainty. It was necessary to consider several other effects to assign realistic errors. One was the statistical validity of the formal errors. Another was the cumulative influence of all modeling errors and editing decisions. Yet another was the magnitude of specific, identifiable systematic errors that could have distorted the results.

The cumulative effect of modeling errors and data editing can be examined by comparison of results from independent analyses. Intercomparison of different radio source catalogues addresses the problem of mismodeling and consequent systematic errors. Aspects that can be probed by catalogue comparisons include:

1. extension of the data set used in analysis;
2. analysis differences at different centers;
3. agreement between subsets of data;
4. intentional perturbations of modeling.

A number of such comparisons involved the ICRF, associated catalogues from the identical data base that were created for test purposes, and existing radio source catalogues. Existing catalogues included the 1994 and 1995 IERS realizations of the celestial reference frame (IERS94 and IERS95) and the radio-optical reference frame of Johnston et al. (1995).

From a consideration of all the known error sources, the subgroup concluded that a realistic error estimate for the invariant source positions could be made by inflating the formal errors by a factor of 1.5 followed by a root-sum-square increase of 0.25 mas. For the most frequently observed sources the 0.25 mas was the dominant error. The errors of the arc sources were also increased by 0.25 mas.

To address the question of whether adopting a unique 'error calibration law' for all sources eliminated or at least minimized systematic effects, a combined frame was made by adopting a 6-parameter comparison model (three angles, drifts in right ascension and declination as functions of declination, and a bias in declination). Individual VLBI frames submitted by GSFC, JPL, NOAA and USNO to the IERS in 1995 (Charlot, 1995) were then included in this combination. For some sources position uncertainties obtained from the catalog combination were larger than predicted, so for these sources the uncertainty of each coordinate was set to the larger of the inflated or comparison value.

5. Categorization of sources

To be most useful in defining the ICRF a source should ideally show no variation in position in the dataset, have sufficient data to support the absence of variation, and not have shown unexplained differences in position between realizations of equivalent validity. Several quality levels can be established for each of the 608 sources in the ICRF catalogue. These were based on one of three sets of criteria:

1. quality of data and observation history
2. consistency of coordinates derived from subsets of data
3. repercussions of source structure.

To qualify for the list of sources that was used to orient the ICRF with respect to the IERS celestial reference system, a source had to meet these three conditions. Some criteria that can be used to identify 'problem sources' are related to their internal structure and their variation with time. If the structure varies with time, the effective centroid can move. Even with static structure, the baselines and their time-varying orientation can induce apparent shifts in position. A simple way to assess the time–variability is to estimate rates of change of right ascension and declinations with time. A more direct assessment of the effects of extended source structure was provided by using externally generated source maps to calculate structure corrections to VLBI observables. A number of such maps were available from the work of Fey with the VLBA.

Sources failed to be defining sources for any of the following reasons:

1. Arclength formal error >1 mas
2. <20 observations
3. <2–year span of data
4. 3σ or >0.5 mas discrepancy between catalogues
5. excessive structure
6. large, significant apparent motions
7. arc source, i.e., position adjusted for each session.

The sources then fell into three categories: 212 defining sources that failed none of the criteria, 294 candidate sources that failed some or all of the criteria, and 102 sources with identified excessive position variation, either linear or random. Figure 1 shows histograms of the inflated errors of the 212 defining sources. Candidate sources potentially could be designated defining sources in future

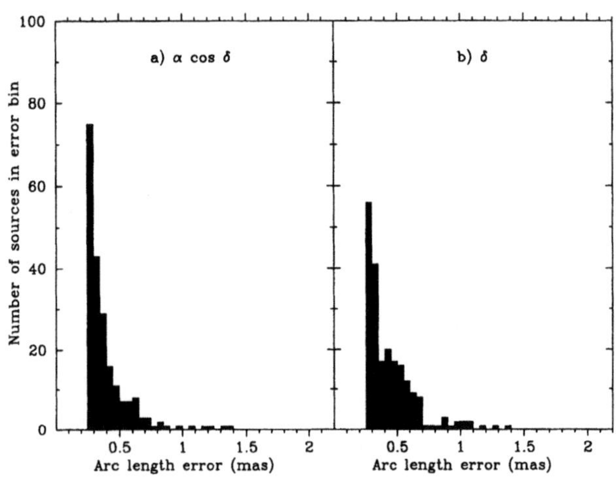

Figure 1. Histograms of position errors for defining sources.

realizations of the ICRF as more data become available or analysis improves. The third category of other sources does include sources that may be useful for purposes such as radio-optical frame ties. While only the defining sources have a formal role in the ICRF, the positions of all sources are consistent with the ICRF.

6. Orientation of the ICRF

The final stage in the ICRF realization was the rigid rotation of the relative positions to the ICRS maintained by the IERS. The ICRF was aligned to the ICRS by rotating it onto the last realization of the IERS celestial reference frame, RSC(IERS) 95 C 02 (IERS, 1996). Radio source coordinates in RSC(IERS) 95 C 02 were evaluated by combining individual extragalactic reference frames submitted to the IERS in 1995. The coordinates adopted for a set of 236 defining sources aligned the axes of RSC(IERS) 95 C 02 to the IERS celestial reference system.

Due to the model adopted in the compilation of RSC(IERS) 95 C 02, the frame was affected by the deformations coming from the individual catalogues. The improvements in the models and procedures applied in the ICRF solution resulted in a frame free from deformations but slightly misaligned with respect to RSC(IERS) 95 C 02. In the procedure applied to rotate the ICRF solution to IERS frame, care was taken not to transfer to the former the deformations of the latter.

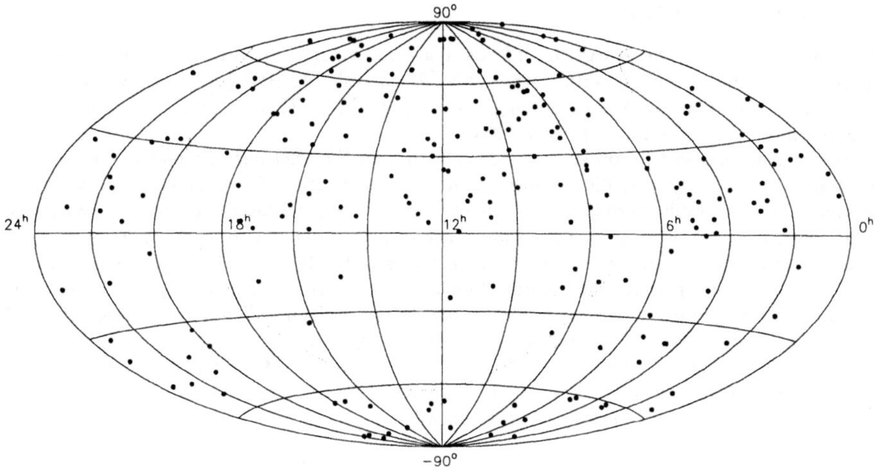

Figure 2. Areal distribution of 212 defining sources.

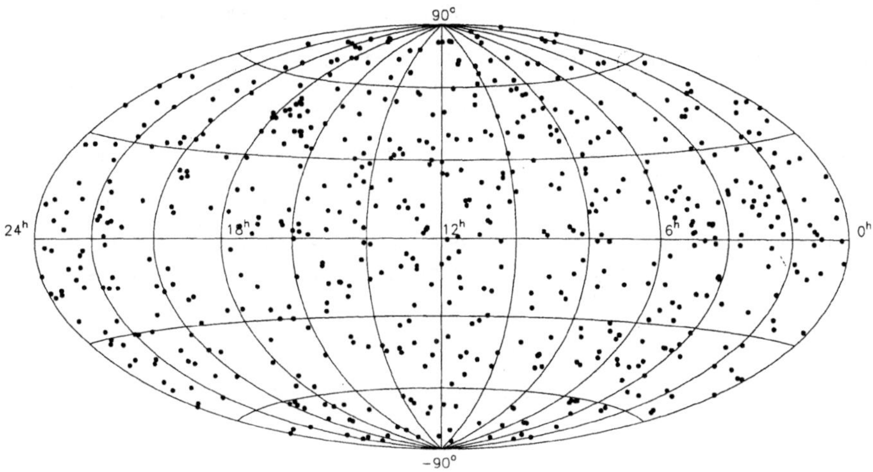

Figure 3. Areal distribution of all 608 ICRF sources.

7. The ICRF catalogue

Figure 2 shows the distributions of objects on the sky for the 212 defining sources, and Figure 3 for all the 608 sources. It can be seen that there is moderately even distribution of all sources, but the Southern hemisphere is deficient in defining sources. This is caused by the small number of VLBI stations in the South and limited observing programs. While all sources are given ICRF designations, it should be emphasized that the quality and intended use of the three categories are quite different. The best astrometric quality resides in the defining sources and those candidate sources with the smallest errors. The positions of the 'other' sources should be used carefully and only where less accuracy can be tolerated.

8. Evolution of the ICRF

The current realization condenses the information from a particular VLBI dataset spanning a defined interval and reflects a certain state of VLBI analysis. As time progresses we expect the realization of the ICRF to evolve, although changes in the ICRF catalogue will be infrequent compared to past practice in VLBI astrometry.

Current sources need to be observed periodically to track their behavior. Second, as observations accumulate, it should be possible to move candidate sources up or down the scale of usefulness. However, it is conceivable, perhaps even probable, that an identical categorization of sources from an analysis using twice as long an interval would show sources changing categories in unpredictable ways. The problem of position variation may be solved in the future if the application of source structure information permits the identification and use of truly kinematically stable points in the sky. This remains to be demonstrated. Unlike stellar catalogues, however, the original VLBI observations should always be accessible for improved analysis de novo.

Despite the burden of maintenance, the ICRF realized by VLBI astrometry is a great step forward. Compared to stellar realizations it is intrinsically simpler, much more accurate, more stable, and less susceptible to systematic deformations.

References

Arias, E.F., Charlot, P., Feissel, M., Lestrade, J.-F. (1995) *The Extragalactic Reference System of the International Earth Rotation Service, ICRS*, A&A **303**, 604-608

Charlot, P. (1995) IERS Technical Note No.19, Observatoire de Paris

Herring, T.A. (1992) *Modeling atmospheric delays in the analysis of space geodetic data*, Symposium on Refraction of Transatmospheric Signals in Geodesy [Eds. J C. De Munk & T.A. Spoelstra], Netherlands Geod. Comm., Delft, 157-164

IERS (1996) 1995 IERS Annual Report, Observatoire de Paris, II-19.

IERS (1997) *Definition and Realization of the International Celestial Reference System by VLBI Astrometry of Extragalactic Objects*, [Eds. C. Ma & M. Feissel] IERS Technical Note No.23, Observatoire de Paris

Johnston, K.J., Fey, A.L., Zacharias, N., Russell, J.L., Ma, C., de Vegt, C., Reynolds, J.E., Jauncey, D.L., Archinal, B.A., Carter, M.S., Corbin, T.E., Eubanks, T.M., Florkowski, D.R., Hall, D.M., McCarthy, D.D., McCulloch, P.M., King, E.A., Nicolson, G., Shaffer, D.B. (1995) *A radio reference frame*, AJ **110**, 880-915

Ma, C., Sauber, J.M., Bell, L.J., Clark, T.A., Gordon, D., Himwich, W.E., Ryan, J.W. (1990) *Measurement of horizontal motions in Alaska using very long baseline interferometry*, J. Geophys. Res. **95**, 21991-22011

McCarthy, D.D. [Ed.] (1992) *IERS Standards 1992*, IERS Technical Note No.13, Observatoire de Paris

McCarthy, D.D. [Ed.] (1996) *IERS Conventions 1996*, IERS Technical Note No.21, Observatoire de Paris

OPTICAL COUNTERPART OF ICRF: HIPPARCOS

L. LINDEGREN
Lund Observatory
Box 43, SE-22100 Lund, Sweden

AND

M.A.C. PERRYMAN
Astrophysics Division, ESTEC
2200AG Noordwijk, The Netherlands

1. The Hipparcos Catalogue

The Hipparcos reference frame is defined by the positions and proper motions of $\sim 10^5$ stars given in the Hipparcos Catalogue (ESA, 1997). It is the optical counterpart of the International Celestial Reference Frame defined by the radio positions of extragalactic sources. The linking of the Hipparcos frame to the extragalactic frame is described by Kovalevsky et al. (1997) and Kovalevsky (1997). In the present review we consider only the intrinsic properties of the Hipparcos reference frame.

A complete description of the contents of the Hipparcos Catalogue is contained in Volume 1 of ESA (1997). Recent overviews have been given by Perryman et al. (1997) and Perryman (1997). The principal astrometric data provided are the barycentric coordinates (α, δ) of the stars, the trigonometric parallaxes (π) and the proper motion components ($\mu_{\alpha*} \equiv \mu_\alpha \cos\delta$, μ_δ). These parameters are given in the main catalogue for 117 955 entries along with estimated standard errors and correlation coefficients, all data being referred to the common reference epoch T_0 = J1991.25 on the Terrestrial Time (TT) scale. Typical ranges of the astrometric standard errors and some other characteristics of the catalogue are shown in Table 1.

The main astrometric data were obtained through a rigorous combination of the results of two independent analyses of the satellite observations, performed by the FAST and NDAC consortia. The parallel reduction scheme included a comprehensive system of cross-checking and validation at several levels of the processing to ensure the highest quality of the end product. The standard errors given in the catalogue are believed to be reliable, and generally representative of the true (external) errors to within a factor 1.0–1.2. Nevertheless, the short duration of the mission—compared with the time base of ground-based astrometric catalogues—and other specific features of the observations make it necessary to interpret the data with due respect for the potential complications caused, in particular, by double and multiple stars (Section 2.3). Flags and pointers in the main catalogue give additional indications of the quality of the solutions and may refer the user to the various annexes for details on specific objects.

TABLE 1. Main characteristics of the Hipparcos Catalogue. The V magnitude and standard errors (s.e.) are median values for the whole catalogue, with the 10th and 90th percentiles given in parentheses.

Measurement period	1989.85–1993.21	Magnitude V	8.44 (6.76/9.88) mag
Catalogue epoch T_0	J1991.25(TT)	S.e. in $\alpha*$ at T_0	0.87 (0.53/1.70) mas
Nominal reference system	ICRS	S.e. in δ at T_0	0.72 (0.51/1.38) mas
coincidence with ICRS at T_0	±0.6 mas	S.e. in π	1.10 (0.71/1.91) mas
residual rotation wrt ICRS	±0.25 mas/yr	S.e. in $\mu_{\alpha*}$	1.02 (0.60/1.98) mas/yr
Number of entries	118 218	S.e. in μ_δ	0.85 (0.57/1.56) mas/yr
with associated astrometry	117 955	Ratio external error/s.e.	$\sim 1.0 - 1.2$
with associated photometry	118 204	Level of systematic errors	< 0.1 mas

J. Andersen (ed.), Highlights of Astronomy, Volume 11A, 287–291.

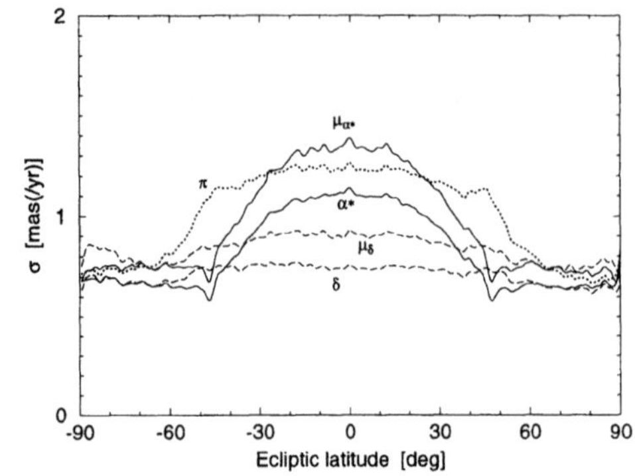

Figure 1. *Median standard errors of the astrometric parameters (epoch J1991.25) as function of ecliptic latitude.*

The five astrometric parameters α, δ, π, $\mu_{\alpha*}$, μ_δ together with the (astrometrically undetermined) radial velocity component define the rectilinear space motions of the stars relative to the solar system barycentre. To within observational uncertainties, this standard model applies to single stars as well as to several categories of non-single stars, such as the components of wide pairs with long orbital periods and the photocentres of close binaries whose absolute orbits have long periods or small amplitudes. The standard model was adopted for 100 038 catalogue entries regarded as single stars, and for the resolved components of 13 211 additional entries, with details given in the Double and Multiple Systems Annex C. Of the remaining 4706 entries with accepted astrometric solutions, most are astrometric binaries with significantly curved proper motion or even complete orbital solutions for the motion of the photocentre (Lindegren et al. 1997). For these objects the astrometric parameters given in the main catalogue refer to the mean motion over the ~3 year measurement period, or (for orbital solutions) to the motion of the centre of mass of the system. Parts G, O, V and X of the Double and Multiple Systems Annex give details of these solutions.

2. Properties of the Hipparcos Reference Frame

In this section we review some of the unusual characteristics of the Hipparcos reference frame and address the important question of the systematic accuracy of the data.

2.1. ACCURACY VERSUS POSITION ON THE SKY

From ground-based astrometry we are used to regard the measurements in α and δ as virtually independent, sometimes produced by different instruments, and usually affected by different kinds of errors. With Hipparcos all measurements are one-dimensional, produced by a single instrument scanning the sky in a complex pattern. Moreover, the scanning law is quasi-symmetric with respect to the ecliptic so that the main variation of the standard errors over the sky is a function of ecliptic latitude (Figure 1), and the principal axes of the error ellipses are on the average oriented along the ecliptic coordinates. As a consequence of the scanning law the distribution of errors is distinctly different in the various regions depending on the ecliptic latitude β:

- in the ecliptic region, $|\beta| < 30°$, the density of observations was rather low, with scans predominantly in the north–south direction, leading to error ellipses elongated along the ecliptic;
- in two narrow bands around $|\beta| \simeq 47°$ the density of observations was very high with scans predominantly in the east–west direction, leading to error ellipses slightly compressed along the ecliptic;
- in the polar regions, $|\beta| > 60°$, the density was moderately high with a rather uniform coverage in different directions, leading to almost constant and isotropic errors.

This pattern is roughly the same for the proper motions as for the positions at the reference epoch, and it therefore persists also when the positions are propagated to other epochs.

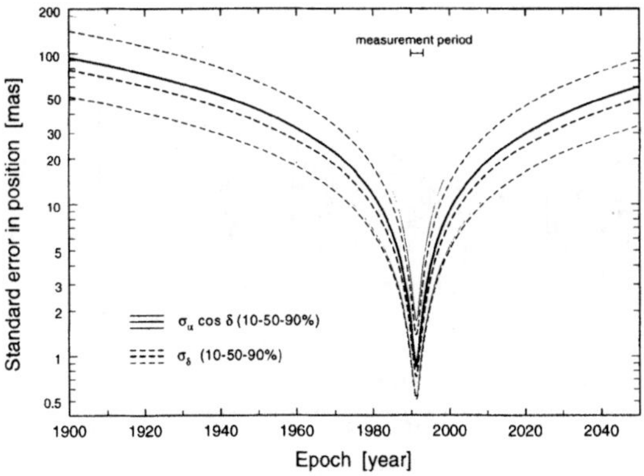

Figure 2. *The accuracy of individual stellar positions at various epochs is here represented by the 10th, 50th and 90th percentiles of the standard errors in right ascension and declination.*

2.2. POSITIONAL ERROR VERSUS TIME

The positional reference frame is defined to sub-milliarcsec accuracy only within a few years of the catalogue epoch J1991.25 and degrades rapidly with the time difference from that epoch (Figure 2). This is a direct consequence of the short measurement period of $\sim$3 years. However, even after several decades the individual accuracies are at least as good as in the best ground-based optical catalogues. Thanks to the expected low level of systematic errors (Section 2.4) and the high density of stars, the statistically defined reference frame is likely to be superior to all existing ground-based optical frames independent of epoch. While the systematic character of the Hipparcos reference frame therefore cannot be improved by incorporating old catalogues, the accuracies of individual objects may in some cases be improved by such a process.

2.3. ROLE OF ASTROMETRIC BINARIES

An important consequence of the presence of astrometric binaries is that the quasi-instantaneous proper motion measured by Hipparcos may be significantly different from the long-term motion measured on ground. This is obvious in the cases where Hipparcos measured a significantly non-uniform motion (Double and Multiple Systems Annex G), but probably applies also to several thousand (as yet) unrecognised astrometric binaries, which in the Hipparcos data analysis appeared to be perfectly normal single stars. When the Hipparcos data are extrapolated to epochs several decades before or after the measurement period, the distribution of positional errors will have an extended, non-Gaussian tail largely consisting of such astrometric binaries. This effect was not taken into account in constructing Figure 2, as the statistics of the phenomenon are not well known. Aspects of the problem have been discussed by Wielen (1997).

2.4. GLOBAL COHERENCE BY DESIGN

Several key features of the Hipparcos mission were specifically aimed at providing an optical reference frame of unprecedented coherence on a global scale, i.e. a system of positions and proper motions free of internal distortion:

- observations from space virtually eliminated all effects of refraction, instrumental flexure and thermal distortion;
- all-sky visibility permitted a direct linking of the stars over the whole celestial sphere observed with a single instrument;
- the two viewing directions of the satellite, separated by a large and suitable angle, resulted in a 'rigid' connection of one-dimensional measurements in different parts of the sky;

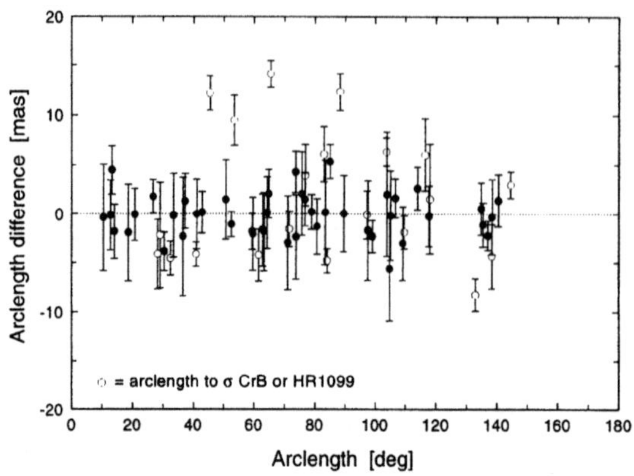

Figure 3. *Differences in arclength between Hipparcos and VLBI observations of 12 radio stars. This is a test of the internal consistency of the Hipparcos reference frame. Apart from arclengths involving σ CrB (HIP 79607) or HR 1099 (HIP 16846) the arclength differences are well centred with a scatter in perfect agreement with the combined standard errors of the Hipparcos and VLBI data.*

- the continuous scanning of the satellite provided multi-epoch measurements in geometrical configurations favourable for the simultaneous determination of the five astrometric parameters of each star;
- critical geometrical calibrations of the instrument were included in a self-consistent manner in the global analysis of the measurements;
- the high redundancy factor of the observations (with typically ~ 6 independent measurements per astrometric parameter) provided high sensitivity to deviations from the standard model of uniform motion and allowed problematic objects to be diverted away from the main processing chain where they did not contribute to the frame definition.

The global analysis of the Hipparcos measurements was thus designed to provide a globally coherent reference frame, which was however decoupled from existing frames and *a priori* inertial directions. Its subsequent linking to the ICRF was considered a completely independent problem.

2.5. TESTS OF GLOBAL COHERENCE

The coherence of the Hipparcos reference frame can be tested by direct comparison with an independent reference frame of similar (or higher) accuracy, provided the catalogues have some sources in common. However, suitable comparison data are scarce, and for this reason one may also have to resort to indirect evidence. We describe briefly the results of one direct and four indirect tests.

Comparison with VLBI: 12 radio stars measured by VLBI as part of the extragalactic link programme (Kovalevsky et al. 1997) have positional accuracies at J1991.25 in the range from 0.3 mas to a few mas and may thus provide a stringent, if somewhat restricted test of the Hipparcos frame. In order to make the test independent of the frame orientation, arclengths were computed for all 66 possible object pairs both from the Hipparcos Catalogue and from the VLBI positions, and the arclength differences plotted against the arclength (Figure 3). The comparison shows a very good agreement between the Hipparcos and VLBI arclengths, on all angular scales, if allowance is made for a possible radio/optical offset in two of the objects.

FAST/NDAC comparisons: Although based on the same satellite data, the global reductions by the two consortia FAST and NDAC were to a large extent independent and used different models for the satellite motion, instrument calibrations, etc. Consistency between the consortia results was a necessary (but not sufficient) condition for global coherence of the combined catalogue. Extensive analyses of the FAST/NDAC differences did not reveal any large-scale systematic differences exceeding about 0.1 mas or 0.1 mas/yr.

Inclusion of bias parameters: It is possible to test for certain types of systematic errors by performing special solutions in which additional unknowns (bias parameters) are introduced. *A priori*, these parameters are expected to come out as zero, within their estimated standard errors. Tests of this kind are described in ESA (1997), Volume 3 (Section 20.4). The general conclusion was that systematic terms exceeding 0.1 mas are rare and affect mainly extreme cases such as very red stars ($V - I > 2$).

Parallax zero point: Of the five astrometric parameters, the parallax is normally the one most easily checked against independent information such as photometric or spectroscopic distances. This could be used to estimate the global zero-point error (z) of the Hipparcos parallaxes and the ratio (k) of the external to nominal parallax errors (Arenou et al. 1995). The results from several tests on different stellar samples indicated that $|z| < 0.1$ mas and $k \simeq 1.05$. Since the (absolute) parallaxes and the positions and proper motions were obtained in the same global reduction, and the absoluteness of the parallaxes depends directly on the accuracy of the wide-angle measurements, it is difficult to imagine that significant large-scale distortion of the reference frame could result from the same measurements that appear to give such accurate parallaxes.

Determination of PPN parameter γ: The bending of starlight in the solar gravitational field is proportional to $\frac{1}{2}(1+\gamma)\cot(\theta/2)$, where γ is the PPN parameter (= 1 in General Relativity) and θ is the angle between the star and the Sun. Hipparcos measured the light bending for $47° < \theta < 133°$, where it amounts to 2–9 mas, and gave $\gamma = 0.997 \pm 0.003$ (Frœschlé et al. 1997). The uncertainty corresponds to a positional shift of 3 to 14 μas at an angular scale length of the order of 90°. This suggests that the level of systematic errors in the Hipparcos measurements, at these scales, is very small indeed.

In summary the direct comparison with VLBI confirms global coherence on a level of about 1 mas. Several indirect tests based on many more stars suggest that the actual level of large-scale distortion is considerably smaller, i.e. of the order of 0.1 mas or less. This does not preclude a higher level of systematic errors in small areas of the sky or for extreme cases such as very red stars.

3. Availability of Data

The Hipparcos Catalogue is available as a 17-volume publication, ESA SP–1200. This includes the main Hipparcos Catalogue (as well as the Tycho Catalogue), the Double and Multiple Systems Annex, the Variability Annex, identification charts, light curves, a full sky star atlas, and extensive documentation of the contents of the catalogue and how it was constructed. All mission products, including intermediate astrometric and photometric information, are also provided on six ASCII CD-ROMs included within the 17-volume publication. On-line access to the catalogue data is provided through several services of the CDS, Strasbourg, and its mirror sites at ADC (Goddard) and NAO (Tokyo), and through the ESA Hipparcos WWW page.

References

Arenou, F., Lindegren L., Frœschlé, M., et al. (1995) Zero point and external errors of Hipparcos parallaxes, A&A 304, 52

ESA (1997) The Hipparcos and Tycho Catalogues, ESA SP–1200

Frœschlé, M., Mignard, F., Arenou, F. (1997) Determination of the PPN parameter γ with the Hipparcos data, Proc. *HIPPARCOS Venice '97*, ESA SP–402, p. 49

Kovalevsky, J. (1997) Linking Hipparcos to ICRF, this volume

Kovalevsky, J., Lindegren, L., Perryman, M.A.C., et al. (1997) The Hipparcos Catalogue as a realisation of the extragalactic reference system, A&A 323, 620

Lindegren, L., Mignard, F., Söderhjelm, S., et al. (1997) Double star data in the Hipparcos Catalogue, A&A 323, L53

Perryman, M.A.C. (1997) The Hipparcos Catalogue contents, Proc. *HIPPARCOS Venice '97*, ESA SP–402, p. 1

Perryman, M.A.C., Lindegren, L., Kovalevsky, J., et al. (1997) The Hipparcos Catalogue, A&A 323, L49

Wielen, R. (1997) Principles of statistical astrometry, A&A 325, 367

THE TYCHO CATALOGUE: EXTENSION OF OPTICAL REFERENCE FRAME

E. HØG
Copenhagen University Observatory
Juliane Maries Vej 30, DK-2100 Copenhagen Ø, Denmark

Abstract. The Tycho Catalogue contains astrometric and photometric data for the one million brightest stars on the sky. About 900 000 stars are high quality astrometric reference stars. Photometry in a blue and visual band is provided.

Two extensions of the catalogue are under way. One is the extension to one magnitude fainter stars by a second reduction of the original Tycho observations, resulting in a catalogue with a total of about three million stars. The other extension is the derivation of more accurate proper motions by means of the Astrographic Catalogue positions as first epoch. This latter work will be carried out first for the one million stars, and later for all the three million, resulting in the Tycho Reference Catalogue (TRC).

1. The Tycho Catalogue (TYC)

The Tycho Catalogue contains astrometric and photometric data for the one million brightest stars in the sky, based on observations with the Hipparcos satellite star mapper. The median astrometric standard errors (in position, parallax, and annual proper motion) are typically around 7 mas for stars brighter than $V_T \sim 9$ mag, and approximately 25 mas for $V_T \sim 10.5$ mag, at the catalogue epoch (J1991.25). The regional systematic errors are $\sim$ 1 mas(/yr). The catalogue is available in machine-readable form as part of the final Hipparcos mission products (ESA 1997), including a detailed description of the Tycho Catalogue contents (in Volume 1) and construction (in Volume 4). Shorter descriptions are given by Høg (1997a, 1997b) and Høg et al. (1997).

2. The second Tycho Catalogue (TYC 2)

The original Tycho observations comprising 1000 Gbits of data are now subject to a second processing (see Høg 1997a, Section 6). This should result in a second Tycho Catalogue (TYC 2) of 3 million stars, including the one million in the present Tycho Catalogue. Expected standard errors at V=11.0 and 12.0 are respectively about 50 and 100 mas for the position components, and 0.10 and 0.20 mag for visual magnitudes. The reduction is being carried out at Copenhagen in collaboration with Heidelberg, with planned completion in 1999.

The second Tycho processing of all raw counts is based on the available satellite attitude, the Tycho Catalogue of one million stars, and a newly constructed Tycho Input Catalogue (TIC2) of 4.3 million stars. The TIC2 production made use of a new astrometric reduction of GSC (GSC1.2 by Röser *et al.* 1996) and the Astrographic Catalogue. The USNO A1.0 catalogue (Monet *et al.* 1996) delivered many of the positions, proper motions and colour indices. The TIC2 is as complete as possible to about $V = 12.5$ mag and has a position accuracy about 0.25 arcsec at the observation epoch. The magnitudes in TIC2 obtained from GSC have been corrected by means of the Tycho Catalogue photometry (Fabricius 1997). The processing is expected to give better astrometric and photometric values for the fainter half of the present one million stars of the Tycho Catalogue. The major reason is that an astrometric estimation based on very few photons does not achieve the Cramér-Rao limit, cf. Yoshizawa & Andreasen & Høg (1985). An estimation based on a superpositioning of all photons from many transits will come much closer to that limit. Furthermore, many

J. Andersen (ed.), Highlights of Astronomy, Volume 11A, 292–293.

transits of the faint stars were below the limit of signal-to-noise ratio 1.5 used for the detection in the first Tycho reduction and were thus completely lost.

3. Tycho Reference Catalogue (TRC)

The Tycho Catalogue constitutes an astrometric reference frame with twice as many stars as PPM (Röser & Bastian 1991) which is the best frame up to now with high star density, cf. Table 2 by Høg (1995). The 30 mas standard error of positions for TYC at the epoch 1990 is ten times smaller than that of PPM and the systematic errors are 100 times smaller. The accuracy is maintained for one or two decades from 1990 if PPM proper motions are used. Even better proper motions will be derived for all Tycho stars by means of positions from the Astrographic Catalogue, reduced to the Hipparcos reference system. This Tycho Reference Catalogue project was proposed by Röser & Høg (1993) and pilot project results are described by Kuzmin *et al.* (1997). The expected standard errors of the proper motion components are about 2.5 mas/yr.

The Astrographic Catalogue will be transferred to the Hipparcos reference system using the PPM catalogue, the ACRS (Corbin & Urban 1991) and the Hipparcos and Tycho Catalogues. A direct use of the Hipparcos Catalogue only is not feasible due to the small number of Hipparcos reference stars per plate. Therefore, the much more numerous Tycho Catalogue stars are better suited to derive certain systematic plate-specific errors of the Astrographic Catalogue even though the Tycho Catalogue proper motions are too inaccurate to transfer the positions properly to the AC epoch. This work is being carried out at Moscow, in collaboration with Copenhagen, Heidelberg and Lund. A first preliminary version of TRC with one million stars should appear in late 1997. After completion of the second Tycho processing a TRC with 3 million stars will be produced.

References

Corbin, T.E., Urban, S.E. 1991, Astrographic Catalogue Reference Stars, U.S. Naval Observatory, Washington

ESA 1997, The Hipparcos and Tycho Catalogues, ESA SP-1200

Fabricius, C. 1997, in Proceedings of the HIPPARCOS Venice '97 Symposium, ESA SP-402

Høg E. 1995, in: E. Høg and P.K. Seidelmann (eds.), Astronomical and Astrophysical Objectives of Sub-Milliarcsecond Optical Astrometry, p. 317, Kluwer Academic Publ., IAU Symposium No. 166

Høg, E. 1997a, The Tycho Catalogue: Astrometry and Photometry, in Proceedings of the HIPPARCOS Venice '97 Symposium, ESA SP-402

Høg, E. 1997b, in Proceedings of IAU Joint Discussion No. 14

Høg, E. *et al.* 1997, Astron. Astrophys. 323, L57

Kuzmin, A., Bastian, U., Høg, E., Kuimov, K., Röser, S. 1997, Tycho Reference Catalogue: Pilot Project Results, in Proceedings of the HIPPARCOS Venice '97 Symposium, ESA SP-402

Monet, D., *et al.* 1996, USNO-A V1.0, U.S. Naval Observatory, Washington

Yoshizawa, M., Andreasen, G.K., Høg, E., 1985, A&A, 147, 227.

Röser, S., Bastian, U. 1991, PPM Star Catalogue, Astronomisches Rechen-Institut, Heidelberg. Spektrum Akademischer Verlag, Heidelberg, Berlin, New York

Röser, S., Høg, E. 1993, In: Workshop on Databases for Galactic Structure. Ed.: A.G. Davis Philip, B. Hauck and A.R. Upgren. Van Vleck Observatory Contr. No.13, 137. L. Davis Press, Schenectady, N.Y.

Röser, S., Morrison, J. Bucciarelli, B., Lasker, B. and McLean, B. 1996, In: IAU Symposium No. 179. Baltimore, USA, August 26-30, 1996. (in press)

PLACING SCHMIDT ASTROMETRY ON THE ICRS

B. BUCCIARELLI AND M.G. LATTANZI
OATo
Strada Osservatorio 20, Pino Torinese TO, Italy

AND

B.M. LASKER
STScI
3700 San Martin Dr., Baltimore, USA

1. Overview

The sky-limited, large-area photographic surveys which are being used for the construction of the second generation Guide Star Catalog (GSC-II) will be for many years to come the largest astronomical database available. The survey plates are typically placed on a 5-degree grid, with each plate area of about 42 square degrees, for a total of more than 1700 plates covering the whole sky. The expected data volume is about 2 Tbytes per bandpass. The final product will consist of an all-sky catalog of positions, proper motions, magnitudes and at least one color to a minimum of V=18. Among the scientific motivations behind such an endeavor let us mention operation of ground-based telescopes and future space missions, observation planning, statistical investigations (see, e.g., Lasker et al. 1995, Cannon 1997)

To fully exploit the astrophysical content of this database, both astrometric and photometric properties of the photographic plates must be calibrated with high precision. In particular, the very important task of placing GSC-II astrometry into the ICRS reference system has to be tackled.

Is is well known that Schmidt plates are astrometrically difficult to treat, essentially because metric distortions increase rapidly at a radius of 2.7° from the plate center outward. Therefore, a classical parametric approach to the plate solution, e.g., using simple polynomials, results in systematic errors of the order of 1 arcsecond or more at the plate edges. Characterization of these systematics deals with plate-holder driven mechanical stresses, asymmetry of image growth, focus-vignetting interactions, filter ghosts, diffraction spikes, etc. While improved understanding of these effects does not obviously provide us with a good modelling, the fact that the bulk of the error is stable within a set of similar plates allows it to be removed by subtracting a **mask** (Taff et al. 1990). The remainder of the error, specific to individual plates, can be removed by the use of a small-scale ($\sim 0.2°$) spatial filter against suitable reference stars. Another important finding is a magnitude dependent effect which has been discovered in the GSC-I catalog (Morrison et al. 1996), and which is typical of Schmidt photographic plates (Irwin et al. 1997). This effect has not yet been thoroughly explained, although it is clear that it stems from the performance of the centroiding algorithm on relatively bright objects. The systematic error increases with magnitude separation from the reference stars and radial distance from the plate centers; Morrison et al. (1996) showed that it can be efficiently removed by subtracting a tabular function of Δm, $\mathbf{r}$.

2. New Resources for GSC-II

From the experience gained with the astrometric re-reduction of the first Guide Star Catalog (Taff et al. 1990, Lattanzi and Bucciarelli 1991, Bucciarelli et al. 1994, Bucciarelli et al. 1995), we know that a very accurate and dense set of reference stars is essential to probe the small scale systematics characteristic of Schmidt plates. To this end, the newly released HIPPARCOS and Tycho catalogs are undoubtedly powerful resources.

J. Andersen (ed.), Highlights of Astronomy, Volume 11A, 294–296.

Although the HIPPARCOS average density is only 3 stars per square degree, its unprecedented accuracy (2 milliarcseconds in positions *and* proper motions) makes it unique for investigating plate-dependent systematic error toward the bright end of the photographic magnitude range of both recent and old-epoch surveys.

The density of the Tycho catalog is 25/sq degree, which is very good, with a typical magnitude of ~ 11, and a magnitude limit of 12. Tycho will contribute to the calibration of photometric zero-points errors across the plate, although its limited magnitude range cannot address the problem of magnitude effect. On the other end, the combination of its density and high positional precision will support a detailed investigation of plate-based systematics as function of different parameters (sky region, zenith distance, filters, etc.) at the mean magnitude of Tycho.

It has to be noted however that, because of large proper motion errors, Tycho will be best suited to reduce recent plates (POSS-II and SES surveys), but could not support the reduction of first epoch plates. For these we will need to use a combination of Tycho and the Astrographic Catalogue (AC) as proposed by Röser and Høg (1993). In such a way, it is intended to reduce the proper motion errors of the Tycho Catalogue to a few mas per year, making it possible then to use it successfully to reduce the first epoch plates. A new version of the AC (AC2000), reduced against the Astrographic Catalog Reference Stars (ACRS) and then tied to the HIPPARCOS system, has been announced at this IAU (Urban et al. 1997). Also, the USNO has recently completed the task of combining the AC2000 with Tycho to produce the ACT which is already available on a personal basis. Therefore, the choice of the reference catalog for the first epoch plates is no longer a matter of concern.

Auxiliary catalogs will also be essential for error control and determination of masks and magnitude effects. The most useful ones are the *Carlsberg Meridian Catalogue* (1997) series, which consist of about 1.4×10^5 objects north of $-40°$, with a significant number in the magnitude range $12-15$. In this context it is worth mentioning the availability of a special catalog of stars along the equator, observed with the *Carlsberg Meridian Telescope* in La Palma between 1994 and 1995, as part of the 'Faint Reference Star' observation program. This contains about 10 000 stars uniformly covering a strip along the equator of about 12 hours in right ascension and 12 degrees in declination. A first attempt to reduce POSS-II plates against Tycho and using this CAMC special catalog as external comparison produced very encouraging results, and is presented elsewhere in these proceedings (Bucciarelli et al. 1997).

Another very interesting catalog is FIRST, which is a VLA survey of about 10 000 square degrees about the North Galactic Pole, with magnitude fainter than 18 (White 1997). The FIRST catalog contains about 90 sources per square degree; most of these are optically faint, but about 15 per square degree have counterparts on the POSS plates ($V < 20$). Most of these objects are faint quasars or nearly unresolved galaxies. They make an ideal reference grid for studying residual astrometric errors: they are faint, they have accurate external positions from the FIRST catalog (which has both good precision and systematic errors smaller than 0.1 arcsec), and they are practically all extragalactic objects (so they have no proper motions).

3. Plans for GSC-II astrometry

With the reference catalog material at our disposal, a careful investigation of the systematic errors will be carried out in order to chose the 'optimal' strategy for GSC-II astrometry. In the following we summarize the general approach which will be adopted, based on our previous experience with Schmidt plate astrometry:

- Apply spherical (Schmidt) projection to x,y measurements and compute the refraction correction using the best available code; then fit a quadratic model to the data using Tycho as the reference catalog;
- For each recent-epoch survey, use the ensemble of Tycho reductions to construct mask(s) at mean magnitude of Tycho;
- Use available auxiliary data to map magnitude effects;
- After applying mask and magnitude equation, determine spatial scale of residual systematics and apply a local filtering technique with that scale.

With this technique, the expected adherence to the reference frame to be $\sim 0.15''$, averaged over all the plates. This means that the systematics will be mostly removed, and the astrometric residuals will reflect essentially the plate measuring errors. Of course, this 'recipe' cannot be applied to first epoch plates; in this case either TRC or a less direct method will have to be used. Another concern is the fact that, with the material at hand, knowledge of magnitude effects in the range 13 to 18 is unsatisfactory. Therefore, the use of short-exposure plates and other indirect methods is indeed required. A lot of testing has still to be done in this respect, but we are confident of succeeding in bridging this gap. In conclusion, high quality GSC-II astrometry is a challenge which can be successfully tackled, thanks especially to the first class astrometric catalogs now available.

Acknowledgement

We wish to thank J. Morrison for useful discussions during this work.

References

Bucciarelli, B., Doggett, J.B., Sturch C.R., Lasker, B.M., Lattanzi, M.G., and Taff, L.G. (1995) GSC 1.2 - A Sub-plate Reduction of the GSC, Proceedings of *Astronomy from Wide Field Imaging*, IAU Symp. 161, 277

Bucciarelli, B., Sturch, C.R., Lasker, B.M., Lattanzi, M.G., Girard, T.M., Platais, I., and van Altena, W.F. (1995), A Test of the Astrometric Quality of the Southern Guide Star Catalogue 1.2, Proceedings of *Astronomical and Astrophysical Objectives of Sub-Milliarcseconds Optical Astrometry*, IAU Symp. 166, 375

Bucciarelli, B., Morrison, J.E., McLean, B.J., and Sturch, C.R., (1997), Astrometry of POSS-II plates using Tycho, these Proceedings

Cannon, R.D. (1997), Optical Reference Frame: high density requirements, these Proceedings

Irwin, M.J., Morrison, L.V., and Argyle, R.W. (1997) Transferring the Hipparcos frame to Schmidt Surveys, these Proceedings

Lasker, B.M., McLean, B.J., Jenkner, H., Lattanzi, M.G., and Spagna, A. (1995), Potential Applications of GSC-II for GAIA Operations, *Proceedings of the Workshop on Future Possibilities for Astrometry in Space, Cambridge, OK, 19-21 June 1995*

Lattanzi, M.G. and Bucciarelli B. (1991), A non-deterministic approach to Schmidt plate Astrometry, A&A, **250**, 565

Morrison, J.E., Röser, S., Lasker, B.M., Smart, R.L., and Taff, L.G. (1996) A Schmidt Plate Coma-Like Term, AJ, **111**, 1405

Röser, S., and Høg, E. (1993), Tycho Reference Catalogue (TRC): A catalogue of Positions and Proper Motions of 1 Million Stars, *Proceedings of a Workshop on Databases for Galactic Structure, Swaaathmore College, PA, May 17-19, 1993*

Taff, L.G., Lattanzi, M.G., and Bucciarelli, B. (1990) Two Successful techniques for Schmidt Plate Astrometry, ApJ **358**, 359

Urban, S.E., Corbin, T.E., and Wycoff, G.L. (1997) AC2000: The Astrographic Catalogue on the Hipparcos System, these Proceedings

White, R. (1997), private communication.

HIGH DENSITY REQUIREMENTS

RUSSELL CANNON
Anglo-Australian Observatory
PO Box 296, Epping, NSW 2121, Australia

1. Specification and Significance

I will consider here some practical astrometric requirements, from the perspective of an optical astronomer working on objects much fainter than the limits of the primary catalogues, i.e., from 15 mag down to 20 mag and fainter. I take 'high density' to mean surface densities of order 100 or more stars per square degree; these two criteria are roughly equivalent (Bahcall & Soneira, 1980). The positions of such objects are routinely available to an accuracy of around 0.5–1.0 arcsec, e.g. from various digitised versions of the Schmidt Telescope photographic sky surveys. However, these sky surveys themselves readily yield *internal* accuracies at the 0.1 arcsec level, as demonstrated by repeated measures of the same plates. The problem is that the discrepancies can be ten times larger than this when different machines or plates are used, when a wide magnitude range is covered, when relative positions are compared over distances of a degree or more, or when absolute positions are needed for comparison with, say, radio catalogues. The challenge now is to improve this accuracy; we should to be able to specify the absolute positions of all well-defined optical objects to better than 250 mas, anywhere in the sky, at any epoch and regardless of magnitude or colour.

1.1. CONNECTING THE REFERENCE FRAMES

A basic problem is that the although reference frame for bright stars, down to $\sim$ 11 mag, is now very well-defined, thanks to Hipparcos and Tycho, it is difficult to connect fainter optical images to this frame. The images on deep sky survey photographs, such as those taken on the Oschin, ESO and UK Schmidt (UKST) telescopes, begin to develop diffraction spikes and other complex structure for stars brighter than $\sim$ 15 mag. Thus there is a problem in relating the astrometry for faint target stars to the bright reference stars; systematic errors arise at the level of several tenths of an arcsec, and since these depend on the detailed shapes of the outer parts of bright star images, they vary across plates, from plate to plate and from machine to machine.

1.2. SUPERNOVA 1987A

A good example of the current situation is afforded by SN 1987A. The early images from the Hubble Space Telescope showed that the supernova was surrounded by a spectacular ring of material about 1 arcsec in diameter, ejected by the progenitor star at an earlier stage of its life. When the supernova reappeared as a radio source in 1990, the first radio maps indicated that the centroid of the emission was located about 0.5 arcsec north of the supernova (Staveley-Smith et al., 1992). Did this mean that the radio emission was coming from part of the pre-supernova ring, or was it really coming from the central supernova remnant? Clearly these were two completely different possibilities, with very different astrophysical implications and explanations. The situation was only resolved after a special campaign to improve the registration of the optical and radio reference frames in this particular direction, using pre-release data from Hipparcos, specially-measured photographic plates and two radio reference stars (Reynolds et al., 1995); this enabled the images to be registered to $\sim$ 70 mas and showed clearly that the radio emission was centred on the supernova, and that the peak radio brightness lay inside the ring, as expected if the emission arises from the expanding ejecta.

J. Andersen (ed.), Highlights of Astronomy, Volume 11A, 297–299.

1.3. MULTI-OBJECT WIDE-FIELD SPECTROSCOPY

Different problems arise in the case of multi-object spectroscopy, as with the new Two-degree Field (2dF) instrument on the Anglo-Australian Telescope (e.g., Cannon, 1997). Here the requirement is to produce positions for 400 faint target objects (magnitude ≥ 18) to a relative accuracy better than 0.25 arcsec across 2°, and to relate these to a few brighter guide stars at $B \sim 15$. At present, errors at the 0.5 arcsec level are not uncommon, leading to significant loss of light with the 2 arcsec diameter optical fibres used in 2dF. There is a one-off need to determine the geometry of the 2dF optics and robot positioner, which requires a few dense sets of reference stars accurate to ~ 100 mas. For the guide stars, the main current problem is proper motions; in the past, errors at the 0.7 arcsec level have not been uncommon, although this should be solved soon when the UKST Second Epoch Survey is completed and fully digitised. The third problem is to map the Schmidt sky survey photographs well enough, especially the non-linear distortions near the plate corners, so that relative astrometry accurate to 150 mas can be achieved routinely, anywhere on the sky.

2. Sources of error

Several separate effects combine to produce the errors in positions derived from Schmidt photographs. Three produce large-scale distortions across the field: the optical characteristics of the telescope itself, including its achromatic corrector and coloured glass filters; non-linear distortions arising from the elastic deformation of the flat glass photographic plates, needed to fit the convex spherical shape of the focal surface; and differential atmospheric refraction. Atmospheric refraction itself involves at least three further separate effects: a large-scale distortion across the field which depends on zenith distance and azimuth; a differential effect for long exposures, due to changes in the above parameters and which also depends on the precise location of the guide star used (see Wallace & Tritton, 1979); and chromatic effects, since each star image is in effect an extremely low dispersion spectrum.

All of these effects are convolved in forming the actual star images, whose structure varies across the field. Additional asymmetrical effects arise from the diffraction spikes caused by the plate holder supports inside the telescope, and from internal reflections arising in the corrector lens (which are generally axially symmetrical, although they suffer vignetting near the edges of the field) and also in the filter, which are eccentric since the flat glass filters are not parallel to the curved photographic plate (see UKST Handbook, 1983, for a full description with illustrations). In addition to all of the above, there can be systematic errors in image shape due to tracking errors or wind shake.

Thus the structure of Schmidt telescope images varies with position in the field, with magnitude and with colour, and every plate is different since 'seeing', tracking, zenith distance and exposure time are all variable. In principle many of these effects could either be modelled or removed empirically, but in practice the photographic detector introduces further problems since it has limited dynamic range, so that the central regions of bright star images become saturated, and it is intrinsically non-linear. These last two factors mean that information has been lost and they limit the extent to which the data can ever be recovered from a single given plate.

Since the image shapes are very complex, especially for bright stars, the positions determined will also depend on the characteristics of each measuring machine (the effective size and shape of the pixel) and on the algorithm used to determine image position (centroid, equal intensity contours or whatever).

One further error can be introduced if incorrect reference star positions are used to reduce the data. Forcing the digitised data to fit erroneous positions will lead to distortions of the coordinate frame, especially if only a small number of reference stars are used.

The result of several of the above effects is to produce large scale 'swirling' patterns in the positional errors on Schmidt plates. These were first derived by the HST Guide Star group (Taff, Lattanzi & Bucciarelli, 1990) and have subsequently been found by other groups (e.g. Irwin, 1994). The patterns are revealed when data for a very large number of fields are combined, and appear to be constant for a given combination of telescope, plateholder and measuring machine, so that they can be corrected by using an empirical template or look-up table.

That these effects are partially due to vignetting of the out-of-focus filter ghosts of bright star images has been demonstrated convincingly by Morrison et al. (1996), in the case of the UKST(J) survey. They find a strong coma-like term when positions of stars in successive magnitude ranges are compared (using the old Astrographic Catalogue positions as a reference frame, and again combining the data for large numbers of plates); the effect starts suddenly at a radius of 2.7° and increases rapidly with radial distance thereafter, following the known pattern of vignetting in the UKST.

3. Cures

Since the astrometric errors increase rapidly towards the edges of Schmidt fields are strongly magnitude-dependent for over-exposed bright stars, and often include significant proper motion terms, there is one obvious quick solution for specific projects requiring accurate astrometry: obtain a new, short-exposure Schmidt plate, centred on the field of interest. This is not however a suitable long-term strategy, since (a) the big Schmidts are expensive to run and may not even be taking plates within a few more years, and (b) the data do already exist on the various all-sky surveys, provided we have sufficient knowledge and adequate tools to extract the information.

It seems that the largest current sources of error arise in tying the deep survey photographs, with good stellar images in the range $15 < B < 20$, to bright ($\sim$ 10 mag) reference stars with badly over-exposed images. Thus a global cure could be effected with a special astrometric survey, designed to bridge the gap from 10 mag to 15 mag. Such a survey could either be done using short-exposure Schmidt plates (but not films), or as a separate exercise using an astrographic camera or transit circle equipped with CCDs. All three possibilities are currently under discussion by various groups; although in principle any one of the three should be adequate to solve the problem, it is highly desirable that at least two of them are carried forward, both to validate the data and because of the long-term funding uncertainties currently affecting most such projects. In due course, the identification of faint radio galaxies and quasars should provide a consistency check, to verify that the bright and faint optical reference frames have been properly connected.

An alternative 'internal' approach would be to use detailed pixel maps of many stellar images and fit them to a standard profile for each plate, after allowing for all known instrumental effects. This would however require more computer power, and better understanding of the image structure than is available today.

So far as the UKST is concerned, a pilot study is underway to check that a Short Astrometric Survey (SAS) would be both necessary and sufficient to solve the problem. There are already indications that this is true: comparison of many deep UKST plates of the same field shows that the scatter in residuals is small for stars at all magnitudes, so that a given exposure is repeatable and hence in principle correctable (Hawkins, private communication), while the large systematic errors seen for bright images on deep plates decrease rapidly with increasing magnitude (Morrison, private communication) and are much smaller on short-exposure plates (Monet, private communication).

Acknowledgements

I have had helpful discussions about Schmidt telescope astrometry with many people and particularly want to thank the following: Barry Lasker, Brian Maclean and Jane Morrison (STScI); Mike Hawkins and Harvey MacGillivray (SuperCOSMOS, ROE); Mike Irwin and Leslie Morrison (APM and CMT, RGO); Dave Monet (PMM, USNO); Karl Glazebrook and Ian Lewis (2dF, AAO); and Bea Bucciarelli and Mario Lattanzi (Torino). This paper was prepared during sabbatical visits to the Space Telescope Science Institute (Baltimore) and the European Southern Observatory (Garching).

References

Bahcall, J. N. & Soneira, R. M., 1980, *ApJS*, **44**, 73.
Cannon, R. D., 1997. In *Wide Field Spectroscopy*, eds. E. Kontizas, M. Kontizas, D. H. Morgan & G. P. Vetollani, pub. Kluwer, p.33.
Irwin, M. J., 1994, *IAU Commission 9 Working Group Newsletter*, ed. H. T. MacGillivray, **No. 5**, 25.
Morrison, J. E., Röser, S., Lasker, B. M., Smart, R. L. & Taff, L. G., 1996, *AJ*, **111**, 1405.
Reynolds, J. E. and 18 other authors, 1995, *A&A*, **304**, 116.
Staveley-Smith, L. and 9 other authors, 1992, *Nature*, **355**, 147.
Taff, L. G., Lattanzi, M. G. & Bucciarelli, B., 1990, *ApJ*, **358**, 359.
UK Schmidt Telescope Unit, 1983, *Handbook of the UK Schmidt Telescope*, Royal Observatory, Edinburgh.
Wallace, P. T. & Tritton, K. P., 1979, *MNRAS*, **189**, 115.

EXTENSION OF THE OPTICAL REFERENCE FRAME: GROUND BASED

N. ZACHARIAS
U.S. Naval Observatory, nz@pyxis.usno.navy.mil
3450 Mass.Av.NW, Washington DC 20392, USA

1. Introduction

The International Celestial Reference Frame (ICRF) is realized by the positions of 608 compact extragalactic radio sources (Ma & Feissel 1997) with milliarcsecond (mas) and sub–mas accuracy, all being on the same system, the International Celestial Reference System (ICRS). The Hipparcos Catalogue (ESA 1997) is the practical realization of the ICRF at optical wavelengths, giving accurate positions ($\approx$ 1 mas) at the mean epoch of 1991.25 and proper motions ($\approx$ 1 mas/yr) for 117 995 stars. This is about 2.5 stars per square degree, most being in the 7 to 9 magnitude range and a few as faint as 12. However, for many astronomical applications the Hipparcos Catalogue is not dense enough and does not reach faint enough magnitudes.

The Tycho Catalogue (ESA 1997) provides accurate positions ($\approx$ 25 mas) for about one million stars to magnitude 11, but lacks precise proper motions. Ground–based observations are an efficient way to complement the Hipparcos mission and to extend the optical reference frame to fainter magnitudes and yield a denser grid of astrometric standard stars. New reductions of early epoch photographic data will be used to provide highly accurate proper motions for stars to about magnitude 12. Current and new observational projects will extend the optical reference frame to even fainter magnitudes.

2. AC reductions, proper motions for Tycho stars

One major step for a densification of the optical reference frame is the reduction of the Astrographic Catalogue (AC) data to the Hipparcos system. For a general overview of the AC and Carte du Ciel (CdC) data, which were taken with the AC data to provide an atlas of the sky, see Eichhorn (1974). The AC covers the entire sphere and provides over 4.6 million positions of stars to about 12th magnitude. The AC 2000 (Urban *et al.* 1998) is the result of a complete re–reduction of the AC data performed at the U.S. Naval Observatory. The ACT contains accurate proper motions derived from the combination of AC 2000 and Tycho data. The ACT is now available on CD–ROM (send a request to seu@sicon.usno.navy.mil). The Tycho Reference Catalog (TRC) is a similar project with contributions from institutes in Moscow, Heidelberg, Copenhagen and Lund (see Høg 1997, this JD), which is expected to be completed by the end of 1997. These AC reductions allow the derivation of proper motions for almost all of the million Tycho stars with an accuracy of 2–4 mas/yr. Currently there is no recent epoch, high precision astrometric catalog available to fully utilize the over 3 million other AC positions. The Tycho 2 project (see Høg 1997, this JD) aims at deriving positions on the 50 mas and 100 mas level for 2 million stars of magnitude 11 and 12, respectively, by a re–analysis of the Tycho space mission data.

3. Re–measuring of old epoch photographic plates

Many zones of the AC project would benefit from a re–measurement. The IAU has set up a working group to explore the options. Of high priority is the Vatican zone ($55° \leq \delta \leq 64°$), with poor published measurements and early plates (around 1900) that are still in good shape. Bordeaux Observatory (Requième 1997, *priv.com.*) has started to measure the Bordeaux Zone CdC plates.

J. Andersen (ed.), Highlights of Astronomy, Volume 11A, 300–303.

Hamburg Observatory plans to re–measure the AGK2 plates (epoch ≈1930) which are kept there under excellent conditions (de Vegt 1988). Only one of the two exposures of about 186 000 stars were originally measured. The new measurements would give results for at least 10 times as many stars, to the limiting magnitude of about 12, with an estimated accuracy for the final catalog positions of 100 mas, resulting in a proper motion accuracy of at least 1.7 mas/yr.

4. Current and near future observational programs

TABLE 1. Current and near future observational programs. The star selection column indicates star list observing (L), selected regions (S) and all stars (A) within the given area and magnitude limits. The approximate standard error of a catalog position, σ_{cat}, is given for *well exposed stars*, which usually means stars of 2 or more magnitudes brighter than the limiting magnitude of the project indicated in the mag.range column.

project name	place	number of stars	mag. range	star sel.	Dec.zone (area)	σ_{cat} (mas)	status remark
automated transit circles:							
CAMC	La Palma	0.1 M	7...14	L	$-40 \leq \delta \leq 90$	50	active, CMC 9
CMASF	El Leoncito		7...14	L	$-90 \leq \delta \leq 30$		1997 →
scanning telescopes:							
AMC	Nicolaev		...15	S	North		start 1998
FASTT	Flagstaff	0.7 M	10...18	S	$-1.7 \leq \delta \leq 1.7$	40	1996 for SDSS
MERIDIAN2000	Bordeaux		9...16	A	$11 \leq \delta \leq 17$		1997–2000
	Sao Paulo		9...15		$-90 \leq \delta \leq 30$		just started
NAOJ	Tokyo	1 M	12...16	S	$-30 \leq \delta \leq 45$		1997–2001
SDSS	Apache Point	100 M	10...23	A	North gal.cap	≈100	start 1998
photographic astrograph:							
FON	Kiev	30 M	...16	(A)	$-23 \leq \delta \leq 90$	200	start 1982
NPM	Lick, Flagstaff		8...18	A	$-23 \leq \delta \leq 90$	150	re-meas. 1997 →
SPM	San Juan, Yale		...18	A	$-90 \leq \delta \leq -23$		2nd epoch ?
CCD astrograph:							
UCA test	Washington	0.3 M	7...16	S	$29 \leq \delta \leq 33$	20	1997
UCAC-S	CTIO, Wash.	40 M	7...16	A	$-90 \leq \delta \leq 2$	20	1998–1999
Schmidt surveys		≥500 M	11...20	A	$-90 \leq \delta \leq 90$	≤250	1950–2000

The Carlsberg Automatic Meridian Circle (CAMC) is the most productive instrument in its class, with the 9th catalog in press (CMC9 1997). Due to the list–mode observing technique, only a relatively small number of stars can be observed each year. However, in selected areas, e.g. around extragalactic reference frame sources, a dense grid of stars to about 14th magnitude has been established. A similar instrument has started test observations on the Southern Hemisphere in El Leoncito, Argentinia (Muinos 1997, *priv.com.*).

Scanning transit telescopes utilize a CCD–camera and integrate light as the sky drifts by due to the diurnal motion of the Earth. The USNO Flagstaff Astrometric Scanning Transit Telescope (FASTT) (Stone *et al.* 1996) did a pioneering job, producing a catalog of star positions along equatorial regions. The Bordeaux meridian circle (Requième 1997, *priv.com.*) and the National Astronomical Observatory of Japan (NOAJ) transit circle (Yoshizawa 1997, *priv.com.*) have recently been upgraded with a CCD–camera and observations are made in scanning mode. A similar instrument is being set up in Sao Paulo, Brazil.

The precision of individual wide–angle observations with modern automated transit circles is about 100 mas for well exposed stars near the zenith, and a catalog accuracy of about 50 mas is reached with multiple observations. For scanning transit circles the wide–angle positional accuracy has been about 150 mas but will be about 40 mas in differential observing with the use of Tycho reference stars. The scanning telescope curved-stellar-path-problem at higher declinations can be overcome; however, active driving along the sky and a rotation of the detector are then required, thus introducing other problems. The Sloan Digital Sky Survey (SDSS) instrument (Richmond 1993) will operate this way, and estimates of its astrometric capabilities range from 50 to 200 mas.

The following photographic astrometric surveys are ongoing or have been recently finished. First data (single overlap) of the FON project (Yatskiv 1997, *priv.com.*) will soon be published. Observations have been made with astrographs on blue sensitive plates by various institutes in the former Soviet Union. The Lick Northern Proper Motion (NPM) survey (Klemola *et al.* 1987) is complete, however only a fraction of the imaged stars have been measured so far (Klemola *et al.* 1993). The Southern Proper Motion (SPM) survey (Girard *et al.* 1997) is lacking the second epoch plates for a large fraction of the sky. Although modern astrographs are capable of an accuracy of under 100 mas for a single observation using photographic plates (Zacharias *et al.* 1994), there are currently no new such surveys in preparation.

The USNO CCD Astrograph Catalog South (UCAC–S) project (Gauss *et al.* 1996) will start in early 1998 to observe the entire Southern Hemisphere with a 4k CCD camera on a 2–meter focal length state–of–the–art astrograph. A positional accuracy of 15 mas has been achieved from guided exposures in test observations with that telescope (Zacharias *et al.* 1997). A 1.3–meter telescope for wide–field CCD imaging is being constructed at USNO, Flagstaff, which will have a wide–field, astrometric capability for faint stars ($\approx 13...22^m$). Similarly, mosaic imaging cameras at major observatories are now planned or already in test operation, which are capable of mapping a significant fraction of the sky to very faint magnitudes ($\geq 22^m$).

5. Reduction of Schmidt plate surveys

A re–reduction of the Space Telescope Science Institute (STScI) Guide Star Catalogue (GSC), version 1.2, has improved its positional accuracy to 250 mas at the plate centers and 500 mas at the plate edge (Röser *et al.* 1997). The version 1.3, with the same data on the Hipparcos system, will become available soon. Although not an astrometric catalog, the GSC has been and will be used for astrometry of faint stars (to 16^m) until it is superseded by e.g. the UCAC.

A complete digitization of Schmidt plate surveys (to $\approx 20^m$) is underway at the Royal Greenwich Observatory, Univ. of Minnesota, Royal Observatory Edinburgh, STScI, and the USNO Flagstaff. A recent review of the Schmidt plate surveys and scans can be found elsewhere (Lasker 1993). USNO, Flagstaff was first in finishing a global catalog (Monet 1996) from the Schmidt plate survey scans. A similar project, the GSC II is in progress with participation from STScI, Torino Observatory, ESA, and the GEMINI consortium. Reference stars fainter than 12^m have proven to be indispensable for astrometric Schmidt plate reductions. A positional accuracy of at least 150 mas seems feasible for well exposed stars from Schmidt plates, provided enough faint reference stars are available and the plates are measured to a linear accuracy of at least a micrometer (Murray & Corben 1979).

References

Carlsberg Meridian Catalogue La Palma Number 9 (1997)

de Vegt, C. (1988) in *Mapping the Sky*, IAU Sympos.133, Eds. S.Debarbat, J.A.Eddy, H.K.Eichhorn & A.R.Upgren, Kluwer Acad.Publ., Dordrecht, p.211

Eichhorn, H. (1974) *Astronomy of Star Positions*. Fred.Ungar Publ.Co., New York

ESA (1997) *The Hipparcos and Tycho Catalogues*. European Space Agency, **SP–1200**

Gauss, F.S., Zacharias, N., Rafferty, T.J., Germain, M.E., Holdenried, E.R., Pohlman, J.W. & Zacharias, M.I. (1997) A new astrometric survey of the southern hemisphere, BAAS **28**, No.4, 1282

Girard, T.M., Platais, I., Kozhurina-Platais, V., van Altena, W.F., & Lopez, C. E. (1997) The Southern Proper Motion Program I, *submitted* to AJ

Klemola, A.R., Jones, B.F., & Hanson, R.B. (1987) AJ **94**, 501

Klemola, A.R., Jones, B.F., & Hanson, R.B. (1993) National Space Science Data Center/ADC Catalog No.A1199

Lasker, B.M. (1993) Photographic Surveys and their Digitization, in *Workshop on Databases for Galactic Structure*, eds. A.G.D.Philip, B.Hauck, & A.R.Upgren, L.Davis Press, Schenectady, NY, p.77

Ma, C. & Feissel, M. (1997) Definition and Realization of the International Celestial Reference System by VLBI Astrometry of Extragalactic Objects, *IERS Technical Note* **23**

Monet, D. 1996, The A1.0 and SA1.0 catalogs on CD-ROM, U.S. Naval Observatory

Murray, C.A. & Corben, P.M. (1979) MNRAS **187**, 723

Richmod, M. (1993) The Sloan Digital Sky Survey, in *Workshop on Databases for Galactic Structure*, eds. A.G.D.Philip, B.Hauck, & A.R.Upgren, L.Davis Press, Schenectady, NY, p.207

Röser, S., Morrison, J., Bucciarelli, B., Lasker, B. & McLean, B. (1997) Contents, test results, and data availability for GSC 1.2, in proceedings IAU Symposium No.179, Baltimore, USA, *in press*

Stone, R.C., Monet, D.G., Monet, A.K.B., Walker, R.L., Ables, H.D., Bird, A.R., & Harris, F.H. (1996) AJ, **111**, 1721

Urban, S.E., Corbin, T.E., Wycoff, G.L., Martin, J.C., Jackson, E.S., Zacharias, M.I., & Hall, D.M. (1998) The AC 2000: The astrographic Catalogue on the System Defined by the Hipparcos Catalogue, *in preparation*

Zacharias, N., de Vegt, C., Winter, L., & Weneit, W. (1994) Some results of CCD-camera based astrometric plate measurements, Proceedings of the IAU Sympos. 161, p.285 Eds.: MacGillivray et al., Kluwer, Dordrecht
Zacharias, N. (1997) Astrometric Quality of the USNO CCD Astrograph (UCA), AJ, **113**, 1925

EXTENSION OF THE OPTICAL REFERENCE FRAME:SPACE BASED

S. RÖSER
Astronomisches Rechen-Institut, s19@ix.urz.uni-heidelberg.de
Mönchhofstr. 12-14, D-69120 Heidelberg, Germany

1. Introduction

At its 23rd General Assembly, Kyoto 1997, the IAU decided that the International Celestial Reference Frame (ICRF) should be defined by the positions of 608 compact extragalactic radio sources (Ma & Feissel 1997), with the Hipparcos Catalogue (ESA 1997) being its practical realization at optical wavelengths. Although the Hipparcos instrument was unable to observe the defining sources of the ICRF directly in the optical regime, it is linked to the ICRF via a number of auxiliary observations (Kovalevsky *et al.* 1997). The Hipparcos Catalogue is the first celestial reference frame which is constructed from observations not made from the surface of the Earth.

Hipparcos represents a major break-through in astrometry at the end of this century and it increases the number of stars in the reference system by a factor of 30 compared to its predecessor, the FK5 (Fricke *et al.* 1988). However, the 120 000 reference stars of Hipparcos are mostly in the range between the 7th and 9th magnitudes. This fact asks for extension to fainter magnitudes and more objects. The short observational period of Hipparcos determines the accuracy with which proper motions could be determined. At its mean epoch, 1991.25, the Hipparcos frame coincides with the ICRF to better than 0.6 mas in all three axes, and deviates from inertiality by not more than 0.25 mas/yr in all three axes (ESA 1997). Mainly, because of the proper motions' accuracy, the coincidence with ICRF will be not better than 5 mas in 2010. This asks for maintenance and improvement of the Hipparcos system.

2. Space missions, overview

Hipparcos has proven that space-based astrometry is, in many aspects, superior to measurements from ground. Consequently, Hipparcos is not the end of the road, but rather the beginning of a new and exciting branch of astrometry. One of the important features of Hipparcos is one-dimensional observations on a great circle using the closure conditions after each revolution. It is therefore quite obvious to use the improved resolution and accuray of an interferometer in space for further progress. And it is for this reason that all the proposed new space missions select a Fizeau interferometer as the instrument.

GAIA (Global Astrometric Interferometer for Astrophysics) (Lindegren & Perryman 1996) is the most ambitious of all the new missions. According to latest estimates it should be able to observe 10^9 objects brighter than V=20 and is aiming at an accuracy of 10 μas in position and parallax and 20 μas/yr in proper motion at V=15. GAIA will be able to directly observe the defining sources of ICRF – mostly 18th magnitude – which will limit the deviation from an inertial frame to less than 10 μas/yr. GAIA is proposed as an ESA mission with a possible launch in 2009 and final results in 2018 at the earliest.

FAME (Fizeau Astrometric Mapping Explorer) (Seidelmann *et al.* 1996) is under study at the USNO and CfA, Cambridge, USA. It is aiming at an accuracy of 50 μas at V=10. A possible launch date for FAME could be in the first half of the next decade.

LIGHT (Light Interferometer satellite for the studies of Galactic Halo Tracers) (Yoshizawa *et al.* 1997) is proposed by NAO (Japan) and others. It will have an interferometer onboard operating in visual and an infrared channels (at 2.4 μm). Detailed performance analysis for LIGHT is not available at the time of writing: the reader is referred to Yoshizawa *et al.* 1997. Launch of LIGHT could be between 2007 and 2010.

J. Andersen (ed.), Highlights of Astronomy, Volume 11A, 304–306.

DIVA (Double Interferometer for Visual Astrometry) (Röser *et al.* 1997a), see also Röser *et al.* 1997b in this volume, is a mission proposed to the German Aerospace Center (DLR). It has a good chance of being selected as the next German small-satellite mission with a launch in 2003.

3. The DIVA mission

DIVA (see Fig. 1) is a Fizeau interferometer on a small satellite of about 1 m diameter and a mass of 100 kg. It has two small interferometer apertures of $5\times5\,\mathrm{cm}^2$ each, separated by a baselength of 10 cm. A beam combiner allows quasi-simultaneous observations of two lines of sight separated by 60°. The limiting magnitude of DIVA is about V=15 for spectral types earlier than M0, but drops to about V=17.5 for stars later than M5. At least 4 million stars will be measured, giving a complete sky survey down to V=12.5. After two years of mission the astrometric accuracy of 10th magnitude stars will be 0.35 mas for positions and parallaxes, and 0.5 mas/yr for proper motions.

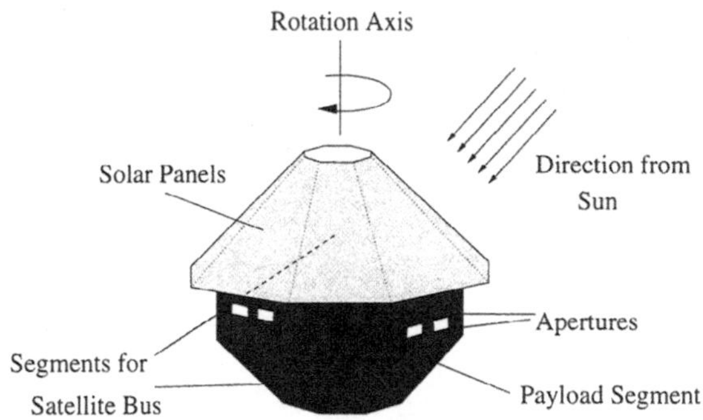

Figure 1. A schematic view of the DIVA spacecraft.

DIVA will be able to observe a few QSOs and AGNs down to its limiting magnitude. Estimates for 3C273 show that an accuracy of 2.6 mas for the position and 3.6 mas/yr for the proper motion are to be expected for this quasar. So, the link of the DIVA system to ICRS must be performed in a similar manner to Hipparcos, with most emphasis to the VLBI observations. Supposing that at least the same number of radio stars as with Hipparcos will be observed, and considering the improved accuracy of the DIVA observations, a coincidence with ICRF of ±0.3 mas can be expected for the mean epoch of DIVA observations around mid-2004. This, together with the link of Hipparcos to ICRF, yields a new representation of the ICRF in the optical regime which does not deviate from inertial by more than $50\,\mu$as/yr.

In summary, we can state, that a small, cheap and fast mission like DIVA will be sufficient to improve the Hipparcos system and extend it to fainter and 40 times more stars. This improved system can be available from 2007 onwards, provided that the implementation of the DIVA mission can be accomplished as planned.

References

ESA (1997) *The Hipparcos and Tycho Catalogues*. European Space Agency, **SP–1200**

Fricke, W., Schwan H., Lederle, T. *et al.* (1988). Fifth Fundamental Catalogue (FK5). Part I. The Basic Fundamental Stars. *Veröff. Astron. Rechen-Institut Heidelberg*, Nr.32.

Kovalevsky, J. *et al.* (1997) The Hipparcos Catalogue as a realisation of the extragalactic reference system, *A&A* **323**, 620

Lindegren, L. & Perryman, M.A.C. (1997) GAIA: Global Astrometric Interferometer for Astrophysics, *A&AS* **116**, 579

Ma, C. & Feissel, M. (1997) Definition and Realization of the International Celestial Reference System by VLBI Astrometry of Extragalactic Objects, *IERS Technical Note* **23**

Röser, S., Bastian, U., de Boer, K.S., Høg, E., Röser, H.P, Schalinski, C., Schilbach, E., de Vegt, Ch., Wagner, S. (1997a) DIVA - Towards Microarcsecond Global Astrometry, *ESA-SP-402*, (in press)

Röser, S., Bastian, U., de Boer, K.S., Høg, E., Schilbach, E., de Vegt, Ch., Wagner, S. (1997b) DIVA - a small satellite for global astrometry and photometry, *(this volume)*.

Seidelmann, P.K. *et al.* (1996) A Fizeau Optical Interferometer Astrometric Satellite, in: Perryman, M.A.C. & van Leeuwen, F. (eds.), *ESA-SP-379*, 187
Yoshizawa, M., Sato, K., Nishikawa, J., Fukushima, T., Miyamoto, M. (1997) Optical/Infrared Astrometric Satellite Project LIGHT, *ESA-SP-402*, (in press)

LINK OF THE HIPPARCOS CATALOGUE TO THE ICRS

J. KOVALEVSKY
OCA/CERGA
UMR CNRS 6527
Av. Copernic, 06130–Grasse (France)

Abstract. In compliance with the 1991 IAU resolutions on reference frames, the merged Hipparcos Catalogue had to be aligned in such a way that it represents at optical wavelengths the International Celestial Reference System (ICRS). This was realized as a combination of particular solutions obtained from observations of radio stars by radio interferometry, observation of quasars relative to Hipparcos stars, photographic plates including galaxies and Hipparcos stars, star catalogues referred to extragalactic objects, HST observations, and comparison of Earth orientation parameters obtained by VLBI and ground–based optical observations reduced with Hipparcos positions and proper motions. The individual results generally agree within 10 mas in orientation at epoch and 1 mas/yr in spin of the system. Two methods were used to combine the results and yielded uncertainties of 0.6 mas in orientation and 0.25 mas/yr in spin.

1. Introduction

A set of recommendations within IAU resolution A4 of 1991 (Bergeron, 1992) on celestial reference systems and frames called for the definition of a celestial reference system that would have no global rotation with respect to a set of distant extragalactic objects. This system, called the International Celestial Reference System (ICRS, see Arias et al., 1995) was recognized by the IAU Working Group on Reference Frames and materialized in the form of a catalogue of positions of 608 objects (Ma and Feissel, 1997) adopted by the IAU as the International Celestial Reference Frame (ICRF) during this General Assembly. This catalogue was prepared in time to be available when the merged Hipparcos Catalogue was completed as a combination of the two catalogues prepared by the two data reduction consortia (FAST, see Kovalevsky et al., 1992 and NDAC, see Lindegren et al., 1992). The merging procedure is described by Arenou (1997) and the provisional astrometric parameters obtained were made available to all participants of the preparation of the link. The ensemble of the work done in this respect is described in Kovalevsky et al., 1997a.

2. Methods and observations

Let the extragalactic reference frame ICRF be represented by the triad of unit vectors $E = (\mathbf{x}_E, \mathbf{y}_E, \mathbf{z}_E)$ and the Hipparcos provisional reference frame by $H = (\mathbf{x}_H, \mathbf{y}_H, \mathbf{z}_H)$. The relation between them can be represented by a time dependent vector

$$\mathbf{V}(t) = \varepsilon + (t - t_0)\omega$$

where ε is the rotation vector at the mean epoch of the Hipparcos Catalogue, $t_o = 1991.25$ and called here *orientation*. While ω is a vector representing the yearly variation, called here *spin*.

The principle of the determination of either of these vectors was first given by Froeschlé & Kovalevsky (1982) for most of the possible cases. It was applied on eleven different sets of data:

J. Andersen (ed.), Highlights of Astronomy, Volume 11A, 307–309.

1. VLBI observations of 12 radio-emitting stars by several networks (Lestrade et al., 1995) provided both ε and ω with uncertainties of the order of 0.5 mas for the components of ε and 0.3 mas/yr for those of ω.
2. Connected radio interferometer MERLIN (Morrison et al., 1997) observed 13 radio stars, but with insufficient spread in time to get reliable results in ω. So, only the components of ε were given within 2.6 mas standard error.
3. VLA observations, mostly done around 1983-86 (Florkowski et al., 1985) and with some additional data in 1994-95, gave standard uncertainties of the order of 5 mas on the components of ε.
4. Optical positions of compact sources derived from CCD observations of 78 selected sources (Johnson et al., 1995 and Zacharias et al., 1995) gave an overall precision of the same order of magnitude as 3 above.
5. About 80 observations of separations of 46 Hipparcos stars to extragalactic objects were made on the Hubble Space Telescope. These observations were insufficient in number, so that the uncertainties (5 mas in ε and 2 mas/yr in ω) were rather disappointing.
6. The Lick proper motion survey referred to galaxies allowed the determination of ω. Two solutions were obtained respectively by I. Platais and his co-workers and S. Röser. The results for ω were scattered, apparently due to strong magnitude errors.
7. A similar work on the results of the Yale-San Juan Southern Proper Motion Program led to values that were about 0.5 mas/yr from the mean (Platais et al., 1997), a very satisfactory result, even if the internal errors have been very underestimated.
8. A similar overall precision on ω was obtained with the catalogue of faint stars KSZ (Kislyuk et al., 1997) using 415 stars from 154 areas on the sky.
9. A special programme of determination of proper motions from widely separated in time photographic plates (70 to 100 years) using 88 Hipparcos stars in 13 fields (Tucholke et al., 1997) yielded actual uncertainties of the order of 0.5 mas/yr in ω.
10. A similar programme based on 24 fields taken at the Tautenburg Schmidt Telescope with time intervals of 20 to 40 years gave better results and better uncertainties (Hirte et al., 1997).
11. Finally, a comparison was made between the Earth rotation parameters determined with VLBI in the ICRS system and those obtained from optical observations rediscussed with Hipparcos data. Many uncertainties are present in this method that could not be evaluated and whilst the results obtained are rather good in x and y components of ω, the results in ε were not satisfactory (Vondrák et al., 1997).

3. Combination of results

The eight results for ω and six results for ε were discussed as described in Lindegren and Kovalevsky (1995) and in Kovalevsky et al., 1997a. Two different methods were used, either separating ε and ω which are only slightly correlated, or in treating all unknowns together. Actually, both methods differed also slightly in the weights attributed to each result.

The solutions obtained in ε differed by less than 0.05 mas in x and z components and 0.16 in the y component. Similarly, the solutions for ω differed by less than 0.08 mas/yr in x and z and 0.13 mas/yr in y. The weighted standard errors were also quite similar. This led to the adoption of values for ε and ω with respectively standard uncertainties of $\pm$ 0.60 mas in all components of ε and $\pm$ 0.25 mas/yr in all the three components of ω. These values are to be compared with $\pm$ 0.50 mas and $\pm$ 0.15 mas/yr as mean values obtained by both solutions for the standard errors. So, a rather large security factor has been retained. In this manner, the interval of confidence includes the spread of values obtained by various solutions with different choices of weights.

In addition, an external check was obtained using a statistical model of stellar motion (Kovalevsky et al., 1997b) for a determination of the components of ω in the galactic plane. The results obtained depend on distance, galactic latitude and colour, but they are all compatible within their own errors with the uncertainty in ω stated above.

So, in conclusion, the rotation components that resulted from this work and that were applied to the preliminary merged catalogue to obtain the final Hipparcos Catalogue and the associated uncertainties appear to be consistent and represent what was achievable.

References

Arenou, F., 1997, The Hipparcos and Tycho Catalogues, ESA Publ. SP-1200, **3**, 369–385

Arias, E.F., Charlot, P., Feissel, M. and Lestrade, J-F., 1995, A&A, **303**, 604–608

Bergeron, J., 1992, *Proceedings of the twenty first General Assembly of the IAU*, Kluwer Acad. Publ., Dordrecht, 41–64.

Froeschlé, M. and Kovalevsky, J., 1982, A&A, **116**, 89–96

Hirte, S., Schilbach, E. and Scholz, R-D., 1997, A&A Suppl., in press

Johnston, K.J., Fey, A.L., Zacharias, N. et al., 1995, Astron. J., **110**, 880-915

Kislyuk, V.S., Rybka, S.P., Yatsenko, A.I. and Kharchenko, N.V., 1997, A&A, **321**, 660–664

Kovalevsky, J., Falin, J-L., Pieplu, J-L. et al., 1992, A&A, **258**, 7–17

Kovalevsky, J., Lindegren, L., Perryman, M.A.C. et al., 1997a, A&A, **323**, 620–633

Kovalevsky, J., Lindegren, L. and Perryman, M.A.C., 1997b, *The Hipparcos and Tycho Catalogues*, ESA Publ. SP-1200, **3**, 410 and foll.

Lestrade, J-L., Jones, D.L., Preston, R.A. et al., 1995, A&A, **304**, 182–188

Lindegren, L., Hoeg, E., van Leeuwen, F. et al., 1992, A&A, **258**, 18-30

Lindegren, L. and Kovalevsky, J., 1995, A&A, **304**, 189–201

Ma, C. and Feissel, M., 1997, IERS Technical Note 23, Observatoire de Paris

Morrison, L.V., Garrington, S.T., Argyle, R.W. and Davis, R.J., 1997, *Hipparcos Symposium*, Venice, ESA Publ. SP-402, 143–146

Platais, I., Kozhurina-Platais, V., Girard, T.M. et al., 1997, *Hipparcos Symposium*, Venice, ESA Publ. SP-402, 153–156

Tucholke, H-J., Brosche, P. and Odenkirchen, M., 1997, A&A Suppl., **122**, 433-437

Vondrák, J., Ron, C. and Pešek, I., 1997, A&A, **319**, 1020–1024

Zacharias, N., de Vegt, C., Winter, L. and Johnston, K.J., 1995, Astron. J., **110**, 3093–3106

LINKING THE DYNAMICAL REFERENCE FRAME TO THE ICRF

E.M. STANDISH
Caltech/JPL
JPL 301-150; Pasadena, CA 91109 USA
ems@smyles.jpl.nasa.gov

Abstract.

The latest JPL planetary and lunar ephemerides, DE405, are referenced to the International Celestial Reference Frame (ICRF) with an accuracy that approaches 1 *mas* for the four innermost planets, the sun, and the moon. This has been accomplished mainly by 18 VLBI observations of the Magellan Spacecraft in orbit around Venus. The ephemeris of Jupiter, however, is not well-determined since the various observations are not consistent within each other. The outer four planets continue to rely almost entirely upon optical observations; their ephemeris uncertainties lie in the 100–200 *mas* range.

1. Introduction

The JPL Planetary and Lunar Ephemerides are now numbered in the 400's, indicating that they are referenced to the ICRF. This reference system has a number of advantages for the ephemerides: 1) the most accurate angular observations are now VLBI measurements of spacecraft near planets with respect to the ICRF, 2) there is also an accurate ephemeris–ICRF frame-tie determination, 3) the ICRF itself is accurate ($< 0\overset{''}{.}001$), 4) the ICRF is stable (there are to be no future net rotations), and 5) timing and polar motion information, distributed by the IERS, is now referenced to the ICRF frame. The only disadvantage, a relatively minor one, is that most of the angular measurements fit by the ephemerides are with respect to the optical reference system, so that an optical–ICRF frame-tie must be applied.

This paper shows the newer ephemeris observations which have enabled the adjustment onto the ICRF. The inner four planets, sun and moon are discussed first, since that system is now well determined in all aspects: relative positions, relative distances, mean motions, and orientation. Then, Jupiter is discussed; there are some remaining inconsistencies between the varied observations which have yet to be resolved. The outer four planets are mentioned at the end.

2. The 4 Inner Planets, Sun and Moon : relative angles, distances, and inertial mean motions

Ranging measurements determine accurately not only the relative angles and distances between the measured planets, but also the bodies' mean motions with respect to inertial space. This has been discussed elsewhere (see e.g., Standish & Williams, 1990). The result is that for the sun, moon, and inner four planets, the relative angles are known well below 1 *mas* and the distances are known below the 1 km level. The inertial mean motions are known to an accuracy of about 20 *mas/cty*; this relatively large uncertainty come from the uncertainties in the masses of the asteroids which perturb the inner solar system, as discussed by Williams (1984).

Ranging, however, will not determine the orientation of the ephemerides with respect to an outside reference frame.

J. Andersen (ed.), Highlights of Astronomy, Volume 11A, 310–312.

3. Orientation onto the ICRF

There are two sets of data which align the ephemerides onto the ICRF:

• The frame-tie determination of (Folkner *et al.*, 1993): The radio sources of the ICRF are connected to the radio antennas via VLBI observations; those antennas are connected to LLR telescopes via geodetic ground surveys; the LLR telescopes are connected to the moon via LLR observations; the sun and planets are connected to the moon through the lunar dynamics. This set of connections is used as an observation of the sun and planets, taken wrt the ICRF. The estimated uncertainty is about 3 *mas*.

• VLBI observations of spacecraft either approaching or orbiting a planet: The VLBI observations w.r.t. the ICRF radio sources are combined with the spacecraft ephemeris to give the planet in the radio frame. These observations are almost purely one-dimensional, measuring along the baseline between the two antennas. They are almost entirely in right ascension when taken between Goldstone and Madrid since those sites are at comparable latitudes. Of these, there were 6 observations of Magellan orbiting Venus, all in 1994, with uncertainties from 1.8 to 4.0 *mas*. There are also 12 observations of Magellan taken between Goldstone and Canberra from 1990 to 1994 with uncertainties from 1.1 to 3.2 *mas*; they split nearly 50-50 between right ascension and declination. Two observations of the Phobos spacecraft, approaching Mars in 1989, have uncertainties of 3.3 and 6.2 *mas*.

The above observations serve to orient the whole inner system of DE405 onto the ICRF with an accuracy of about 1 *mas*. A confirmation of this estimate came in July 1997 with the arrival of the Pathfinder spacecraft at Mars. The ephemeris error, determined by JPL's navigation, was about 1.1 km. With Mars at a distance of 1.3 AU, this down-track error is the equivalent of 1.2 *mas*.

4. Problems with Jupiter

The situation with the ephemeris of Jupiter is less satisfying. There is a wide variety of observations of Jupiter, but they are not all consistent with each other. Figures 1-3 show the residuals vs. DE405. Transit observations of Ganymede and Callisto taken at La Palma, are represented by stars, each star showing the mean of the residuals taken over an opposition of Jupiter; the open boxes show the opposition means of transit observations of the limbs of Jupiter, taken from Washington.

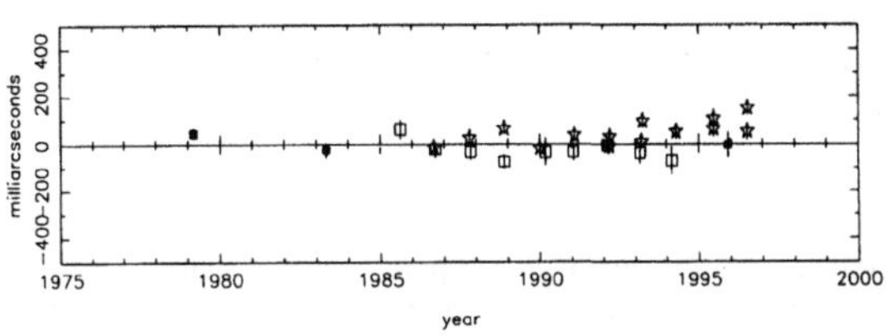

Figure 1. Right Ascension residuals of Jupiter

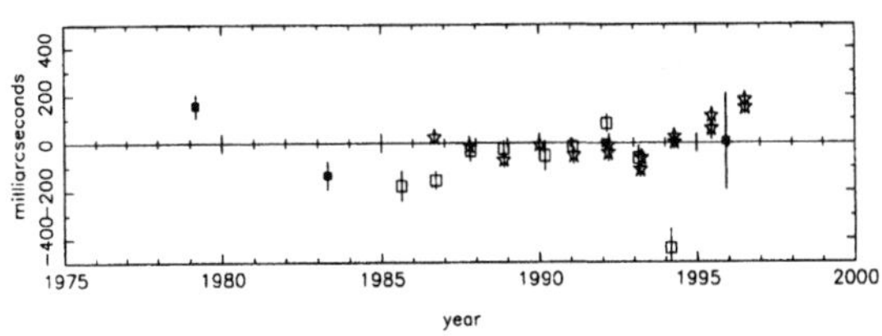

Figure 2. Declination residuals of Jupiter

The Voyager 1 point, shown as a filled square in 1979, is from JPL Navigation's Orbit Determination Program's fitting to the standard tracking data, range and doppler; the VLA observation, filled square in 1983, measured the thermoelectric emission of Jupiter w.r.t. the radio source background; the Ulysses point, filled square in 1992, is from an ODP fit to range, doppler, and VLBI; and the Galileo ODP, filled square in the end of 1995, is a fit to range and doppler during approach.

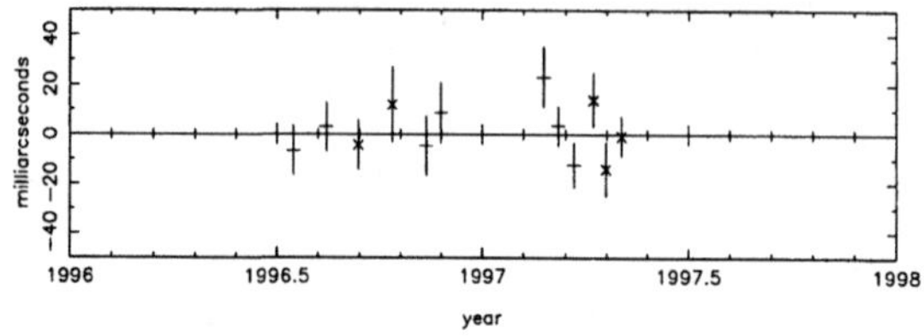

Figure 3. VLBI residuals of Jupiter

Figure 3 shows the Galileo VLBI, taken while in orbit around Jupiter; the ×'s mark the Goldstone–Canberra observations and the –'s are from Goldstone-Madrid. (Note the difference in scales between the figures.) These VLBI observations anchor the ephemeris firmly in 1996-97.

One can see from the figures that while the right ascension residuals seem to be fairly well-behaved, except for a slight positive trend from La Palma, the declination residuals show a more serious problem. Both the Voyager 1 and VLA points are nearly 3σ away from zero and there is a definite positive trend in the La Palma opposition means. Since the 1996-1997 VLBI points are so dominant, there is no chance of flattening the La Palma points. Further, there is no explanation for the poorly fitting Voyager 1 and VLA residuals. If one were to increase the weight of these two observations during the fitting process, the La Palma points would be even further away from the axis. The uncertainty in the Jupiter ephemeris is therefore estimated as a tentative 50 *mas* during the latter half of the 1990's; it could be worse at other times.

5. The outer four planets

The ephemerides of the outer four planets continue to be markedly less well-determined than those of the inner planets. The orbital periods are long; they cover decades, during which observational techniques have changed and during which the planets move through different zones of the optical catalogues. Not surprisingly, the ephemerides of Saturn through Pluto have uncertainties in the 100–200 *mas* range.

6. Summary

The inner four planets, sun, and moon have ephemerides which are well-determined with respect to the ICRF at the 1 *mas* level. Jupiter's ephemeris is fit to a variety of observations, but problems still exist; the uncertainty is estimated at 50 *mas* during the present few years, but probably worse at other times. The ephemerides of Saturn through Pluto have uncertainties in the 100–200 *mas* range.

References

Folkner,W.M., Charlot,P., Finger,M.H., Williams,J.G., Sovers,O.J., Newhall,X X, and Standish,E.M.: 1993, "Determination of the extragalactic frame tie from joint analysis of radio interferometric and lunar laser ranging measurements", *Astron. Astrophys.*, **287**, 279–289.

Standish,E.M. & Williams,J.G. : 1990, "Dynamical Reference Frame in the Planetary and Earth-Moon Systems", in "Inertial Coordinate System on the Sky" (J.H.Lieske and V.K.Abalakin, eds.) Kluwer Academic Publishers, Dordrecht, 173-181.

Williams,J.G.: 1984, "Determining Asteroid Masses from Perturbations on Mars", Icarus 57, 1-13.

LINKING THE FK5 TO THE ICRF

F. MIGNARD & M. FROESCHLE
OCA/CERGA
UMR CNRS 6527
Av. Copernic, 06130-Grasse (France)

Abstract. The Hipparcos optical reference frame is compared to the basic FK5 in order to determine the orientation at $T_0 = 1991.25$ and the global spin between the two frames. The components of the spin are significant and suggest a correction the IAU76 value of the precession constant and to a possible non-precessional motion of the equinox of the FK5. The regional errors are analysed with harmonic functions and found to be as large as 150 mas in position and 3 mas/yr in proper motion.

1. The FK5 catalogue

The FK5 Catalogue has provided until the advent of the ICRF a practical realization of a dynamical frame through the position and proper motion of carefully measured stars. It represents a revision of the FK4 and results from the determination of systematic and individual corrections to the mean positions and proper motions of the FK4, the elimination of the error in the position and motion of the FK4 equinox, and the introduction of the IAU (1976) system of astronomical constants (Fricke et al., 1988). Fricke (1982) has derived the following correction in mas,

$$E(T) = (525 \pm 45) + (12.75 \pm 1.5)(t - 1950.0) \ ,$$

to the zero point of the right ascension. One must bear in mind that any error in the above rate propagates directly into a systematic effect in the system of proper motion and into a defect of inertiality.

The expected positional accuracy of the 1535 bright stars is 30 mas at the mean epoch of the catalogue and 0.6 mas/yr for the proper motion of the Northern hemisphere and 1.0 mas/yr for the Southern. By projecting the FK5 positions to the Hipparcos epoch J1991.25, this leads to an expected error in right ascension and declination of 40 to 60 mas, according to the hemisphere.

The 1535 stars of the FK5 have been observed successfully by Hipparcos and their positions are known in 1991.25 with an accuracy of typically 0.4 ± 0.1 mas in declination and 0.6 ± 0.2 mas in right ascension, with in the latter case a marked dependence with the declination. The corresponding figures for the proper motions are 0.7 ± 0.2 mas/yr and 0.55 ± 0.15 mas/yr.

The Hipparcos Catalogue has been constructed in such a way that the reference frame materialized by the positions coincides with the International Celestial Reference System, within the measurement errors. The Hipparcos positions and proper motions provide a realisation of an optical non-rotating reference frame with axes pointing to fixed extragalactic directions to within 0.6 mas for the alignment and 0.25 mas/yr in the rate of rotation, resulting in a similar uncertainty in the intertiality of the system of proper motion (ESA 1997). In terms of consistency, the optical reference frame defined by the Hipparcos positions and proper motions is believed to have a global accuracy of about 0.1 mas and 0.1 mas/yr, with no regional distortion.

2. The comparison with Hipparcos

All the multiple stars and objects with questionable solutions in Hipparcos have been excluded from the analysis and the selection has ended up with 1233 reliable solutions. The reason for excluding

J. Andersen (ed.), Highlights of Astronomy, Volume 11A, 313–316.

the double stars and the astrometric binaries, is not in general the quality of the solution, but the fact that the Hipparcos and FK5 astrometric solutions may correspond to two different points on the sky and that the Hipparcos proper motion determined over a timespan shorter than the orbital period does not represent the systematic motion of the barycenter. The comparison is performed at $T_0 = 1991.25$, and for each star one has computed the positional differences $(\alpha_F - \alpha_H)\cos\delta$ and $\delta_F - \delta_H$ and the proper motion differences, $(\mu_{\alpha*})_F - (\mu_{\alpha*})_H$, $(\mu_\delta)_F - (\mu_\delta)_H$.

2.1. GLOBAL ROTATION

These differences have been projected on a set of orthogonal vectorial harmonics independently for the positional and proper motion differences. The first degree of these harmonics represents the pure rotation, while the harmonics of higher degree account for the zonal differences at decreasing wavelengths with increasing degree. The value of the three angles of the global orientation and of the components of the spin vector are given in Table 1 with their respective uncertainties. The orientation is given at the Hipparcos epoch J1991.25 and at J2000. The standard errors are similar in both cases and there is no need to consider the systematic difference of the order of 0.6 mas between the Hipparcos frame and the ICRS. The sign conventions are such that the rotation represents the orientation of the Hipparcos triad with respect to the FK5 triad or equivalently the three consecutive rotations to align the FK5 triad to the Hipparcos axes.

TABLE 1. *Global orientation and spin between the Hipparcos and FK5 Catalogues at J1991.25.*

	Orientation (mas)				Spin(mas/yr)		
	J1991.25	J2000	σ			σ_{HIP}	σ_{ICRF}
ϵ_x:	−17.3	−19.9	2.3	ω_x:	−0.30	0.10	0.27
ϵ_y:	−14.3	− 9.1	2.3	ω_y:	+0.60	0.10	0.27
ϵ_z:	+22.9	+16.8	2.3	ω_z:	+0.70	0.10	0.27

The global rotation has no particular physical meaning, giving the relative orientation of the FK5 frame with respect to the ICRF, insofar as it can be represented by the Hipparcos Catalogue. The three angles of rotation are consistent with the IAU recommendations stipulating that the direction of the Conventional Celestial Pole relative to that of the FK5 should be within 50 mas (Arias et al., 1995). The uncertainty of the orientation with respect to the Hipparcos frame or to the ICRF are similar for the orientation.

The components of the spin are linked to the lack of inertiality of the FK5 frame with respect to the non–rotating extragalactic frame. The uncertainty with respect to Hipparcos is the formal result of the fit, while we have taken into account the uncertainty of the link of 0.25 mas/yr to determine the true default of inertiality of the FK5 proper motion system. It was combined quadratically, although the difference between the Hipparcos frame and the ICRS should be considered as a systematic bias. Various levels of filtering have been tried to determine this rotation, which proved very stable against the different selections. Comparisons with the bright end of the PPM Catalogue has yielded comparable values for the spin, irrespective of whether the FK5 stars were removed or not. The rotation between the two Catalogues is thus real and can be traced back to the construction of the FK5. We suggest a tentative interpretation based on a correction to the precession constant, Δp, and a combination of a non–precessional motion of the equinox of the FK5, ΔE, and a correction, $\Delta\lambda$, to the planetary precession, although the latter is likely to be negligible. With,

$$\begin{aligned} \omega_y &= -\Delta p \sin\epsilon \\ \omega_z &= \Delta p \cos\epsilon - (\Delta\lambda + \Delta E) \end{aligned}$$

which gives,

$$\begin{aligned} \Delta P &= -1.5 \pm 0.7 \text{ mas/yr} \\ \Delta\lambda + \Delta E &= -2.7 \pm 0.7 \text{ mas/yr.} \end{aligned}$$

The correction to the precession constant is not compatible with the recent determinations based on the VLBI or LLR which are both close to −3 mas/yr. On the other hand, the correction to the non–precessional motion of the equinox is in the range of the uncertainty of ±1.5 mas/yr obtained by Fricke.

2.2. ZONAL DIFFERENCES

The regional differences are shown in Fig. 1 for the positions at 1991.25 and in Fig. 2 for the components of the proper motion. The magnitude of the systematic differences are much larger than the expected accuracy of the FK5, which strictly speaking applies to small fields. Large systematic differences are conspicuous around the South celestial pole and more generally more significant in the Southern hemisphere than in the Northern. The two plots have a rather similar appearance, indicating that most of the zonal errors in the positions come from the propagation to 1991.25 of the FK5 proper motions. A correction table will be made available and endorsed by the newly appointed Working Group on Reference Systems.

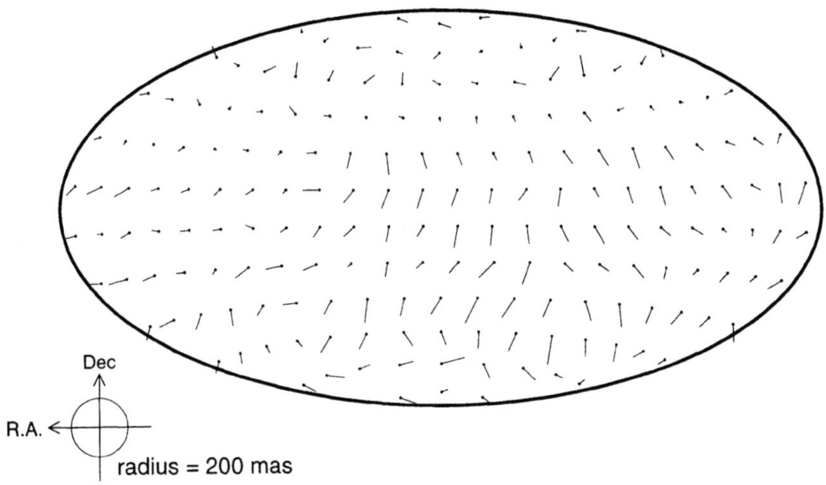

Figure 1. Differences FK5 - HIP for the positions at 1991.25. Right ascension runs right to left from 0 to 360 deg.

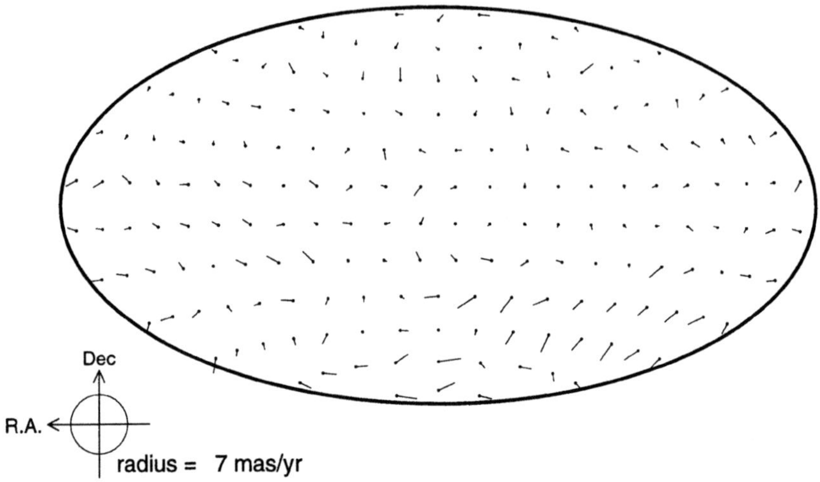

Figure 2. Differences FK5 - HIP for the proper motions. Right ascension runs right to left from 0 to 360 deg.

References

Arias, E.F., Charlot, P., Feissel, M., Lestrade, J.F., 1995, A&A, 303, 604-608.
ESA, 1997, The Hipparcos and Tycho Catalogues, SP-1200, Vols. 1–4.
Fricke, W., 1982, A&A, 107, L13-L16.
Fricke, W., Schwan, H., Lederle, T., 1988, Veröf. Astr. Rechen-Institut, No. 32.

MAINTENANCE OF THE ICRF: RADIO

ALAN L. FEY
U.S. Naval Observatory
Washington, DC 20392–5420 USA

1. Introduction

Extragalactic radio sources are assumed to be very distant and thus should exhibit little or no detectable proper motions. A reference frame defined by the positions of extragalactic radio sources may be said to be a quasi–inertial frame (i.e. a frame whose basis is inertial) with little or no time dependency. Unfortunately, although extragalactic sources are good as ICRF objects, most extragalactic sources display intrinsic structure on angular scales larger than the accuracy of their position estimates. Temporal variations of the intrinsic structure of these objects will result in apparent motion when observations are made at several epochs. Because the underlying physics of extragalactic sources is not as well understood as that of the stars which define the stellar reference frame, we can only describe with certainty what the radio sources did during the particular interval of time covered by previous observations. We cannot predict theoretically what behavior can be expected in the future. It is therefore necessary to regularly measure the structure of ICRF sources, and monitor their changes, in order to maintain the ICRF and make it more useful to astronomers.

2. Frame maintenance and improvement

2.1. CONTINUED VLBI OBSERVING AND ANALYSIS

The National Earth Orientation Service (NEOS)[1] has been appointed by the International Earth Rotation Service (IERS) as the VLBI Coordinating Center charged with observations, analysis, and other related activities toward the maintenance of the ICRF.

NEOS is committed to making the observations required to maintain the ICRF. Observations will be made using the NEOS array of telescopes, including both regular NEOS observing for Earth orientation parameters estimation and special sessions organized specifically for reference frame maintenance. Observations for the new Continuous Observations of the Rotation of the Earth (CORE) program, when operational, will be included as well. The Very Long Baseline Array (VLBA) telescope of the National Radio Astronomy Observatory (NRAO) will also continue to be used for ICRF maintenance, at least down to its southern declination limit.

2.2. ACCURACY OF THE ICRF

Factors limiting the accuracy of the ICRF include tropospheric modeling, the paucity of data on some sources (particularly in the southern hemisphere), and intrinsic source structure. Considerable progress has been made toward understanding atmospheric effects by estimating gradients in the troposphere directly from the VLBI data. The resulting effect is a systematic shift of up to about 0.5 mas in source declinations from analyses with and without gradient estimates. The effect is much larger than the position formal errors, and is caused by the greater troposphere thickness nearer the Earth's equator (MacMillan & Ma 1997). While the effect is not large in absolute terms, it is systematic and would distort the celestial reference frame if ignored. Special southern hemisphere experiments involving NEOS antennas and/or selected VLBA antennas are being scheduled

[1]NEOS is a cooperative effort between the Goddard Space Flight Center (GSFC), the U.S. Naval Observatory (USNO), and the National Oceanographic and Atmospheric Administration (NOAA) and serves as the IERS Sub–bureau for Rapid Service and Predictions.

J. Andersen (ed.), Highlights of Astronomy, Volume 11A, 317–319.

to obtain additional data on poorly observed southern sources. The VLBA is also being used extensively in the northern hemisphere. Observations obtained with the VLBA accumulate quickly due to the larger number of antennas compared to the NEOS array. As observations accumulate, it should be possible to move candidate sources up or down the scale of usefulness. The remaining factor limiting the accuracy of the ICRF is intrinsic source structure and will be the focus óf the remainder of this paper.

2.3. INTRINSIC SOURCE STRUCTURE

Until recently, the intrinsic structure of the majority of the ICRF sources has been mostly unknown. The surveys of Fey, Clegg & Fomalont (1996) and Fey & Charlot (1997) show that most sources, when examined in detail, exhibit spatial structure on milliarcsecond scales. Their results show that the variation of intrinsic structure from source to source can be quite extreme, ranging from relatively compact naked–core objects, to compact double sources, to complex core–jet objects. The situation is exacerbated by the fact that compact extragalactic radio sources are known to have variable intensity and to have frequency– and time–dependent intrinsic structure. Consequently, unknown and/or unmodeled source structure effects may be introduced into the astrometric solution.

Charlot (1990) has modeled the effects of radio source structure on measured VLBI group delays and delay rates. Fey & Charlot (1997) calculated structure corrections based on the Charlot (1990) analysis using source models derived from VLBA observations of 169 extragalactic sources. Results of these calculations show that intrinsic structure contributions to the measured bandwidth synthesis delay are significant, ranging from maximum corrections of only a few picoseconds for the most compact sources to maximum corrections of several nanoseconds for the most extended sources. Fey & Charlot (1997) found a correlation between the compactness of the sources and their position formal uncertainties indicating that the more extended sources have larger position formal errors. They also define a source "structure index" based on the median of the calculated structure corrections. They suggest that this index can be used as an estimate of the astrometric quality of the sources.

As mentioned in §1, temporal variations of the intrinsic structure of extragalactic sources will result in apparent motion when observations are made at several epochs. The position variations for some sources show clear correlations with changes in intrinsic structure and, to some extent, the position changes can be derived from the structure, but there is no strong evidence of any regularly repeated behavior. Fig. 1 shows the time evolution of the right ascension of 0923+392 (4C 39.25). The astrometric time series agrees very well with the differential motion of the brightest image component measured relative to an assumed stationary image component in VLBI images. Essentially, the brightest component of this source has traversed a distance of ~ 2 mas in the span of about 17 yrs.

3. The Future

New VLBI observations of ICRF sources will be made and added to the current database. Sources will be monitored for variability and/or structural changes. The current realization of the ICRF condenses the information from a particular VLBI dataset spanning a defined interval of time and reflects a certain state of VLBI analysis. As new observations are added and the modeling is improved, we expect the realization of the ICRF to evolve, although changes in the ICRF catalogue will be infrequent compared to past practice in VLBI astrometry. The problem of position variation may be solved in the future if the application of source structure information permits the identification and use of truly kinematically stable points in the sky.

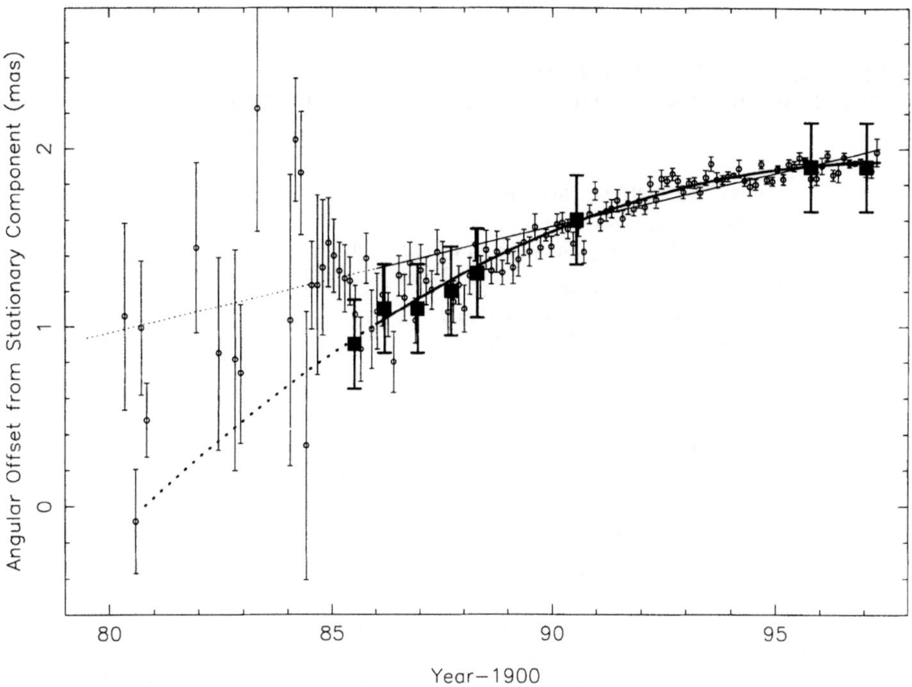

Figure 1. The astrometric right ascension $\alpha \cos \delta$ of 0923+392 (4C 39.25) offset from the position of an assumed stationary image component. Superposed are the angular separations in right ascension between the positions of the brightest image component at $\lambda = 3.6$ cm (large solid squares) and the position of the stationary image component for VLBI images from several epochs. Also shown are linear and quadratic weighted least-squares fits to the astrometric positions. The dashed portions of these curves represent extrapolations.

References

Charlot, P. (1990) Radio–Source Structure in Astrometric and Geodetic Very Long Baseline Interferometry, AJ, **99**, 1309–1326

Fey, A.L., Clegg, A.W. & Fomalont, E.B. (1996) VLBA Observations of Radio Reference Frame Sources. I., ApJS, **105**, 299–330

Fey, A.L. & Charlot, P. (1997) VLBA Observations of Radio Reference Frame Sources. II. Astrometric Suitability Based on Observed Structure, ApJS, **111**, 95–142

MacMillan, D.S. & Ma, C. (1997) Atmospheric gradients and the VLBI terrestrial and celestial reference frames, Geophys. R. Lett., **24**, 453–456

THE ACCURACY OF THE ICRF: AN INTERCOMPARISON OF VLBI ANALYSIS SOFTWARE

C.S. JACOBS AND O.J. SOVERS
Jet Propulsion Laboratory (JPL)
4800 Oak Grove Dr., Pasadena, CA 91109-8099

D. GORDON AND C. MA
Goddard Space Flight Center (GSFC)
Code 926, Greenbelt, MD, 20771

AND

A.-M. GONTIER
Observatoire de Paris
61 Avenue de l'Observatoire, F-75014 Paris France

1. Introduction

As discussed in other papers in this volume, the IAU XXIII General Assembly adopted a new fundamental celestial reference frame: the International Celestial Reference Frame (ICRF) based on VLBI observations of extragalactic radio sources (Ma *et al.*, 1997). It is approximately 300 times more accurate than its predecessor, the FK5. At present, no other technique has produced a more accurate celestial frame than VLBI. Since no other astrometric technique provides an external standard of accuracy, the VLBI claim of a great leap forward in accuracy must be verified by internal consistency tests. This paper addresses one aspect of internal consistency: the ability of independent VLBI software packages to reproduce a celestial frame without significant loss of accuracy. This is no small task since the software packages are large – involving on the order of 100 000 lines of code. What does VLBI software do? Aside from routines designed to collect the data and extract raw observables which will not be considered here, its principal task is to model the differential group delay and phase delay rate of radio signals received at two widely separated antennas (Sovers, Fanselow & Jacobs, 1998). The software then refines this model via a least squares adjustment of relevant physical parameters which describe station locations, source positions, clock offsets, atmospheric refraction, tidal effects, etc. In the early 1990s, studies revealed that differences in software implementation and analyst's choices of model options were one of the largest contributors to differences in independent calculations of VLBI celestial frames. These differences were of comparable size to the formal uncertainties of the celestial frame's source positions.

2. Comparisons of VLBI analysis software

In order to isolate the cause of these differences, we compared three software packages: JPL's MODEST (Sovers, Fanselow & Jacobs, 1998), GSFC's CALC/SOLVE (Ma, 1978) and Paris Observatory's GLORIA (Gontier, 1992). Because almost all available high quality astrometric VLBI data was already set up for analysis by GSFC, their software was chosen to produce the ICRF. Because of the complexity of the software, the first step taken was to compare the calculations of each VLBI model component from each software package. This comparison was limited to data taken on 27 March 1992 with Deep Space Network antennas in Goldstone, CA and Madrid, Spain, and a second experiment which included a southern hemisphere station: Tidbinbilla, Australia to Goldstone, CA on 22 March 1992. After considerable effort, we reached our goal of 1 ps and 1 fs/s *wrms* agreement in delay and delay rate, respectively (< 0.01 mas). The geometric component (station locations, source positions, relativity) of the delay models agreed to much better than 1 ps.

J. Andersen (ed.), Highlights of Astronomy, Volume 11A, 320–321.

The solid Earth tide models limited the accuracy to the stated levels. Antenna axis offset and troposphere gradient models were not tested.

3. Celestial frame comparisons and analyst choices

Once the model implementations were verified, we constructed a full VLBI reference frame using the GSFC and JPL software (time constraints prevented Paris Observatory from producing a frame). The initial comparison was based on a limited data set of just 11,600 delay/rate observation pairs taken between Aug. 1988 and Feb. 1993 using single baselines from JPL's Deep Space Network. This restricted data set allowed the GSFC and JPL analysts to carefully match *a priori* model parameters and modeling options. Analyst choices to remove outlier data points were not forced to agree *a priori*, but, in fact, editing choices differed for only 1% of the observations (mostly due to an error in the exchange of raw data). The 267 sources in the resulting frames agreed to 0.15 and 0.21 mas *wrms* in right ascension ($\alpha \cos \delta$) and declination, (δ), respectively. These differences were slightly smaller than the formal uncertainties of the individual frames and are thought to be due to differences in the number of parameters used by GSFC and JPL analysts to describe clock and troposphere behavior. There was insufficient time to match exactly the details of these parameterizations. Coordinate biases were 0.01 and 0.02 mas in $\alpha \cos \delta$ and δ, respectively. Rotational alignment was $\leq$0.1 mas in each of three orthogonal directions. Lastly, we made a frame comparison using a much more extensive data set of $> 10^6$ delay/rate observation pairs spanning 1978 to 1997, which included multiple baseline sessions using several global networks of stations. This extensive data set largely overlaps the data set used to construct the ICRF. However, its large size precluded the careful matching of modeling options, data editing, and parameterization choices that were possible with the smaller data set. Despite these limitations, the advantages of averaging over a large number of observations prevailed. The 400 sources common to the resulting GSFC and JPL frames agreed to 0.07 and 0.10 mas *wrms* in $\alpha \cos \delta$ and δ, respectively. Coordinate biases were $\leq$0.005 mas and rotational offsets were $\leq$0.08 mas about all three axes. This agreement is exclusive of the 0.5 mas zonal differences caused by an analyst's choice of whether or not to estimate troposphere gradients (Ma *et al.*, 1997).

4. Conclusions

The newly adopted ICRF claims to be accurate to $\approx$ 0.3 mas. Such an accuracy requires that the analysis software used to construct the ICRF be free from any errors that could potentially degrade the frame at a significant level. To that end, we compared JPL's MODEST, GSFC's CALC/SOLVE, and Paris Observatory's GLORIA software in order to verify that they reproduce each other's calculations. Agreement was achieved to a level of $\leq$0.01 mas thereby ensuring that model implementation is not a significant error. Comparison of reference frames derived from GSFC and JPL VLBI software indicate that analyst modeling option choices, data editing, and model parameterization using VLBI software may contribute 0.1 to 0.2 mas of uncertainty to the newly adopted ICRF. This partially accounts for the 0.25 mas noise floor used in deriving the "realistic" errors reported for the final ICRF source coordinates.

References

Gontier, A.-M., (1992) "Orientation de la Terre par Mesure VLBI", Ph. D. dissertation, Observatoire de Paris, Paris, France.

Ma, C., (1978) "VLBI Applied to Polar Motion, Relativity and Geodesy" Ph. D. dissertation, University of Maryland; also NASA TM 79582, GSFC, Greenbelt MD.

Ma, C., Arias, E.F., Eubanks, T.M., Fey, A.L., Gontier, A.-M., Jacobs, C.S., Sovers, O.J., Archinal, B.A., and Charlot, P. (1997) in *IERS Technical Note 23*, C. Ma and M. Feissel, eds., Observatory of Paris, Paris, France. See also: Ma, C., *et al.* (1998) "The ICRF Based on VLBI Observations of Extragalactic Radio Sources", in prep. *A&A*.

Sovers, O. J., Fanselow, J.L., and Jacobs, C.S., (1998) "Astrometry and Geodesy with Radio Interferometry: Experiments, Models, Results", *Rev. Mod. Phys.*, **70**.

MAINTENANCE OF THE ICRF: OPTICAL

N. ZACHARIAS
U.S. Naval Observatory, nz@pyxis.usno.navy.mil
3450 Mass.Av.NW, Washington DC 20392, USA

1. Introduction

1.1. THE ICRF

The International Celestial Reference Frame (ICRF) is presently realized by the positions of 608 compact, extragalactic radio sources (Ma & Feissel 1997) with milliarcsecond (mas) and sub–mas accuracy, all being on the same system, the International Celestial Reference System (ICRS). The Hipparcos Catalogue (ESA 1997) is the practical realization of the ICRF at optical wavelengths, giving high accuracy positions (≈1 mas) and proper motions (≈1 mas/yr) for 117 955 stars (ESA 1997). The link between the radio and optical frames has been achieved by various methods (Kovalevsky *et al.* 1997). The accuracy of this link is estimated to be 0.6 mas at the mean epoch of Hipparcos (1991.25) for the orientation, and 0.25 mas/year for the system rotation.

1.2. WHY DO WE NEED TO MAINTAIN THE OPTICAL SYSTEM ?

The optical frame maintenance, particularly its tie to the defining radio frame, is needed even more than the maintenance of the radio frame itself. Due to the large distances, the frame defining radio sources are basically fixed in the sky, while the optical frame is based on stars displaying significant individual proper motions. Without further observations, the accuracy of the radio frame will remain constant, whereas the accuracy of the optical positions will degrade due to the proper motion error propagation.

1.3. THE PLAN

The original plan was to wait ≈ 10 years after the Hipparcos mean epoch and then repeat a broad international effort to establish another link of the radio and optical systems at that later epoch. That would improve the system rotation link significantly when combined with the original link effort used for the Hipparcos project. However, some radio–optical link projects have not been completed in time to be fully included in the Hipparcos link program. Results are coming in right now which will help to maintain the alignment of the optical to the radio frame.

2. Radio stars

The largest weight in the current link is provided by a few (12) radio stars which were observed optically by Hipparcos and at radio wavelengths by VLBI (Lestrade 1995) on the 1 mas level. Other radio star programs were carried out with MERLIN (Morrison *et al.* 1997) and the VLA (de Vegt & Johnston *et al.* 1998), which involves more stars (≈50), however with lower precision for a single source (≈30 mas). These new observations of the radio stars will improve their radio positions and proper motions with respect to the system defining extragalactic sources. No new independent optical observations are being made. Thus, the existing Hipparcos positions and proper motions for the radio stars will be used for an improved radio–optical reference frame link.

In the more distant future, new high precision, optical observations of radio stars are required for the radio–optical reference frame link, because the accuracy of the Hipparcos positions will

J. Andersen (ed.), Highlights of Astronomy, Volume 11A, 322–324.

degrade with time due to proper motion error propagation. The Hipparcos system, realized by the sum of all Hipparcos stars, is much more accurate than the positions and proper motions of individual stars. Optical observations of radio stars referenced to many other Hipparcos stars will be required on the mas level, which currently can best be provided by optical interferometers. The Navy Prototype Optical Interferometer (NPOI) (Armstrong *et al.* 1998) will shortly be able to perform large angle astrometry on the few mas level for stars as faint as 9th magnitude. Besides observing the accessible radio stars, this instrument will also be used to find any possible local frame distortions in the Hipparcos system and to resolve double stars.

3. The Extragalactic Link

There are three different methods utilizing extragalactic sources for the radio–optical reference frame link. The Northern (NPM) and Southern (SPM) Proper Motion surveys (Lick, Yale), the Bonn, Potsdam and the KSZ (Kiev) program determine *absolute proper motions* with respect to galaxies without providing much information about the *orientation* link. These data have been used for the current Hipparcos system rotation link. No similar, major, new program is planned. The second method measures separations between selected pairs of Hipparcos stars and extragalactic sources by HST. No major new data is expected soon.

The third method involves direct observations of optical counterparts of the extragalactic radio sources, which represent the reference frame, in order to obtain a position at a particular epoch. Several groups around the world have been and are participating in this effort, with the Hamburg/USNO program making the biggest impact as far as accuracy and number of observed sources are concerned. See (Kovalevsky *et al.* 1997) for a review of all frame link investigations, which were completed in time for the Hipparcos link program.

3.1. THE HAMBURG/USNO EXTRAGALACTIC PROGRAM

The Hamburg/USNO program aims at radio and optical observations of about 400 compact, extragalactic sources globally to establish a radio–optical reference frame link (Johnston *et al.* 1988). Only a relatively small fraction (20%) of the data could be reduced in time (Zacharias *et al.* 1995) to be included in the current Hipparcos link. A multi–step procedure is used to relate the faint (≈ 17 to 20 mag) optical counterparts of the compact optical radio sources to the bright (≈ 7 to 9 mag) Hipparcos stars.

The first step is the wide–field astrograph program, which also determines the epoch of the reference frame orientation link. Mostly between 1980 and 1993 observations have been performed with the Hamburg astrograph for the Northern, and the USNO astrograph at Black Birch, New Zealand, for the Southern Hemisphere. Over 400 fields globally have been mapped on photographic plates, which were then measured at Hamburg Observatory. From these plates secondary reference stars with 50 mas accuracy have been obtained in a one square degree field centered on each source. In addition the Lick astrograph was used for over 100 fields.

The second step is the determination of each reference frame source position in the secondary reference star systems. Most of that data were obtained by CCD imaging with the 0.9–meter telescopes on Kitt Peak and Cerro Tololo (over 600 fields). Some faint sources required additional CCD observations made with 4–meter telescopes. Subframes of those images in FITS and GIF format have been made available on the Web at `ftp://rorf.usno.navy.mil/OID/oid.html`. For general information and the radio data see also `http://maia.usno.navy.mil/rorf/rorf.html`.

On the average a total standard error of 35 mas per source position has been obtained for the 78 sources used for the Hipparcos link. Two error contributions dominate the overall accuracy of this program. Firstly, the field distortion pattern (FDP) of the 0.9–meter telescopes is insufficiently known, with poorest image quality at the CTIO instrument. Unfortunately, each observing run had its own FDP. The other major error contribution is the relatively poor limiting magnitude (13 to 14) of the wide–field photographic work, providing relatively few secondary reference stars at high galactic latitude fields. Recent observations with wide–field CCD cameras at the Hamburg and USNO astrographs significantly improved the FDP determination. The Hamburg/USNO observing program is now complete and the final publication is in preparation.

3.2. THE UCAC-S PROJECT

The USNO CCD Astrograph Catalog South (UCAC–S) project (Gauss *et al.* 1996) will map the entire Southern Hemisphere to about 16th magnitude. A catalog positional accuracy of 20 mas is expected for stars in the magnitude range of 7 to 14. A direct tie to the Hipparcos and Tycho stars is part of the program, and additional long exposures at the astrograph and other telesocpes will provide a strong link to the counterparts of extragalactic radio sources ($\approx$ 10 mas per source). This will provide another radio–optical *orientation* link at a much later epoch than for the original Hipparcos Catalogue link effort. The resulting improvement over the current Hipparcos *system rotation* link from the UCAC–S project alone is estimated to be significant (factor 1.5 to 2).

4. Structure of optical counterparts

Structure analysis on the sub–mas level of the reference frame sources is now being performed regularly at radio wavelengths (Fey & Charlot 1997). At optical wavelengths no comparable resolutions can be achieved in the foreseeable future. A few, relatively nearby, compact galaxies and BL Lac reference frame objects show optical structure on the arcsecond level. Observations by HST can reveal structure on the 0.1 mas level. However, relevant for the reference frame tie are the internal structures of the dominating, compact cores of these sources. Even currently planned space interferometer missions with mas resolution will not be able to observe most of these sources because of their faintness at optical wavelengths. Imaging of the ICRF sources in the infra–red has just started (USNO project). Currently, the best option to reveal structure problems of a source is by comparing radio interferometric observations made at different frequencies. Also, the radio–optical *position residuals* for each individual source will reveal problematic candidates, which should then be excluded from the frame link process. These problematic candidates, however, are of astrophysical interest and justify high resolution follow–up observations.

References

Armstrong, J. T., Mozurkewich, D., Rickard, L. J, Hutter, D. J., Benson, J. A., Bowers, P. F., Elias, N. M. II, Hummel, C. A., Johnston, K. J., Buscher, D. F., Clark, J. H. III, Ha, L., Ling, L.-C., White, N. M., & Simon R. S. (1998), The Navy Prototype Optical Interferometer (NPOI), ApJ, accepted

de Vegt, C., Johnston, K.J. *et al.* (1998) VLA radio star observations for the reference frame link, *in preparation*

ESA (1997) *The Hipparcos and Tycho Catalogues*. European Space Agency, **SP–1200**

Fey, A.L. & Charlot, P. (1997) ApJS **111**, 95

Gauss, F.S., Zacharias,N., Rafferty,T.J., Germain,M.E., Holdenried,E.R., Pohlman,J.W. & Zacharias,M.I. (1997) A new astrometric survey of the Southern Hemisphere, BAAS **28**, No.4, 1282

Johnston, K.J., Russell,J., de Vegt,C., Hughes,J., Jauncey,D., White,G., & Nicolson,G. (1988) In *The Impact of VLBI on Astrophysics and Geodesy*, IAU Symposium No.129, edited by M.Reid and J.Moran, Kluwer, Dordrecht, p.317

Johnston, K.J., Fey,A.L., Zacharias,N., Russell,J.L., Ma,C., de Vegt,C., Jauncey,D.L., Reynolds,J.E., Archinal,B.A., Carter,M.S., Corbin,T.E., Eubanks,T.M., Florkowski,D.R., Hall,D.M., McCarthy,D.D., McCulloch,P.M., Nicolson,G.D., Shaffer,D.B., Sovers,O.J., & Jacobs,C.S. (1995) AJ **110**, 880

Kovalevsky,J., Lindegren,L., Perryman,M.A.C., Hemenway,P.D., Johnston,K.J., Kislyuk,V.S., Lestrade,J.F., Morrison,L.V., Platais,I., Roeser,S., Schilbach,E., Tucholke,H.-J., de Vegt,C., Vondrak,J., Arias,F., Gontier,A.M., Arenou,F., Brosche,P., Florkowski,D.R., Garrington,S.T., Preston,R.A., Ron,C., Rybka,S.P., Scholz,R.-D., & Zacharias,N. (1997) The Hipparcos Catalogue as a realisation of the extragalactic reference system, A&A **323**, 620

Lestarde, J.F., Jones, D.L., Preston, R.A., *et al.* (1995) A&A **304**, 182

Ma, C. & Feissel, M. (1997) Definition and Realization of the International Celestial Reference System by VLBI Astrometry of Extragalactic Objects, *IERS Technical Note* **23**

Morrison, L.V., Garrington, S.T., Argyle, R.W. & Davis, R.J. (1997) *Future control of the Hipparcos frame using MERLIN*, in Hipparcos Venice '97, ESA SP-402, 143-145

Zacharias, N., de Vegt, C., Winter,L., & Johnston,K.J (1995) AJ **110**, 3093

Zacharias, N. (1997) Astrometric Quality of the USNO CCD Astrograph AJ **113**, 1925

MAINTENANCE OF ICRF: COMPLEX OPTICAL OBJECTS

CHARLES E. WORLEY
U.S. Naval Observatory, Washington D.C., USA

Abstract. Heretofore, double and multiple stars have presented a tolerable but manageable problem in most astrometric programs. However, the drive to ever greater precision, with detections in the milli–to–micro arcsecond range of these complex objects, over a wide range of wavelengths, presents new and as yet unsolved problems in both the astrometry and photometry of such systems.

This discussion is aimed at a better delineation of the implications of the problem.

J. Andersen (ed.), Highlights of Astronomy, Volume 11A, 325.

MAINTENANCE OF THE ICRS - STATISTICAL TREATMENT

H. SCHWAN & R. WIELEN
Astronomisches Rechen-Institut
Moenchhofstrasse 12-14, D-69120 Heidelberg, Germany

1. Introduction

In this paper we briefly describe the influence of astrometric binaries on derived proper motions and we draw some conclusions with respect to the reference systems. For more details the reader is referred to Wielen (1997) and Wielen et al. (1997).

Increased measurement accuracy determines not only the derived parameters with much more precision, but it may even allow one to describe new phenomena, which were of no importance before. This happened in the case of the highly precise Hipparcos measurements (ESA, 1997) which were able to determine stellar proper motions within less than four years with an accuracy comparable with the best proper motions available from ground-based observations as given in the FK5 (Fricke et al. 1988). The basic difference between the Hipparcos and the FK5 proper motions showed up within our task of validating the Hipparcos proper motions during the reduction process. Our main result was that the dispersion of the residual differences between both proper motions was about three times as large compared with the error estimates expected from the data given in the catalogues. Since it was rather unlikely that one of the two catalogues had so strongly underestimated its errors, we looked for other reasons to explain the discrepancy. For many of the stars with the largest differences between the Hipparcos and FK5 proper motions, information on their duplicity was found in the literature. We thus suspected that the influence of astrometric binaries to be responsible for the discrepancy.

2. Mean and instantaneous proper motions

With its high measurement accuracy of better than 1 mas for the bright FK5 stars, Hipparcos was able to determine stellar motions with a precision of 1 mas/yr from observations during less than four years. A star's motion within this short time interval may, in the case of a double star, be completely different from the motion of its center of mass (Fig. 1). The observations used to construct the FK5, on the other hand, are less precise by a factor of about one hundred and in most cases they can therefore not resolve the wavy motion of a double star. The FK5 proper motions, derived from observations over about 250 years, reflect the star's mean motion in that time interval, and it will therefore be near to the motion of the center of mass, except for double stars with very long orbital periods. We believe that this basic difference in the time interval used – yielding 'instantaneous' Hipparcos and 'mean' FK5 proper motions – is responsible for the unexpectedly high dispersion of the differences between the Hipparcos and FK5 proper motions.

3. Comparison between Hipparcos and FK5 proper motions

We have computed the differences between the proper motions in the Hipparcos and the FK5 catalogue. In the following we have eliminated all stars for which we could find any information on suspected duplicity or for which Hipparcos itself has found indications of a non-linear motion. A subsample of 1202 stars was left from the 1535 FK5 stars. The FK5 proper motions were put onto the Hipparcos system by the following procedure. First some stars with highly discordant differences were eliminated, then the global rotation between both proper motion systems were determined and eliminated, and finally the remaining regional systematic differences were determined

J. Andersen (ed.), Highlights of Astronomy, Volume 11A, 326–328.

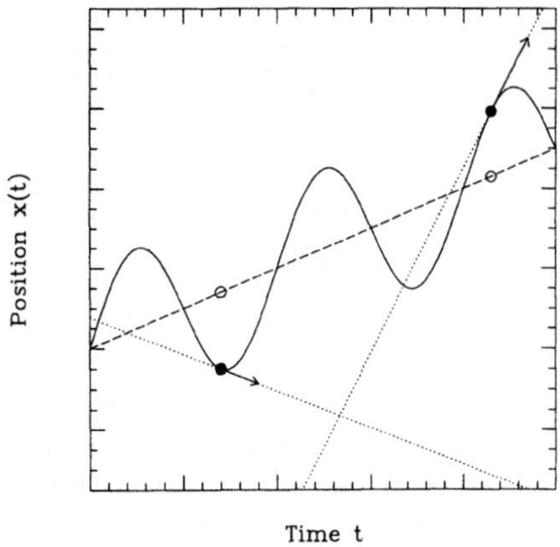

Figure 1. *Wavy motion of the photo-center of an astrometric binary (solid curve) around the linear motion of its center of mass (dashed line). Two instantaneous positions and proper motions are indicated (filled dots; arrows). The linear predictions based on the instantaneous values are shown as dotted lines.*

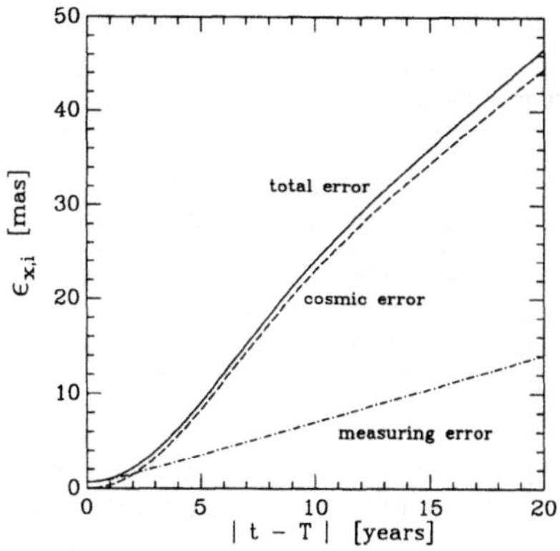

Figure 2. *Mean error $\epsilon_{x,p}$ of a predicted instantaneous position $x_p(t)$ at an epoch t, based on a linear extrapolation using an instantaneous position and an instantaneous proper motion observed at epoch T (e.g. the Hipparcos epoch). The total mean error (solid) is the combination of the cosmic error (dashed) and the measuring error (dashed dotted).*

and eliminated from the FK5 proper motions with the aid of the analytical method described by Bien et al. (1978). An iteration of this process has yielded no additional systematics.

For the the dispersion $D_{\mu,obs}$ of the residual differences Hipparcos minus FK5 we obtained (averaged over $\mu_\alpha \cos\delta$ and μ_δ) :

$D_{\mu,obs} = 2.38$ mas/yr.

The error estimates given in the two catalogues yield expected mean errors (again averaged over both coordinates) of 0.77 mas/yr for the FK5, and 0.68 mas/yr for the Hipparcos proper motions. Taking also into account the average mean error 0.28 mas/yr of the derived systematic corrections we compute the total expected mean proper motion error as

$D_{\mu,exp} = 1.07$ mas/yr.

Thus, the real dispersion of the observed differences exceeds the dispersion expected from the estimated proper motion accuracy in the two catalogues by a value $D_{\mu,cos}$ which is

$D_{\mu,cos} = (D^2_{\mu,obs} - D^2_{\mu,exp})^{\frac{1}{2}} = 2.13$ mas/yr.

This is about three times the errors estimated in the Hipparcos or the FK5. We explain this discrepancy with the presence of unresolved astrometric binaries in our sample. Hipparcos has measured for many of them 'instantaneous' proper motions which may deviate significantly from the mean FK5 proper motions (Fig. 1). We denoted this contribution in the observed dispersion as 'cosmic error'. This is not a real error, it merely results from our insufficient knowledge of a possible binaries.

4. Supporting arguments for the presence of cosmic errors

If our hypothesis is true that unresolved binaries produce the unexplained contribution in the observed dispersion, then the effect should depend on the star distance such that it decreases with increasing distance since the orbital displacement, measured in angular measure, becomes smaller. This effect is indeed significantly present as shown in Table 1. In various distance intervals are give the observed dispersion (resulting from comparing the published proper motions), the expected dispersion (resulting from the error estimates in the catalogues and including the uncertainty of the system) and the cosmic error (quadratic difference of both entries).

Another argument for the presence of the cosmic errors in the Hipparcos proper motions is given in Table 2. For the 1202 'apparently' single stars and the 95 so-called g and $\dot{g}$ stars (stars for which significant deviations from linear motion were detected from the Hipparcos reduction procedure) we give the cosmic errors in proper motion (first line) and in position at the mean

TABLE 1. Cosmic errors in proper motion as a function of the parallax

	Parallax p [mas] all	≥ 20	20 - 10	10 - 5	< 5
$D_{\mu,obs}$	2.38	2.94	2.50	2.05	1.58
$D_{\mu,exp}$	1.07	1.01	1.06	1.11	1.11
$D_{\mu,cos}$	2.13	2.76	2.26	1.72	1.12

TABLE 2. Cosmic errors in proper motion and position at the Hipparcos and FK5 epoch, for the 'single' and the g , $\dot{g}$ stars

'type' of star number of stars	'single' 1202	$g, \dot{g}$ 95
$D_{\mu,cos}[mas/yr]$	2.13	9.24
$D_{pos,cos}(T_{FK5})[mas]$	77.5	388.2
$D_{pos,cos}(T_{Hip})[mas]$	45.4	59.8

FK5 epoch (about 1950, last but one line) and the mean Hipparcos epoch (about 1991.25, last line). Since the cosmic errors are dramatically larger for the g and $\dot{g}$ stars, where Hipparcos had already found strong indication for orbital motion, we may conclude that unresloved binaries are also responsible for the cosmic error in our 'single' star sample. Comparing the dispersions in position at the FK5 and Hipparcos epoch makes obvious that the cosmic errors are primarily in the Hipparcos proper motions because their contribution over approximately 40 years epoch difference produces a significantly larger dispersion at the FK5 epoch than in the case of the FK5 proper motions which contribute to the dispersion at the Hipparcos epoch.

5. Mathematical treatment of the cosmic errors

The correct treatment of the cosmic errors makes use of correlation functions which describe the statistical correlation between the orbital displacements in position and motion relative to the mean motion. Since it is far beyond the scope of this brief presentation we refer the reader to Wielen (1997) for an introduction to this new field of 'statistical astrometry'. The correlation functions have been used by Wielen et al. (1997) to estimate the cosmic errors in a predicted Hipparcos position as a function of the time (Fig. 2). It is obvious that the cosmic errors dominate over the measuring errors, except for epochs very near to the mean Hipparcos epoch.

6. Conclusions for the Reference Frame

From 1998 onwards the International Celestial Reference Frame in optical wave lengths will be represented by the positions and motions of the stars in the Hipparcos catalogue. Since the orbital elements of the unresolved binaries in that catalogue, which are responsible for the cosmic errors described above, are distributed randomly in space and time, the cosmic errors will not introduce systematic errors into the frame. One has, however, to be aware that the precision of the positions and, in particular, of the proper motions, may be inferior to the estimates given in the catalogue. Since the cosmic errors dominate the error budget (except near the mean Hipparcos epoch) we suggest the use of stars at preferentially larger distances from the Sun for the maintenance of the reference frame, because the cosmic errors are less important for the distant stars (see Table 1).

References

Bien, R., Fricke, W., Lederle, T. & Schwan, H., (1978) *Methods for the comparison of systems of star positions to be applied in the construction of the FK5*, Veroeff. Astron. Rechen-Inst., Heidelberg, No. 29, Verlag G. Braun, Karlsruhe

ESA, 1997, The Hipparcos Catalogue, ESA SP-1200

Fricke, W., Schwan, H., Lederle, T. et al., (1988) Fifth Fundamental Catalogue (FK5), Part I; Veroeff. Astron. Rechen-Inst., Heidelberg, No. 32, Verlag G. Braun, Karlsruhe

Wielen, R., (1997) *Principles of statistical astrometry* A&A, 325, 367–382

Wielen, R., Schwan, H., Dettbarn, C., Jahreiss, H. & Lenhardt, H. (1997) *Statistical astrometry based on a comparison of individual proper motions and positions of stars in the FK5 and in the Hipparcos catalogue* in Hipparcos-Venice '97, ESA SP-402, 727-732

ABSTRACTS OF POSTER PAPERS

1. The International Celestial Reference System (ICRS)

The International Celestial Reference System, ICRS

M. Feissel, F. Arias
Bureau Central de l'IERS, Observatoire de Paris, 61 Av. de l'Observatoire, 75014 Paris , France
<feissel@obspm.fr>

Taking advantage of the participation of worldwide Very Long Baseline radio Interferometry (VLBI) in its activities, the International Earth Rotation Service (IERS) has defined and realized an extragalactic celestial reference system, proposed for adoption by IAU as the International Celestial Reference System (ICRS). Its origin is at the barycentre of the solar system. The directions of its axes are fixed with respect to the distant quasars to better than ±20 microarcseconds; they are aligned with those of the FK5 within the uncertainty of the latter (±80 milliarcseconds). The ICRS is realized by estimates of the coordinates of a set of quasars, the International Celestial Reference Frame (ICRF). The Hipparcos catalogue of directions and proper motions of galactic objects, and the Jet Propulsion Laboratory ephemerides of the solar system bodies are referred to the ICRS.

Reference:

Arias E. F., Charlot P., Feissel M., Lestrade J.-F., 1995, The Extragalactic Reference System of the International Earth Rotation Service, ICRS, Astron. Astrophys., 303, pp. 604–608.

2. Accuracy of the International Celestial Reference Frame

Internal analyses of the International Celestial Reference Frame (ICRF)

F. Arias, C. Mercuri-Moreau, A.-M. Gontier, M. Feissel, P. Charlot
Bureau Central de l'IERS, Observatoire de Paris, 61 Av. de l'Observatoire, 75014 Paris , France
<felicitas@obspm.fr>

The celestial reference system of the IERS and its materialization by the International Celestial Reference Frame (ICRF), is proposed to the IAU for adoption as the conventional celestial system, under the name International Celestial Reference System (ICRS). The ICRF includes about 200 extragalactic radio emitting objects (mostly quasars) selected from a list of 600 objects that are monitored by VLBI. The selection process that was implemented by the IAU Working Group on Reference Frames involved a number of criteria. This presentation concentrates on statistical analyses of systematic and random errors based on comparison of catalogues.

Modeling Radio Source Structures for the Maintenance of the Celestial Reference Frame

P. Charlot[1], O. J. Sovers[2]
[1] Observatoire de Paris - CNRS/URA 1125, France <charlot@obspm.fr>
[2] Jet Propulsion Laboratory, Pasadena, CA <ojs@logos.jpl.nasa.gov>

One of the largest error sources in astrometric VLBI analyses is caused by internal milliarcsecond-scale structures of extragalactic radio sources. Such errors show up, in particular, through systematic trends in the time variation of the estimated source coordinates, likely caused by time variation of the structural features. These trends, however, are difficult to model because some radio sources are known to eject fast moving components on time scales of months, while maps are seldom available more frequently than once or twice a year. To overcome this limitation, we have devised a scheme that permits calculation of structure delay and delay rate corrections at any epochs, based

J. Andersen (ed.), Highlights of Astronomy, Volume 11A, 329–338.

on interpolation between maps. Tests of the validity of this interpolation scheme, implemented in the JPL VLBI analysis software MODEST, have been carried out for the extended source 3C273, using VLBI observations collected during the years 1984–1988 and dual-frequency (S and X) models of the structure of 3C273 generated twice a year during this period. By comparing monthly positions of 3C273 estimated, both with and without structure corrections, we demonstrate that such processing largely accounts and removes the effects of varying source structure in astrometric VLBI analyses.

On the consistency of the ICRF and the reference frame realized by RSC(GAO UA)97 C 01

O.Molotaj[1], V.Tel'nyuk-Adamchuk[1], Ya.Yatskiv[2]
[1]Kyiv Taras Shevchenko Univ., Ukraine <vtel@aoku.freenet.kiev.ua>
[2]Main Astronomical Observatory, National Academy of Sciences, Ukraine <yatskiv@mao.kiev.ua>

Using the method of arcs between the radio sources (RS) developed by Yatskiv and Kuryanova, the Kyiv series of compiled catalogues of RS positions has been continued. This method provides the possibilities of deriving the uncertainty of source catalogues and of detecting inconsistencies in the separation of RS positions. A comparison between the constructed catalogue RSC(GAO UA)97 C 01, on the one hand, and the reference frames ICRF94 and RSC(WGRF)95 R 02, on the other hand, has been carried out. Systematic differences and the consistency of these frames are discussed.

Fluctuation of Extragalactic Reference Frame due to Gravitational Lensing in Our Galaxy

Mizuhiko Hosokawa[1], Toshio Fukushima[2], Kouji Ohnishi[3]
[1] Communications Research Laboratory, Japan <hosokawa@crl.go.jp>
[2] National Astronomical Observatory, Japan <toshio@spacetime.mtk.nao.ac.jp>
[3] Nagano National College of Technology <ohnishi@ge.nagano-nct.ac.jp>

When we observe the extragalactic objects through our galaxy, their positions will be fluctuate due to the gravitational lensing of matter in our galaxy. The positional shift will set a limit to the accuracy of the extragalactic reference frame and the variation of that will cause the aging of the frame. We calculate the expected values of these shifts on the stars and MACHOs in our galaxy. Their typical values are 10 microarcsecond(μas) and several μas/y, respectively.

The optical depth of such fluctuations are also calculated as a function of the positional shift or its variation. It is found that the fluctuations at 10 μas level would occur rather rarely. In such cases, they are considered to be induced mainly by the nearest single source to the line of sight and would be correctable to some extent. At the same time, the detection of these effects will be useful for probing the components of our galaxy.

Consistency of IERS global references

P. Sillard[1], F. Arias[2], M. Feissel[2], Z. Altamimi[1], C. Boucher[1], D. Gambis[2]
[1] Institut Géographique National, ENSG, 77455 Marne la Vallée Cedex 2, France <sillard@ensg.ign.fr>
[2] Bureau Central de l'IERS, Observatoire de Paris, 61 Av. de l'Observatoire, 75014 Paris <feissel@obspm.fr>

Classically, general spatial references involve three main components that have to be considered as a whole: the Celestial Reference System, the Terrestrial Reference System and the link between them, the Earth's Orientation. The International Earth Rotation Service (IERS) is in charge of maintaining mutually consistent numerical realizations of these three components: the International Celestial and Terrestrial Reference Frame (ICRF, ITRF) and Earth Orientation Parameters (EOP) time series. These are constructed by combining results from VLBI, GPS, SLR, LLR and DORIS,for which the Terrestrial Reference Frame, the EOP series and, in the case of VLBI, the Celestial Reference Frame are simultaneously available.

Currently the IERS EOP series, the Celestial and the Terrestrial Frames are combined separately. These combinations involve the estimation of rotation angles from each individual solution to the IERS one that should be in agreement with each other, although they are obtained from different subsets of the same global solution. We give some realistic present-day evaluations of the consistency of the ICRF, ITRF and IERS EOP. This approach as proved successful to identify and correct sources of inconsistencies between the various individual analyses, a necessary step in view of the future implementation of the simultaneous computation of a global set of spatial references.

3. Linking Hipparcos to the ICRF

Future control of the Hipparcos frame using MERLIN

L.V. Morrison[1], S.T. Garrington[2], R.W. Argyle[1], R.J. Davis[2]

[1] Royal Greenwich Observatory <merlp@ast.cam.ac.uk>

[2] Nuffield Radio Astronomy Laboratories, Jodrell Bank <stg@jb.man.ac.uk>

The equatorial cartesian axes of the Hipparcos Catalogue have been fixed relative to those of the International Celestial Reference Frame (ICRF) with an accuracy of ±0.6 milliarcseconds (mas) at the epoch 1991.25 and with a constraint of ±0.25 mas/yr on their rotation. The maintenance of the Hipparcos frame will depend on periodic checks of its alignment to the ICRF. One of the most effective ways of doing this will be to compare the radio positions of Hipparcos stars with the Hipparcos optical positions brought forward with Hipparcos proper motions.

¿From the radio positions of 11 Hipparcos radio stars, the MERLIN connected-element radio interferometer confirmed that the alignment of the Hipparcos frame at epoch ~1994 was within 2.3 mas of the ICRF. Ten years from the mean epoch of Hipparcos (2001) the misalignment of the Hipparcos frame to the ICRF could reach 3 mas. By doubling the number of measurable radio stars to ~20 and improving corrections for atmospheric phase errors, MERLIN will be capable of testing the alignment of Hipparcos to the ICRF with an accuracy of about ±1 mas. We discuss some new MERLIN positions of radio stars, which is the next step towards the realization of this objective.

The orientation of the Hipparcos celestial system with respect to ICRS as induced from the Earth orientation parameters

J. Vondrák[1], C. Ron[1]

[1] Astronomical Institute, Prague, Czech Republic <vondrak@ig.cas.cz, ron@ig.cas.cz>

The new solution of Earth orientation parameters is worked out in the Hipparcos reference frame, based on the observations of latitude/universal time made in the years 1899.7–1992.0 with nearly fifty astrometrical instruments at about thirty observatories. The individual values of polar motion, celestial pole offset and UT1–TAI are determined at 5-day intervals. These parameters are used to determine the time-dependent orientation of the Hipparcos reference system with respect to extragalactic system ICRS. The orientation and slow rotation of the Hipparcos reference system is usually defined by six independent parameters. We make use of only the values of the celestial pole offset, that are virtually independent of the terrestrial reference system. Therefore, we obtain only four (out of the six) parameters, defining the orientation and slow rotation of the Hipparcos system around the x and y axes.

Investigation of the relative orientation of reference frames: HIPPARCOS- FK5–ICRF

I.I. Kumkova, A.M. Kolomietz

Institute of Applied Astronomy, Russian Academy of Sciences, 8 Zhdanovskaya st.,St.Petersburg, 196042 Russia <kumkova@isida.ipa.rssi.ru>

Three main reference frames used at present in astronomy are considered. These are: FK5 – a classical fundamental reference frame being the only one in use during the last decade; HIPPARCOS – the most modern, precise and complete optical frame; and ICRF – International Celestial Reference Frame – recommended to be the new IAU system. These reference systems are based on different types of objects (FK5 and HIPPARCOS on optical stars, and ICRF on extragalactic radio sources), and are obtained by different techniques.

These differences lead to the need to determine the relative orientation of their frames. In the work presented here, we investigate the relative orientation of FK5 – HIPPARCOS – ICRF. The comparison between radio and optical frames was made possible by the availability of photographic astrometry of extragalactic radio sources made with respect to the FK5 and HIPPARCOS catalogues. Values of rotation angles are estimated from this comparison. New results for the determination of relative orientation parameters for HIPPARCOS – FK5 – ICRF are presented and discussion and comparison with the results of the other authors are made.

Determinations of high precision proper motions for 32 Hipparcos stars from photographic plates

Shuhua WANG, Zhenghong TANG, Wenjing JIN, Jiaji WANG
Shanghai Observatory, Chinese Academy of Sciences, China <jwj@center.shao.ac.cn>

Seventy exposures on 32 photographic plates of four areas of the sky – around the radio source 0316+413; the extragalactic nebula NGC7302; the Pleiades astrometric standard region, and so on – taken with the 40*cm* astrographic refractor at SheShan(Zô-Sè) station of Shanghai Observatory, are reduced. Using the PPM catalogue as the initial reference catalogue and the central overlapping method with the iteration process, we obtain high-precision positions and proper motions for 32 Hipparcos stars.

Due to the large epoch differences of up to 86 years, the mean internal consistencies of the proper motions for 32 stars are ± 0.59 mas/yr and ± 0.53 mas/yr in right ascension and declination. While the positional accuracies in RA and Dec are ± 23 mas and ± 18 mas, respectively. Comparisons between our results and those of other catalogues are made.

Residual rotation of the Hipparcos Reference Frame, or a new class of Galactic interior motion ?

Masanori Miyamoto, Zhu Zi
National Astronomical Observatory, Mitaka, Tokyo <zhuzi@c00.cc.nao.ac.jp>

The HIPPARCOS Reference Frame has been constructed on the extragalactic reference frame, ICRS. The nominal value of the proper motion deviation from inertial is 0.25 mas/yr, which means that the overall pattern analysis of HIPPARCOS proper motions should reveal the new class of the galactic interior motion other than the classical galactic rotation, such as the non-axisymmetric stream of stars, the warping motions, and the oscillatory motions of the Galaxy.

We have analysed HIPPARCOS proper motions, on the basis of the 3-D Ogorodnikov-Milne model (Miyamoto and Soma 1993). Two populations of stars, K-M giants and O-B stars, have been examined. We have found statistically meaningful shear and vorticity, including the classical galactic rotation. If we accept these motions, then we need a new concept of the galactic interior motion. However, the possibility of a remaining rotation of the HIPPARCOS Reference Frame larger than 0.25 mas/yr should be considered. To examine the remaining rotation, we have carefully investigated the systematic difference between the HIPPARCOS and NPM proper motion systems.

Effect on EOP of the systematic differences between the ICRS and the previous reference system

Tongqi XU, Peizhen LU
Shanghai Observatory, Chinese Academy of Sciences, China <lpz@center.shao.ac.cn>

Since 1984 the FK5 fundamental catalogue and the IAU 1976 astronomical constants have been used in astronomical calculations. The changes brought about by this system directly affected the Earth Orientation Parameters(EOP) determined from observations made by optical instruments. The effects of these changes on the EOP were calculated in the 1990s. Following the adoption of the ICRS for the Hipparcos Catalogue, the effects on EOP will have to be recalculated. The purpose of this paper is to calculate and analyse the systematic differences between the ICRS system and the previous system in order to establish a uniform system of EOPs since 1957.

Mykolayiv AMC positions of faint stars in selected fields around extragalactic radio sources for linking the optical and radio reference frames

A. Kovalchuk, G. Pinigin, Yu. Protsyuk, A. Shulga
Mykolayiv Astronomical Observatory, Ministry of Science and Technology of Ukraine
<root@mao.nikolaev.ua>

The principal features and technical possibilities of the Mykolayiv Automatic Meridian Circle (AMC) permit it to be used efficiently for improving the accuracy of the link between the optical (stellar) and radio (extragalactic radio sources - IERS) reference frames. A technical description of the AMC was presented at IAU Symposium 166. From 1996, the observation of reference stars in fields around 400 IERS sources in declination zone +90 to -20 and in magnitude range 12-14 was started with the AMC. The reference stars were selected from the Guide Star Catalogue and the measured positions are derived in the Hipparcos frame.

Additionally, Tycho stars were observed in the selected fields. The observations were carried out in short strips using a CCD camera. At present, about 2000 strips have been observed around 200 IERS sources containing 60 000 stars from the GSC, TC, and 2500 strips with HC stars. The observation and current reduction are described. It needs about three years of observation to complete the catalogue with an accuracy of 0."02–0."05.

4. Linking the dynamical frame to the ICRF

Link between the dynamical reference system and the ICRS, with Hipparcos minor planets

J.-F. Bange[1], A. Bec-Borsenberger[2], M. L. Bougeard[34]

[1] Bureau des Longitudes, URA 707 CNRS, Paris [bange@bdl.fr]

[2] Bureau des Longitudes, URA 707 CNRS, Paris [borsen@bdl.fr]

[3] UFR Maths., Université des Sciences, LYON I

[4] URA 1125 CNRS, Observatoire de Paris [bougeard@bdl.fr]

We used the observations of eighty-eight minor planets by Hipparcos to position the dynamical reference system with respect to the ICRS. The dynamical reference system is materialized by ephemerides of Bureau des Longitudes or ephemerides of JPL and the ICRS frame at optical wawelength is given by Hipparcos data. The data reduction consortia FAST and NDAC gave two sets of data for the observed positions of minor planets. We will present the results of the link obtained from these two sets using algorithms based on the least absolute value and the Huber M-estimators. Accuracy will be discussed.

Future study of minor planets' motion with respect to the ICRF

O.P. Bykov

Pulkovo Astronomical Observatory, Russia <bykov@venvi.usr.pu.ru>

To study the motions of the Minor Planets in the system of the ICRF one needs to know the accuracy the modern CCD observations are able to give.

The CERES software package created at the Institute of Theoretical Astronomy (Russia) was applied to calculate (O–C) differencies for more than 1000 numbered Minor Planets observed irregularly and quasi- simultaneously in 1995-96 with CCD as well as photographic techniques at nearly 70 observatories.

The accuracy of observations was estimated by analysing the dispersion of the average (O–C) for each type of observation obtained by each telescope. The star catalogue GSC was used by practically all the observers for the astrometric reduction of their CCD–scans. A representative sample of the considered observations was statistically reliable. To eliminate errors of the theory, short arc observations were made during one or several adjacent nights.

On the whole, the CCD observations of the numbered Minor Planets published in the Minor Planet Circulars from 06.1995 to 06.1996 are more accurate than the photographic ones.

Results of internal and external accuracy will be given for each observatory.

Linking Hipparcos with DE200/DE403 reference frame by photoelectric observations of the occultation of ϵ *Gem* by Mars (1976 April 8)

G. Krasinsky

Inst. of Applied Astronomy, Russian Acad. of Sciences <kra@ipa.rssi.ru>

There are two ways to link radio and optical realizations of ICRS. In the direct method, when coordinates of common sources are compared, a number of well-known problems arise. Now that very accurate ephemerides DE200/DE403 are linked to the radio system, an alternative approach may be realized in which inner planets are used as intermediate targets. In the optical band milliarcsecond accuracy is reached in photoelectric observations of occultations of stars by planets. Thus, it appears possible to tie Hipparcos with the frame of DE ephemerides, and thus with the VLBI radio frame.

We have tried this approach by making use of 16 published timings of contacts made during the occultation of the bright Hipparcos star ϵ *Gem* by Mars in 1976. Normal places for the differences in right ascension and declination were constructed which give the following residuals for DE403 (in the sense Mars minus star) :

$d\alpha = -60 \pm 2\, mas,\ d\delta = -7 \pm 9\, mas$ (Hipparcos coordinates)

$d\alpha = -45 \pm 2\, mas,\ d\delta = 103 \pm 9\, mas$ (FK5 coordinates)

Though with these observations it is possible to determine only two linear combinations of three unknowns, the residuals show that the system of Hipparcos needs corrections up to 60 mas. Even with a single additional observation of this type, it would be possible to determine the orientation of the Hipparcos frame with the milliarcseond accuracy. The large number of stars in Hipparcos makes reasonable the task to search for such events in near future.

5. Extending the Hipparcos frame to fainter magnitudes

CCD positions for extragalatic radio sources

M. Assafim[1], R. Vieira Martins[2], A.H. Andrei[2]
[1] Depto. Astronomia - Obs. Valongo, Rio de Janeiro, Brazil <massaf@on.br>
[2] Observatorio Nacional, Rio de Janeiro, Brazil <rvm@on.br and <at1@on.br>

We discuss a method to obtain on local area in the sky, precise positions for the stars of the Guide Star Catalog (GSC), referred to more precise catalogues. The corrections to the right ascensions and declinations are carried out in the tangent plane in a fashion similar to that of standard photographic astrometry.

We applied this method for CCD observations made at the 1.60m Telescope of the Laboratorio Nacional de Astrofisica, Itajuba, Brazil. We observed one hundred extragalatic radio sources uniformly distributed over $-80° < \delta < +20°$ declination, from the IAU list of objects scheduled to materialize the International Celestial Reference System (ICRS). The positions are referred to the Carlsberg Series (CAMC), PPM, ACRS and Hipparcos and Tycho catalogues. A set of fiducial sources was selected by means of their positional arclength differences, to link and compare the optical results and the VLBI radio source positions. Relative orientation angles between the optical and the radio frames were found within 0".02 precision. Comparison of positions on both frames for all radiosources has shown a dispersion consistent with the errors in the reference catalogues used.

Transferring the Hipparcos frame to Schmidt surveys

M.J. Irwin[1], L.V. Morrison[2], R.W. Argyle[3]
[1] Royal Greenwich Observatory, Cambridge, UK. <mike@ast.cam.ac.uk>
[2] Royal Greenwich Observatory, Cambridge, UK. <merlp@ast.cam.ac.uk>
[3] Royal Greenwich Observatory, Cambridge, UK. <rwa@ast.cam.ac.uk>

A series of faint astrometric standards obtained on the Carlsberg Meridian Telescope and a zonal astrometric sequence obtained by Stone for the Sloan Digitial Sky Survey, have been used to investigate the magnitude-dependent astrometric errors on sky survey Schmidt telescope photographic plates. In conjunction with the PPM and Tycho astrometric catalogues this has enabled us to characterise and quantify problems related to transferring the Hipparcos frame at V $\approx$ 9th magnitude to the limit of sky survey plates, V $\approx$ 21st magnitude.

As expected, the generally asymmetric image profiles on Schmidt plates lead to pronounced magnitude-dependent astrometric distortion terms of up to $\approx$1 arcsec in size. We demonstrate that most of this distortion is caused by the complex image structure of saturated stars, mainly caused by the appearance of diffraction spikes and low light level halos due to multpiple reflections in the telescope optics. Fainter than V $\approx$ 14–15th magnitude, the magnitude-dependent astrometric distortion rapidly declines to a level of order $\approx$0.1 arcsec.

USNO-A and the ICRF

D. Monet[1] and T. Corbin[2]
[1] U.S. Naval Observatory <dgm@nofs.navy.mil>
[2] U.S. Naval Observatory <tec@sicon.usno.navy.mil>

USNO-A is a catalog of 488 006 860 sources detected on both the red and blue first-epoch Schmidt survey plates (POSS-I in the north, SRC-J+ESO-R in the south). The currently available release, USNO-A1.0, is tied to the system of GSC1.1 because that was the only option. As soon as the Hipparcos and Tycho catalogs are released, the task of compiling USNO-A2.0 will be begin, and the expected release date is summer of 1998. Because of its depth, density, and accuracy, USNO-A is a useful catalog for a variety of astrometric applications. The astrometric accuracy is typically 0.25 arcseconds, and the photometric accuracy ranges from 0.2 to 0.4 magnitudes (depending on declination and distance from the nearest available calibration data), and objects as faint as (O=21,E=20) are included.

Please consult the USNO Web page *http : //www.usno.navy.mil/pmm/* for further details and ordering information.

The U.S. Naval Observatory Pole–to–Pole Program

T.J. Rafferty[1], E.R Holdenried[2]

[1] U.S. Naval Observatory, Washington DC <tjr@sicon.usno.navy.mil>

[2] U.S. Naval Observatory, Washington DC <erh@sicon.usno.navy.mil>

Between the years 1985 and 1996 the U.S. Naval Observatory carried out a program of absolute transit circle observations covering the entire sky. Two transit circles were involved, one located in Washington, DC (USA) and the other in Blenheim (New Zealand). Over 660 000 observations were made, primarily of International Reference Stars (IRS), as well as all the major planets (except Pluto) and nine minor planets. Over 37 000 observations were obtained of day-time objects including the Sun, Mercury, Venus, and Mars. The final objective is a self-consistent, all-sky catalog of absolute positions which can be linked by rigid rotations to the dynamical system.

The optical coordinates of 150 galactic radio sources

N.G. Rizvanov, V.N. Kitkin, L.I. Rakhimov, I.A. Dautov

Engehardt Astronomical Observatory <rng@astro.ksu.ras.ru>

Using the astrograph Zeiss 400/2000 at the Zelenchuk mountain station, plates of 150 galactic radio-sources were taken. A new method of reduction of the wide-angle plates was developed. Through their reduction, the optical coordinates of the target objects were obtained in the systems of the catalogues PPM and ACRS. These optical coordinates are compared with the radio-coordinates. In the future the radio stars' coordinates will be reduced on the system of the Hipparcos Catalogue.

The accuracy of the optical coordinates corresponds to the typical accuracy obtained by the photographic method.

Linking QSOs to the Carlsberg AMC

R.L. Smart, B. Bucciarelli, R. Casalegno, G. Chiumiento, G. Massone, F. Marocco and M.G. Lattanzi

Osservatorio Astronomico di Torino, Pino Torinese, Italy <smart@to.astro.it>

Procedures used to directly determine the positions of twelve radio sources in the frame of the Carlsberg Automatic Meridian Catalogues are discussed. We show that the precision and accuracy of these positions is solely limited by the CAMC and that this will remain the case when the CAMC is rereduced using the results from the Hipparcos mission.

We discuss how this data will help the ultimate tie of the Hipparcos coordinate system to the extragalactic reference frame and other ongoing projects at Torino to address this problem.

AC 2000: The Astrographic Catalogue on the Hipparcos System

S.E. Urban, T.E. Corbin, G.L. Wycoff

U.S. Naval Observatory, Washington, D.C. <seu@pyxis.usno.navy.mil>

The United States Naval Observatory has completed new reductions of the Astrographic Catalogue data. This catalog, called AC 2000, contains 4 660 000 stars distributed globally. Since the density of Hipparcos stars is inadequate to reduce the AC data on a plate-by-plate basis, the measures have been initially reduced using the Astrographic Catalog Reference Stars (ACRS). All positions have been converted to the Hipparcos system using systematic differences computed for those stars in common between Hipparcos and the AC.

Since each participating observatory used different instruments in photographing and measuring the plates, data from each were reduced as if they comprised an independent catalog. Measured images from each plate have been analyzed for scale, rotation, non-perpendicularity of measuring axes, differences in scale between measuring axes and plate tilt. Presence of radial and non-radial optical distortions, coma and magnitude equation were investigated and corrected, where applicable. Every attempt has been made to apply all published errata to the data, and a detailed search for previously unknown typographical errors was performed. All data have been combined and the final catalog contains positions, magnitudes, epochs, error estimates, and cross reference information. All positions are on the Hipparcos system for equinox J2000.0.

A CD-ROM containing the catalog and the reduction details is being distributed at this meeting. For additional information visit WWW site *http://aries.usno.navy.mil/ad/ac.html.*

Instrumentation for the U.S. Naval Observatory CCD Astrograph

T.J. Rafferty[1], M.E.Germain[2], N. Zacharias[3]

[1] U.S. Naval Observatory, Washington DC <tjr@sicon.usno.navy.mil>

[2] U.S. Naval Observatory, Washington DC <meg@crux.usno.navy.mil>

[3] with Universities Space Research Assoc., Washington DC <nz@pyxis.usno.navy.mil>

The U.S. Naval Observatory CCD Astrograph will start an observing program in mid-1997 on Cerro Tololo (CTIO) in Chile to produce a high density, high accuracy, astrometric catalog of the southern hemisphere stars down to 16th magnitude. The program will be done using a robotic, refracting telescope with a 8-inch five-element red-corrected lens. A Kodak 4k x 4k (9 micron pixels) CCD camera will allow a one square-degree field of view, which is large enough to provide the necessary reference stars. The dome rotation, setting and clamping the X-Y slide for the guidescope, setting and clamping the focus, telescope setting, and use of a Hartman screen for determining the focus will be done automatically. The system is controlled via an embedded single-board computer. A host PC sends commands to the embedded computer, receives status information, controls the camera, saves the images to disk, and does the on-line reduction of the previous CCD frame.

Extension of the ICRF: US Naval Observatory CCD Astrograph Catalog South (UCAC-S)

N. Zacharias[1], T.J. Rafferty[2], M.E. Germain[3]

[1] with Universities Space Research Assoc., Washington DC <nz@pyxis.usno.navy.mil>

[2] U.S. Naval Observatory, Washington DC <tjr@sicon.usno.navy.mil>

[3] U.S. Naval Observatory, Washington DC <meg@crux.usno.navy.mil>

The U.S. Naval Observatory CCD Astrograph Catalog project is about to start in July 1997 at Cerro Tololo (CTIO), Chile, to cover the entire Southern Hemisphere in a 2–fold overlap down to 16th magnitude. The program should be completed with 2 years of observing. The goal is to produce a high density, high accuracy, astrometric catalog in the Southern Hemisphere.

A 4k CCD camera will be used in combination with the 5–element, 206mm lens in a 579–642nm bandpass covering 61' at 0.9"/pixel. The Kodak chip is of high cosmetic quality with no dead pixels in the light sensitive area. Tycho stars will be used for preliminary reductions, and a direct tie to the Hipparcos stars is feasible with block adjustment techniques. A catalog accuracy of 20 mas per coordinate is expected for stars in the 7 to 14 magnitude range.

In addition to the program fields, longer exposures around extragalactic reference frame sources, calibration fields and minor planet observations are planned.

6. Extending the Hipparcos frame to the Infrared

First step of an Infrared Reference Catalogue based on the FK5 and the creation of an IR Reference System after Hipparcos

A.S. Kharin

Main Astronomical Observatory Ukrainian National Academy of Sciences <kharin@mao.kiev.ua>

Recently, an IR reference catalogue (CPIRSS) was compiled at the USNO (R.Hindsly and R. Harrington, 1994, AJ 207). The second version available on CD-ROM includes 37700 stars from the basic and supplement FK5, IRS, ACRS and WCC catalogues which were also observed by IRAS and were included in IRAS PSC. This catalogue is not homogeneous in accuracy and, as with optical catalogues, it may be naturally divided into two hierarchical catalogues. The first of these is the IR analogue of the FK5, and the second, based on the IRS, ACRS and WCC stars, is the analogue of the photographic reference catalogues. But after the FK5 is replaced by Hipparcos, the first step of the IR hierarchical coordinate system must be also recompiled, based on the IR stars from the Hipparcos catalogue.

We found from direct comparison of the HIC with the IRAS PSC that more then 25 000 HIC stars have IR counterparts. Therefore, these IR stars with positions and proper motions in the Hipparcos Catalogue must represent the first stage of a new IR coordinate system. Evidently, the first stage of the new IR catalogue, based on the Hipparcos Catalogue, will be ten times denser and practically two orders of magnitude more precise than that based on the FK5. The stars in the second stage of the CPIRSS should be adjusted to the Hipparcos (ICRF) reference frame, and they must represent the second stage of the new IR coordinate system.

A Milli–arcsecond Astrometric Interferometer: Mitaka Optical and Infrared Array (MIRA) Project

J. Nishikawa[1], K. Sato[1], T. Fukushima[1], Y. Machida[1], T. Miyazawa[2], Y. Honma[2], M. Yoshizawa[1], and M. Miyamoto[1]
[1] National Astronomical Observatory, Japan <nisikawa@optik.mtk.nao.ac.jp, toshio@spacetime.mtk.nao.ac.jp, koichi@optik, machida@optik, yoshzwms@cc.nao.ac.jp>
[2] Hosei Univ., Faculty of Engineering, Japan
<miyazawa@ksg.sc.hosei.ac.jp, honma@ksg.sc.hosei.ac.jp>

A series of interferometer projects, called MIRA, is in progress in NAO. We are now making experiments of stellar fringe detection and fringe tracking at Mitaka optical and infrared array called MIRA–I. MIRA–I has two 25 cm coude telescopes with a 4 m N–S baseline. Recently we started upgrading the interferometer to achieve 10 milliarcsec astrometry by using vacuum delaylines and temperature stabilized optics, which is called MIRA-I.2.

We are planning the next version of Mitaka array as MIRA–II. We use 30 cm siderostats and 20 cm beam reducing telescopes there. Four fixed astrometric stations with from 16 m to 32 m baselines and three movable ones from 8 m to 256 m are placed in a T–configuration array. We hope to construct an interferometer array with 1.5 m siderostats and about 1.4 km baselines at a good site like Mauna Kea, which we call MIRA–III. It will have some fixed and movable stations in a Y–configuration array. MIRA–II and MIRA–III aim at 1mas astrometry and 1mas or more resolution imaging. Wavelength between 400 nm and 2500 nm will be used.

MIRA–II data will maintain the Hipparcos catalog and MIRA–III would be able to make global astrometry of many quasars and optical to near–IR sources, thus contributing to the new ICRF.

Astrometry with Optical and Infrared Interferometers

K. Sato, J. Nishikawa, M. Yoshizawa, T. Fukushima, Y. Machida, M. Miyamoto
National Astronomical Observatory, <koichi@optik.mtk.nao.ac.jp nisikawa@optik.mtk.nao.ac.jp, yoshzwms@cc.nao.ac.jp toshio@spacetime.mtk.nao.ac.jp machida@optik.mtk.nao.ac.jp>

We are promoting the construction of the ground–based optical and infrared interferometers, called MIRA Project(I, II, and III), at the National Astronomical Observatory of Japan. The outline of the MIRA Project is given in the paper by Nishikawa et al. (this JD). Siderostats are used in order to carry out precise astrometric observations. The aperture will be large enough to reach a limiting magnitude fainter than 12 or 13 mag at K–band.

The scientific impacts of the astrometry with MIRAs are discussed in this paper. It is very important to link the optical positions of stars to the radio reference frame determined by VLBI through the observations of quasars and stars with the interferometers. It is also interesting to compare the positions of central stars of maser sources to individual maser-spot positions of VLBI. With K–band, the proper motions of the red supergiants within 20 kpc from the sun can be observed even through the galactic disk because the interstellar absorption at K–band is very small. The dynamics of the bulge and disk of the Galaxy studied by using these observations will provide a new global kinematic scope to the formation and evolution of the Galaxy.

7. Astrolabe catalogues

Second Catalogue of Stars with the Photoelectric Astrolabe in San Juan

W.T. Manrique[1], Lu. Lizhi[2], R. Perdomo[3] Wang Zezhi[2], R.A. Podesta[1], Zeng Zhifang[2]

E.L. Actis[1], Zeng Fanmiao[2], E. Alonso[1]
[1] Observatorio Astronomico Felix Aguilar, Argentina <postmaster@unsjfa. edu.ar>
[2] Beijing Astronomical Observatory, China <nrd@bao01.bao.ac.cn>
[3] Observatorio Astronomico de La Plata, Argentina <perdomo@fcaglp.unlp. edu.ar>

According to the cooperation between Beijing, San Juan and La Plata Astronomical Observatories, the photoelectric astrolabe mark II(PA II) of Beijing Astronomical Observatory was moved and installed at San Juan Observatory in Argentina in January, 1992 for the observation of the catalogue of stars in the southern hemisphere.

On the basis of the observations in San Juan made with the instrument from Feb. 23, 1992 to January 12, 1997, residuals of 10 930 stars were formed from about 390 000 observations of stars made on 1458 days. The mean accuracy of a star position derived from these residuals is 0".045. Using these data, the Second Catalogue of 5195 stars (CPASJ2) has been compiled. The mean accuracies are ±3.1 ms and ±0.059" in right ascension and declination, respectively. The magnitudes of stars are from 2.0 to 11.5. The declinations are from –3° to –60°. The mean epoch is 1994.8. Finally, systematic corrections (CPASJ2–FK5) are given.

Preliminary results from the new astrolabe in Bucharest

Radu Popescu, Petre Popescu, Petre Paraschiv
Astronomical Institute of Romanian Academy <ppopescu@imar.ro>

The Danjon astrolabe transferred from the Royal Observatory of Brussels to Bucharest has been upgraded. The wollaston micrometer was replaced by an illuminated reticle that serves for the determination of astrolabe tilt versus the local vertical (the mercury pool). The transparent prism was replaced with a reflecting one, cut from a Zerodur block; it contains two pieces glued at the contact surfaces. The left piece has the reflecting surface tilted at 45 degrees to the optical axis of the astrolabe, and is the autocollimating mirror. The right piece has the reflecting surface tilted at 22.5 degrees to the astrolabe objective, and serves for star observations. The camera used is a COHU 4710 (borrowed from Paris Observatory, DANOF). The images are obtained in the CINE–CCD mode. The software was written in Pascal, based on advice received from Dr. F. Chollet. In the first step we have determined the inner parameters of the new astrolabe. We hope to observe by this technique stars fainter than 8 mag with a new intensified CCD camera.

JD8

Stellar Evolution in Real Time

Chairpersons and Editors: **E. F. Guinan & R. H. Koch**

JOINT DISCUSSION 8: STELLAR EVOLUTION IN REAL TIME, PART I

EXPLICIT EVIDENCES FOR STELLAR EVOLUTION

E. F. GUINAN
University of Villanova, Villanova, PA 19085
guinan@ucis.vill.edu

AND

R. H. KOCH
210 Roberts Road, Ardmore, PA 19003, USA
rhkoch@mail.sas.upenn.edu

JD8 assembled for half-day sessions on August 22 & 23, 1997 in a satisfactory meeting hall. Between 350 and 400 delegates – about 50% larger than anticipated – attended most of the sessions.

The theme of the meeting developed from initial conversations between R. H. Koch and Y. Kondo and was extended by E. F. Guinan. Opinions from the officers of Commissions 27 and 42 were favorable to seeking IAU approval and eventually the co-sponsoring Commissions came on board.

In essence, all the organizing workers recognized that there did not exist a broad recognition that we really do know specific stars whose evolution we have actually perceived and that there are also statistical interpretations (not the conventional isochrones in an HRD or CMD) for ensembles of stars which lead to the same conclusion.

The presentation compiled here is really a distillation of short offerings from the JD participants provided against a deadline which was regrettably close at hand. A full presentation of the meeting Proceedings will appear in *The New Astronomy* series of Elsevier. E. R. Jewell prepared the final files.

J. Andersen (ed.), Highlights of Astronomy, Volume 11A, 341.

JOINT DISCUSSION 8: STELLAR EVOLUTION IN REAL TIME, PART II

EXPLICIT EVIDENCES FOR STELLAR EVOLUTION

This Part of the report contains the summaries of the most affirmative conclusions about recognized evidences for stellar evolution. We would certainly prefer to know that we have actually discovered this or that piece of evidence but it can be equally valuable to know that the present state of knowledge forbids a conclusion of the sort which we wish. Not surprisingly, there are instances for which we are lucky to be able to make only a limiting statement about the matter. So *affirmative* here really means that a conclusion - positive or negative - has actually been attained. To the *layman* in stellar evolutionary studies it may be surprising that there exists such a diversity of evolutionary criteria - radiometric, spectrometric, orbital dynamical, spin dynamical - all with a defined time base available at the present day. It may be hoped that some of the conclusions summarized here will make their way routinely into undergraduate texts in the forseeable future and that tools such as the HRD or CMD will not appear as only static snapshots.

The convenors of the JD8 are also well aware that some specific subjects are actually missing from the following highlights and expect to cover these gaps in the full presentation of the Proceedings. The editors also note that the summaries of the impressive presentations by N. Mowlavi and K. Nomoto did not arrive by press time; these too will appear in the final volume. In what follows, the summaries are arrayed by advancing stellar evolution as closely as the variety of masses, the single or double star condition, and current knowledge permit.

J. Andersen (ed.), Highlights of Astronomy, Volume 11A, 342.

WATCHING STARS EVOLVE: RAPID EVOLUTIONARY CHANGES OF THE POST-AGB STAR – FG SGE

E. F. GUINAN
Villanova University
guinan@ucis.vill.edu

YU. S. EFIMOV
Crimea Astrophysical Observatory
efimov@elvis.astro.nwu.edu

AND

S.J. MARGHEIM
Villanova University
guinan@ucis.vill.edu

The variable post-AGB star FG Sge has provided an unique opportunity to study phenomenon which may be characteristic of the post-PN stage of stellar evolution. FG Sge is the central star in the 36-arcsec PN He 1-5 that over the last 50 yrs has been undergoing rapid evolutionary changes. Most notable are changes in temperature, luminosity, chemical composition and pulsation period. Starting in mid-1992, the star has undergone several large dimming events from which it has yet to recover. These events closely resemble the dimming events commonly observed in R CrB stars which result from dust formation in ejected envelopes. From all the evidence, it now appears that FG Sge has evolved into a R CrB star.

In this paper we first summarize the recent photometric history of the FG Sge from UBVRI photometry obtained during 1990-1997. We then combine these data with earlier photometry to define the star's light bevavior over the last 100 yrs. The photometry is also combined with available spectroscopic polarimetric, and IR observations. Particlar attention is given to comparing FG Sge's post-dimming behavior with contemporaneous observations of R CrB itself, also done on the same photometric system.

J. Andersen (ed.), Highlights of Astronomy, Volume 11A, 343.

A STUDY OF THE YOUNG STAR V1331 CYG IN A COMPACT STAR-FORMING REGION

ALISHER HOJAEV
High Altitude Maydanak Observatory, Ulugbek Astronomical Institute Tashkent, Uzbekistan
AND
ALEXANDER ZHELEZNYAK
Astronomical Observatory, Kharkov University, Kharkov, Ukraine

The peculiar T Tauri type star V1331 Cyg = LkH 120, located in the dark cloud Lynds 984, is a FU Orionis pre-outburst candidate (McMuldroch *et al.*, 1993). This star embedded in circumstellar bright nebulosity is also surrounded by a helix-shaped nebula originated from the star. We obtained a series of speckle images of V1331 Cyg on the standard *VBR* system on July 18-20, 1994 using the Zeiss1000 1.0 m telescope of the High Altitude Maydanak Observatory in Uzbekistan. We also used the results of Johnson's system photometry of V1331 Cyg, made simultaneously at Maydanak Observatory with a 0.6 m reflector. Images were computer processed. Surface photometry of the star and surrounding nebula was made and isophotes were constructed. Fourier analysis of light curves for V1331 Cyg has shown 36.2 day periodic process (Melnikov, priv. comm.) which could be due to the presence of a circumstellar disk. Morphologic analysis revealed a complex fine structure of nebula, consisting of many bright knots and obviously show a jet-like outflow. According to spectral observations (Chavarria, K. 1981; Penston & Keavey 1977) the line profiles of V1331 Cyg are P Cyg type only; there are no detections of anti-P Cyg profiles during the last 25 years observations. This indicates a stable mass loss (on average, $10^{-7} M_\odot$/yr) with mean velocities about -410 km/s.

J. Andersen (ed.), Highlights of Astronomy, Volume 11A, 344.

AN ANALYSIS OF ALMAGEST MAGNITUDES FOR THE STUDY OF STELLAR EVOLUTION

J. B. HEARNSHAW
University of Canterbury, Christchurcy, NZ
j.hernshaw@phys.canterbury.ac.nz

Visual magnitude data in Ptolemy's *Almagest* are analysed by comparing them with modern photoelectric magnitudes on the Pogson scale, after taking extinction into account. This result is used to transform *Almagest* magnitudes to the Pogson scale.

A study is made of changes in the vusual magnitudes of supergiant stars in the *Almagest* between classical and modern times (an interval of nearly 19 centuries). No evidence is found for any secular changes in the mean brightness of these stars, contrary to the conclusion of Mayer (Observatory 104, 77, 1984). However, it is shown that the Almagest magnitudes for stars in the Milky Way are on average 0.34 Pogson magnitudes too faint and about the Almagest supergiants are affected by this. Finally, some evidence is cited for the visual atmospheric extinction in classical times being significantly less than today. If this is the case, and is not taken into account, it would mimic an apparent supergiant brightening of about 0.1 magnitudes over this time interval.

J. Andersen (ed.), Highlights of Astronomy, Volume 11A, 345.

EVOLUTIONARY CHANGES IN LUMINOUS STARS

E. ZSOLDOS
Konkoly Observatory, Hungarian Academy of Sciences, H-1527 Budapest, Hungary

The light curves of luminous stars often show spectacular secular changes which can be connected to stellar evolution. Such events are, e.g. the outbursts of P Cygni in the 17th century and η Carinae in the last century. Both stars belong to the Luminous Blue Variables, but these changes are not restricted to blue stars. The light curve of HR 8752 (V509 Cassiopeiae) shows a certain similarity to that of the former two stars. When it was first catalogued in the middle of the 19th century, it had been a 6^{m} star. During 100 years the star showed a secular brightening of 1^{m}. A similar yellow hypergiant, ρ Cassiopeiae produced at least two outbursts this century, though both have smaller amplitudes than it is in the case of the LBVs. Moreover, these yellow variables also have an apparently secular colour change: the $B-V$ colour of HR 8752 is decreasing while that of ρ Cassiopeiae is increasing. In both cases evolutionary changes are possible but one cannot exclude other causes.

Besides these well studied stars there are several other yellow hypergiants with promising light curves. One of the most interesting cases seems to be R Puppis, which was discovered to be variable in the last century, but then did not show any appreciable change in the following 70–80 years. In the late 1970s, however, it began to vary once more.

J. Andersen (ed.), Highlights of Astronomy, Volume 11A, 346.

LONG-TERM PERIOD AND AMPLITUDE VARIATIONS IN β CEPHEI STARS

M. JERZYKIEWICZ
Wrocław University Observatory, Wrocław, Poland
mjerz@astro.uni.wroc.pl

We first consider all β Cephei stars for which long-term behavior of their pulsation periods is known from observations. As it turns out, these objects fall into two categories: one consisting of stars with complex period variations, and the other, with a constant and non-negative rate of period change, $\dot{P}$ = const ≥ 0. The problem is that stars with $\dot{P}$ close to that predicted for the hydrogen-shell burning phases of evolution are over-represented. We then compare the five β Cephei stars which exhibit long-term variations of the pulsation amplitudes and note that in each case the cause may be individual. In one star, 16 (EN) Lac, two different effects are responsible for the observed variation of the amplitudes.

An IAU travel grant is gratefully acknowledged. This work was supported by KBN grant 2 P03D 016 11.

J. Andersen (ed.), Highlights of Astronomy, Volume 11A, 347.

LONG-TERM VARIATIONS IN THE β CEPHEI STARS 16 (EN) LAC AND ν ERI

M. JERZYKIEWICZ AND A. PIGULSKI
Wrocław University Observatory, Wrocław, Poland
(mjerz,pigulski)@astro.uni.wroc.pl

Using all available data, we investigate long-term variations of the amplitudes of the three strongest pulsation modes of 16 (EN) Lac and the period of the strongest mode of ν Eri. We show that in case of 16 (EN) Lac there are feasible explanations of the observed variations. The explanations involve non-linear interaction between pulsation modes (the first two observed modes) and beating between two non-interacting modes (the third observed mode). In case of ν Eri, however, we can only indicate the difficulty of accounting for the long-term period variation.

MJ gratefully acknowledges an IAU travel grant. This work was supported by KBN grant 2 P03D 016 11.

J. Andersen (ed.), Highlights of Astronomy, Volume 11A, 348.

MODE AND PERIOD CHANGES IN PULSATING STARS NEAR THE MAIN SEQUENCE: δ SCUTI STARS

M. BREGER
Institut für Astronomie, Türenschanzstr. 17, A-1180 Wien, Austria
breger@astro.univie.ac.at

Reported mode changes in δ Scuti stars cna usually be explained by insufficient data to obtain multiperiodic solutions and by the effect of time-variable amplitudes. The extensive data on 4 CVn obtained by the Delta Scuti network are used to illustrate that the large changes in the appearance of the power spectra in the years from 1966 to 1996 can be ascribed to amplitude variability.

The period changes of δ Scuti stars have been collected or redetermined from the available observations and are compared with values computed from evolutionary models. For the radial pulsators of Pop.I, the observations indicate (1/P)dP/dt values around 10^{-7} year^{-1} with equal distribution between period increases and decreases. The evolutionary models, on the other hand, predict that the vast majority should show increasing periods a factor of about 10 smaller than observed. Arguments are given why the rate of evolution for these relatively unevolved stars cannot yet be deduced from the observed period changes.

J. Andersen (ed.), Highlights of Astronomy, Volume 11A, 349.

EVIDENCE OF ANGULAR MOMENTUM LOSS IN THE ECLIPSING BINARY VW CEPHEI

M. J. DEVITA AND D. H. BRADSTREET
Eastern College, St. Davids, PA
mdevita@eastern.edu

E. F. GUINAN
Villanova University guinan@ucis.vill.edu

AND

Z. GLOWNIA
Palomar College glownia@minaka.sdsu.edu

VW Cep is one of the brightest and longest observed short-period (P = 6.67 hours) W UMa type binaries. It consists of a G5V and K0V components in contact with their Roche surfaces. We investigated complex period changes based upon eclipse timings from the past 70 years. In addition to the well-known 30 year light time effect due to the presence of a third star in the system, we find evidence for a long term decrease in the orbital period of dP/dt = -0.02 sec/yr. This decrease in period could arise from angular momentum loss from the binary or mass exchange between components. From these timings we have refined the properties of the tertiary component and redetermined its mass and orbital parameters. After correcting the O-C's for the third body and the steady decrease in Keplerian period, we uncovered small systematic deviations in the residual O-C's. We are studying these second order period variations, possibly arising from mass loss in winds and/or mass flow between components, to determine if they correlate with known cycles of spot and surface activity.

J. Andersen (ed.), Highlights of Astronomy, Volume 11A, 350.

EVOLUTIONARY CHANGES IN THE ECLIPSING BINARY β LYRAE

P. HARMANEC
Astronomical Institute, Acad. of Sciences of the Czech Republic, 251 65 Ondřejov, Bohemia, Czech Republic

An attempt at the first direct test of the theory of large-scale mass transfer between the components of a binary system is discussed. It is argued that β Lyr is the only binary in the sky for which such a test is conceivable.

J. Andersen (ed.), Highlights of Astronomy, Volume 11A, 351.

LIGHT CURVE VARIATIONS OF THE OLD DISK ECLIPSING BINARY R CANIS MAJORIS

Y.W. KANG
Sejong University
kangyw@astro.sejong.ac.kr

AND

E. F. GUINAN
Villanova University
guinan@ucis.vill.edu

R CMa is one of the best observed low mass Algol-type systems, believed to be a semidetached system composed of a nearly spherical F0V primary and a less massive highly distorted K3 IV-V star. Most of the published light curves have asymmetries of one kind or another. Analysis of the individual light curves, dating back to over 75 years, yielded somewhat different orbital solutions depending on the epoch. These inconsistencies are caused by obvious differences in the shape of the light curves obtained at different epochs and by the different models used to find solutions.

A series of light curves of R CMa (obtained at different epochs) have been analyzed by a new version of the Wilson-Devinney computing program to find a homogeneous solution. All parameters were adjusted to fit a light curve of each epoch. Then we found the weighted mean value of each parameter. These mean values were used as initial parameters and the potential and radius of each component were re-adjusted for each epoch to see the variation of radii for over 30 years. Several radial velocity curves were used to find spectroscopic parameters. The photometric parameters were combined with the corresponding spectroscopic parameters to calculate an absolute dimension of R CMa.

Finally, we examined the *IUE* spectra of R CMa for existence of circumstellar gas in the system and evidence of magnetic activity on the cooler component. We also discuss the evolution of the system and possible observable *real time* evolutionary processes.

J. Andersen (ed.), Highlights of Astronomy, Volume 11A, 352.

ANGULAR MOMENTUM LOSS AND TRANSFER IN CLOSE BINARIES: EFFECTS ON A HUMAN TIME-SCALE?

C. MACERONI
Rome Astronomical Observatory, I-00040 Monteporzio C. (RM), Italy; CNRS Institut d'Astrophysique de Paris, France
maceroni@coma.mporzio.astro.it

The orbital periods of binaries are known to great accuracy, their changes produce an easily detectable cumulative effect and many systems have been observed for more than a century. In tidually locked late-type binaries the orbital period changes are often related to structural or evolutionary changes. The study of the orbital period secular evolution can therefore provide information on phenomena taking place on timescales very short when compared to the typical stellar evolutionary scales, but still much longer thant the human lifetime. This paper focuses on the dynamical evolution due to magnetic braking in late-type close binaries and on the detectability of angular momentum transfer among the stellar layers.

J. Andersen (ed.), Highlights of Astronomy, Volume 11A, 353.

SYNCHRONIZATION ANOMALIES IN CLOSE BINARIES AS A TEST OF RAPID EVOLUTIONARY PROCESSES

A. GIMÉNEZ
LAEFF(I.N.T.A.-C.S.I.C.), Apartado 50727, E28080 Madrid, Spain
ag@laeff.esa.es

AND

A. CLARET
Instituto de Astrofisica de Andalucia, C.S.I.C., Apartado 3004, E-18080 Granada, Spain
claret@iaa.es

Close binaries with late-type components generally present rotational velocities synchronized with their orbital period. This is a well known effect of tidal evolution in binary systems which is particularly efficient in stars with convective envelopes. Nevertheless, some cases are known where the angular velocities of the componenent stars are clearly different from each other or even from that of the orbit. Good examples are the evolved systems TZ For and α Aur.

We have made a detailed study of the tidal evolution history of these binaries through integration of the relevant differential equations using the formalisms by Zahn as well as Tassoul. Our results show that a satisfactory explanation of the synchronization status of the component stars can be achieved when the influence of the rapid expansion phases, suffered by the stars after leaving the main sequence and starting core He combustion, in the actual rotational velocities is taken into account.

J. Andersen (ed.), Highlights of Astronomy, Volume 11A, 354.

REAL TIME EVOLUTION OF EVOLVED STARS

J. R. PERCY
Erindale Campus, Univ. of Toronto, Mississauga, ON, Canada L5L 1C6
jpercy@erin.utoronto.ca

We briefly discuss the methods of observing stellar evolution in real time, and of interpreting these observations. Then we discuss recent results on RR Lyrae stars, Population I and II Cepheids, RV Tauri stars, and related objects.

J. Andersen (ed.), Highlights of Astronomy, Volume 11A, 355.

EVIDENCE FOR STELLAR EVOLUTION IN MIRA VARIABLES

PATRICIA A WHITELOCK
South African Astronomical Observatory, P O Box 9, Observatory, 7935, South Africa
paw@saao.ac.za

After briefly reviewing our understanding of Miras and their evolutionary status, three aspects of real-time evolution in these and related stars are examined:

1. Chemical changes (O-rich to C-rich) due to third dredge-up,
2. Period changes due to the effects of the helium shell-flash,
3. The existence of 'fossil' dust and gas shells.

Studies of resolved gas and dust shells are highlighted as particularly interesting. They will enable us to examine the mass-loss histories of many late-type stars over the last ten thousand years or so. Such observations have only recently become technically feasible and they are expected to provide important new insight into the late stages of stellar evolution.

J. Andersen (ed.), Highlights of Astronomy, Volume 11A, 356.

SEARCH FOR SIGNS OF A NEW OUTBURST IN THE QUIESCENT STATE OF CH CYGNI

D. KOTNIK-KARUZA AND R. JURDANA-ŠEPIĆ
University of Rijeka, Faculty of Education, Department of Physics, Omladinska 14, HR-51000 Rijeka, Croatia

A review of our investigations of CH Cygni in its inactive state during the period 1987-1989 has been given. Optical spectra reveal the development of physical conditions in the photosphere of the cool component as well as changes in the accretion complex around the hot component. The apparently quiescent star was actually evolving in trainsition between adjacent outbursts.

J. Andersen (ed.), Highlights of Astronomy, Volume 11A, 357.

SAO 244567 (HEN1357) : A POST-AGB STAR WHICH HAS TURNED INTO A PLANETARY NEBULA WITHIN THE LAST 20 YEARS

M. PARTHASARATHY
Indian Institute of Astrophysics, Bangalore, India
partha@iiap.ernet.in

SAO 244567 (Hen1357) (IRAS 17119-5926) is a high galactic latitude (l = 331°, b = -12°) early type star, originally classified as a B or A type H-alpha emission line star by Henize (1976). It is an IRAS source with far infrared colours similar to planetary nebulae. The IUE ultraviolet spectra obtained during the last eight years show that the central star is rapidly evolving. It is found that the central star of this young PN has faded by a factor of 3 within the last seven eight years. The terminal velocity of the stellar wind has decreased from -3500 km/sec in 1988 to almost zero in 1994. In 1988 the C IV (1550Å) line which was a P-Cygni profile with strong absorption component had almost vanished by 1994. The CIII] 1909A emission strength increased markedly within 4 years from 1988 to 1992. The optical spectra obtained since 1990 shows very clearly only the nebular spectrum which is very similar to that of low excitation planetary nebula. The optical spectrum of SAO 244567 obtained in 1971 shows that it was a post-AGB B 1 or B2 supergiant at that time. This result shows that SAO 244567 has turned into a planetary nebula within the last 20 years. Recently Bobrowsky (1994) obtained narrowband optically resolved images in both H-beta and [OIII] 5007A with the HST planetary camera which revealed a well resolved nebula of size 2 seconds of arc. In this paper we discuss the recent new results.

J. Andersen (ed.), Highlights of Astronomy, Volume 11A, 358.

RECENT DEVELOPMENTS IN THE YOUNG PLANETARY NEBULA HEN-1357

M. BOBROWSKY
CTA Inc., Rockville, MD

Observations over the past four decades have revealed significant changes in the spectrum of Hen-1357. Here we present HST images and spectra showing the most recent developments. In 1950, Henize saw only Hα in emission; but more recent observations by Parthasarathy *et al.* in 1992 showed strong forbidden lines consistent with a young planetary nebula. The spherically aberrated 1992 HST images, in which Bobrowsky first optically resolved the nebula, showed a compact nebula surrounding the central star. Nebular gas appeared most strongly concentrated in an ellipse with its major axis subtending $1\overset{''}{.}6$ from NW to SE. If this ellipse is actually a circular ring viewed obliquely, then our line of sight is inclined from the symmetry axis by 5°. Above and below the ring of gas are two bubbles containing lower-density gas. At the tip of each bubble, there is a hole where the gas inside the bubbles has broken through and is now escaping. The windblown appearance is consistent with the blueshifted Si IV (1394-1403Å) and Al III (1855-1863Å) doublets observed by Parthasarathy *et al.* that indicated the presence of a strong stellar wind. The N V (1239-1243Å) to C IV (1548-1551Å) ratio has increased in recent years, consistent with a young nebula becoming increasingly ionized. Our recent (1996) spectra reveal additional developments that show the real-time development of this young nebula. Finally, the new HST Planetary Camera images of the nebula show detailed structure indicating a much more complex object than previously known, including the presence of a companion star $0\overset{''}{.}3$ from the central star.

J. Andersen (ed.), Highlights of Astronomy, Volume 11A, 359.

SAKURAI'S OBJECT: CONSTRAINTS FROM ITS VARIABILITY?

A. GAUTSCHY
Astronomisches Institut der Univ. Basel, Switzerland
gautschy@astro.unibas.ch

H.W. DUERBECK
Space Telescope Science Institute, Baltimore, USA
duerbeck@stsci.edu

A.M. VAN GENDEREN
Leiden Observatory, Leiden, Netherlands
genderen@strw.LeidenUniv.nl

AND

S. BENETTI
European Southern Observatory, Santiago, Chile
benetti@eso.org

The peculiar outburst of the star baptized Sakurai's Object (SO) is a conceivable example of a late He shell flash in a post-AGB object. The new source of nuclear energy forces such objects toward high luminosities and eventually low effective temperatures; they cross the HR diagram in a comparable fashion as FG Sge did in the past – i.e., they move noticeably on the HR diagram on human timescales. From monitoring campaigns of SO during the last year, first estimates of its cooling rate were derived and in particular cyclic light variability was established. We present first results from attempts to model stellar envelopes appropriate for SO. As we hypothesize the light variability to be attributable to stellar pulsations, we aim at constraining the basic stellar parameters based on stability analyses of our envelope models. Radial, nonadiabatic stability computations provided predictions of the modal content which should be observable as SO evolves. The particular components in such mode spectra of SO as they are to appear in the coming years should indeed help to constrain basic stellar parameters such as mass and luminosity.

J. Andersen (ed.), Highlights of Astronomy, Volume 11A, 360.

SAKURAI'S OBJECT, A LATE HE-FLASH

F. KERBER, H. GRATL AND S. KIMESWENGER
Institut für Astronomie, Universität Innsbruck, Technikerstr. 25, A-6020 Innsbruck, Austria

AND

M. ROTH
Las Campanas Observatory, Carnegie Institution of Washington, Casilla 601, La Serena, Chile

We report on dramatic change in Sakurai's object (RA: 17h52m32.7s, DEC: −17d41'07", J2000.0). Discovered by the Japanese amateur astronomer Y. Sakurai in Feb. 1996, this object was first classified as a slow nova (Nakano S., Sakurai Y., Hazen M., McNaught R.H., Benetti S., Duerbeck H.W., Leibundgut B., 1996, IAU Circ. 6322). Further study revealed a spectrum rich in C, N and O-lines, but poor in hydrogen, as well as an old planetary nebula surrounding the star; it was then realized that this might be a star undergoing a late He-flash (Benetti, S. et al. 1996, IAU Circ. 6325). Despite the fact that 10 to 20 % of all low mass stars will experience a late He-flash (Iben I. Jr., MacDonald J., 1996), it is an extremely rare observational event, due to its short duration. The most recent example is V605 Aql/A 58 in 1919 (Eitter W.C., 1987, ESO Messenger, 14: see also Clayton C.G., De Marco O., 1997, AJ for a review).

In March 1997 we discovered dramatic change in the optical spectrum of Sakurai's object using the 2.5 m Du Pont telescope at Las Campanas, Chile. The dominating features now are the prominent bands of CN (bandheads at 388.8 and 421.6 nm) and C_2 (Swan-Bands at 439.2, 473.7, 516.5, 558.5 and 565.5 nm)(Kerber F., Gratl, H., Roth, M., 1997, IAU Circ. 6601) that were not evident a year ago(Duerbeck, H.W., Benetti, S., 1996, ApJ 468, L111).

Another dramatic change was the development of a strong IR excess (Kimeswenger S., Gratl,H., Kerber, F., Fouque, S. Kohle, S. Steele., 1997, IAU Circ. 6608). The optical data were obtained at UTSO's 60 cm, Las Campanas, and ESO's Dutch 90 cm. The NIR data are from the DeNIS project operated at ESO's 1 m. The IR-measurements are from ISOCAM, obtained on Feb. 25th, 1997. The energy distribution found in 1996 could be explained by a 7000 K blackbody in agreement with the observed spectrum, resembling a F-type supergiant. The 1997 energy distribution can no longer be described by a single Planck curve, and a second component of standard grain distribution (Mathis J.S., Rumpl W., Nordsiek K.H., 1977, ApJ 217) with a temperature of up to 1700 K is required as is confirmed by the ISO data. The specific output of this component is about 30 times larger than that of the stellar source. We conclude that this is the signature of very hot dust that has recently formed in an extended, optically thin shell around the star.

J. Andersen (ed.), Highlights of Astronomy, Volume 11A, 361–362.

For two reasons the dust observed in Sakurai's object must have formed most recently:

(1) Sakurai's object was not detected as an IRAS source in the 80's;

(2) The location of the dust excludes an age older than the He-flash because dust so close to a PN central star would have been destroyed long ago.

Both formation of molecules and dust are compelling evidence for an episode of mass loss from Sakurai's object. While rapid change is expected for such a star, Sakurai's object literally outpaces all predictions made by current theory (Iben I. Jr., MacDonald J., 1996).Discretionary observing time on ISO has been granted in order to follow the evolution of the freshly formed dust.

This work is financially supported by the Austrian *Fonds zur Förderung der wissenschaftlichen Forschung*, Project P-11675.

THE REMARKABLE EVOLUTION OF THE POST-AGB STAR FG SGE

JOHANNA JURCSIK
Konkoly Observatory, Hungarian Academy of Sciences, H-1527 Budapest, Hungary
AND
BENJAMIN MONTESINOS
LAEFF(I.N.T.A.-C.S.I.C.), Apartado 50727, E28080 Madrid, Spain

FG Sagittae is one of the most important key objects of post-AGB stellar evolutionary studies. As a consequence of a final helium shell flash, this unique variable has shown real evolutionary changes on human time scales during this century. The observational history was reviewed in comparison with predictions from evolutionary models.

The central star of the old planetary nebula (He1-5) evolved from left to right in the HR diagram, going in just hundred years from the hot region of exciting sources of planetary nebulae to the cool red supergiant domain just before our eyes becoming a newly-born post-AGB star. The effective temperature of the star was around 50,000 K at the beginning of this century, and the last estimates in the late 1980s give 5,000-6,500 K. Recent spectroscopic observations obtained by Ingemar Lundström show definite changes in the nebular line intensities. This fact undoubtedly rules out the possibility that, instead of FG Sge, a hidden hot object would be the true central star of the nebula. Consequently, the observed evolutionary changes are connected with the evolution of a single star.

Due to mixing processes accompanying the rapid evolution, the surface chemical composition of FG Sge also showed drastic changes. In the early 1970s, an enhancement in the abundances of s-process elements was observed, then in the 1980s FG Sge became a carbon star. In 1992 a 4 mag dimming of the object was observed which was later identified as an R CrB type decline. Since then, FG Sge has been one of the most active members of this class of variables. Though FG Sge is not so hydrogen deficient as a typical R CrB star, both its photometric and spectroscopic behaviour support the R CrB classification. Witnessing the birth of a new R CrB star reveals the evolutionary state of these variables as born-again AGB objects.

During the redward evolution FG Sge became unstable against radial pulsation, thus the pulsational properties also provide useful information on the changes in the surface physical parameters. Investigating the changes in the pulsational behaviour coincidental with the occurrence of the decline episodes also help to understand the R CrB phenomenon. The direct observational evidence of chemical evolutionary processes (nucleosynthesis, mixing, dust condensation) and of different type mass loss episodes (planetary nebula, mass ejection connected with the final flash and with the R CrB behaviour) also give the opportunity to study such phenomena which are closely related with stellar evolution.

J. Andersen (ed.), Highlights of Astronomy, Volume 11A, 363.

EVOLUTION OF NOVA 1934 DQ HERCULIS OVER 40 YEARS

E.S.DMITRIENKO
Crimean Astrophysical Observatory
dmitrien@crao.crimea.ua

We present the results of the analysis of the photometrical behaviour of Nova 1934 DQ Her for a time interval about 40 years based on our data and those published by different authors. Our observation were done in UBVRI in 1982-1997 during apptoximatly 170 nights. They give the possibility of investigating different types of the activity of the old nova with amplitudes of brightness variability from several tenths of magnitudes to 1^m. The light oscillations on a time-scale of either several days or several dozens of days would be conditioned by the change of the velocity of the accretion from the disc on the white dwarf caused by the intrinsic disc instability or by the irregular receiving of matter from the secondary component of the system – a red dwarf. The cyclic change of the light of DQ Her on a time-scale of about several years could be the reaction of the accretion disc in response to the activity of the secondary star.

J. Andersen (ed.), Highlights of Astronomy, Volume 11A, 364.

RAPID EVOLUTIONARY CHANGES IN THE WR BINARY HD 5980

V.S. NIEMELA, R.H. BARBÁ AND N. I. MORRELL
FCAG, Univ. of La Plata, Argentina
virpi@fcaglp.fcaglp.unlp.edu.ar

Observations of the Wolf-Rayet eclipsing binary HD 5980 in the Small Magellanic Cloud show how one component of this binary system unexpectedly underwent an eruption transforming its spectral type from a hot WNE to a cooler and luminous WN11 type. Radial velocity analysis of the preoutburst spectra indicates that both components of the binary are massive stars of WN type. The component that erupted is found to be the more massive one. Differential analysis of the H Balmer and HeII Pickering emission line strengths indicates that the hydrogen content changes with the cooling of the spectrum. The maximum light of the eruption of HD 5980 had a duration of a few months, but the slow change from a WNE spectrum to WNL type took about 10 years. The present spectrum of HD 5980 is described. The environment and orbital parameters of this massive binary suggest that the lower metal content of the Small Magellanic Cloud changes the spectral appearance of WR type stars, as compared with their galactic counterparts.

J. Andersen (ed.), Highlights of Astronomy, Volume 11A, 365.

THE CRAB NEBULA AND OTHER ANOMALOUS ASTROPHYSICAL ARTHROPODS

VIRGINIA TRIMBLE
Physics Dept., Univ. of California, Irvine, CA 91697, USA
and Astronomy Dept., Univ. of Maryland, College Park, MD 20742, USA

1. Supernovae and Their Remnants

The Crab Nebula is expanding and its pulsar is slowing down, both on time scales that indicate some very interesting real-time stellar evolution must have occurred about 1000 years ago, as indeed we know from Chinese and other ancient records. The Crab was, in fact, the first SNR to be confidently identified as such and to have its expansion rate measured and the first pulsar to have its first and second time derivatives measured. Among the interesting related details are (a) the current expansion is a bit faster than the average since the 1054 explosion, indicating outward pressure from the relativistic electrons and magnetic field responsible for the synchrotron emission, (b) some of the central synchrotron emission features wiggle and oscillate around at much larger speeds than the expansion but in a non-secular way, (c) the braking index of the pulsar is about 2.5 rather than 3.0 (as predicted for a pure magnetic dipole emitter), (d) energy is conserved (it wasn't obviously so before 1968), in that the pulsar is losing rotational kinetic energy fast enough to keep up the supply of optical and X-ray emitting electrons, and (e) glitches in the slowing of the pulsar (for which data go back to a 1964 balloon flight) indicate complex coupling of crust to interior, presumably a magnetic one.

Many other supernova remnants show net outward motion (for instance as revealed by radio images of Cas A) on time scales like their ages. Some also have slower moving features, presumably a result of pre-SN mass loss. The very interesting case of SN 1987A (which has both of these and other activity as well, but no pulsar yet) is presented here in Nomoto's paper.

Other recent advances in our understanding of supernovae include (a) two- and three-dimensional calculations of core collapse and neutrino propagation in evolved massive stars, which indicate that neutrino-driven turbulence, Rayleigh-Taylor instabilities, and convection can probably eject the outer layers of the star (thus solving the longstanding problem that one-dimensional models usually didn't explode) and (b) possible identification of SN Ia progenitors in the form of supersoft X-ray sources (see paper by Kahabka in proceedings of JD18), whose accreting stars are quite massive white dwarfs that do not seem to expel as much material as they accrete (thus solving the ancient problem that we knew SN Ia progenitors had to be white dwarfs driven above the Chandrasekhar limit somehow, but could not find any real binary or single stars that would do it).

J. Andersen (ed.), Highlights of Astronomy, Volume 11A, 366.

RADIO SUPERNOVAE AS DIRECT EVIDENCE OF STELLAR EVOLUTION IN REAL TIME

S.D. VAN DYK
Visiting scientist, UCLA, USA
vandyk@astro.ucla.edu

M.J. MONTES AND K.W. WEILER
Naval Research Lab, USA
mmontes@moon.nrl.navy.mil, kweiler@sne.nrl.navy.mil

R.A. SRAMEK
NRAO/VLA, USA
dsramek@zia.aoc.nrao.edu

AND

N. PANAGIA
STScI/ESA
panagia@stsci.edu

The radio emission from supernovae provides a direct probe of a supernova's circumstellar environment, which presumably was established by mass-loss episodes in the late stages of the progenitor's presupernova evolution. The observed synchrotron emission is generated by the SN shock interacting with the relatively high-density circumstellar medium which has been fully ionized and heated by the initial UV/X-ray flash. The study of radio supernovae therefore provides many clues to and constraints on stellar evolution. We will present the recent results on several cases, including SN 1980K, whose recent abrupt decline provides us with a stringent constraint on the progenitor's initial mass; SN 1993J, for which the profile of the wind matter supports the picture of the progenitor's evolution in an interacting binary system; and SN 1979C, where a clear change in presupernova mass-loss rate occurred about 10^4 years before explosion. Other examples, such as SNe 1994I and 1996cb, will also be discussed.

J. Andersen (ed.), Highlights of Astronomy, Volume 11A, 367.

JOINT DISCUSSION 8: STELLAR EVOLUTION IN REAL TIME, PART III

IMPLIED EVIDENCES

It is inevitable that understanding of some indicators of stellar evolution is not so definite as we could wish. This can be equally true of observational material and of the beginning and intermediate stages of theoretical development. Even though these matters may be arguable and untidy, they are not negligible and, in fact, represent obvious recognized opportunities for the near future. It is possible that these matters are really the most valuable consequence of the meeting. Subject matter of this kind was presented at the JD8, in both invited and contributed forms, and their summaries follow here in the same approximation to stellar evolution stages that characterized Part II.

J. Andersen (ed.), Highlights of Astronomy, Volume 11A, 368.

STELLAR PROPERTIES OF THE CLASSICAL LBV R127 – IMPLICATIONS FOR MASSIVE STELLAR EVOLUTION

P.A. CROWTHER AND L.J. SMITH
Dept of Physics & Astronomy, UCL, Gower Street, London, WC1E 6BT, U.K
pac@star.ucl.ac.uk

We present a new analysis of R127 (HDE 269858) – the classical Luminous Blue Variable in the LMC – during various phases of its spectral evolution over the past two decades, ranging from 9,000K to 28,000K. High quality CTIO, ESO and AAT spectroscopy are utilised to trace the variation in stellar properties for R127 with spectroscopic phase. R127 is found to possess almost identical stellar properties to AG Carinae during each stage, with a comparable helium enriched surface composition (H/He=2 by number). Results obtained from recent HST/FOS spectroscopy also indicate near identical nebular conditions to AG Carinae, with comparable electron densities and temperatures. In contrast with expectations, nebulae represent lightly processed material, suggestive of H-rich red supergiant envelopes. We discuss the importance of our results for LBVs, and the wider picture of massive stellar evolution within different environments.

J. Andersen (ed.), Highlights of Astronomy, Volume 11A, 369.

THE PERIOD CHANGES OF YY ERIDANI

C.-H. KIM AND J. H. JEONG
Dept. of Astron. & Space Science, Chungbuk Nat'l. Uni., San 48, Gaeshindong, Cheongju 361-763, Korea
kimch@astro.chungbuk.ac.kr, jeongjh@astro.chungbuk.ac.kr
O. DEMIRCAN AND Z. MUYESSEROULU
Ankara Uni. Obs., Dept. of Astron., Tandogan, Ankara, Turkey
demircan@dione.astro.science.ankara.edu.tr, muyes@dione.astro.science.ankara.edu.tr
AND
E. BUDDING
Central Insitute of Technology, P.O. Box 40740, Wellington, New Zealand
budding@ee.cit.ac.nz

A total of eighteen times of minimum lights for YY Eri were determined from relatively new or unpublished photoelectric observations collected from Korea and Turkey. All minima available to us were intensively analyzed to deduce the character of period variation of YY Eri. It is either formed by a sinusoidal variation superimposed on an upward parabola, or a set of abrupt changes. The abrupt changes appeared to have alternatively occured in the pattern of two increases following one decrease, which may be an indication of sinusoidal variations rather than real sudden changes of period. Upward parabolic variation can be due to a secular period increase caused by mass transfer from less massive to more massive component. The sinusoidal character can arise from a third body or from a strong magnetic activity cycle. Long term sinusoidal light level variation in the light curves supports the cyclic magnetic activity effect on the orbital period. However, the third body hypothesis can not be ruled out by the present data.

J. Andersen (ed.), Highlights of Astronomy, Volume 11A, 370.

V392 ORIONIS: OBSERVATIONS AND EVOLUTIONARY STATE OF A LOW-MASS SYSTEM

S. NARUSAWA
Nishi-Harima Astronomical Observatory, Japan
narusawa@nhao.go.jp
A. YAMASAKI
National Defence Academy, Japan
yamasaki@cc.nda.ac.jp
AND
Y.NAKAMURA
Fukushima University, Japannakamura@cosmos.educ.fukushima-u.ac.jp

Although the evolution of binary systems has been qualitatively interpreted with the evolutionary scenario, the quantitative interpretation of any observed system is still unsatisfactory due to the difficulty of the quantitative treatment of mass and angular momentum transfer/loss. To reach a true understanding of the evolution of binary systems, we have to accumulate more observational evidence. So far, we have observed several binaries that are short-period and noncontact, and found the existence of extremely small-mass systems. In the present paper, we study another short-period (P=0.659d), noncontact, eclipsing binary system, V392 Ori. We have made photometric and spectroscopic observations of V392 Ori. The light curves are found to vary, suggesting the existence of circumstellar matter around the system. Combining the photometric and spectroscopic results, we obtain parameters describing the system; we find the mass of the primary component is only $0.6 M_\odot$ – undermassive for its spectral and luminosity class A5V, suggesting that a considerable amount of its original mass has been lost from the system during the course of evolution. The low-mass problem is very important for investigation of the evolution of close binary systems: large mass loss within and/or after the main-sequence will have a significant influence on the future evolution of binary systems.

J. Andersen (ed.), Highlights of Astronomy, Volume 11A, 371.

THE UNUSUAL SHORT PERIOD BINARY V342 AQUILAE; DOES IT BREAK THE EVOLUTIONARY THEORY MOLD?

C. N. HARTMAN
The Catholic University of America and Goddard Space Flight Center
chartman@iue.gsfc.nasa.gov

R. S. POLIDAN
Goddard Space Flight Center

A. D. WELTY
Space Telescope Science Institute

R. WADE
Pennsylvania State University

P. B. ETZEL
San Diego State University

AND

F. C. BRUHWEILER
The Catholic University of America

Analysis of the eclipsing (period=3.39 days) binary V342 Aquilae at visible wavelengths reveals that the system is heavily enshrouded with gas, masking the identity of the underlying stars. In the ultraviolet V342 Aquilae shows six times more attenuation at phases outside of eclipse than most Algols. It is likely, then, that V342 Aquilae is the most enshrouded short period Algol system known. Our ultraviolet spectra show very pronounced Fe II absorption lines arising from ground and metastable levels, indicating an extensive circumstellar gas envelope. The strength of this absorption shows both an orbital and a cycle-to-cycle variability. The eclipse spectra display very strong emission from lines such as C II, Si IV, and C IV , with a striking similarity to the interacting binary TT Hydrae (B9.5V+K0III). We have deduced that the system consists of a late B star with a late GIII-IV companion and a mass ratio on the order of 0.5. A detailed description of the observations, as well as deduced geometrical and physical properties of the system are presented. The apparent conflict of these observations with presently understood theory of binary evolution will be discussed.

J. Andersen (ed.), Highlights of Astronomy, Volume 11A, 372.

SEMI-EMPIRICAL DETERMINATION OF THE POSITION OF RRAB STARS ON THE H–R DIAGRAM

J. JURCSIK AND G. KOVÁCS
Konkoly Observatory, H-1525 Budapest XII, Box 67, Hungary

In a recent series of papers, it was shown that the metallicity and the absolute V, I and K mean magnitudes of RRab stars can be accurately calculated using only their periods and the low order Fourier parameters of the V light curves. The application of these formulae in conjunction with the synthetic colours of the model atmospheres and with the theoretical pulsation periods enables us to determine all the basic physical parameters (*i.e.* $\log L$, $\log T$, $\log M$ and [Fe/H]) of any variable with an accurate light curve. As a result, we might trace evolutionary tracks on the H–R diagram by plotting the stars with the same mass and metallicity. From a present sample of nearly 300 stars, signs of differences in the evolutionary states can be suspected but not yet proven.

The accumulation of large numbers of precise light and colour curves, on the one hand, will further improve the empirical formulae and therefore the accuracy of the derived parameters and, on the other hand, will increase our chances of constructing better-populated empirical evolutionary tracks.

J. Andersen (ed.), Highlights of Astronomy, Volume 11A, 373.

OBSERVATIONAL CONSTRAINTS ON EVOLUTION OF DWARF NOVAE IN AND BELOW THE PERIOD GAP

D. NOGAMI, T. KATO, H. BABA AND S. MASUDA
Kyoto Univ., Japan
nogami, tkato, baba, masuda@kusastro.kyoto-u.ac.jp

There is a gap in the distribution of the orbital periods of cataclysmic variables (CVs) between 2 and 3 hours. The period gap is explained at present by cessation of the mass transfer making CVs fainter and preventing them from being discovered. After restarting the mass transfer, CVs have been believed to evolve with the orbital periods becoming shorter, as angular momentum is released by gravitational wave radiation. In this view, the mass transfer rate depends almost only on the orbital period.

However, reconsideration of these views is urged by a series of recent discoveries of ER UMa stars (a subclass of SU UMa stars having enormous mass-transfer rates), dwarf novae bridging "classical" SU UMa stars and ER UMa stars, and the first in-the-gap dwarf nova, PG 1510+234. These objects suggest two possibilities; 1) (a part of) SU UMa stars undergo large secular changes of the mass-transfer rate, and 2) there exist parameters overlooked but strongly influential in the evolution of CVs.

J. Andersen (ed.), Highlights of Astronomy, Volume 11A, 374.

MULTIPLE TIME SCALES IN CATACLYSMIC VARIABLES: OBSERVATIONS VS. MATHEMATICAL MODELS

I. L. ANDRONOV
Department of Astronomy, Odessa State University
root@astro.odessa.ua

Theoretical models and observational evidence for various processes in magnetic cataclysmic variables are briefly reviewed. Among them: modulation of the accretion rate by the magnetic field of the white dwarf; excitation of the orientation change of the magnetic axis of the white dwarf with respect to the secondary; structure of the accretion column and its instability; mass and angular momentum transfer; magnetic activity of the secondary; high/low luminosity state transitions; QPO's, "shot noise" and "red noise" in polars, intermediate polars and nova-like objects.

J. Andersen (ed.), Highlights of Astronomy, Volume 11A, 375.

MAGNETOROTATIONAL MECHANISM OF SUPERNOVA EXPLOSIONS – RESULTS OF 2D SIMULATIONS

S.G. MOISEENKO
Space Research Institute, Moscow, Russia
moiseenko@mx.iki.rssi.ru

Results of 2D numerical simulation of the magnetorotational mechanism of a supernova explosion are presented. Simulation has been done for the real equations of state and neutrino energy losses have been taken into account. Simulation has been done on the basis of an Implicit Lagrangian scheme on a triangular grid with grid reconstructuring. It is shown that, due to differential rotation of the star, a toroidal component of the magnetic field appears and grows with time. Rotational momentum transfers outwards as the toroidal component grows with time. With the evolution of the process, part of the envelope of the star is ejected. The amounts of the thrown-off mass and energy are estimated. The results of the simulation could be used as a possible explanation for the supernova explosion picture.

J. Andersen (ed.), Highlights of Astronomy, Volume 11A, 376.

JOINT DISCUSSION 8: STELLAR EVOLUTION IN REAL TIME, PART IV

MISCELLANY

For a number of interesting contributed papers at the JD8 it was not obvious at the meeting or later that their contents and conclusions are germane to the main theme of secular evolution evidences. The editors have decided that these papers should be listed by title only. They appear below in alphabetized order by (first or only) author.

J. Andersen (ed.), Highlights of Astronomy, Volume 11A, 377.

WINDS AND MASS-LOSS FROM EVOLVED, LOW-GRAVITY COOL STARS

K. G. CARPENTER
Laboratory for Astronomy and Solar Physics, NASA-GSFC, USA
hrscarpenter@tma1.gsfc.nasa.gov

AND

R. D. ROBINSON
Astronomy Programs, Computer Sciences Corporation, @NASA-GSFC, USA
hrsrobinson@stars.gsfc.nasa.gov

The character of the winds and the amount of mass-loss from stars during the red giant and supergiant stage of evolution are critical to the progression of that evolution. We therefore summarize here the results of a series of programs utilizing the Goddard High Resolution Spectrograph (GHRS) on the Hubble Space Telescope (HST) to study the winds and mass-loss from the surface of evolved, low-gravity cool stars. We discuss the structure of the chromospheres in which the base of these stellar winds lie, the velocity of the wind and its acceleration with height, and what these spectra tell us about the mass-loss from these stars. Because of its relevance to this Joint Discussion, we will also discuss in some detail what we have learned about real-time variability of these phenomena and the implied somewhat erratic nature of the mass-loss process in at least some of these stars.

J. Andersen (ed.), Highlights of Astronomy, Volume 11A, 378.

R CORONAE BOREALIS STARS: LONG-TERM PHOTOMETRIC & SPECTROSCOPIC STUDIES

P.L. COTTRELL, L. SKULJAN, P.M. KILMARTIN AND C. GILMORE
Mt John Univ Obs, Univ. of Canterbury, NZ
p.cottrell,l.skuljan,p.kilmartin,a.gilmore@phys.canterbury.ac.nz

AND

W.A. LAWSON
Dept of Phys, ADFA, AU
wal@phadfa.ph.adfa.oz.au

For more than a decade we have been able to acquire and analyse a significant amount of photometric data of the highly variable R Coronae Borealis (RCB) stars. This has made been possible by a photometric service observing programme instigated at the Observatory. These photometric data have been combined with less extensive spectroscopic coverage, particularly of the decline phase of these stars. These have been supplemented by observations obtained at Mount Stromlo and Siding Spring Observatories for a radial velocity study. Significantly more spectroscopic observations are now being acquired with the development of a new medium resolution spectrograph at Mount John University Observatory.

In this poster we will present recent photometric and spectroscopic results for a number of the RCB stars in our sample. This observational and analysis work can be used to provide further insight into the nature of these stars, their likely progeny and progenitors and the processes that are involved in the formation and evolution of the obscuring dust clouds which cause the decline phase.

J. Andersen (ed.), Highlights of Astronomy, Volume 11A, 379.

OBSERVATION OF THE CRAB PULSAR WITH BEPPOSAX: PHASE RESOLVED SPECTROSCOPY AND STUDY OF THE PULSAR PROFILE.

G. CUSUMANO, C. MACCARONE, T. MINEO, A. SEGRETO AND B.SACCO
IFCAI ,Palermo , Italy
cusumano@ifcai.pa.cnr.it

E. MASSARO
IOA institute of Rome University, Italy
massaro@astrom.astro.it

AND

L. NICASTRO
I. Te.S.R.E. - C.N.R, Bologna , Italy
nicastro@tesre.bo.cnr.it

The Crab Pulsar(PSR 0531+21) was observed by the four Narrow Field Instruments on board the Italian-Dutch satellite BeppoSAX on August and September 1996, during the Science Verificatio Phase. The fine time resolution (15 μs) and the high statistic of the data provided phase histogram of very good quality, well suited for phase resolved spettroscopy over the entire energy band (0.1-300 keV) covered by BeppoSAX payload. In this contribution we present preliminary results of the spectra distribution of the peaks and the inter peak region. Moreover, we also carried out a detailed analysis of the P1 shape and the behaviour of the P2/P1 ratio with the energy.

J. Andersen (ed.), Highlights of Astronomy, Volume 11A, 380.

OUTFLOWING ENVELOPES FROM STARS AT ARBITRARY OPTICAL DEPTHS – A NEW APPROACH TO THE PROBLEM

A.V. DORODNITSYN
Space Research Institute, Moscow, Russia
dora@mx.iki.rssi.ru

We have considered a stationary outflowing envelope accelerated by the radiative force in arbitrary optical depth case. Introduced approximations provide satisfactory description of the behavior of the matter flux with partially separated radiation at arbitrary optical depths. The obtained system of differential equations provides a continuous transition of the solution between optically thin and optically thick regions. We analytically derived approximate representation of the solution at the vicinity of the sonic point. Using this representation we numerically integrate the system of equations from the critical point to the infinity. Matching the boundary conditions we obtain solutions describing the problem system of differential equations. The theoretical approach advanced in this work could be useful for self-consistent simulations of massive star evolution with mass loss.

J. Andersen (ed.), Highlights of Astronomy, Volume 11A, 381.

CCD OBSERVATIONS AND ANALYSIS OF THE W UMA TYPE BINARY SS ARIETIS

WONYONG HAN
Korea Astronomy Observatory
CHUN-HWEY KIM AND JAE WOO LEE
Department of Astronomy & Space Science, Chungbuk National University
AND
HO-IL KIM AND WOO-BAIK LEE
Korea Astronomy Observatory

The *BVR* CCD observations of W UMa-type eclipsing binary SS Arietis were made for ten nights from November 1996 to December 1996 at the Sobaeksan Astronomy Observatory. From the observed light curves, nine new times of minimum lights were derived from the Kwee and van Woerden's method. Improved light elements for this system were determined from these minimum lights with all the published minima. The analysis of the times of minima of SS Ari confirms that the orbital period of SS Ari has been suffering from a sinusoidal variation as the suggestions of other previous investigators (Kaluzny & Pojmanski 1984, Demircan & Selam 1993). The calculation shows that the cyclic period change has a period of about 56.3yrs with an amplitude of about $0.^{d}052$. The period variation has been discussed in terms of two potential mechanisms : 1) the light-time effect due to a hypothetical third body and 2) deformations in the convective envelope of a magnetically active component. The *BVR* light curves of SS Ari observed for about one month showed the existence of cycle to cycle light variations. Long-term light changes of SS Ari are discussed in terms of the period variation of the binary system.

J. Andersen (ed.), Highlights of Astronomy, Volume 11A, 382.

OPTICAL AND NEAR INFRARED OBSERVATIONS OF POST-AGB STARS

T.FUJII
Dept. of Astronomy, Univ. of Tokyo
fujii@mtk.ioa.s.u-tokyo.ac.jp

T.ONO
Nishi-Harima Astronomical Observatory
ono@nhao.go.jp

Y. NAKADA
Kiso Observatory, Univ. of Tokyo
nakada@kiso.ioa.s.u-tokyo.ac.jp

AND

M.PARTHASARATHY
Indian Institute of Astrophysics, Bangalore
partha@iiap.ernet.in

The post-AGB stars rapidly change their spectral energy distribution on their way from the asymptotic giant branch(AGB) to planetary nebula stage. They are generally surrounded by cool and extended dust shells emitting largely in the infrared wavelengths.

A selected sample of the post-AGB candidates were observed using the CCD and IR camera at Kiso and Nishi-Harima Astronomical Observatory. They were classified based on their broad band spectra from B to K' band and their evolutionary stages have been discussed using the simple model of the dust shell around a star.

The light variation of some of these objects were monitored and their light curves are also presented.

J. Andersen (ed.), Highlights of Astronomy, Volume 11A, 383.

PROPERTIES OF THE SUBGIANTS IN VARIOUS TYPES OF ECLIPSING BINARY SYSTEMS

V. G. KARETNIKOV
Astronomical Observatory, Odessa State University
root@astro.odessa.ua

The radius and temperature excesses are found to be largest for the subgiant stars in DS systems and in low-mass and long-period SD systems, being the smallest for the subgiant stars in AR systems and in low-mass and short-period SD-systems. Subgiant stars in AR and in more massive SD systems exhibit even the temperature deficiency which is the largest for the primaries of the AR systems. Necessity is shown to divide the class of SD systems into several groups according to masses and orbital periods while studying their properties.

J. Andersen (ed.), Highlights of Astronomy, Volume 11A, 384.

THE Hγ LINE SPECTRUM OF INTERMEDIATE POLARS

Y. KIM
Dept. of Astronomy and Space Science, Chungbuk National University, Cheongju, Chungbuk 361-763, Korea
ykkim@astro.chungbuk.ac.kr

AND

K. BEUERMANN
Universitäts-Sternwarte, Geismarlandstrasse 11,37083 Göttingen, Germany
beuermann@usw050.dnet.gwdg.de

By using of a model for the propagation of the emitted X-rays through the infalling material and the reemission of the energy deposited by photoabsorption in the optical (and UV) spectral range (Kim & Beuermann, 1995) we calculated Hγ line spectrum of intermediate polars. Photoabsorption of X-rays by the infalling material is the dominant process in forming the observed energy-dependent rotational modulation of the X-ray flux. X-ray and optical modulations are sensitive to model parameters in different ways. In principle these dependences allow us to obtain improved insight into the geometry of intermediate polars. Some results of our calculation will be presented in this paper in comparison with the Hβ spectrum calculated by Kim & Beuermann (1996).

J. Andersen (ed.), Highlights of Astronomy, Volume 11A, 385.

SEARCH FOR SIGNS OF A NEW OUTBURST IN THE QUIESCENT STATE OF CH CYG

D. KOTNIK-KARUZA AND R. JURDANA ŠEPIĆ
Department of Physics - Faculty of Education, Rijeka, Croatia
kotnik@mapef.pefri.hr

Our high-resolution optical spectroscopic survey of the symbiotic star CH Cyg, extending over the quiescent period 1987 - 1989 between two outbursts, shows that the spectrum did not completely lose its symbiotic character. Moreover, signs of radiation of the hot source were present, not only as remnants from the previous outburst, but also as an indication of a new one. Some of the evidences which support this idea are following:

- Variation of excitation and vibration temperatures of neutral metals and TiO moleculs respectively
- Untypical microturbulent velocity dependance of the excitation potential of the lower state for Fe I
- Double-peaked Balmer profiles of Ha and Hb with temporary disappearance of the central absorption
- Presence of forbbiden emission lines [FeII] [SII] [NeIII] together with OIII with complex profile and variable intensity.

J. Andersen (ed.), Highlights of Astronomy, Volume 11A, 386.

THE 283-DAY PERIODICITY IN THE O-C DIAGRAM OF BETA LYRAE

J. M. KREINER AND G. PAJDOSZ
Mt Suhora Observatory, Cracow Pedagogical University, Poland
sfkreine@cyf-kr.edu.pl

A detailed analysis of an extensive collection of primary visual, photographic and photoelectric minima of Beta Lyrae was performed. After removing the effect of the increase in orbital period we applied the Fourier analysis to the O-C residuals. The most prominent periodicity is 283 days with amplitude about 0.05 days. It confirms the period found in the photometric light curves of Beta Lyrae [Van Hamme, Wilson and Guinan (1995)]. The 283-d quasi-variation of the period may be explained by assymetries of the light curve near the minima caused by some instabilities in disc and/or hot spot(s) brightness.

J. Andersen (ed.), Highlights of Astronomy, Volume 11A, 387.

DETECTION OF SUPERHUMPS DURING THE OUTBURST OF BC UMA

CHATIEF KUNJAYA
Bosscha Observatory, Indonesia

AND

TAICHI KATO AND RYUKO HIRATA
Kyoto University, Japan

BC UMa was observed during its rare outburst in 1994. The superhumps with period 0.0619 day are clearly apparent, accompanied by secondary superhumps. This confirms the classification of BC UMa as an SU UMa type dwarf nova. The maximum brightness is 12.5 mag, which is about six magnitude brighter than quiescence phase, and it makes BC UMa be classified as a TOAD (Tremendous Outburst Amplitude Dwarf-nova). The superoutburst's magnitude was fainter than the preceeding normal outburst in September 1990. This faint superoutburst is uncommon among SU UMa stars and also among TOADs. The period analysis of the lightcurve taken after the end of the outburst yields the period of 0.0618 day which is shows that late superhumps is in operation.

J. Andersen (ed.), Highlights of Astronomy, Volume 11A, 388.

ON THE PROFILES AND THE POLARIZATION OF RAMAN SCATTERED EMISSION LINES IN THE SYMBIOTIC STARS: II

K.W. LEE
Kyungpook National Univ., Taegu 702-701, South KOREA
kwlee@vega.kyungpook.ac.kr

AND

H.-W. LEE
Seoul National Univ., Seoul 151-742, South KOREA
hwlee@astro9.snu.ac.kr

A Monte Carlo code is used to calculate the profiles and the polarization of Raman scattered O VI lines($\lambda\lambda$6827, 7088) in symbiotic stars. Various wind models around the cool giant are investigated including disk type and bipolar wind models. We obtain a large degree of polarization in the blue part and the red-asymmetric profile, which is generic to the Raman scattered feature in symbiotic stars.

We also investigate the effect of the orbital motion of the hot component around the cool giant, where the orbital velocity is comparable to the wind terminal velocity. In the case of small inclination angle of the orbital plane, we obtain the polarization flip around the red wing of the scattered feature and the triple peak structure in the polarized flux, which is characteristic of many symbiotic systems including RR Tel.

Brief observational consequences are discussed and it is concluded that spectropolarimetry may provide a powerful diagnostic about the physical conditions of symbiotic stars.

J. Andersen (ed.), Highlights of Astronomy, Volume 11A, 389.

THE CHEMICAL EVOLUTION OF THE MAIN COMPONENT OF THE BINARY SYSTEM υ SAGITARRII

V. LEUSHIN
Special Astrophysical Observatory of RAS, Russia
leushin@sci.lpi.msk.su

V. CHUVENKOV
Rostov State University, Russia
chuvenkov@phys.rnd.runnet.ru

AND

L. SNEZHKO
Special Astrophysical Observatory of RAS, Russia

A model of internal structure and evolution of the peculiar binary system υ Sgr is presented. The model corresponds well to the observed chemical composition of the main component atmosphere (10^{-4} H, 0.844 He, 0.013 C, 0.042 N by mass). It is supposed that about 5 million years ago the main component passed the stage of hydrogen nuclear burning in the shell over the helium core where the helium-carbon reactions took place. Because of convective mixing, the synthesized carbon diffusing into the hydrogen burning zone was catalyzing the reactions of the CN – cycle. This has resulted in anomalies in the chemical composition, particularly high nitrogen abundance in the layer observed now as the atmosphere of the main component since external layers were thrown off during the evolution. Following the calculated results, the quantitative restrictions of temperature and density in the layers and values of mixing parameters are obtained. It is shown that the best agreement with observations exists if the mass of the matter penetrating from the zone of helium-carbon reactions into the helium layer is 0.25 of the helium-carbon core mass. Moreover, the ratio of mass concentration He/C in this matter should equal 2, and the mass share diffusing into the layer of hydrogen burning should be in the range 0.25 - 0.30 of the layer mass.

J. Andersen (ed.), Highlights of Astronomy, Volume 11A, 390.

THE NATURE OF THE CHEMICAL ANOMALIES OF THE SIRIUS A

V. LEUSHIN
Special Astrophysical Observatory of RAS, Russia
leushin@sci.lpi.msk.su

On the basis of the analysis of the observed equivalent line widths of FeI, FeII, and TiII in the spectra of Sirius A and Omicron Peg and calculation of abundances of these elements, the oscillator strengths of the lines used are refined. With the improved oscillator strengths, the iron and titanium abundances in the atmosphere of Sirius A are obtained with a higher accuracy than previously:

lgN(FeI) $= 7.899 \pm 0.011$,

lgN(FeII) $= 7.908 \pm 0.010$,

lgN(TiII) $= 5.30 \pm 0.02$.

The improved accuracy allowed one to conclude that the surface magnetic field is absent in the atmosphere of Sirius A: H =3D 0 ± 100 Gs. The equivalent widths of 7 helium lines of the red region of the spectrum are measured. The calculations of these lines (with allowance for their blending with lines of other elements) show normal helium abundance. Following a comparison of features of Sirius A and Omicron Peg, a suggestion is advanced on possible reasons for the existence of Am stars.

J. Andersen (ed.), Highlights of Astronomy, Volume 11A, 391.

THE VARIATIONS IN LIGHT CURVES OF CONTACT BINARY AU SER

LI ZONG-YUN DING YUE-RONG
Department of Astronomy, Nanjing University, Nanjing 210093, China
email: zyli@netra.nju.edu.cn

Photometry of AU Ser, a contact binary in poor thermal contact, was made using the 1-meter reflector of the Yunnan Observatory, China, in 1991,1992 and 1995. Our observations, together with Binnendijk's(1972), show variations in light curves similar to ones of VW Cep:

1) *Changes in sign and value of O'Connell effect.* According to Binnendijk(1972), the maximum following primary minimum was higher than the other maximum, i.e., the O'Connell effect was positive. In our 1991 and 1992 light curves it became negative. After three years, however, the situation has reversed again.

2) *Variations of the difference between minimum depths.* The difference in 1991 was smaller than that in 1969 and 1970. It got smallest in 1992 and largest in 1995. The difference increases by about 0.3 mag in three years, which is hard to explained.

3) *The phase shift.* The observed minimum phase in 1991, 1992 and 1995 have progressively displaced forward with respect to computed minimum phase according to Binnendijk's ephemeris. The O-C diagram seems to indicate a sudden decrease of orbital period after 1970. Also, short period oscillations have been found at frequencies of 0.0003, 0.005 and 0.0075 Hz. Flickering, with an average amplitude of about 0.05 mag, can be seen in the light curves. The power spectral analysis suggest the existence of oscillations. It is unusual that such oscillations exist in a contact binary.

J. Andersen (ed.), Highlights of Astronomy, Volume 11A, 392.

X-RAY EMISSION DIFFERENCES IN SNR MSH14-63

FANGJUN LU, MEI WU AND TIPEI LI
High Energy Astrophysics Lab, Institute of High Energy Physics, Academia Sinica, Beijing 100039,P.R. China

AND

XUEJUN SUN
High Energy Astrophysics Lab, Institute of High Energy Physics, Academia Sinica, Beijing 100039,P.R. China and Beijing Astronomical Observatory, Academia Sinica, Beijing 100080, P.R. China

In the paper we report the discovery of X-ray emission differences in SNR MSH14-63 based on the ROSAT PSPC observation. The structures of MSH14-63 are different in different energy bands. These images along with the radial brightness distributions in these energy bands show the existence of a region which only emit X-ray photons harder than 1.0keV. Though weak, there actually exist large scale X-ray emissions beyond the bright rim in many parts of the northeast remnant component, which is in conflict with the Sedov phase assumption used in previous research. These new soft X-ray features provide astronomers more information to study its physical characteristics.

J. Andersen (ed.), Highlights of Astronomy, Volume 11A, 393.

NEW CCD OBSERVATIONS OF ECLIPSING BINARY VW CEPHEI

ADBDULRAHMAN ALI S. MALAWI
Astronomy Dept., Faculty of Science, King Abdul Aziz University

New CCD observations of the eclipsing binary VW Cephei were made on the 9th and 10th of October 1996 and on the 11th of Nov. 1996. The light curves and short time-scale variations are discussed. Photometric elements were determined from analyses of the light curves. These new solutions are also discussed.

J. Andersen (ed.), Highlights of Astronomy, Volume 11A, 394.

DISCOVERY OF INFRARED STARS IN GLOBULAR CLUSTERS IN THE MAGELLANIC CLOUDS AND THEIR LIGHT VARIATIONS

S. NISHIDA, T. TANABÉ, S. MATSUMOTO, T. ONAKA AND Y. NAKADA
University of Tokyo, Japan
snishida@mtk.ioa.s.u-tokyo.ac.jp, ttanabe@mtk.ioa.s.u-tokyo.ac.jp
K. SEKIGUCHI
National Astronomical Observatory of Japan, Japan
AND
I. S.GLASS
South African Astronamical Observatory, South Africa

A systematic near-infrared survey was made for globular clusters in the Magellanic Clouds. Two infrared stars were discovered in NGC419 (SMC) and NGC1783 (LMC). NGC419 and NGC1783 are well-studied rich globular clusters whose turn-off masses and ages are estimated $M_{\rm TO} \sim 2.0\ M_{\odot}$ and $\tau \sim$1.2 Gyr for NGC419, and $M_{\rm TO} \sim 2.0\ M_{\odot}$ and $\tau \sim 0.9$ Gyr for NGC1783, respectively.

The periods of the infrared light variations were determined to be 540 d for NGC419IR1 and to be 480 d for NGC1783IR1, respectively. Comparison of the measurements with the period$-K$ magnitude relation for carbon Miras in the LMC by Groenewegen and Whitelock(1996) revealed that the K magnitudes of the infrared stars were fainter by about $0.3 - 0.8$ magnitude than those predicted by the $P - K$ relation. This deviation can be explained if the infrared stars are surrounded by thick dust shells and are obscured even in the K band. The positions of NGC419IR1 and NGC1783IR1 on the $P - K$ diagram suggest that AGB stars with the main sequence masses of about 2 $M_{\odot}$ start their heavy mass-loss when $P \sim 500$ d.

J. Andersen (ed.), Highlights of Astronomy, Volume 11A, 395.

ON THE SHORT TIME SCALE EVOLUTIONARY HISTORY OF THE CONTACT BINARY VW CEPHEI

I. PUSTYLNIK
Tartu Observatory, EE2444, Estonia

We study the short-time evolutionary history of the well-known contact binary VW Cep. Our analysis is based partly on the numerous *UBV* light curves obtained at Tartu Observatory, *IUE* spectra, and samples from the published data. Special attention is given to the effects of asymmetry of the light curves. A higher degree of asymmetry outside the eclipses along with the significant displacements of the brightness maxima in respect to the elongation phase is interpreted as evidence that a considerable portion of the flaring source is concentrated close to the neck connecting the components. We discuss the nature of asymmetry in terms of possible mass exchange and the flare activity and compare the results of our model computations with the record of orbital period variations over the last 60 years.

J. Andersen (ed.), Highlights of Astronomy, Volume 11A, 396.

EXPLORATION OF STELLAR PROCESSES THROUGH INTERSTELLAR ABUNDANCE STUDIES

U. J. SOFIA
Villanova University
usofia@ast.vill.edu

AND

D. M. MEYER
Northwestern University
meyer@elvis.astro.nwu.edu

Interstellar abundances studies can provide answers to some questions about stellar processes, and they can lead to new questions about others. We will discuss some recent interstellar abundance determinations and their implications for stellar studies.

Although Cd and Sn are chemically similar to elements which are moderately to heavily depleted from the gas-phase interstellar medium (ISM), they appear to have solar-like abundances in interstellar clouds. This is likely the result of these elements being depleted from an ISM which has been enriched with Cd and Sn since the formation of the sun. These two elements are mainly produced by s-process neutron capture in intermediate to low mass AGB stars. The present ISM abundances of Cd and Sn can therefore provide information about s-process yields and the cumulative injection of this material into the ISM over the past 4.5 Gy.

Recent studies of C and O abundances in the ISM have shown that the C/H and O/H are constant out to approximately 1500pc from the sun. Local B star abundance studies, however, find that C/O is not constant, thereby indicating that the ISM is not well mixed or homogeneous. We currently do not have an explanation for this apparent inconsistency between stars and the ISM.

J. Andersen (ed.), Highlights of Astronomy, Volume 11A, 397.

SOLITONS IN THE CLOSE BINARY SYSTEMS

KENJI TANABE
Okayama Univ. of Science
tanabe@das.ous.ac.jp

Propagation of the surface waves of the lobe-fillng components of close binary systems is investigated theoretically. Such waves are considered to be analogous to the gravity waves of water on the earth. As a result, the equations of the surface wave in the rotating frame of reference are reduced to the so-called Korteweg-de Vries(KdV) equation and non-linear Schroedinger(NLS) equation according to its "depth". Each of these equations is known to have the solution of soliton. When this soliton is sent to the other component of the binary system through the Lagrangian point,it can give rise to the flare activity observed in some kinds of close binary systems.

J. Andersen (ed.), Highlights of Astronomy, Volume 11A, 398.

A PECULIAR VARIABLE STAR HD229221 – FROM 1994 TO 1997

JUN-JIE WANG AND JING-YAO HU
Beijing Astronomical Observatory, Chinese Academy of Sciences
wangjj@bao01.bao.ac.cn, hjy@bao01.bao.ac.cn

From 1994 to 1997, we observed the star HD229221 several times spectroscopically by using the Beijing Astronomical Observatory 2.16 m telescope and the spectrograph Universal.

From the spectra we clearly see that the H-α spectral feature of this star underwent a peculiar variation, from a single emission line to one composed of three components, i.e. emission+absorption+filled emission. In our paper, based on an analysis of its spectral line variation and some physical parameters (e.g. luminosity, age, etc.), we discuss the possible reasons that might cause this peculiar variation.

J. Andersen (ed.), Highlights of Astronomy, Volume 11A, 399.

THE LUMINOSITY FUNCTION OF THE MAIN SEQUENCE STARS IN THE SOLAR NEIGHBORHOOD

YANG TING-GAO
Haanxi Astronomical Observatory, Academia Sinica, Shaanxi 710060
SHU CHENG-GANG, FU CHENG-QI AND JIANG DONG-RONG
Shanghai Astronomical Observatory, Academia Sinica, Shanghai 200030
AND
PENG QIU-HE
Department of Astronomy, Nanjing University, Nanjing 210093

The Third Catalogue of Nearby Stars (CNS3) was analyzed. A study of the stellar luminosity function in the solar neighborhood with CNS3 was described. The luminosity function for main sequence stars derived from CNS3 was compared with that from CNS2 and that based on the method of photometric parallaxes. The results from CNS3 for stars with $M_v < 15.5$ were well defined. Luminosity functions for the giants and A, F, G, K and M type main sequence stars are also given, based on CNS3.

J. Andersen (ed.), Highlights of Astronomy, Volume 11A, 400.

MODELLING OF LIGHT CURVES OF ALGOL-TYPE SYSTEMS WITH ACCRETION DISKS: WW AND

S. ZOLA
Astronomical Observatory, Jagiellonian Unversity, ul. Orla 171, 30-244 Cracow, Poland
sfzola@cyf-kr.edu.pl
and
Mt. Suhora Observatory, Pedagogical University, ul. Podchorazych 2, 30-084 Cracow, Poland

Olson's light curves of the Algol-type binary WW And, showing evidence of harbouring an accretion disk, were re-analysed. Two approaches were attempted to find a solution: the first one, the Roche geometry model assuming the system consisting of two stars, and the second one, including also the existence of an accretion disk around the hotter component. In both cases the Monte Carlo method was employed to search for the best solution. We found that adding an additional parameter, namely the third light, the data can be satisfactorily reproduced with only a marginal underfilling of the Roche lobe of the cool star, for the Roche geometry. A thorough search within the semidetached configuration, incuding an accretion disk around the hotter component, was performed. Although such a fit is slightly worse (as measured by the *rms* error) than that for the Roche geometry, it can explain the observed spectroscopic features. Absolute parameters of the components of this interacting binary are presented for both models.

J. Andersen (ed.), Highlights of Astronomy, Volume 11A, 401.

JOINT DISCUSSION 8: STELLAR EVOLUTION IN REAL TIME, PART V

CONCLUSIONS

There is no reason to believe that the prospects expressed in Part III are the ultimate ones for development of evidences about stellar evolutionary information. Instead, (and just as for all other domains of science) technology advances and the availability of funds and manpower will remain important discriminants for new knowledge. As it happened, a presentation of some significant and *imminent* and distant hardware programs was given at the JD8 and it has been chosen (out of sequence with its order in Kyoto) to end this report.

J. Andersen (ed.), Highlights of Astronomy, Volume 11A, 402.

STELLAR EVOLUTION IN REAL TIME: CURRENT AND FUTURE EXPECTATIONS

YOJI KONDO
NASA Goddard Space Flight Center

1. Introduction

Directly measuring stellar evolution may be accomplished in at least two distinctly different ways. The first approach will be observing it in real time as the phenomenon is taking place; the examples will include explosive events in recurrent novae, stel- lar oscillations in Cepheid and RR Lyrae variables, and mass flow in and out of interacting binaries. The second approach will be to observe a number of stars that are believed to belong to the same class at different stages of their evolutionary paths.

In this talk, I will focus on the first method of direct observations. The most direct method will probably be to observe images of the events that we wish to investigate. The same astronomical phenomena can and should also be studied using high-time and spectroscopic resolution data; however, this is already being done with increasingly better spectral-and-temporal resolutions and will probably be reported elsewhere in this Joint Discussion. It is also noted that over the past two decades multiwavelength observations, using various telescopes on the ground and in space, have come to play a crucial role in astronomy, although I am not addressing this subject in this paper. I will talk primarily about imaging using interferometry that will soon become possible due to the advances in technology.

2. Observing Stellar Evolution in Real Time with Interferometry

2.1. GROUND-BASED TELESCOPES

1) An array of five 1-meter telescopes at Mount Wilson with a maximum separation of 150 meters is being constructed at Mount Wilson under the leadership of Harold McAlister of the Center for High Angular Resolution Astronomy (CHARA) at Georgia State Uni- versity. Adaptive optics will be used for the mirrors. [As a point of interest, the 100-inch telescope at Mt. Wilson is presently used successfully with adaptive optics]. An approxi- mate resolution of 10×10^{-9} radian will have an image resolution of 10×10^{11} cm at a distance of 10 light years. The base can be ex- tended to a 1,000 meters if desired. The first light is expected before the end of this century, which will come in three years.

2) Interferometry technology is also being applied to the Keck 10-meter telescope at Mauna Kea in Hawaii and to the 5-meter reflector at Palomar Mountain. As presently envisaged, the baselines for those two projects will be shorter than that for the CHARA array; their mirrors will however have greater apertures, which should be an asset in achieving high angular resolution.

J. Andersen (ed.), Highlights of Astronomy, Volume 11A, 403.

2.2. TELESCOPES IN FREE ORBIT

There currently are three projects under active planning at NASA.

1) *Space Interferometry Mission (SIM)* with five 40-cm class telescopes on a 10-meter beam (baseline) is being planned for early next century.

2) *Deep Space Three Mission*, which involves an array of free-orbiting telescopes, will test the technology crucial to future interferometry in space; for this technology to work, we will need to know the positions of each telescope to within a few percent of the wavelength at which observations are being made. This technology will be crucial in realizing *astronomically-long baseline* interferometry in the future; attainable resolution will become virtually unlimited with it.

3) *Terrestrial Planet Finder (TPF) Mission* on a 50-to-75 meter platform will be sent to beyond Jupiter's orbit. This is to avoid the background noise from the zodiacal light within our solar system. [Not much can be done about any zodiacal light background at the other end.]

4) We should perhaps mention that the *New Generation Space Telescope (NGST)* with an aperture of 8 meters will improve the resolution currently attained with the 2.4-meter Hubble Space Telescope by more than a factor of 3.

2.3. INTERFEROMETRIC TELESCOPES ON THE MOON

Even before the technology being tested in *Deep Space Three* (B: 2) becomes available, we can consider the construction of an array of interferometric telescopes on the Moon; our natural satellite is essentially an immense optical bench in vacuum. Using innovative technologies (Chen, Kondo & Oliversen), the cost of placing a one-meter class telescope on the lunar surface can now be brought down to the range of NASA's MIDEX projects, i.e., to the several times $10M range. Once reusable launch vehicles (RLVs) reduce the cost of placing payloads in low Earth orbit by a factor of ten to a hundred, the cost of getting telescopes to the lunar surface can decline correspondingly, making the concept of interferometry from the Moon even more economical.

The natural curvature of the Moon will give us a baseline of about a kilometer. However, this baseline can be extended significantly using various methods, including the use of struts, lunar crater limbs, and optical relays. In this way, we can eventually attain a resolution of 10×10^{-12} radian or higher, which should give us a $10 \times 10^{+9}$ cm resolution or better for objects at a distance of some 1,000 light years. (As a point of comparison, the solar diameter is greater than $10 \times 10^{+11}$ cm.)

See: Chen, P., Kondo, Y. & Oliversen, R., in JD-22 "Astronomy from the Moon", *Highlights of Astronomy* for the 1997 IAU General Assembly.

JD9

Future Large Scale Facilities in Astronomy

Chairperson: **H. Butcher**

MANUSCRIPT NOT RECEIVED

JD10

Low Luminosity Stars

Chairperson and Editor: **J. J. Binney**

JD 10: LOW-LUMINOSITY STARS

SCIENTIFIC ORGANIZING COMMITTEE
T. Axelrod, J.J. Binney (Chairman), A.S. Burrows, G.S. da Costa, M. Grenon, T. Nakano, M.A.C. Perryman, H.B. Richer, J.A. Sellwood

Most of the mass near the Sun takes the form of stars less luminous than the Sun. A central problem of Galactic astronomy is the determination of how much mass these stars contain.

The only thoroughly reliable way of detecting mass is through its gravitational field. Classically this has been done by studying the Galaxy's rotation curve and the Oort limit, both of which remain active areas of research. In the last few years an exciting new way of probing the Galaxy's gravitational field has been opened up by large surveys for microlensing events. These surveys are yielding important information about the numbers of low-mass objects both in the disk and above it. Another truly dynamical probe for low-luminosity objects is provided by studies of clusters. JD 10 reviewed what we have learned about the density of low-luminosity stars from each of these approaches.

An important orthogonal approach to determining the density of low-luminosity stars involves seperately determining the luminosity function and the mass-luminosity relation for stars of a given spectral class (e.g. main-sequence stars) and then combining them to obtain the mass function. The development of infrared detectors and the refurbishment of the The Hubble Space Telescope have recently yielded important advances in each of these areas. A major difficulty with this line of research is the fact that, in the neighbourhood of $M = 0.08\,\mathrm{M}_\odot$, the main-sequence mass-luminosity relation is (i) steep and non-linear, and (ii) dependent not only on the masses of the faintest stars, but also on their ages, metallicities, rotation rates, binarity etc. Moreover, cool low-luminosity stars have extremely complex spectra, which are difficult to simulate with radiative transfer calculations. Similar difficulties are encountered in the determination of the mass-luminosity relation for white dwarfs. JD 10 reviewed the current status of these very difficult problems.

Programme

Disk mass from large-scale dynamics	J.A. Sellwood
The Oort limit	J.J. Binney
Results of microlensing surveys	D. Bennett
Prospects for microlensing surveys	A. Udalski
Dynamical constraints on the mass function	N.W. Evans
Main-sequence $L(M, \mathrm{age}, Z)$ from models	A. Burrows
The mass–luminosity relation from binaries	D. Latham
Spectra of M dwarfs	Takashi Tsuji
Ground-based surveys for low-luminosity stars	H.R.A. Jones
Space-based surveys for low-luminosity stars	A. Gould
Very low-luminosity objects in star-forming regions	M. Tamura
Star formation: can there be a break in the IMF near $M = 0.1\,M_\odot$?	T. Nakano
White-dwarf cooling curves and searches for white dwarfs	M.A. Wood
Contribution of white dwarfs to cluster masses	T. von Hippel
Low-luminosity stars: past and future	I.N. Reid

J. Andersen (ed.), Highlights of Astronomy, Volume 11A, 409.

DISK MASS FROM LARGE-SCALE DYNAMICS

J. A. SELLWOOD
Rutgers University
Dept. of Physics & Astronomy, PO Box 849, Piscataway, NJ 08855, USA

1. Introduction

The radial distribution of mass in a disk galaxy is strongly constrained by its rotation curve. The separate contributions from the individual stellar populations and dark matter (DM) are not easily disentangled, however, especially since there is generally no feature to indicate where the component dominating the central attraction switches from luminous to dark matter. Here I summarize three recent thesis projects at Rutgers University which all suggest that DM has a low density in the inner parts of bright galaxies, and that most of the mass therefore resides in the disk. In addition, I present some preliminary work on the Milky Way. If we are able to determine the M/L of a typical disk stellar population, it should provide a useful constraint on the numbers of low mass stars.

2. Masses of galactic disks

Palunas & Williams (1997) have I-band surface photometry and 2-D kinematic maps of 76 southern disk galaxies. They calculate the gravitational field in each case, assuming a constant M/L for the disk and bulge, and find that the shape of the inner rotation curve is very well reproduced by the light distribution for $M/L_I \sim 2 - 3$ in most cases (for $H_0 = 75$ km s^{-1} Mpc^{-1}). Mass discrepancies indicative of the dark halo contribution become noticeable only in the outer parts of some galaxies.

Weiner (1997) models the 2-D flow pattern of gas in the barred galaxy NGC 4123 by calculating gas flows in a model potential derived from the I-band photometry. He finds that a $M/L_I \sim 2.5$ is again required for the predicted flow pattern to match that observed. As a result, almost all the mass in the inner galaxy must be distributed as the luminous component, else the rotation curve in the outer disk would be too high.

Debattista & Sellwood (1997) use fully self-consistent simulations of barred disks embedded in halos to verify Weinberg's (1985) prediction that the rotation rate of the bar should quickly decrease through dynamical friction. In even a moderate density halo, the bar slows dramatically as it loses angular momentum and the ratio of corotation radius to bar semi-major axis quickly becomes inconsistent with observed properties of barred galaxies. Only if the central halo density is low, and the disk is massive, can rapid braking be avoided.

3. The Milky Way

Both Spergel et al. (1996) and Freudenreich (1997) have proposed models for the 3-D luminosity density in the Milky Way that are based on the all-sky multi-color photometry obtained by the DIRBE instrument on the COBE satellite. Both groups take extinction by dust into account, but in different ways, and then fit parameters of an idealised model to the projected sky brightness. While the volume luminosity density distributions are different in the two models, the integrated luminosities for the Milky Way in the different IR color bands (see Mulhotra et al. 1996) are very similar. Freudenreich gives a useful conversion to bolometric luminosity for the whole galaxy: $L_{bol} = 2.3 \times 10^{10}\ L_\odot$ ($\pm 20\%$).

We can use these models to derive a mass density for the light distributions, provided a constant M/L assumption is valid. I make this assumption in the present work, but it should be borne in mind that localized excesses of supergiant stars may cause features in the IR light distribution that

J. Andersen (ed.), Highlights of Astronomy, Volume 11A, 410–411.

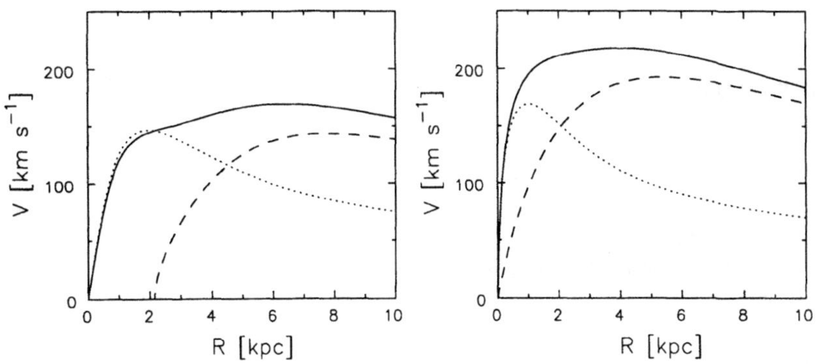

Figure 1. Axially symmetrised "rotation curves" derived from the photometric models of the Milky Way proposed by Freudenreich (left) with $M/L_{bol} = 2$ and Spergel et al. (right) $M/L_{bol} = 3$. The dotted (dashed) curve shows the bar/bulge (disk) contributions, the full curve is their sum.

have no counterparts in the mass; one such feature may be the IR bright ring at 3 kpc where the molecular gas density is high, which could be responsible for the apparently short scale length for the disk. It is also questionable whether the disk and bulge/bar should be assigned the same M/L.

I determine my single free parameter, the M/L for the Milky Way, by requiring that the surface density of luminous mass at the Sun be approximately 50 $M_\odot$ pc^{-2}, in accordance with recent determinations (Kuijken & Gilmore 1991). A safer normalization can be made by modelling the gas flow in the barred region (Englmaier & Gerhard, this meeting) and it is encouraging that they also find that the mass of the inner galaxy is dominated by luminous material.

Both models have two stellar components, the bulge/bar and a disk. The rotation curve obtained by solving for the gravitational field is shown in Figure 1; the inner "rotation curve" of these barred galaxy models is simply derived from the azimuthal average of the central attraction. Freudenreich's model S has a hole in the disk; setting $M/L_{bol} = 2$ gives a surface density of 46 $M_\odot$ pc^{-2} at $R = 8$ kpc. The model preferred by Spergel et al. has not yet been published, but Binney, Gerhard & Spergel (1997) give one fit which has almost the same total light as Freudenreich's, even though the disk in this model does not have a hole. The circular velocity curve shown in Figure 1(b) results from setting $M/L_{bol} = 3$ to give a local surface density of 49 $M_\odot$ pc^{-2}. The central brightness, extrapolated inwards from 3 kpc, is very high in this model and clearly overstates the disk light in this region; a greater fraction of the light probably should be attributed to the bar.

4. Conclusions

Both studies of external galaxies, as well as of the Milk Way, suggest that the disk component contains most of the mass in the bright inner parts of high luminosity galaxies. The data seem to suggest a $M/L_I \sim 2 - 3$ for the stellar population in typical galaxy disk. Very preliminary fits of constant M/L models for the Milky Way suggest $M/L_{bol} \sim 2 - 3$ (a value that also indicates little DM in the inner Galaxy).

This work was supported by NSF grant AST 93/18617 and NASA Theory grant NAG 5-2803.

References

Binney, J., Gerhard, O. E. & Spergel, D. (1997) *MNRAS*, **288**, 365
Debattista, V. P. & Sellwood, J. A. (1997) *ApJ Lett*, to appear
Freudenreich, H. T. (1997) astro-ph/9707340
Kuijken, K. & Gilmore, G. (1991) *ApJ Lett*, **367**, L9
Mulhotra, S., Spergel, D. N., Rhoads, J. E. & Li, J., (1996) *ApJ*, **473**, 687
Palunas, P. & Williams, T. B. (1997) *ApJ Suppl*, to appear
Spergel, D. N., Mulhotra, S. & Blitz, L. (1996) In *Spiral Galaxies in the Near-IR*, ed. D. Minnitti & H-W. Rix (Berlin: Springer) 128
Weinberg, M. D. (1985) *MNRAS*, **213**, 451
Weiner, B. (1997) PhD Thesis, Rutgers University

THE OORT LIMIT

J.J. BINNEY
Theoretical Physics
Keble Road, Oxford OX1 3NP, U.K.

1. Introduction

During galaxy formation non-baryonic dark matter is not expected to pack down into a thin disk. Hence the mass-to-light ratio $\Upsilon_{\rm d}$ of the Galactic disk is a key number as the mass-to-light ratio of a purely baryonic system. If $\Upsilon_{\rm d}$ proves to be large, it suggests that the large values of Υ that are detected elsewhere in the Universe are due to dark baryonic matter rather than some exotic form of matter. Conversely, a small value $\Upsilon_{\rm d}$ lends credence to the idea that most of the mass in the Universe is non-baryonic by suggesting that baryonic systems are well endowed with luminous stars.

2. Basic formulae

In cylindrical polar coordinates (R, ϕ, z) Poisson's equation may be written

$$4\pi G\rho = \frac{1}{R}\frac{\partial v_{\rm c}^2}{\partial R} - \frac{\partial K_z}{\partial z}, \tag{1}$$

where $v_{\rm c}$ is the circular speed and K_z is the vertical component of the gravitational field. The term $\partial v_{\rm c}/\partial R$ can be evaluated in terms of the Oort constants and proves to be negligible. The Jeans equation governing the vertical distribution of a tracer population of number density ν can be written

$$\nu K_z - \frac{\partial}{\partial z}(\nu\langle v_z^2\rangle) = \frac{\partial}{\partial R}(\nu\langle v_R v_z\rangle) + \frac{\nu}{R}\langle v_R v_z\rangle \tag{2}$$

Near the plane we expect that $\langle v_R v_z\rangle \simeq \alpha\langle v_R^2 - v_z^2\rangle z/R$ with $0 < \alpha < 1$ and $\nu\langle v_i^2\rangle \sim {\rm e}^{-2R/R_{\rm d}}$, so the r.h.s. of (2) evaluates to $-(2\alpha/R_d - 1/R)\langle v_R^2 - v_z^2\rangle\nu z/R$. For plausible values of α and $R_{\rm d}$ ($\alpha = 0.5, R_{\rm d} = R_0/3$) the ratio of the r.h.s of (2) to a typical term on its l.h.s. is r.h.s./l.h.s. $\sim hz/R_0^2$, where $h \simeq 300\,{\rm pc}$ is the scale height of the disk. This argument suggests that the r.h.s. of (2) can be neglected. Combining the resulting approximate equation with (1) and integrating the result through the disk we obtain successively

$$4\pi G\rho = -\frac{\partial}{\partial z}\left(\frac{1}{\nu}\frac{\partial}{\partial z}(\nu\langle v_z^2\rangle)\right) \quad \text{and} \quad 2\pi G\Sigma(z) \simeq -\frac{1}{\nu}\frac{\partial}{\partial z}(\nu\langle v_z^2\rangle), \tag{3}$$

where $\Sigma(z) = \int_{-z}^{z} {\rm d}z\,\rho$.

The bad news is that these equations require differentiation of the measured quantities ν and $\langle v_z^2\rangle$. In the approximation in which motions in and perpendicular to the plane are decoupled, the good news is that *both* $\nu(z)$ and $\langle v_z^2\rangle(z)$ are determined by a single unknown function, the distribution of tracers with respect to vertical energy $E = \frac{1}{2}v_z^2 + \Phi$. Kuijken & Gilmore (1989) and Créze et al. (1997) exploit this connection: they adopt a parametric form of $\Phi(z)$ and use it and the measured form of $\nu(z)$ to determine $f(E)$, from which they can predict the distribution of velocities at any z. The parameters in $\Phi(z)$ are chosen to optimize the fit between the predicted and measured velocity distributions.

Kuijken & Gilmore applied this technique to a sample of 512 K-dwarfs that were identified in a photometric survey of fields near the SGP. Since such tracer stars lie in a cone with the Sun at

J. Andersen (ed.), Highlights of Astronomy, Volume 11A, 412–415.

the apex, there are few stars near $z = 0$, and the sample cannot constrain ρ_0. Kuijken & Gilmore (1991) concluded that the sample most srongly contrains

$$\Sigma(1.1\,\mathrm{kpc}) = 71 \pm 6\,\mathrm{M}_\odot\,\mathrm{pc}^{-2}. \tag{4}$$

The samples of Créze et al. are formed by stars with $M_V < 2.5$ in the Hipparcos Catalogue. Depending on the distance out to which they include stars, their samples contain 3 000 – 2 300 stars. Since they require v_z but do not have radial velocities for most stars, their velocity data are confined to the proper motions of the 17% of the samples that lie at $|b| \leq 10°$. They conclude that $\rho_0 = 0.076 \pm 0.015\,\mathrm{M}_\odot\,\mathrm{pc}^{-3}$.

Bahcall (1984) and Bahcall, Flynn & Gould (1992) chose rather to parametrize $f(E)$. Specifically, they assumed that the tracer population was isothermal: $f \propto \mathrm{e}^{-E/\sigma^2}$. Indeed, they assumed that *every* mass-bearing population is isothermal with some measured velocity dispersion σ_i. Then $\rho_i(z) = \Upsilon_i \nu_i(z) = \rho_{i0}\mathrm{e}^{-\Phi/\sigma_i^2}$ and Φ has to solve the non-linear o.d.e.

$$\frac{\mathrm{d}^2\Phi}{\mathrm{d}z^2} = 4\pi G \sum_i \rho_i = 4\pi G \sum_i \rho_{i0}\mathrm{e}^{-\Phi/\sigma_i^2}. \tag{5}$$

This equation was repeatedly solved for assumed values of the parameters ρ_{i0} until the predicted density law of a given tracer, $\nu_i(z) = (\rho_{i0}/\Upsilon_i)\mathrm{e}^{-\Phi/\sigma_i^2}$, agreed with the measured density profile.

This technique was applied to old samples of F and K giants by Bahcall (1984), and to a new sample of K giants by Bahcall et al. (1992). In each study at least one population required an unexpectedly large value of Υ. Different solutions $\Phi(z)$ were obtained depending on the population that was assigned the anomalous value of Υ. Bahcall et al. (1992) concluded that

$$\Sigma(1.1\,\mathrm{kpc}) = 83\,\mathrm{M}_\odot\,\mathrm{pc}^{-2} \quad \text{and} \quad \rho(0) = 0.26\,\mathrm{M}_\odot\,\mathrm{pc}^{-3} \quad \text{with errors} \sim \pm 0.1\,\mathrm{dex} \tag{6}$$

This estimate of $\Sigma(1.1\,\mathrm{kpc})$ agrees with that of Kuijken & Gilmore within the errors, while the Bahcall et al. value of $\rho(0)$ is incompatible with that of Crezé et al. Since the Bahcall et al. samples, like those of Kuijken & Gilmore, are confined to small patches of the sky, while those of Crezé et al. are all-sky, the newer value, $\rho(0) \simeq 0.076\,\mathrm{M}_\odot\,\mathrm{pc}^{-3}$ is surely to be preferred.

3. Estimating Υ_d

Once $\Sigma(1.1\,\mathrm{kpc})$ has been estimated, two things remain to be done before Υ_d can be evaluated: (i) estimate how much of $\Sigma(1.1\,\mathrm{kpc})$ is attributable to the disk rather than any spheroidal halo; (ii) estimate the disk's surface luminosity density L_d.

Kuijken & Gilmore argued that, near $z = 1.1\,\mathrm{kpc}$, Φ could be approximated by the first two terms in its Maclaurin series: $\Phi(z) = Kz + Fz^2$. Physically, K quantifies the mass within $|z|$ of the plane while F quantifies the equatorial halo density. The Kuijken & Gilmore sample imposes a single constraint on K and F: $K_z = K + 2z_0F \simeq$ const. with $z_0 \simeq 1.1\,\mathrm{kpc}$. If the halo were spherical, the Galaxy's circular-speed curve would impose a second constraint: $0.094K + F \times 1\,\mathrm{kpc} = 15 \pm 3\,\mathrm{M}_\odot\,\mathrm{pc}^{-2}$. Combining these constraints Kuijken & Gilmore concluded that $\Sigma_\mathrm{d} = 48 \pm 9\,\mathrm{M}_\odot\,\mathrm{pc}^{-2}$.

Dehnen & Binney (1997) reached an essentially identical conclusion by a rather different route: by an analysis of global Galactic mass models similar to that of Caldwell & Ostriker (1981), they concluded that (4) implies that $\Sigma_\mathrm{d} = 50 \pm 2\,\mathrm{M}_\odot\,\mathrm{pc}^{-2}$ unless either R_d is unexpectedly short ($\lesssim R_0/4$) or the halo is highly flattened ($b/a \leq 0.3$).

The interstellar medium is thought to contribute $\Sigma_\mathrm{ISM} \simeq 13\,\mathrm{M}_\odot\,\mathrm{pc}^{-2}$ to the local surface density (Bahcall et al. 1992), so dynamical estimates of the local disk mass imply that stars contribute $\Sigma_* \simeq 37\,\mathrm{M}_\odot\,\mathrm{pc}^{-2}$ to the local surface density.

From K-band *SPACELAB* data Kent et al. (1991) infer that the Galaxy's surface luminosity density at R_0 is $I(R_0) = 70\,\mathrm{L}_\odot\,\mathrm{pc}^{-2}$ for a solar K-band absolute magnitude $M_K(\odot) = 3.41$. By comparison, from *COBE/DIRBE* data Malhotra et al. (1996) find that the Galaxy has $M_K = -24.06$, or $L_K = 9.7 \times 10^{10} L_\odot$, 44% larger than the value $L_K = 6.7 \times 10^{10} L_\odot$ inferred by Kent et al. Thus it seems unlikely that $I(R_0)$ is significantly smaller than $70\,\mathrm{L}_\odot\,\mathrm{pc}^{-2}$. In view of the estimate given above of Σ_* we conclude that the local K-band mass-to-light ratio is $\Upsilon_K \lesssim 0.52$.

Bruzual & Charlot (1993) estimate that a system in which the star-formation rate declines exponentially with time constant $\tau = 5\,\mathrm{Gyr}$ and the IMF has Salpeter's form down to $0.1\,\mathrm{M}_\odot$, has

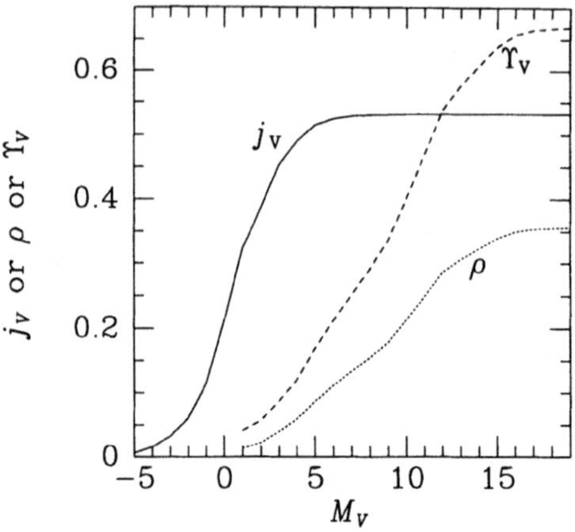

Figure 1. The full curve shows the V-band luminosity (in $10^{-1} L_\odot\,\mathrm{pc}^{-3}$) in stars brighter than M_V. The dotted curve shows the mass (in $10^{-1}\,\mathrm{M}_\odot\,\mathrm{pc}^{-3}$) in these stars. The dashed curve shows the corresponding values of Υ_V.

$\Upsilon_K \simeq 0.74$ after 8 Gyr and $\Upsilon_K = 0.87$ after 12 Gyr. Thus the local mass-to-light ratio is lower than a Salpeter IMF implies, and the IMF should flatten significantly above $0.1\,\mathrm{M}_\odot$.

Recently Gould et al. (1996) used HST to determine the luminosity function and the vertical density profile of low-mass stars. Their luminosity function agrees with that of Stobie et al. (1989) in turning down sharply faintward of $M_V = 12$, with the result that they concluded that M stars contribute only $\Sigma_\mathrm{M} \simeq 13.5 \pm 3\,\mathrm{M}_\odot\,\mathrm{pc}^{-2}$ to the local surface density. When this figure is added to the surface density in stars other than M stars, $\sim 15\,\mathrm{M}_\odot\,\mathrm{pc}^{-2}$, we obtain $\Sigma_* \simeq 28.5\,\mathrm{M}_\odot\,\mathrm{pc}^{-3}$, which is smaller than the corresponding dynamical estimate.

The luminosity functions of Stobie et al. and Gould et al. both rely on photometric distances. At $M_V > 12$ luminosity functions that are based on trigonometric parallaxes of high-proper-motion stars lie consistently above those based on photometric parallaxes and therefore predict that there is more mass in M stars. For example, the luminosity function that Jahreiß and Wielen (1997) obtain from Hipparcos parallaxes of stars within 25 pc of the Sun is flat from $M_V \simeq 12$ to $M_V \simeq 16.5$. When this luminosity function is combined with older data for luminous stars and the main-sequence mass-luminosity relation of Kroupa et al. (1993), one finds that stars brighter than $M_V = 18$ have a cumulative mass-to-light ratio $\Upsilon_V = 0.62$ – see Fig. 1. Observations of external galaxies that are similar to our own (de Jong 1996) suggest that the local disk has $V - K \simeq 2.8$. If the disk is this red, then $\Upsilon_V = 0.62$ corresponds to $\Upsilon_K = 0.17$, which is smaller even than the work of Malhotra et al. would suggest. Hence even with the larger number of low-mass stars that are found by parallax-based luminosity functions, the local mass-to-light ratio is surprisingly small.

In summary: estimates of the local luminosity surface density combined with dynamical estimates of the local mass surface-density yield a mas-to-light ratio Υ_K that is smaller than expected from a population of appropriate age with Salpeter IMF right down to $0.1\,\mathrm{M}_\odot$. On the other hand, counts of faint stars yield an even smaller value of Υ_K than is observed. This suggests that there are more low-luminosity stars than even parallax-based surveys find, but fewer than a Salpeter IMF predicts.

References

Bahcall J.N., 1984, ApJ, 287, 926
Bahcall J.N., Flynn C., Gould A., 1992, ApJ, 389, 234
Caldwell J.A.R., Ostriker J.P., 1981, ApJ, 251, 61
Créze M., Chereul E., Bienaymé O., Pichon C., 1998, A&A submitted
Dehnen W., Binney J.J., 1997, MNRAS in press
de Jong R.S., 1996, A&A, 313, 377
Gould A., Bahcall J.N., Flynn C., 1996, ApJ, 465, 759
Jahreiß, H., Wielen, R., 1997, in "Hipparcos – Venice '97", ed. B. Battrick, ESA, Noordwijk, p. 675
Kent, S.M., Dame, T.M., Fazio, G., 1991, ApJ, 378, 131

Kroupa, P., Tout C.A., Gilmore G., 1993 MNRAS, 262, 545
Kuijken K., Gilmore G., 1989, MNRAS, 239, 650
Kuijken K., Gilmore G., 1991, ApJL, L367, L9
Malhotra, S., Spergel D.N., Rhoads J.E., Li J., 1996, ApJ, 473, 687
Stobie R.S., Ishida K., Peacock J.A., 1989, MNRAS, 238 709

DYNAMICAL CONSTRAINTS ON THE MASS FUNCTION

N. W. EVANS
Theoretical Physics, 1 Keble Rd, Oxford, OX1 3NP, UK

1. Introduction

The aim of this paper is to summarise the dynamical constraints on the contribution of low luminosity stars. Important restrictions come from disk stability and the Oort Limit. These are reviewed elsewhere in these Proceedings in the contributions from Sellwood and Binney. Here, our attention is concentrated on two further pieces of evidence – constraints from (1) the stochastic heating of wide binaries and from (2) microlensing.

2. Wide Binaries

The physical principles underlying the constraint from the stochastic heating of wide binaries are easy to understand. Consider two stars in circular orbit with a combined mass of 1 $M_{\odot}$. Suppose they have a relative velocity of $\sim 0.1\,\mathrm{km\,s^{-1}}$. Perturbers encounter the binary at a speed some two orders of magnitude greater than this. As the binary gains energy, it expands and eventually it disrupts. The more perturbers, the fewer wide binaries. Therefore, the goal is to constrain the mass and the density of the unknown perturbers using the observed distribution of binaries and models of their evolution. This is a subject that starts with observational work by Bahcall & Soneira (1981) that appeared to detect a cut-off in the separation of wide binaries at about 0.1 pc. Retterer & King (1982) calculated that a binary with a semi-major axis of 0.1 pc will disrupt in roughly a Galactic age (10 Myrs). Bahcall, Hut & Tremaine (1985) then extended these arguments to include a population of dark matter perturbers of unknown mass. By fixing the density of the perturbers (at the value suggested by the kinematics of vertical tracer populations), they found the characteristic mass of the perturbers must be less than two solar masses. This work stimulated a major effort by Martin Weinberg and co-workers (1987, 1990), who provided a substantial theoretical framework with which to analyse the problem. They argued, first, that the observed break in the semi-major axes is consistent with being a sampling or selection effect, and, second, that no sharp cut-off is expected anyhow. As this spoilt the original claim, Weinberg (1987) then looked to see if a dynamical constraint on low mass objects could be restored somehow. Encounters are catastrophic if a single event disrupts the wide binary. Encounters are diffusive if many events are needed to disrupt the binary. Encounters with stars and low mass objects predominate and are typically diffusive, whereas encounters with Giant Molecular Clouds are catastrophic. Given that the dominant physical processes are diffusive, it is natural to treat the problem with a Fokker-Planck equation. Explicit forms for the diffusion coefficients are available, as are the Greens functions (Weinberg 1990). Given any assumed form for the binary birth-rate, the endpoint of the population can be determined. Of course, the binary birth-rate is not well-understood observationally, so it is not clear exactly what should be assumed for this unknown. However, a reasonable assumption is that the dependence is separable in energy and time, with a power-law dependence on energy (or equivalently semi-major axis). Weinberg actually examined models in which the binaries are created continuously and are created in bursts. The original idea of Bahcall, Hut & Tremaine (1985) was to use the shape of the distribution in semi-major axes – a sharp cut-off. However, Weinberg's analysis make it clear that this is untenable. Diffusion smoothes things out, so sharp cut-offs are not expected. The other possibility is a limit based on overall normalisation. For a particular model, the fraction of total number of binary pairs ever created may be computed from the binary birth rate. If we observe n_w

J. Andersen (ed.), Highlights of Astronomy, Volume 11A, 416–418.

wide binaries in a given volume of space containing n_c of stars in total, then we have the obvious limit that $n_c > 2n_w/f$. However, even here, it is sad to report that the best fit model does not give an interesting limit on the mass function of the dark objects. Weinberg (1990) found he could constrain the masses of perturbers in excess of twenty solar masses in some model forms for the binary birth rate, but not the low mass end of the mass function.

Given the recent resurgence of interest in careful astrometry, it is clearly worthwhile trying to build better catalogues of wide binaries. Whether the stochastic heating of wide binaries will ever give definite dynamical constraints on the mass function is less clear – the limitations seem partly to do with the scaling of the diffusion coefficients with the masses of the deflectors. Low mass deflectors are simply much less effective at heating than high mass ones.

3. Microlensing

Let me now pass to microlensing where the prospects are more upbeat! But first let me issue some cautionary words. Any inference concerning the mass function of deflectors depends on assumptions as regards the unknown distribution of deflectors along the line-of-sight and the unknown velocity distributions of the deflectors and sources. One way of gaining a better understanding of the uncertainties is to set up thought experiments. Suppose an observer and a source are separated by a single species of deflectors of known mass and of a known triaxial Gaussian velocity distribution with known dispersion. Suppose that all the microlensing observables – the optical depth, the rate and the histogram of events with respect to timescales – are known to arbitrary precision. Can the density of the deflectors be inferred uniquely? Another way to phrase this problem is, given a configuration that reproduces the observables, is it possible to change the density distribution along the line-of-sight without changing the observables? It is straightforward to show that densities which have no effect on all three of the microlensing observables – known as *microlensing konus densities* – do exist by expanding the quadratures as Fourier series. A microlensing konus density contains both positive and negative densities. It is useless in itself, but can be added to any model that reproduces the observables to generate further models that do so. In other words, it is a neat way to parametrise the degeneracy of the problem. So, the result of the first thought experiment is *the first microlensing theorem of ignorance.* Even given perfect accuracy, it is not possible to restrict the density of deflectors along the line-of-sight from microlensing data alone. Let us now pass to a second thought experiment. A microlensing group monitors the optical depth, rate and timescale histogram of a single species of deflectors of known density possessing a triaxial Gaussian distribution of velocities. The semi-axes of the velocity dispersion tensor are unknown. Can the microlensing group even in principle infer the mass of the species? This problem is also straightforward to solve - the rate and the timescale histogram depend on the unknowns not separately but only through the combination $M/\sigma_{\rm T}^2$, where M is the characteristic mass and $\sigma_{\rm T}^2$ is the square of the tangential velocity dispersion. Only this combination is knowable from the microlensing observables. To find the mass of the species is not possible, unless the tangential velocity dispersion is accurately known. So, any determination of the mass function of the deflectors must necessarily depend on uncertain assumptions as to the density distribution and the velocity distribution of the deflectors. The result of the second thought experiment is *the second microlensing theorem of ignorance.* It is not possible to infer the characteristic mass of deflectors from microlensing data alone. These theorems of ignorance tell us that only by combining the microlensing data-set with assumptions regarding the structure of the Milky Way can useful constraints on the density and mass of low luminosity stars be obtained.

There are two data-sets available thanks to the tenacity of the observers (e.g., Alcock et al. 1997a,b) . Let me first consider microlensing towards the Large Magellanic Cloud. This is less useful for the purposes of investigations into low luminosity stars, so I wish to deal with it very quickly. There is genuine uncertainty as to where the lenses lie along the line-of-sight. If we assume that the lenses are in the halo and have a roughly isotropic velocity dispersion, then we arrive at the conclusion that roughly a third or half of the halo is built of objects of around half a solar mass or so. But it is not clear that these assumptions are correct. In particular, the lenses may lie in the warped and flaring Milky Way disk (Evans et al. 1997) or in an intervening stellar population (Zaritsky & Lin 1997). Until the location of the lenses is known, it is not possible to come to any reliable conclusion regarding the masses of the lenses. The data-set towards the Galactic Centre is simpler to analyse, at least given a working model of the Galactic Bulge. There were two early analyses of the data-set of the nine OGLE events – one by Evans (1995) using an axisymmetric bulge and spheroid, and a second by Zhao, Spergel & Rich (1995) using a triaxial bar. Both came to the conclusion that mass functions dominated by brown dwarfs – such as Salpeter laws extrapolated down 0.01 $M_\odot$ – generate too many short timescale events to be consistent with the data-set. Han & Gould (1996) combined the OGLE data-set with the 45 events in the first year Macho data-set and produced an interesting discussion of the mass function of the deflectors. They used a double exponential disk, together with an analytic bulge model that is a reasonable approximation to the COBE infra-red light. The velocity distribution of disk objects is assumed to be Gaussian about the rotation velocity. The bulge is assumed to be non-rotating. The distribution of velocities is a triaxial Gaussian with components deduced from the flattening via tensor virial theorem. Han & Gould considered two possible mass functions for the deflectors The first is a Gaussian in the logarithm of the mass, with some unknown mean and unknown dispersion in logarithmic mass. The second is a power-law mass function of unknown index which is assumed to be valid down to some unknown lower cut-off. For each model, Han & Gould found the best fits to the unknowns by performing a maximum likelihood test. In the best fitting power-law, the mass function goes like mass$^{-2.1}$ down to a lower mass limit of 0.04 $M_\odot$. This is of course very close to a Salpeter mass function. The best-fitting Gaussian mass function is centered around 0.1 $M_\odot$ and has a broadish width, extending well into the brown dwarf régime. It seems that the lenses must have a mass function that is different to that of the local disk stars found by Kroupa, Tout & Gilmore (1993). It also seems from Han & Gould's analysis that a significant fraction of the events are caused by sub-stellar masses. This is rather surprising given the earlier analysis by Evans (1995) and by Zhao (1995) on the admittedly much smaller OGLE data set. One possible explanation is the neglect of over-all bulge rotation by Han & Gould. This causes the transverse velocity of the lenses to be reduced, and hence given events are produced by lower masses. Han & Gould make one further interesting point – there is a tail of long timescale events that are poorly fit by all the models. This led Han & Gould to speculate that they might be caused by a kinematically cold population with a low scale height. These would be observed near the Sun for sources near Baade's Window. Such a population would have a low transverse speed and so cause very long timescale events.

4. Conclusions

Microlensing is a more powerful method than stochastic heating of wide binaries for constraining the contribution of low luminosity stars. Han & Gould's (1996) nice analysis points the way forward. It would be very interesting to repeat it with the larger data-set soon to become available and with state-of-the-art models of the Galactic Bulge. The conclusion reached on the mass function clearly depend on the model adopted – and, in particular, on the velocities in the model. As readers of the Proceedings of Joint Discussion 15 will find out, such Bulge models will soon be available thanks to the work of my Oxford colleagues.

References

Alcock C. et al. 1997a, ApJ, 479, 119
Alcock C. et al. 1997b, ApJ, 486, 697
Bahcall J.N., Soneira R., 1981, ApJ, 246, 122
Bahcall J.N., Hut P., Tremaine S.D., 1985, ApJ, 290, 15
Evans N.W., 1995, ApJ, 440, L13
Evans N.W. et al., 1997, Nature, submitted.
Han C., Gould A., 1996, ApJ, 467, 540
Kroupa P., Tout C., Gilmore G., 1993, MNRAS, 262, 545
Retterer J.M., King I., 1982, ApJ, 254, 214
Weinberg M., 1990, In Baryonic Dark Matter, eds. D. Lynden-Bell, G. Gilmore, (Kluwer Academic: Dordrecht), 117
Weinberg M., Shapiro S.L., Wasserman I., 1987, ApJ, 312, 367
Zaritsky D, Lin D.N.C, 1997, ApJ, submitted.
Zhao H.S., Spergel D.N., Rich R.M., 1995, ApJ, 440, L13

THE MASS-LUMINOSITY RELATION FROM BINARIES

DAVID W. LATHAM

Harvard-Smithsonian Center for Astrophysics

60 Garden Street, Cambridge, Massachusetts, 02138, USA

What is known about the masses of main-sequence stars from the analysis of binary orbits? Double-lined eclipsing binaries are the main source of very precise stellar masses and radii (e.g. Andersen 1997), contributing more than 100 determinations with better than 2% precision over the range 0.6 to 20 $M_\odot$. For lower-mass stars we are forced to turn to nearby systems with astrometric orbits (e.g. Henry et al. 1993). Not only is the number of good mass determinations from such systems smaller, but also the precision is generally poorer. We are approaching an era when interferometers should have a major impact by supplying good astrometric orbits for dozens of double-lined systems. Already we are beginning to see the sorts of results to expect from this (e.g. Torres et al. 1997).

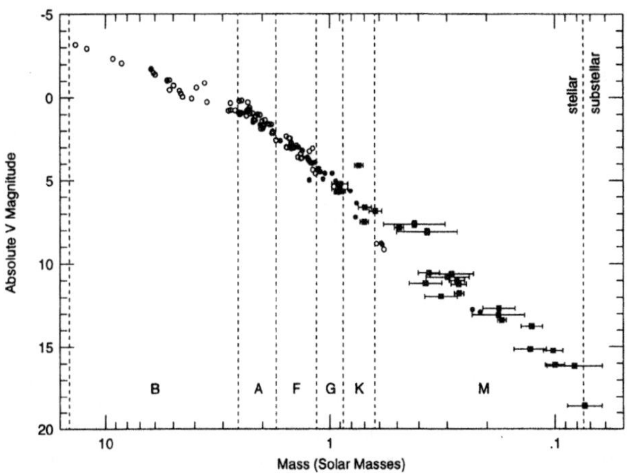

Figure 1. Mass vs. absolute *V* magnitude for eclipsing binaries (circles) and nearby astrometric binaries (squares)

Figure 1 is an updated version of a diagram presented by Henry et al. (1993, their Figure 2). It shows the general run of mass determinations from about 10 $M_\odot$ down to the substellar limit near 0.075 $M_\odot$. Ninety of the points in Figure 1 are for eclipsing binary masses from Andersen's review (1991) and are plotted as open circles. The results for eclipsing binaries published since 1991 are plotted as 30 filled circles, adopting the same limit of 2% for the mass precision. In most cases the uncertainties are similar to the size of the symbols. Especially noteworthy is the pair of new points for CM Draconis (Metcalfe et al. 1996) with masses near 0.25 $M_\odot$. Together with the points for YY Geminorum near 0.6 $M_\odot$, these are the only M dwarfs that have precise mass determinations. For the most part we are forced to rely on nearby stars with astrometric orbits, to fill in the M dwarf region of the diagram. We have used filled squares in Figure 1 for 29 such systems from Henry et al. (1993), updated using 14 new parallaxes from Hipparcos and 4 from the new Yale Parallax Catalog (1995). Gliese 508 is not included, because it is now known to be a triple, while Gliese 67AB, 570BC, and 623AB are not included because there are not yet any direct measurements of the *V* magnitude difference for these systems.

J. Andersen (ed.), Highlights of Astronomy, Volume 11A, 419–420.

For eclipsing binaries it is normally possible to derive good values for the luminosities, if suitable photometry is available for deriving a reliable effective temperature, because the stellar radius comes directly out of the orbital solution. However, for the cool stars the bolometric corrections are rarely well determined, and the usual practice is to quote absolute magnitudes rather than luminosities. I have chosen to use the absolute magnitude in the V band for the vertical axis in Figure 1.

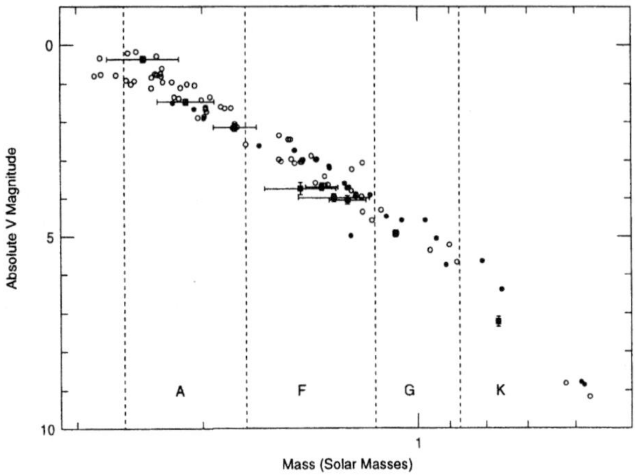

Figure 2. Mass vs. absolute V magnitude for Hyades binaries (filled squares) and eclipsing binaries (circles)

Astronomy is now entering an era when interferometers should be able to provide excellent astrometric orbits for binaries with much shorter periods than traditionally has been possible. At the same time, new techniques have been developed for deriving double-lined spectroscopic orbits for binaries where the two sets of lines are never well separated (e.g. Andersen 1997). Together these two developments promise a rich harvest of new mass determinations. A good example is the Hyades cluster, where recent work (Torres et al. 1997) has provided masses and orbital parallaxes (and therefore luminosities) for three binaries. Figure 2 shows these new results (filled squares) plotted together with the results for eclipsing binaries. The agreement is good, although the uncertainties for the Hyades binaries are still larger than one would want. However, the spread among the points for the eclipsing binaries is often larger than the uncertainties due to observational errors. This illustrates a point emphasized by Andersen (1991, 1997) that detailed analyses of the individual stars in eclipsing binary systems must include the effects of metallicity, helium abundance, and age, in order to take full advantage of the precision now available in the masses and radii, when testing stellar models and evolution theory.

We thank Todd Henry for useful discussions and for providing revised results from Henry et al. (1993) in advance of publication.

References

Andersen, J. 1991, A&AR 3, 91

Andersen, J. 1997, in Fundamental Stellar Properties: The Interaction between Observation and Theory (IAU Symp. No. 189), eds. T.R. Bedding et al. (Dordrecht: Kluwer Acad. Publ.), p. 99

Henry, T.J., McCarthy, D.W., Jr. 1993, AJ 106, 773

Metcalfe, T.S, Mathieu, R.D., Latham, D.W., Torres, G. 1996, ApJ 456, 356

Torres, G., Stefanik, R.P., Latham, D.W. 1997, ApJ, 485, 167

GROUND-BASED SURVEYS FOR LOW-LUMINOSITY STARS

HUGH R.A. JONES
Institute of Astronomy, The University of Tokyo, 2-21-1 Osawa, Tokyo 181
Astrophysics Group, Liverpool John Moores University, Byrom Street, Liverpool

1. Introduction

Most of what we know about the luminosity and mass function of low-mass stars has been derived from analysis of ground-based photometric and astrometric surveys. A major uncertainty in the interpretation of such surveys lies in the appropriate correction for unresolved binaries and in the use of a reliable colour-magnitude relation. Various new surveys are now uncovering large numbers of objects which should rapidly ameliorate these ambiguities.

The recent discovery of brown dwarfs and planets has highlighted the importance of low-luminosity stars. Once we understand their spectra and colours and have statistically significant samples, it will be important to use them to reliably constrain theories of star formation and chemical evolution as well as to use them as a probe of galactic structure and to reliably ascertain their contribution to baryonic dark matter. In this brief review I shall focus on ground-based field searches. Hambly (1997) reviews the very successful searches that have recently been carried out in nearby clusters.

Until the 1970s, most work on the faint-star luminosity function was based on proper-motion searches primarily by Dutch researchers, in particular Luyten (e.g. 1938, 1979). The last 20 years have seen a switch from astrometric to photometric surveys employing plate-scanning machines or CCD cameras. Combined with substantial increases in computing power these technologies have led to the discovery of large numbers of very low-luminosity stars.

2. Problems & Progress

Surveys for low-mass stars have usually being made in two near-optical bands. However low-mass stars release most of their energy in the infrared, furthermore large numbers of low-mass stars are only found in the faintest bins. In addition the near-optical relationships between colour, luminosity and mass are not smooth and are not well calibrated. This has lead to newer surveys moving to longer wavelengths and obtaining more calibration data. A problem which all surveys have to overcome is Malmquist bias. Unless corrected for unresolved binary systems beyond the designated survey volume will be erroneously included as single stars because of their apparent brightness. The treatment of Malmquist bias and other biases are described in detail by Stobie, Ishida & Peacock (1989) and Kroupa, Tout & Gilmore (1993).

A major aim of most surveys is to probe the stellar luminosity and mass functions with good statistics. Early work broadly supported the Salpeter hypothesis of a power-law. As statistics improved, e.g., Reid & Gilmore (1982) structure became apparent. A number of studies have improved the statistics and extended such findings to fainter stars, though, the broad trend of a maxima in the photometric luminosity function ($\psi_{\rm phot}$) around $M_{\rm v} \sim 12$ ($M_{\rm bol} \sim 10$) with a decline at fainter magnitudes has remained. However, this apparent lack of faint stars is not seen in luminosity functions constructed from nearby stars samples ($\psi_{\rm near}$), typically within 5 or 8 pc, which show a much flatter distribution. One explanation for this might be a local overdensity. However the stars which define $\psi_{\rm near}$ have a velocity dispersion of ~ 50 *kms* which means that within 4 Myr they would have dispersed over ~ 200 pc. Dahn et al. (1986) suggested that the apparent deficit of faint stars in the photometric surveys might arise from the non-detection of low-mass companions, which in studies of nearby stars can be resolved as separate from bright primaries. Kroupa and collaborators (e.g. 1993, 1995) and Mera et al. (1996) have used the binary fraction of 49 % and the companion mass

J. Andersen (ed.), Highlights of Astronomy, Volume 11A, 421–422.

ratios based on the G and K dwarf sample of Duquennoy & Mayor (1991) to model the contribution of unresolved binaries to $\psi_{\rm phot}$. Using a model where both primary and secondary are drawn from the same initial mass function they find that unresolved binaries are sufficient to solve the deficit of faint objects in $\psi_{\rm phot}$. However Reid & Gizis (1997) have failed to re-produce these results and contest that the Duquennoy & Mayor sample is appropriate for use with M dwarfs. Reid & Gizis's 8-pc sample of M dwarfs and the work of Mazeh et al. (1992) indicates that the binarity fraction and distribution fraction for M dwarfs is quite different from that of G and K dwarfs. Rather than a smooth tendency to produce large binary mass ratios, Reid & Gizis find that M dwarfs show a strong propensity toward near-equal mass binary systems. Reid & Gizis model unresolved binarity using a binary fraction of 36% and find that the shape of $\psi_{\rm phot}$ is altered little by correcting for unresolved binaries. They propose that the discrepancy between $\psi_{\rm phot}$ and $\psi_{\rm near}$ is rectified by using an $M_{\rm v}(V-I)$ relation which is constructed using only objects with high quality photometry. The clear clarification of this issue requires

- M dwarf multiplicity to be thoroughly quantified for a large sample and
- further work on colour-luminosity-mass relations.

Even without these essential large scale observational efforts, research would benefit from the use of the same nearby star sample and from using bolometric luminosities rather than $M_{\rm v}$ magnitudes. Although it is easier to make accurate measurement in one photometric band, disk low-luminosity stars show very significant colour variations which are not yet understood in terms of their spectra, age or metallicity (e.g. Tinney et al. 1995). Furthermore the slow cooling (>1 Gyr) of very low-mass objects onto the main sequence means that analyses need to take age into account. The necessary observational work is already being tackled by a number of large-scale projects. New surveys are now producing large numbers of nearby M dwarfs and brown dwarf candidates which are expected to clarify the appropriate corrections to apply to $\psi_{\rm phot}$.

TABLE 1. Some of the new candidates which have been discovered. References are DEN97 - Delfosse et al. 1997a, KBS97 - Kirkpatrick et al. 1997, O97 - Oppenheimer 1997, F97 - Fleming 1997, D97 - Delfosse et al. 1997b, R97 - Ruiz 1997, T96 - Thackrah 1996 and H97 - Hawkins et al. 1997.

Technique & Survey	Recent Highlights	Reference
Large area IR	Brown dwarf + 3 excellent candidates	DEN97, KBS97
e.g. DENIS and 2MASS	+ many M dwarfs within 25 pc	
Coronographic	Brown dwarf + ∼5 M dwarfs within 25 pc	O97
ROSAT all-sky	55 M dwarfs with 25 pc	F97
Radial velocities	lots of planets + 12 M dwarfs within 9 pc	D97
Photographic searches	Brown dwarf + many candidates	R97, T96, H97
	+ many M dwarfs within 50 pc	

References

Delfosse et al., 1997a, A&A, in press
Delfosse et al., 1997b, Cool Stars and the Sun No. 10, ASP Conf. Ser., in press
Duquennoy A., Mayor M., 1991, A&A, 248, 485
Fleming T., 1997, Cool Stars and the Sun No. 10, ASP, in press
Hambly N.., 1997, Extrasolar planets and brown dwarfs, ASP Conf Ser., in press
Hawkins M.R.S., Ducourant C., Jones H.R.A, Rappaport M., 1997, MNRAS, in press
Kirkpatrick J.D., Beichman C.A., Skrutskie M.F., 1997, ApJ, 473, 311
Kroupa P., Tout C.A., Gilmore G., 1993, MNRAS, 262, 545
Kropua P., 1995, ApJ, 453, 350
Kropua P., 1995, ApJ, 453, 358
Luyten W.. 1938, Observatory Publications, University of Minnesota Press, 2, 5, 15
Luyten W.. 1979, LHS catalogue, University of Minnesota Press
Mazeh T., Goldberg D., Duquennoy A., Mayor M., 1992, ApJ, 401, 265
Mera G., Chabrier G., Baraffe I., 1996, ApJ, 459, L87
Oppenheimer B., 1997, Extrasolar planets and brown dwarfs, ed. Rebolo R., ASP Conf Ser., in press
Reid N., Gilmore G., MNRAS, 201, 73
Reid I.N., Gizis J.E., 1997, AJ, 113, 2246
Ruiz M., 1997, http://www.eso.org/outreach/press-rel/pr-1997/pr-07-97.html
Stobie, Ishida, Peacock, 1989, MNRAS, 238, 709
Thackrah A., 1996, PhD thesis, University of Edinburgh
Tinney C.G., Reid I.N., Gizis J., Mould J.R., 1995, AJ, 110, 3014

VERY LOW-LUMINOSITY OBJECTS IN STAR-FORMING REGIONS

MOTOHIDE TAMURA
National Astronomical Observatory of Japan, Osawa 2-21-1, Mitaka, Tokyo 181, Japan
YOICHI ITOH AND YUMIKO OASA
University of Tokyo, Osawa 2-21-1, Mitaka, Tokyo 181, Japan
ALAN TOKUNAGA
Institute for Astronomy, University of Hawaii, 2680 Woodlawn Drive, Honolulu, HI96822, USA
AND
KOJI SUGITANI
Nagoya City University, Mizuho-ku, Nagoya 467, Japan

Abstract. In order to tackle the problems of low-mass end of the initial mass function (IMF) in star-forming regions and the formation mechanisms of brown dwarfs, we have conducted deep infrared surveys of nearby molecular clouds. We have found a significant population of very low-luminosity sources with IR excesses in the Taurus cloud and the Chamaeleon cloud core regions whose extinction corrected J magnitudes are 3 to 8 mag fainter than those of typical T Tauri stars in the same cloud. Some of them are associated with even fainter companions. Follow-up IR spectroscopy has confirmed for the selected sources that their photospheric temperature is around 2000 to 3000 K. Thus, these very low-luminosity young stellar sources are most likely very low-mass T Tauri stars, and some of them might even be young brown dwarfs.

1. Introduction

One of the most important issues about the IMF is its slope, particularly at its lower-mass end extending towards substellar region. In addition, whether it generally turns over and whether its slope differs among various star-forming regions are important questions to be explored. This issue is directly related with the formation mechanisms of brown dwarfs. A variety of techniques at various wavelengths such as X-ray, optical, and infrared surveys have been applied to search for low-mass objects in many star-forming regions, but infrared studies have two advantages. First, low-mass objects in regions of star-formation will be in the contraction phase and radiate their energy mostly at infrared wavelengths. Second, at infrared wavelengths, the radiation from the sources is less affected by the cloud and circumstellar matter extinction than at optical wavelengths. In addition, recent great advances in infrared array detectors enable us to conduct deep and wide-area near-infrared surveys. Armed with these advantages, we have conducted deep near-infrared surveys of nearby molecular clouds of Taurus, Chamaeleon, and Perseus. In this contribution, we report on the results of the surveys of the Taurus and Chamaeleon dark clouds, which resulted in the discovery of a number of very low-luminosity young stellar objects (YSOs), and the results of our follow-up spectroscopy and high-resolution imaging of the newly detected very low-mass YSOs.

2. Observations and Results

We have conducted a JHK survey of a part of the Taurus molecular cloud complex with a completeness limit of J=15.2 mag (an area of 1.5 square degrees covering most of the Heiles Cloud 2). We have detected a number of low-luminosity sources with near-infrared excesses on the J-H/H-K diagram, which we ascribe to the emission from the circumstellar matter similar to that of optical T Tauri stars (TTSs). However, their luminosity is 3 to 6 mag fainter than those of typical TTSs

J. Andersen (ed.), Highlights of Astronomy, Volume 11A, 423–424.

in Taurus. These sources are too faint to be detected with the previous YSO searches including optical Hα, *IRAS*, and X-ray surveys. Their masses estimated from the combination of the J-band fluxes and the evolutionary tracks for low-mass YSOs (D'Antona & Mazzitelli 1994), assuming ages of typical TTSs in the cloud, are near the stellar/substellar boundary and some of them are indeed in the substellar region. The luminosity function of these low-luminosity YSOs does not appear to have any cut-off at its low-end. See Itoh, Tamura, & Gatley (1996) for the details of this observation.

In the core region of the Chamaeleon molecular cloud I complex, we have conducted a higher-sensitivity JHK survey of the cloud core. With a completeness limit of J=17.1 mag, we have identified 13 YSO candidates within the central 30 square arcminutes. Some of them are of very low-luminosity, even 8 mag fainter than the typical TTSs in the same cloud. The luminosity function of the YSOs seems to be bimodal in this region, and the fainter group of YSOs shows an increase of counts toward the detection limit (Oasa, Tamura, & Sugitani 1997).

3. Follow-Up Observations and Discussion

The photometric approach described above is very useful to identify the very low-luminosity YSOs with IR excesses. However, in order to correctly estimate the mass of the detected YSOs, it is necessary estimate the age of the sources. If the age of the sources is older than our assumed ages of typical TTSs, the low-luminosity of the detected could be explained by their older ages. One of the best ways for determining both masses and ages simultaneously is determining the effective temperature and luminosity of the sources, then placing them in the HR-diagram and comparing with the theoretical YSO tracks. In order to determine the effective temperature, we have conducted near-infrared spectroscopy of the selected low-luminosity sources detected in our Taurus survey with CGS4 at the 3.8-m UKIRT and with OASIS at the Okayama 1.88-m telescope (Itoh, Tamura, & Tokunaga 1997). First, we have confirmed that most of our low-luminosity sources are not simple background giants/dwarfs and their spectra are more or less similar to those of TTSs. We have then used the Na/Ca line ratio to determine that their photospheric temperature is around 2000 to 3000 K. The results as well as the photometric results are therefore consistent with the interpretation that the very low-luminosity YSOs detected in our near-infrared surveys are very low-mass TTSs and that some of them might even be young brown dwarfs.

We have also been conducting high-resolution imaging of the detected low-luminosity YSOs and have discovered that some sources appear to be associated with even fainter companions. The companions also have IR excesses. If these two sources have the same age, the fainter companion is less massive than the primary low-luminosity YSO, which is strong evidence that the nearby star-forming region is the site of brown dwarf formation. Brown dwarfs appear to form in both isolated mode and as companions, although the relative frequencies are not clear. Both extensive deep and wide area surveys and high-resolution observations will be necessary for a more statistical study. The simultaneous-band infrared imager for unbiased surveys (SIRIUS) and the infrared coronagraphic imager with adaptive optics (CIAO; Tamura et al. 1997) for the Japanese Subaru 8-m telescope, both now under development, will be a most powerful tool for this study.

References

D'Antona, F., & Mazzitelli, I. 1994, ApJS, 164, 349
Itoh, Y., Tamura, M., & Gatley, I. 1996, ApJ, 465, L129
Itoh, Y., Tamura, M., & Tokunaga, A. 1997, in preparation
Oasa, Y., Tamura, M., & Sugitani, K. 1997, in preparation
Tamura, M. et al. 1997, in Extrasolar planets and brown dwarfs, ed. R. Rebolo, ASP Conf Ser., in press

STAR FORMATION: CAN THERE BE A BREAK IN THE IMF NEAR 0.1M$_\odot$?

TAKENORI NAKANO
Nobeyama Radio Observatory, National Astronomical Observatory
Nobeyama, Minamimaki, Minamisaku, Nagano 384-13, Japan

1. Introduction

The initial mass function of stars (IMF) at small masses depends on several factors. First, it depends on the mass function of cloud cores in which stars form. Second, there must be a lower limit to the core mass for contraction; very small mass cores may not contract even if they exist. This must affect greatly the IMF near its lower end. Third, not all core matter may become stars; we must determine the stellar mass $M_\star$, or the star formation efficiency $M_\star/M_{cc}$, as a function of the mass of the cloud core, M_{cc}. In this paper we discuss the second and third points.

2. Critical Core Mass for Contraction

There is a widely accepted assumption that low-mass stars form in magnetically subcritical cloud cores and high-mass stars form in magnetically supercritical ones (*e.g.*, review by Shu, Adams, & Lizano 1987); the magnetic force in magnetically supercritical cores is not strong enough to support the cores against the self-gravity, and subcritical cores can be in magnetohydrostatic equilibrium only when their expansion by the magnetic force (diluted by the gravity) is controlled by the external magnetic field and/or external pressure. However, Nakano (1998) showed that cloud cores, which are generally observed as portions of a molecular cloud having higher column density than the surrounding, are magnetically supercritical.

For a magnetically supercritical cloud core there is a critical value P_{cr} for the surface pressure P_s, above which no magnetohydrostatic equilibria exist and the core collapses, given by (Nakano 1998)

$$\frac{P_{cr}}{k} \approx 7.1 \times 10^4 \frac{1}{a_{eff}^4}\Big(\frac{M_{cc}}{1M_\odot}\Big)^{-2}\Big(\frac{C_{eff}}{0.19\,\mathrm{km\,s^{-1}}}\Big)^8 \mathrm{cm^{-3}\,K}, \tag{1}$$

where k is the Boltzmann constant, C_{eff} is the effective sound velocity in the core including the effect of turbulence, and $a_{eff} \approx 1 - (\Phi/\Phi_{cr})^2$ is the effective coefficient for the gravity diluted by the magnetic force; Φ and $\Phi_{cr} \approx 2\pi G^{1/2} M_{cc}$ are the magnetic flux and the critical magnetic flux, respectively, of the core, G is the gravitational constant, and $\Phi < \Phi_{cr}$ for a magnetically supercritical core.

Because the surface pressure P_s on the cores is nearly equal to the mean pressure in the cloud in which the cores are embedded unless they are located close to the cloud surface, we have

$$\frac{P_s}{k} \approx \frac{G\Sigma^2}{k} \approx 8.0 \times 10^4 \Big(\frac{A_V}{3\,\mathrm{mag}}\Big)^2 \mathrm{cm^{-3}\,K}, \tag{2}$$

where Σ and A_V are the column density and the extinction at visible, respectively, of the cloud.

Although C_{eff} can hardly be less than $0.19\,\mathrm{km\,s^{-1}}$ corresponding to a temperature of 10 K even without turbulence, cloud cores of $M_{cc} \gtrsim 0.3M_\odot$ can contract if they are embedded in ordinary molecular clouds with $A_V \sim 10$ mag as seen by comparing equations (1) and (2).

J. Andersen (ed.), Highlights of Astronomy, Volume 11A, 425–426.

3. Star Formation Efficiency

Dynamical contraction of a cloud core is highly nonhomologous. A small protostellar core in quasiequilibrium forms first, and this core grows gradually by inflow of matter from the outer part of the cloud core. The mass of a forming star is fixed when the mass inflow is stopped.

As a mechanism of stopping accretion onto protostars many people considered the reversion of inflows by stellar winds (*e.g.*, review by Shu *et al.* 1987). However, only a part of the inflow is reversed as seen from the shapes of bipolar outflows.

Nakano, Hasegawa, & Norman (1995) proposed a different mechanism that determines the mass of the star. When a significant fraction of the core matter is blown off by the mass outflows from the forming star, the remaining matter is no longer gravitationally bound unless the initial core mass is much larger than the critical mass for contraction, and disperses. In this way the forming star stops growing. With some simple modifications to the theory of Nakano *et al.* (1995) the star formation efficiency is given by

$$\frac{M_\star}{M_{\rm cc}} \approx 0.043\left(\frac{\dot{M}_{\rm w}/\dot{M}_{\rm I}}{0.1}\frac{V_{\rm w}}{200\,{\rm km\,s^{-1}}}\right)^{-1/2}\left(\frac{\Omega_{\rm w}}{4\pi}\right)^{1/2}\left(\frac{1-\dot{M}_{\rm w}/\dot{M}_{\rm I}}{0.9}\right) \times\left(\frac{M_{\rm cc}}{100M_\odot}\right)^{1/6}\left(\frac{n_{\rm cc}}{10^4\,{\rm cm^{-3}}}\right)^{1/12}+\frac{M_0}{M_{\rm cc}}, \tag{3}$$

where $V_{\rm w}$ is the velocity of the wind near its base, $\dot{M}_{\rm I}$ is the mass inflow rate to the star, $\dot{M}_{\rm w}$ is the mass loss rate by the wind, $\Omega_{\rm w}$ is the solid angle of the outflows seen from the star, $n_{\rm cc}$ is the mean density of the cloud core, and M_0 is the mass of the star when it begins to drive the wind. For $M_\star \gg M_0$ the star formation efficiency is rather insensitive to $M_{\rm cc}$ and $n_{\rm cc}$ and is 10 % on the order of magnitude.

4. Onset of Mass Outflows

Winds from protostars are essential in determining the stellar mass except for very high mass stars where expansion of HII regions may be important (Nakano *et al.* 1995). At what mass do protostars begin to drive winds? Although the mechanism of driving bipolar outflows from young stellar objects is not well known, we have an example of the outflows from a seemingly very low mass protostar. André, Ward-Thompson, & Barsony (1993) estimated the mass of the protostar driving the bipolar flows in VLA 1623 as less than $0.4M_\odot$. They also argued a possibility that the mass is less than $0.03M_\odot$. Because brown dwarfs have effective temperature as high as 3000 K (*e.g.*, Hayashi & Nakano 1963; D'Antona & Mazzitelli 1985, 1994), which is nearly equal to that of T Tauri stars, the ionization fraction in the atmosphere is not much different from that of T Tauri stars. In addition mass accretion is going on at the surface. Therefore, we can expect magnetic activities and mass outflows from mass-accreting stars with mass much smaller than $0.1M_\odot$, i.e., M_0 in equation (3) may be much smaller than $0.1M_\odot$.

5. Summary

We have found that cloud cores of mass as low as $0.3M_\odot$ can contract. We can expect mass outflows even from protostars of $M_\star \ll 0.1M_\odot$. The star formation efficiency is of the order of 10% at $M_\star \gg M_0$. Therefore, we can expect formation of stars whose mass is significantly smaller than $0.1M_\odot$. Thus, from the theory of star formation we have found no clear evidences for the existence of a break in the IMF at $M_\star \approx 0.1\,M_\odot$ unless there is a break in the mass function of cloud cores at $M_{\rm cc} \sim 1M_\odot$.

References

André, P., Ward-Thompson, D., Barsony, M. 1993, ApJ, **406**, 122.
D'Antona, F., Mazzitelli, I. 1985, ApJ, **296**, 502.
D'Antona, F., Mazzitelli, I. 1994, ApJS, **90**, 467.
Hayashi, C., Nakano, T. 1963, Prog. Theor. Phys., **30**, 460.
Nakano, T. 1998, ApJ, in press.
Nakano, T., Hasegawa, T., Norman, C. 1995, ApJ, **450**, 183.
Shu, F.H., Adams, F.C., Lizano, S. 1987, ARAA, **25**, 23.

WHITE DWARF COOLING CURVES AND SEARCHES FOR WHITE DWARFS

MATT A. WOOD
Florida Institute of Technology
Department of Physics and Space Sciences
Melbourne, FL, 32901-6988, USA
wood@kepler.pss.fit.edu

1. White Dwarf Cooling Curves

White dwarf (WD) evolution is essentially a cooling problem in two parts. Degenerate electrons in the core provide the support while non-degenerate C and O ions provide the thermal reservoir (Koester & Chanmugam 1990, Kawaler 1996). The rate of energy loss through the non-degenerate envelope of H and He—segregated by gravitational settling—is meager enough that the vast majority of WDs have not had time to cool to undetectability in the ~10 Gyr age of the Galactic disk. Because WDs are a remarkably homogeneous class with a mass function sharply peaked about $0.6M_{\odot}$, and because the cooling physics is relatively straightforward to calculate, the observed deficit of WDs below $M_V = 16.5$ [$\log(L/L_{\odot}) \approx -4.5$, $T_{\rm eff} \approx 4000$ K; see Liebert et al. 1988 (LDM), Oswalt et al. 1996 (OSWH), and Leggett et al. 1997] provides constraints on the age and star-formation history of the Galactic disk (Winget et al. 1987 (WEA97), Wood 1992 (W92), Yuan 1992, Hernanz et al. 1994).

The maximum fractional surface layer masses that can remain following post-AGB evolution are $\log(q_{\rm H}) \sim -4$ and $\log(q_{\rm He}) \sim -2$ (for a 0.6 $M_{\odot}$ remnant—the values are inversely correlated with remnant mass), although we know from WD asteroseismology that surface H layer masses range from this value to smaller than $\log q_{\rm H} \sim -6$ (e.g., Fontaine & Wesemael 1997). Some 75% of WDs in the McCook & Sion (1987) catalog reveal pure-H photospheres (spectral type DA). The remainder are broadly classified as non-DAs, with most showing show pure-He spectra (DB) or continuous spectra (DC), and with a handful that show mixed compositions (DAB), or C (DQ) or metal lines (DZ). See Bergeron et al. 1997 for a thorough discusion of the chemical evolution of cool WDs.

As a hot WD, cooling is dominated by neutrino processes which can "out shine" the photon luminosity by $\lesssim 100$ near $T_{\rm eff} \sim 10^5$ K ($\log(L/L_{\odot}) \sim 2$, again for a 0.6 $M_{\odot}$ model), which results in a core temperature inversion, and which accelerates the cooling (Fig. 1). Neutrinos become unimportant below $\log(L/L_{\odot}) \sim -1.6$ ($T_{\rm eff} \sim 20,000$ K); at this point, the models are closest to the Mestel (1952) ideal, with nearly isothermal cores and radiative envelopes. A significant surface convection zone develops near $T_{\rm eff} \sim 10^4$ K which persists from there on, reaching a maximum fractional depth in mass of $\log(q_{\rm cz}) \approx -6$ and potentially mixing with the subsurface He layer. Cool WDs ($T_{\rm eff} \lesssim 7000$ K, $\log(L/L_{\odot}) \lesssim -3.5$) undergo a crystallization phase transition of their C-O interiors. The release of latent heat associated with this first-order phase transition slows the cooling temporarily, but quantum effects in this phase act to lower the heat capacity and accelerate cooling. Based on the spindle form of the phase diagram of the C-O binary ionic mixture (e.g., Segretain et al. 1994, Isern et al. 1997), Hernanz et al. (1994 and references therein), have suggested that the solidification phase transition should be accompanied by a global redistribution that effectively moves O inward and C outward, and that acts as an additional energy source comparable in magnitude to the latent heat released. Detailed evolutionary models incorporating this process have recently been published by Salaris et al. (1997). The models shown in Fig. 1 include the latent heat release but phase separation is not implemented. The age-luminosity relation must be extrapolated below $\log(L/L_{\odot}) \sim -5.5$.

J. Andersen (ed.), Highlights of Astronomy, Volume 11A, 427–429.

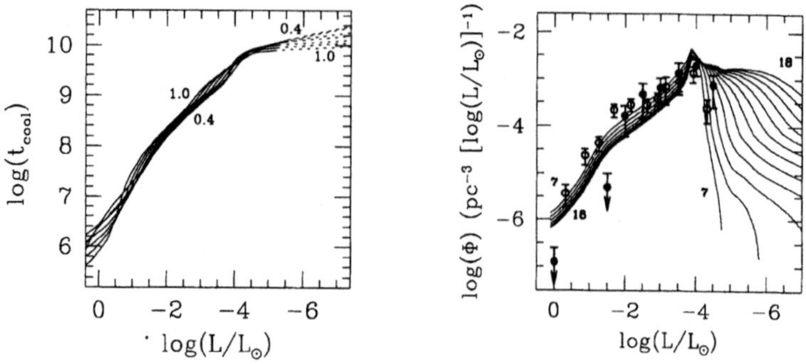

Figure 1. Left panel shows cooling curves of C-O core DA models with masses ranging from 0.4 to 1.0 $M_{\odot}$, as indicated. Extrapolations are shown as dashed lines. Right panel shows integrated WD LFs from Wood (1995) shown over the data of LDM (open circles) and OSWH (filled circles). Ages range from 7 to 18 Gyr, as indicated.

2. Searches for White Dwarfs

Most hot WDs have been found through colorimetric surveys searching for faint blue objects [Green et al. 1986 (PG); Kondo et al. 1984 (Kiso); Demers et al. (1986) Lamontagne et al. 1997 (MCT); Stobie et al. 1997 (EC); Cristiani et al. 1995 (HBQS)]. Cool WDs are faint and therefore must be close if they are to be detected. Their photometric colors are similar to K dwarfs and therefore the majority of cool WDs have been found through proper motion surveys (e.g., Luyten 1978, 1979 (etc.); Giclas et al. 1971, 1989; Ruiz 1996 and references therein) with extensive spectroscopic followup work (e.g. Oswalt et al. 1991, 1993). Chuck Claver and collaborators (1998, in preparation) are using a MgH and CaH filter to separate WD candidates from the lower MS in a photometric CCD survey; spectroscopic followup observations are being made using WIYN Hydra. Recently, WDs have been detected in open and globular clusters with HST (Richer et al. 1995, 1997; Cool et al. 1996; Renzini et al. 1996).

3. White Dwarfs and the Age of the Galactic Disk

Much of the recent excitement surrounding WDs centers on their use as age indicators for the Galactic disk (WEA87, W92, OSWH), and the MACHO collaboration result indicating that up to 50% of the mass of the halo may be composed of WDs (Alcock et al. 1993, 1997). Using a sample of common proper motion binary WDs and updated evolutionary models (Wood 1995), OSWH find a most likely age of ~9.5 Gyr, where the data of LDM fit with the same models suggest an age ~2 Gyr younger (see Fig. 1). Monte Carlo simulations (Wood & Oswalt 1998) suggest that sample-to-sample variations are large, and that inferred ages should be considered lower limits. The 10 Gyr WD ages compare favorably with the 8-12 Gyr Hubble expansion ages and the recent 12 Gyr globular cluster age estimates (e.g., Chaboyer et al. 1997).

Alcock et al. (1993, 1997) report a significant population of $\sim 0.5\ M_{\odot}$ WDs in the galactic halo. In order for this population to exist in kinematically significant numbers and not have been detected via other means, it must have been the result of an enormous star formation burst ~15 Gyr ago with an IMF sharply peaked near $M_{\rm MS} \sim 3 M_{\odot}$ (Tamanaha et al. 1990, Ryu et al. 1990, Adams & Laughlin 1996, Chabrier et al. 1996). Adopting such an IMF, Fields et al. (1997) modeled schematically the formation of baryonic galactic halos by following the dynamics, chemical evolution, etc. on a hierarchy of mass scales, finding a best-fit model that yielded halos that were 40% baryonic and dominated by WDs. However, Fields et al. only considered the global metallicity Z. Gibson & Mould (1997) revisited the problem following the individual elemental abundances and found that all such models overproduce C and N relative to O by a factor of 10 or more. In addition, Kawaler (1996) notes that up to 12 faint halo WDs should be visible in the Hubble Deep Field, but Flynn et al. (1996) found no objects with $V - I > 1.8$. To date, we must conclude that no completely satisfying explanation of the origin of the MACHO collaboration objects exists.

References

Adams, F.C. & Laughlin G. 1996, ApJ, 468, 586

Alcock, C., et al. 1993, Nature, 365, 621

Alcock, C., et al. 1997, ApJ 486, 697

Bergeron, P., Ruiz, M.-T., and Leggett, S.K. 1997, ApJS, 108, 339

Chaboyer, B., Demarque, P., Kernan, P.J., & Krauss, L.M. 1997, ApJ, in press.

Chabrier, G., Segretain, L. & Méra, D. 1996, ApJ, 468, L21

Cristiani, S., et al. 1995, A&A Suppl. Ser., 112, 347

Cool, A.M., Piotto, G., & King, I.R. 1996, ApJ, 468, 655

Demers, S., Kibblewhite, E.J., Irwin, J.J., Nithakorn, D.S., Béland, S., Fontaine, G., & Wesemael, F. 1986, AJ, 92, 878

Fields, B.D., Mathews, G.J. & Schramm, D.N. 1997, ApJ, 483, 625

Flynn, C., Gould, A., & Bahcall, J.N. 1996, ApJL, 466, L55

Fontaine, G. & Wesemael, F. 1997, in White Dwarfs, eds. I. Isern et al. (Kluwer: Dordrecht) p. 173

Giclas, H.L., Burnham, R., Thomas, N.G. Lowell Proper Motion Survey, Northern Hemisphere Catalog (Flagstaff: Lowell) (1971).

Giclas, H., Burnham, R., Thomas, G. Lowell Proper Motion Survey, Southern Hemisphere Catalog. Lowell Obs. Bull. 164, Vol. VIII, p.89 (1978).

Gibson, B.K. & Mould, J.R. 1997, ApJ, 482, 98

Green, R. G., Schmidt, M., & Liebert., J. 1986, ApJS, 61, 305

Hernanz, M., García-Berro, E., Isern, J., Mochkovitch, R., Segretain, L., & Chabrier, G. 1994, ApJ, 434, 652

Isern, J. Mochkovitch, R. García-Berro, E. & Hernanz, M. 1997, ApJ, 485, 308

Kawaler, S.D., Novikov, I., & Srinivasan, G. 1997, Saas-Fee Adv. Course 25 (New York: Springer) eds. G. Meynet & D. Schaerer

Kawaler, S.D. 1996, ApJL, 467, L61

Koester, D., & Chanmugam, G. 1990, Rep. Prog. Physics, 53, 837

Kondo, M., Maehara, H., & Noguchi, T., 1984, Ann. Tokyo Astron. Obs., 20, 130

Lamontagne, R., Wesemael, F., Fontaine, G., Demers, S., Bergeron, P., Irwin, M.J., and Kunkel, W.E. 1997, in White Dwarfs, eds. I. Isern et al. (Kluwer: Dordrecht) p. 143

Leggett, S.K., Ruiz, M.-T., & Bergeron, P., 1998, ApJ, in press.

Liebert, J., Dahn, C. C., & Monet, D. G. 1988, ApJ, 332, 891

Luyten, W.J. 1978, Proper Motion Survey with the 48-inch Schmidt Telescope: L Double Stars with Common Proper Motion (Minneapolis: University of Minnesota)

Luyten, W. J. 1979, The LHS Catalogue, Second Edition (Minneapolis: University of Minnesota)

McCook, G.P., & Sion, E.M. 1987, ApJS, 65, 603

Mestel, L. 1952, MNRAS, 112, 583

Oswalt, T.D., Sion, E.M., Hintzen, P.M. & Liebert, J. 1991, in White Dwarfs, eds. G. Vauclair & E.M. Sion (Kluwer: Dordrecht), p. 379

Oswalt, T.D. et al. 1993, in White Dwarfs: Advances in Observation & Theory, ed. M. Barstow (Kluwer: Dordrecht) p. 419

Oswalt, T. D., Smith, J. A., Wood, M. A., & Hintzen, P. M., 1996, Nature, 382, 692

Renzini, A., et al. 1996, ApJ, 465, L23

Richer, H.B., et al. 1995, ApJ, 451, L17

Richer, H.B., et al. 1997, ApJ, 484, 741

Ruiz, M.-T. 1996, AJ, 111, 1267

Ryu, D., Olive, K. A., & Silk, J. 1990, ApJ, 353, 81

Salaris, M., Dominguez, I., García-Berro, E., Hernanz, M., Isern, J., & Mochkovitch, R. 1997, ApJ, 486, 413

Sion, E.M. 1986, PASP, 98, 821

Segretain, L., Chabrier, G., Hernanz, M., García-Berro, E., Isern, J., & Mochkovitch, R. 1994, ApJ, 434, 641

Stobie, R.S., et al. 1997, MNRAS, 287, 848

Tamanaha, C.M., Silk, J., Wood, M.A., and Winget, D.E. 1990 ApJ, 358, 164.

Winget, D. E., Hansen, C. J., Liebert, J., Van Horn, H. M., Fontaine, G., Nather, R. E., Kepler, S. O., & Lamb, D. Q. 1987, ApJ, 315, L77

Wood, M. A. 1992, ApJ, 386, 539

Wood, M.A. 1995, in *White Dwarfs*, eds. D. Koester and K. Werner (Berlin: Springer-Verlag), p. 41

Wood, M.A. & Oswalt T.D. 1998, ApJ, in press.

Yuan, J. W. 1992, A&A, 261, 105

CONTRIBUTION OF WHITE DWARFS TO CLUSTER MASSES

TED VON HIPPEL
Department of Astronomy
University of Wisconsin
Madison, WI 53706, USA
ted@noao.edu

1. Introduction

The study of cluster white dwarfs (WDs) has been invigorated recently by the Hubble Space Telescope (HST). Recent WD studies have been motivated by the new and independent cluster distance (Renzini et al. 1996), age (von Hippel et al. 1995; Richer et al. 1997), and stellar evolution (Koester & Reimers 1996) information that cluster WDs can provide. An important byproduct of these studies has been an estimate of the WD mass contribution in open and globular clusters. The cluster WD mass fraction is of importance for understanding the dynamical state and history of star clusters. It also bears an important connection to the WD mass fractions of the Galactic disk and halo. Current evidence indicates that the open clusters (e.g. von Hippel et al. 1996; Reid this volume) have essentially the same luminosity function (LF) as the solar neighborhood population. The case for the halo is less clear, despite the number of very good globular cluster LFs down to nearly 0.1 solar masses (e.g. Cool et al. 1996; Piotto, this volume), as the field halo LF is poorly known. For most clusters dynamical evolution should cause evaporation of the lowest mass members, biasing clusters to have flatter present-day mass functions (PDMFs) than the disk and halo field populations. Dynamical evolution should also allow cluster WDs to escape, though not in the same numbers as the much lower mass main sequence stars. The detailed connection between cluster PDMFs and the field IMF awaits elucidation from observations and the new combined N-body and stellar evolution models (Tout, this volume). Nevertheless, the WD mass fraction of clusters already provides an estimate for the WD mass fraction of the disk and halo field populations. A literature search to collect cluster WDs and a simple interpretive model follow. This is a work in progress and the full details of the literature search and the model will be published elsewhere.

2. Observations

Using the NASA ADS abstract service I searched the literature on white dwarfs in globular and open clusters through July 31, 1997. I included cataclysmic variables (CVs) and other types of binary systems where the authors specifically discussed the WD nature of one of the binary components. My literature search covered 82 globular cluster references and 29 open cluster references. In assessing whether an object was a cluster WD I assessed both the likelihood of cluster membership and the likelihood that the object was a WD. For the globular clusters I used relaxed criteria in both categories, allowing photometric properties to be sufficient if the reported WDs had appropriate magnitudes and colors, and if there were few, or no, field stars with the same magnitudes and colors. For the open clusters, where field contamination is much more severe, I required that the authors used proper motions or some other criteria to evaluate membership and that the resulting membership probability was at least "probable".

3. Discussion

At present there are reliable estimates for the total WD populations for only two clusters, the Hyades and M4. The Hyades is near enough, and its population has been well enough studied, that

J. Andersen (ed.), Highlights of Astronomy, Volume 11A, 430–432.

essentially all the cluster WDs have been identified. This includes 7 single WDs and 3 WDs in binaries. The combined total mass is 6.4 $M_\odot$ (Boehm-Vitense 1993; White et al. 1993; Wegner et al. 1989). Errors on individual masses are generally $\leq$ 5%, so the expected error of the total mass is smaller than 5%. More importantly, it is unlikely that the current Hyades census is missing a significant number of cluster WDs (Reid, private communication). The total mass of the Hyades is $410 - 480 M_\odot$ (Reid 1992). The WD mass fraction in this young (~0.7 Gyr, Weidemann et al. 1992) cluster is thus 0.013 - 0.016.

For the globular cluster M4 the situation is very different. Observations would be needed to V $\geq$ 31 over the entire cluster field to count all the WDs. Currently this is impossible, so Richer et al. (1995) used the number of Horizontal branch stars and their evolutionary lifetimes as compared to the lifetime of the cluster WDs (essentially the cluster age) to estimate the number of cluster WDs. They calculate that M4 contains 2×10^4 WDs. Among the $\geq$ 200 that Richer et al. (1997) find in M4, they determine a mean mass of $0.51 \pm 0.03 M_\odot$. Since the observable WDs are younger than the cluster, and therefore from lower mass progenitors, I adjusted the mean cluster WD mass to $0.55 M_\odot$. Sigurdsson (1993) used dynamical models to estimate that the mass of M4 is $\sim 10^5 M_\odot$. The WD mass fraction for this old cluster (~12 Gyr, e.g. Chaboyer et al. 1997) is thus 0.11 ± 0.05, where I have guessed that the dominant sources of uncertainty yield a precision of 50%.

The WD mass fractions for the Hyades and M4 supports a picture where WDs account for ~1% by mass of 1 Gyr populations and ~10% by mass of 10 Gyr populations. How reasonable is such an interpretation? To address this I have created a simple model of cluster formation and evolution. I assume clusters are created in a single burst star formation event and that their mass functions can be characterized by a single power law, $N \sim M^\alpha$, over the mass range $0.1 \leq M/M_\odot \leq 80$. Current HST work on globular clusters (e.g. Piotto, this volume) is roughly consistent with the single power law model, at least below the present-day turn-off mass, ~0.85 $M_\odot$. The timescales of stellar evolution are from the parameterizations of Eggleton, Fitchett, & Tout (1989). Evolving stars up to either 5 or $8 M_\odot$ (Koester & Reimers 1996) are converted to WDs via the initial mass – final mass relation as parameterized by von Hippel et al. (1997) and based on Weidemann & Koester (1983). All gas ejected from evolving stars and all neutron star and black hole remnants are assumed to leave the cluster. These models do not include cluster binary stars. Open clusters are known to have a large number of binaries while globular clusters have binary fractions typically $\leq$ 5% (e.g. Richer et al. 1997). The challenge in comparing these models to the clusters is to observationally correct for binaries in the open clusters. This has been done for the Hyades. Finally, I do not model cluster kinematical evolution. Thus clusters which have undergone significant dynamical evolution should show marked differences from these simple models. Mass segregation can be simply corrected for since King models for positions near a few core radii (Cool, Piotto, & King 1996; Piotto, this volume) have consistently shown that the PDMFs in those regions are expected to be very similar to the global MF. Most globular cluster observations have been taken at just such intermediate cluster radii. Stellar evaporation, on the other hand, would pose a problem for the models. I note, however, that globular clusters contain many stars down to at least $0.2 M_\odot$, so I do not expect globular clusters to have lost a significant fraction of their WDs.

Figure 1 compares the simple model to the data for M4 and the Hyades. The M4 WD mass fraction is consistent with a Salpeter IMF, while the Hyades WD mass fraction is consistent with a somewhat steeper IMF. The Hyades is likely to have seen a considerable amount of dynamical evolution (Weidemann et al. 1992), however, and it is difficult to know how to make the appropriate correction to the stellar number counts. Regardless, it appears that the M4 and Hyades WD mass fractions are reasonable, and probably typical for clusters of their age. The general picture is thus that the old stellar populations of the Galactic disk, bulge, and halo should contain ~10% of their stellar mass in the form of WDs. Ultimately, precise comparisons between the field and cluster MFs for both the disk and halo would test whether the clusters we see today are typical of those which formed the field populations.

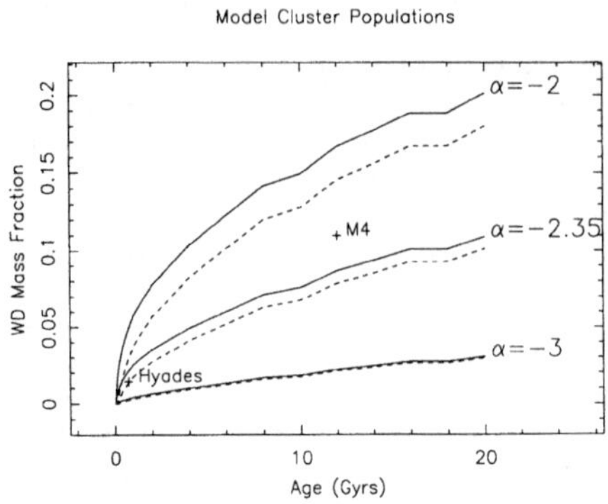

Figure 1. The WD mass fraction for the model clusters as a function of the IMF slope, for $M_{up} = 5$ (dashed lines) and 8 (solid lines) $M_{\odot}$. Also plotted are the WD mass fraction values for M4 and the Hyades.

References

Boehm-Vitense, E. 1993, AJ, 106, 1113
Chaboyer, B., Demarque, P., Kernan, P.J., & Krauss, L.M. 1997, ApJ, in press
Cool, A.M., Piotto, G., & King, I.R. 1996, ApJ, 468, 655
Eggleton, P.P., Fitchett, M.J., & Tout, C.A. 1989, ApJ, 347, 998
Koester, D., & Reimers, D. 1996, A&A, 313, 810
Reid, N. 1992, MNRAS, 257, 257
Renzini, A., et al. 1996, ApJ, 465, L23
Richer, H.B., et al. 1995, ApJ, 451, L17
Richer, H.B., et al. 1997, ApJ, 484, 741
Sigurdsson, S. 1993, ApJ, 415, L43
von Hippel, T., Bothun, G.D., & Schommer, R.A. 1997, AJ, 114, 1154
von Hippel, T., et al. 1996, AJ, 112, 192
von Hippel, T., Gilmore, G., & Jones, D.H.P. 1995, MNRAS, 273, L39
Weidemann, V., Jordan, S., Iben, I., & Casertano, S. 1992, AJ, 104, 1876
Weidemann, V., & Koester, D. 1983, A&A, 121, 77
Wegner, G., Reid, I.N., & McMahan, R.K. 1989, in White Dwarfs, ed. G. Wegner, (Springer-Verlag: Berlin), 378
White, S.M., Jackson, P.D., & Kundu, M.R. 1993, AJ, 105, 563

LOW-LUMINOSITY STARS: PAST AND FUTURE

I. N. REID
105-24, California Institute of Technology
Pasadena, CA 91125

The expression "low luminosity stars" is a descriptive category which spans a wide range of objects, from the oldest stellar remnants in the Galaxy (Wood) through failed stars (Jones, Gould) to the enigmatic MACHOs, discernible not in themselves but only through their effects on others (Bennet). All of these have attracted considerable attention in recent years, and significant progress has been achieved in each case, even if our understanding has failed to keep up with observations. However, it is in the area of brown dwarfs where the most dramatic results have been obtained. The existence of such objects has been predicted theoretically for well over thirty years, but predictions can fail. Thus the discovery of Gl229B (Nakajima et al, 1995), an unambiguous substellar-mass object, followed by the detection of lithium in Teide 1 and Calar 3 in the Pleiades (Rebolo et al, 1996) mark a turning point in studies of brown dwarfs. The issue now is not whether they exist, but what are their properties as a class. With that in mind, a number of obvious questions arise.

First, what is the most effective method of finding brown dwarfs? Both brown dwarfs and VLM stars have effective temperatures below 2500K. Hence, many surveys concentrated on searching for objects which were very red at near-infrared wavelengths, and the blue JHK colours exhibited by Gl 229B came as something of a surprise. This should not have been the case, of course, since Tsuji (1964) predicted the presence of methane in such cool atmospheres well before the term 'brown dwarf' was invented. Current theory, as summarised in these sessions by Tsuji and Burrows, predicts that, as one moves to lower temperatures, grain formation initially drives (J-K) redward, before grains settle below $\tau \sim 1$ and CH_4 absorption sets in. Throughout, however, the optical-to-IR colours, straddling the peak in the emergent flux distribution, become progressively redder (Reid, 1994). Those colours therefore offer the most effective means of identifying both GD 165B-like and Gl 229B-like sources - an argument reinforced by the first results from the IJK DENIS survey (Delfosse et al, 1997).

Second, given that one can construct a statistically well-defined sample of VLM dwarfs and brown dwarfs, how does one derive the mass function, $\Psi(M)$, from the luminosity function, Φ(M)? Clearly, the mass-luminosity relation is the key. As pointed out originally by D'Antona and Mazzitelli (1983), features in Φ(M) may stem from features in $\frac{dM}{dL}$, not $\Psi(M)$, but, despite the heroic efforts summarised here by Latham, the mass-luminosity relation remains poorly-constrained observationally below 1 $M_\odot$. It is particularly disconcerting that the one eclipsing system with $M < 0.5M_\odot$, CM Dra, is offset from the mean relation defined by astrometric systems. Follow-up observations of any eclipsing M-dwarfs found by MACHO, OGLE et al should be given the highest priority.

Third, how universal is $\Psi(M)$? Even without an absolute definition of $\Psi(M)$, it is apparent. particularly from open cluster work, that there are variations. The Pleiades appears to have $\Psi(M) \propto M^{-1}$, matching results for local field stars (which data, as Henry and Becklin points out, may still be incomplete); Hillenbrand (1997), however, finds that the Orion core stars are closer to a Miller-Scalo function, peaking at $\sim 0.2M_\odot$; and some young open clusters (eg IC 2391) seem to lack any late-type M-dwarfs. Are these differences real, or selection effects? Note that, unlike field-star surveys, selection effects in cluster surveys predominantly lie in including field stars, and overestimating $\Psi(M)$.

Fourth, are there significant differences between $\Psi(M)$ for single stars (and primaries) and companions? Imaging surveys for companions to white dwarfs (Becklin & Zuckerman) and M-dwarfs (Henry, Oppenheimer), and radial velocity surveys of nearby stars all detect many M-dwarf companions, but few brown dwarfs - contradicting one's expectations given a mass function rising

J. Andersen (ed.), Highlights of Astronomy, Volume 11A, 433–434.

towards lower masses. Does this reflect significant differences in the physics of formation, or are selection effects still sufficient to account for this result?

Clearly, many other questions remain, notably whether brown dwarfs (or white dwarfs) have anything to do with MACHOs. These are issues for the next three years - or more.

References

D'Antona, F. & Mazzitelli, I. 1983, Astr. Ap. 127, 149-152
Delfosse, X. et al, 1997, Astr. Ap., in press
Hillenbrand, L.A., 1997, AJ 113, 1733-1754
Nakajima et al, 1995, Nature 378, 463-465
Rebolo, R. et al, 1996, ApJ 469, L53-56
Reid, I.N. 1993, Science with astronomical near-IR surveys, ed. N Epchstein
Tsuji, T. 1964, Ann. Tokyo Obs. Ser. II, 9, 1-24

A DEEP LARGE AREA SEARCH FOR LOW LUMINOSITY STARS

HUGH R.A. JONES
Institute of Astronomy, University of Tokyo, 2-21-1 Osawa, Tokyo 181
Astrophysics Group, Liverpool JM University, Byrom Street, Liverpool L3 3AF

AND

MIKE R. S. HAWKINS
Royal Observatory Edinburgh, Blackford Hill, Edinburgh EH9 3HJ

In a recent survey for faint red stars from a digital stack of Schmidt plates a number of candidate objects were identified. Parallax's for three of these objects have been reported showing them to have luminosities which interpreted within the available evolutionary models indicate them to be good brown dwarf candidates. Here we examine spectra of these objects and others from the plate stack. Using standard spectral indices we find that for a given spectral type their spectra are more consistent with the Pleiades brown dwarfs (PPL 15, Teide 1 and Calar 3) than with standard late-type M dwarfs. Our interpretation is that this is due to their selection by $R_F - I_N$ colours which at values > 3 preferentially selects objects with relatively low gravities. For late-type M dwarfs and brown dwarfs low gravities are expected to be a reliable indication of youth. We also notice that the stack objects generally have strong FeH absorption for their spectral type. Current model atmospheres suggest that FeH strongly increases in strength toward lower metallicities and lower temperatures. We believe that this is not consistent with the available observational evidence from late-type M dwarfs. It is possible that solid Fe is forming in the low temperature atmospheres relatively depleting FeH strengths toward lower temperatures. We find some evidence that for dwarfs at low temperatures dust formation is less prevalent in lower gravity objects suggesting that dwarfs at low temperatures stronger FeH may be an indication of youth. In addition to the spectral evidence the three stack objects whose parallax's have been measured show small tangential velocities which is a further indication of youth.

We have constructed preliminary models for the number of brown dwarfs and M dwarfs expected from the digital stack survey. We find that most of the objects are brown dwarfs less than 1 Gyr old and because of this we make models using a time-dependent scale height. M dwarfs will dominate over brown dwarfs until around $I = 18$ for $\alpha = -2.35$ and $I = 20$ for $\alpha = 0.00$. For the case of $\alpha = 0.00$ the survey would be expected to detect few brown dwarfs. Noting the large uncertainties in constructing reliable models we believe our findings are more consistent with $\alpha = -2.35$. Overall their low luminosities, peculiar spectra, low tangential velocities and the expectation of detecting young brown dwarfs suggest that the stack objects are young brown dwarfs some of which may have very low masses. The main problem with such an interpretation is that their position in the HR diagram does not easily fit with the Pleiades sequence of brown dwarfs or the predictions of evolutionary models. However for late-type M dwarfs there is evidence that evolutionary models and synthetic spectra do yet give consistent results. The clear resolution of these issues probably requires further parallax measurements and follow-up observations of the new objects together with considerable work on high-temperature dust opacities and their inclusion in evolutionary models.

J. Andersen (ed.), Highlights of Astronomy, Volume 11A, 435.

WIDE BINARIES: PROBES OF THE GALAXY'S DARK MATTER CONTENT

TERRY D. OSWALT, J. ALLYN SMITH* AND MATT A. WOOD
Florida Institute of Technology
Department of Physics and Space Sciences
Melbourne, Florida, 32901-6988 USA
oswalt@tycho.pss.fit.edu

Photometric parallaxes derived from B, V, R, I photometry of wide binaries with white dwarf components were used to set a firm lower limit to the age of the Galactic disk (see Oswalt et al. 1996 and the article by Wood elsewhere in these proceedings). Here we show the distribution of projected semi-major axes derived from new B, V, R, I data for over 200 wide pairs.

Like clusters, wide binaries are gradually disrupted, primarily by the gentle but cumulative effects of encounters with the denser parts of giant molecular clouds (see Wasserman & Wienberg 1991 for a review). A sharp cut-off in semi-major axes > 0.1 pc was one of the predictions of competing models. Wood & Oswalt (1992) showed that additional orbital expansion occurs in wide pairs which have experienced isotropic post-MS mass loss. Can we see the signatures of these events in the wide binary sample?

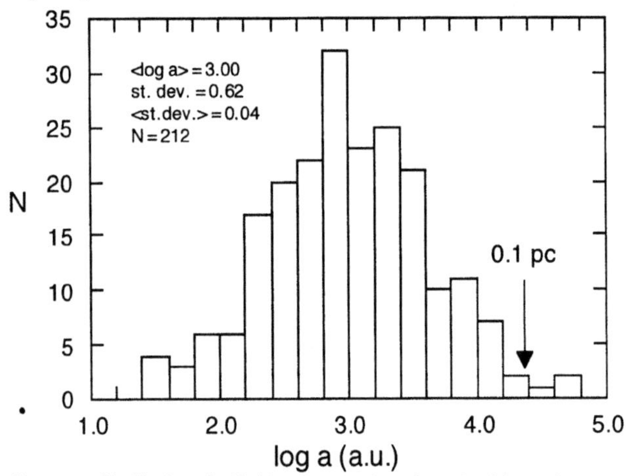

Figure 1. Frequency distribution of wide binary separations determined from photometric parallaxes.

Figure 1 shows that typical pairs are separated by $\approx 10^3$ a.u., with very few wider than 0.1 pc. However, the distribution in log(a) is nearly Gaussian, with no evidence for a sharp cut-off at the widest separations. A detailed analysis and comparison to the log(a) distribution for un-evolved pairs will be published elsewhere.

References

Oswalt, T.D. et al. 1996, Nature (letters), 382, 692.
Wasserman, I., & Weinberg, M.D. 1991, ApJ 382, 149.
Wood, M.A., & Oswalt, T.D. 1992, ApJ 394, L53.

*Now at the University of Michigan.

J. Andersen (ed.), Highlights of Astronomy, Volume 11A, 436.

DWARF CARBON STAR MODEL ATMOSPHERES AND SYNTHETIC SPECTRA

[1]U. G. JØRGENSEN, [1]A. BORYSOW, [1]S. HÖFNER AND [2]R. F. WING
[1]*Niels Bohr Institute, Copenhagen University Observatory, Denmark*
[2]*Astronomy Department, Ohio State University, USA*

Dwarf carbon (dC) stars are believed to be main-sequence stars which have received carbon-rich material from a former AGB carbon star companion. We have computed model atmospheres, synthetic spectra, and colours for dC stars and compared them with observations. Our models are computed with an updated version of the MARCS code (Jørgensen et al. 1992, A&A 261, 263) and include atomic and molecular line blanketing, grain formation, and collision-induced absorption (CIA) processes (Borysow et al. 1997, A&A 324, 185). The models assume hydrostatic equilibrium, which for giant stars (i.e., high luminosity and low gravity) would be inconsistent with the inclusion of dust. Typical gravities of dC stars are, however, considerably larger than the acceleration due to radiation pressure even on dust grains with a relatively high absorption coefficient.

For the cooler solar metallicity models, grain formation can be substantial, but it depends strongly on the assumed type of dust. In a test model of T_{eff} = 2800 K, C/O = 1.17, and log(g) = 5, 1% of the free carbon (i.e., that not bound in CO molecules) condensed when the dust was assumed to be amorphous carbon, whereas as much as 30% condensed when nano-diamonds (of the extra-solar form which are abundant in carbonaceous chondrite meteorites and are believed to have formed in stellar outflows) were allowed for. This difference in degree of condensation is due to the very small diamond opacity compared to that of amorphous carbon. In both cases, however, the main effect of the dust was a 400 K heating of the surface layers. The dust itself is not clearly visible in the synthetic spectrum, but the heating and correspondingly reduced molecular partial pressures cause a weakening of spectral features in the 3μm region (due mainly to reduced C_2H_2 abundance in the upper layers) and reduced CO fundamental and first overtone band intensities.

For the low metallicity models, collision-induced absorption of H_2-H_2 complexes dominates the opacity. The continuous CIA is strongest at intermediate infrared wavelengths (typically around 2 to 5μm at temperatures of our interest) and causes the near infrared colours to become more blue. For a test model of T_{eff} = 4000 K, $Z = 10^{-4} Z_\odot$, and C/O = 1.17, the effect of CIA was a reduction in the value of the synthetic H−K and J−H colours of 0.15 and 0.3 mag., respectively.

The computed features due to C_2, CO, and CN are much weaker in dC stars than in giants of the same T_{eff}, and the H^- opacity minimum hump is almost absent. Polyatomic molecules, on the other hand, form more readily in the dwarfs than in the giants, due to the higher gas density.

The solar metallicity dwarf models occupy the same region of the JHK colour-colour diagram as the giant carbon stars, but the low-metallicity giant and dwarf models are well separated. There is a good agreement between synthetic and observed colours for our giant models of all metallicities, but the computed JHK colours for the low-metallicity dwarfs are bluer than the observed dC stars. Ironically, we are able to fit the observed colours of the dwarfs with less complete opacities in our models. However, our models with what we consider the best available opacity data can fit the colours of the dwarf carbon stars if the stars for example are surrounded by a dusty envelope or disk.

This work was supported by the Danish Natural Science Research Council, NASA Astrophysics Theory Program, and the Austrian Science Fund FWF.

J. Andersen (ed.), Highlights of Astronomy, Volume 11A, 437.

PARALLAXES USING INFRARED ARRAYS

R. M. BROCKIE[1], H. R. A. JONES[2,3], M. WELLS[4] AND A. J. LONGMORE[4]
[1]*Institute of Astronomy, University of Edinburgh,* [2]*Institute of Astronomy, University of Tokyo,* [3]*Astrophysics Group, Liverpool JM University,* [4]*Royal Observatory Edinburgh*

1. Abstract

We have recently carried out an infrared parallax programme on the UKIRT conducted with a 256^2 array. Our observations were made in the near-infrared K-band which is ideal for observing red stars and brown dwarf candidates and has the advantage that differential colour refraction is negligible. We find a parallax of $\pi_{\rm abs} = 156.0 \pm 13.3$ mas for the archetypal late-type M dwarf vB 10 and $\pi_{\rm abs} = 11.1 \pm 7.5$ mas for PC 0025+0447 which we conclude is too distant to be a clear case of a brown dwarf and is probably a late-type M dwarf close to the hydrogen-burning limit. Recent improvements in telescopes and instrumentation working in the infrared should enable accuracies of ~ 1 mas to be obtained for future infrared parallax programmes.

2. Accuracy of Future Observations

The accuracies for the parallax measurements in our programme are of order $\delta\pi \sim 10$ mas. The introduction of larger detectors in conjunction with tip/tilt or low–order correction adaptive optics systems promises to allow even greater accuracy to be attained in the K–band.

One can estimate the benefits such a system will deliver. A factor of two increase in accuracy will come from smaller image sizes with a tip/tilt system. A second factor of two can be obtained due to improved contrast for stars fainter than the background. A third factor of two comes from using a 1024^2 array. This will have 16 times more pixels than the 256^2 array in IRCAM3, but if a tip/tilt system is being used, the requirement of fully sampling the point spread function means that only 4 times the area on the sky can be viewed. Thus, we will have roughly 4 times as many reference stars, but we expect the accuracy to increase as $\sqrt{n_{\rm stars}}$. This larger number of reference stars will also reduce the residuals when constructing the global reference frame. Overall, we expect about an order of magnitude of improvement and that accuracies of $\delta\pi$ = 1–2 mas are possible given current innovations making infrared-based parallaxes fully competitive with optically-based parallaxes for red objects.

As they become available we hope to include brown dwarf candidates and other red interesting objects discovered from large area surveys such as 2MASS, DENIS, EROS and SLOAN. Based on the case of the brown dwarf Gl 229B (R_{F675W} - J_{CIT} = 10.6, Oppenheimer 1997, in preparation) field brown dwarfs will be too faint for parallax observations to be made at visible wavelengths and infrared parallax programmes such as this will be essential.

The move to conduct parallax programmes at infrared wavelengths will enable observers not to be concerned critically about airmass when making measurements. This will make it considerably easier for parallax programmes to be conducted on common–user telescopes using queue–based scheduling.

J. Andersen (ed.), Highlights of Astronomy, Volume 11A, 438.

SPECTRA OF M DWARFS – DIMMED BY DUST

T. TSUJI, K. OHNAKA, W. AOKI AND H. R. A. JONES
Institute of Astronomy, The University of Tokyo
2-21-1 Osawa, Mitaka, Tokyo, 181 Japan

1. Introduction

Spectra of M dwarfs are rich in atomic and molecular lines. These spectra provide such basic information as $T_{\rm eff}$ (or radius), log g (or mass), surface chemical composition, and something more (e.g. activity) if properly interpreted. It is recognized, however, that spectra of M dwarfs are already dimmed by the dust formed in their photospheres (Tsuji et al. 1996a) and this effect, which has been overlooked until recently, should be taken into account in any interpretation and analysis of the spectra of very low mass objects (VLMOs) including late M dwarfs and brown dwarfs.

2. Dusty Model Atmospheres of M Dwarfs and Brown Dwarfs

In cool and dense photospheres of VLMOs, the thermodynamical condition for condensation is well fulfilled. We first assume that the dust forms in the local thermodynamical equilibrium (LTE): This assumption is of course not trivial and we will return to this point later (Sect.4). Then, we consider condensation of such grains as Al_2O_3 (corundum), Fe (iron), and $MgSiO_3$ (enstatite) formed from the most abundant refractory elements (Fe, Mg, Si, Al). These condensates are effective sources of opacity due to their large extinction coefficients, which can be evaluated by the Mie theory (we assumed grain radius of 0.1μm). The volatile molecules such as CH_4, H_2O, NH_3, H_2S, PH_3 remain as important sources of opacity, together with the collision-induced dipole transitions of H_2.

Then, we constructed non-grey dusty model in radiative-convective equilibrium within the framework of the classical assumptions, and extended our grid of model atmospheres to $T_{\rm eff}=$ 800K (solar composition, log g = 5.0, and $v_{\rm micro}$ = 1km s^{-1}). The effects of dust appear in models of $T_{\rm eff}=$ 2,800K and lower: dust (especially iron) effectively blocks the radiation and the photosphere is heated appreciably. For this reason, molecular bands such as H_2O are much weaker than in dust-free model of the same $T_{\rm eff}$. Also, optical spectra suffer large extinction by the dust.

3. Evidences for Dust in Obseved Spectra and Colours

We have already shown evidence for dust in late M dwarfs (Tsuji et al. 1996a), but a more clear case is shown in Fig.1a for the brown dwarf candidate GD 165B discovered by Becklin & Zuckerman (1988): The spectra observed by Jones et al.(1994) can best be explained by the predicted spectrum by our dusty model (solid line) but not at all by our dust-free model (dashed line). Further evidence for dust in late M dwarfs is also found in the weakening of TiO bands (Jones & Tsuji 1997). Somewhat ironically, however, our dusty models could not explain the spectra of the genuine brown dwarf Gl 229B discovered by Nakajima et al.(1995) just when we have completed our dusty models, although our earlier model without dust (Tsuji & Ohnaka 1995) could (Tsuji et al 1996b).

In the $(I-J, J-K)$ diagram shown in Fig.1b, $J-K$ of M dwarfs (Leggett 1992) remains nearly constant up to M6V but shows reddening at M7V and later including GD 165B (Tinney et al. 1993). This trend is well explained by our dusty models (solid line) but not by dust-free models (dashed line). Recently, three brown dwarf candidates, one of which (DBD J1228-1547) is confirmed to be brown dwarf, have been found by the DENIS survey (Delfosse et al. 1997), and they appear in Fig.1b near GD 165B which we already know to be well dusty. Thus, the agreement between the observed colours of late M dwarfs (and brown dwarf candidates) and the predicted

J. Andersen (ed.), Highlights of Astronomy, Volume 11A, 439–440.

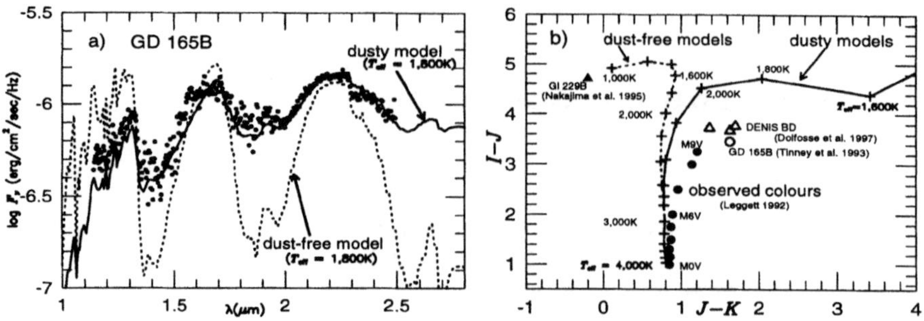

Figure 1. **a)** The spectrum of GD 165B (dots: Jones et al. 1994) is compared with the predictions by dusty (solid line) and dust-free (dashed line) models ($T_{\rm eff}$= 1,800K), and shows definite evidence for dust in this brown dwarf candidate. **b)** Predicted loci based on dusty (solid line) and non-dusty (dashed line) models on the $(I-J, J-K)$ diagram. Observed colours are shown for M dwarfs (filled circles), GD 165B (open circle), DENIS brown dwarf candidates (open triangles), and GL 229B (filled triangle). Note that dusty and non-dusty cases can clearly be separated.

ones based on our dusty models can be regarded as evidence for dust in most late M dwarfs and brown dwarf candidates. On the other hand, the predicted locus based on our dust-free models turns to the blue at about $T_{\rm eff}$= 1,600K and explains the observed colous of Gl 229B quite well. Thus there should be two kinds of brown dwarfs which can clearly be separeted on $(I-J, J-K)$ diagram: blue brown dwarfs of non-dusty type (e.g. Gl 229B) and red brown dwarfs of dusty type (e.g. DBD J1228-1547). However, M dwarfs later than about M7V are probably all dusty and red.

4. Discussion and Conclusions: how dust forms in stellar photosphere

The presence of dust in the photospheres of M dwarfs later than M7V and hot brown dwarfs can be regarded as well established. The fact that the majority of VLMOs are dusty implies that dust exists in the photospheres of these objects throughout their lifetime (or as long as the Hubble time in the case of M dwarfs). This should be possible if dust formation and destruction are repeating forever, that is the dust is in detailed balance with the gas. This means that the dust-gas mixture is in LTE, and our assumption that dust forms in LTE (Sect.2) is justified by the very fact that dust is present in a large number of VLMOs.

The case of the very cool brown dwarf Gl 229B is clearly different. Probably grain size turns to be larger than the critical radius, and dust may begin to grow (that is an irreversible process starts). Then, the dust can no longer be in detailed balance with the gas and will finally segregate from the gas (or dust clouds will form). Volatile gas will then appear in the observable layer in the photosphere and this case explains the observation of Gl 229B (Tsuji et al. 1996b). Thus, the presence of two kinds of brown dwarfs (Sect.3) can be explained. However, where and why the transition from the dust-gas detailed balance regime to dust-gas segregation phase occurs is unknown. Probably, some mechanism to prevent dust-growth is working until a certain phase. Certainly, more work is needed to fully understand the spectra of M dwarfs and brown dwarfs, and to better determine their basic properties. For this purpose, dust is again an important ingredient, since the Universe is everywhere dusty and M dwarfs are no exception.

References

Becklin, E. E. and Zuckerman, B. (1988) Nature, **336**, 656
Delfosse, X. et al. (1997) A&A, in press
Jones, H. R. A., Longmore, A. J., Jameson, R. F., Mountain, C. M. (1994) MNRAS, **267**, 413
Jones, H. R. A. and Tsuji, T. (1997) ApJL, **480**, L39
Leggett, S.K. (1992) ApJS, **82**, 351
Nakajima, T., Oppenheimer, B. R., Kulkarni, S. R., Golimowski, D. A., Matthews, K. and Durrance, S. T. (1995) Nature, **378**, 464
Tinney, C. G., Mould, J. R. and Ried, I. N. (1993) AJ, **105**, 1045
Tsuji, T. and Ohnaka, K. (1995) in *Elementary Processes in Dense Plasmas* eds. S. Ichimaru and S. Ogata, Addison-Wesley, p.193
Tsuji, T., Ohnaka, K. and Aoki, W. (1996a) A&A, **305**, L1
Tsuji, T., Ohnaka, K., Aoki, W. and Nakajima, T. (1996b) A&A, **308**, L29

OPTICAL AND INFRARED SPECTROSCOPY OF CM DRACONIS

S. VITI[1], H. R. A. JONES[2,3], F. ALLARD[4], A. SCHWEITZER[5] P. HAUSCHILDT[6]
[1]*Department of Physics and Astronomy, University College London,* [2]*Institute of Astronomy, University of Tokyo,* [3]*Astrophysics, Liverpool JM University,* [4]*Department of Physics, Wichita State University,* [5]*Landessternwarte Heidelberg,* [6]*Department of Physics and Astronomy, The University of Georgia*

1. Abstract

We compare observations of the binary system CM Draconis with synthetic spectra. Spectroscopic observations from 0.40 to 2.41μm, combined with photometry and the accurately known surface gravity enable us to estimate the temperature and metallicity using detailed spectra synthesis. We find discrepancies between the analysis of the infrared and optical spectrum: while the optical spectral energy distribution (SED) yields a metal-rich solution with $T_{\rm eff}$ = 3000K, the infrared SED yields around 3200K and $-0.8 \leq$ [M/H] ≤ -0.6 compatible with the high space motion of the system. The low-metallicity characteristics of the infrared SED could be real and is partly supported by the detailed analysis of the atomic lines in the optical region. Although, the known incompleteness of the TiO and H_2O line lists in the models may cause substantial systematic errors.

2. Background

CM Draconis (hereafter, CM Dra) is the lowest mass main-sequence eclipsing binary known. It has a short period of 1.268 days (Metcalfe et al. 1996, ApJ, 456, 356), a very large space velocity (163 km s^{-1}) and low flaring rate which suggests it belongs to Population II (Lacy 1977, ApJ, 218, 444). Because it is one of the faintest, smallest and least massive main-sequence binaries so far known, it is a prime calibrator of what happens to very low mass stars at the bottom of the main sequence. Some of the fundamental physical properties of the components of CM Dra have already been accurately determined. Most important are the individual masses of $M_{\rm A} = 0.2307\pm 0.0010$ and $M_{\rm B} = 0.2136\pm 0.0010\ M_\odot$ and radii of $R_{\rm A} = 0.2516\pm 0.0020$ and $R_{\rm B} = 0.2347\pm 0.0019\ R_\odot$ (Metcalfe et al., 1996), and surface gravities of log g = 4.999 $\pm$ 0.007 for CM Dra A and log g = 5.028 $\pm$ 0.007 for CM Dra B (Jones et al. 1996, MNRAS, 277, 767). The precision of these values far exceeds those known for any M dwarf and means that CM Dra is an excellent system for comparison with model calculations. The spectra taken for our study are shown below in Figure 1.

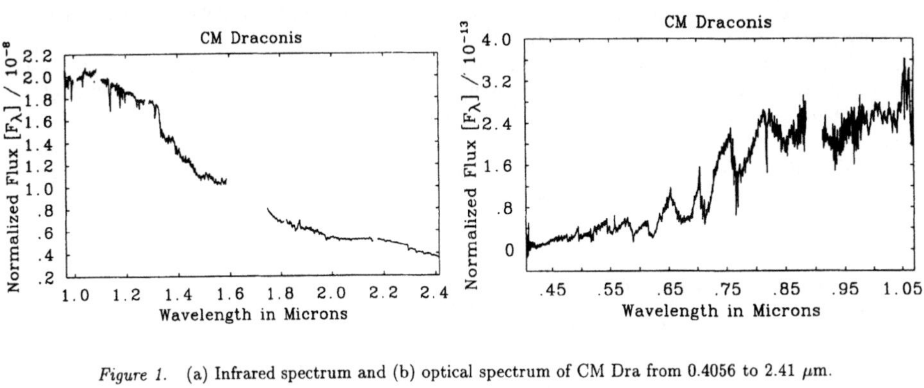

Figure 1. (a) Infrared spectrum and (b) optical spectrum of CM Dra from 0.4056 to 2.41 μm.

J. Andersen (ed.), Highlights of Astronomy, Volume 11A, 441.

NEW MEMBERSHIP CRITERIA FOR 437 FLARE STARS IN THE PLEIADES

G. SZÉCSÉNYI-NAGY
Eötvös University, Department of Astronomy, Budapest, Hungary
szena@ludens.elte.hu

AND

E. SCHILBACH, S. HIRTE AND R.-D. SCHOLZ
Astrophysikalisches Institut Potsdam, Potsdam, Germany
eschilbach@aip.de, shirte@aip.de and rdscholz@aip.de

1. New Proper Motion Determinations and Photometry

Most of the flare stars (FSs) discovered in that field are very faint and are missing from all previously published lists of Pleiades cluster members. Reliable membership probabilities (MPs) were only determined for one third of these objects while the rest were supposed to be more or less probable members with a few exceptions. A recent proper motion survey of that field (Schilbach et al. 1995) based on plates taken with the Tautenburg Schmidt telescope and extended to an area of 16.5 square degrees provided new cluster MPs and also photometric data even for a number of the lowest luminosity stars (in fact the faintest object measured has $V = 18.26$). Of the 520 known FSs in the field 437 (85%) were successfully identified, their J2000.0 coordinates, high precision proper motion components and apparent photographic R magnitudes (and for the majority of them their V and B magnitudes too) determined. Based on these data new cluster MPs have been deduced which show that a considerable number (40%) of the so called Pleiades FSs are not members of that cluster at all. This conclusion follows the tendency already found: when we involve more and more faint FSs into the investigations the percentage of non-members monotonically increases (cf. Haro et al. 1982 and references therein).

2. The Most Important Results

1. The number of FSs with well determined MPs has been increased by 137.5 percent in the so called Pleiades field (to 85% of the catalogued objects). 2. The statistics is even more complete in the Largest Common Field (LCF, for more details see Szécsényi-Nagy 1994) of the Pleiades (it involves 407 FSs i.e. 91% of the LCF FSs). 3. The relative frequency of cluster member FSs amongst all FSs discovered in the region is between 0.45 and 0.50. 4. Low activity FSs (with one or two very low amplitude flares observed) have even lower mean MPs (30% to 33%). 5. At least 40% of all FSs catalogued here are background or (less probably) foreground stars (which are obviously not more than one tenth of the non-members). 6. Amongst cluster member FSs no one which is fainter than absolute R magnitude 11.2 is known. This means that 70 Myr old FSs are not active anymore below this brightness (and the corresponding mass) limit.

References

Haro, G., Chavira, E. & Gonzalez, G. 1982, Bol. Obs. Ton. 3, No. 1
Schilbach, E., Robichon, N., Souchay, J. & Guibert, J. 1995, A&A 3, 6
Szécsényi-Nagy, G. 1994, IAU Symp. 161, 61

This research has been supported partly by the FHHER (grants: AMFK-418/96 *and FKFP* – 0201/1997).

J. Andersen (ed.), Highlights of Astronomy, Volume 11A, 442.

COMPLETING THE LOCAL SAMPLE WITH TYCHO

R. L. SMART, R. PANNUNZIO, M. G. LATTANZI
Osservatorio Astronomico di Torino, Italy

AND

B. MCLEAN
Space Telescope Science Institute, USA

The Tycho Catalogue contains proper motions and low precision parallaxes with a median precisions of 25 mas, while having high precision B and V magnitudes (median precision of 0.07 and 0.06 respectivly). The number of high parallax stars is quite unexpected, with over 30 previously unreferenced, non-Hipparcos, stars with a parallax >500 mas and over 2000 with a parallax >250 mas. While many of these are probably mistakes, the completeness limit of Tycho and the precise magnitudes provide us with a way to intelligently build a complete picture of the nearby star field. Catalogues of nearby stars provide us with many fundamental parameters in the field of stellar astronomy, but the completness of known stars within say even 8pcs is still underestimated by about 30%, i.e., 50-60 systems (T.J. Henry, CSSS 8, ASP Conf Series 64, 1994).

In a new program at Torino we are using the Tycho Catalogue to intelligently build a list of possible candidates for a 10pc volume. Figure 1 shows that Tycho will contain basically all ZAMS to spectral type M3 within 10 pcs, M5 to 4pcs and M6 to 2pcs. A comparison of the photometric and Tycho distances provides a list free of giants and systems where binarity has affected the color. This list is then added to the Torino Parallax program where we expect to provide more precise parallaxes.

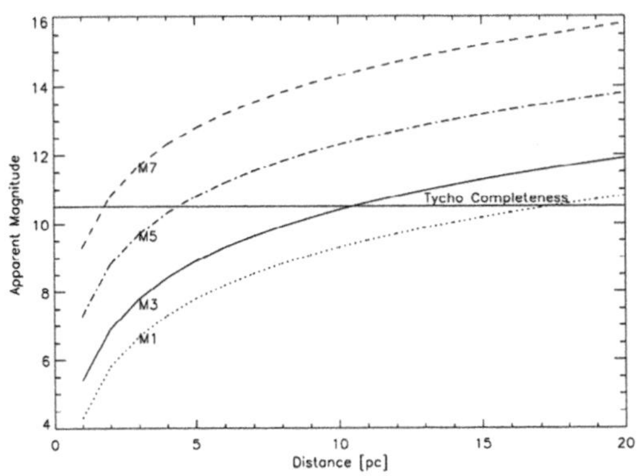

Figure 1. Change of observed magnitude with distance under the assumption of negligible absorption. The straight line indicates the approximate completeness limit of the Tycho catalogue.

J. Andersen (ed.), Highlights of Astronomy, Volume 11A, 443.

JD11

Redshift Surveys in the 21st Century

Chairpersons and Editors: **J. Huchra & A.P. Fairall**

INTRODUCTION

A.P.FAIRALL
Department of Astronomy, University of Cape Town
Rondebosch, 7700. South Africa

As we approach the 21st Century, enormous technical advances are being made in the acquisition of galaxy redshifts. This Joint Discussion offers a perspective on those advances and the associated redshift surveys. It is representative, but by no means all inclusive.

My introduction offers some historical background, against which the new methods and achievements can be measured.

Redshift surveys tells us about the spatial distribution and evolution of galaxies. Yet the study of large-scale structures in the distribution of galaxies goes back further than one might imagine. The first accurate description of the Virgo Supercluster was given by John Herschel in the mid-19th Century. To paraphrase Sir John, "Virgo is the central condensation of a roughly spherical cluster of nebulae – our system lies outside the denser part of the cluster, but is involved with its outlying members – forming an element of some one of its protuberances or branches".

Victorian science might have beaten Hubble had more people believed John Herschel – and had he himself more confidence in his interpretation. Few were prepared to accept the 'Island Universe' theory that it implied, and strong criticism followed. Moreover, Herschel was very reluctant to go against the beliefs of his revered father, William Herschel, who, though once believing in the 'Island Universes', had in later life accepted that all nebulae were somehow gaseous condensations in our system. Consequently, even John Herschel's textbook still stated the older, incorrect interpretation.

Richard Proctor, a popularizer of astronomy, was one of those critics of Herschel's, Virgo 'Supercluster'. Yet Proctor also produced plots of Herschel's nebulae over the sky. One of these is quite remarkable; in the Centaurus region, the galaxies are clearly seen to form long filaments surrounding empty voids – much as we have seen in modern days 'redshift slices'. The fabric of the Universe, on its largest scale, had been revealed – long before the first redshift had been obtained. John Herschel had seen it too, and had made the correct guess as to the geometry of the Virgo Supercluster – but the general criticism of the 'Island Universe' theory and his deep respect for his father — kept it for science to rediscover a hundred years later.

The very first galaxy 'redshift' (actually a slight blueshift) was obtained early this century – Vesto Slipher extracted a radial velocity for the Great Nebula in Andromeda in 1913. Yet the observation was made on the basis that spiral nebulae might be solar systems in formation. Slipher worked for Percival Lowell and carried out instructions; Lowell's obsession with Mars had let him to consider looking for other solar systems. Yet Slipher was to persevere, obtaining redshifts of other galaxies – even though some of the exposures took nights to accumulate sufficient signal. By 1914, he presented some 15 redshifts to a meeting of the American Astronomical Association, for which he received a standing ovation. As we know, his early redshifts were later to be interpreted by Hubble as the linear recession of galaxies.

Many redshift pioneers followed. I would particularly like us to remember Fritz Zwicky (for whom I worked as a graduate student) who was the first to obtain redshifts of a cluster of galaxies – the Coma Cluster – and consequently the first to call attention to the presence of 'dark matter'. Also Gerard de Vaucouleurs who from 1953 championed the existence of the Local supercluster – the rediscovery of Herschel's great concentration.

Improved technology greatly speeded up redshift production. The remarkable labyrinth of large-scale structures was revealed as we saw the distribution of galaxies in the three dimensions of redshift space. The early plots of Joeveer and Einasto were confirmed by the statistically controlled redshift surveys of the Center for Astrophysics. The rest is well known. Whereas at the start of the century, a single redshift took many nights of exposure, we now have the prospect of state-of-the-art

J. Andersen (ed.), Highlights of Astronomy, Volume 11A, 447–448.

Figure 1. Sir John Herschel's telescope was located at the Cape of Good Hope during 1834-38, thereby enabling him to complete the global survey of nebulae that led to the 'New General Catalogue'

instruments that will gather more than a thousand a night as the new century begins.

References

Proctor, R., 'The Universe of Stars', *Longmans, Green and Co.*, 1878.
Smith, R., 'The Expanding Universe', *Cambridge University Press*, 1982.

THE SLOAN DIGITAL SKY SURVEY

MASATAKA FUKUGITA
Institute for Cosmic Ray Research, University of Tokyo
Tanashi, Tokyo, 188 Japan

Abstract

The concept and the present status of the Sloan Digital Sky Survey are described, with emphasis on the instrumentation aspect of the project.

1. Introduction

The Sloan Digital Sky Survey (SDSS) is a project undertaking a photometric survey of a half of the northern sky (π ster) and a follow-up spectroscopic survey of over one million astronomical objects, one million galaxies and a hundred-thousand quasars, complete within precisely defined selection criteria. This project aims at producing a large homogeneous data sample of the northern sky with high accuracy photometry. The data base should provide a basis for quantitative cosmology in the 21st century.

The most important feature of the project lies in a careful design of the system and a strategy to precontrol the quality of data and to enable us to understand the selection effects. No surveys are free from selection biases, and SDSS is not an exception. The important requirement is to know quantitatively what is observed and what is missed in the survey.

The photometric survey will be made in five colour bands to a limiting magnitude of $r' \simeq 22.5$ mag. We expect the photometric catalogue to contain 5×10^7 galaxies and a comparable number of stars. The follow up spectroscopy is wide band extending from 3900Å to 9100Å with a moderate dispersion of $\lambda/\Delta\lambda \sim 2000$. The spectroscopic sample will include 10^6 galaxies to $r' \simeq 18.2$ mag and 10^5 quasars to $g' \simeq 19.5$ mag.

We have been building a dedicated wide-field telescope of 2.5m diameter (f/5), a large format mosaicked CCD camera that fits the focal plane diameter of 65 cm, and two double spectrographs fed with 640 fibres altogether. The photometric data will be calibrated using an ancillary 60cm telescope installed for this purpose, and all detected objects will be catalogued. Target selection is made based on this catalogue and selected objects are then subjected to spectroscopic studies. A fast computing facility and supporting software are also an important and challenging ingredient of this project.

The system parameters were determined to minimize the cost of instrumentation under the requirement that 10^6 spectroscopic objects are obtained in 5 years, and under the assumption of the availability of 2048×2048 thinned CCD chips with 24 μm pixels (physical size is 4.9 cm square), the state-of-the-art technology at the time of the system design. These two constraints fix the basic parameters of the instruments.

This survey is being carried out by a collaboration of University of Chicago, Princeton University, Institute for Advanced Study, Fermilab, JPG (a group of Japanese astronomers), Johns Hopkins University, University of Washington and US Naval Observatory. We refer to Gunn & Weinberg (1995) for an earlier publication that describes this project.

2. Telescope

It is crucial to realize a wide field of view for efficient photometric imaging. It is known that a wide, flat field of view can be achieved with a Ritchey-Chrétien design with two hyperboloids with

J. Andersen (ed.), Highlights of Astronomy, Volume 11A, 449–459.

the same curvature (Bowen & Vaughan 1973). The most successful telescope of this type is the du Pont 2.54m at Las Campanas Observatory, which has a 1.5° field of view (FoV) at f/7.5. The SDSS telescope has a diameter of 2.5m with a faster focal ratio f/5 and a FoV of 3°. Our survey strategy requires very low image distortion and uniform image scale across the field. The two mirrors (2.5m f/2.25 primary and 1.08m secondary) have 6th and 8th order polynomial corrections of the order up to a few wavelengths to offset a positive curvature introduced by a Gascoigne corrector that corrects for astigmatism. Our requirements are fulfilled by preparing two aspheric correctors, one conventional Gascoigne type, and the other negatively and more strongly figured which is placed close to the focal plane. The mirror structure function has been tested using the null lens technique with the aid of a computer generated hologram, and verified that it is equivalent of an atmosphere that adds 0.2 arcsec FWHM to otherwise perfect seeing. Two more correctors are prepared, one optimized for the imaging camera and the other for the spectrographs. The 3° field is 65 cm in diameter and the net focal surface of the camera is still slightly quartic in the radius. 27% of the incoming beam is obscured by the secondary mirror and baffles.

The image scale is $16.''5$ mm^{-1}, or $0.''4$ per 24 micron CCD pixel. We want a $3''$ fibre aperture for efficient spectroscopy of galaxies at our limiting magnitude; this aperture corresponds on the focal plane to 180 μm, for which good quality fibres that do not cause serious f degradation are available.

The telescope is of an altitude-azimuth design, which allows for a light weight supporting structure, resulting in low inertia, low friction and mechanical simplicity. It is housed in a small shelter which rolls away during observations, and has independently tracking wind and light baffles.

The observing site, Apache Point Observatory (APO), is situated on Sacramento Peak, New Mexico (32°46′50″N, 105°49′12″W), at 2800 m altitude, one kilometre away from National Solar Observatory. A 3.5m telescope of Astronomical Research Consortium already exists at APO. The Observatory is 30 km east of Alamogordo. We expect the median seeing to be about $0.8''$ in good photometric conditions. Surface brightness of the dark night sky is $V = 21.7$ mag (arcsec)$^{-2}$.

3. Survey Strategy

We want to target the largest possible contiguous volume of the Universe. A naive solution is to choose a 120° cone centred on the north Galactic pole, i.e., the $b > 30°$ region. This is not optimal for the following reasons: (i) Galactic extinction is not minimum for all the region of $b > 30°$. (ii) This choice requires the telescope to be inclined down to 25°, which causes a significant differential chromatic refraction for fibre-fed spectroscopy. We have thus decided to squeeze the region to 110° in the north-south direction but 130° in the east-west. The central meridian is 12^h20^m with the survey centre at $\delta = +32.8°$. The selective extinction is reduced to $E(B-V) = 0.08$ at the survey boundary. With this choice we need to incline the telescope only to 35°, for which the loss of light due to differential chromatic refraction is at most 16% at the end of the spectrum. An extra advantage of this choice is that it makes the prime time of the observation earlier in the year, which is favourable from the weather point of view at APO, and slightly extends the observing season.

The photometric survey will be carried out with the drift scan or time-delay-and-integration (TDI) mode (Schneider et al. 1989) at the sidereal rate of 15 arcsec/sec. Off the equator, the telescope must be tracked in great circles. The effective exposure time is 55 sec. An important advantage of this technique is that we can save the readout time of CCD, which is comparable to the exposure time for our shallow imaging. Another advantage is that we can obtain excellent flat-fielding, since objects directly pass through 2048 pixels: extra flat-fielding is required only for one direction. The cost is that specifications of the optics and camera are significantly more demanding. As we discuss later we have already solved this problem for the camera. For the telescope optics, calculations show that the requirement is well satisfied.

We cover the survey area with 45 stripes of 2.5° wide and 130° long. Because the CCDs are not contiguous in the focal plane, one stripe is filled with two swaths of TDI scan. This requires 780 hr, or about 1.5 yr taking account of the fraction of clear, dark, and photometric nights. With 55 sec exposure, we expect $S/N = 5$ images at $r' = 22.8$ mag and $S/N = 170$ at $r' = 18.2$ mag, the limit of the spectroscopic survey.

We note that the survey efficiency is given by

$$\epsilon = \Omega D^2 q, \tag{1}$$

where Ω is the solid angle of FoV, D the diameter of the telescope and q is the quantum efficiency of the detector. The SDSS ensures $\epsilon = 4.8$ m^{2}deg^2 compared to 0.30 m^{2}deg^2 for 48″ Schmidt photography. This gain enables us to carry out multi-colour photometry simultaneously; we decided to carry out five colour photometry. This not only brings us important information on the stellar content for a huge number of galaxies, but also provides the essential means for the quasar target selection for a wide range of redshift.

When we cannot observe the north pole in autumn, we intend to carry out a survey of the southern sky. We plan to repeat scans of long thin strips (probably three strips): i.e., the region of 200 square degrees will be surveyed 35-45 times. This makes the limiting magnitude 2 mag deeper and should bring us new insight on variable objects.

The strategy for spectroscopy is more conventional. We take 1800 3° fields, chosen with the aid of a tiling algorithm to minimize the loss of targets. Each field contains 640 objects and average exposure time is 45 min. With an appropriate overhead, we need 2000 hr or about 3.8 yr for this survey. We expect 100 galaxies per square degree at $r' = 18.2$ mag, and hence one million galaxies for π steradian of the sky. $S/N \geq 13/\text{Å}$ is expected for the galaxies at our limiting magnitude.

An option for the spectroscopic survey is that we may select 0.1 million bright red galaxies and 0.9 million normal galaxies. The depth of the survey would be significantly increased with such a selection.

We will need to interchange quickly the imager and the spectrograph when the weather condition changes. So, these instruments are built in the form of cartridges.

4. Imaging Camera

The camera should fit the focal plane with a diameter of 65 cm. We decided to make an array of $5\times6 = 30$ CCDs mosaicked on the focal plane. Extra space is used to arrange 24 smaller CCD chips for astrometry and automated focusing. A schematic view is given in Fig. 1. The central 30 CCD chips constitute the main photometric camera where 5 chips in one column are housed in one dewar. The u' imaging that requires the most stringent optics is placed in the central row. For g' and r' this requirement is less stringent and they are placed in the two edges. The space of the adjacent CCD in the row direction is slightly smaller than the size of the chip, so that it results in an 8% overlap for two swaths.

All photometric CCD chips have 2048×2048 imaging pixels procured from Scientific Imaging Technologies (SITe, formally Tektronics). The CCD for u' imaging is a thinned back-illuminated chip with UV antireflection coating which attains UV sensitivity of 40% on average. We use thinned back-illuminated chips with normal antireflection coating for the $g'r'i'$ colour bands. Typical quantum efficiency from 5000Å to 7000Å is 80%. For z' band we use thick front-illuminated devices without coating. The filters are described in the next section, but here we only mention that they are plano-concave lenses serving as field-flatteners to match a slightly convex surface of the chip. The front side of the corrector (made of quartz) is coated differently in 5 strips to match the colour bands.

One of the major challenges for constructing this camera was to satisfy the stringent requirement for the accuracy of the position of the CCD chips. The direction of the rows of CCD must agree precisely with that of the tracking: we achieved the rotation of the chip with respect to the tracking line only $< 3\mu$m from the centre to the edge, which satisfies well the requirement. The departure of the CCD surface from the focal plane is only 15μm, much smaller than 24μm×5 for the f/5 beam, image degrading being negligible. We also demand stability less than 2 μm (30 mas) across the whole camera during the operation.

The auxiliary 24 chips are 2048×400 thick front-illuminated devices with r' and neutral density filters, and are divided into an upper and a lower group housed by two dewars. 22 chips are for astrometry and two chips are for focusing. These chips allow observation of the bright stars in astrometric star catalogues to tie the SDSS system into these catalogues (see below).

A number of technically demanding requirements are satisfied by mounting the camera directly on a thick (about 5cm) quartz corrector, which provides a very stiff support. The filters are cemented directly onto the backside of the corrector to reduce the number of surfaces that would cause a loss of light. Holes are drilled on the corrector and screw anchors are embedded into these holes to support

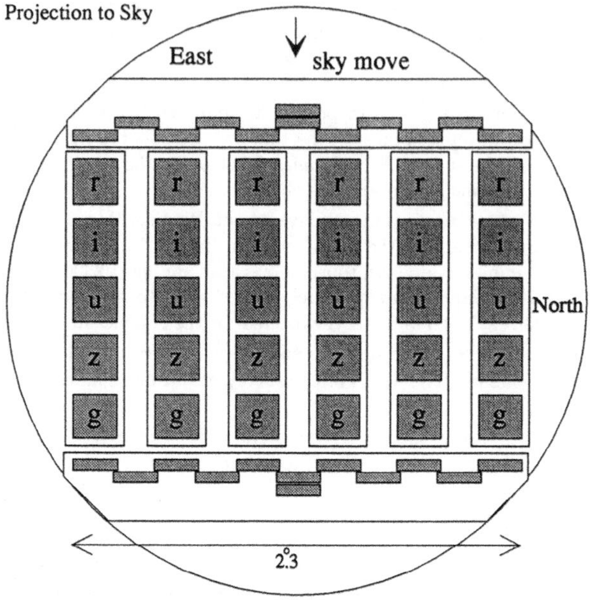

Figure 1. Schematic view of the photometric imager.

both optical benches, on which CCD chips are mounted, and dewars that contain electronics and liquid nitrogen containers.

High modularity of electronics is another notable feature of this camera. The camera contains altogether about 80 channels. 42 clock driver and 42 preamp cards (each contains two channel processors) are packed in the dewars to minimize noise. The signal from preamps goes into dual slope integrator and then digitized with a 16 bit ADC on signal-bias chain cards (also 42 units). The control signal, received by 8 bus-receiver cards, drives in-dewar clock drivers. The clock of the CCD is semi-synchronous (for some of the CCDs it is a factor 2 faster for technical reasons). One pixel is read in 24μs and it takes 52 sec per frame including overscan columns. The output of the on-chip charge amplifier is 1 μV per electron and the readout noise is typically 5 e. The photometric chips are cooled to $-80°$C by LN_2 to suppress the dark current.

5. Photometric System and Photometric Calibration

Our photometric system is five colour bands (u', g', r', i', z') that divide the entire range from the atmospheric ultraviolet cutoff at 3000 Å to the sensitivity limit of silicon CCDs at 11000 Å into five essentially non-overlapping pass bands. The SDSS system is substantially different from the Thuan-Gunn (1976) photometric system. The most salient feature is the very wide bandpasses used, even significantly wider than that of the standard Johnson-Morgan-Cousins system to ensure high efficiency for faint object detection.

The division of the pass bands is designed to exclude the strongest night-sky lines of O I $\lambda5577$ and Hg I $\lambda5460$. The u' band response is designed so that the bulk of the response is shortward of the Balmer discontinuity; this produces a much higher sensitivity (combining with g') to the Balmer jump at the cost of lower total throughput. The i' band terminates at 8500Å before a rapid increase of OH brightening. The filters are made from colour glass elements and a multi-layer interference film coating as detailed in Table 1. The transmission is $> 90\%$ except for the u' band.

For the calibration of the photometric system we have installed a 61 cm monitor telescope

TABLE 1. Elements of the SDSS filters

filter	glass	coating
u'	1mm UG11 + 1mm BG38 + 3mm quartz	suppression 6600-8200Å
g'	2mm GG400 + 3mm BG38	short-pass, cut off at 5500Å
r'	4mm OG550 + 1mm BK7	short-pass, cut off at 7000Å
i'	4mm RG695 + 1mm BK7	short-pass, cut off at 8500Å
z'	4mm RG830 + 1mm BK7	

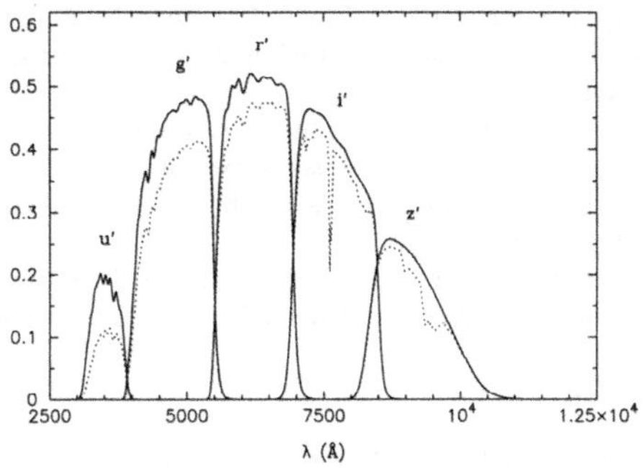

Figure 2. Response function of the SDSS photometric system (solid curves are without atmospheric extinction; dashed curves are at 1.2 airmass).

(Ritchey-Chrétien, f/10), and the precise definition of the SDSS photometric system is given by the response function of this monitor-telescope detector system. A thinned UV-antireflection coated 2k×2k chip is installed in this detector (FoV 27′ × 27′). The system response functions are presented in Fig. 2 and the characteristic numbers are given in Table 2. The 2.5m survey telescope uses the same filters, but their response is slightly different from that of the monitor telescope. Instrumental magnitudes obtained with the survey telescope are then corrected into the monitor telescope magnitude. Details are given by Fukugita et al. (1996).

Another new feature is the use of the AB magnitude, which can directly be translated into a flux in physical units, with the additional advantage of making the translation from spectroscopic flux to photometric wide band flux straightforward. The AB system of Oke & Gunn (1983), designated as AB_{79}, is now somewhat outdated; we have developed the AB_{95} system to meet the demand (Fukugita et al. 1996). In the new scheme we have retained the definition:

$$AB_\nu = -2.5 \log f_\nu \,(\mathrm{erg\ s^{-1}cm^{-2}Hz^{-1}}) - 48.60 \,, \tag{2}$$

where f_ν is the flux per unit frequency from the object. The broad band magnitude is written

$$m = -2.5 \log \frac{\int d(\log\nu) f_\nu S_\nu}{\int d(\log\nu) S_\nu} - 48.60, \tag{3}$$

where S_ν is the response function including atmospheric extinction. A transformation into the conventional zero point (zero mag for A0 stars) is approximately $u' = u + 0.93$, $g' = g - 0.09$, $r' = r + 0.16$ and $i' = i + 0.40$, where primed refer to AB magnitude. To utilize the AB system, it is essential to know the accurate system response. A spectrophotometer was built specifically to measure the quantum efficiency of the CCD.

Photometric Calibration

We take as the fundamental photometry calibrators two spectrophotometric standard stars (metal deficient F subdwarf) BD+17°4708 and BD+26°2606 (Oke & Gunn 1983), which are given

TABLE 2. Characteristics of the response function at 1.2 air mass

	u'	g'	r'	i'	z'
λ_{eff} (Å)	3557	4825	6261	7672	9097
FWHM (Å)	599	1379	1382	1535	1370

accurate spectrophotometry with respect to α Lyr. We want to add one more metal deficient F subdwarf BD+21°607 (Kent 1985) so that two stars are always accessible at any time, although this star lacks spectrophotometry. Some 100 stars ($V \leq 14$ mag) have been selected as candidates for primary standard stars which are to be calibrated against the fundamental triad to set up the network of standard stars over the sky.

The monitor telescope is used to observe several primary standard stars every hour to determine atmospheric extinction. In addition, we observe 'secondary standard stars' ($13 < V < 15$ mag) located in a few patches set in the survey strips with almost one hour interval to tie the survey magnitude with monitor telescope magnitude.

Our goal of accuracy is 0.01 mag for $g'r'i'$ and 0.02 for $u'z'$ for the primary standard star network, and 0.02 for $g'r'i'$ and 0.03 for $u'z'$ for the stellar objects in the survey.

6. Astrometric Calibration

The purpose of astrometry is two-fold: (i) to locate the fibre positions for drilling plates for spectroscopy, and (ii) for a good catalogue. For (i) an accuracy budget of < 270 mas is required not to degrade the spectroscopic survey.

We will use the Tycho Catalogue for astrometry, which will be supplemented with the proper motion data from Twin Astrograph Catalogue of USNO. The stars that appear in Tycho or TAC are as bright as $V = 12$ mag, which saturates the photometric camera. The astrometric camera installed in the imager assembly, however, can observe these bright stars, and the positions derived from these stars are used to give the positions of fainter 'secondary standard' stars of $14 < V < 17$ mag, imaged in the main camera.

We expect the error budget to be: errors in primary standards (40 mas); seeing in primary standards (25 mas); seeing in secondary standards (42 mas); shot noise in secondary standards (20 mas), and uncorrelated tracking errors (55 mas). These altogether yield 90 mas as the total error in the position.

7. Photometric pipelines

All data processing must be fully automated to deal with huge data volume. This software is called a "pipeline" and there are many pipelines in the survey system. For instance for the photometric data reduction, the camera data are cut into "frames" and each set of five frames sent to the "frames pipeline" to detect and measure objects. To obtain the point spread function and flat-field vectors for a whole run, which are used by the frames pipeline, a reduced amount of data from the camera is sent to the "postage stamp pipeline". In this way, the code is parallelized, and the frames can be reduced in any order. The "postage stamp pipeline" also receives data from "monitor telescope pipeline" (where data from the monitor telescope are processed) and the "astrometric pipeline" (where data from the astrometric camera are processed), and forward them to the frames pipeline for calibrations.

The output of the frames pipeline is written into the "operational data base" and these data are sent to the "target selection pipeline" to give coordinates for plate drilling. The "science data base", which will be used for science, is also derived from the operational data base.

Apparently the most demanding element is the frames pipeline, which is a package for automated surface photometry, but which is significantly more sophisticated than existing software. The main part consists of "correct frame" (flat-fielding; interpolating bad columns or bleed trails of saturated pixels), "bright star subtraction", "find object", and "measure object" modules. The find object module is basically peak finding after smearing the image with the point spread function. A verification with simulated data ensures almost complete detection of normal objects to $r' = 22.5$ mag. What deserves some discussion is the choice of photometric parameters for galaxies derived by the measure object module. It is particularly difficult to measure a total flux for galaxies; isophotal magnitudes are clearly inappropriate for our purpose. We measure magnitudes at many apertures, and in addition measure the Petrosian magnitude (Petrosian 1976) as a pseudototal magnitude. Our trial parameters are the radius r_P at which surface brightness is 1/8 the average, and the flux encircled within $2r_P$. For the exponential profile this gives virtually the total magnitude, and for the de Vaucouleurs profile 92% of flux is included.

Target selection

We select objects that are subject to spectroscopy from the photometric catalogue. This process crucially determines the selection function of the spectroscopic sample. There are severe requirements from both science goals and practical feasibility; some compromises are required.

For the galaxies, we plan to use the Petrosian magnitude in the r' band for the flux limit, but we also apply the constraint that average surface brightness (or $3''$ aperture magnitude) be brighter than some limiting value to avoid observing objects for which we cannot obtain a redshift. This parameter introduces complication into the selection function but is necessary in order not to waste fibres.

We can take advantage of multi-colour photometry for quasar target selection, with which one expects 90% completeness and an efficiency of 65%. While we have set up a default selection criteria in multi-colour space, we need extensive testing of our tactics before the survey begins.

We also allocate a small number of fibres for stars, blank sky, calibrations and for the purpose of data quality assurance. Our provisional partition of fibres is 470-500 for galaxies, 100-110 for quasars, 5 for stars, 10–25 for blank sky and 5 for other purposes. All selection criteria are set in the target selection pipeline.

Adjacent fields of exposures necessarily have some overlaps. We can utilize this overlap to cover more galaxies in dense regions with appropriate tiling. Allowing for some regions with three fields overlapping, we can take virtually all galaxies to $B = 19$ for the APM catalogue (Maddox et al. 1990), if infinitesimally small spacing were allowed for adjacent fibres. In practice, however, we cannot allocate two fibres closer than 3.2mm in the focal plane, which corresponds to 55 arcsec. With this constraint 95.6% of the APM objects are taken as spectroscopic targets with 610 fibres.

In our project we plan to have an extended test period to fix all detailed parameters and tactics, no changes being allowed after the start of the survey. Target selection must be particularly carefully tested and verified.

8. Spectrograph

The requirements on the spectrograph are (i) a resolution $\lambda/\Delta\lambda \sim 2000$, corresponding to $\Delta\sigma \sim 140$ km s^{-1}, a typical velocity dispersion for normal galaxies, to avoid the loss of S/N due to contamination of absorption features by continuum light. (ii) Wavelength coverage extends from 3900 to 9100Å. The minimum requirements are to cover Ca II H+K and Hα for all redshift up to $z = 0.25$; the above range extends a little further. G band and Mg b features are also covered. We miss [O II]λ3727 for nearby galaxies ($z < 0.05$), which, however, are a small fraction of the sample. (iii) we need > 550 fibres for the survey to be finished in 5 years.

These requirements are satisfied by constructing two double spectrographs each with dichroic splitter and 2k×2k CCD chips. The schematic design is shown in Fig. 3: A is a slit head with a slit 4.9 inches long and with 320 fibres installed perpendicular to the sheet, emitting f/4 light; B is a collimator of the Schmidt design (without a corrector) with a 6° FoV, movable with a DC servo motor. A 150 mm diameter beam is reflected from the collimator; C is a dichroic splitter, blue light (3900-6100Å) being reflected, while red light (5900-9100Å) going straight. Almost no loss of light is expected for reflection and 90% efficiency for transmission; D stands for two right angle grisms with grating of ruling densities of 640 (blue) and 440 (red) lines/mm. Zero angular deviation is at 4960Å and 7400Å . The use of a transmitting (rather than reflecting) grating allows us to install cameras close to the system pupil, so that the size of the camera can be reasonably small. The efficiency is 50-60% for all wavelengths. The flexure of the spectrograph is designed to be < 0.1 pixel for one hour exposure.

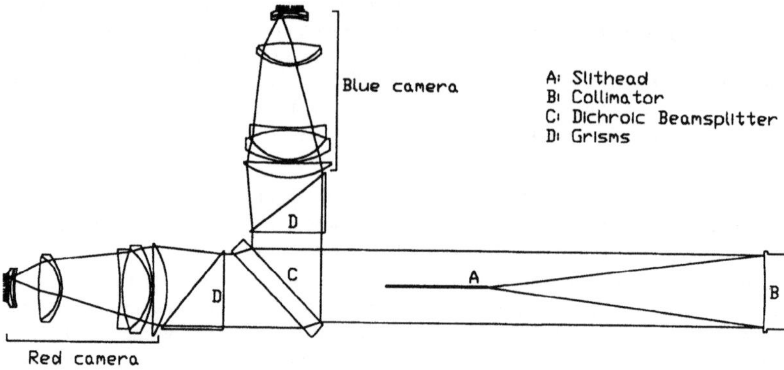

Figure 3. Schematic view of the spectrograph. See explanations in the text.

Behind the optics are two CCD cameras. In order to exploit the advantage of 2k×2k CCD chips with 24μm pixel, we need a wide field camera with a FoV of 16.5°. The focal length is 240 mm with f/1.3. The camera is of Epps' design; it is similar to that of LRIS at Keck (Oke et al. 1995) but with smaller aperture and larger FoV. The optics is composed of one singlet, one triplet and one doublet Petzval lenses plus a flattener. All but one of the surfaces are spherical and the asphere on the backside of the doublet is only mild. The detector is a thinned back-illuminated CCD with normal antireflection coating. The readout electronics is the same as that for the imager. The output flux spans 72 μm, i.e., 3 pixels, and hence the demagnification of the spectrograph is modest, only a factor of 2.5. We place individual fibres every 360μm (twice the inner aperture of fibres) at the slit head to avoid cross talk. So each camera can image >320 spectra. The net image quality is 20μm, which is about 1 pixel. The total throughput is expected to be 15-20% including the quantum efficiency of CCD and telescope losses.

We adopt the old technology of manual plugging of fibres (Limmongkol et al. 1993). A plug plate of diameter 81 cm is plugged with 640 fibres and 10 coherent fibre bundles, the latter used for a TV guider mounted on the spectrograph. There is one more larger hole with a wide-field fibre bundle for sky monitoring. Holes are drilled by applying a bending moment to the edge of the plate to match the best focus surface. We have confirmed that the engineering accuracy of the hole position is 6-10μm (100-170 mas) and the deviation from the focal plane is 60μm rms (the maximum is 200μm at the centre where the optics is the best). We require the total de-centreing error to be smaller than 500 mas, and this accuracy is sufficient taking account of errors from astrometry and differential refraction. Fibres with a core diameter of 180 μm are made of silica and UV enhanced with a total flux loss of 10% rms (max 13%). The output beam is f/4 for the f/5 input beam. Fibres of 2 m long are bundled in units of 20 to ease handling. These fibres are plugged in day time, and installed on a cartridge, which contains a plug plate holder and a slit head of the spectrograph. We will prepare 10 cartridges which can be changed quickly from one to another. We do not need to mind which fibres are plugged into which holes. All connections are mapped by illuminating fibres from the spectrograph side using a fibre mapper.

9. Spectroscopy

Spectra of each target will be obtained with three 15 minutes exposures. This exposure time will be modified according to airmass, Galactic extinction and sky brightness to maintain the homogeneity of the spectroscopic sample. The overhead is expected to be 18 min per target field, including wavelength calibration and flat-fielding exposures each time.

The data reduction is made by the "spectroscopy pipeline" which is divided into two parts. The first is the "2D spectro pipeline", which processes the raw 2D spectral frames taken with three exposures into single 1D calibrated spectra for each object. The process includes bias subtraction, flat-fielding, wavelength calibration and sky subtraction. We use modified IRAF routines. The second ("1D spectro pipeline") processes the 1D spectra to extract scientific information. After merging red and blue halves of spectra together and masking pixels contaminated with sky emission lines, emission lines are identified and classified. This is followed by an analysis of absorption features with the cross-correlation technique (Tonry & Davis 1979). We also attempt to classify the spectra using a set of template spectra, and carry out a principal component analysis. Spectrophotometry with an accuracy of $\sim$ 15% is also one of the aims of the analysis. We must process 5000 spectra per day.

10. Data Base

A conspicuous feature of the SDSS is the huge amount of data it produces. The photometric imager produces data at a rate of 4.6 MB/sec and 12 TB altogether by the end of the survey. The astrometric camera also yields 3.7 MB/sec, but these data will not be archived. Postage stamp images of bright stars and quartile data of image frames, used to generate flat-field vectors, amount to 99 kB/sec (254 GB altogether) and 10 kB/sec (25 GB), respectively. The spectroscopic survey yields 153 kB/sec or 360 GB. We need to keep all data in the operational data base. We use the commercial object-oriented data base system Objectivity.

TABLE 3. Science data base

Level	data	storage
Level 1	object list (w/ spectroscopy)	2.5 GB
Level 2	object list (all)	250 GB
Level 1	atlas image	700 GB
	spectra (1D)	20 GB
	4×4 pixel all sky map	50 GB

From these data bases we intend to produce the "science data base" which contains the data fully calibrated, indexed and read only, open to the public. We are planning a three level structure referring the frequency of use and amount of storage required, as presented in Table 3. The most memory consuming is the archive of atlas image, small regions that comprise all detected objects and also those where other surveys (e.g., X-ray, radio, IR surveys) found objects.

The requirement is that we must be able to sort out objects with desired properties, such as "blue galaxies fainter than 21 mag which are located within 10$''$ of quasars of $z > 2.5$", as well as allowing easy access without any special facilities. Our simulations show that a brute force approach does not work any more, if a data base comprises 100 GB or more. We plan to use an object-oriented data base with a hierarchical structure, such as ODBMS which is commercially supported. Progress in data processing is rapid, and we still have some time to choose before the survey starts.

11. Science

The science that can be contemplated with the SDSS data base is immense, and I do not intend to describe it here. Rather I mention only a few basic features anticipated for the study of galaxies, large-scale structure and quasars.

The depth of the survey reaches $z \simeq 0.2$, and the median of the redshift distribution is $z \simeq 0.1$ (see Fig. 4). The high redshift tail is cut off by the selection criterion for surface brightness compared to what is expected without this selection. The survey is dense and contiguous for the northern sky. We can, in principle, obtain power spectrum to $\lambda \geq 500$ Mpc, which overlaps with the region surveyed by COBE, to good accuracy. What we found crucial during our preliminary study is the necessity of accounting for Galactic extinction. Patchy Galactic extinction induces mock large-scale power in excess of the real power spectrum beyond 250 Mpc. This means that we cannot obtain meaningful power spectrum for $\lambda \geq 250$ Mpc unless the Galactic extinction is accurately known. In our survey, we plan to cross correlate an extinction map obtained from differential extinction of

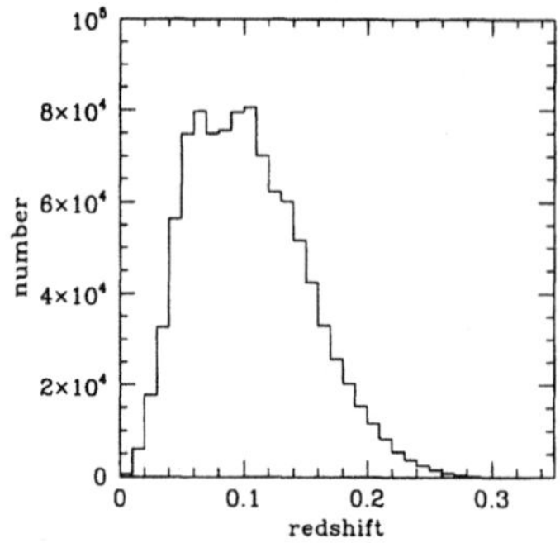

Figure 4. Redshift distribution of one million galaxies expected with the selection criteria discussed in the text. In this example a half-light surface brightness cut is imposed at < 22 mag $(\text{arcsec})^{-2}$ in addition to the Petrosian-magnitude flux limit.

hot halo stars with those from galaxy number counts, star colour map and also with those from external sources, e.g., from HI and infrared data. If we succeed to control the extinction effects down to 0.05 mag in r', we can obtain a reliable power spectrum up to 400 Mpc or so.

The quasar survey is another important item. With a "normal selection" we expect to have 84,000 quasars and 250 with $z > 4$, median redshift being $z \sim 1$. The spectra will contain ~200,000 Lyman α absorbers with $2.3 < z < 3.0$ and ~2000 CIV absorbers with $2.8 < z < 5.0$. About 400 damped Lyman α absorbers will be obtained for $z > 3.4$. We also plan to produce a "high z quasar sample" using different selection criteria, resulting in 4500 $z > 4$ quasars and hopefully 300 $z > 5$ quasars. Roughly 400 lens candidates will be contained in the quasar sample, although we will need a separate follow-up study to confirm each lens.

12. Present Status and Schedule

The imaging camera has been completed and shipped to APO (October 1997). One of the spectrograph is finished except for the dewar in the red channel. The optics has been verified to work as designed, and readout is being tested. The blue camera is also being tested now, and will be installed in a month or so. All parts of the second spectrograph are ready except for a collimator that is not verified so far, but they yet have to be assembled. We except one spectrograph will be ready for verification in early 1998. The telescope still need more work. Basically all components are in hand, but the present image rotator is not adequate, and we have to wait until summer of 1998 for a new image rotator to be installed. The work on the telescope control will also take some more time. We expect, however, first light on the imager in the early summer of 1998. We plan to start the verification of the imager and tests for photometry with the telescope parked shortly after the first light. The monitor telescope system is in a commissioning phase, and expected to start data taking from the spring of 1998 (this work must be done in advance of the main survey). Spectroscopic first light will be achieved only in 1999. So, the year 2000 is perhaps a reasonable estimate for the survey to start.

I would like to thank all my colleagues working for this Project, in particular M. Carr, M. Doi, J. Gunn, T. Ichikawa, S. Kent, G. Knapp, R. Lupton, T. Mckay, E. Mannery, G. Pauls, M. Sekiguchi, W. Siegmund, D. Schneider, A. Szalay, A. Uomoto, D. Weinberg, N. Yasuda and D. York for discussions and information. I am grateful to G. Knapp and E. Turner for comments improving the manuscript.

References

Bowen, I. S. & Vaughan, A. H. 1973, *Applied Optics*, **12**, 1430
Fukugita, M., Ichikawa, T., Gunn, J. E., Doi, M., Shimasaku, K. & Schneider, D. P. 1996, *AJ*, **111**, 1748
Gunn, J. E. & Weinberg, D. H. in *Wide Field Spectroscopy and the Distant Universe*, ed by S. J. Maddox and A. Aragón-Salamanca (World Scientific, Singapore, 1995), p.3
Kent, S. M. 1985, *PASP*, **97**, 165
Limmongkol, S., Owen, R. E., Siegmund, W. A. & Hull, C. 1993, in *Fiber Optics in Astronomy II*, ASP Conference Series vol. 37, p. 127
Maddox, S. J., Sutherland, W. J., Efstathiou, G. & Loveday, J. 1990, *MNRAS*, **243**, 692
Oke, J. B. & Gunn, J. E. 1983, *ApJ*, **266**, 713
Oke, J. B., Cohen, J. G., Carr, M. Cromer, J., Dingizian, A., Harris, F. H., Labrecque, S., Lucinio, R., Schaal, W., Epps, H. & Miller, J. 1995, *PASP*, **107**, 375
Petrosian, V. 1976, *ApJ*, **209**, L1
Schneider, D. P., Schmidt, M. & Gunn, J. 1989, *AJ*, **98**, 1507
Thuan, T. X. & Gunn, J. E. 1976, *PASP*, **88**, 543
Tonry, J & Davis, M. 1979, *AJ*, **84**, 1511

THE CNOC2 FIELD GALAXY REDSHIFT SURVEY

H.K.C. YEE, M.J. SAWICKI, R.G. CARLBERG, H. LIN
University of Toronto
Dept. of Astronomy, Toronto ON M5S 3H8, Canada
S.L. MORRIS
DAO/HIA, 5071 Saanich Rd, Victoria BC V8X 4M6 Canada
D.R. PATTON, G.D. WIRTH
University of Victoria
Dept. of Physics & Astronomy, Victoria BC V8W 3P6, Canada
C.W. SHEPHERD
University of Toronto
Dept. of Astronomy, Toronto ON M5S 3H8, Canada
E. ELLINGSON
University of Colorado
CASA, CB 389, Boulder CO 80309, USA
AND
D. SCHADE, R. MARZKE
DAO/HIA 5071 Saanich Rd, Victoria BC V8X 4M6 Canada

1. Introduction

Fundamental to our understanding of the universe is the evolution of structures, from galaxies to clusters of galaxies to large-scale sheets and filaments of galaxies and voids. The investigation of the evolution of large-scale structure not only provides us with the key test of theories of structure formation, but also allows us to measure fundamental cosmological parameters. The CNOC2 (Canadian Network for Observational Cosmology) Field Galaxy Redshift Survey is the first large redshift survey of faint galaxies carried out with the explicit goal of investigating the evolution of large scale structure. This survey also provides the largest redshift and photometric data set currently available for the study of galaxy population and evolution at the moderate redshift range between 0.1 and 0.6. In this paper we describe the scope and technique of the survey, its status, and some preliminary results.

2. Observational Strategy

The CNOC2 survey is being conducted using the MOS arm of the MOS/SIS spectrograph at the Canada-France-Hawaii Telescope. The primary goal of the survey is to obtain a large enough sample of galaxies with high quality spectroscopic and photometric data for the purpose of studying the large scale structure at the redshift of ~ 0.35. We need to obtain a sample of galaxies comparable to the large nearby redshift surveys (e.g., Geller & Huchra 1989; Shectman et al. 1996). This requires obtaining redshifts for the order of 10^4 galaxies covering an area of sky subtending well over 10 h^{-1} Mpc with velocity measurements accurate to better than 100 km s^{-1}.

To avoid being dominated by a small number of large structures, the survey covers four widely separated regions, called *patches*, on the sky. The patches are distributed so that two regions can be observed at any time of the year. Each patch is a mosaic of 20 MOS fields, covering $\sim$ 1400 sq arcmin. The total area of the survey is $\sim$ 1.5 sq degrees. Each spectroscopically defined MOS field is $\sim 9' \times 8'$, with about $15''$ overlapping area with the adjacent fields. The overlapping of the

J. Andersen (ed.), Highlights of Astronomy, Volume 11A, 460–463.

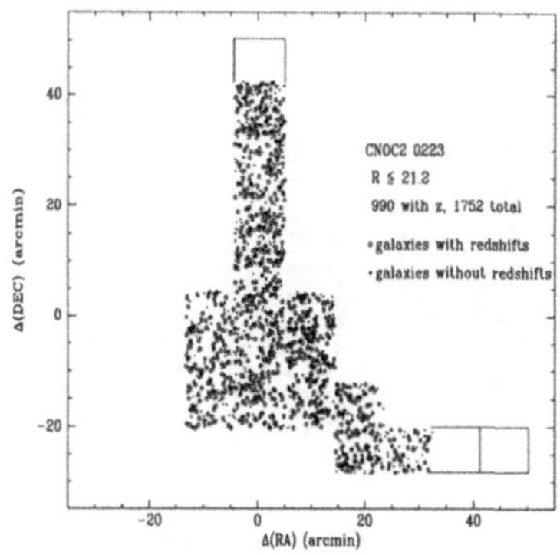

Figure 1. Field layout in a patch. Currently, 17 of 20 fields in 0223+00 have been observed, with a total of 1310 redshifts. Symbols denote galaxies in a $R \leq 21.2$ sample.

fields minimizes the systematic uncertainties in photometric calibrations from field to field. The field layout, in a L shape with a central block, is shown in Figure 1. The maximum dimensions of the patches span $\sim 80'$ NS, and $63'$ EW. The patches are chosen to avoid bright (< 12 mag) stars, low-redshift (e.g., Abell) clusters, and other known low redshift bright objects. They have galactic latitudes between 45° and 60° , chosen to avoid excessive Galactic extinction (< 0.05 mag in A_B), yet close enough to the galactic plane to provide sufficient number of stars to overcome the star-galaxy classification problem for the MOS images which have variable point-spread-function across the field.

We utilize the observational technique developed for the CNOC1 Cluster Redshift Survey (Yee, Ellingson, & Carlberg 1996; hereafter, YEC). Here, we describe briefly the procedure and some improvements. For the first 3 runs, two different CCDs with 15μ/pixel were used. However, for the remaining 4 runs, from which most of the data were (or will be) obtained, a STIS CCD, which has higher quantum efficiency (QE) and superior blue response, but with a larger 21μ pixel size, was used. Here, for brevity, we describe primarily the observational parameters for runs using the higher QE STIS CCD.

Images in I, R, Gunn g, B, and U for each field are obtained using the imaging mode of MOS with integration times of 6 to 15 min, obtaining average 5σ detection magnitudes for R and B of 24.0 and 24.6, respectively. The B and U images for all the fields are taken with the blue sensitive CCD. The photometry is reduced and multicolor catalogs produced in real time at the telescope. Using the catalogs, multislit masks are designed using a computer program which allows one to prioritize the sample in various ways and optimize the number of slit placements. The slits have a width of $1''\!.3$ and a minimum length of $11''$. The masks are cut using LAMA, a computer controlled laser cutting machine, and are available for the spectroscopic observation within as little as 3 hours after the direct imaging observation. The primary spectroscopic sample is the union set of galaxies with $R < 21.5$ or $B < 22.5$. This allows us to construct unbiased R and B samples from the redshift catalogs.

The spectroscopy observations are done using the B300 grism providing a resolution of ~15Å. A band limiting filter is used to shorten the spectrum so that more slits can be placed on each mask, with typical numbers ranging from 85 to 110. The wavelength coverage of the filter is 4300Å to 6300Å, allowing for the unbiased coverage of important spectroscopic features within the target redshift range of $0.10 < z < 0.55$ for the survey.

Two masks are used for each field. Mask A covers primarily galaxies brighter than R=20.0, with a total integration time of 40 min. For mask B, which has an integration time of 80 min, the highest priorities are assigned to galaxies with R between 20 and 21.5. Fainter objects are designed into the masks whenever possible. The masks are designed so that about 25% of the galaxies observed spectroscopically are common to both masks. These redundant observations serve two purposes: first, for the fainter objects, the summed spectra provide improved signal-to-noise ratio; second, objects with redshifts determined independently from both masks provide a rigorous empirical

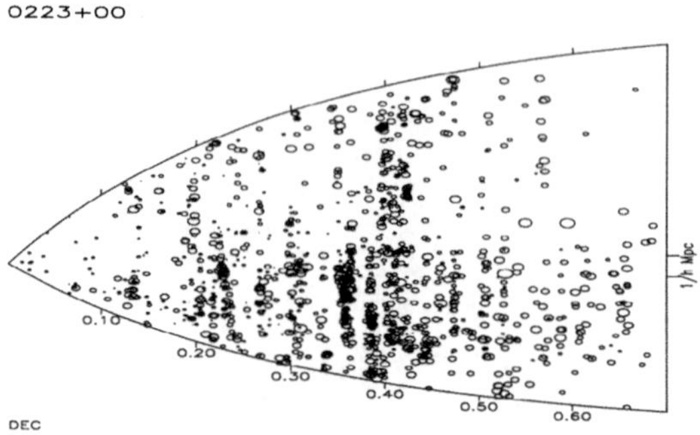

Figure 2. Wedge Diagram for the patch 0223+00. The tick marks on the vertical axis denote 1 h^{-1} Mpc, while the horizontal axis is marked by redshift. Note the sheet-like structures spanning across the whole field in the Dec axis. The uneveness of the density of points across the structures is due to an unequal numbers of fields projected.

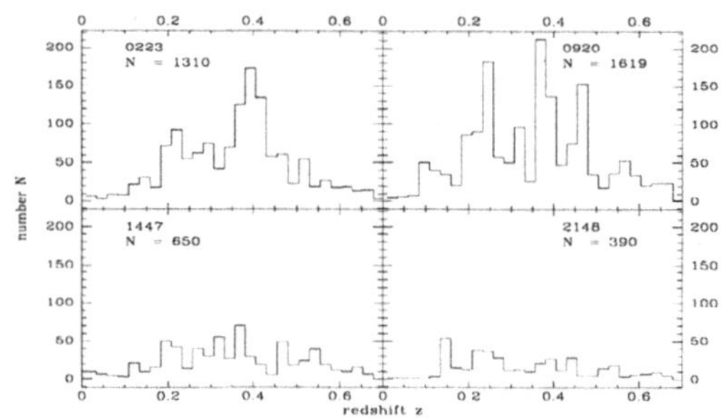

Figure 3. Histograms of the z distributions from the 3969 redshifts currently available in the 4 patches

check for the velocity accuracy and the probability of catastrophic redshift errors. The two-mask strategy also allows us to compensate for the under-sampling of close pairs that occurs when only one mask is used.

The total time required, using the more efficient CCD and including overhead, for the 5 colors and 2 spectroscopic masks per MOS field is about 3.5 hrs, yielding typically 80 to 90 redshifts. Thus, the complete survey of 80 fields requires about 27 clear nights, providing about 6500 redshifts with a mean z of 0.35. At the intended survey spectroscopic limit of R=21.5, the cumulative sampling rate for the redshift sample is about 0.50 of the photometric sample, while the differential completeness is about 0.25.

3. Catalog Creation

The data reduction procedure follows basically that outlined in YEC for the CNOC1 survey. Photometry is performed using an improved version of PPP (Yee 1991). Velocities are determined using cross-correlation. With 5 photometric colors we are also able to verify the redshift assignment.

Because the survey uses a sparse sampling strategy, accurate weighting factors for each galaxy are essential for proper interpretation. Although this has been discussed in detail in YEC where magnitude, geometric, and color weights were presented, additional effort has been put into better defining the selection function due to the limited redshift range over which the survey is sensitive. Photometric redshifts will be used to estimate the survey completeness as a function of redshift, and simulations of spectra will be used to investigate how the redshift identification success rate depends on variables such as spectral signal-to-noise ratio, galaxy type, redshift, and surface brightness. The objects from all the fields in each patch are merged into a single list after correcting for distortion in the MOS image, creating a final catalog which includes photometry, redshift, and selection weights for all objects brighter than $R = 24$.

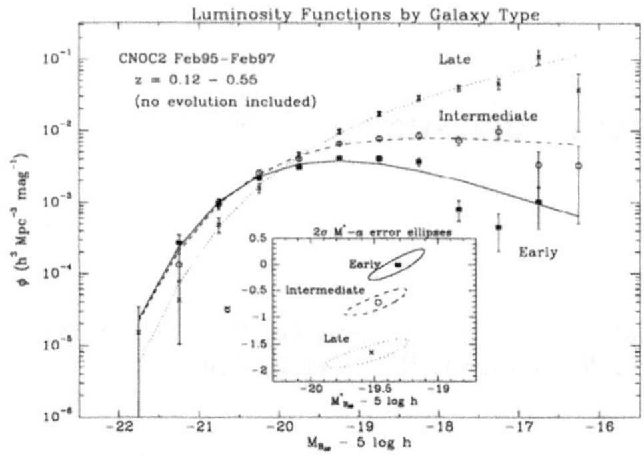

Figure 4. Preliminary luminosity function of galaxies of different population types as chosen by $B - R$ colors.

4. Survey Status

The survey is presently (09/97) 80% complete, and will be completed by the spring of 1998. The current reduced sample contains ~4000 redshifts, with an additional ~1000 redshifts to be obtained from the data currently being reduced. From the redundant observations the typical rms uncertainty of the velocity determination is about 75 km s^{-1} in the rest frame.

Figure 1 illustrates the sky distribution of galaxies for the redshift sample in the 17 fields currently available in the 0223+00 patch. Figure 2 shows the wedge diagram of the 0223+00 patch. Note the many sheet-like structures in redshift space spanning the whole NS width of the patch of ~10–15 h^{-1} Mpc. The redshift distributions of the current data set in the 4 patches are shown as histograms in Figure 3.

Figure 4 shows some preliminary LFs obtained from a subsample of 2768 galaxies with $R < 21.2$. The galaxies are divided into early, intermediate, and late types based on $B-R$ colors. Note the clear difference in the LFs for galaxies with different colors. The CNOC2 sample is an order of magnitude larger than any other redshift surveys in this redshift range (e.g., CFRS: Lilly et al. 1995; Autofib: Ellis et al. 1996). This large sample size, in combination with extensive *UBgRI* photometry (plus *K* for some fields), will permit precise measurements of the LF and its evolution with redshift for different galaxy populations.

Currently, various projects are being carried out using this large data base. We will be able to determine the galaxy correlation function in redshift and morphological bins, estimate the evolution of the pairwise velocity distribution, and measure the bias of galaxy clustering relative to mass clustering. The data are also being used to create a complete catalog of groups and pairs of galaxies, from which we can study the dynamics and evolution of galaxy groups and the redshift dependence of the merger rate. Line index measurements and principal component analysis of the spectra will be carried out, allowing us to conduct a detailed investigation of the star formation history of the galaxies. In addition, we will also derive quantitative morphological parameters of the galaxies, allowing us to examine the relationship between morphology, environment, and galaxy evolution. We will be able to obtain well-calibrated photometric redshifts for about 20,000 galaxies down to $R \sim 22.0$, which will further improve our studies of galaxy clustering and populations, and also extend the results to higher redshift.

References

1. Ellis, R.S., et al. 1996, MNRAS, 280, 235
2. Geller M. & Huchra, J.P. 1989, Science, 246, 897
3. Lilly, S., Le Fèvre, O., Crampton, D., Hammer, F., & Tresse, L. 1995, ApJ, 455, 50
4. Shectman S. et al. 1996, ApJ, 470, 172
5. Yee, H.K.C., Ellingson, E., Carlberg, R.G. 1996, ApJS, 102, 269 (YEC)
6. Yee, H.K.C. 1991, PASP, 103, 396

LIQUID-MIRROR TELESCOPE SURVEYS

P.HICKSON
University of British Columbia, Dept. Physics & Astronomy
2219 Main Hall, Vancouver, B.C. V6T1z4 Canada

Abstract. Recent advances in the technology of rotating liquid-mirrors now make feasible the construction of large optical telescopes for dedicated survey programs. Two three-metre-class astronomical telescopes have been built and a six-metre telescope is under construction. These instruments observe in zenith-pointing mode, using drift-scanning CCD cameras to record continuous imaging of a strip of sky typically 20 arcmin wide. This enables them to observe of order 100 square degrees of sky with an integration time of a few minutes per night. Data can be co-added from night to night in order to increase the depth of the survey. Liquid-mirror telescopes are particularly well suited to surveys using broad or intermediate bandwidth filters to obtain photometric redshifts and spectral energy distributions for faint galaxies and quasars.

1. Introduction

The idea of using a uniformly rotating liquid surface to focus reflected light is an old one [1]. The potential of liquid mirrors for astronomy was realized at least as early as 1850, with the first example being constructed around 1872 [2]. Well-documented experiments were conducted by Wood [3] who achieved considerable success with a 0.5-m diameter mercury mirror. Using a mechanical bearing, the telescope was able to resolve double stars of 3 arcsec separation. Unfortunately, it was regarded more as a curiosity than a scientific tool. Fifteen years ago, the idea was revisited by Borra [4]. His group at Laval University introduced technological innovations, such as the use of air bearings and thin mercury layers, culminating in a 1.5-m mercury mirror capable of producing diffraction-limited images [5]. The subsequent development, at the University of British Columbia (UBC), of designs and fabrication techniques for large liquid mirrors [6] lead to several fully-functional three-metre-class liquid-mirror telescopes (LMTs) [7-10].

Uniform rotation of a liquid in a homogeneous gravitational field naturally produces a parabolic surface. The focal length f is simply related to the angular velocity w and gravitational acceleration g by $f = g/2w^2$. An advantage of a liquid-mirror is that the parabolic surface is accurately maintained by fluid equilibrium despite small deformations of the supporting structure due to mechanical or thermal effects. This greatly simplifies the design and construction of liquid-mirrors, resulting in a cost which is almost two orders of magnitude lower than that of a comparable glass mirror.

A key to the success of mercury mirrors is the thin oxide layer which forms on the surface within a few hours. Virtually transparent, this monomolecular layer is very effective at damping waves and stabilizing the surface. The reflectivity of mercury is between 76% and 78% throughout the visible spectrum. Although the reflectivity of fresh aluminum is higher (typically 88% in the visible and 98% in the infrared), weathering degrades aluminum coatings, making mercury mirrors competitive in performance at visible and near-infrared wavelengths.

Because the rotation axis must be vertical, these mirrors typically observe in zenith-pointing mode. A prime-focus corrector lens typically provides an instantaneous field of view of order 0.5 deg. Because of the Earth's rotation, images track across this field at the local siderial rate. By aligning the CCD columns with the direction of image motion and continuously reading the CCD at the appropriate rate (drift scanning [11]), this motion is compensated and sharp images result. Except at the equator, there are small departures from linearity in both the direction and speed of

J. Andersen (ed.), Highlights of Astronomy, Volume 11A, 464–467.

the images which, if uncorrected, result in degradation of image quality in proportion to the solid angle subtended by the detector. Recent corrector designs incorporate geometric distortion which cancels these effects and produce high-quality images [12].

2. The UBC-NASA Multiband Survey

The NASA Orbital Debris Observatory (NODO) employs a 3-m diameter mercury mirror. It is located at an altitude of 2750 m near Cloudcroft, New Mexico – a dark, dry site having an average of 120 photometric nights per year. The primary purpose of the NODO is to observe space debris in low Earth orbit. However, during the hours when the debris cannot be illuminated by the sun, the telescope may be used for astronomical observations. Since 1996, we have used it to survey a 20-square degree region of the sky at high galactic latitude through a series of intermediate band-width filters ([13]). When completed, the survey will provide fluxes in 33 contiguous wavelength bands in the range 455 - 948 nm for all detected objects. The data are calibrated by means of spectrophotometric standard stars throughout the survey area established by observations at Kitt Peak National Observatory. The resulting spectral energy distribution (SED's) have resolving power $R \sim 44$ and can provide photometric galaxy redshifts with typical accuracy of $\delta z \simeq 0.01$ [14]. The characteristics of the survey are summarized in Table 1.

The primary scientific goals of this study are to measure the luminosity function, clustering properties and morphological and spectral evolution of galaxies to a redshift $z \sim 0.5$. The survey will be unique in providing flux-calibrated SED's for a large sample of objects. These can be used to investigate, for example, star formation rates for galaxies in a range of environments. We anticipate that redshifts will be obtained for over 10^4 galaxies. These will allow us to reduce contamination by foreground and background galaxies when searching for clusters and groups. Physical properties of these systems can then be investigated and compared with theory. For example, one can measure the frequency of close pairs and thus estimate the variation of the merger rate with redshift [15,16].

A particular advantage of drift-scanned images is the high spatial uniformity which results because pixel sensitivities are averaged over the entire length of the CCD. For the UBC-NASA survey, systematic background variations, after flat-field correction, are less than 0.02%. Because of this the survey is well-suited to the detection of low-surface brightness galaxies. Such galaxies may comprise a high fraction of the population, particularly at low luminosities, so it is particularly important not only to detect them, but also to characterize the selection biases [17]. In a parallel study, broadband observations are being made of the survey region in order to detect and study the structure of objects of low surface brightness.

The survey should also provide a relatively large sample of QSOs, and other objects having strong emission lines, which are quite conspicuous in the multiband data. Because of the wide wavelength range of the survey, it should be relatively free from redshift biases. Simulations indicate that typical QSOs can be reliably identified, and distinguished from stars, when the signal-to-noise ratio of the data is ~ 10 or more [18]. The high success rate indicated by the simulations means that statistical studies can proceed without the need for a large amount of follow-up spectroscopy.

The NODO telescope performs very well. At times of good seeing, the resolution of the drift-scanned images is limited to 1.0 arcsec FWHM by optical distortion (the corrector lens was not designed for drift scanning because a video camera is used to observe space debris). The corrector is presently being upgraded to remove the distortion. The telescope can be used even in relatively high winds without significant image degradation. The telescope is operated and maintained by a minimal staff, at a cost substantially lower than those of conventional observatories.

3. The Large Zenith Telescope

The latest addition to the LMT family is the 6-m Large Zenith Telescope (LZT) under construction near Vancouver. Its dual goals are to develop and test technology needed for the next-generation LMTs, and to extend the multiband drift-scan surveys to fainter magnitudes. The LZT is scheduled to begin observations in early 1998. It is designed to achieve subarcsecond image quality over an 0.3-degree filed of view. Equipped with a thinned 2K $\times$ 4K CCD camera, it should reach a magnitude limit of $m_{AB} \sim 24$ in a simple night. This will allow it to study the faint-blue galaxy population [20], and achieve sample sizes of order 10^6. The fainter magnitude limit will also facilitate studies of

TABLE 1. UBC-NASA Multiband Survey

ra range (2000):	12 00 00 - 18 00 00
dec range (2000):	32 52 00 - 33 08 00
central dec (2000):	33 00 00
dec width (arcmin):	16.00
ra length (arcdeg):	75.48
area (sq deg):	20.13
pixel size (arcsec):	0.598
wavelength range (nm):	455 - 948
spectral resolving power:	44
spatial resolution (arcsec):	1.5
astrometric accuracy (arcsec):	1.0
magnitude limit (AB):	20.4

large-scale structure and the topology of the galaxy distribution, providing a clear discrimination between competing models of structure formation [21].

4. Future Developments

Technical solutions employed in the LZT will facilitate the construction of larger liquid mirrors. The air bearing developed for the LZT has sufficient capacity for mercury mirrors of 10-m diameter. The modular construction techniques employed can be readily extended to larger mirrors. A 30-m LMT is not implausible. Alternatively, arrays of 10-m-class telescopes could be contemplated.

A limitation of current LMTs is the need to observe near the zenith. Refracting correctors can provide an instantaneous field of view of about one degree, but larger fields are possible with reflecting correctors. The well-known three-mirror designs of Paul [22] and Baker [23] can provide good image quality over a field of several degrees [24]. Another approach would be the use of adaptive correctors. These contain movable mirrors whose shape can be adjusted to correct aberrations over a range of field angle. Such a corrector could in principle track objects as far as 15 degrees from the zenith [25].

The performance of three-metre LMTs has been proven, and they are regularly used for astronomical observations. The LZT will soon begin operation, providing multi-band data for a million galaxies. Larger LMTs are possible and can provide a cost-effective approach to obtaining large apertures. The potential impact of LMTs, particularly for survey programs, is likely to be substantial.

References

1. Gibson BK, 1991, J Ray Astr Soc Canada, 85, 158
2. Skey H, 1874, Nature, 10, 274
3. Wood, RW, 1909, Ap J, 29, 164
4. Borra EF, 1982, J Roy Astr Soc Canada, 76, 245
5. Borra EF, Content R, Girard L, Szapiel S, Tremblay LM & Boily E, 1992, Ap J, 393, 829
6. Hickson, Gibson BK & Hogg D, 1993, Pub Astro Soc Pacific, 105, 501
7. Hickson P, Borra EF, Cabanac R, Content R, Gibson BK & Walker GAH, 1994, Ap J, 436, L201
8. Sica RJ, Saragoytchic S, Borra EF, Girad L, Argall S, Sarrow CT & Flatt S, 1995, Appl Opt, 34, 6925
9. Wurker RF & Wong A, preprint
10. Potter AE & Mulrooney MK, 1997, Adv Space Res, 19, 213
11. McGraw JT, Angel JRP & Sargent TA, 1980, Proc Soc Photo-Opt Instr Eng, 264, 20
12. Hickson P & Richardson EH, 1997, in preparation
13. Hickson P & Mulrooney MK, 1997, Ap J Suppl, in press
14. Hickson P, Gibson BK & Callaghan KC, 1994, MNRAS, 267, 911
15. Zepf S & Koo DC, 1989, Ap J, 337, 34

16. Woods D, Fahlman GG & Richer HB, 1995, Ap J, 454, 32
17. Impey C & Bothun G, 1997, Ann Rev Astr Ap, 35, 267
18. Peri F, Iovino A & Hickson P, 1997, *Science with Liquid-Mirror Telescopes*, in press
19. Hickson P, Walker G, Chapman S, de Lapparent V, Borra EF & Cabanac R, 1997, in preparation
20. Elis RS, 1997, Ann Rev Astr Ap, 35, 389
21. Brandenberger R, 1997, private comm.
22. Paul M, 1935, Rev d'Opt, 14, 169
23. Dimitroff GZ & Baker JG, 1945, *Telescopes and Accessories*, Blakiston, NY, p. 107
24. Wilson RN, *Reflecting Telescope Optics I*, Springer-Verlag, Berlin, p. 222
25. Borra EF, Moretto G & Wang, 1995, Astr. Ap. Suppl., 109, 563, in press

THE DEEP PROJECT

DAVID C. KOO
UCO/Lick Observatory, University of California,
Santa Cruz, CA 95064, USA

Abstract.
DEEP is a multi-institutional program designed to undertake a major new spectroscopic survey of faint field galaxies with the Keck II 10-m telescope. The scientific goals are broad and include exploring galaxy formation and evolution, mapping the large scale structure at moderate to high redshifts, and constraining the nature and distribution of dark matter and cosmology. Besides the primary goal of securing large numbers of redshifts (10,000$^+$) to very faint limits of $I \sim 23$, DEEP intends to acquire spectra of high enough quality and spectral resolution to extract rotation curves, velocity dispersions, age estimates, and chemical abundances for a brighter subset of galaxies. A new imaging spectrograph for Keck called DEIMOS has been specifically designed to achieve these goals and is currently scheduled for completion by the end of 1998. DEIMOS will provide an overall gain for multi-object spectroscopy of about 7x compared to the current low-resolution spectrograph (LRIS). While awaiting for DEIMOS to be operational, the interim DEEP science programs have been diverse, but largely concentrated on spectroscopy of faint galaxies observed with HST, especially in the "Groth Strip" and Hubble Deep Field (HDF) and its flanking fields. Recent highlights include redshift and kinematic studies of compact galaxies, high redshift ($z \sim 3$) galaxies, and distant spirals.

1. Introduction

As we approach the 21st century, the suite of new 8-10 m class optical telescopes promises to offer spectroscopic surveys of galaxies to unprecedented faintness. Equally tantalizing are the new opportunities to combine such deep ground-based spectra with deep space-based measurements of high spatial resolution, such as size, shape, and morphology, from the Hubble Space Telescope (HST), as well as in new wavebands, such as can be expected from SIRTF and AXAF. Besides supplying redshifts that are so critical to convert, e.g., magnitudes to intrinsic luminosities, and to obtain the lookback time, the high S/N and high spectral-resolution from 8-10 m class telescopes provide three new, powerful diagnostics for the analysis of distant galaxies, namely internal velocities (and hence masses when size is used), abundances, and age estimates. These parameters are clearly important and independent probes of galaxies in the early universe and provide via mass, e.g., more direct links to the outputs of cosmological simulations of galaxy formation and evolution. Finally, because both galaxy evolution and their large scale patterns are very likely to be complex, surveys with large samples will be needed to address these problems.

2. What is DEEP?

To meet the aforementioned challenge, the Deep Extragalactic Evolutionary Probe (DEEP: more details at website **http://www.ucolick.org/ ~deep/home.html**) has been initiated as a project designed to gather spectral data for over 10,000 faint field galaxies (Mould 1993, Koo 1995) using the Keck II 10-m telescope and a new instrument (DEIMOS). Depths are expected to be $I \sim 23$ or fainter, corresponding to median redshifts $z \sim 0.8$. DEEP will thus be at least 10x larger and one magnitude deeper than the pioneering and important Canada France Redshift Survey (CFRS: see proceedings). Based on our experience with Keck so far, redshifts are expected for

J. Andersen (ed.), Highlights of Astronomy, Volume 11A, 468–472.

the vast fraction of this sample (~80%, which is comparable to the 75% reached in CFRS for $21 < I < 22$). The redshifts will be used not only to study the galaxies themselves, but also to map the clustering patterns and global velocity distortions and motions of distant galaxies. Since the large-scale structure and its evolution to redshifts $z \sim 1$ are quite sensitive to the amount and nature of dark matter as well as cosmology (Jenkins et al. 1997), DEEP is expected to place powerful independent constraints on both.

A distinguishing aspect of DEEP is that the survey also aims to gather internal kinematic data in the form of rotation curves or linewidths. Such measurements are of special interest because, if the system is bound and if sizes and inclinations can be independently measured or estimated, the inferred gravitational potential depth can be used to estimate the actual masses of distant galaxies. Tracking how the volume density of galaxies with different masses (as well as luminosities, sizes, structure, etc.) vary with lookback time will provide an entirely new dimension to study the rate of galaxy merging, luminosity evolution, disk and bulge structure formation, etc. Finally, if the mass, luminosity, and number evolution of galaxies can be better understood, the large sample of DEEP will provide an unprecedented opportunity to measure the geometry of the universe not only through the evolution of large scale structure but also via the classical volume test. Such volume tests are more sensitive than size or luminosity tests by factors of $\sim(1+z)^2$ and $\sim(1+z)$, respectively.

3. What is DEIMOS?

DEIMOS (DEep Imaging Multi-Object Spectrograph) is a new spectrograph for Keck II (more information is at website **http://www.ucolick.org/ ~loen/Deimos/deimos.html**). DEIMOS will provide both a wide-field direct imaging capability as well as multi-object spectroscopy via the use of focal-plane masks with multiple slitlets. Besides being optimal for DEEP, this Nasymth instrument will be ideal to tackle a broad range of targets that are densely populated in a small region of sky, including dense star fields, stars in globular clusters, extragalactic globular clusters, clusters of galaxies, etc. It is expected to be commissioned by late 1998 and operational by 1999.

Although DEIMOS was originally planned to be a four camera system, and then a two camera system, DEIMOS is currently being constructed with only one camera. This camera will include a CCD mosaic of 8K x 8K comprised of eight 2K x 4K CCDs with pixels of 15μ. Not only will this camera accommodate a total slit length of over 16 arcmin (or roughly 75 slitlits) and an imaging area of of roughly 76 square arcmin at 0.12 arcsec per pixel, but also it will allow the spectra to span a large range in a single exposure with moderately high spectral resolution. The current design includes a low resolution mode with a 600 l/mm grating that provides 0.65Å per pixel over, e.g., the nominal range of 4300Å to 9500Å and a high resolution mode with a 1200 l/mm grating that provides 0.33Å per pixel over, e.g., 6800Å to 9400Å. The wavelength useable for spectroscopy spans from 4300Å to 11000Å while the imaging mode will reach bluer limits and will have standard B, V, R, I, and two user filters. After including the silver-coated Keck II telescope, the end-to-end throughput is expected to be 40% for imaging and nearly 30% for spectroscopy. Present plans provide for 10 pre-fabricated slit masks for each camera. Compared to Keck's current Low Resolution Imaging Spectrograph (LRIS: Oke et al. 1995), DEIMOS has a total throughput (sensitivity, spectral range, resolution, area) that is roughly 7x greater.

4. Summary of Early DEEP Projects

While waiting for the completion of DEIMOS so that the major DEEP survey of $10{,}000^+$ galaxies can begin, we have undertaken a number of smaller, pilot-style projects with the existing LRIS to determine what is feasible with Keck and thus to help define the DEEP survey. To maximize the scientific returns for our relatively small samples, we have chosen to focus our efforts in fields where relatively deep HST WFPC2 images already exist. Such HST images are crucial not only for determining the morphology of our galaxy samples, but also for obtaining reliable measures of structure, including size and inclination, which are needed to convert kinematic observations from Keck into direct measures of mass. Overall, our early results have been encouraging and reassure us that our major DEIMOS survey is feasible.

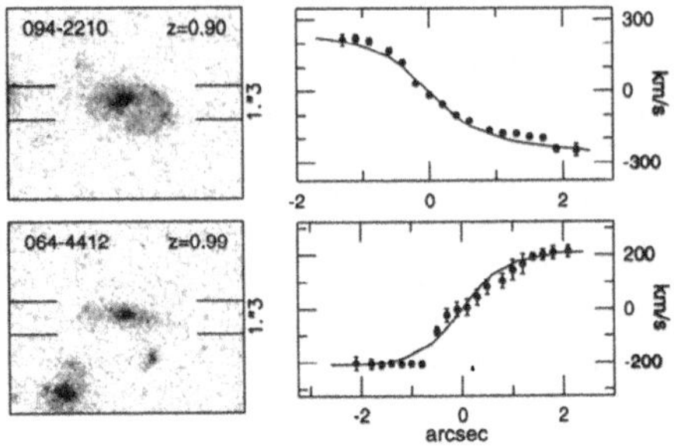

Figure 1. Fig. 1. Examples of the rotation curves measured for two high redshift galaxies (Vogt et al. 1996).

4.1. SURVEY DEPTH AND COMPLETENESS

Our faintest redshift samples are too small to obtain any detailed assessment of the expected completeness in securing reliable redshifts as a function of total I band magnitudes, size, surface brightness, strength of emission lines, redshift, etc. Based on a representive sample of 33 galaxies that reached fainter than $I \sim 24$ in the Groth Strip, we find that our exposures of 1.5 to 3.5 hours yielded a completeness of 100% to the $I \sim 22$ limit of CFRS and over 80% for the entire sample (Koo et al. 1996). A large fraction of all successful redshifts greater than $z \sim 0.9$ are based on the detection of the [OII] doublet at 3727Å , which can easily be hidden by the atmospheric "A" band absorption at 7600Å and the dense forest of strong night sky OH lines. New redshifts from our most recent observing runs indicate that our completeness is fully consistent with those found above, with over 80% completeness to $I \sim 23$. Thus the DEEP survey will be able to achieve such completeness for 1.5 to 2 hour exposures for roughly four sets of targets (i.e., about 300 total targets) for each clear night. Thus a program to target 20,000 galaxies with a final yield of 16,000 galaxies to $I \sim 23$ can be completed with roughly 70 clear nights. Assuming a 30% loss due to weather and other adverse conditions, a 100 night program would be needed for such a survey.

We note that these faint and relatively high redshifts also provide a crucial set of tests and calibration targets for the use of broadband photometry to estimate redshifts. Within our DEEP team, such photometric redshifts have been studied by Koo 1985; Connolly et al. 1995; Connolly et al. 1997; Brunner 1997 and shown to be one of the most promising and reliable techniques to push beyond the limits of spectroscopy. Such photometric redshifts, which can be estimated with relatively inexpensive multicolor images, are expected to play an important role in DEEP. As one example, we plan to use photometric redshifts to improve probes of mass that rely on weak lensing distortions. As another, we intend to maximize the number of galaxies from a given redshift interval in the selection of the DEEP sample used to track the evolution in large scale structures. A demonstration of the power of photometric redshifts for survey pre-selection is the use of U band drop-outs to define the spectroscopic samples that led to the discovery of large scale structures at redshifts $z \sim 3$ by Steidel et al (1997).

4.2. ROTATION CURVES OF DISTANT SPIRALS

One of the major themes of DEEP is the kinematics of distant galaxies. As seen in Fig. 1, we have clearly demonstrated the feasibility of discerning emission-line rotation curves of likely spirals seen to redshifts near $z \sim 1$ for galaxies as faint as $I \sim 22$ with one to two hour exposures. To ensure alignment with the apparent major axis, the slitlets have been milled with tilts. Based on a total sample of 16 galaxies, we find little evidence for any major change (< 0.6 mag) in the zero-point of the optical Tully-Fisher relation (Vogt et al. 1996; Vogt et al. 1997). The DEEP results are in stark contrast to claims for more extensive evolution of 1.5 mag (Rix et al. 1997) to 2.5 mag (Simard and Pritchet 1997) for very blue galaxies.

4.3. EMISSION LINEWIDTHS

Although rotation curve measures are possible for extended galaxies, the vast majority of very faint galaxies are in fact too small to yield more than linewidths as kinematic data. Except for very bright galaxies that yield absorption linewidths, emission lines are used. Though winds, dust obscuration, and poor representation of the gravitational potential by the luminous star formation regions may all invalidate the use of the emission line widths for probing the gravitational potential, especially for very compact, starbursting systems, the available data to date on such extreme galaxies show a fairly tight ratio of 0.7 +/- 0.1 between such optical line-width measures and radio measures of HI motions (Telles and Terlevich 1993). The HI kinematics presumably sample most of the true potential.

We have thus proceeded with the assumption that linewidths, after an upward correction of 40%, are meaningful measures of the true potential. Then when combined with HST size data, we are able to obtain masses. At least for blue compact galaxies, we find that luminosity alone can be a very poor gauge of their masses, i.e., the M/L ratio can vary enormously (Guzmán et al. 1996). Indeed, for some very compact galaxies brighter than the Milky Way, we needed the the High Resolution Echelle Spectrograph (HIRES: see Vogt et al. 1994) to resolve velocity widths smaller than 30 km-s^{-1} (Koo et al. 1995). Fortunately for the DEEP program, the vast majority of faint galaxies appear to have linewidths that can be resolved at the nominal DEIMOS spectral resolution of $\sigma \sim 40$ km-s^{-1}. This was empirically verified for a sample of compact galaxies, and thus most likely to have the smallest masses (Phillips et al. 1997; Guzmán et al. 1997). The key point is that the lack of correlation between luminosity and mass, i.e. stable M/L, demonstrates the usefulness, necessity, and promise of kinematics as an important new dimension to discern the evolution of different galaxy populations.

4.4. VERY HIGH REDSHIFT $Z \sim 3$ GALAXIES

One of the major advances with Keck has been the dramatic demonstration by Steidel et al (1996a) that very high redshift ($z \sim 3$) galaxies, originally selected on the basis of broadband colors, can be routinely confirmed spectroscopically. The DEEP team has extended the pioneering efforts of Steidel et al. (1996b) in the Hubble Deep Field (HDF) by pushing the faintness limits beyond $I \sim 25$, using redder "dropouts" to reach higher redshifts and higher levels of completeness, and adopting higher spectral resolutions to improve kinematic measurements (Lowenthal et al. 1997). Based on the evidence so far, we suggest that the high redshift galaxies may also be small mass systems that become dwarfs today or that later merge to form more massive galaxies instead of being only the cores of massive galaxies. Regardless of the eventual understanding of their nature, the success of these surveys now open another possible avenue for the DEEP survey with DEIMOS, namely an exploration of not only the $z \sim 1$ universe, but also the very high redshift regime with 1000's of galaxies at $z \sim 3$ or beyond, where galaxy formation and pre-merger subcomponents are likely to be in their earliest phases.

5. Summary and Future Plans

Our various pilot programs with LRIS clearly show the need and power of kinematics for galaxy surveys, the feasibility of reaching $I \sim 23$ or fainter with respectable completeness in one to two hour exposures with Keck, and the accessiblity of the very high redshift universe. Thus the definition, feasibility, and value of the DEIMOS DEEP program is no longer merely in the speculative proposal stage. The selection of the targets has not yet been fully settled, but is likely to be dominated by the need for very large numbers of galaxies in the large scale structure program. Simulations to optimize the sample definition have not yet been completed, but the basic survey is anticipated to consist of redshift slices near $z \sim 1$, with the spectroscopic sample chosen from photometric redshifts. The final choice of fields for the DEIMOS large-scale structure survey has not yet been settled, but is very likely to be comprised of four or five regions of roughly 16'x120' each and accessible from spring to fall. The final choices may be dictated by the availability of deep, wide-field multicolor images being collected by several different groups.

6. Acknowledgements

The DEEP scientific program was initiated and enjoyed financial support (NSF AST 91-20005) from the Berkeley Center for Particle Astrophysics (CfPA), an NSF Science and Technology Center aimed to explore the nature of dark matter, as well as an NSF grant, AST 95-29098. The scientific DEEP program includes teams from several institutions, including a core group comprised of Garth Illingworth, S. M. Faber, and the author at UC Santa Cruz, Marc Davis at UC Berkeley, Richard Kron at the Univ. of Chicago/Fermi Lab, Alex Szalay at Johns Hopkins, and Chuck Steidel and Roger Blandford at Caltech. Except for management, the bulk of the hard work has been undertaken by a large pool of talented junior astronomers: Duncan Forbes (now at Birmingham), Rafael Guzmán (now at Yale), Jesus Gallego (U. of Madrid), James Lowenthal (now at U. of Mass., Amherst), Drew Phillips, Nicole Vogt, Luc Simard, and Vicki Sarajedini (all at UC Santa Cruz), Matthew Bershady (now U. of Wisconsin), Andy Connolly and Robert Brunner (Johns Hopkins), and Marianne Takamiya and Brad Holden (Univ. of Chicago).

The instrument (DEIMOS) is being funded largely by NSF through its Academic Research and Infrastructure Program and by the Keck foundation (CARA). The instrument effort is being led by S. M. Faber with support from Richard Stover (UC Santa Cruz) for the CCD effort; David Cowley who is Project Manager; and a team of dedicated technical experts and staff among the UC Observatories that is transforming design goals into a working instrument.

References

Brunner, R. J. 1997, Ph.D. thesis, Johns Hopkins Univ.
Connolly, A. J., et al. 1995, *AJ*, **110**, 2655
Connolly, A. J., et al. 1997, *ApJ*, **486**, L11
Guzmán, R. et al. 1996, *ApJ*, **460**, L5
Guzmán, R. et al. 1997, *ApJ*, **489**, 559
Jenkins, A., et al. 1997, *ApJ*, submitted (Astro-ph 9709010)
Koo, D. C. 1985, *AJ*, **90**, 418
Koo, D. C. 1995, *Wide Field Spectroscopy and the Distant Universe*, eds. S. J. Maddox and A. Aragón-Salamanca, p. 55
Koo, D. C. et al. 1995, *ApJ*, **440**, L49
Koo, D. C. et al. 1996, *ApJ*, **469**, 535
Lowenthal, J. D. et al. 1997, *ApJ*, **481**, 673
Mould, J. 1993, *ASP Conf. Ser.*, **43**, 281
Oke, J. B. et al. 1995, *PASP*, **107**, 375
Phillips, A. C. et al. 1997, *ApJ*, **489**, 543
Rix, H.-W., et al. 1997, *MNRAS*, **285**, 779
Simard, L., and Pritchet, C. 1997, *ApJ*, submitted
Steidel, C. C., et al. 1996a, *AJ*, **112**, 352
Steidel, C. C., et al. 1996b, *ApJ*, **462**, L17
Steidel, C. C., et al. 1997, *ApJ*, in press
Telles, E., & Terlevich, R. 1993, *Ap&SS*, **205**, 49
Vogt, N. P. et al. 1996, *ApJ*, **465**, L15
Vogt, N. P. et al. 1997, *ApJ*, **479**, L121
Vogt, S., et al. 1994, *Proc. SPIE*, **2198**, 362

REDSHIFT SURVEYS WITH 2DF

MATTHEW COLLESS*
Mount Stromlo & Siding Spring Observatories, Australia

AND

BRIAN BOYLE†
Anglo-Australian Observatory, Australia

1. Introduction

This IAU Joint Discussion proposes to address the subject of redshift surveys in the 21st century. This paper, however, deals with two major new redshift surveys that those involved sincerely hope will be completed in the 20th century. Nonetheless, these surveys are relevant to the topic of the meeting, as they clearly foreshadow the scope and style of redshift surveys, if not in the coming millennium, at least in the coming decade.

The surveys are being carried out with the new Two Degree Field (2dF) facility on the Anglo-Australian Telescope (AAT), a 400-fibre multi-object spectrograph with the capability, as described in Section 2, to increase the size of redshift surveys by an order of magnitude over current best efforts. The main scientific goals, survey strategy and some preliminary results from the 2dF Galaxy Redshift Survey are outlined in Section 3, while Section 4 similarly describes the 2dF QSO Redshift Survey. Further information can be found on the WWW at `http://www.aao.gov.au/2df/` for the 2dF facility, at `http://msowww.anu.edu.au/~colless/2dF/` for the galaxy survey and at `http://www.aao.gov.au/local/www/rs/qso_surv.html` for the QSO survey.

2. The 2dF Facility

The 2dF facility on the AAT consists of several major components, shown schematically in Figure 1: a new prime focus optical assembly, comprising an atmospheric dispersion compensator and field corrector; a pair of field plates, each with 400 positionable fibres, which can be tumbled between the focus location and a configuration location; a robotic fibre positioner which can rapidly re-configure the 400 fibres; two spectrographs, each dealing with the light from 200 fibres; and a top-end ring assembly to carry these various components.

2dF obtained its first scientific data in November 1996. For the next nine months it operated with only 200 fibres and one spectrograph while the system was tested and commissioned and the second spectrograph completed. In September 1997 the full 400 fibres were installed (see Figure 2) and for the first time 400 spectra were obtained simultaneously using the two-spectrograph system.

On behalf of the 2dF Galaxy Redshift Survey Team: Matthew Colless (MSSSO), Richard Ellis (IoA), Joss Bland-Hawthorn (AAO), Russell Cannon (AAO), Shaun Cole (Durham), Chris Collins (LJMU), Warrick Couch (UNSW), Gavin Dalton (Oxford), Simon Driver (UNSW), George Efstathiou (IoA), Simon Folkes (IoA), Carlos Frenk (Durham), Karl Glazebrook (AAO), Nick Kaiser (IfA), Ofer Lahav (IoA), Ian Lewis (AAO), Stuart Lumsden (AAO), Steve Maddox (IoA), John Peacock (ROE), Bruce Peterson (MSSSO), Will Sutherland (Oxford), Keith Taylor (AAO)

† *On behalf of the 2dF QSO Redshift Survey Team:* Brian Boyle (AAO), Scott Croom (Durham), Lance Miller (Oxford), Mike Read (ROE), Tom Shanks (Durham), Robert Smith (IoA)

J. Andersen (ed.), Highlights of Astronomy, Volume 11A, 473–481.

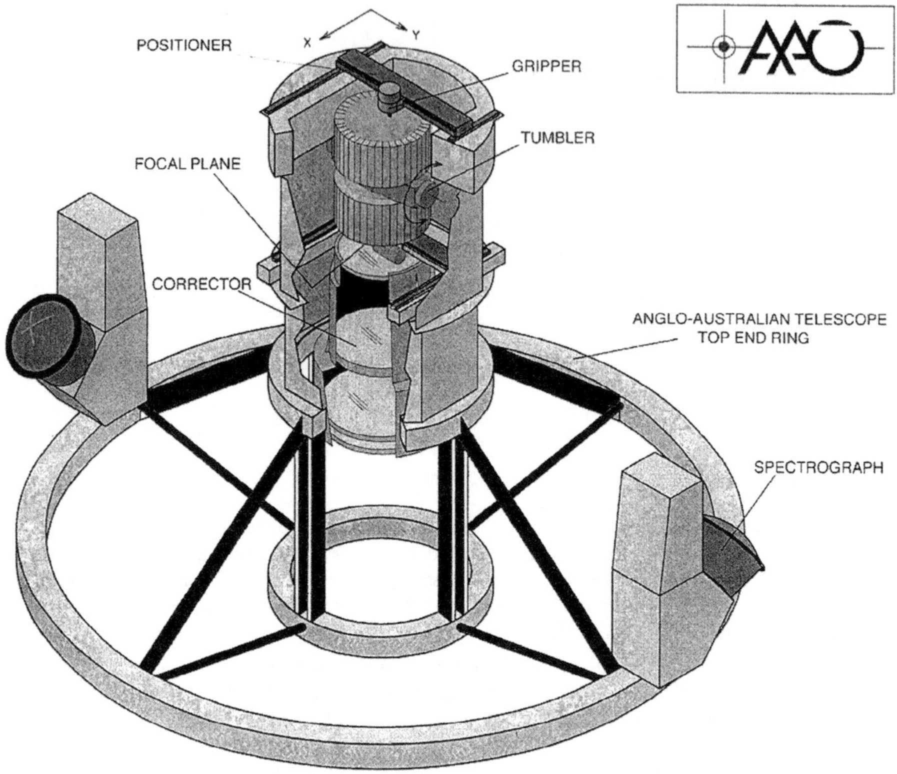

Figure 1. Schematic of the 2dF facility.

3. The 2dF Galaxy Redshift Survey

3.1. GOALS OF THE SURVEY

The observational goal of the 2dF Galaxy Redshift Survey is to secure high quality spectra and redshifts for 250,000 galaxies brighter than b_J=19.5 (extinction-corrected), with a deeper extension to R=21 making best use of good conditions. The brighter galaxies cover an area of 1700 square degrees selected from both the southern galactic hemisphere APM galaxy survey and an extension into the north galactic hemisphere equatorial region. The arrangement of 2dF survey fields is shown in Figure 3.

We aim to use the survey to address a variety of fundamental problems in galaxy formation and cosmology. The main scientific goals include:

1. Accurate measurements of the power spectrum of galaxy clustering on scales of up to a few hundred Mpc, allowing direct comparisons with microwave background anisotropy measurements of fluctuations on the same spatial scales.

2. Measurements of the distortion of the clustering pattern in redshift space, providing constraints on the cosmological density parameter Ω and the spatial distribution of dark matter.

3. A determination of variations in the spatial and velocity distributions of galaxies as a function of luminosity, type and star-formation history, providing important constraints on models of galaxy formation.

4. Measurements of the evolution of the galaxy luminosity function, clustering amplitude and star formation rates out to a redshift of $z\sim0.5$. This will be accomplished by combining the bright

Figure 2. One of the two fieldplates, showing 400 fibres positioned in the focal plane.

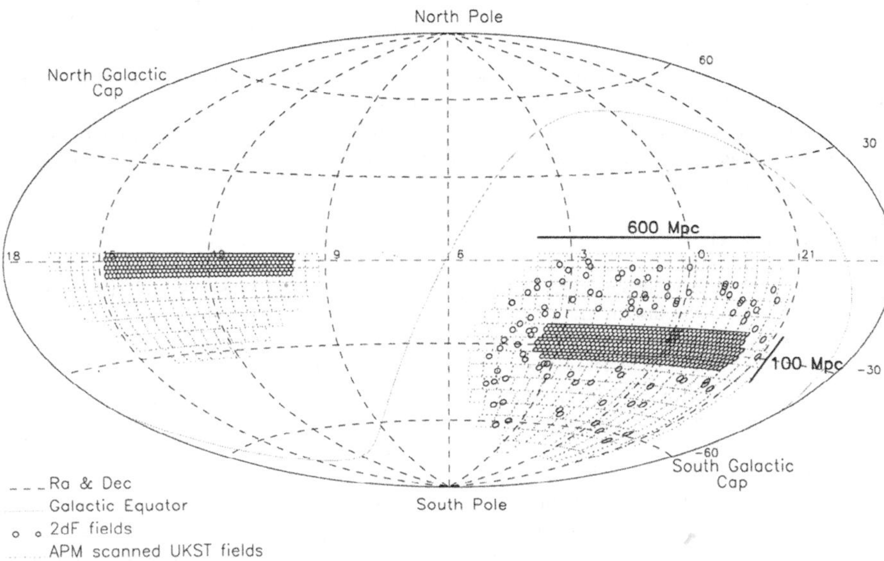

Figure 3. The 2dF survey fields (small circles) superimposed on the APM survey area (dotted outlines of Sky Survey plates). There are approximately 140,000 galaxies in the 75°×12.5° South Galactic Hemisphere (SGH) strip centred on the South Galactic Pole, 70,000 galaxies in the 65°×7.5° North Galactic Hemisphere (NGH) equatorial strip, and 40,000 galaxies in the 100 random 2dF fields scattered over the entire southern region of the APM galaxy survey.

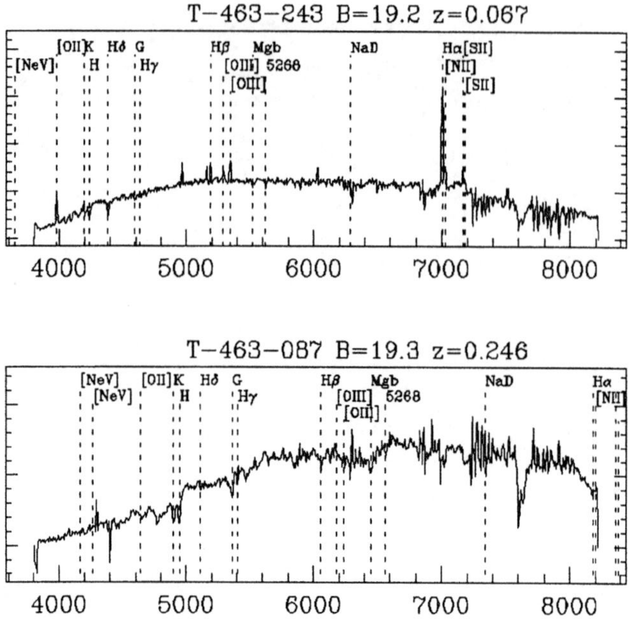

Figure 4. Example spectra from the survey: a b_J=19.2 emission-line galaxy at z=0.067 and a b_J=19.3 absorption-line galaxy at z=0.246.

b_J<19.5 survey with the faint R<21 survey, which is comparable in size to the largest current local surveys.

5. Investigations of the morphology of galaxy clustering and the statistical properties of the fluctuations—for example, whether the initial fluctuations are Gaussian as predicted by inflationary models of the early universe.

6. A study of clusters and groups of galaxies in the redshift survey, in particular the measurement of infall in clusters and dynamical estimates of cluster masses at large radii.

7. Application of novel techniques to classify the uniform sample of 250,000 spectra obtained in the survey, thereby obtaining a comprehensive inventory of galaxy types as a function of spatial position within the survey.

3.2. SURVEY STATUS

As of mid-1997 we have taken nearly 2000 redshifts for the survey. These were obtained in commissioning time on the instrument using only one of the two spectrographs and 200 fibres. Later this year we will begin observations with all 400 fibres and the survey will enter production-line operation. We expect to complete the survey observations by the end of 1999.

Two example spectra taken in the commissioning period are shown in Figure 4. One is a b_J=19.2 emission-line galaxy at z=0.067 and the other is a b_J=19.3 absorption-line galaxy at z=0.246. The quality target for the survey spectra is a S/N of at least 10 per 2-pixel resolution element (FWHM≈9Å); most spectra will easily exceed this target.

This minimum S/N permits reliable automatic spectral classification and redshift measurement. Employing both a standard cross-correlation and line-fitting code and a new code which uses Principal Component Analysis and χ^2-fitting to simultaneously classify the spectrum and measure its redshift (Glazebrook et al. 1997), we find we achieve a very high level of reliability. A comparison of the redshifts obtained from these codes with redshifts determined semi-manually via visual inspection (Figure 5) shows a very low level of failures in the automatic algorithms, even with the variable-quality commissioning data.

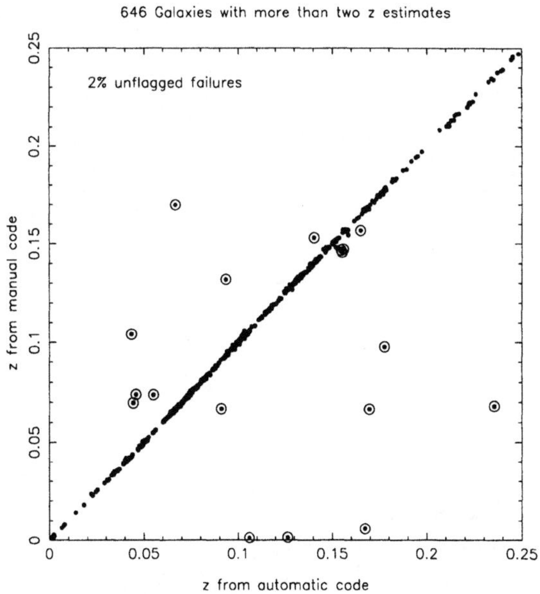

Figure 5. A comparison of redshifts determined automatically and via visual inspection of the spectra.

3.3. PRELIMINARY RESULTS

Figure 6a is a cone plot, combining NGH and SGH strips, showing the galaxies with redshifts measured to date—this is less than 1% of the full sample. For comparison, Figure 6b shows a simulated cone plot of a single 2° declination slice in the redshift survey; even this slice, which contains 34,000 galaxies, represents less than 15% of the survey.

Although so far we have observed too few fields to effectively address questions of large-scale structure, we do have a sufficiently large sample of redshifts to begin to look at the properties of local galaxies. Figure 7 shows a preliminary determination of the galaxy luminosity function (LF) at a mean redshift of $\bar{z}$=0.1. This LF is in good agreement with the LFs from the Autofib survey (Ellis et al. 1997) and the ESO Slice Project (Zucca et al. 1997), and gives an overall galaxy density about 50% higher than that obtained in the Stromlo-APM survey (Loveday et al. 1992). It is important to note that we are reaching more than 5 magnitudes below L^* even with this small sample, and that we should be able to tightly constrain the faint end of the local luminosity function for each morphological or spectral type with the full sample.

4. The 2dF QSO Redshift Survey

4.1. GOALS OF THE SURVEY

The observational goal of the QSO redshift survey is to obtain spectra for ~30000 B<21 QSOs in two declination strips at the South Galactic Pole and in an equatorial region at the North Galactic cap. The QSO survey will probe the largest scales in the universe ($10\,h^{-1}$ Mpc $< r <$ $1000\,h^{-1}$ Mpc) over a wide range in redshift space ($0.3 < z < 2.9$), and is thus complementary to the galaxy survey. The primary scientific aims of the survey are:

1. To obtain the primordial fluctuation power spectrum out to COBE scales.

2. To determine the rate of QSO clustering evolution in the non-linear and linear regimes, and hence obtain new limits on the value of Ω and b (Croom & Shanks 1996).

3. To apply geometric methods to measure Λ (Ballinger et al. 1996).

4. To identify large statistical samples of unusual classes of QSOs (e.g. BALs) or absorption line systems (e.g. damped Lyα systems).

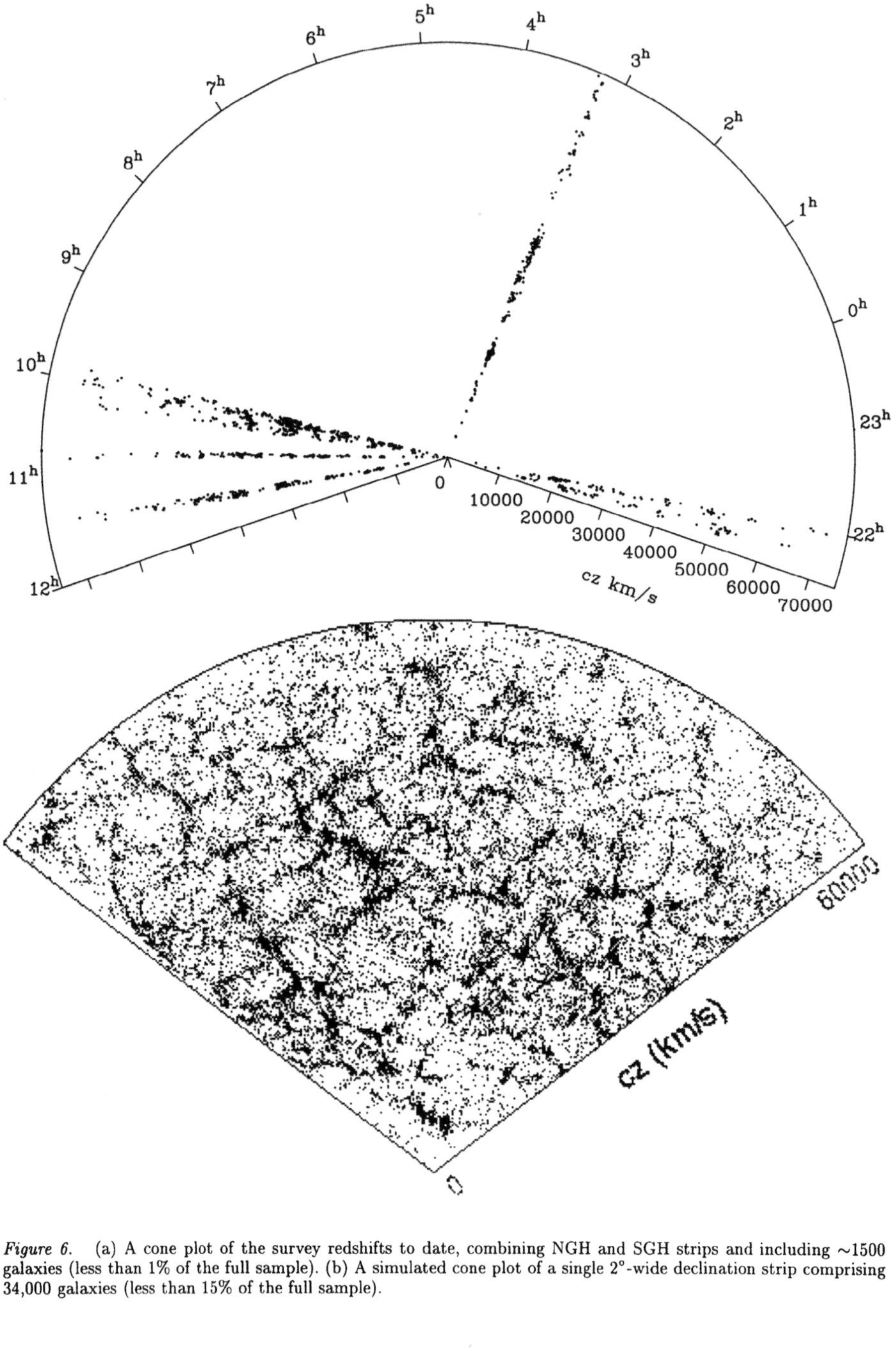

Figure 6. (a) A cone plot of the survey redshifts to date, combining NGH and SGH strips and including ~1500 galaxies (less than 1% of the full sample). (b) A simulated cone plot of a single 2°-wide declination strip comprising 34,000 galaxies (less than 15% of the full sample).

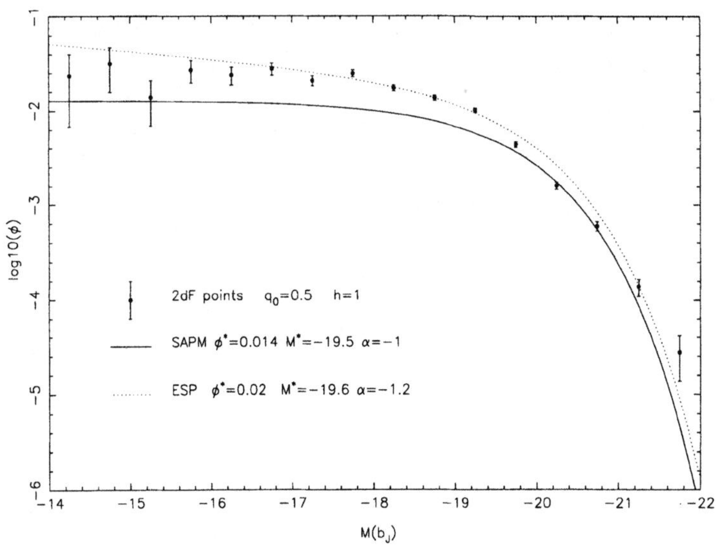

Figure 7. A preliminary galaxy luminosity function from the 2dF survey, compared to previous determinations by the APM-Stromlo and ESO Slice Project surveys.

Observations that radio-quiet QSOs exist in average galaxy cluster environments ($r_0 \sim 5\,h^{-1}$ Mpc) at low-to-moderate redshifts (Smith et al. 1995 and references therein) demonstrate that radio-quiet QSOs can be used to provide important information on the structure of the Universe at the largest scales.

4.2. SURVEY STRATEGY AND STATUS

The 2dF QSO survey covers an area of 740 deg^2 comprising two 75° × 5° strips on the sky (small areas surrounding bright stars have been excluded from the analysis). QSO candidates were selected from APM measurements of UK Schmidt U, J and R plates/films, on the basis of their anomalous position in the stellar $U-B/B-R$ colour–colour diagram. The vast majority (>85%) of the candidates are selected solely on the basis of their $U-B$ colour (sensitive to QSOs with $z<2.2$), but the additional use of the R magnitude minimises contamination from Galactic stars and extends the redshift range over which QSOs can be selected to $z\sim2.9$. The initial UVX-selected catalogue is shown in Figure 8. In total over 150 U, J and R plates (comprising 30 UKST fields) were used to compile the input catalogue. Great care was taken to keep photometric variations at <0.1 mag level over the entire catalogue (Smith 1997).

The QSO 2dF redshift survey fields form part of the area of sky surveyed by the galaxy 2dF survey, and so the objects in the QSO and galaxy 2dF redshift surveys are being observed simultaneously with 2dF. This leads to significant gains in both efficiency (the combined survey can be carried out in 20 nights less than if the surveys had been separate) and completeness (the QSO survey will pick up compact blue galaxies, the galaxy survey will identify extended low-redshift QSOs).

To date (October 1997), almost 800 QSO candidates in the 2dF QSO survey have been observed, comprising over 400 QSOs. Most of these have been observed with 2dF, but over 100 bright ($B<18$) QSOs have been identified with the UKST FLAIR system and 20 QSOs in close pairs (separations less than 20 arcsec) have been discovered with the MSSSO 2.3m telescope. Representative spectra from the QSO survey are shown in Figure 9.

We hope to complete the QSO 2dF redshift survey over the next two years, providing an invaluable resource with which to study the large-scale structure of the Universe out to $z\sim3$. Progress of the survey can be monitored through the WWW site (see Introduction), and by-products of the survey (CCD calibration, input catalogue) will be available at regular intervals.

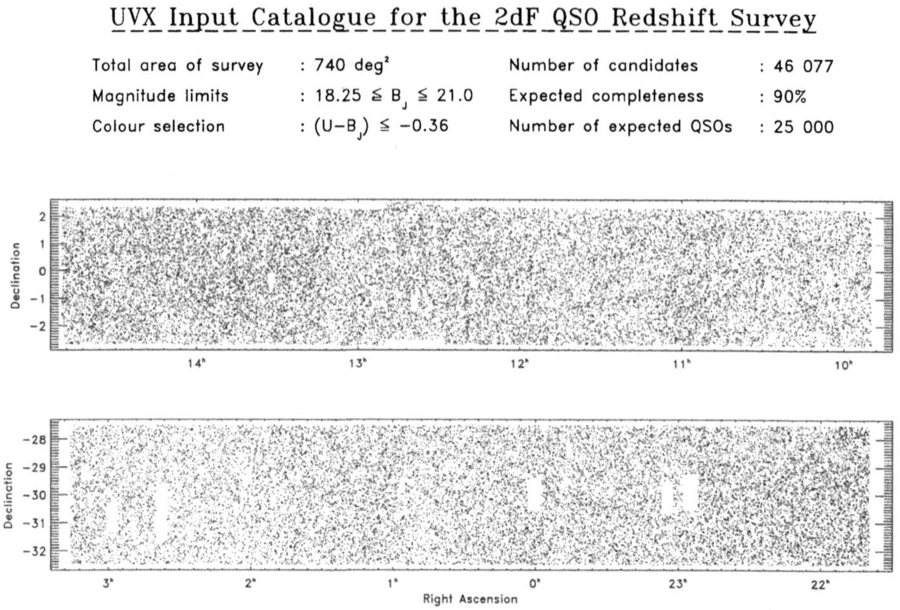

Figure 8. The UVX-selected objects for the QSO 2dF input catalogue. The RA and Dec of all UVX-selected objects are shown in the two 75° × 5° strips of the catalogue. Increasing contamination by Galactic sub-dwarfs at lower Galactic latitudes gives an obvious gradient (<50%) in number density along the survey strips. Completeness estimates are based on our success at recovering previously identified QSOs.

References

1. Ballinger W.E., Peacock J.A., Heavens A.F., 1996, MNRAS, 282, 877
2. Colless M.M., 1996, Wide Field Spectroscopy and the Universe, *Wide Field Spectroscopy*, eds Kontizas M., Kontizas E., Kluwer, pp.227–240
3. Croom S.M., Shanks T., 1996, MNRAS, 281, 893
4. Ellis R.S., Colless M.M., Broadhurst T.J., Heyl J.S., Glazebrook K., 1996, MNRAS, 280, 235
5. Glazebrook K., Offer A.R., Deeley K., 1997, MNRAS, in press
6. Loveday J., Peterson B.A., Efstathiou G., Maddox S.J., 1992, ApJ, 390, 338
7. Smith R.J., 1997, PhD Thesis, University of Cambridge
8. Smith R.J., Boyle B.J., Maddox S.M., 1995, MNRAS, 277, 270
9. Zucca E., Zamorani G., Vettolani G., Cappi A., et al., 1997, A&A, in press

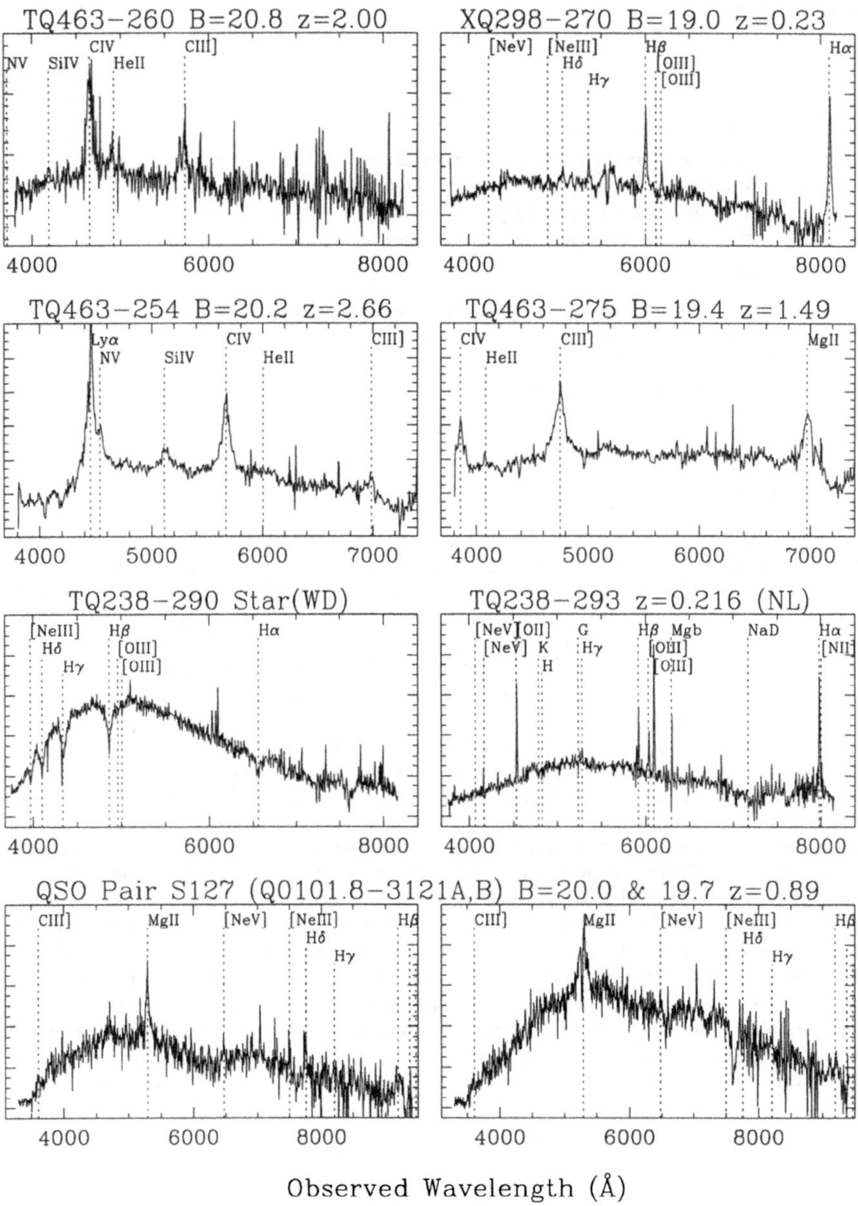

Figure 9. Representative spectra from the 2dF QSO redshift survey. The first six spectra were obtained with the 2dF and comprise 4 QSOs, a white dwarf and a narrow-emission-line galaxy. The latter two classes of object are some of the typical 'contaminants' of the UVX-selected sample. The bottom pair of objects are a close pair of QSOs (20 arcsec separation) observed with the MSSSO 2.3m.

MULTICOLOR SURVEY FOR HIGH REDSHIFT QUASARS BY KISO SCHMIDT TELESCOPE

TOMOHIKO YAMAGATA
Elementary and Secondary Education Bureau
Ministry of Education, Science, Sports and Culture
Kasumigaseki, Chiyoda Tokyo 100, Japan
Electric mail: yamagata@omega.mtk.ioa.s.u-tokyo.ac.jp

AND

KIMIAKI KAWARA, TSUTOMU AOKI AND YOSHIAKI SOFUE
Institute of Astronomy, University of Tokyo
Mitaka Tokyo 181, Japan
Electric mail: kkawara@ioa.s.u-tokyo.ac.jp

1. Introduction

Several groups have been successful in the survey of high-redshift ($z > 4$) quasars. For example, Schneider et al. (1994) and Schmidt et al. (1995) were successful in detecting 90 quasars with $2.75 < z < 4.75$, $R \leq 21$ for $62deg^2$ using Palomar Transit Grism. Among them, there are 9 quasars with $z > 4.0$. Storrie-Lombardi et al. (1996) were detected 31 quasars with $z \geq 4.0$, $m_r \leq 19$ for $2500deg^2$ by APM photographic multicolor survey using UK Schmidt telescope. Survey observations of quasars up to now have to select one of the following conditions; wide field surveys using photographic Schmidt plate which cover large areas but do not go faint enough, or deep surveys using large telescopes which go faint magnitude but which cover relatively small areas.

In this situation, we decided to begin a deep and wide-field multicolor survey for high-redshift quasars using the Kiso Schmidt telescope with a wide-field CCD camera. The Kiso Schmidt telescope has a corrector of $105cm$ in diameter and a $150cm$ main mirror. In order to utilize wide-field of the telescope, wide-field CCD cameras are being developed at the Kiso Observatory. As a first step a new large formatted CCD camera (hereafter referred to as 2K-CCD) is commissioning in a few month. The CCD chip used in this camera was provided by SITe and has 2048×2048 pixels which gives us a field of view of $50' \times 50'$ at the prime focus of the telescope. The coverage of the field is expected to be upgraded soon to a $2° \times 2°$ with mosaic of 2×4 CCD chips as a next generation imaging system of the Kiso Observatory. The standard $UBVRI$ Johnson-Cousins color bands are adopted for 2K-CCD system. We modified the standard system dividing I band into I' and Z bands shown in Figure 1b so as to be appropriate to detect quasars of redshift up to 6.0 (see discussion in section 3). In this paper, we present the result of the test observation of known quasars with the existing CCD camera, and discuss the performance of the survey by simple model of spectral energy distribution (hereafter referred to as SED) of high-redshift quasars.

2. Observation of Known Quasars

In order to examine the feasibility of multicolor CCD survey with the Kiso Schmidt telescope, the existing CCD Camera system (hereafter referred to as 1K-CCD) of the Kiso Observatory was used to observe VRI colors of 4 known quasars with redshift $z > 4$. The camera was made of Tektronics CCD chip with 1024×1024 pixels giving a field of view of $12'.5 \times 12'.5$ with $0.37''/pixel$ at the prime focus of the Kiso Schmidt telescope. Figure 1a shows VRI response function for the 1K-CCD system. We used VRI color bands for the test observation since the Lyman alpha line emitted at $4.0 < z < 5.0$ is redshifted to $6000\mathring{A}$ or more and drops into the R band. Therefore

J. Andersen (ed.), Highlights of Astronomy, Volume 11A, 482–486.

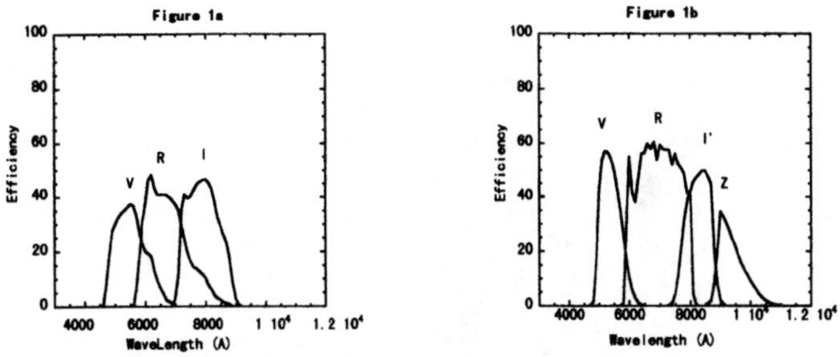

Figure 1. (a) *VRI* response function for the 1K-CCD system. *R* band of 1K-CCD is slightly bluer and narrower than the standard system. (b)$VRI'Z$ responce function for modified 2K-CCD system. The standard *UBVRI* Johnson-Cousins color bands are adopted for original 2K-CCD system. We modified the standard system dividing *I* band into I' and *Z* bands so as to be appropriate to detect quasars of redshift up to 6.0.

quasars with $4.0 < z < 5.0$ is expected to be distinguished from field stars using 2-color diagram of $(V - R)$ vs. $(R - I)$. Table 1 summarizes the coordinates and redshifts of the observed quasars. We obtained *VRI* images around 4 quasars. The standard IRAF system was used for flat-fielding and dark-subtracting of CCD data. Source detection and photometry are performed by the automated detection package developed by Yamagata (1986) and Yamagata and Yoshii (1991). The output from the detection package is compared to the CCD image, and then extended sources and artifact such as cosmic ray hits are removed. The sky brightness is assumed to be $21mag/deg^2$ for all the bands. The results are shown in Figure 2, which demonstrates that quasars at $z > 4$ can be well distinguished from field stars by the observation using Kiso Schmidt telescope.

TABLE 1. Observed Quasars

Object	α	δ	z	r
PC1247+3406	12^h49^m	$+33°50'$	4.897	20.4
PSS0248+1802	2^h49^m	$+18°\ 3'$	4.43	18.4
PC0751+5623	7^h56^m	$+56°19'$	4.281	19.9
PC0307+0222	3^h10^m	$+\ 2°31'$	4.379	20.4

3. Performance Test by Model SED of Quasars

The model colors of quasars of various redshifts were calculated using simulated spectral energy distribution (SED) of quasars. We used a composite SED of quasars constructed from HST data by Zheng et al. (1997) as a SED at the rest frame. SED's for various redshifts were derived by redshifting the SED from the rest frame. The correction of Lyman absorption due to intergalactic HI clouds were executed. The intergalactic Lyman absorption as a function of redshift and wavelength were calculated from the estimation of the optical depth due to Lyman line and Lyman continuum by Yoshii and Peterson (1994). Figure 3 shows the Lyman absorption as a function of the rest frame wavelength for various redshifts. The colors of quasars of various redshits were calculated by integrating model SED's by convoluting response function of 1K-CCD or 2K-CCD systems. Calibration of colors were done using the flux standard of Hynes and Latham (1975). The colors of stars of various types were also calculated from the stellar SED's in the spectrophotometric atlas of Gunn and Stryker (1983). Figure 4 shows the simulated two-color diagram ($(V - R)$ vs. $(R - I)$) of redshifted quasars on 1K-CCD system. Black dots show colors of stars in the atlas of Gunn and Stryler, and blue dots indicate those of redshifted quasars calculated from the model SED's. Red dots in the figure show calculated colors of the actual quasars, PC1247+3406 ($z = 4.897$) and PC0307+0222 ($z = 4.379$). The SED of PC1247+3406 is from Schneider et al.(1991) and that

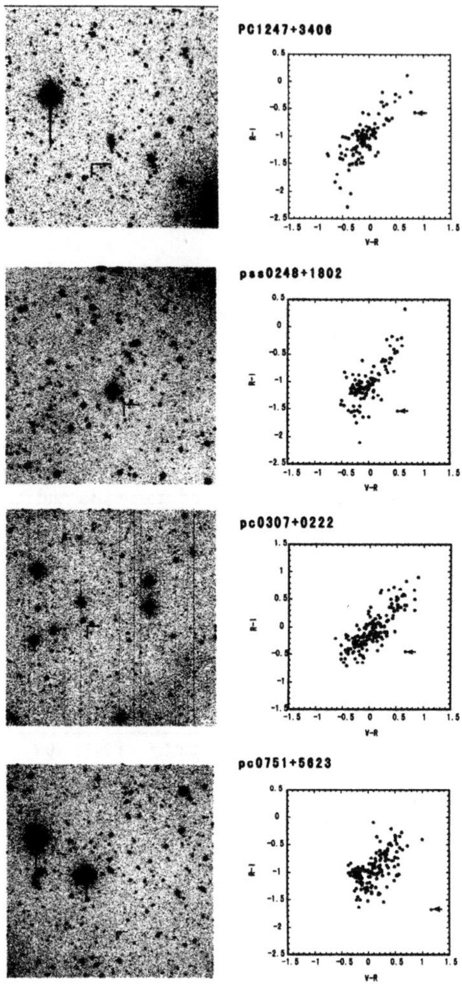

Figure 2. Left panels show R-band images of observed fields. Quasars are indicated by lines. Right are two-color $((V-R)$ vs. $(R-I))$ diagrams of observed images of each field. Arrows indicate quasars. These diagrams demonstrate that quasars at $z > 4$ can be well distinguished from field stars by CCD color observation using the Kiso Schimidt telescope.

of PC0307+0222 is from Schneider et al.(1989). The simulated colors mimic the actual colors of quasars reasonably well. The trend of redder $(V-R)$ color with increasing redshift is shown. Smith et al.(1994) have presented similar models but for $(B-R)$ vs. $(R-i)$. Our $(R-I)$ colors of quasars with redshift $z > 5$ are redder due to R filter of 1K-CCD being slightly bluer and narrower than the standard system.

Figures 5a and 5b show the predicted two-color diagrams on 2K-CCD system. Figure 5a is that of $(V-R)$ vs. $(R-I')$, and Figure 5b is of $(R-I')$ vs. $(I'-Z)$. As shown in Figure 5a, $(V-R)$

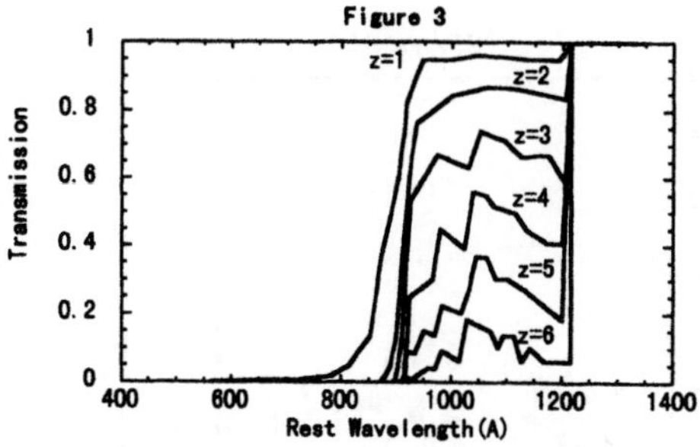

Figure 3. The Lyman absorption as a function of the rest frame wavelength for various redshifts. The intergalactic Lyman absorption as a function of redshift and wavelength were calculated from the estimation of the optical depth due to Lyman line and Lyman continuum by Yoshii and Peterson (1994).

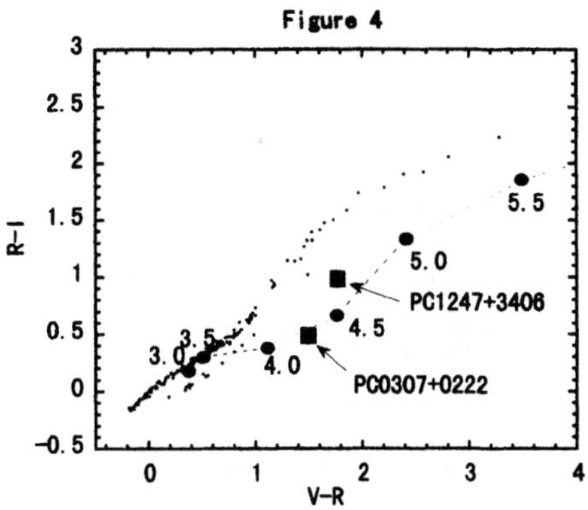

Figure 4. The simulated two-color diagram ($(V-R)$ vs. $(R-I)$) of redshifted quasars on 1K-CCD system. Dots are colors of stars in the atlas of Gunn and Stryler(1983), and solid circles are those of redshifted quasars calculated from the model SED's. Squares in the figure indicate calculated colors of the actual quasars, PC1247+3406 ($z = 4.897$) and PC0307+0222 ($z = 4.379$). The SED of PC1247+3406 is from Schneider et al.(1991) and that of PC0307+0222 is from Schneider et al.(1989). The simulated colors mimic the actual colors of quasars reasonably well.

colors of quasars of redshift $z > 5$ is greater than 3.6, therefore it is practically difficult to detect those quasars with V band unless they are very bright. In this case, two-color diagram of $(R - I')$ vs. $(I' - Z)$ can be used as shown in Figure 5b.

4. Concluding Remarks

We have verified the availability of the deep and wide-field multicolor CCD imaging method for searching high-redshift quasars using Kiso Schmidt telescope. The 2K-CCD camera system is now at the debugging stage at the Kiso observatory, and is expected to be put in service by the winter of 1997. The survey observation will be started soon after it is available.

References

Gunn, J.E., and Stryker,L.L. (1983) *ApJS*, **52**, 121
Haynes.,D.S., and Latham,D.W. (1975) *ApJ*, **197**, 593
Irwin,M., McMahon,R.G., and Hazard, C. (1991) *ASPCS*, **21**, 117
Schneider,D.P., Schmidt, M., and Gun, J.M. (1989) *AJ*, **98**, 1507

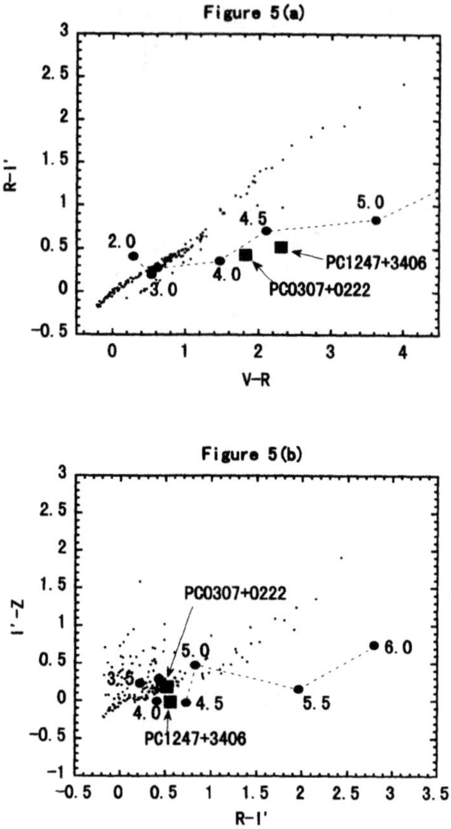

Figure 5. (a)The predicted two-color diagram of $(V-R)$ vs. $(R-I')$ on 2K-CCD system. (b)The predicted two-color diagram of $(R-I')$ vs. $(I'-Z)$ on 2K-CCD system.Figure 5a indicates that $(V-R)$ colors of quasars of redshift $z > 5$ is greater than 3.6, therefore it is practically difficult to detect those quasars with V band unless they are very bright. In such a circumstance, two-color diagram of $(R-I')$ vs. $(I'-Z)$ is useful as shown in Figure 5b.

Schneider,D.P., Schmidt, M., and Gun, J.M. (1991) *AJ*, **102**, 837
Schneider,D.P., Schmidt,M., and Gun,J.M. (1994) *AJ*, **107**, 1245
Schmidt,M., Schneider,D.P., and Gun,J.M. (1995) *AJ*, **110**, 68
Smith,J.D., Thompson,D., and Djorgovski,S. (1994) *ApJ*, **107**, 24
Storrie-Lombardi,L.J., McMaHon, R.G., Irwin, M.J., and Hazard,C. (1996) *ApJ*, **468**, 121
Yamagata, T. (1986) *Annals Tokyo Astron. Obs.*, **2nd Ser., 21**, 31
Yamagata, T. and Yoshii, Y. (1991) *AJ*, **103**, 117
Yoshii,Y., and Peterson, B.A. (1994), *ApJ*, **436**, 551
Zheng,W., Kriss,A.G., Telfer, R.C., Grimes,J.P., and Davidsen,A.F. (1997) *ApJ*, **475**, 469

THE 2MASS REDSHIFT SURVEY

J. HUCHRA & E. TOLLESTRUP
Harvard-Smithsonian Center for Astrophysics, Cambridge, MA USA
S. SCHNEIDER & M. SKRUTSKI
University of Massachusetts, Amherst, MA USA
AND
T. JARRETT, T. CHESTER & R. CUTRI
IPAC, Caltech, Pasadena, CA, USA

1. Introduction

With the current convergence of determinations of the Hubble Constant (e.g. *The Extragalactic Distance Scale,* 1997, Livio, Donahue and Panagia, eds.) to values within ±25% rather than a factor of two, and the clear possibility of determining q_o using high redshift supernovae (Garnavich et al. 1998), the major remaining problem in observational cosmology is the determination of Ω — what is the dark matter, how much is there, and how is it distributed?

The most direct approach to the last two parts of the question has been to study galaxy dynamics, first through the motions of galaxies in binaries, groups and clusters, and in the last decade and a half, driven by the observation of our motion w.r.t. the Cosmic Microwave Background (CMB) and the notion that DM must be clumped on larger scales than galaxy clusters if Ω is to be unity, through the study of large scale galaxy flows.

The ratio of the mass density to the closure mass density, Ω, is thought by most observers to be ~0.1–0.3, primarily based on the results of dynamical measurements of galaxy clusters and, more recently, gravitational lensing studies of clusters. In contrast, most theoretical cosmologists opt for a high density universe, $\Omega = 1.0$, based on the precepts of the inflation scenario, the difficulty of forming galaxies in low density models given the observed smoothness of the microwave background radiation, and the observational evidence from the matching of the available large scale flow measurements (and the absolute microwave background dipole velocity) to the local density field. However this last result is extremely controversial—matching the velocity field to the density field derived from IRAS (60μ) selected galaxy samples yields high Ω values (e.g., Dekel et al. 1993) but matching to optically selected samples yields low values (Hudson 1994; Lahav et al. 1994; Santiago et al. 1995). On small scales, the high Ω camp argues that the true matter distribution is much more extended than the distribution of galaxies, so the dynamical mass estimates are biased low.

A subplot of the Omega problem is conclusively finding the cause of the microwave background dipole. The leading candidate is, of course, the gravitational attraction of nearby density enhancements (or "repulsion" by voids). The current debate centers on two still unresolved issues, the actual motion of the galaxy/local group with respect to reference frames of nearby and moderately distant galaxies and clusters (Rubin et al. 1976; Lauer & Postman 1994, Postman & Lauer 1995; Courteau et al. 1993; Riess, Press & Kirshner 1995), and the cause of those motions (Lynden-Bell et al. 1988; Dressler & Faber 1990; Mathewson & Ford 1994; Huchra 1996; Scaramella et al. 1994; Kraan-Korteweg et al. 1996). The Great Attractor, the Hydra-Centaurus supercluster, Abell 3627, and the Shapley Supercluster all remain candidates, while Virgo is only slowly being understood and the lack of strong motions towards Pisces-Perseus is an unsolved mystery. Because of incompleteness and the very sparse sampling through the galactic plane we still do not have a complete (or even coherent) picture of even the very local universe and galaxy motions nearby (e.g. Shaya, Tully and Pierce 1992).

J. Andersen (ed.), Highlights of Astronomy, Volume 11A, 487–491.

2. Existing Surveys

Existing samples are limited. The two major large area/all-sky samples used to study the galaxy density field are relatively small and suffer significant biases. The Optical Redshift Survey (Santiago et al. 1995) consists of about 8000 optically (blue) selected galaxies, diameter limited, from the merge of three catalogs based on visual examination of photographic plates (the UGC, ESO and a merge of the VV and ESGC catalogs in the southern equatorial strip). By definition, the catalog has a very strong bias with galactic latitude due to extinction, moderately large errors in the "by eye" diameter determinations (20-30%), and only poorly determined uniformity across the sky due to the melting of these three catalogs which used differing photographic material (never mind just photographic!). Due to selection in the blue, the catalog underestimates the clustering of early type, generally more massive galaxies.

The IRAS Catalogs (1.936 Jy, Strauss et al. 1992; 1.2-Jy, Fisher et al. 1995; QDOT, Rowan Robinson et al. 1990, Lawrence et al. 1998; PSC-Z, Saunders et al. 1998), while only weakly biased with galactic latitude, select galaxies bright at 60μ — i.e. luminous, dusty spirals, with lots of star formation — and completely miss early type galaxies which dominate rich clusters and the small and intermediate scale clustering seen in most redshift survey maps. Good arguments have been made (e.g. Huchra 1996) that IRAS galaxies are not likely to be the best tracers of mass in the Universe. Both the largest optical and IR *all sky* samples contain only of order 10^4 galaxies, and analyses of these samples are limited to small distances ($\leq$ 10,000-12,000 km/s) and low resolution (1000 km/s or poorer) in the determination of the density field.

Similarly with regard to the flow field, the available datasets are small, crude and limited to small distances. The Mark III catalog (Willick et al. 1997) has ~3500 objects with relative distances measured using several different techniques, but only samples the velocity field out to about 4500 km/s, and again with poor spatial resolution. Deeper catalogs such as the SN catalog of Reiss, Press and Kirsher, while using a high quality distance indicator, have less than 50 objects. Perhaps the deepest survey, the Abell cluster survey of Postman and Lauer, has a few hundred objects and surveys the velocity field only very coarsely.

Both the determinations of the galaxy velocity field and the galaxy density field need to be significantly improved before a conclusive answer is reached.

3. The 2 Micron All-Sky Survey

To attack this problem as well as other problems in galactic structure, unusual objects, etc., a proposal was developed here in 1988 (Fazio, Huchra and others) to do a near infrared all sky survey from an explorer class satellite (NIRAS). While the satellite was not selected, the scientific program developed into today's groundbased 2 Micron All-Sky Survey (2MASS). 2MASS will cover the whole sky and produce large, photometrically accurate catalogs of stars ($\sim 10^8$ to K_S=14.5) and galaxies ($\sim 10^6$ to K_S=13.5). The goal of the survey is to provide catalogs with a uniformity over the sky (variation of photometric zeropoint with position) of $\sim$ 1% or better and with completeness and reliability uniform to 1% or better above $|b| = 30$, and as good as we can make it at low latitudes, including external photometric verification.

Depending on the weather, 2MASS will be completed, in both hemispheres, five years from the start of collecting survey data, or summer of 2002. This is the **only** planned catalog with (1) full sky coverage, including both celestial hemispheres and the galactic plane, (2) precision photometry and astrometry, (3) large numbers of objects, and (4) coverage at wavelengths where galactic extinction is minimal. Survey operations started at the end of April 1997, and currently we are running slightly ahead of schedule with 30% of the northern tiles observed in good to photometric weather.

More information on 2MASS can be obtained at the 2MASS web site:
`http://www.ipac.caltech.edu/2mass/`

4. The 2MASS Redshift Survey

We intend to use the 2MASS galaxy catalog to provide both a significantly improved density field map and an improved velocity field map. The 2MASS Redshift Survey (2MRS) will use the 2MASS catalog as a basis for a whole sky redshift survey of ~250,000 selected galaxies to a limiting K_s magnitude of 13.5 (approximately the combined completeness and reliability limit of the survey), and second to combine the accurate J, H, and K_s band photometry for the edge-on spiral galaxies in our survey with 21-cm line width measurements from Arecibo, Parkes and the GBT and with optically measured rotation curves to measure IR Tully-Fisher distances to a large sample of galaxies ($\geq$ 10,000).

There are several major advantages in the use of the 2MASS galaxy catalog: (1) By definition, we are much less sensitive to extinction both in our galaxy and in others. (2) The near IR is the where the peak of the energy distribution is for low mass stars, the objects that contribute most to the mass of any individual galaxy. (3) The survey processing will provide not only magnitudes but also sizes, inclinations, colors and accurate coordinates.

The 2MRS consists of two phases. The **primary** survey we wish to complete at or near the time of completion of the photometric survey (and the primary survey we are proposing for here) includes all galaxies brighter than $K_s \sim 12.2$ ($\sim$ 150,000 galaxies over the whole sky.) The **secondary** survey, which we will only start on if enough observing time is available, is one-in-ten of the remaining galaxies to $K_S = 13.5$, or about another $\sim$ 100,000 galaxies over the whole sky.

2MRS builds on the narrower CfA (Geller and Huchra 1989; Huchra et al. 1990) and SSRS2 (da Costa et al. 1994) redshift surveys, as well as the ORS and IRAS 1.2-Jy surveys already completed. The 2MASS Redshift Survey (2MRS) will sample the galaxy distribution both to a factor of 3-5 times deeper and over a factor of more than two more of the sky ($\sim$90% of 4π steradians) than the early optical surveys. The median redshift of 2MRS will be $\sim$ 0.08 (compared to about 0.025 for the CfA survey), with dense sampling of the regions inside that redshift and reasonable sampling of the density field out to a redshift of 0.1. This median is computed from the real data — K^* from our preliminary surveys of the A262 and Hercules fields is $\sim$ -24.0$\pm$0.15 (H=100 km/s/Mpc) for K selected galaxies. At our limit of 13.5, the sample depth estimated from K^* is a little over 300 Mpc or a little over z=0.1. We expect about half the galaxies in the survey to be inside z$\sim$0.07.

There are **no** *other comparable surveys underway or planned.* The PSC-Z survey is the only all-sky redshift survey yet to be released, and it (1) only contains 13,000 objects and (2) is still 60μ based, so biased towards dusty galaxies with star formation. Sloan (SDSS), 2DF, LCRS, 15R all cover much smaller areas of the sky — combined, perhaps 4 steradians — so are not very useful for large scale flow studies or the determination of the masscon(s) responsible for our CMB motion. Most of the area we are going to survey, even in the north, is not covered by other surveys, even SDSS. For the first phase of the 2MASS survey, most of the high latitude galaxies to our bright K_s limit have already been observed and the additional redshifts needed are small in number. The goal of the 2MASS redshift survey is to increase, by at least one order of magnitude, the size of the whole-sky sample available for the analysis of the local density field. This is not the goal of SDSS or any other currently planned survey,

5. Northern Primary Survey

The primary program at FLWO will be to provide a large fraction of the redshifts for the 2MRS galaxies brighter than K_S=12.2 and above δ=0°. This sample will be about 60,000 galaxies in total, and $\sim$ 35,000-40,000 of them already have redshifts — the CfA Redshift Catalogue lists redshifts for $\sim$43,200 galaxies above δ=0°, and this does not include the redshifts from the 15R survey and PSC-Z surveys which should be released before 2MRS is completed.

When the 2MRS is completed, we believe we will have taken a very large step toward solving the Omega problem and will have a density map that should determine the source of our motion with respect to the CMB. The data set will be very densely sampled with accurate positions, magnitudes and redshifts. Given the expected number density of objects inside z=0.07, we will be able to measure density enhancements accurate to $\sim$10% of the mean density, on scales of $\sim$ 10 Mpc (h=1). This admittedly crude estimate is based on an effective survey volume, V $\sim 10^6$ Mpc^3, and a mean number density for phase 1 of about 0.1 galaxy /Mpc^3 (for the sample of 150,000 redshifts), which implies that a 10 Mpc cube will contain on average 100 galaxies. This is comparable to the density accuracy in the CfA redshift survey inside z=0.03 (and only over 1/4 the sky), comparable to the ORS survey inside z=0.02, and a factor of 10 better than the IRAS surveys. By definition, even without redshifts, the 2MASS galaxy sample can and will be used to calculate the JHK band dipoles, and will provide a strong and simple test of the relation between the local galaxy density field and the CMB dipole — on the scale probed by 2MASS, in any reasonable cosmological scenario, there should be convergence between the gravity/light vector and the CMB frame.

As part of the program, we will also measure IRTF distances to all the edge on spirals in the sample. Even though classification of the galaxies in the normal Hubble sequence is difficult in the IR, any galaxy with an axial ratio of more than 5-to-1 is almost certainly a spiral. For many galaxies in the northern hemisphere inside about 8,000 km/s, HI velocity widths already exist from the surveys of Giovanelli and Haynes. We will obtain velocity widths for the remaining galaxies from Arecibo, the GBT, or using long-slit spectroscopy. We will observe edge on galaxies which we cannot observe/detect at Arecibo or Green Bank to obtain rotation curves. Again, to provide an order of magnitude improvement over existing compilations (Mark III) we need to obtain ~20,000 distances for galaxies inside z~0.05, more or less uniformly distributed over the sky.

These distance measurements, inside z=0.05, assuming a scatter of 0.30 magnitudes at H, which is slightly pessimistic (Freedman 1990; Pierce and Tully 1992), will measure the flow field at z=0.03 accurate to 200 km/s on a scales of 25 Mpc. This will easily allow a determination our motion inside the Lauer and Postman sample volume, and, via analyses such as POTENT (Dekel et al. 1993; Dekel 1994) a derivation of the inferred density field for comparison to the observed density field. This comparison will provide a strong test of the hypothesis that light traces mass on very large scales.

We also plan to produce intermediate results. 2MASS, like CfA, is being done in strips across the sky. This is a result of the fact that the most efficient tiling algorithms for surveys approximate strips on the sky. This provides moderately large contiguous areas and volumes to study. Several programs we intend to complete include (1) a significantly improved near-IR field galaxy luminosity function, (2) measurement of the standard galaxy clustering statistics ($\xi(r), \xi(r_p, \pi), power spectrum, etc.$, but from an IR selected (and thus early type galaxy weighted) sample, (3) tests of the somewhat discrepant flow field results (e.g. Strauss and Willick, 1995) between Virgo and Pisces-Perseus, (4) maps of large scale structure through the galactic plane (e.g. Marzke, Huchra and Geller 1996), and (5) studies of the colors of galaxies as a function or morphological and spectroscopic type. 2MASS provides both excellent photometry and positions, something in fact found in very few optical surveys — of all the large optical redshift surveys, only LCRS has a photometric sample, and even then in only one filter.

References

Aaronson, M., Huchra, J., Mould, J., Tully, R. B., Fisher, J. R., van Woerden, H., Goss, W. M., Chamaraux, P., Mebold, U., Siegman, B., Berriman, B. & Persson, S. E. 1982, ApJS 50, 241
Bothun, G., Aaronson, M., Schommer, R., Mould, J., Huchra, J. & Sullivan, W. III. 1985, ApJS 57, 423
Courteau, S. et al. 1993, ApJ 412, L51
da Costa, L., et al. 1994, ApJ 424, L1
Davis, M. & Peebles, P. J. E. 1983, Ann. Rev. A&Ap. 21, 109
Dekel, A. 1994, Ann. Rev. A&Ap. 32, 371
Dekel, A. et al. 1993, ApJ 412, 1
Dressler, A. & Faber, S. 1990, ApJ 354
Fisher, K., Huchra, J., Strauss, M., Davis, M., Yahil, A. and Schlegel, D. 1995, ApJS 100, 61
Freedman, W. 1990, ApJL 355, L35
Garnavich, P. et al. 1998, ApJL 493, 53.
Gavazzi, G. & Boselli, A. 1996, Astro. Lett. & Comm. 35, 1
Geller, M. & Huchra, J. 1989, Science 246, 897
Geller, M. & Peebles, P. J. E. 1973, ApJ 184, 329
Guzzo, L. 1996, in Mapping, Measuring & Modelling the Universe, eds. P.Coles & V.Martinez, APS.
Huchra, J. et al. 1990, ApJS 72, 433
Huchra, J. 1996, in The Big Bang & Diffuse Background Radiation, IAU Symposium 168, M. Kafatos & Y. Kondo, eds., p 143.
Hudson, M. J. 1994, MNRAS 266, 468
Kraan-Korteweg, R., et al. 1996, Nature 379, 519
Lahav, O. et al. 1994, ApJ 423, L93
Lauer, T. & Postman, M. 1994, ApJ 425, 418
Lawrence, A. et al. 1998, MNRAS, in press. (QDOT)
Lynden-Bell, D. et al. 1988, ApJ 326, 19
Marzke, R., Geller, M., daCosta, L. & Huchra, J. 1995, AJ, 110, 477
Marzke, R., Huchra, J. & Geller, M. 1994, ApJ 428, 43
Marzke, R., Huchra, J. & Geller, M. 1996, AJ 112, 1803
Mathewson, D., & Ford, V. 1994, ApJ 434, L39
Perlmutter, S. et al. 1998, ApJL in press (Astroph 9712212).

Pierce, M. & Tully, R. B. 1992, ApJ 387, 47
Postman, M., & Lauer, T. 1995, ApJ 440, 28
Riess, A. Press, W. & Kirshner, R. 1995, ApJ 445, L91
Rowan-Robinson, M. et al. 1990, MNRAS 247,1
Rubin, V. C., et al. 1976, ApJ 208, 662
Shaya, E., Tully, R. B. & Pierce, M. 1992, ApJ391, 16
Santiago, B. et al. 1995, ApJ 446, 457
Saunders, W. et al. 1998, in preparation.
Scaramella, R. et al. 1994, ApJ 422, 1
Strauss, M. & Willick, J 1995, Physics Reports 261, 271
Strauss, M., Huchra, J., Davis, M., Yahil, A., Fisher, K. & Tonry, J. 1992, ApJS 83, 29
Willick, J., Courteau, S., Faber, S., Burstein, D., Dekel, A., and Strauss, M. 1997, ApJS, 109, 333
Zurek, W. et al. 1994, ApJ 431, 559

THE VIRMOS PROJECT

D. MACCAGNI
IFCTR/CNR, Milano, Italy, dario@ifctr.mi.cnr.it

O. LE FÈVRE
LAS, Marseille, France, lefevre@cosobs.obspm.fr

G. VETTOLANI
IRA/CNR, Bologna, Italy, vettolani@astbo1.bo.cnr.it

D. MANCINI
OAC, Napoli, Italy, mancini@astrna.na.astro.it

J.P. PICAT
Observatoire Midi-Pyrénées, Toulouse, France, picat@srvdec.obs-mip.fr

J.G. CUBY
ESO, Garching, Germany jcuby@eso.org

Y. MELLIER
IAP, Paris, France mellier@iap.fr

AND

A. MAZURE
LAS, Marseille, France, mazure@astrsp-mrs.fr

Large and deep spectroscopic samples of galaxies are essential to study galaxies and large scale structure evolution out to look-back times $\sim$ 10% the current age of the universe. Keeping this scientific and observational goal in mind, we designed and are presently building two wide-field imaging spectrographs to be installed at the Nasmyth foci of the ESO-VLT Unit Telescopes 3 and 4.

VMOS is a multi-object spectrograph with a field of view of over 200 arcmin^2, and a multiplexing gain between 200 and 840 (depending on the spectral resolution), working in the wavelength range 370 nm$\leq \lambda \leq$1000 nm. It will become operational in the year 2000.

NIRMOS is its near-IR counterpart, working in the wavelength range 800 nm$\leq \lambda \leq$1800 nm (with imaging capabilities extended into the K band). It has a field of view of 192 arcmin^2 and a multiplexing gain of 190 at $R = 2500$. It will become operational one year after VMOS.

We show that with these two instruments it will be possible to measure the redshifts of $\sim 10^5$ I$\leq$ 24 galaxies, of which a sizable fraction will be at $z > 1$, in a reasonable amount of telescope time.

J. Andersen (ed.), Highlights of Astronomy, Volume 11A, 492.

THE LARGE SKY AREA MULTI-OBJECT FIBRE SPECTROSCOPY TELESCOPE (LAMOST)

YAOQUAN CHU
Center for Astrophysics
University of Science and Technology of China
Hefei, Anhui 230026, China

Abstract

In China, we will start to build a new spectroscopic survey telescope. It will be a meridian reflecting Schmidt telescope. The optical system is horizontal. The primary mirror is spherical and segmented. It faces towards the north. In its spherical center there is a plane mirror which is segmented too. The different shape of reflecting Schmidt plate is realized by active optics. The clear aperture is 4 m and f-ratio is 5. The angular field of view is 5 square degree, 4000 objects can be observed simultaneously by using optical fibre. The science goal is to make a 20000 square degrees spectroscopic survey, including ten million galaxies and stars, many variable objects and the identification of numerous objects found by radio, x-ray, IR and other survey in these sky areas.

J. Andersen (ed.), Highlights of Astronomy, Volume 11A, 493.

JD12

Electronic Publishing: Now and the Future

Chairperson: **A. G. Hearn**

Editor: **P. B. Boyce**

ELECTRONIC PUBLISHING: NOW AND THE FUTURE

A. G. HEARN
Sterrekundig Instituut, Utrecht, The Netherlands

Electronic publishing of journals has already arrived, and is established. The Astrophysical Journal and Letters, Astronomy and Astrophysics Supplements and New Astronomy have been publishing a parallel electronic edition since the beginning of 1997. Astronomy and Astrophysics will follow in 1998, and the Monthly Notices of the Royal Astronomical Society are being developed. The American Astronomical Society has been the leader in developing electronic publishing in astrophysics. They have led not only by being first, but also by the high standards of their electronic publication.

One of the stories that school children in Great Britain learn at an early age is about King Canute II. Canute was a Danish king who ruled part of what is now Great Britain from 1016 to 1035. He is famous because he sat by the sea as the tide was coming in and commanded the tide to go back. It did not.

Electronic publication is going to bring many changes to the way research is done and published. There are obvious advantages and disadvantages. There are certainly problems which we have not yet realized. The purpose of this Joint Discussion is to stimulate a discussion of what the international astrophysical community would like from electronic publishing. Please do not say that you want electronic publishing to go away, because then you will be behaving like King Canute. But the publishers of journals and the bodies responsible for maintaining the quality of publication are feeling their way into completely new territory. To exploit electronic publishing to the full,and to minimize the disadvantages, discussion in the community is needed.

Electronic publishing will bring great connectivity into journals. While looking at one paper, a click on the mouse will immediately connect with another paper published in another journal. Large tables, theoretical or observational, will be read in the same way from databases. An electronic publication will longer be static. It can be modified to refer to papers published later, it can contain movies or computer programs that can be executed online.

Electronic publishing of a journal is different from a preprint server. A journal, be it on paper or electronic, has a refereeing system. In many cases this process improves the readability and consistency of papers. A preprint server is automatic and uncontrolled. Will preprint servers eventually destroy the journals? A somewhat surprising result of the Joint Discussion was a poll on the question of should peer review be preserved. Everybody present voted yes, with no one voting no and no abstentions.

The most worries were expressed over archiving and the function of the libraries. Who should be responsible for preserving the electronic archive? One can still read Newton's Principia after three hundred years. Will we be able to read a comparable piece of work published electronically after three hundred years? We cannot buy a record player any more to play the old gramophone records that went at 78 rpm. Traditionally the keeper of the written archives has been the libraries. What is going to be the function of a library in the age of electronic publishing, and can they afford the extra costs that it brings?

The other main worry is the networks. The internet networks are already very slow at some times. This can be helped by mirror copies of the journal on different continents. In most countries, the scientific user of the internet does not pay the bill for the network connections. But as networks expand in capacity, the unit cost has been decreasing rapidly. Look at the cost of international telephone calls.

How will all this impact on developing countries? Officially almost all countries have email connections. But in practice in some countries it does not work at all. Will this increase the

J. Andersen (ed.), Highlights of Astronomy, Volume 11A, 497–498.

gulf between scientists working in the developed and developing countries? One cannot restrain developments worldwide because some countries cannot keep up, but certainly in that development one must consider how the scientists in those countries can best be helped.

ELECTRONIC PUBLISHING IN ASTRONOMY

PETER B. BOYCE
American Astronomical Society
2000 Florida Ave., NW, Washington, DC 20009

Abstract. Links are important. The providers of electronic astronomical data are linked together the way no other science is. Electronic journals are part of this distributed, linked resource. Permanence is important. The AAS journals use a new publishing process which ensures the ability to maintain permanent access to our electronic journals. Process is important. With the right process, we have been able to add links automatically to our journal; links to references, links to citations (where the electronic material exists) and internal links for ease of navigation within the journal. Readers like this. Links are important.

1. Introduction

As of late 1997, astronomy still leads other sciences in the quality and number of electronic journals. Beginning in 1992, the American Astronomical Society (AAS) has pioneered the successful electronic publishing efforts in astronomy. Boyce and Dalterio (1996), Boyce (1996), and Boyce, Owens and Biemesderfer(1997) have summarized the AAS electronic publishing program. Calvin (1996) has given the AAS electronic journals a positive review. To date, we have found no other electronic journal which provides the number of features found in the electronic version of the AAS's Astrophysical Journal (ApJ), and no other science provides as much as forty percent of their literature in a good electronic format; HTML presentation with copious links to references and citations, backed by a robust, electronic archival version. By 1998 we expect that 75 percent of the astronomical peer reviewed literature will be on line in such a format as described below.

The Letters Section of the ApJ has been available electronically since mid 1995,[1] and the main journal joined it on line in January 1997. The Astronomical Journal and the Publications of the Astronomical Society of the Pacific will have electronic versions by January 1998. All three journals are now published by the University of Chicago Press. The Elsevier electronic journal, New Astronomy,[2] started in late 1995 and includes all the features of the ApJ plus the inclusion of video clips which will also become part of the AAS journals in January 1998. Astronomy and Astrophysics Supplements have been on line since 1997 and the main Astronomy and Astrophysics journal is scheduled to be on line in 1998.

2. Links

What makes a good electronic journal? After two years of development and a further two years of publishing, it is clear which features are appreciated by readers. By far the most important feature is the presence of links, both internal, for navigation within the document and external, to the references and citations which can be retrieved by the reader instantly. The internal links make it possible to start browsing an article by looking at the figures, and then to jump right to the text where the figure is discussed. Or a reader can start with the reference list, find a reference and jump into the middle of the article where the reference was cited.

We produce two versions of the journal in HTML format, designed for on-screen browsing and effective linking. Special symbols and mathematics are inserted as graphic images using a specially

[1] http://www.journals.uchicago.edu/ApJ/
[2] http://www1.elsevier.nl/journals/newast/

J. Andersen (ed.), Highlights of Astronomy, Volume 11A, 499–503.

designed library of GIF images which are inserted in the proper places. Although the Astrophysical Journal requires the use of about 1,200 special characters, the processing has been automated to the point where this poses no particular problem. One of our HTML versions includes the whole article in one file. In the other version, the article is broken into smaller chunks, allowing a user with lower bandwidth connections to only download the abstract, or to look at only the references or the figures. Of course, each chunk has the same navigational links. About 20 percent of the HTML readers use the chunky version of the articles.

However, readers need to do more than just read articles on the screen. They want to be able to print out articles of particular interest to them. For this, we produce a page image version of the journal using Adobe's PDF format. Many other publishers have begun publishing PDF articles as their only electronic version. They lack the links, machine readable tables and other electronic-only features which our readers find indispensable. An analysis of our access site shows that readers access the HTML version six times more often than the PDF version, a ratio which indicates that readers browse the material in HTML and only use the PDF version for printing out the articles of interest.

The real strength of the electronic journals in astronomy lies not in the articles themselves, but in the links to the outside references as well as to the articles which cite the one being read. Readers say these external links are the single most important feature, but they do not come automatically. First, the reference material has to be available on line. Second, the information has to be identified using a uniform naming scheme so that we know how to access the specific information. Third, to ensure longevity of the links, they have to be made using the uniform names coupled with name resolvers which translate the logical names to the actual "hard coded" URLs. Each information provider must run a name resolver. Fourth, the linking system has to be designed so that the links may be added automatically during the production process. Adding links by hand, indeed doing any action by hand is prohibitively expensive.

Astronomy is fortunate that much of the required material was already available on line. The abstracts of the core journals in astronomy already existed in the NASA-supported Astrophysics Data System(ADS), a searchable database of abstracts.[3] At the same time as the AAS started on the electronic publishing project, the ADS began to accumulate the full text page images of the material in addition to the abstracts. Full text presentations of much of the older astronomical literature are now available on line with accompanying lists of references and citations.

Astronomy was also fortunate to have several data centers in which published tabular data and catalogs were available in electronic form. Ten years ago, the data centers had already devised a system of for assigning logical names to published articles. It was a simple matter to expand the use of these names, known first as "refcodes" and now called "bibcodes," to the electronic journals. The material in the data centers was organized by astronomical object name and provided pointers to the paper literature. With the appearance on the scene of the electronic journals it was simple to tie into the existing system, providing another entry point by which to link to abstracts of articles and to the full text page images at the ADS. By officially linking together these internationally distributed electronic resources, and smoothing a few rough edges at the interfaces, astronomy now has a full fledged, linked data system which operates in a collaborative mode.

To recount, astronomers worldwide now have access to much of the core literature, along with published data tables and data on astronomical objects. All of the material is linked together through a common naming convention and using name resolvers to ensure robust links. We call the whole collection, the Urania resource.[4] Astronomy now has a very powerful electronic information resource. As she looks down from Olympus, the Muse of astronomy has to be proud that we chose her name for this research tool.

3. Permanence

The inherently impermanent nature of electronic information is a major drawback for scholarly material, much of which is expected to retain value decades and even centuries into the future. Garrett and Waters (1996) authored a report on archiving Digital Information which lays out the problem. Many of us still remember eight track tapes and 45 rpm records, yet it is virtually

[3]http://adswww.harvard.edu
[4]http://www.aas.org/Urania

impossible to play music on either of those media today. In the scientific world, much information stored on 7-track tapes was simply abandoned because it was too difficult to read the tapes. Changes in technology make our storage media obsolete. But, software advances and changes in storage format also render material inaccessible after a relatively short lifetime. In either case, with today's pace of change, scholarly material with expected lifetimes of more than five years must be designed with the archival storage requirements in mind.

Electronic archiving implies that users have continued access to the material. For simple page image documents with no links, this is still fairly simple. But, based upon the reception given the electronic ApJ by the community, storing the pages without also storing the links, machine readable data tables, video clips and other electronic-only enhancements will not suffice. Keeping electronic material accessible implies that the material will have to be managed during its lifetime. Translations to new storage media and new formats will have to be done. We are not talking just about saving the material, we are also envisioning updating the material to incorporate new technology, keeping the material available to one new generation of Web browsers after another, or even making the material accessible as we inevitably move beyond browsers.

How do we do this? The answer lies in preparing the original material in an open, standard format which incorporates logical markup. At present publishers have generally agreed to use the Standard Generalized Markup Language (SGML). The ApJ archival database, which the public never sees, is composed of manuscripts coded in SGML. From this database, we can automatically derive both HTML screen versions and the PDF version.

Updating the public versions now becomes an automatic, almost trivial, process. We have updated the HTML versions of the electronic ApJ three times since its inception. Originally we presented tables as graphics. All the electronic issues now support the HTML table specification, even though we started publishing before the specification was developed. Our HTML versions now take advantage of the expanded HTML character set, reducing the number of character graphics we have to use. Perhaps the most noticeable change has been the incorporation of the citation list with every article. All this, plus a host of smaller improvements, have been done by modifying the translation program which produces the HTML version, and reconstituting the whole corpus of publicly available material. The success of this translation leads us to believe that we can do the same with the complete archival database when the need arises, translating it to whatever new standard is appropriate. As of late 1997, XML appears to be emerging as the new standard. By planning for a robust, long-lived archival format from the beginning we have taken a large step toward ensuring the effective access to our scholarly journal far into the indefinite future.

As we have demonstrated, maintenance of an electronic archival journal is far different from simply putting the material into a storage medium and leaving it forever. Maintaining successful and effective access over decades requires active management of a well prepared database. Since the archive management function is tied so closely to the original preparation of the database, it seems logical that the responsibility for maintaining the journal archives should rest with the publisher.

How to pay for the necessary active management is a problem. Commercial publishers are loathe to take on the responsibility because they envision an increasingly heavy burden to maintain material which is used less and less frequently. The AAS is a non-profit publisher with some moral obligation to maintain the availability of the research corpus in astronomy. Still, we have to worry about funding the necessary management activity. We have chosen to establish a journal archive fund into which we will put an amount of current revenue sufficient to enable us to completely translate our archival database every five years. We fully expect that the growth in computing capability will outpace the growth of the size of the journal archive. Our experience leads us to expect that, if we maintain our own archive, we can do so at less than one percent of the cost of the journal per year. Whereas, if any other organization tried to take over archival maintenance it would cost substantially more.

4. Process

Producing an effective electronic journal requires approaching the task with a new mind set. Publishers have been putting words on paper for hundreds of years. But an effective electronic journal has to go beyond the printed page. The reality of that statement has been demonstrated by the community's acceptance of, and demand for, links, machine readable tables, videos and other material which can not be rendered on paper. Yet, most publishers are still concentrating on creating a typeset paper page first and generating an electronic rendition after the fact; a process which includes all the costs of the traditional journal plus additional costs for the electronic version.

Since the UCP and the AAS have been partners from the beginning in the development of our electronic publishing process, we have been able to retool the entire publishing process and to take advantage of some of the economies which an electronic environment can provide. Our process now focuses first on producing the electronic database of manuscripts in a robust, richly-tagged, SGML format. Then we derive both the paper and public electronic versions from the electronic archival database.

We are realizing a number of advantages. We accept electronic manuscripts and translate them into SGML automatically. We use carefully designed, electronic tools to copyedit on the screen. We use electronic transmission of material. We create links automatically and check that they work by using tools maintained at the ADS. In short, we have abandoned the traditional paper-based process while still keeping the rigor of our peer review and the high quality copyediting which has been a hallmark of the ApJ for 100 years. There are some cost advantages to our new process, too. We are still in the development phase, and it is too early to know how much we are going to save in the long run. But, we do know that we are producing both paper and electronic versions (including the contribution to the electronic archive maintenance fund) for less than we were producing paper three years ago.

The new electronic process makes it possible to shorten production time. Our prime example is the electronic ApJ Letters. By typesetting them in house, we are able to produce and post an electronic version without page numbers within two weeks from when the scientific editor accepts the paper and six to eight weeks before the paper issue date. We are posting individual papers as they are completed and only collecting them into volumes when the paper version is ready.

5. Summary

Astronomy is well ahead of most other disciplines in providing accessible, interconnected electronic information. The scholarly journals, while important, make up only a part of this information resource. The power of the full system lies in combining the electronic journals with the bibliographic databases, electronically accessible historical literature and a full range of observational data ranging from the raw observations to the published catalogs. Yet, all of us associated with the electronic delivery of information feel we have barely begun to understand how to make the most effective use of the new technology. The future will surely see impressive new developments which will help accelerate the pace of discovery.

6. Acknowledgments

Evan Owens of the University of Chicago Press and Chris Biemesderfer of ferberts associates are responsible for making the AAS program a success. Our work has been supported in part by a grant from the National Science Foundation and a cooperative agreement with the National Institutes of Standards and Technology.

References

Boyce, P.B. and H. Dalterio (1996), Physics Today, Vol. 49 pp. 42
Boyce, P.B. (1996), Computers in Physics, Vol 10, pp. 216
Boyce, P.B., C. Biemesderfer and E. Owens (1997) Ser. Rev., Vol 23, No. 3, pp.1
Calvin, W. H. (1996), Science Surf, 96, 2 http://WilliamCalvin.com/column02.htm
Garrett and Waters (1996) http://lyra.rlg.org/ArchTF/

Comments

GENOVA: A comment on a question about the cost of the efforts asked of authors in electronic publishing. The possibility of checking the contents is increased - e.g., electronic tables are better checked than by the referee, tagged object names, having been checked, will be better recognized ten years from now than in the present situation.

BOYCE: You are right that better tools for the author will relieve some of the burden of preparing electronic manuscripts. I should add that in the electronic era, where we want to take advantage of the automation, we, as authors, have to be more precise about references, tables, links, etc.

EICHHORN: PASA intends to build links into reference lists. They plan to ask the authors to provide the links. When the ApJL appears in electronic form, will the page numbers be known?

BOYCE: No, it will not. This means that we will need to have different identifiers for articles before pagination. When the papers are posted one by one the page numbers are not known. The publisher will have to supply a document ID. It should be published with the paper, and there will be tools to provide that information upon query as well. There had to be a correspondence map to equate the various article IDs.

SEIDELMANN: What is the future of costs for the libraries? Currently costs are going up, so subscriptions must be cancelled. What is the future of libraries?

BOYCE: Our experience shows that by retooling the production process cost savings can be achieved. Other publishers seem to be charging more for the electronic journals, but they have not retooled their process yet. In astronomy the combination of pressure from the low- priced AAS journals and the preprint archives should help to contain costs.

CHRISTENSEN-DALSGAARD: Comment 1. This is a comment relative to publishers being responsible for the historical record. Libraries now collect past issues and are willing to handle electronic journals more than eight years old. Comment 2. As a society space is in a unique relation to standards. You have mentioned SGML, HTML and XML. There are more, such as those for access and query: Z39.50, for storing information: VICML, and NYTIME.

BOYCE: Concerning comment 1, the publishers are better able to handle the links but there are cost problems for commercial publishers. Concerning comment 2, We have tried Z39.50 access for remote queries to our journal, but that standard is not implemented with access control so we can no longer use it. The situation is changing so fast that many of the older standards are, for the moment, irrelevant. It is a problem, but we have had to make our choices based on things which we know will provide a graceful path for migration into the future. The critical item is to use open and public formats, and protocols.

HJELMING: Your talk deals with the present and the future. What is being done to include the many decades of journals in electronic form?

BOYCE: The ADS is doing that for major journals as you will hear about in a coming talk.

DE GEUS: Regarding the archivability, the commercial publishers take this issue very seriously. We don't consider it to be entirely the role of publishers, but libraries will need (and want) to continue the ability to archive materials. We talk to many libraries to find out what their wishes are in this respect, and try to provide as good a service as possible.

SORENSEN: I am worried about your comment that the publishers need to be the keepers of information in the electronic community. I fear for the continuity and access implications. Also, you have not addressed the cost implication of maintaining a fully transparent system as new technologies become available.

BOYCE: We have done two things to make archiving affordable. First, we prepare from the beginning for the ability to update and move to new technology in a way which does not use a lot of resources, i.e., automatically. Second, we have established a fund for archiving into which we put a small portion of the present income. This will allow us to update the archive completely every five years.

ABT: With regard to finances, traditionally most of the cost of producing a journal is in type setting. By convincing authors to submit electronically we provide savings to finance the on-line editions. All manuscripts submitted electronically are used in that form. However, receipt of electronic manuscripts is often not trouble- free, so we hesitate to ask volunteer science editors and referees to do the extra work. But the techniques currently being explored is to post the manuscripts on the Web with passwords so that the scientific editors and referees can call them up. Peter, how long do you think that CD-ROMS will still be useful?

BOYCE: CD ROMS will lose their attraction very rapidly. I put the lifetime at a few years at most.

ARCHIVING ELECTRONIC PUBLICATIONS – A LIBRARIAN'S POINT OF VIEW

UTA GROTHKOPF
European Southern Observatory

Abstract. Archiving electronic journals is complex. There are many questions; Who archives? What is archived? How many copies should be archived. How can one ensure the integrity of electronic documents? How can access be maintained? Technology will provide some answers. Electronic archives will require periodic refreshing and migration to new technology. Many factors remain unknown at this time.

1. Introduction

Today's scientific literature is not limited to one single medium; various media have coexisted for some time. In recent years, an increasing number of publications have been published electronically. Electronic format alters the traditional way of publishing, retrieving, obtaining and preserving information, and it modifies the structure, composition and features of documents. Electronic publications are available in digital format and therefore require the use of a computer in order to access and use them. They can be delivered to users on physical media (diskette, CD-ROM, tape etc.), over networks (via electronic mail, ftp etc.) or may have to be accessed actively by users (typically on the World Wide Web). They can be made available in various formats with identical information and functionality in parallel. Various types of information (text, images, graphics, sound, software programs etc.) can be combined in one multimedia product. Embedded software may allow users to interactively modify underlying data or launch virtual experiments. Networked electronic publications can be accessed and used from anywhere at any time by several simultaneous users; interconnected documents provide links to other electronic resources. Depending on whether or not they can be modified after publication, electronic publications are dynamic or static documents.

In order to preserve the scientific and cultural knowledge of today, electronic publications have to be archived just as print publications have been. However, archiving electronic media is considerably different from archiving print publications. Up to now, archiving has been relatively easy and straight-forward. Libraries subscribe to journals and buy books, and after publications have arrived in our libraries, they are cataloged, classified, and technically prepared, and finally they are placed on the library shelves where they are available for users. References to print publications follow generally agreed-upon rules for citations that uniquely describe the source, and users can trust that a print document has not been changed after publication. Libraries own the publications they have bought, and unless it is decided to re-sell or remove the items, they are kept in the libraries indefinitely. Users can rely on a decentralized system of libraries where they can consult both current and historical publications whenever needed without additional costs.

In contrast, archiving electronic publications is much more complex. Up to now, no particular institution has been responsible for long-term preservation of electronic publications, hence future access to electronic documents is still uncertain. As of today, users don't know how they will be able to retrieve, obtain, and use electronic publications even a few years from now. Once archiving institutions have been selected, they must solve a large number of political, administrative and technological questions. What exactly has to be archived? Who provides and maintains the necessary tools and technological equipment? Who pays the personnel? Who will have access to the

J. Andersen (ed.), Highlights of Astronomy, Volume 11A, 504–510.

archives and at which costs? Is the technological infrastructure capable of meeting the demand for access to documents via networks?

Archives must guarantee that information resources can be identified, accessed, retrieved and used over time, and in the interest of the general public they must ensure access is available at reasonable costs. Archiving therefore is one of the most important and urgent problems arising from electronic publications and requires thoughtful solutions.

2. Archiving electronic publications

2.1. WHO ARCHIVES?

At present, originators (i.e., authors and publishers) are responsible for preserving the electronic documents they publish. There is a danger that publishers will not be willing or able to archive documents without commercial value. Therefore, in the future institutions whose interest in archiving goes beyond immediate economic consideration should be responsible. Electronic archiving is more expensive than print archiving, and small libraries may not be able to do it. Large institutions and national or state libraries, which can keep up with changing technology, remain the possible candidates. Legal deposit regulations, which currently exist for print media, could be extended so that also electronic publications are required to be deposited in national libraries. Legal deposit regulations in the United States have already been changed accordingly, so that a copy of all electronic publications must be deposited in the Library of Congress.

According to the Task Force on Archiving of Digital Information, institutions "must meet or exceed the standards for archival certification" (Garrett and Waters, 1996) if they want to become certified digital archives. Minimum requirements include the guarantee to provide access to electronic publications, even if the initial vendor has disappeared and the original system configuration is no longer available, as well as the provision of "a critical fail-safe system [...] in order to actively and aggressively rescue electronic publications which are in acute danger". Exact definitions to determine which institutions are eligible for becoming digital archives will vary in different countries. However, guidelines and criteria should be established in order to provide a general framework for international standards.

2.2. WHAT SHALL BE ARCHIVED?

Documents always have been subject to appraisal in order to determine which material had to be preserved. Archiving institutions need to develop standards that define what exactly constitutes an "electronic publication" as opposed to electronic communication, databases and other ephemeral information, and which electronic publication fulfills the quality criteria for archiving. In addition to contents-related quality criteria, further factors may influence the decision to archive electronic media, for instance, functionality, usability, technological standard and adequacy of maintenance.

2.3. HOW MANY COPIES SHALL BE ARCHIVED?

For networked electronic publications, it may be attractive to archive only one copy of a document which can be accessed remotely, but this approach bears many risks with regard to loss or damage of that master copy. Redundant copies of electronic documents should be stored in various places in order to avoid loss or corruption of archived material. Providing multiple copies in a distributed network also accelerates access times as users can log into a server that is geographically close to them or to which they maintain particularly fast links.

2.4. STORING ELECTRONIC PUBLICATIONS

Currently, the majority of electronic publications are static documents, i.e., they are not changed once they are released. Users may be able to change underlying data interactively according to their needs, launch virtual experiments from embedded software or add comments to networked documents, thus creating a dialog with the authors and other readers. However, the original document remains unchanged. Archiving institutions will have to decide whether or not comments and other additions should be preserved together with the publications. A subset of networked electronic publications are dynamic documents which can be modified after publication. Archiving all versions of dynamic publications is difficult. Probably only "snapshots" of the available data can be taken at regular intervals. Alternatively, the entire record of changes will have to be logged in order to be able to reproduce all changes carried out after release of the original document.

Electronic publications can be stored on a variety of physical media. If high user demand can be expected, archiving institutions may decide to provide publications online on networked servers in order to reduce access times. Otherwise, offline storage might be more appropriate (and cheaper). Theoretically, electronic publications provided on networks can be accessed from anywhere at any time. Documents remain interconnected, i.e., they can contain active hypertext links to other digital documents and vice versa. In practise, however, global accessibility requires stable and fast telecommunication networks worldwide which are not yet in place.

2.5. TECHNOLOGY REFRESHING AND MIGRATION

Electronic publications are not inseparably tied to a particular physical format, but can be copied to new storage media with little or no loss. However, experience about the long-time durability of the media is not yet available. An even greater danger than deterioration of the storage media lies in the technological obsolescence of the reading devices which are necessary to use electronic publications. Technology is changing rapidly; hence, in the digital environment "preservation means copying, not physical preservation" (Graham, 1994). This process usually is called technology refreshing; in an even broader concept, the Task Force on Archiving of Digital Information refers to it as "migration" which is defined as "a set of organized tasks to achieve the periodic transfer from one hardware/software configuration to another or from one generation of computer technology to a subsequent one" (Garrett and Waters, 1996). The purpose of migration is to retain the ability for clients to use digital objects in the face of constantly changing technology. Copying should take place at regular intervals in order to avoid unexpected and possibly unbridgeable gaps in the availability of necessary hardware and software. Information should be encoded in a system-independent format, for instance SGML, as the ability to migrate and refresh will be much greater for such a richly tagged format than it is for a page image format like PDF or Postscript.

2.6. ARCHIVAL DATA

Just like printed documents, electronic publications must be retrievable in order to be used. Structured descriptive information about electronic documents, so-called metadata, must be mapped to retrieval systems (e.g., library catalogs). Metadata typically are contained in the documents themselves or accessible through them; they must include formal data (author, title, release date etc.) as well as contents-related data (keywords, time and geographic coverage). In addition, the reference system must provide a variety of further information, for instance about modes of access (WWW, telnet, ftp), tools necessary to use the electronic material (e.g., particular software required at the user's site), and computer file characteristics (size, format). The authenticity (identity) of electronic publications needs to be easily recognizable in order to determine which version of the document is being used. Electronic time stamping of documents (i.e., creating a permanent, indelible mark in the digital file that indicates the version), digital watermarking technologies (i.e., software that allows encoding an identification into a document that can be located by WWW search engines), digital fingerprints (i.e., creating hash codes that reflect every bit in a record) and other authentication methods are hoped to help assure the authenticity of electronic publications. Additional information should be available about embedded links to other digital objects, access rights and restrictions, as well as data to prove the integrity of documents.

2.7. INTEGRITY OF ELECTRONIC DOCUMENTS

A critical distinction between printed and electronic publications is the ease with which their integrity can be determined. The Task Force on Archiving of Digital Information defined five components that constitute the integrity of digital documents: content (intellectual substance contained in information objects), fixity (content fixed in a discrete object as opposed to continuously updated documents), reference (reliable systems for locating and citing), provenance (a record of the document's origin and chain of custody), and context (a document's interaction with elements in the wider digital environment) (Garrett and Waters, 1996). Printed publications do not create doubt about their integrity; we take for granted that they have not been changed since they were published. Electronic publications, in contrast, can be changed in various ways: accidentally (for instance during copying to newer devices); on purpose, with well-meant intention (as is the case with dynamic documents); or on purpose, without well-meant intention, i.e., through fraud (Graham, 1994). Mechanisms to prove the integrity of data and information systems are not only necessary in order to trace fraud, but also to make sure the documents have been moved to a new storage medium without loss.

3. Libraries and Archiving

At first glance, electronic publications seem to be not only an alternative format for scientific documents, but also the solution to many problems libraries face today (Lehmann, 1996): they offer accelerated access to literature for several simultaneous users over networks regardless of their actual location; the growing need for expensive storage space on shelves is diminished; the problem of deterioration of books due to acid paper is solved. On the other hand, digital publications raise a large number of new questions and problems, many of which are not yet solved. The traditional librarian's role involved retrieving, obtaining, making accessible and archiving information needed by our users. As archiving is undergoing vast changes, this also has a major impact on the other classical library services.

3.1. RETRIEVING ARCHIVED ELECTRONIC PUBLICATIONS

If electronic publications are received offline, i.e., on physical media, they will be treated in libraries similar to print media. Items will be cataloged and shelved and will be retrievable through the library catalog. Networked electronic media have to be cataloged, too, and library records must be enhanced so that all available archival information can be included in catalog entries. The Dublin Core Metadata Element set provides a resource description that will improve access to information on networks.[1] If national libraries become digital archives, electronic publications will appear in the respective national bibliographies which will make their retrieval easier than it is today (Deutsche Forschungsgemeinschaft, 1995). However, the ability to retrieve networked electronic publications in distributed systems depends not only on their appearance in library catalogs and bibliographies, but also on a reliable global naming and address system. Current network addresses still depend on the Uniform Resource Locator (URL). But URLs are easily mistyped and can lead to error messages when trying to access networked documents. Name resolvers which are already used by some electronic astronomical journals allow a unique name to be assigned to an online publication that will not be changed even if the document is moved to another server. The name resolver will keep track of the actual location of the requested file and translate the name into the current network address.

3.2. OBTAINING ARCHIVED ELECTRONIC PUBLICATIONS

In the digital environment, journal subscriptions and book purchases as we know them from the print environment are replaced by the concept of granting access to publications just for a given time. The library does not "own" the items, but "leases" access to the documents. The detailed access and usage conditions are described in contracts between publishers and libraries. These contracts, typically called license agreements, regulate, for instance, for how long the library has access to the publication, what users and librarians are allowed to do with it, how many simultaneous users are granted access, and how often the publication can be viewed, printed or otherwise used. License agreements must also determine what happens after the contract terminates. Since it is the nature of leased material that it remains physically with the provider, many librarians fear that they will be left without anything when a leasing contract ends.

As a result, publications requested by our users more and more often will not be available in our libraries, but will have to be obtained from external sources. No standard mode for access to back volumes of publications is yet being applied, and terms and conditions vary from publisher to publisher. It is possible, however, that libraries will have to pay twice (or more often) for the same

[1] http://www.oclc.org:5046/research/dublin_core/

product – once while a contract lasts, and again after cancellation or termination of a title if a user needs a particular article from the timeperiod when the library subscribed.

3.3. ACCESSING ARCHIVED ELECTRONIC PUBLICATIONS

Once a requested electronic publication has been located, accessing and using it can be cumbersome. Electronic publications require that certain hardware and software be available at the user's site. At present, a large variety of physical storage media for electronic publications (CD-ROM, diskette, tape etc.) exist and will continue to exist, since newly evolved technologies do not always replace previous ones, but co-exist with them. A library that owns or leases electronic publications therefore will have to provide the necessary reading devices for each storage medium to be used. Electronic publications in HTML format on the World Wide Web not only require networked terminals at the user's site, but also often make it necessary that a particular version of the browser and other software be installed. Networked documents depend on a stable and reliable infrastructure. At present, bandwidth does not keep up with demand. Even in technologically developed countries, accessing networked documents can be painfully slow, and the situation is even worse in countries with less stable infrastructure. It is not yet clear who will be able to pay for new versions of software and hardware that are necessary at ever shorter intervals, and who will ensure reading devices for offline electronic publications are available in our libraries many years from now.

Further important changes in the electronic environment refer to copyright law and interlibrary loan. Copyright law is meant to grant the owners of creative and scientific works specific rights. For instance, those who want to use copies of a work must have the copyright holder's permission and usually have to pay a certain amount to the owner. In order to balance the rights of copyright-holders and those who want to use the information, there are some exceptions to this rule. Interlibrary loan (ILL) is one such exception. With regard to journal articles, this name is somewhat confusing, because libraries today do not usually lend bound volumes of journals, but rather send photocopies of individual articles. However, the concept of interlibrary loan has remained the same up to now – a library that needs a publication owned by another library may ask for a photocopy, provided the principle of fair use (or fair dealing) is adhered to, i.e., the copy is meant for personal, scientific or educational, but not commercial, use. Misuse of intellectual property seems to be all too easy in the electronic environment; therefore many publishers currently are trying to diminish some of the rights users had with print publications, in particular the right of interlibrary loan. But "there is a misconception that fair dealing does not apply to electronic databases. Fair dealing takes no regard of the medium" (Oppenheim, 1997). In the interest of the scientific and general public, interlibrary loan and the fair use concept must be maintained with regard to electronic publications.

Following a phase of free access for test purposes, we now witness a commercialization of electronic information resources on the Internet. Publishers are concerned about their earnings if they are not able to trace who has accessed, printed, or otherwise used the publications they market. Most probably, their concerns will vanish soon once reliable Electronic Copyright Management Systems (ECMS) become available that will be able to automatically tag copyrighted works and monitor their use (Oppenheim, 1997). There is another reason for publishers not to worry. Those libraries that can afford it will continue to order needed publications from commercial document delivery services (from where publishers can be sure to receive royalty revenues), as in many countries commercial services are quicker than interlibrary loan due to work overload at the lending library. However, those libraries whose budgets are too limited for purchasing articles from commercial services must be able to rely on ILL, or their users will be entirely excluded from access to electronic information resources.

3.4. LIBRARIES IN THE ELECTRONIC ENVIRONMENT

One of the most important tasks of librarians always has been to provide access to information resources. This role will not change in the electronic environment. Users need not worry about the question from where a publication can be obtained; they can rely on a system of libraries to obtain needed information for them. In the print as well as in the electronic environment, librarians assist users in searching and retrieving information. They mediate between authors/readers, publishers, digital archives and other parties involved, provide appropriate tools capable of matching user needs with the available information, and deliver current as well as historic publications to the users.

4. Conclusion

Many of the complex, interrelated factors that influence archiving of electronic publications are not yet known. Decisions made now will have an enormous impact on the future availability of scientific literature. Archiving electronic publications requires thoughtfulness, vision, and longterm commitment from all parties involved. A stable and reliable infrastructure must exist. International collaboration is essential in order to develop standards that ensure preservation of the scientific and cultural record throughout time and regardless of national boundaries. The current increasing commercialization of information access must not turn information into a privilege of the rich. In the interest of the scientific and general public, archiving institutions must guarantee future access to and use of electronic publications at reasonable costs.

Acknowledgements

My sincere thanks go to Ellen Bouton (National Radio Astronomy Observatory, Charlottesville) and Sarah Stevens-Rayburn (Space Telescope Science Institute, Baltimore) who provided very helpful comments and suggestions. I also wish to thank Peter Boyce (American Astronomical Society, Washington, DC) for sharing his expertise and knowledge with me.

References

Deutsche Forschungsgemeinschaft (DFG), Bibliotheksausschuß, 1995: Elektronische Publikationen im Literatur- und Informationsangebot wissenschaftlicher Bibliotheken. *Zeitschrift für Bibliothekswesen und Bibliographie (ZfBB)*, **Vol. 42, no. 5**, 445-463. Electronic version at http://www.dfg.de/foerder/biblio/web.urz.uni-heidelberg.de/epub.html

Garrett, J. and D. Waters (co-chairs), 1996: Preserving digital information. Report of the Task Force on Archiving of Digital Information, commissioned by The Commission on Preservation and Access and The Research Libraries Group. Electronic version as of May 1, 1996 at http://www.rlg.org/ArchTF/

Graham, Peter S., 1994: Intellectual preservation: electronic preservation of the third kind. Washington, DC: Commission on Preservation and Access. Electronic version at http://aultnis.rutgers.edu/texts/cpaintpres.html

Lehmann, K.-D., 1996: Making the transitory permanent: the intellectual heritage in a digitized world of knowledge. *Daedalus*, **Vol. 125, no. 4 (Fall)**, 307-329

Oppenheim, Charles, 1997: Copyright in the electronic age. In: *UNESCO World Information Report*. Unesco, Paris. Electronic version at http://www.unesco.org/cii/wirerpt/chap26.htm

Comments

CORBIN: In the past we relied on lists such as the Durchmusterungs, HD, NGC for identifications. Obviously new developments such as 500,000,000 stars and their need to cross identify objects at various wavelengths make these obsolete. For a while stellar astronomers thought to form an all inclusive identifier, based on something like Guide Star Catalog, but this has been abandoned. From your comments I conclude that the direction in the future will be to rely on cross-identifications in databases rather than an all-inclusive catalog. Do you agree?

GROTHKOPF: In a world of dstributed systems it is essential that information can be retrieved easily. No single library can obtain and store all available information resources, but we must be able to point our users to the source, for instance on the World Wide Web, from where the information actually is available. Identifying and retrieving information sources is becoming one of the core tasks of librarians.

CORBIN: Yes, and so authors give the best possible coordinates for a new object as part of the identifier. Coordinates are the glue that will hold the system together - especially at low galactic latitudes.

ANONYMOUS: The original vision of the Web was of quick individual publication. I see a chance that the astronomical community will fragment into closed interest groups unless something is done to counter this trend. There is a chance that the Web will limit data flow rather than encourage it.

CHRISTENSEN-DALSGAARD: I was slightly puzzled by the last part of your talk regarding the use of IL as it seems contradictory to the general idea of locating the source and then deal directly with the place storing the document.

GROTHKOPF: We are witnessing an ever increasing commercialization of information. More and more often, only prosperous organizations can afford to pay for expensive electronic information resources, and scientists from less fortunate institutions or developing countries might be excluded from access to large parts of information. It is extremely important that essential library fu;nctions, for instance obtaining publication on behalf of our users through interlibrary loan, continue to be permitted in the electronic environment.

DE GEUS: How do you envision the archives by National Libraries to work? Through subscriptions?

GROTHKOPF: National libraries and publishers need to collaborate very closely in order to guarantee the long-term availability of publications. Legal deposit regulations should be changed so that electronic publications have to be deposited in national libraries just like print publications. They could be treated as another mirror site.

ROTS: As to archiving journals: why would not the present model of library services work, provided publishers could be made to agree on a standard journal format? Small libraries would subscribe to a limited nuber of journals, put them on-line for their users and provide name solving routines. Requests for documents not in the library's holdings would be passed on to a larger library with which the originating library has an ILL subscription, as well as outside name solving services.

It would appear to me that such a model would solve the monetary issues, ensure archiving and furnish quick response to requests for the most frequently accessed journals.

GROTHKOPF: Archiving electronic publications requires continuous updating of the necessary technology. I doubt that small specialized libraries will be able to bear the immense costs involved in keeping the technology up-to-date, checking the integrity of documents after they were moved to new media, etc. If archiving, including technology refreshing, was done by libraries individually, it would not only mean a duplication of work, time and money invested, but would also lead to a variety of technically different archives for the same publication, making it even more difficult to determine the identity of publications.

ELECTRONIC PREPRINTS

HARRY E. PAYNE
Space Telescope Science Institute
3700 San Martin Drive, Baltimore, MD 21218, USA

Abstract. Paper preprints are in decline in astronomy, while the use of electronic preprints is on the rise, mainly *via* the LANL `astro-ph` preprint archive. I discuss the decline of the paper preprint, some preprint servers available on the Web and some general electronic preprint issues, the `astro-ph` archive and its response to these issues, and alternatives for the future of electronic preprints.

1. What are Preprints For?—Moving from Paper to Electronic

Preprints are an integral part of the astronomical research culture, serving two important purposes: as rapid communications of results, and as advertisements for the authors and their institutions. The interval of months or years between having a paper accepted for publication and having the printed journal reach subscribers could be dramatically shortened—in the electronic era, however, that interval itself should nearly be eliminated. Authors want to be sure that the senior astronomers who will influence tenure decisions are aware of their work, and the institutions want to be seen as scientifically productive.

Paper preprints have an impact that cannot be matched by asking someone to download and print an electronic preprint. And, there is some suspicion that senior astronomers do not download preprints. On the other hand, reproducing and distributing paper preprints has an identifiable cost, is slower, and greatly reduces the chance of discovery by someone whose name or library is not on the distribution list. So, while paper preprints are an effective way of reaching the intended audience, they exclude other audiences. In addition to impact *vs.* cost, the forces changing the balance between paper and electronic preprints are the move to electronic journals and the success of the Los Alamos National Laboratory (LANL) preprint archive,[1] claimed by its creators to be a desirable replacement for the refereed literature.

Paper preprints are in decline. In Figure 1a, I have plotted the rate at which paper preprints were added to two independent (but largely overlapping) lists: my institute's STEPsheet[2] (Space Telescope Exhibited Preprints) and the National Radio Astronomy Observatory's RAPsheet[3] (Radio Astronomy Preprints) lists, maintained by the librarians at those institutions. I have also shown the total number of articles published in the ApJ, as determined by querying the NASA Astrophysics Data System[4] (ADS). In 1996, ApJ articles made up more than 40% of the total number of articles published in ApJ, A&A, AJ, MNRAS, and PASP (Abt 1997), i.e., the main astronomical journals. The curves track each other extremely well, except since about 1994, when the article count is rising slowly and the preprint rate is declining sharply. A number of paper preprint series have disappeared—some were announced, others just stopped arriving. My own institution has twice discussed eliminating paper preprints, motivated by the potential distribution cost reduction.

The LANL `astro-ph` archive is taking up the slack, and more, caused by the decline of paper preprints, as shown in Figure 1b. The `astro-ph` submission rate has doubled in the last year and a half, after doubling during the year and a half before that. Cross-postings and updates are not

[1]http://xxx.lanl.gov/
[2]http://sesame.stsci.edu/lib/stsci-preprint-db. html
[3]http://www.nrao.edu/aoclib/rapsheet.html
[4]http://adswww.harvard.edu/

J. Andersen (ed.), Highlights of Astronomy, Volume 11A, 511–515.

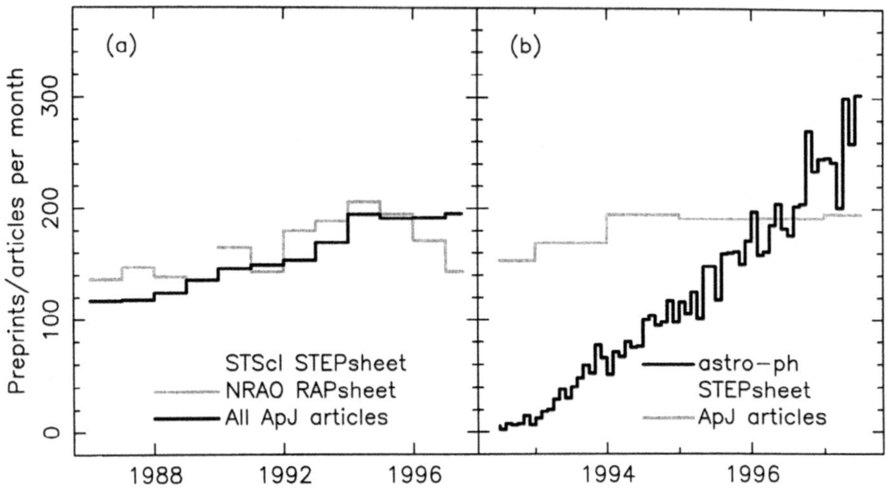

Figure 1. (a) A comparison of the rates at which paper preprints are added to STScI and NRAO preprint lists with publication rates in all sections of the ApJ. (b) The rate of LANL `astro-ph` submissions, compared to the STScI and ApJ data shown in (a).

counted in these numbers. At the moment, the area under this curve represents a small fraction of the literature—negligible compared to the ADS. But if the rate of growth continues, then most new journal articles will be found in the `astro-ph` archive.

2. Electronic Preprint Servers: What is Available Now?

Putting aside the exchange of preprints via e-mail, and focusing on preprint services available on the World Wide Web, we find that the field is dominated by servers or archives in various disciplines of physics and mathematics, with a sprinkling of computer science, biology, philosophy, and, of course, astronomy services. Prominent physics servers include the Los Alamos National Laboratory (LANL) preprint archive, the Stanford SPIRES high-energy physics database,[5] maintained by SLAC librarians and containing over 345,000 entries, and the CERN preprint server.[6] The American Physical Society preprint server[7] provides an experimental platform for simplifying the communications between the author, referees, and the editorial office of the journals. The American Mathematical Society preprint server[8] collects only bibliographic information, including an abstract, at their site, but authors are encouraged to provide URLs linking to full text versions of their preprints. Interesting features of the AMS server include asking authors to classify their contribution into one of nearly 100 standard subject classifications (e.g., Number theory or Differential geometry), and explicitly expiring links after two years.

In astronomy there is the `astro-ph` archive at LANL, a number of institutional, departmental, or group servers, and little in between. Of the 90+ preprint services listed by AstroWeb,[9] most refer to departmental pages that actually have no links to preprints or no recent preprints. Exceptions include servers at the Australia Telescope National Facility[10] and the McDonald Observatory Péridier Library.[11] The Princeton radio pulsar group maintains a directory of pulsar-related preprints[12]—

[5]http://www-spires.slac.stanford.edu/find/hep
[6]http://preprints.cern.ch/
[7]http://publish.aps.org/eprint/docs/
[8]http://e-math.ams.org/preprints/preprints-home.html
[9]http://www.stsci.edu/astroweb/cat-preprint.html
[10]http://wwwatnf.atnf.csiro.au/Preprints/.
[11]http://abstract.as.utexas.edu/abstracts.html
[12]http://pulsar.princeton.edu/rpr.shtml

authors are asked to submit links to their own preprints. The Astrophysics Data System does not do preprints, other than to index the `astro-ph` archive.

Among the technically more interesting services is NCSTRL[13] (pronounced "ancestral"), the Networked Computer Science Technical Reports Library, which consists of "a distributed network of interoperating servers." Collections of computer science reports maintained independently by the participating institutions can be searched as if they were a single collection. Bibliographic records describing the items in each collection are exchanged between participating sites, so that each may host its own search engine. "NCSTRL-lite" sites consist of a simple ftp repository, with bibliographic records held at a central indexing site. This service is making use of important developments in the digital library field, and provides an interesting approach to distributed document collections of any type.

3. Electronic Preprint Issues

The purpose of the scientific literature is communication, and works that are published in electronic form but cannot be found do not advance that communication. Discoverability is everything. But as a paradigm for the future of publishing, having authors put up contributions on their personal, departmental, or institutional Web pages and waiting for readers to come to them is not working, because of the lack of discoverability. Using a global search engine, like AltaVista,[14] turns up so much junk that nothing discovered this way has a scholarly aura, and no search engine targeted at distributed scholarly works exists in astronomy (or any other discipline). A distributed system with participation by institutions rather than individuals, like NCSTRL,[15] would probably result in a better maintained, more homogeneous set of preprints, simply because of the extra attention they would get from their local librarian or site maintainer.

Preprints often change in wholesale ways, not amenable to corrigenda the way published papers are, and we expect them to go through several versions up to the published paper, which we then expect to be fixed. But it can be almost impossible to tell the lineage of an electronic document. And there is no system connecting preprints and papers. The librarians at my institution spend an hour or two with each issue of the major astronomical journals, trying to match the table of contents with their paper preprint list. The matching must be able to deal with re-ordered or different author lists, changed titles, and papers appearing in journals other than the one to which the preprint claims to have been submitted.

Authors relinquish their copyright when they publish, and they are entitled to keep their privately published electronic preprint only if the publisher does not object. In astronomy, the copyright holders have not yet decided how strenuously to object, but expiring electronic preprints does not appear to be common.

4. LANL's `astro-ph` Archive

Astronomy has been fortunate in that timing and our relatively small size allowed winners to emerge early in the transition to electronic services. The Electronic ApJ[16] is a model for other professional societies in how an electronic journal should work. The Astrophysics Data System[17] is another treasure. And the LANL `astro-ph` preprint archive is the dominant source of pre-publication and unrefereed literature. The LANL archives are committed to the rapid exchange of information that is scholarly, but not refereed. They are very serious about being a preprint archive—preprints never expire. They are opposed to having the intellectual content retreat into the holdings of the journals, from which they can be retrieved only for a fee. Authors are encouraged to submit the final, published version of the paper, with an entry in the bibliographic record that gives the journal reference, and leave it there forever. That is the "subversive" nature of the archive. They challenge the existence of journals and the refereeing process.

The `astro-ph` archive is having a major impact on astronomy. Over 3000 astronomers subscribe to the `astro-ph` notification service, which brings a daily e-mail message containing the titles and

[13] http://www.ncstrl.org/
[14] http://altavista.digital.com/
[15] http://www.ncstrl.org/
[16] http://www.journals.uchicago.edu/ApJ/journal/
[17] http://adswww.harvard.edu/

abstracts of new submissions. The ApJ Letters started encouraging authors to submit accepted papers to `astro-ph` in 1996, and `astro-ph` contribution numbers have become a *de facto* bibliographic reference. In at least one case, the availability of preprints from the LANL archive has caused a library to abandon the collection and cataloging of preprints. And `astro-ph` is having an effect on the research process: a provocative `astro-ph` contribution can create a flurry of activity where problems are hashed out on an impressive timescale of days to weeks.

The dominance of `astro-ph` solves the discoverability problem, they have effectively solved the version tracking problem, with the major exception of tracking all the way to the published paper, and there is a philosophical difference over the copyright problem. The failure of the `astro-ph` archive to link up with the astronomical literature is its greatest shortcoming. You can still retrieve the text of the first `astro-ph` submission,[18] from April 1992, even though a search through the ADS[19] shows that it was published in August of 1992. The `astro-ph` preprint has no links under the "Refers to"[20] and "Cited by"[21] buttons, whereas the ADS[22] entry for the corresponding paper lists 24 references[23] and 62 citations.[24]

It may be because astronomy is so far out in front in the move to electronic publishing that the LANL shortcomings are so apparent. And that is a shame, since we have chosen to support a service which directly threatens our world-class electronic journals.

5. The Future

There are a couple of ways that the relationship between electronic preprints and the refereed literature could go. One is for electronic preprints to actually replace the journals. It would be straightforward to enhance the LANL server, say, to provide a mechanism for readers to record their comments as a running commentary, as a substitute for anonymous refereeing. The problems of the missing links could be fixed by checking references and citation against the ADS, contributions could be broken down by category, and our professional organizations could step in with some guarantee of permanence.

Another path for electronic preprints is a closer integration with the journals. That is the goal of our NASA AISRP[25] grant-funded electronic preprints project,[26] headed by Bob Hanisch (STScI). We would like to track a preprint through the publication process, so that it could automatically be identified with the published paper. If an author were able to tell the journal that this manuscript corresponded to a particular preprint (only the author can know, for sure), then, when the paper appears, the journal could notify some tracking authority to update the bibliographic record for that preprint with a link into the electronic journal. The system is a name resolver for manuscripts, returning the URL of the preprint until the paper is published, and the URL of the electronic journal article afterwards. At that point the preprint becomes indiscoverable, and can be removed. To make a complete system, we would collect and index metadata for each preprint, as in the LANL and NCSTRL systems. We would not archive most preprints, but instead rely on a distributed network of departmental or institutional preprint servers, but with centralized metadata and search engine.

If the journals were willing to accept an electronic preprint as a submission, or to produce an electronic preprint from a submission, then the journals could match preprints and papers without input from authors, and simplify distributing copies to the referees. Following the NCSTRL model, it should also be possible to decentralize the metadata, search engine, and name resolver, allowing each site a rapid local search limited to its own collection, to enhance the local preprint server. Or perhaps these elements should be merged into the ADS. Proper attention to the choice of metadata elements should eventually be rewarded by greater interoperability with other on-line document resources, in other disciplines.

[18] http://xxx.lanl.gov/abs/astro-ph/9204001
[19] http://adswww.harvard.edu/
[20] http://xxx.lanl.gov/refs/astro-ph/9204001
[21] http://xxx.lanl.gov/cits/astro-ph/9204001
[22] http://adsabs.harvard.edu/cgi-bin/bib_query?1992ApJ...395L..83N
[23] http://adsabs.harvard.edu/cgi-bin/nph-ref_query?bibcode=1992ApJ...395L..83N&refs=REFERENCES
[24] http://adsabs.harvard.edu/cgi-bin/nph-ref_query?bibcode=1992ApJ...395L..83N&refs=CITATIONS
[25] http://www.hq.nasa.gov/office/oss/aisr/aisrp.html
[26] http://doright.stsci.edu/Epreps/

Acknowledgements

I would like to thank Helmut Abt, Gene Avrett, Ellen Bouton, Andy Fruchter, Evan Owens, and Archie Warnock, for helpful conversations and contributions, and especially my STScI colleagues Sarah Stevens-Rayburn and Edwin Huizinga for both discussions and a critical reading of the manuscript.

References

Abt, H. A. 1997, "Is the Astronomical Literature Still Expanding Exponentially?" submitted to PASP

Comments

EICHHORN: Preprints probably cannot replace the journals unless they provide the same connectivity as the journals, data centers and abstract services provide. Since they do not seem to intend to do that, they will not have the influence on astronomy that the E-journals have.

HJELMING: The LANL pre-print service and "personal publication" by email both address the fact that most papers are not worth permanent publication. The institutional publishers, librarians, etc. do not address this. They (LANL, Web publication) fulfill a perceived need that current institutions of publishing do not yet address.

FALCLEE: You mentioned two reasons for sending out preprints. 1. Advertizing. The first reason may disappear soon, since journals publish more quickly every year. The second reason will, however become more important since the journals grow bigger each year. Preprints allow you to target a very specific audience. We have had a very successful experience with launching a subject specific newsletter on the Galactic Center (GCNEWS). This newsletter distributes preprints (abstracts and lists) to a small but well selected list of readers - the majority of active researchers working in that field. This allows quick and efficient communication between scientists, helping, not threatening the role of major journals.

PAYNE: This service would be enhanced by the preprint to paper tracking I discussed.

BOYCE: A comment on unsolicited public comments serving as peer review. It hasn't worked. You get comments from people who don't contribute much to the field and the knowledgeable people don't have the time to read everything, much less comment.

As for Galactic Center and Be Star newsletters - this is one way a system "overlaid" on the current journal/preprint system can help sift out the relevant articles from the overwhelming mass of information available electronically.

CONNECTIVITY IN THE ASTRONOMY DIGITAL LIBRARY THROUGH THE ADS

GUENTHER EICHHORN, MICHAEL J. KURTZ, ALBERTO ACCOMAZZI
AND
CAROLYN S. GRANT, STEPHEN S. MURRAY
Smithsonian Astrophysical Observatory Cambridge, MA 02138

Abstract. The Astrophysics Data System (ADS) is an integral part of the Astronomy Digital Library and the collaboration Urania. It provides access to 1 million references and connects these references with many other information centers and their data, such as on-line journals, object databases, and scanned journal articles. This article describes some of the features and links between the ADS and other on-line services.

1. History

Astronomy was always at the forefront of networking technology. With the Astrophysics Data System (ADS) we had a head start in distributed data system. The Classic ADS was the first distributed data system in Astronomy, before the World Wide Web (WWW) became available (Eichhorn, 1994). However, it was based on proprietary technology and therefore never became widely used. When the WWW became popular, the ADS was the first astronomy system to make its services available over the Web (Eichhorn et al., 1995). Other data centers soon followed.

Individual services generally allowed their holdings to be retrieved through specialized query forms, depending on the nature of the data. These ranged from text based queries to the ADS abstract service to queries by astronomical object name in the SIMBAD database to positional queries in catalog databases.

It soon became clear that in order to make these new services really useful, they needed to be connected. One way to connect the different systems is through the published literature. By correlating data sets with the articles in which they were published, it became easy to search for information through natural language queries. This was made possible through the ADS abstract service.

A second way to connect data systems is via object names. This allows the user to find information about specific objects. This type of search was made possible by the SIMBAD database at the Centre de Données astronomiques de Strasbourg in Strasbourg, France (Genova et al., 1997).

By directly connecting to the SMBAD database (and later also to the NASA Extragalactic Database (NED)) and then combining both types of queries in one system, the ADS provided the capability to find published literature for a specified topic or for a specified astronomical object through one system. The published literature in turn was linked to the on-line information at various information centers and to the object databases.

2. Connectivity

The distributed nature of the on-line digital library allows the different information providers to concentrate on their specific tasks, while utilizing the expertise of other systems by directly linking to their on-line information. The ADS provides an easy way for other systems to link to the published literature and, through the ADS links, to other information. The ADS allows other systems to directly link to any of its references or articles. By following a link to an ADS reference, the user then immediately has access to other information linked to this reference. Figure 1 shows

J. Andersen (ed.), Highlights of Astronomy, Volume 11A, 516–519.

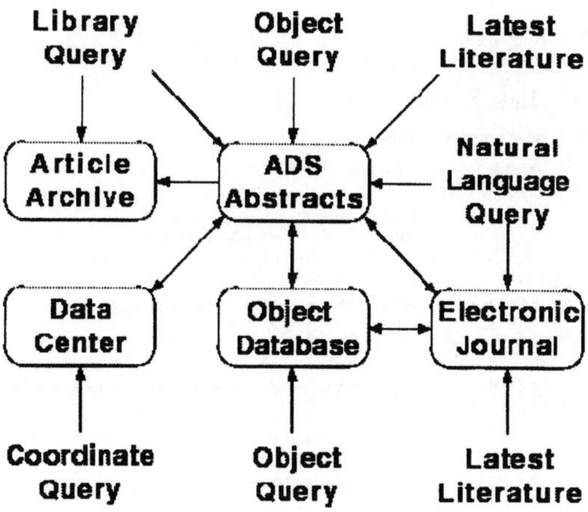

Figure 1. Connectivity in the Astronomy Digital Library.

schematically how interconnected the parts of the astronomy library are and the types of queries that can come in to different parts of the system.

One extremely important property of any on-line system is that it is open to connections from the outside. There are some on-line journals for instance where the user has to log in and then follow internal links to the final article. There is no provision for outside systems to link directly to individual articles in that system. Such closed systems cannot be part of a global digital library since other information providers cannot point to information in these systems directly. It is essential that any information provider who wants to be a part of the global digital library be open to connections from the outside.

3. ADS Query Example

The ADS query page[1] allows users to search the ADS database by author, object name, title, and/or abstract text words. More detailed descriptions of the ADS abstract and article service are available from (Eichhorn et al., 1995), (Accomazzi et al., 1996), (Eichhorn et al., 1996).

TABLE 1. List of data links with anchor letter.

Anchor	Data Item	Data Provider
A	NASA/STI abstract	ADS
C	Citations for current article	ADS
D	Data available	Data centers
E	Electronic on-line article	On-line journals
F	Full text of article (Postscript or PDF)	ADS, on-line journals
I	Author information	ADS
L	Library catalog entry	Library of Congress
M	Mail ordering of articles	Publishers, Service Organizations
N	NED objects	NED
O	Original author abstract	ADS
P	PDS on-line data	Planetary Data System
R	References in article	ADS, on-line journals
S	SIMBAD objects	CDS
T	Table of contents for current volume	ADS

[1] http://adsabs.harvard.edu/abstract_service.html

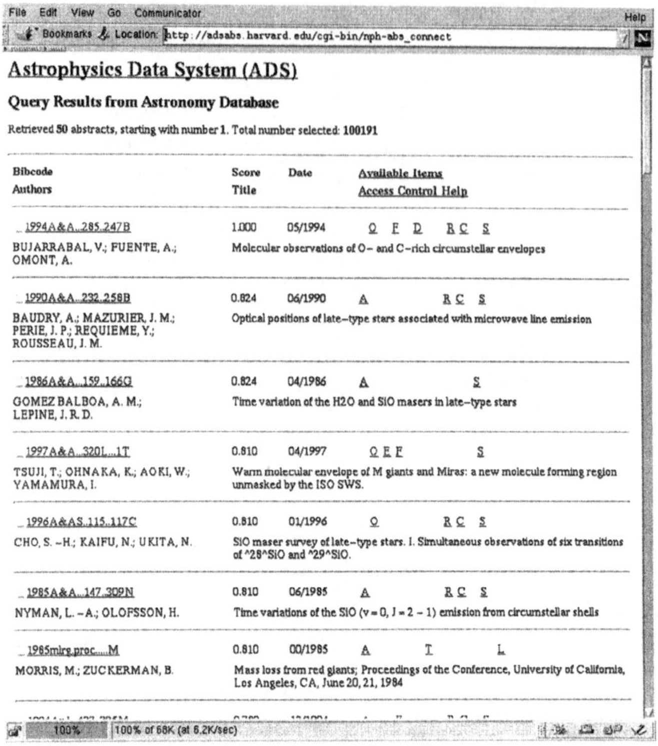

Figure 2. ADS results page.

Figure 2 shows the result of such a query. For each reference, the title, author list, publication date, bibliographic code (see below), and a list of links are returned. The links, anchored on different letters, provide access to additional information for each reference. Table 1 shows the list of possible links and their anchor letters, together with the organizations that have this information on-line. These links are an important part of the ADS and provide the user with the capability to easily move to different sets of information. Most of these links use the bibliographic codes to identify the reference for which the link is specified.

4. Bibcodes

The bibliographic codes (or bibcodes) are by now used by most information systems in astronomy to identify published articles. A first definition of these codes is available from (Schmitz et al., 1995). These bibcodes can be constructed from a regular journal reference. They are 19 digits long and have the following format:

YYYYJJJJJVVVVMPPPPA

YYYY	Publication year
JJJJJ	Abbreviation for the journal (e.g. ApJ, AJ, MNRAS, Sci, PASP, etc.)
VVVV	Volume number
M	To indicate "special" issues (such as "L" for Letters, "P" for pink pages)
PPPP	Page number
A	First letter of the first author's surname.

The fields are padded with periods (.) so that the code is always 19 characters long. The journal abbreviation is left-justified and padded right. The volume and page numbers are right-justified and padded left.

New journal abbreviations have to be unique, and follow existing naming conventions. A list of journal abbreviations already in use is available at our website.[2]

A summary of naming conventions is being prepared and will be on-line at the ADS website soon.

The references at the end of this article include the bibliographic codes and hyperlinks to the ADS abstracts in the on-line version. A description of the encoding is available.[3]

5. Conclusion

Linking different systems together is becoming more and more common. There are many new ways available now to connect different information sources. This new way of presenting information is already having a profound influence on how information is made available and how scientists use this information, and it is certain to continue to do so.

6. Acknowledgment

This work is funded by NASA grant NCCW-0024.

References

1996adass...5..558A: Accomazzi, A., Grant, C.S., Eichhorn, G., Kurtz, M.J., Murray, S.S., 1996, Astr. Data Analysis. Softw. and Syst. V, ASP Conf. Ser., **Vol. 101**, 558

1994ExA.....5..205E: Eichhorn, G, 1994, Exp. Astr, **Vol. 5**, 205

1995adass...4...28E: Eichhorn, G., Murray, S.S., Kurtz, M.J., Accomazzi, A., Grant, C.S., 1995, Astr. Data Analysis. Softw. and Syst. IV, ASP Conf. Ser., **Vol. 77**, R.A. Shaw, H.E. Payne, and J.J.E. Hayes, eds., 28

1995VA.....39..217E: Eichhorn, G., Accomazzi, A., Grant, C.S., Kurtz, M.J., Murray, S.S., 1995, Vistas in Astr., **Vol. 39**, 217

1996adass...5..569E: Eichhorn, G., Accomazzi, A., Grant, C.S., Kurtz, M.J., Murray, S.S., 1996, Astr. Data Analysis. Softw. and Syst. V, ASP Conf. Ser., **Vol. 101**, 569

1997BaltA...6..192G: Genova, F., Bonnarel, F., Bartlett, J. G., Dubois, P., Egret, D., Jasniewicz, G., Lesteven, S., Ochsenbein, F., Wenger, M. 1997, Baltic Astr., **Vol. 6**, 192

1995ioda.book..259S: Schmitz, M., Helou, G., Dubois, P., LaGue, C., Madore, B., Corwin Jr, H.G., Lesteven, S., 1995, Inf. and On-line Data in Astr., D. Egret and M.A. Albrecht, Eds., Kluwer Acad. Publ.

[2] http://adsabs.harvard.edu/abs_doc/journal_abbr.html

[3] http://adsdoc.harvard.edu/abs_doc/bibcodes_help.html

ELECTRONIC PUBLISHING: THE NEW ROLES OF CDS

F. GENOVA, J.G. BARTLETT, F. BONNAREL, P. DUBOIS, D. EGRET, P. FERNIQUE, G. JASNIEWICZ, S. LESTEVEN, R. MONIER, F. OCHSENBEIN AND M. WENGER
Centre de Données astronomiques de Strasbourg (CDS)
Observatoire de Strasbourg, URA CNRS 1280, Université Louis Pasteur, 11 rue de l'Université, 67000 Strasbourg, FRANCE

Abstract. The Centre de Données astronomiques de Strasbourg (CDS) has dealt with bibliographic information for many years. References of publications, published observational data related to objects, data tables, nomenclature, have been homogenized and organized into information retrieval systems: SIMBAD, the reference database for the identification and bibliography of astronomical objects; the catalogue service and the VizieR catalogue Browser, for data tables; the Dictionary of Nomenclature of Astronomical objects, which is now maintained by the CDS. Evolution in recent years has brought the Data Centers closer to the publishing process. General standards for astronomy, for the description of references and of data tables, have been proposed and implemented. Data tables from papers are now directly published in electronic form, and distributed on-line by the Data Centers. The emergence of fully electronic publication paves the way to innovative new services, linking the journals to other sources of informations (data bases, tables, then images, data archives), and making use of new methods for information retrieval. This also has an effect on the publishing process, with the possibility to implement new checks and links from text to other kinds of information (from objects names, positions, etc.). The CDS will bring some of the key features in the evolution towards a fully linked astronomy information system, in close collaboration with the journal editors, the ADS, the other Data Centers, and the data providers.

1. Introduction

The *Centre de Données astronomiques de Strasbourg* was founded in 1972, by the French *Institut National des Sciences de l'Univers* (INSU, which was then the INAG - *Institut National d'Astronomie et de Géophysique*). At that time, it was called the *Centre de Données Stellaires*, and its objectives were:

- to collect 'useful' data about astronomical objects, in electronic form;
- to improve these data by critical evaluation and combination;
- to distribute the results to the international community; and
- to conduct research programs using these data.

For this purpose, the CDS was from the beginning imbedded in a research environment at the *Observatoire de Strasbourg*. During the first years, it dealt solely with stellar data (hence Stellar Data Center) to study the galactic structure.

The CDS aims were somehow extended in 1983, when it began to deal with all astronomical objects (except the solar system ones), and was then renamed *Centre de Données astronomiques de Strasbourg*. The CDS objectives can now be summarized by: collect, homogenize, distribute, preserve astronomical information, for the usage of the astronomical community all over the world. It is worth noting that the primary important functions defined at the very beginning, have been kept in the somehow more general present definition of the CDS activities: to deal with information in electronic form, to develop and implement expertise about this information, and to serve the international scientific community in astronomy.

Another important point is that the CDS does not deal directly with observational data, but rather builds 'metadatabases', containing high-level selected, organized, homogenized information.

J. Andersen (ed.), Highlights of Astronomy, Volume 11A, 520–525.

In the following, the CDS activities linked to the bibliography, and their evolution with the rapid development of electronic journals, will be described.

2. Bibliographic information services at CDS

A general description of the CDS services can be found e.g. in Egret *et al.* (1995) and Genova *et al.* (1996), or at the CDS website,[1] which is regularly updated and gives access to the on-line WWW services. In the following, one will mainly describe the service contents linked to the bibliography. Bibliographic information are displayed in two of the main CDS services: SIMBAD and the catalogue/VizieR service.

2.1. SIMBAD

SIMBAD is the reference database for the identification and bibliography of astronomical objects (outside the solar system). It contains identifications, 'basic data', bibliography links, and some measurements, for more than 1,500,000 objects (September 1997). This information is extracted from published papers and selected catalogues, through a large set of collaborations. The scanning of bibliography is a very large effort, involving several French institutes: 90 journals are systematically screened, in partnership with the *Institut d'Astrophysique de Paris*, and the Paris (DASGAL) and Bordeaux Observatories. References, object names, 'basic data' such as position, object type, magnitude, and so on are systematically recorded by experienced staff which scan the published papers (Laloë *et al.*, 1993; Laloë, 1995). In parallel, long tables in electronic form are now more and more often entered by semi-automatic procedures.

In practice, when dealing with a published paper, the SIMBAD 'bibliographers' will for instance:

- find the objects cited in the paper;
- interpret the origin of the acronyms mentioned in the paper (an object called "M 13" is not always from the Messier catalogue!);
- identify new lists of objects in the paper, then create the corresponding acronym, and describe it in the *Dictionary of Nomenclature of Astronomical Objects*;
- identify errors on object names in the paper, and collect this information in a note appended to the reference in SIMBAD.

The *Dictionary of Nomenclature of Astronomical objects*, developed for many years by M.-C. Lortet and her collaborators (Lortet *et al.*, 1994), is now fully maintained by the CDS, in collaboration with the Paris Observatory (DASGAL), as a by-product of SIMBAD bibliography scanning. Older lists cited in recent papers are also tracked back during the process. An on-line version, updated every week, is available in the CDS WWW service, together with the *Specifications concerning designations for astronomical radiation sources outside the solar system*, from the Task Group on Astronomical Designations of IAU Commission 5 which provides basic advices in this topic, and the *Advance registry for acronyms* service.

2.2. THE CATALOGUE/VIZIER SERVICE

This service contains large catalogues and published tables, which are particularly important as 'metadata', since they contain information calibrated and homogenized by the authors. This is a co-operative action between the CDS and the other Data Centers, ADC (GSFC), INASAN (Moscow), NAOJ (Tokyo) and Beijing, with also a regional center for India at IUCAA in Pune. The Data Centers share the same data sets and collaborate in the building up of the contents.

The key to this cooperation is a general standard for the description of the tables, proposed by CDS (Ochsenbein 1994)[2] and now shared by the Data Centers and several major journal editors (*Astronomy and Astrophysics*, the *American Astronomical Society* on its CD-ROM with the tables from the *Astrophysical Journal*, the *Astronomical Journal*, and the *Publications of the Astronomical Society of the Pacific*, *Soviet Astronomy*). The standard description is both fully readable by the user, and an important tool for electronic checks, transformation and exchange of data. It describes all the files which constitute the catalogue, and the byte-per-byte contents of each file. Tools have

[1] http://cdsweb.u-strasbg.fr/CDS.html
[2] http://vizier.u-strasbg.fr/doc/catstd.htx

been developed to manage the units, check that the contents of the tables is coherent with their description (e.g., that declination values are between −90 deg and +90 deg), and transform the tables to FITS format.

In practice, the staff activities for the catalogue service are the following:

- create or improve the description of a table;
- check the consistency of data in a table; and
- discuss with the author when discrepancies are found.

In September 1997, the catalogue service contains more than 2,350 catalogues, of which more than 1,800 are available on–line as full ASCII or FITS file. Several hundreds tables published in journals are added each year, mostly by agreement with the journal editors as explained below. Nearly 1,450 tables are also available through the VizieR browser, developed in 1995 in collaboration with ESA–ESRIN as a follow–up of the ESIS Catalogue Browser, and now fully maintained by the CDS (Ochsenbein, 1996). The standard description is used in the pipeline to include the ASCII tables in the Sybase VizieR database, which makes it very easy to include new catalogues and tables in the system (just build the standard description). Tables in VizieR can be queried by constraining any of their field, or for data around one position, or around one object (the position being then given by SIMBAD).

Plans are now to include logs of ground–based and space–borne observatories in the system, and to build active links from the VizieR access tool to observatory on–line archives.

3. Recent evolutions and perspectives

As shown in the previous section, the CDS has been dealing with information from the bibliography from the very beginning, and SIMBAD (like the NED database for extragalactic objects) and the catalogue service are high value–added services to retrieve checked, homogenized, standardized bibliographic information.

The recent, rapid evolution towards electronic publication has brought the Data Centers much closer to the publishing process. From 1993 on, the CDS has been building on–line tables (which are in general not printed any more) for *Astronomy and Astrophysics* and the *Supplement Series* (Ochsenbein and Lequeux, 1995). On–line abstracts are also prepared for these journals, to which active links to references, objects (from SIMBAD), and soon images (from ALADIN – Bonnarel *et al.*, 1997), are progressively added. In parallel, and by agreement with the *American Astronomical Society*, the CDS is also installing the tables from the AAS CD-ROMs on–line. In addition, the Russian Center is now building the tables from *Soviet Astronomy*, by agreement with the journal editor. As explained above, due to the agreement of all the partners on a common standard, all these tables can be put together in a dataset common to all the Data centers, and can easily be exchanged and transformed. The tables are accessible to users from all over the world thanks to the collaboration between the Data Centers, and information retrieval services such as the VizieR Catalogue Browser also constitute a new, powerful way to access published information.

A very appealing opportunity offered by the WWW, is the possibility to build links between distributed, heterogeneous on–line services. The user can already navigate from one object in SIMBAD, to the references citing the object, then to the ADS services for each reference and to the full paper from the editors when available; or from one SIMBAD object, to the *Dictionnary of Nomenclature* information about one of the object names, to the list of origin of this name in the catalogue service, to the reference containing the list, etc... Reciprocally, from one reference found for instance in the ADS, one can go to the list of SIMBAD objects from this reference, and then to SIMBAD for each of these objects. Or the ADS can be queried by an object name, and then the list of references is retrieved from SIMBAD and NED (through a client/server procedure) and processed by the ADS. The links between bibliographic information and databases are build in particular by regular exchange of information between CDS and ADS (Eichhorn 1997).

Another important tool for navigation among on–line bibliographic information is the bibcode/refcode, a 'de facto' standard first defined by the CDS and NED, then extensively used by the ADS, and shared by the journal editors. One reference is uniquely described by 19 characters, easily readable (at least for the references from journals) (Schmitz *et al.*, 1995).

In the frame of the rapid development of electronic bibliographic services, more and more linked to the Data Center information, the CDS now hosts the European mirror of the ADS NASA bibliographic database (the CfA is implementing a mirror copy of SIMBAD for US usage), and the European mirror of the electronic *Astrophysical Journal*. The possibility of hosting the long term archive of the electronic *Astronomy and Astrophysics* is being discussed with the journal editors.

This evolving context offers good opportunities to develop innovative services. Recognition of object names in the published texts, the possibility of linking the text of journals to databases, images (e.g. through the ALADIN service), etc., is certainly very appealing. The first attempts were done in the electronic version of *New Astronomy* (by visual detection and individual check), and in the CDS bibliographic service (for objects in the keyword list, by semi–automatic detection and check). CDS is developing automatic procedures to recognize object names, and one can also recommend the editors to ask the authors to tag the object names themselves when submitting a paper in electronic form. The experience with the scanning of the objects in published papers for SIMBAD, is that the different methods will be complementary, and that a final check by an expert will probably remain necessary for completeness and correctness, mainly because of the complexity of astronomical nomenclature, and also unfortunately because of the carelessness of too many authors in these matters.

In parallel, a method of organizing sets of documents by the similarity of their contents, by using neural network techniques, is being developed (Kohonen maps, Lesteven *et al.*, 1996). The neural network is built from the keywords attached to the papers. The common list of keywords shared by *Astronomy and Astrophysics*, the *Astrophysical Journal*, and the *Monthly Notices of the Royal Astronomical Society*, together with the fact that *Astronomy and Astrophysics* and the *Astrophysical Journal* provide CDS with electronic reference information including the keywords, has proven to be a very important tool to build a service common to several journals.

4. A very practical wish–list

An IAU Joint Discussion is certainly a good opportunity to present a wish–list, so here is a list of very practical points (some of them discussed in the various Commission 5 meetings during this General Assembly), which could help the future development of fully linked astronomy, from observational data to published papers. Many of these points refer to 'de facto' exchange standards, which already play a tremendous role in the existing links and collaborations.

- to have more journals collaborate in the *Astrophysical Journal*, *Astronomy and Astrophysics* and *Monthly Notices of the Royal Astronomical Society* common list of keywords;
- to have more journals collaborate to the on–line published table service;
- to get tags of object names in the text of published papers, from the authors;
- to add the origin of observations (number of observation run, etc.) as an additional 'keyword' to the published papers, to open the possibility of building a direct link between one observation and the resulting published research results. This should also imply the definition of a 'standard description' to identify observations (observation # N with one instrument in one ground–based or space–borne observatory);
- to get the recognition of the bibcode as the 'de facto' standard for describing references in astronomy, and to improve the procedures to maintain this standard.

IAU Commission 5, as shown in the Joint Discussion, and in the meetings during this General Assembly, and the Urania initiative, are certainly good forums in these matters. Commission 5 proposes for instance that the bibcode standard be under the responsibility of its Working Group on Information Handling.

5. Conclusions

For the future, one can certainly expect more links between distributed services. The users can already use the first new paths and new tools to access published information, as shown above. An important objective is now to include access to the observatory archives (i.e. to observational results) in the landscape of the fully linked astronomy.

The management of links (in particular keeping track of their evolution) is certainly a key question here. The CDS has developed a solution to manage its own set of heterogeneous, distributed services and mirror sites, the GLU (Générateur de Liens Uniformes – Wenger *et al.*, 1996), which avoids storing hard-coded URLs, permitting the resolution of URL names, and which might be used in a wider context.

The coming of electronic publication already produces an important evolution of the journal editor procedures, as explained elsewhere in these Proceedings. The refereeing process is also be expected to evolve since new checks are and will be possible (e.g. the new checks operational already for the table contents). Clearly also, the internal Data Centers procedures will have to be adapted, towards more automated processes, but preserving steadily the expert intervention, and the building of high value-added 'meta-information'. The Data Centers will certainly continue to jump at the opportunity to give access to the wealth of new information available, and to build up innovative information retrieval services. They have some of the keys for the future development of new links, with the 'metadata' services as bridges between different types of information, and information retrieval tools that can be linked to distant services. They will also continue to keep on other essential functions of the Data Center 'mission', push and participate to the definition of standards, and to the development of international cooperation.

References

Bonnarel, F., Ziaeepour, H., Bartlett, J.G., Bienaymé, O., Crézé, M., Egret, D., Florsch, J., Genova, F., Ochsenbein, F., Raclot, V., Louys, M. and Paillou, Ph. (1997) The Aladin Interactive Sky Atlas, *IAU Symp. 179, New Horizons from Multi-Wavelength Sky Surveys*, Kluwer Academic Publishers.

Eichhorn, G., Kurtz, M.J., Accomazzi, A., Grant, C.S., Murray, S.S. (1997), Connectivity in the Astronomy Digital Library through the ADS, this volume.

Egret, D., Crézé, M. , Bonnarel, F., Dubois, P., Genova, F., Jasniewicz, G., Heck, A., Lesteven, S., Ochsenbein, F. and Wenger, M. (1995) A global perspective on astronomical data and information: the Strasbourg astronomical data centre (CDS), *Information & On-line Data in Astronomy*, Egret & Albrecht (Eds.), Kluwer Academic Publishers, pp. 163–174.

Genova, F., Bartlett, J. G., Bienaymé, O., Bonnarel, F., Dubois, P., Egret, D., Fernique, P., Jasniewicz, G., Lesteven, S., Monier, R., Ochsenbein, F. and Wenger, M. (1996) CDS as an Astronomical Information Hub, *Vistas in Astronomy*, **40**, pp. 429–437.

Laloë, S., Beyneix, A., Borde, S., Chagnard-Carpuat, C., Dubois, P., Dulou, M.R., Ochsenbein, F., Ralite, N. and Wagner, M.J. (1993) Updating the bibliography in SIMBAD, *CDS Inform. Bull.*, **43**, pp. 57–63.

Laloë, S. (1995) The updating of the bibliography in the SIMBAD database, *Vistas in Astronomy*, **39**, pp. 259–270.

Lesteven, S., Poinçot, Ph., Murtagh, F. (1996) Neural networks and information extraction in astronomical information retrieval, *Vistas in Astronomy*, **40**, pp. 395–400.

Lortet, M.C., Borde, S., and Ochsenbein, F. (1994) Second Reference Dictionary of Nomenclature of Celestial Objects, *Astron. Astrophys. Suppl.*, **107**, pp. 193–218.

Ochsenbein, F. (1994) Adopted standards for catalogues at CDS, *CDS Inform. Bull.*, **44**, pp. 19–28.

Ochsenbein, F. (1996) VizieR, the new catalogue interface at CDS , *CDS Inform. Bull.*, **48**, pp. 47–50.

Ochsenbein, F. and Lequeux J. (1995) The A & A tables and abstracts: An example of collaboration between data centres and editors, *Vistas in Astronomy*, **39**, pp. 227–234.

Schmitz, M., Helou, G., Dubois, P., LaGue, C., Madore, B., Corwin, H.G. Jr and Lesteven, S. (1995) NED and SIMBAD conventions for bibliographic reference coding, *Information & On-line Data in Astronomy*, Egret & Albrecht (Eds.), Kluwer Academic Publishers, pp. 259–270.

Wenger, M., Fernique, P., Genova, F., Bartlett, J.G., Bienaymé, O., Bonnarel, F., Dubois, P., Egret, D., Jasniewicz, G., Lesteven, S., Monier, R. and Ochsenbein, F. (1996) SIMBAD on the Web and Links to other Services, *BAAS*, **189**, #06.02.

Comments

BOYCE: I am impressed with two things you said. 1. The degree to which you have to automate your functions and 2. The advent of new tools, such as Kohonen maps, which are new tools available only in the electronic environment.

GENOVA: 1. Yes, it is very important to automate the functions as much as possible, and to concentrate the work of experienced staff on value-added activities. This is also the only way to keep pace with the increasing volume of published literature.

2. There will certainly be many new tools thanks to the electronic environment, and this might increase the acceptability of electronic publishing for the scientific community.

HJELMING: The comment about supplying accurate postitions of X-Ray sources is related to the management of misinformation. Often errors are large and poorly described. Dealing with this, and changes, is critical. Often only the scientist can make the judgement among diverse data with errors.

GENOVA: Dealing with errors and changes is certainly a very important task for the data centers. For instance, SIMBAD is continuously updated for cross-identification purposes, and notes and errors are added to the references. Active links can help to give access to checked and/or updated information.

CHEUNG: The Astrophysics Data Facility (ADF) at NASA's GSFC is actively working on the problem of correcting designations of objects identified from NASA missions to astronomical nomenclature, and building tools to help the researcher do object cross identification. See the NASA project (http://amase.gsfc.nasa.gov) homepage.

ELECTRONIC PUBLISHING. IMPACT ON DEVELOPING COUNTRIES

LESLEY ONUORA
The Astronomy Centre, University of Sussex, Brighton BN1 9QL
(formerly of the University of Nigeria, Nsukka, Nigeria)

Abstract. The developing countries at the low end of the scale face almost insurmountable problems in accessing electronic information. In Nigeria, only two of 40 universities have e-mail. Although electronic access to current information could revoutionize the educational situation at these places, the communication infrastructure does not exist, nor does the necessary funding.

The technological development and telecommunications infrastructure of countries which may be considered to be in the developing world varies greatly, some countries being almost as well equiped as any country in the developed world, while others have almost no access to electronic media. Some of the more electronically advanced countries such as India, Mexico and Brazil have successfully developed their own software packages (Report of Unesco Expert Conference on Electronic Publishing in Science, 1996).

However, attention needs to be drawn to countries at the extreme end of the range since their situation is so different. Apart from the problem of lack of access to electronic publications of all types and no access to email, there is also often a problem of poor communications in general. The ease of communication which is taken for granted by scientists in the developed world is just an unimagineable dream with no prospect of being obtainable in the forseeable future.

This is the situation at the University of Nigeria, and at many other universities in Nigeria and probably many other African countries. Nsukka is in a very rural area far from large cities and areas of commerce. There is therefore little pressure from private businesses for improvements in communications and the liklihood that the private sector may provide some of the necessary funding for internet facilities is remote.

The regional map of world wide internet connections (the Internet Society), available on the world wide web, shows that Africa in general puts a blight on the map with only a few countries having access to the internet, in contrast to the rest of the world. This situation probably reflects the unique economic and political problems of African countries. Even so the map tends to paint a rather rosier picture than reality. According to the map Nigeria appears to have email, but in fact only two of its over 40 universities have access. However, even for these two exceptions, connections are intermittent and unreliable. This point, that email may be available but not reliable or readily accessible, was highlighted at the Sixth UN/ESA Workshop on Basic Space Science held in Bonn in 1996 and probably applies also to some Asian countries. The basic problems in Nigeria are that the telecommunications system needs to be upgraded, and there is gross underfunding of universities. Satellite communications are used over long distances, but more locally connections are by overhead cables. These are easily damaged and often not repaired promptly. The problem is not really one of expertise. A level of competence in computing is already available in Nigeria which could easily be built on and developed through further training.

To put the situation in perspective, communications of any form are a major problem at Nsukka. The telephone rarely works and even ordinary mail is unreliable. Courier services, which are prohibitively expensive, are the only sure way of communicating. Individual staff have to pay personally for any form of communication because of the poor financial situation of the universities. Thus even if email became available, the cost would be a non trivial problem. Increasingly research papers have to be published in local journals because the cost of mailing them overseas cannot be borne. This is very detrimental for research in an area like astronomy as there are few people with sufficient interest or expertise in astronomy in the country to either referee the papers, or indeed read them.

J. Andersen (ed.), Highlights of Astronomy, Volume 11A, 526–527.

Accessibility to computers is restricted to the elite, who can afford their own personal computers, or to the few research groups who have succeeded in obtaining an external grant. Naturally enough, astronomy is not seen as a priority area for Nigeria by the international agencies that award grants.

CD-ROM readers provided by the World Bank are available in the University Library but are not readily accessible to the large number of students and staff in the university. Books and journals are also scarce and are largely out of date. In astronomy most of the books available in the library are at least 20 years old. Preprints from observatories and donations (some arranged through the IAU) are relied on for research, as well as photocopies obtained by anyone who is able to travel out of Nigeria. Despite this, for the present printed material is still far more accessible than any electronically published material and this is likely to remain the situation at Nsukka for the forseeable future. The only way that change could come about would be if the necessary infrastructure were put in place by the government, and if funding of universities were given a higher priority.

It is clear that being on-line would revolutionize the situation at a place like Nsukka which is currently so isolated. Reliable and readily available access to email and the world wide web would solve many of the problems of scientists in developing countries, and CD-ROMs would provide an invaluable resource for book-starved students. It would also reduce the need for observational facilities to carry out astronomical research because of the extensive data archives which are available in electronic form. The continuing trend in developed countries to rely more and more on electronic communication and publication is however at present worsening the situation for scientists in developing countries who are without access, and who are falling further and further behind and becoming increasingly cut off from current research. At the same time their colleagues in the developed world, who take for granted email, access on the web to research results months before they come out in print, access to data banks, etc., often do not comprehend their plight.

The developed world cannot be held back by developing countries, this is clear. It has rightly been said that the developing world must be encouraged to catch up. However, scientists in places like Nsukka do need the support and understanding of their more fortunate colleagues and need to be remembered when decisions are being taken about producing publications in "electronic only" form. Some of them have struggled long and hard to try to convince their governments of the needs of astronomy and science. Government backing and national policies are needed to bring about the necessary changes. The present piecemeal acquisition of technology by the elite alone will not solve the problems.

Comments

EICHHORN: Will the new satellite telephone communications technology change the situation?

ONUORA: Not right away, since the main problem is lack of funding.

GENOVA: An example is the cancellation of Internet access in a lab of FSU because of costs.

SMITS: Nigerians visiting South Africa have no background in computing because of the lack of facilities at their home institute (University of Nigeria). We teach them how to use UNIX, Web browsers and email but when they leave South Africa they have no contact with us once back in Nigeria.

JD13

Detection and Study of Planets Outside the Solar System

Chairperson: **A. J. Penny**

MANUSCRIPT NOT RECEIVED

JD14

The First Results of Hipparcos and Tycho

Chairperson and Editor: **C. Turon**

JOINT DISCUSSION 14
THE FIRST RESULTS OF HIPPARCOS AND TYCHO

Monday 25 August 1997

Supporting Division: I
Supporting Commission: 24

Co-supporting Divisions: IV, V, VII
Co-supporting Commissions: 5, 8, 19, 20, 25, 26, 27, 29, 33, 37, 42, 45.

Chairperson:
C. Turon, DASGAL, URA CNRS 335, Observatoire de Paris, 92195 Meudon Cedex, France.

SOC: L. Blitz (USA), M. Feast (South Africa), G. Gilmore (United Kingdom), M. Grenon (Switzerland), C. Jaschek (Spain), M. Lattanzi (Italy), A. Maeder (Switzerland), J.E. Norris (Australia), M.A.C. Perryman (The Netherlands), C. Turon (Chairperson, France).

In August 1997, at the time of the IAU General Assembly in Kyoto, the Hipparcos and Tycho data had been widely available. They have been published in June 1997, and Principal Investigators had access to the data related to their proposals during 1996 and early 1997.

The Joint Discussion aims at presenting the Hipparcos and Tycho astrometric and photometric results; their quality and reliability; their calibration and the systems to which they have been linked; and the results obtained for double and multiple stars, for variable stars, for minor planets and satellites.

It is also one of the first occasions to widely present the new and exciting scientific results obtained from these data in a variety of domains: reference frames and the methods used to link the Hipparcos frame to ICRF (presented in detail in JD 7 on 'The new International Celestial Reference Frame' chaired by Leslie V. Morrison); impact of Hipparcos and Tycho data on astrometric observations and reductions; impact of Hipparcos parallaxes on our knowledge of nearby stars (stellar interiors, stellar ages and masses, luminosity function); impact of Hipparcos and Tycho data on galactic structure, kinematics and evolution, on the calibration of distance scale primary indicators; including some highlights about new determination of the distance, three-dimensional structure, HR diagram, chemical composition and internal dynamics of very nearby galactic open clusters; discovery of new double stars, impact on duplicity statistics; discovery of new variable stars, impact on studies of stellar variability and their applications; impact of Hipparcos and Tycho astrometric and photometric observations on studies of minor planets and satellites.

J. Andersen (ed.), Highlights of Astronomy, Volume 11A, 533–535.

Session 1 : Hipparcos and Tycho results

Talks

Posters

Session 2 : Astrometry

Posters

Session 3 : Impact on stellar physics

Talks

Posters

Session 4 : Impact on galactic physics

Talks

Posters

Session 5 : Impact on distance scale primary indicators

Talks

Posters

Session 6 : Astrophysics and microarcsec astrometry

Talks

Posters

HIPPARCOS ASTROMETRIC RESULTS

J. KOVALEVSKY
OCA/CERGA
UMR CNRS 6527
Av. Copernic, 06130–Grasse (France)

Abstract. The astrometric results of Hipparcos include the positions at epoch (1991.25), the proper motion in the new IAU extragalactic reference system (ICRS), and parallaxes for about 118 000 stars. One dimensional positions are also given for 48 asteroids and 3 satellites. Due to the non-isotropy of the scanning law, the uncertainties are position dependent. Some indications of the remaining correlations are given. Various tests and comparisons show that systematic errors in parallax, if any, are not larger than 0.1 millisecond of arc.

1. DESCRIPTION OF THE CATALOGUE

The Hipparcos catalogue, now published as ESA publication SP 1200 (July-September 1997) and in electronic form, in particular from the CDS, provides astrometric data for 117 955 stars almost uniformly distributed over the sky. It includes a survey of about 52 000 *bright* stars complete to a limiting visual magnitude V such that, if b is the galactic latitude :

$V \leq 7.9 + 1.1 sin\,|b|$ for spectral types earlier or equal to G5
$V \leq 7.3 + 1.1 sin\,|b|$ for spectral types later than G5

The other 66 000 stars were selected for their astrophysical or astrometric interest or in order to ensure a minimum density in star poor regions.

2. THE ASTROMETRIC PROPERTIES OF THE CATALOGUE

They can be summarized by the few following features.

- The positions and the proper motions are referred to the ICRS (International Celestial Reference System; Ma and Feissel, 1997) which is fixed with respect to distant extragalactic radio-sources. This link has been realised from a discussion of eleven different determinations using radio-interferometry, photographic plates, star catalogues linked to galaxies, HST observations, and analysis of Earth rotation determined by Hipparcos stars. The method used and the results obtained by each team are described in Volume 3 of the catalogue (ESA 1997) and with many additional details in Kovalevsky et al (1997). The remaining offset at epoch (J1991.25) with ICRS axes has an uncertainty at one sigma level of 0.6 mas. The uncertainty in stopping the spin of the catalogue axes is 0.25 mas/yr (1 sigma). This means that in using the catalogue for galactic dynamics, one has to enter this latter uncertainty in the error budget.
- The median precision of positions at epoch for stars with Hipparcos magnitudes $H_p < 9$ mag is 0.77 mas in RA and 0.64 mas in declination.
- The median precision of proper motions for the same selection of stars is 0.88 mas/yr in RA and 0.74 mas/yr in declination.
- The median precision of parallaxes for the same selection of stars is 0.97 mas. About 21 000 stars have a relative uncertainty in parallaxes smaller than 10%, out of which, 3000 are double or multiple systems.

J. Andersen (ed.), Highlights of Astronomy, Volume 11A, 536–538.

- There are 23 882 solved or suspected double stars. The astrometric parametre uncertainties are generally somewhat poorer than for single stars.
- In addition, results of the observations of 48 asteroids and 3 planetary satellites (JII - Europa, SVI Titan and SVIII Iapetus) are given as a one dimensional quantity: the projection of the position of the object on a reference great circle for each observed transit. The observations as reduced by FAST and NDAC are both given without merging, because the reduction procedures are not identical. The standard uncertainties of each data point range between 5 and 30 mas, depending upon the magnitude of the object. In total, 2887 NDAC and 2657 FAST asteroid positions are provided. Almost all FAST positions are in addition complemented by a determination of the magnitude. Practically only NDAC provided the 107 positions of the planetary satellites.

3. PROPERTIES OF THE UNCERTAINTIES

The scanning law of the satellite was not isotropic: all the scans had to be inclined by at least 43° to the ecliptic in order to minimize the solar strayglight. The consequences are that parametres in declination are better determined than in right ascension. In addition, in the mean, the standard errors are larger by a factor of about 1.6 in proper motions in RA and in parallaxes for stars around the ecliptic than around the ecliptic poles. This factor is smaller than 1.1 for proper motions in declination. At intermediate latitudes, the situation is of course somewhere in-between (see figure 1).

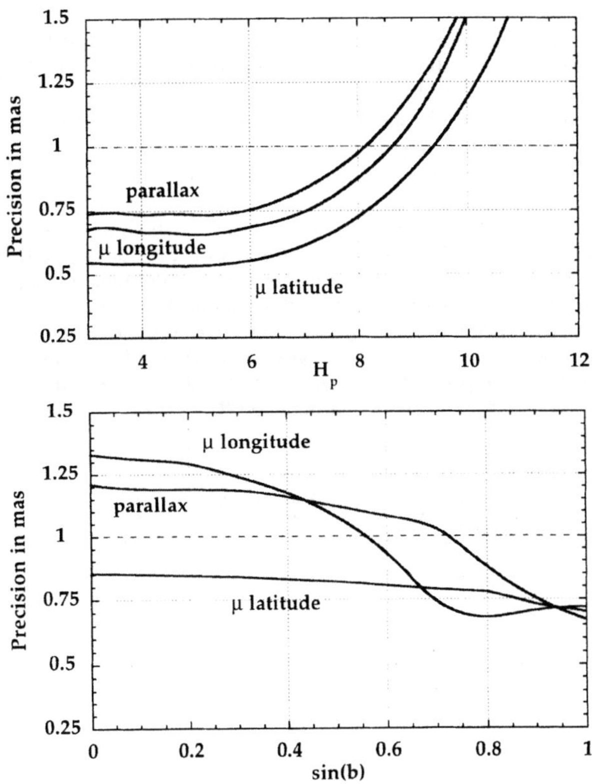

Figure 1. Mean precision of parallaxes and proper motion for single stars as functions of Hipparcos magnitudes (above) and the ecliptic latitude (below).

The uncertainties also increase with magnitude. They are approximately constant up to magnitude $H_p = 6.5$ for which $\sigma_\pi \cong 0.75$ mas; $\sigma_{\alpha \cos\delta} = 0.65$ mas/yr and $\delta_{\mu\delta} = 0.55$ mas/yr. Then they steadily increase to the double of these values for $H_p = 10$.
Because of the scanning symmetry about the plane of the ecliptic, the correlations are small when expressed in ecliptic coordinates between l and b and between the components of proper motions (< 0.2) but reach 0.4 when one uses equatorial coordinates. This is an artefact, the increase results only from breacking down the symmetry. Correlations between α and $\mu_{\alpha\cos\delta}$ or δ and μ_δ may be larger than 0.5, but again is an artefact whenever the actual mean observation time is far from the adopted J1991.25 epoch. It can be reduced to a minimum by referring stars to their mean observation time.
The only systematic effect concerns the correlation between the right ascension and the parallax. It reaches -0.4 around RA = 6h and +0.4 around RA = 18h for latitudes between ±60°. This effect is well understood as a consequence of an uneven distribution of observations in time due to payload failures, provoking an unbalanced distribution of longitude differences with the Sun.

4. SYSTEMATIC ERRORS

Several methods were used to test whether Hipparcos catalogue has significant biases, particularly in the parallaxes.

- A comparison of Hipparcos with the ground based 88 optical parallaxes determined by the USNO 61-inch reflector gave a median difference of +0.2 ±0.35 mas and parallaxes determined by VLBI showed similar identity.
- The average parallax of 38 non double stars in the Magellanic Clouds was found to be -0.1 ±0.23 mas.
- Comparing between distances of moving open clusters as determined from classical ground based astronomy and the distance of some of their members provided by Hipparcos yielded a mean difference of 0.04 ±0.06 mas.
- Cluster distances determined by multiband photometry were compared with Hipparcos parallaxes. One has obtained -0.05 ±0.05 mas.

These results all converge in estimating that if there is zero-point error in Hipparcos parallaxes, it is certainly below 0.1 mas.
A similar analysis cannot be made for proper motions, mainly because of regional errors present in all ground-based catalogues. However a comparison with FK5 proper motions, permitted to point out one difficulty: the Hipparcos proper motions are quasi-instantaneous proper motions, in contrast with FK5, based upon 50 or more years of observations. If a star is an unresolved double, the Hipparcos proper motion is polluted by the effect of the orbital motion on the photocentre, while this effect is smoothed out in FK5. So in some cases, the uncertainty of the Hipparcos proper motion, considered as the proper motion of the barycentre of the system, is underestimated. But by no means does this imply a systematic kinematic effect.

References

ESA 1997, The Hipparcos and Tycho Catalogues, ESA Publ. SP-1200.
Kovalevsky, J., Lindegren, L., Perryman, M.A.C. et al., 1997, A&A, **323**, 620–633.
Ma, C. and Feissel, M., 1997, Technical Note 23, Observatoire de Paris, June 1997.

THE HIPPARCOS DOUBLE AND MULTIPLE STAR SOLUTIONS

F. MIGNARD
OCA/CERGA
UMR CNRS 6527
Av. Copernic, 06130–Grasse (France)

Abstract. The Hipparcos Catalogue provides general astrometric and photometric information on double and multiple stars in specific fields of the main Catalogue and detailed data on the components in the various sections of a dedicated annex : the Double and Multiple Systems Annex (DMSA). Overall statistics of these solutions are presented for the 13211 entries of this annex and the different types of solutions are outlined.

1. THE HIPPARCOS DOUBLE STARS

It was known well before the Hipparcos mission that a significant fraction of the observing programme will have to deal with double and multiple systems, a circumstance that seriously complicated the data treatment. When a Hipparcos target was a single object, the standard astrometric model in the data reduction assumed a uniform rectilinear motion in space relative to the barycentre of the solar system. In this case only five parameters were necessary to describe the full astrometric information embedded in the Hipparcos measurements : the two angular coordinates to specify the coordinate direction at the Catalogue epoch (T_0 =J1991.25), the two components of the proper motion and the parallax. In the case of a non single object the situation was more complex and additional parameters were required first to recognize that the target comprised at least two objects and to model properly the grid observations in order to achieve a good decoupling between the absolute and relative astrometry and photometry. The principles applied to the detection and the astrometry of the double stars have been presented in detail in the literature (Mignard et al, 1992 , Söderhjelm et al., 1992 and Mignard et al, 1995) and the most updated information is now available in the Hipparcos documentation (ESA, 1997) and for the classification of the double star solutions in Lindegren et al (1997).

TABLE 1. *Content in entries of the different parts of the Hipparcos Catalogue.*

Catalogue	Solutions	Entries	Comments
Main Catalogue	Single star solutions	100 038	6763 suspected non single
DMSA :			
C	Component solutions	13 211	2996 new systems
G	Acceleration solutions	2 622	Likely astrometric binaries
O	Orbital solutions	235	
V	Variable movers	288	
X	Stochastic solutions	1 561	

About 15 percent of the Hipparcos Catalogue refers to objects seen non–single by the satellite, although in the remaining 85 percent there are likely many double and multiple systems not detectable and treated as single stars. Because of the variety of non–single stars, several types of solutions had to be devised to cope with this problem according to the separation, magnitude difference, orbital period. The resolved systems for which both the separation and magnitude difference

J. Andersen (ed.), Highlights of Astronomy, Volume 11A, 539–541.

were measured constitute the largest single group and the solutions are given in term of absolute astrometry of the components in the part C of the DMSA. Most systems (12 005) are double stars and 190 systems were solved for three or four components, corresponding in total to the 13 211 entries of the Catalogue. Most of these systems were solved as fixed components (separation and position angle fixed over the mission) while 1186 include a linear relative motion of the components and for the remaining 202 the two parallaxes were solved independently of each other.

Other non single star solutions are referred to as *acceleration solution* when the absolute motion on the sky was significantly non-linear, *orbital solutions* when one or several orbital elements of the photocenter could be fitted to the data, *variability-induced movers* for which the photocenter exhibited a systematic motion due to the variability of at least one component and the *stochastic solutions* when no acceptable single or double star solutions were found. The respective content and denomination of each part are given in Table 1. Further ground-based observations should provide a better on the still uncertain status (single or multiple) of most of the entries in the part G, V and X of the annex. In addition a special effort should be devoted to the subset of suspected non-singles, in particular with speckle interferometry.

The Double and Multiple star solutions are published in the printed form in various parts of the 17-volume Hipparcos Catalogue. The ASCII files are also available on the set of CD-ROMs published simultaneously, which also includes the intermediate data (abscissae and grid signals for 30 percent of the stars) that could be used for a reprocessing using a different modelling or to take advantage of new ground based data useful to constraint the solution.

2. THE ASTROMETRIC SOLUTION

Among the $\sim$ 11 100 entries of the main Catalogue solved for two components, there are $\sim$ 8100 which were known as non single stars from ground based observations carried out before or during the Hipparcos mission and about 3000 new systems recognized as double from the Hipparcos data. The plots in Figure 1 show the distribution of these systems on a diagram magnitude difference - separation. They reveal a striking difference between the two populations : rather larger separations for the known systems, comprising mainly visual detections, with the exception of the speckle measurements in the left part of the diagram. The Hipparcos newly detected systems, on the contrary are mainly close visual binaries with separation < 0.5 arcsec. The envelope of the distribution follows closely the detection limit of the Hipparcos optical and grid system. This indicates also the level of completeness to be considered in further statistical studies.

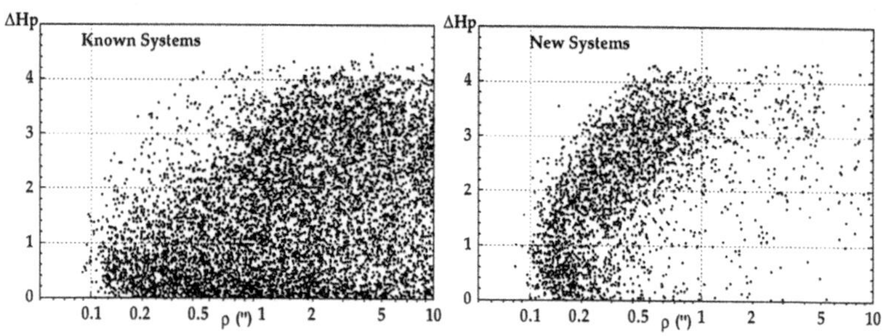

Figure 1. Distribution of the magnitude difference versus separation for the components of the double stars measured by Hipparcos. The left panel refers to systems known from ground based measurements and the right one to the systems discovered by Hipparcos.

The astrometric solutions have been derived by correcting the single star for the effect of the companion allowing to produce the absolute astrometry of either the primary or the photocenter, according to the separation. For the users, the part C of the annex gives the astrometric solution of each components derived from the merging of the FAST and NDAC solution after the two groups agreed on a list of indisputable detected non single stars. A slight possibility of grid step errors remains for few stars, which affect both the separation and the absolute position. For catalogue entries resolved in strictly two components, the relative astrometry and photometry is also given in the main Catalogue and is perfectly consistent with the data of the Annex.

Apart for the relative astrometry, the parallax is the most relevant parameter for the double and multiple systems because of its implication in stellar physics. Because of the complication in the modelling of the double stars, it was reasonable to expect a significant degradation in the quality of the parallax compared to that of the single stars. The plots in Figure 2 compare the relative error in the parallax σ_π/π for the two categories. All the entries (= 96731) not included in the Double and Multiple Systems Annex and with a non-negative parallax make the set of *single stars*, whereas the set *multiple stars* comprises the entries (= 12569) of the DMSA-C with non negative parallax. The difference is noticeable but not very large, given the difference in the modelling. There are 18 per cent of the single stars with $\sigma_\pi/\pi < 0.1$ versus 13 per cent for the multiple and the level of 18 per cent of the multiples is reached at $\sigma_\pi/\pi < 0.125$ instead of 0.10 for the singles. This results from a very fortunate circumstance which made the parallax solution moderately sensitive to the quality of the double star solution, unlike the other astrometric parameters.Thus, one can say, that despite a small difference of quality, the Hipparcos parallaxes are statistically homogeneous irrespective to whether the star was processed as single or not.

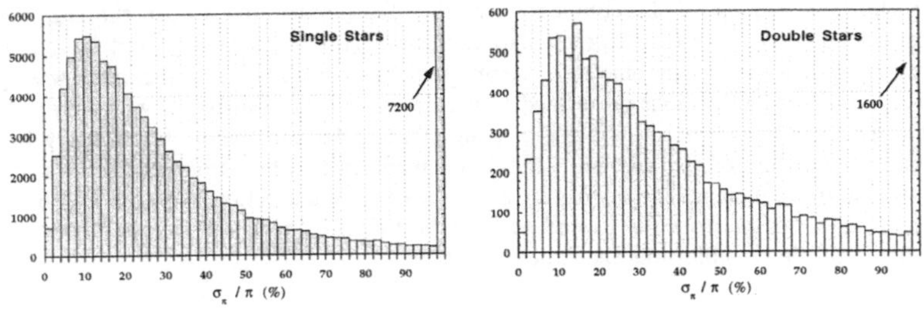

Figure 2. *Relative standard errors of the Hipparcos parallaxes for the single stars and the double and multiple systems of the Annex-C.*

Regarding the relative astrometry of double stars, the separations solved with Hipparcos range from 0.1 to 20 arcsec, with a median of 0.97 arcsec. The distribution of the standard errors in ρ is shown in Fig. 3 for all the separations of the known double stars (left) and the new systems detected by Hipparcos (right). The difference between the two sets is much apparent and linked to the fact that the new double stars have predominantly small separations (then more difficult to resolve) with a median of 0.33 arcsec versus 1.55 arcsec for the complementary set. The medians of the standard errors are respectively 23 and 8 mas.

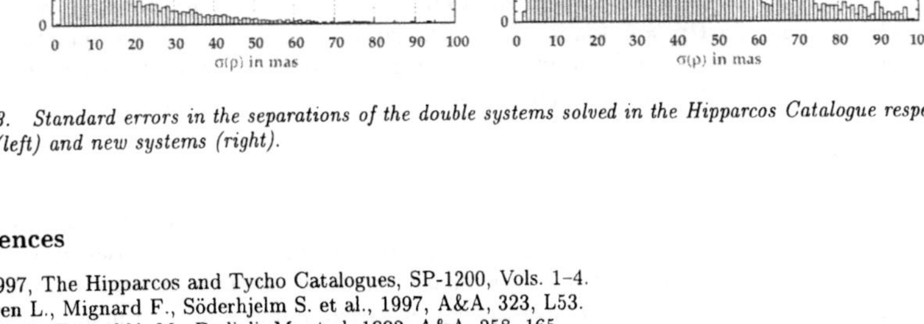

Figure 3. *Standard errors in the separations of the double systems solved in the Hipparcos Catalogue respectively for known (left) and new systems (right).*

References

ESA, 1997, The Hipparcos and Tycho Catalogues, SP-1200, Vols. 1–4.
Lindegren L., Mignard F., Söderhjelm S. et al., 1997, A&A, 323, L53.
Mignard, F., Froeschlé, M., Badiali, M. et al, 1992, A&A, 258, 165.
Mignard, F., Söderhjelm, S., Bernstein, H. et al, 1995, A&A, 304, 94.
Söderhjelm S., Evans, D.W., van Leeuwen F. and Lindegren L., 1992, A&A, 258, 157.

THE STELLAR VARIABILITY FROM HIPPARCOS PHOTOMETRY

M. GRENON
Geneva Observatory
1290 Sauverny, Switzerland

1. Introduction

During the development of the Hipparcos satellite, the opportunity to use the star mapper to perform a whole sky two colour intermediate accuracy photometric survey, and the main-mission detector for high accuracy photometry of target stars was identified. The instrument was optimised and the pre-launch calibrations performed to pre-determine the pass-bands with an accuracy sufficient to cope, in-orbit, with the foreseen aging of the detection chains. Photometry became an important by-product of the mission, very complementary to the astrometric results. It was the first opportunity to monitor the sky during 3 - 4 years without selection biases and with a precision similar to that achieved from the ground with the best classical techniques. A systematic detection of small amplitude variables was the expected return.

2. The correction of the aging effect

The radiation by cosmic and solar particles induced transmission losses mainly in the green to violet part of the spectrum. Since the Hp band is broad, $\lambda\lambda$ 380-900 nm, the effect was a chromatic change of the band, and hence of the Hp magnitude as large as 0.6 and 0.2 mag over the mission for V-I = 0 and 7 mag respectively. Photometric reductions were performed twice a day using a subset of the 29000 standard stars. Zero-point errors on the magnitude scale were then less than 0.001 mag for each data set. Correction of chromatic drifts requires the knowledge of the star colour; consequently error on it induces a linear trend in the Hp time series.

3. The variability detection threshold

The Hipparcos photometry is essentially photon-noise limited. The error on Hp per transit are 3, 5, 8, 15 and 33 mmag for Hp = 3, 5, 7, 9 and 11 mag. The mean number of useful transits, function of the galactic latitude, is 110, with a range of 20 to more than 300. The time sampling is peculiar with groups of 4 to 6 transits separated by 20 and 108 min, repeated 3 to 5 weeks later, an interval depending on the scanning law. This sampling limits the range of detectable periods and introduces severe aliasing problems. Several algorithms developed at Geneva and RGO, allow to classify stars as non-variable, suspected or variable with thresholds depending on the magnitude and the number of transits cf. Eyer et al.(1997), van Leeuwen (1997). The detectable amplitude for variables are 0.8, 20 and 120 mmag for Hp = 4, 8 and 12. Known variables as Ap or BY Dra type may escape the detection if faint, whereas on the brighter side, stars previously accepted as constant appeared variable because of the higher accuracy and the large number of measurements.

4. The variability analysis

According to statistical tests about 11600 stars are variable or possibly variable at the 0.1 % error level. When the ratio of intrinsic scatter (corrected for photon and reduction noises) over the noise exceeds a threshold leaving less than few percent spurious periods, a search for signal periodicity was undertaken. Some 3343 stars were not investigated. A careful analysis of instrumental effects due to mispointing, image overlap, high-background, allowed to clean the time series and remove cases mimicking flares or eclipses. Abbe test was used to detect long term trends true or due to errors on

J. Andersen (ed.), Highlights of Astronomy, Volume 11A, 542–543.

the colour. A series of algorithms optimised for specific light-curve shapes, Fourier transform for symmetric distributions, Renson for possible eclipsing binaries etc. were applied to the final time series. Period search was an iterative process, followed by a semi-automatic curve fitting leading to the period, epoch, light-curve extrema and residuals for all periodic variables. Solutions from Geneva and RGO were then carefully cross-checked and compared to literature data when available at RGO. Due to the time sampling, only the first mode of multiperiodic stars is detectable from Hipparcos data.

5. The Catalogues and Atlases

The data relative to periodic variables, as described in the Guide to the data (ESA 1997), are given in the Hipparcos Variability Annex Part 1. Among 2712 periodic variables, 951 are new discoveries and other 118 were previously NSV suspected variables. 171 solutions were found by combining Hipparcos data with literature information. Folded light-curves are displayed in part A of the Atlas with super-imposed fitted curves in 2287 cases (1189 from Geneva and 1098 from RGO).

The catalogue of unsolved variables (Annex Part 2) contains 5542 entries; among them 1123 were already known variables and 233 were NSV stars. 4086 variables are new and for 1953 of them, the variability status was clear enough to have a star name allocated by the Samus' team at Moscow. About 3000 unsolved variables have still no variability type. For unsolved variables an estimation of the peak to peak amplitude as described by Eyer et al.(1997) is provided. The best defined light-curves are displayed in part C of the Atlas.

Large amplitude irregulars were monitored by AAVSO observers prior and during the mission. The transformation of visual magnitudes into Hp allowed to construct combined light-curves for 274 LPV's and irregulars, with a high time coverage, displayed in part B of the Atlas.

6. New variables

The most frequent new periodic variables are EB, then EA eclipsing binaries, followed by SR and SPB. Eclipsing are mainly B to early G type stars with one component somehow evolved. A bias against long period eclipsing binaries is present. The number of SPB's was increased by a factor 10. ACV and δ Sct variables are the next most frequently detected new variables. Periods were rarely found for multiperiodic main sequence stars nor for late type rotating stars.

Among 4086 new unsolved variables, only 1134 have a variability type. The most frequent types for new variables are SR, I, L then γ Cas. Many B and A stars were classified as L type, waiting for a spectroscopic improvement of their type.

7. The Variability across the HR-diagram

The fraction of variables in the HR-diagram was estimated for parallax stars, showing all red giants to be variable with amplitudes anticorrelated with Teff and surface gravity. Early type stars appear nearly all variable, especially if luminous or B2-B3 main sequence stars. The low main sequence looks constant because the amplitudes are below the detection limit, a simple magnitude effect. Distributions of periods and amplitudes across the HR-diagram are now available.

The distribution of the intrinsic scatter, in bins of spectral sub-type and luminosity class, provides quantitative information on the mean amplitude and its range across the HR-diagram for all stars. Islands of stability are found on both sides of the instability strip: the first from B9 to A7 stars, narrowing towards high luminosity and the second from F2 to G5 dwarfs up to G5 to K2 giants. The level of microvariability across the whole HR-diagram allows to select targets for further study as asteroseismology, or constant comparison stars. The new limits of gaps and instability strips are direct tests of the stellar evolution models.

8. References

Eyer L., Grenon M., 1997, 'Hipparcos Venice' 97', ESA SP-402, 462-473

van Leeuwen F., 1997, 'Hipparcos Venice' 97', ESA SP-402,19-24

ESA, The Hipparcos and Tycho Catalogues, 1997, ESA SP-1200

THE TYCHO CATALOGUE: ASTROMETRIC AND PHOTOMETRIC RESULTS

E. HØG
Copenhagen University Observatory
Juliane Maries Vej 30, DK-2100 Copenhagen Ø, Denmark

Abstract. The Tycho Catalogue contains basic astrometric and photometric data for a million stars on the sky, complete to about $V = 10.5$ mag. The accuracy at the catalogue epoch (J1991.25) of positions and two-colour magnitudes is generally an order of magnitude better than has hitherto been available for the majority of these stars. Results from the production and verification of the catalogue is presented. The catalogue and associated epoch photometry annex are available in machine-readable form as part of the final Hipparcos mission products.

1. The Tycho Catalogue

The Tycho Catalogue of a million stars was obtained from photon counts obtained by scanning with the Hipparcos star mapper, carried out simultaneously with the Hipparcos observations in the adjacent main field of view of the telescope.

The observational capabilities and operational principles of the Hipparcos satellite have been presented in the literature on previous occasions, with the most complete and definitive treatment contained within the published Hipparcos and Tycho Catalogues (ESA 1997). Short descriptions are given in this volume, in Astronomy & Astrophysics by Perryman *et. al.* (1997), Høg *et al.* (1997) and van Leeuwen *et al.* (1997), and at the Venice Symposium (Perryman 1997.)

The principal observational characteristics of the Tycho Catalogue are given in Table 1. Complete details of the data, the reductions, and the properties of the final Tycho Catalogue, are published by ESA (1997), especially in Volumes 1 and 4 to which we refer in the following.

2. Astrometry

The five astrometric parameters, (α, δ), the annual parallax, π, and the proper motion components $\mu_{\alpha*} = \mu_\alpha \cos\delta$ and μ_δ, in angular measure per unit time, are given for almost all stars in the catalogue. The catalogue includes astrometric quality flags, and indicators of variability and/or duplicity – nearly 900 000 of the catalogue entries are classified as 'recommended' reference stars, having good Tycho astrometric quality, and not recognised as double.

The median value of the formal (internal) astrometric standard errors given for each star (in position, parallax, and annual proper motion) are typically around 7 mas for stars brighter than $V_T \sim 9$ mag, and approximately 25 mas for $V_T \sim 10.5$ mag, The external standard errors (i.e. accuracies) are only slightly larger by a factor of 1.1 for bright stars of $V_T < 10$ mag, increasing to a factor of 1.5 larger at the median magnitude $V_T = 10.5$ of the Tycho Catalogue stars. This has been found from a comparison with the much more accurate Hipparcos values for stars common to the two catalogues. The systematic errors are less than about 1 mas(/yr).

3. Photometry

The Tycho Catalogue includes accurate and homogeneous photometric information for each star: two-colour (B_T and V_T) magnitudes derived from the Tycho (star mapper) observations; Johnson V magnitude and $B - V$ colour index, derived from the observed $B_T - V_T$ by simplified transfor-

J. Andersen (ed.), Highlights of Astronomy, Volume 11A, 544–546.

TABLE 1. *Principal observational characteristics of the Tycho Catalogue. ICRS is the International Celestial Reference System.*

Measurement period	1989.85–1993.21
Catalogue epoch	J1991.25
Reference system	ICRS
coincidence with ICRS[1]	±0.6 mas
deviation from inertial[1]	±0.25 mas/yr
Number of entries	1 058 332
based on Tycho data	1 052 031
with only Hipparcos data	6301
Median astrometric standard errors[2]	
$V_T < 9$ mag	7 mas
all stars	25 mas
Median photometric std. errors on V_T	
$V_T < 9$ mag	0.012 mag
all stars	0.06 mag
Sky density[3]	~ 25 deg^{-2}
Limiting magnitude	$V \sim 11.5$
Completeness to ~ 90 per cent	$V \sim 10.5$
Completeness to ~ 99.9 per cent	$V \sim 10.0$
Total number of observations	$\sim 130 \times 10^6$

[1] about all 3 axes
[2] ratio of astrometric external to internal errors at $V_T = 10.5$ is ~ 1.5
[3] depending on galactic latitude

mations reflecting the fact that the spectral type, luminosity class, reddening, etc., are unknown for the majority of Tycho Catalogue objects; and various flags resulting from (preliminary) studies of variability based on an average of 130 transits per star during the 3 year observing period. The details of the Tycho B_T and V_T photometric system, and details of the corresponding transformations to V and $B-V$, are given in the published catalogue, and also described in the context of the Hipparcos main mission photometry by van Leeuwen et al. (1997).

Published mean photometric values were based on all transits for each star, including those transits where the star was not detected because the signal was too faint. These 'censored' observations were taken into account in a dedicated 'de-censoring' processing. In the basic Tycho data reductions only the detections with a signal-to-noise ratio larger than 1.5 in the combined photon counts from the B_T and V_T detectors provided astrometric and photometric estimates. The photometric results for individual transits, including the censored observations, are provided separately in a Tycho Epoch Photometry Annex for specific subsets of stars. These individual magnitude determinations, at the specified measurements epochs, will allow detection of variable stars and further studies of the variability of known variables.

The principal photometric characteristics are as follows. Standard errors of V_T mean magnitudes are for $V_T < 9$ and ~ 10.5 respectively about 0.012 and 0.06 mag. Errors on B_T are typically 10 per cent larger for the same values of B_T. The standard errors of individual transits in the Tycho Epoch Phtometry Annex are typically 10 times larger than for the mean value of a given star.

A histogram of all stars in the Tycho Catalogue as function of V_T is given in Figure 1. The catalogue is generally complete to 99.9 and 90 per cent at respectively $V = 10.0$ and 10.5 mag. But the degree of completeness at the faint magnitudes depends also on the colour index and on the position in the sky, due to the scanning law (see Volume 4, Figures 16.4 and 16.5). The degree of completeness is approximately constant for all stars with a given value of $V_T = V_{T0} - 0.50(B_T - V_T)$ because the Tycho detection was carried out in the added photometric counts from the two channels B_T and V_T. This equation is almost equivalent to $V = V_0 - 0.70(B-V)$ where $V_0 = V_{T0}$, according to the transformation by Equations 1.3.20 in Volume 1.

The distribution of the colour index $B - V$ (Johnson system) is shown in Figure 2 of Høg *et al.* (1997) for different ranges of the galactic latitude. The bimodal structure and other features related to the HR diagramme of field stars in the solar neighbourhood are thus illustrated.

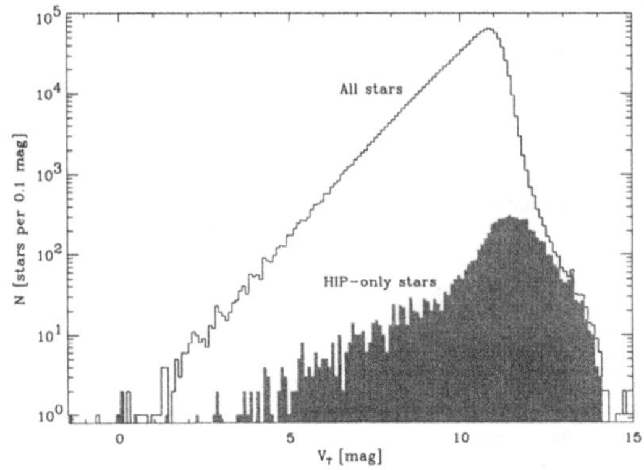

Figure 1. Histogram of all stars (upper curve) in the Tycho Catalogue as function of V_T. For the sake of completeness the catalogue also includes entries (HIP-only) for stars only observed in the Hipparcos main field and thus contained in the main Hipparcos Catalogue.

4. Applications

Many applications of the Tycho Catalogue have already been initiated. The Tycho Reference Catalogue (TRC) and the 2nd Tycho processing are described by Høg (1997b); the latter project is based on the 1000 Gbits of original Tycho photon counts. Furthermore, the *Celestia 2000* is a CD-ROM for catalogue interrogation (see Volume 1), the *Millenium Star Atlas* contained in the Volumes 14–16 will show all Tycho stars. The Tycho Catalogue will be or has been used as an astrometric reference frame for, e.g., space craft control by ESA/ESOC, control of the AXAF satellite, CCD drift scan, CCD astrograph, calibration of the USNO A 1.0 etc., calibration of the GSC 1.1, calibration of the GSC II, calibration of quasar fields, calibration of a multi-fibre spectrograph.

The photometric reference stars of the Tycho Catalogue will be used for calibration of the GSC 1.1, calibration of the USNO A 1.0 etc., calibration of variable star observations by amateurs. Finally, e.g., new nearby stars have been found, kinematic studies will be based on the Tycho Reference Catalogue, and new variable stars will be found from the Tycho epoch photometry.

Acknowledgements. The Tycho Catalogue is a result of the Hipparcos space astrometry mission, undertaken by the European Space Agency. The data reduction was undertaken by the Tycho Consortium (TDAC) in collaboration with the NDAC, FAST and INCA Consortia, many members of which were also members of TDAC. The efforts of the many individuals and organisations participating in the Hipparcos project over many years have been an essential component of the successful completion of the Tycho Catalogue project.

References

ESA, 1997, The Hipparcos and Tycho Catalogues, ESA SP-1200
Høg, E. 1997a, The Tycho Catalogue, in: Proceedings of the HIPPARCOS Venice '97 Symposium, ESA SP-402
Høg, E. 1997b, in: Proceedings of IAU Joint Discussion No. 7
Høg, E. *et al.* 1997, Astron. Astrophys. 323, L57
van Leeuwen, F. *et al.* 1997, Astron. Astrophys. 323, L61
Perryman, M.A.C. *et al.* 1997, Astron. Astrophys. 323, L49
Perryman, M.A.C. (ed.) 1997, Proceedings of the HIPPARCOS Venice '97 Symposium, ESA SP-402
Röser, S., Høg, E., 1993, in: Workshop on Databases for Galactic Structure. Ed.: A.G. Davis Philip, B. Hauck and A.R. Upgren. Van Vleck Observatory Contr. No.13, 137. L. Davis Press, Schenectady, N.Y.

SOME CONSIDERATIONS IN MAKING FULL USE OF THE HIPPARCOS CATALOGUE

A.G.A. BROWN[1], F. ARENOU[2], F. VAN LEEUWEN[3], L. LINDEGREN[4], X. LURI[2,5]
[1] *Sterrewacht Leiden, Postbus 9513, 2300 RA Leiden, The Netherlands*
[2] *Observatoire de Paris, Section d'Astrophysique, 92195 Meudon Cedex, France*
[3] *Royal Greenwich Observatory, Madingley Road, Cambridge, CB3 0EZ, United Kingdom*
[4] *Lund Observatory, Box 43, S-22100, Lund, Sweden*
[5] *Departament d'Astronomia i Meteorologia UB, Avda. Diagonal 647, E08028 Barcelona, Spain*

Introduction. This contribution is intended as a 'rough guide' to the Hipparcos Catalogue for the non-expert user. We shall focus here on the interpretation of the measured parallaxes in terms of distances and luminosities of stars, and on the use of the covariance matrix of the errors on the astrometric parameters. Issues not addressed here are the level at which one may expect systematic errors to occur in the Hipparcos Catalogue, and the correlation of astrometric parameters over a given region of the sky. For more details on the latter two subjects we refer to the documentation of the Hipparcos Catalogue (ESA, 1997) and to Brown *et al.* (1997). We note that at present the Hipparcos Catalogue contains no identified systematic errors.

Correct Use of Trigonometric Parallaxes. Notwithstanding the unprecedented quality of the Hipparcos data, the correctness of the *astrophysical* results is not assured, as the estimation of stellar distances, absolute magnitudes and other physical quantities from trigonometric parallaxes is not a trivial process. The statistical properties of the relationships involved and the effects of sample selection hide several pitfalls that, if not avoided, lead to biased estimates. In the following π and $\pi_{\rm H}$ stand for the true and measured parallax of a star.

Several quantities, such as the stellar distance or the absolute magnitude, have a non-linear dependence, $h(\pi)$, on the parallax. In this case, the expectation value of the function, $E[h(\pi_{\rm H})]$, is in general different from $h(\pi)$, even if the individual Hipparcos parallaxes are unbiased, i.e. if $E[\pi_{\rm H}] \approx \pi$. In other words, $1/\pi_{\rm H}$ is a biased estimate of the star's true distance, and $m+5\log\pi_{\rm H}+5$ is a biased estimate of its absolute magnitude: $E[1/\pi_{\rm H}] \neq 1/\pi$ and $E[m+5\log\pi_{\rm H}+5] \neq m+5\log\pi+5$. The expectation values of distance and absolute magnitude can be calculated. For example, if parallaxes below 0.01 mas, which is the smallest non-zero value of parallax that can be found in the catalogue, are excluded the computed distance may be overestimated by more than 100 per cent when the relative error in π is 100 per cent. Note that if negative parallaxes are not rejected, the bias, although reduced, is still present. Absolute magnitudes computed from observed parallaxes are on the average 0.2 mag too bright when the relative parallax error is about 50 per cent, and 0.6 mag too faint for a 200 per cent relative error.

Both bias and variance are negligible for relative errors better than about 10 per cent. For parallaxes with a higher relative error, one could naively hope to correct the computed distance from the calculated bias, but this is not possible because the bias is a function of $\sigma_{\pi_{\rm H}}/\pi$, where the *real* parallax π is unknown. What is available is *not* the real relative parallax error but the observed one $\sigma_{\pi_{\rm H}}/\pi_{\rm H}$. Given the uncertainty on the true relative error, a bias correction is simply not feasible. On the other hand, such a correction would only have a statistical meaning when applied to a sample, but would be questionable on an individual basis.

Another source of biases are the selection effects in a sample. A well known selection bias is the Malmquist bias. The use of the parallaxes of a truncated sample without caution may lead to similar biases in the derived results. Another type of bias is caused by an indirect truncation of the parallaxes. For instance, suppose that the spatial velocities of a given sample are computed

J. Andersen (ed.), Highlights of Astronomy, Volume 11A, 547–548.

and stars with high spatial velocity are selected. This will select stars with a truly high velocity but also stars with an overestimated distance ($\pi_{\rm H} \ll \pi$): the estimated distances of objects in this subsample will be biased in the mean. Consequently its estimated mean absolute magnitude will be too bright.

Summarizing, one can easily calculate the expected biases for a *given* true parallax. However, one only has the observed values, so the correction will depend on what kind of assumption one makes concerning the true values. In other words, the *distribution* of the true parallaxes has to be known, and this is an astrophysical question, not a statistical one! Hence, one cannot solve this problem just by statistics, but needs also some kind of modeling of the objects or sample under study. The reader is strongly encouraged to perform a detailed analysis of this sort *for each specific case* in order to obtain a correct estimation of any parameter of a star or a sample of stars using trigonometric parallaxes. This means in particular that one should neither ignore the possible biases nor apply blindly 'Malmquist' or 'Lutz-Kelker' corrections.

Use of the Covariance Matrix. One of the unique features of the Hipparcos Catalogue is that not only the standard errors of the five astrometric parameters are provided but also their correlation coefficients. This allows the user to make full use of the information contained in the astrometric parameters. Here we concentrate on using the covariance matrix when interpreting the statistics of a particular data set.

If $\mathbf{x}$ is an observed vector with covariance matrix $\mathbf{C_x}$ then the confidence region around $\mathbf{x}$ is given by: $c = \mathbf{x}'\mathbf{C}_\mathbf{x}^{-1}\mathbf{x}$, where the prime denotes matrix transposition. The distribution of c is described by a χ^2_ν probability distribution, where ν, the number of degrees of freedom, is equal to the dimension of $\mathbf{x}$. In the one-dimensional case this reduces to the well-known Gaussian distribution, where $c = 9$ corresponds to '3σ', the 99.73 per cent confidence level. For other values of ν the value of c will be higher for the same confidence level. It is 11.8 for $\nu = 2$ and 14.2 for $\nu = 3$. Note that the distribution of the errors around $\mathbf{x}$ is described by a multi-dimensional Gaussian and the equation above describes a confidence 'ellipsoid' around $\mathbf{x}$. If the vector $\mathbf{y}$ is derived from $\mathbf{x}$ via some transformation $\mathbf{f}(\mathbf{x})$, the covariance matrix of $\mathbf{y}$ is: $\mathbf{C_y} = \mathbf{J}\mathbf{C_x}\mathbf{J}'$. Here $\mathbf{J}$ is the Jacobian matrix of the transformation from $\mathbf{x}$ to $\mathbf{y}$: $[\mathbf{J}]_{ij} = \partial f_i/\partial x_j$. Thus one can calculate the covariance matrix of any set of variables derived from the observed astrometric parameters.

We now turn to the example of space velocities for cluster stars, specifically the Hyades. For full details we refer the reader to Perryman *et al.* (1997). When deriving space velocities for cluster stars we make use of the observed vector $(\pi, \mu_{\alpha*}, \mu_\delta, V_{\rm R})$, where $\mu_{\alpha*} = \mu_\alpha \cos\delta$, and $V_{\rm R}$ is the radial velocity. This vector is transformed to a space velocity, implicitly invoking a transformation to $(V_{\alpha*}, V_\delta, V_{\rm R})$ ($V_{\alpha*} = \mu_{\alpha*}A_v/\pi, V_\delta = \mu_\delta A_v/\pi$, $A_v = 4.74047...$ km yr s^{-1}). One can show that even in the absence of correlations between astrometric errors, the parallaxes and velocity components $V_{\alpha*}$ and V_δ will in general be correlated. Moreover, because of the position of the convergent point of the Hyades the correlated errors will lead to systematic behaviour of the uncertainties in the sample as a whole. These systematics will be transferred to the space velocities.

In the case of the Hyades both the overall distribution of the velocity residuals (with respect to the mean cluster motion), as well as the correlation of the direction of the residuals with spatial position, can be fully attributed to observational errors. We stress here that ignoring the covariance matrix can easily lead to false interpretation of, for example, kinematic data. For cases other than the Hyades the way in which features due to correlated errors enter may differ. It is important to carry out this kind of analysis and consider the implications for each case individually.

Finally, we refer again to Brown *et al.* (1997) and Perryman *et al.* (1997) for many more details and further references than can be provided here.

References

ESA, 1997, *The Hipparcos and Tycho Catalogues*, ESA SP-1200

Brown, A.G.A., Arenou, F., Lindegren, L., van Leeuwen, F., Luri, X., 1997, in *Hipparcos Venice '97*, ESA SP-402

Perryman, M.A.C., Brown, A.G.A., Lebreton, Y., Gómez, A., Turon, C., Cayrel, G., Mermilliod, J.-C., Robichon, N., Kovalevsky, J., Crifo, F., 1997, *Astron. & Astroph.*, in press

BINARIES IN ACCELERATION AND STOCHASTIC HIPPARCOS SOLUTIONS

F. ARENOU
DASGAL/URA 335 du C.N.R.S. - Observatoire de Paris-Meudon,
F-92195 Meudon Cedex - E-mail: Frederic.Arenou@obspm.fr

The Hipparcos astrometric parameters were obtained through the merging of abscissae data obtained by the FAST and NDAC Data Reduction Consortia. Using a different reduction scheme, other astrometric solutions may be obtained, although not statistically different: a weighted solution using FAST and NDAC astrometric solutions is shown, which gives both astrometric data and associated covariance matrix. This is applied to the acceleration and stochastic solutions. The intersection between merged and weighted solutions is then studied from the point of view of the possible signature of astrometric binaries.

New astrometric binaries were found during the Hipparcos data reduction process and their orbits were determined or improved. Other astrometric binaries may however be detected by searching for a quasi-sinusoidal signature in right ascension and declination. For this purpose the observations of the candidate stars have been combined at a few different epochs, using the Hipparcos astrometric intermediate data.

A simultaneous fit of the astrometric and orbital parameters has then been done. However in most of the cases other ground-based data is needed in order to confirm the suspected binarity and to give a reliable orbit.

APERTURE SYNTHESIS OF HIPPARCOS TRANSIT DATA

C.F. QUIST, L. LINDEGREN, S. SÖDERHJELM
Lund Observatory
Box 43, SE-221 00 Lund, Sweden

The ESA Hipparcos astrometry satellite did not contain an imaging detector (such as a CCD) but used a modulating grid of alternating transparent and opaque bands to modulate the light from a given object. The Fourier coefficients of the modulated detector signal were the basis for the astrometric and photometric processing leading to the Hipparcos Catalogue. These coefficients are included (for a subset of the stars) in the CD-ROM version of the Hipparcos Catalogue as the 'Hipparcos Transit Data'. In this work, the Transit Data are used in a completely different and novel application: the modulation components from a single scan are considered to be equivalent to the interference fringes from an interferometer oriented along the scan direction and having baselines of 0, $\simeq$ 10 and $\simeq$ 20 cm and a wavelength of 550 nm. Using all the scans collected over the mission lifetime, it is possible to use the complex visibilities (containing both amplitude and phase information) calculated from the Transit Data to create aperture synthesis images of various type of objects such as visual or orbiting binary systems. More information, including sample images, can be found in: Quist C.F., Lindegren L., Söderhjelm S., 1997, Using Hipparcos Transit Data for Aperture Synthesis Imaging, Hipparcos Venice '97 Symposium, ESA SP-402 (in press).

J. Andersen (ed.), Highlights of Astronomy, Volume 11A, 549.

HIPPARCOS STARS IN THE GCVS

N.N. SAMUS
Institute of Astronomy, Russian Acad. Sci., 48, Pyatnitskaya, Moscow 109017, Russia

Prof. M. Grenon asked the team of the General Catalogue of Variable Stars to attribute GCVS names to new Hipparcos variables in order to present, in the Hipparcos Catalogue, the information on these stars already with final designations. We received lists of 5665 variables. Of them, 3157 qualified to the standard GCVS criteria and will appear in the 74th Namelist of variable stars.

The main problems in compiling the Namelist were the following:

1. Cases of spurious variability for double and multiple stars and for misidentified stars (wrong colour indices used in reductions). Some wrong identifications come from the SIMBAD data base.

2. Lacking identifications with catalogues of spectroscopic binaries. Some stars attributed to slowly pulsating B variables are in fact ellipsoidal variables. HIP 98954 = V1472 Aql was classified as a semiregular star, but it is a rather unusual, probably eclipsing, star (Samus, 1997), a known spectroscopic-binary M giant with a 200^d period (Lucke and Mayor, 1982).

References

Lucke, P.B. and Mayor, M. (1982) *Astron. & Astrophys.*, **105**, 318
Samus, N.N. (1997) *Inf. Bull. Var. Stars*, No. 4501 (in press)

J. Andersen (ed.), Highlights of Astronomy, Volume 11A, 550.

CPC2 REDUCTION WITH HIPPARCOS AND PROPER MOTIONS IN THE SOUTHERN HEMISPHERE

N. ZACHARIAS[1], M.I. ZACHARIAS[2], C. de VEGT[3], C.A. MURRAY[4]
[1] *U.S. Naval Observatory, 3450 Mass.Av.NW, Washington DC 20392, USA*
[2] *Universities Space Research Association, Washington DC, USA*
[3] *Hamburger Sternwarte, Univ.of Hamburg, Germany*
[4] *12 Derwent Rd, Eastbourne, Sussex BN20 7PH, UK*

The Second Cape Photographic Catalog (CPC2) contains 276,131 stars covering the entire Southern Hemisphere in a 4-fold overlap pattern. Its mean epoch is 1968, which makes it a key catalog for proper motions. A new reduction of the 5687 plates using on average 40 Hipparcos stars per plate has resulted in a vastly improved catalog with a positional accuracy of about 40 mas (median value) per coordinate, which comes very close to the measuring precision. In particular, for the first time systematic errors depending on magnitude and color can be solved unambiguously and have been removed from the catalog. In combination with the Tycho Catalogue (mean epoch 1991.25) and the upcoming U.S. Naval Observatory CCD Astrograph Catalog (UCAC) project proper motions better than 2 mas/yr can be obtained. This will lead to a vastly improved reference star catalog in the Southern Hemisphere for the final Astrographic Catalogue (AC) reductions, which will then provide proper motions for millions of stars when combined with new epoch data. These data then will allow an uncompromised reduction of the southern Schmidt surveys on the International Celestial Reference System (ICRS).

ASTROMETRY OF POSS-II PLATES USING TYCHO

B. BUCCIARELLI[1], J.E. MORRISON[2], B.J. MCLEAN[2], C.R. STURCH[2,3]
[1] *Astronomical Observatory of Torino, Italy*
[2] *Space Telescope Science Institute, USA*
[3] *Computer Science Corporation at STScI*

We present a first attempt to compute an astrometric solution of POSS-II plates using TYCHO stars as the reference catalogue. Also, we have made use of a special catalogue of stars observed with the *Carlsberg Meridian Circle* in La Palma between 1994 and 1995, brought into the HIPPARCOS reference system. Such a catalogue consists of a double strip along the equator, fairly dense (approximately 10 stars per square degree), faint (average magnitude 12.5), with an average error at epoch of ~ 150 *mas* per coordinate. Since the limiting magnitude of TYCHO stars is about 11.5, almost exactly one magnitude less than the CAMC special catalogues' average, the latter has served as a very useful external comparison, although a more interesting magnitude range would have been 14-15. In this respect, we note that the Schmidt plates under study cover a large area (6.4×6.4 degree) and have fairly long exposures (~ 60 *min*); therefore, a possible magnitude dependence of the model needs to be carefully investigated. The results shown in the table above, which were obtained using a 2nd order polynomial in conjunction with a *mask* method using 5 plates only are very promising. The average *external* error over the entire plate area is about 0.2 *arsec* per coordinate, after the CAMC error have been subtracted.

		TYCHO rms (arsec)			CAMC rms (arsec)	
Plate ID	TYCHO stars	Ra	Dec	CAMC stars	Ra	Dec
A094	1019	0.21	0.20	358	0.30	0.30
A096	1057	0.22	0.20	346	0.29	0.32
A0OZ	586	0.18	0.16	358	0.24	0.26
A0NR	509	0.14	0.15	398	0.26	0.25
A15M	379	0.16	0.16	346	0.25	0.27

J. Andersen (ed.), Highlights of Astronomy, Volume 11A, 551.

IMPACT OF THE HIPPARCOS DATA ON THE ASTROMETRIC REDUCTION OF THE OUTER PLANETS

A. FIENGA
Bureau des Longitudes, 77, Av. Denfert-Rochereau 75014 Paris, France

We investigated the improvement of photographic astrometric observations of outer planets by use of Hipparcos Catalogue stars. The position of Jupiter on the plates was determined from measurements of the Galilean satellites, combined with ephemerides accurate to better than 10 mas. In a 6 contants model of photographic plates reduction, when images are near the center of the field, only offsets have to be computed, by considering the scale and orientation known. We applied this idea to the galilean satellites by using 137 photographic plates realised between 1967 and 1974 by D. Pascu, with the Mc Cormick 26-inch refractor, Virginia, and its twin in the USNO, Washington DC. A study of the scale factors reveals an expected anisotropy of the scale factors. One possible hypothesis is that this effect is a consequence of the thermal expansions of the telescopic components. The use of Hipparcos data in the astrometric reduction allows to obtain an external error of around 40 mas on the determination of the observed absolute position of Jupiter. It is 5 times better than the classic computations using photographic or CCD reductions. Furthemore, with such results, badly-known effects can be easly estimated like the anisotropy of scale. At the end, important impacts on the fit of planetary ephemerides to the visual observations are expected.

A PRELIMINARY STUDY ON THE IMPROVEMENT OF PROPER MOTIONS FOR HIPPARCOS STARS BY USING PHOTOGRAPHIC PLATES

W. JIN, Z. TANG, J. LI, S. WANG
Shanghai Observatory, Academia Sinica, 80 Nandan Road, Shanghai 200030, China

The 3435 photographic plates, which were obtained by using the 40cm astrograph at Zô-Sè station of Shanghai Observatory during 1901-1993, have been accumulated for observing open clusters, extragalatic nebulas, radio stars and variable stars etc. These plates cover 764 unoverlapping regions ($2^\circ.3 * 1^\circ.7$) on the sky. Most of the plates distribute along the ecliptic, about 10 000 Hipparcos stars, 27 000 PPM stars and 500 FK5 stars are included in these regions. Among these plates, there are 2718 plates having two observing epochs, and they cover 359 different regions; there are 142 and 92 regions where the intervals of two epochs are about 50 and 70 years, and the numbers of plates in these regions are about 1494 and 1023 respectively. About 372 plates were photographed before 1949, which play an important action for improvement of proper motions of Hipparcos stars.

Four methods used to improve the proper motions for Hipparcos stars were designed. If the 2718 plates mentioned above having two epochs were used for improvement of the proper motions of Hipparcos stars, the stellar number of 299, 1378, 377 and 1664 corresponding to the four methods will be obtained with final precision 0.7 mas/yr. If supplementary observations are made in 1997, the stellar number of 494, 2431, 1534 and 2899 will be obtained with the same precision.

J. Andersen (ed.), Highlights of Astronomy, Volume 11A, 552.

RIGIDITY ESTIMATION OF THE HIPPARCOS SYSTEM IN THE EQUATORIAL ZONE BY 20-TH CENTURY GROUND-BASED OBSERVATIONS

Y. B. KOLESNIK
Institute for Astronomy of the Russian Academy of Sciences, 109017,48 Piatnitskaya St.,Moscow, Russia

The external systematic accuracy of the HIPPARCOS positions and proper motions in the equatorial zone as represented by the subset of its bright FK5 stars is estimated by comparison with modern ground-based catalogues of 1980-1990s as well as with the most precise compiled catalogues made in the 20th century. Significant zonal and regional systematic differences between HIPPARCOS and ground-based catalogues have been detected for the different epochs ranging from 1920 to 1990. The most typical behaviour of the analysed differences at the epoch 1991.25 is almost absence of small-scale deviations and presence of the large-scale features. For the epoch 1920, both large and small-scale discrepancies are more pronounced reaching up to 50 mas. This gives evidence of the inferiour consistency of the HIPPARCOS system with ground-based observations comparing with the best compiled catalogues.

EARTH ORIENTATION PARAMETERS 1899.7–1992.0 IN THE HIPPARCOS REFERENCE FRAME

J. VONDRÁK[1], C. RON[1], I. PEŠEK[2], A. ČEPEK[2]
[1] *Astronomical Institute, Boční II, 141 31 Prague 4, Czech Republic*
[2] *Czech Technical University, Thákurova 7, 166 29 Prague 6, Czech Republic*

The optical astrometry observations of latitude/universal time variations made with 48 instruments at 31 observatories are used to determine the Earth orientation parameters (EOP) since the beginning of the century. The Hipparcos Catalogue is used to bring more than four million individual observations, made in the interval 1899.7–1992.0, into the International Celestial Reference System. The Earth orientation parameters (polar motion, celestial pole offsets and, since 1956.0, also universal time UT1) are determined at 5-day intervals, with average uncertainties ranging from 8 mas (in the eighties) to about 40 mas (in the forties). Making use of very long series of ground-based observations, the solution also leads to the improvement of proper motions of about ten per cent of the observed Hipparcos stars, with precision of $\pm 0.2 - 0.5$ mas/yr. In addition, 474 auxiliary parameters, describing the rheological properties of the Earth and seasonal deviations of the observations at contributing observatories, are found. The new solution provides the EOP series suitable for further analyses, e.g., for studying long-periodic polar motion, length-of-day changes or precession/nutation.

J. Andersen (ed.), Highlights of Astronomy, Volume 11A, 553.

CHECK ON JPL DExxx USING HIPPARCOS AND TYCHO OBSERVATIONS

L.V. MORRISON[1], D. HESTROFFER[2], D.B. TAYLOR[1], F. VAN LEEUWEN[1]
[1] *Royal Greenwich Observtvatory, Cambridge, UK*
[2] *ESTEC, ESA, NL*

Obervations of the positions of Europa (J2) and Titan (S6) by Hipparcos, and Ganymede (J3) and Callisto (J4) by Tycho are analysed to give checks on the latest JPL ephemerides of the planets Jupiter and Saturn.

The observed positions of the satellites are compared with DExxx, using the G5 theory of the Galilean satellites and D.B. Taylor's theory of Titan to calculate their offsets from the barycentres of the two systems.

Each Tycho observation can be resolved into RA and Dec because the slits are inclined to the scan direction. Hipparcos, on the other hand, which has a uni-directional scan, requires a minimum of two (preferably orthogonal) scans to resolve the offsets into RA and Dec. We have subdivided the Hipparcos observations in time and solved for corrections to DExxx at several epochs over the period of the mission.

The Hipparcos observations of J2, and Tycho of J3 and J4, put tight constraints on the orbit of Jupiter, and agree closely with the series of ground-based observations made by the Carlsberg Meridian Telescope. In particular, they help to resolve a problem in the declination of Jupiter associated with larger than expected errors in the tracking of Voyager II and radio positions from the VLA.

J. Andersen (ed.), Highlights of Astronomy, Volume 11A, 554.

HIPPARCOS AND THEORY OF STELLAR INTERIORS

What is being learned ?

A. BAGLIN
DESPA, Observatoire de Paris, URA CNRS 264,
5 place J. Janssen, Meudon CEDEX, France.

1. Introduction

The theory of stellar interiors is generally considered as well settled, and many branches of astrophysics rely on its results. However, the true situation is not so bright. Despite great successes at its beginning, in the sixties, it has to deal with many physical processes, generally poorly understood, which have large influences on the results and the derived quantities, like for instance ages.

After a rapid description of these difficulties, I will recall the major questions of the confrontation between theory and observations, and how HIPPARCOS can and will help. As an illustration, a few preliminary results will be presented, and future prospects foreseen. But, this short review is by no means exhaustive; there are many more results dealing directly or indirectly with stellar structure theory which have already been presented at the Venice Symposium (Perryman *et al.*, 1997a), and many more, I am not aware of, will certainly appear soon.

2. The Present Status of Internal Structure Theory

This question is developed in many text books, and a recent critical review can be found in (Lebreton and Baglin, 1995). Standard stellar modelling rely, up to now, on hydrostatic models neglecting most of the hydrodynamical processes or replacing them by a very crude phenomenology. The microscopic description of the stellar material is generally safer, and has been seriously improved in the recent years. Several observational facts find no explanation in this framework showing that something is missing in the description.

The present major uncertainties concern hydrodynamical processes, like the treatment of the energy transport in convective regions, the extension of the convectively mixed regions, either convective cores or outer convective envelopes, and the transport of chemicals and momentum in radiative zones. Models are still generally ignoring the non axisymetric processes, as for instance rotation, and the hydrodynamical instabilities that they produce; see i.e. Zahn (1995), Schatzman (1995), Roxburgh (1997).

One of the major problem in testing the physical processes by confronting observational data with predicted ones, is that, for an individual object, the number of observables is generally smaller than the number of unknown quantities. Then, this comparison accepts a large variety of solutions, and additional hypothesis are needed. Dealing with 'homogeneous' ensemble of objects, like clusters or binaries, helps to reduce this indetermination. But, increasing the number of observables, for instance measuring radii and eigenmode frequencies, as expected, is the only way to solve the question.

In this comparison, the determination of the total luminosity is of great importance. Up to now, except for a few very nearby objects, the accuracy of direct determinations was too small to be useful, and one had to rely on photometric calibrations. HIPPARCOS, as expected, contributes very significantly in reducing these error bars in the Hertzprung-Russell diagram (Baglin, 1988).

3. The solar-like non evolved objects and the convection parameter.

These stars have convective envelopes and radiative interiors; they are slow rotators, and their evolutionary time scale is long, so that most of them are still on the main sequence. The treatment

J. Andersen (ed.), Highlights of Astronomy, Volume 11A, 555–557.

of the convective transport remains very rough though progress is underway, and up to now the available phenomenological treatments contain an almost free parameter. But the precise slope of the intermediate zero age main sequence ($0.6 \leq M_{\odot} \leq 1$) can help.

Assuming homogeneity of the chemical composition and of the date of formation, ther position in the HR diagram depends essentially on the efficiency of the convective transport in the external outerlayers, which governs the temperature gradient, i.e. the radius and then the effective temperature. In the classical phenomenological formalism, the mixing length parameter α determines the slope of the main sequence (Fernandes, 1997). This slope is now well defined observationally, and common to many different objects, the Hyades and several nearby open clusters (Mermilliod *et al.*, 1997), the field stars and subdwarfs, for which effective temperature and metal content are well constrained by detailed spectrocopic analysis and luminosities are derived from HIPPARCOS. It is compatible with a constant value of α, in the region between 1 and 0.6 $M_{\odot}$, $\alpha = \alpha_{\odot}$, where $\alpha_{\odot}$ is the value derived from the solar calibration, using the same physical description of the interior.

This result is valid for a large interval of metal content Z, but has not been checked for giants. Equivalent results are obtained with the more refined FST description (Canuto *et al.*, 1996).

On can also gain a lot by calibrating nearby visual binaries, for which individual masses, luminosities and effective temperatures, as well as their common metal content Z are measured. As, for these objects the number of unknown quantities is equal to the number of free parameters, a calibration similar to the one conducted for the Sun is possible (Noels *et al.*, 1991); ages, helium contents, and α can be determined. Once more, with the same standard physics (Fernandes *et al.*, 1997), the same value $\alpha_{\odot}$ is obtained, confirming the previous statement.

HIPPARCOS plays a crucial role here in increasing both the accuracy of the masses of already well known systems, and the number of double stars with accurate masses (Martin and Mignard, 1997), (Soderhjhelm *et al.*, 1997).

These results, showing that the phenomenological scaling of the treatment of the superadiabatic layers of outerconvective zones is now settled, allow to go a step further. It becomes now possible to determine the Helium content Y for groups of non evolved objects, as i.e. the Hyades cluster (Perryman *et al.*, 1997b), and several nearby open clusters (Robichon *et al.*, 1997). Ages of objects where Y is known, i.e. old subdwarfs, can also be estimated as a function of Y .

4. Position of the main sequence and metallicity

The drastic reduction of the uncertainty on luminosity allows now to built representative sets of 'well known' object, sorted rigourously to avoid pecularities, and for which detailed spectroscopic analysis reaches its best quality. With these criteria, sets are quite small; i.e., starting wih a sample of more than 160 candidates of main sequence stars of spectral type between F and K, with $-1 \leq [\frac{Fe}{H}] \leq 0.2$, only 38 satisfy the quality criteria necessary to make significant tests in the HR diagram, in particular $\sigma(\pi)/\pi \leq 0.05$ (Lebreton *et al.*, 1997).

For 'solar' metallicity, the width of the main sequence is quite small, in favor of a large value of the galactic enrichment ratio $\frac{\Delta Y}{\Delta Z}$. But a difficulty appears for stars with low metallicity $-1 \leq [\frac{Fe}{H}] \leq -0.45$. Standard models with this metallicity could occupy such a position only if their helium content Y was smaller than the primordial value. To solve this problem, a shift towards lower effective temperature of $\sim 250?K$, or towards higher luminosities of 0.2 mag is needed. And it seems to be definitely too large to be attributed to observational errors.

Going to the closest subdwarfs, $[\frac{Fe}{H}] \leq -1.0$, the sample is even more restricted, but the main sequence is well defined (Pont *et al.*, 1997). The same kind of difficulty arises, with a slightly smaller, but still significant amplitude: the shift in temperature is only of $\sim 150?K$, and in luminosity of 0.15, (Cayrel *et al.*, 1997).

If this effect is real, and not due to some uncertainties in the observed parameters or in the physical description, it looks as if the surface abundances (Z and/or Y) were not representative of the interior abundances. We already know from the Sun, that elements sedimentation seems to occur in radiative zones. But we also know that a complex interplay between sedimentation and mixing of the material is at work, and that its influence on stellar evoution can be important.

The present observed surface mixture would not represent the initial material from which the star was formed; it would be the result of a complex time dependent process, which has now to be understood.

If these results are confirmed, HIPPARCOS will contribute to built this new refined description of stellar interiors.

5. Dwarf A/F stars of the instability strip

For these stars, the main questions are the extension of convective core and the influence of rotation. Both have consequences on the age determinations. Improved luminosities from HIPPARCOS, show that, in the average they are brighter than predicted by the previous calibrations, though a large scatter exists (Liu *et al.*, 1997), (Høg and Petersen, 1997) . It has been also shown that variables, non variables and peculiar stars occupy the same domain of the HR diagram (North *et al.*, 1997).

Unfortunately, the size of the convective core cannot be measured with the HR diagram alone. As most of these objects are pulsating, seismology will be the appropriate tool. But to be applied succesfully, it needs an excellent knowledge of the basic parameters of the object, which is reached presently only for the closest ones (Lebreton *et al.*, 1995), and also a correct treatment of rotation. For instance, the new position of the binary system θ^2 Tau in the Hyades, shifts significantly the predicted eigenfrequencies ($\sim 100 \mu Hz$), and consequently the interpretation of the observed spectrum.

6. Prospects

This very preliminary and uncomplete sketch shows already that the HIPPARCOS mission, combined with progresses in other domains, contribute to the renewal of stellar structure and evolution theory. Convergence of better physics and observations raises new problems to fundamental physics and cosmology. In the domain of astrophysics, it is demanding more to stellar atmosphere theory and data acquisition of spectroscopic parameters which are now limiting the accuracy of the Observation/Theory confrontation.

But, one has to remind that the domain in which HIPPARCOS is largely improving the accuracy on luminosities remains very limited:

$$\frac{\Delta L}{L} \sim 2 \frac{\sigma(\pi)}{\pi} \leq \delta(B.C.), \; if \; \pi \leq 10 \; parsecs$$

where $\delta(B.C.)$ is the uncertainty on the bolometric correction.
This is why one forsees already a need for even higher accuracy parallaxes to obtain higher precision for a few closeby objects, as well as the extension of the present accuracy to a larger sample of stars, and also to rare types of objects.

References

Baglin, A.(1987) in *Scientific aspects of the Input Catalogue Preparation II*, J. Torra and C. Turon eds, p. 103.
Canuto,V.M., Goldman,I., Mazitelli, I.(1996) Astrophys. J., **473** , 550.
Cayrel, R., Lebreton, Y., Perrin, M.N., Turon, C. (1997), *Hipparcos symposium* Venice, ESA publ. SP-402, p. 219.
Fernandes, J. (1997) Ph. D. Thesis, Paris VII University
Fernandes, J., Lebreton, Y., Baglin, A. (1997), Astron. Astrophys, in press.
Høg, E., Petersen, J.O.(1997) , Astron. Astrophys., in press.
Lebreton, Y., Michel, E., Goupil, M.J., Baglin, A., Fernandes, J. (1994) , IAU Symposium 166, p.135.
Lebreton, Y., Baglin, A.(1995) in *Stellar evolution, what should be done?* 32th Liege Inst. Astroph. Coll, p.1.
Lebreton, Y., Perrin, M.N., Fernandes, J. et al. (1997) *Hipparcos symposium* Venice, ESA publ. SP-402, p. 379.
Liu, Y.Y, Baglin, A., Auvergne, M. et al. (1997) *Hipparcos symposium* Venice, ESA publ. SP-402, p. 363.
Martin C, Mignard F. (1997), Astron. Astrophys, in press
Mermilliod, J.C., Turon, C., Robichon, N. et al. (1997), *Hipparcos symposium* Venice, ESA publ. SP-402, p. 643.
Noels , A., Grevesse, N., Magain, P., Neuforge, C., Baglin A., Lebreton, Y. (1991), Astron, Astrophys, **247**, 91.
North, P., Jaschek, C., Egret, D. (1997) *Hipparcos symposium* Venice, ESA publ. SP-402, p. 367.
Perryman M.A.C. et al. (1997a), *Hipparcos symposium* Venice, ESA publ. SP-402,
Perryman, M.A.C., Brown, A.G.A., Lebreton, Y. et al. (1997b), Astron. Astrophys., in press.
Pont, F., Charbonnel, C., Lebreton, Y. et al. (1997), *Hipparcos symposium* Venice, ESA publ. SP-402, p. 699.
Robichon, N., Arenou, F., Turon, C., et al. (1997), *Highlights in Astronomy*, **11**, in press.
Roxburgh, I.W.,(1997) *IAU symposium 185* , in press.
Schatzman, E..(1995) in *Stellar evolution, what should be done?* 32th Liege Inst. Astroph. Coll, p.121.
Soderhjelm, S., Lindengren, L., Perryman, M.A.C. (1997), *Hipparcos symposium* Venice, ESA publ. SP-402,p.251.
Zahn, J.P.(1995) in *Stellar evolution, what should be done?* 32th Liege Inst. Astroph. Coll, p.111.

THE LUMINOSITY CALIBRATION OF THE HR DIAGRAM

A.E. GÓMEZ[1], X. LURI[1,3], M.O. MENNESSIER[2], J. TORRA[3]
AND
F. FIGUERAS[3], F. ROYER[1]
[1] *Observatoire de Paris-Meudon, D.A.S.G.A.L., CNRS URA 335*
Pl. J. Janssen, 92195 Meudon Cedex, France
[2] *Universite de Montpellier II*
Pl. E. Bataillon, 34095 Montpellier Cedex 5, France
[3] *Universitat de Barcelona*
Diagonal, 647, 08028 Barcelona, Spain

An extensive work on the luminosity calibration of the HR diagram has been performed using the LM method based on the Maximum Likelihood estimation (Luri et al. 1996). The method uses all available information: Hipparcos parallax and proper motion data as well as radial velocities and some measured spectroscopic or photometric parameters related to luminosity. On the other hand, the method takes into account the censorship of the sample and the errors of the data. Finally, it is able to identify and separate in a given sample, groups with different luminosity, kinematical or spatial characteristics providing not only the corresponding luminosity calibration relationship, but also improved individual absolute magnitude estimates.

In order to asses how reliable the luminosity calibrations based on the MK spectral classification are, the LM method has been applied to a sample of B to K stars with MK classification selected from the Hipparcos survey. All known binaries, variables and spectroscopically peculiar stars were removed from the sample. The final sample contains 22054 stars. Our results show that the relation between absolute magnitude and luminosity class has a large intrinsic dispersion. As a consequence, the assignation to a star of a mean absolute magnitude given by a MK classification calibration is a rough procedure. It is preferable to use the individual absolute magnitudes estimated using all the available information for each star, like the estimates provided by the LM method (Gómez et al. 1997).

The LM method is well adapted to study the evolutionary state of Barium stars which spread out over a large range in luminosity, from dwarfs to supergiants, most of them showing luminosities of normal giants. From a sample of 297 stars, our results confirm that Ba stars constitute an inhomogeneous group. Five groups were found: supergiants, normal giants, giants concentrated in the clump, dwarfs and a small group of stars with kinematical behaviour of halo stars with a large spread in luminosities. Some of the supergiants could be 'false' Ba stars. Moreover, stars with masses greater than 3 $M_\odot$ are found in contradiction with current theories for the enhancement of s-process elements in Ba stars (Mennessier et al. 1997).

A Period-Luminosity relation in the K band has been obtained for Oxygen-rich Mira variables (103 stars). Two main groups has been found corresponding to old disk stars and to a mixing of thick disk and halo stars, respectively. The two PL relations have similar slopes, the shift being probably due to metallicity effect (Alvarez et al. 1997).

References

Alvarez, R., Mennessier, M.O., Barthès, D., Luri, X., Mattei, J. (1997), *A&A*, in press
Gómez, A.E., Luri, X., Mennessier, M.O., Torra, J., Figueras, F. (1997), *Hipparcos Venice'97*, **ESA SP-402**, in press
Luri, X., Mennessier, M.O., Figueras, F., Torra, J. (1997), *A&AS*, **117** pp. 405–415
Mennessier, M.O., Luri, X., Figueras, F., Gómez, A.E., Grenier, S., Torra, J., North, P. (1997), *A&A*, in press

J. Andersen (ed.), Highlights of Astronomy, Volume 11A, 558.

PARALLAXES AND PROPER MOTIONS OF PROTOTYPES OF ASTROPHYSICALLY INTERESTING CLASSES OF STARS

V. TRIMBLE [1,2], G.H. HERBIG [3], A. KUNDU [2,4]
[1] *Physics Dept., Univ. of California, Irvine CA 92697 USA*
[2] *Astronomy Dept., Univ. of Maryland, College Park MD 20742 USA*
[3] *Inst. for Astronomy, Univ. of Hawaii, Honolulu HI 96822 USA*
[4] *Space Telescope Science Institute, Baltimore MD 21218*

Early release data were obtained under the 1982 announcement of opportunity for about two dozen stars belonging to seven classes whose basic properties or evolutionary status were uncertain at the time. A good many remain so. Many of the stars were too faint and far for parallaxes to be statistically significant, though many of the proper motions are. Two papers presenting the data have been accepted by archival journals (Trimble and Kundu 1997, 1998). Both bear the same title as this abstract. The first deals with the R Coronae Borealis variables, the second with the other six classes (FK Comae Stars, CVs, runaway T Tauri stars, and so forth).

References

Trimble, V. & Kundu, A. 1997. PASP (in press, October)
Trimble, V. & Kundu, A. 1998. AJ (in press, probably January)

J. Andersen (ed.), Highlights of Astronomy, Volume 11A, 559.

THE AGE OF OLD GALACTIC POPULATIONS

M. GRENON
Geneva Observatory
1290 Sauverny, Switzerland

1. Introduction

As a preparation to the HIPPARCOS mission, a large observing programme on NLTT stars (proper-motion $> 0.18''$/yr) was started in Geneva photometry. The original programme consists of 10047 stars brighter than mR = 11.5, or mR = 12.5 if of colour class m. Among them, 7813 targets could be included in the HIPPARCOS programme, selected according to their observability and internal priorities in favour of large parallaxe stars (photometric distances < 100 pc) and high-velocity stars. The bulk of new nearby, halo, mild-metal poor and SMR stars in the HIP Catalogue originates from this proposal (N° 139). No less than 208 new nearby stars with $\pi \geq 40$ mas were discovered south of δ +10°, the closest has π(HIP)= 182 mas. Radial velocities were obtained with CORAVEL at OHP and ESO. Most aspects of the early evolution of the Galaxy may be addressed with this sample. Here we discuss, as examples, the ages of the thick disk and of the galactic bulge.

2. The stellar physical and kinematical parameters

For stars measured with Geneva photometry, the quantities Mv, T_{eff}, and, if T_{eff} > 4800°K, [M/H] are obtained with uncertainties of ± .15 mag, 20°K and 0.05 dex respectively. The Geneva calibration provides [Fe/H] = 0.142 ± 0.008 for Hyades for all T_{eff}. Space velocities and galactic orbits are computed for 4222 F to M stars. 2930 stars have now their kinematical and physical parameters presently defined, [M/H] included. This material, free from bias against metal-rich stars namely, is ideal to address the question: how old are the early galactic populations.

3. The thick-disk age

Because the space velocities and metal-abundances distributions overlap for halo, thick disk (TD) and old thin disk populations, the concept of TD is not unique in literature. Here we consider as typical TD members, stars having, in the solar vicinity, V velocity component from -170 to -50 km/s, Zmax < 2 kpc and [M/H] from -0.90 to -0.55. The comparison of trigonometric and photometric parallaxes reveals binaries undetected by HIPPARCOS which may appear as spurious metal-poor stars. With the condition photometric and HIPPARCOS parallaxes > 10 mas, 198 stars are retained. Their mean [M/H] is -0.67. The HR-diagram of single TD stars shows a well defined isochrone, typical of metal-rich globulars, with Mv(RHB) = 0.75. With similar [M/H], the TD population and 47 Tuc show an age identity from their HR-diagram, i.e. δMv(mid-SGB - RHB) = 3.10 in both cases.

At mid-subgiant branch (SGB), effects as He sedimentation, rotation, C, N, O abundance anomalies cancel, and small errors on E(B-V) have a negligible impact. Mv(SGB) turns out to be the best age estimator for old mild metal-poor to metal-rich stars. Around log(T_{eff}) = 3.735, TD Mv(SGB) = 3.89, hence an age of 14 Gyr is found from Bergbush et al. (1992) isochrones. The fit of the empirical HR-diagram to theoretical isochrones requires a shift of -0.026 in theoretical log(T_{eff}) or 300°K around T_{eff} 5400°K, making the absolute dating still questionable.

J. Andersen (ed.), Highlights of Astronomy, Volume 11A, 560–561.

4. The age and evolution of the bulge

The simplest way to identify bulge-like stars in the solar vicinity is to select high eccentricity SMR stars with [M/H] higher than that of disk stars moving on nearly circular orbits. The majority of the 203 stars with $[M/H] > +0.30$ forms the metal-rich tail of the local disk having reached lately the maximum metallicity observable at solar galactocentric distance. Stars with $V <$ -110 to -50 km/s are members of the inner disk or of the bulge. From [O/Fe] anomalies Barbuy & Grenon (1990) assigned extreme SMRs to the bulge population. The local counterparts of the bulge are moving on trifoil orbits, with $Zmax \leq 1$ kpc and pericentric distances Rp as small as 2 - 3 kpc. The age of old SMRs is deduced from the HR-diagram Mv(HIP) versus T_{eff} from Geneva photometry. When binaries and blue stragglers are taken into account, no noticeable age dispersion is detected for stars with $Rp < 5.2$ kpc. Both from the turn-off location and the SGB luminosity, an age of 10.5 ± 0.5 Gyr is found from Schaerer et al.(1993) isochrones with $Z = 0.04$. SMRs represent few percent of the bulge population; their age corresponds to the epoch of the end of the star formation in the bulge. [O/Fe] excess and [Ba/Fe] deficiency, (Castro et al. 1997) indicate that SN Ia and AGBs played a limited role in the bulge nucleosynthesis.

If few bulge SMRs are seen in the solar vicinity, less metal-rich bulge stars are also scattered out up to the solar orbit. When stars are selected according to $Rp \leq 5.2$ kpc, the metal abundance distribution happens to be the same as that observed in the bulge by Mc William & Rich (1994), i.e. a bell shaped distribution with [M/H] from -0.70 to +0.50, a maximum around -0.10 dex and a tail of more metal-poor stars as seen in the bulge. With the constraints from the SN II, SN Ia and AGB yields, the time scale for the bulge chemical evolution must be < 1 Gyr, thus the onset of star formation may have occured about 11 Gyr ago.

5. Summary

The inner old disk and the bulge appear to be built from the same stellar populations as for their age and chemistry are concerned, but with distinct subsequent kinematical evolution. The orbital diffusion, triggered by a central bar, played a major role on the old stars space distribution. It is why the metallicity gradient from old stars with eccentric orbits (Grenon 1987) tends to flatten or vanish inside the solar orbit, contrarily to the situation in the young disk as observed from B type stars or the gas. The thick disk appears related to the inner halo rather than to the thin disk although some continuity seems to exist between the thick and the thin disk. The time delay between the formation of the so-called thick-disk and the onset of the main burst of star formation in the bulge could be as large as 2 Gyr.

References

Barbuy B., Grenon M. (1990) Oxygen in bulge like SMR stars, in *Bulges of Galaxies*, B.J.Jarvis & D. Terndrup., Garching, ESO-CTIO Conf. Workshop Proc.

Bergbush P.A., VandenBerg D.A., *Astrophys.J.Suppl.Ser.* **81**, pp. 163-220

Castro S., Rich R.M., Grenon M., Barbuy B. (1997) *Astron.J.* **114** , pp. 376-999

Grenon M., (1987) *J.Astrophys.Astr.* **8**, pp. 121-193

Schaerer D., Charbonnel C., Meynet G., Maeder A., Schaller G. (1993) *Astron.Astrophys.Suppl.Ser.* **102**, pp. 339-342

Mc William A., Rich R.M. (1994) *Astrophys.J.Sup.* **91**, pp 749-999

NEW AGES FOR OLD CLUSTERS

I. N. REID
Palomar Observatory
105-24, Caltech, Pasadena, CA 91125, USA

A major result of the Hipparcos mission is the availability of accurate parallaxes for a much-increased sample of nearby subdwarfs. On the basis of those data, both Reid (1997-R97) and Gratton et al (1997-G97) have used main-sequence fitting techniques to obtain new estimates of globular cluster distances, deriving significantly larger moduli (by 0.15 to 0.3 mag) for extreme, metal-poor clusters such as M92 and M15. These re-calibrations are not solely the result of the revised parallax data - both D'Antona, Caloi & Mazzitelli (1997 - DCM) and Reid show that similar results can be derived from pre-Hipparcos subdwarf data if the calibrations are not tied exclusively to HD 103095. The longer distance scale leads to a brighter inferred turnoff, and, combined with the latest models (DCM), ages of only 11 to 13 Gyrs - substantially younger than the 16±2 Gyrs proposed by Bolte & Hogan (1992).

Recently, Pont et al (1997) have suggested that these revised distances reflect systematic bias in the calibration, notably underestimation of subdwarf abundances. Their distance modulus for M92 ($(m\text{-}M)_0$=14.68) is closer to the Bolte/Hogan value (14.65 mag) than those derived by G97 (14.82 mag) or Reid (14.93 mag). While we believe that Pont et al overestimate the bias, particularly where high-resolution spectroscopic analyses are concerned (G97, Reid, preprint), there is no question that significant uncertainties remain in the cluster distance calibration. *Colours*, rather than luminosities, are particularly vulnerable, and a mismatch of 0.01 in (B-V) translates to $\delta(m-M)_0 \sim 0.05$ mag. Amongst the problems are:

1) regardless of the random uncertainties in [Fe/H] for subdwarfs, there are *systematic* differences of up to 0.3 dex between different calibrations (Reid, preprint);

2) the $\delta(B-V)$ correction required to adjust subdwarfs to a mono-metallicity scale is poorly-defined empirically;

3) interstellar reddenings remain uncertain for many subdwarfs and clusters;

4) discrepancies of up to 0.04 mag exist in the colours adopted for individual subdwarfs, and most stars have only BV photometry;

5) there are systematic differences in the (M_V, (B-V)) isochrones predicted by the DCM and Vandenberg models - in *shape* as well as zeropoint. These discrepancies are often ignored.

Overall, the main legacy of Hipparcos should be a better appreciation of the true level of uncertainty in main-sequence fitting. There are over 200 HIP stars with parallaxes measured to $\frac{\sigma_\pi}{\pi} < 15\%$, and falling below the 47 Tuc main-sequence. Accurate photometry and high-resolution spectroscopy of those stars will go a long way toward setting cluster distances on a firmer footing.

References

Bolte, M. & Hogan, C. (1992) Conflict over the age of the Universe, *Nature*, **376**, 399-402

D'Antona, F., Caloi, V. & Mazzitelli, I. (1997) The Universe and globular clusters, *ApJ*, **477**, 519-534

Gratton, R.G., Fusi Pecci, F., Carretta, E., Clementini, G., Corsi, C.E., & Lattanzi, M. (1997), Ages of globular clusters from Hipparcos parallaxes of local subdwarfs *ApJ*, in press

Pont, F., Mayor, M., Turon, C., Vandenberg, D.A. (1997) Hipparcos subdwarfs and globular cluster ages, *Astr. Ap.*, in press

Reid, I.N. (1997) Younger and Brighter, *AJ*, **114**, pp. 161-179

J. Andersen (ed.), Highlights of Astronomy, Volume 11A, 562.

DISTANCE AND AGE OF M92 FROM HIPPARCOS SUBDWARFS

F. PONT[1], M. MAYOR[1], C. TURON[2], D.A. VANDENBERG[3]
[1] *Observatoire de Genève, 1290 Sauverny, Switzerland*
[2] *DASGAL/URA CNRS 335, Observatoire de Paris-Meudon, France*
[3] *Department of Physics and Astronomy, University of Victoria, Canada*

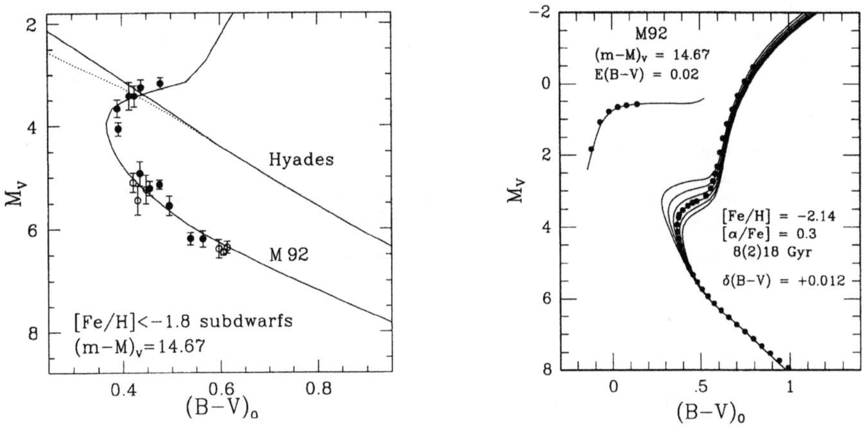

The parallax data gathered by Hipparcos for field subdwarfs allow a much more precise determination of globular cluster distances by direct sequence fitting than was previously possible. We determined the distance and age of the old, representative globular cluster M92 from a set of more than 500 subdwarf candidates with Hipparcos parallaxes. Precise [Fe/H] values were derived using the equivalent width of the CORAVEL cross-correlation function.

Our best estimate of the distance of M92 is $(m-M)_V = 14.67 \pm 0.08$ (including binaries with a statistical correction) or $(m-M)_V = 14.74 \pm 0.08$ (classic treatment, i.e. excluding binaries). The agreement of the cluster sequence with the position of extreme metal-poor field subdwarfs is remarkable [**figure, left**]. The distance found is slightly higher than previously thought. The corresponding ages, derived by comparing the luminosity of the turnoff and subgiant-branch stars with up-to-date evolution models [**figure, right**], are 14±1.2 Gyr or 13±1.2 Gyr respectively, implying a *minimum* age of 13 Gyr for the Universe.

Other authors have claimed that larger distances, and smaller ages, resulted from Hipparcos subdwarf parallaxes. Although there are some differences in the data sets used by each author, the main difference resides in the treatment of systematic and selection biases. We have examined at some length the biases affecting the determination of the mean luminosity of a set of subdwarfs selected in the usual way. By means of Monte Carlo simulations, we show that selection biases act in a direction opposite to the classic Lutz-Kelker bias affecting parallax data, and that they can be dominant. The biases introduced are of the order of 0.1 mag.

It is pointed out that, now that the whole Hipparcos catalogue is available, we shall be able to greatly reduce the systematic biases by refining the fitting procedure. Firstly, by imposing no [Fe/H] limits and fitting the subdwarf locus for all [Fe/H] values at the same time. Secondly, by imposing no parallax error limit and fitting the data in parallax space instead of magnitude space (comparing observed parallaxes directly with parallaxes predicted from the models).

J. Andersen (ed.), Highlights of Astronomy, Volume 11A, 563.

SPECTROSCOPIC RADIAL VELOCITIES: PHOTOSPHERIC LINESHIFTS CALIBRATED BY HIPPARCOS

D. GULLBERG, D. DRAVINS
Lund Observatory, Box 43, SE-22100 Lund, Sweden

Wavelengths of stellar spectral lines depend not only on the star's motion. Until recently, accurate studies of shifts not caused by radial motion were feasible only for the Sun. Solar lineshifts are interpreted as gravitational redshift (636 m/s) and convective blueshifts ($\sim$ 400 m/s; caused by velocity-brightness correlations). In other stars, such effects may be greater (Dravins & Nordlund 1990). Accurate astrometric radial velocities are now available from Hipparcos (Dravins et al. 1997a; 1997b), permitting studies of such shifts also in some other stars.

For such stars in the open clusters of Hyades, Ursa Major and Coma Berenices, a spectroscopic program is in progress, analyzing wavelength shifts in groups of lines with different strengths, excitation potentials, etc., using the ELODIE high-precision radial-velocity instrument (Baranne et al. 1996) at Haute-Provence Observatory.

References

Baranne, A., et al. (1996) A&AS 119, 373
Dravins, D., Nordlund, Å. (1990) A&A 228, 203
Dravins, D., Lindegren, L., Madsen, S., Holmberg, J. (1997a), HIPPARCOS Venice '97, ESA SP-402, p. 733
Dravins, D., Lindegren, L., Madsen, S., Holmberg, J. (1997b) Highlights of Astronomy 11 (this volume)

Further project description: http://www.astro.lu.se/~dainis/

ASTROMETRIC RADIAL VELOCITIES FROM HIPPARCOS

D. DRAVINS[1], L. LINDEGREN[1], S. MADSEN[1,2], J. HOLMBERG[1]
[1] *Lund Observatory, Box 43, SE-22100 Lund, Sweden*
[2] *Copenhagen University Observatory, J. Maries Vej 30, 2100 Copenhagen, Denmark*

Space astrometry now permits accurate determinations of stellar radial motion, without using spectroscopy. Although the feasibility of deducing astrometric radial velocities from geometric projection effects was realized already by Schlesinger (1917), only with Hipparcos has it become practical. Such a program has now been carried out for the moving clusters of Ursa Major, Hyades, and Coma Berenices. Realized inaccuracies reach about 300 m/s (Dravins et al. 1997).

Discrepancies between astrometric and spectroscopic radial velocities reveal effects (other than stellar motion) that affect wavelength positions of spectral lines. Such are caused by stellar surface convection, and by gravitational redshifts. A parallel program (Gullberg & Dravins 1997) is analyzing high-precision spectroscopic radial velocities for different spectral lines in these stars, using the ELODIE radial-velocity instrument at Haute-Provence.

References

Dravins, D., Lindegren, L., Madsen, S., Holmberg, J. (1997) *ESA SP-402*, p. 733
Gullberg, D., Dravins, D. (1997) *Highlights of Astronomy* **11** (this volume)
Schlesinger, F. (1917) *AJ* **30**, 137

Further project description: http://www.astro.lu.se/~dainis/

J. Andersen (ed.), Highlights of Astronomy, Volume 11A, 564.

HELIUM, [Fe/H] ABUNDANCES AND THE HR (LOG T_{EFF}, M_{BOL}) DIAGRAM WITH HIPPARCOS DATA OF THE FOUR NEAREST OPEN CLUSTERS: HYADES, COMA BERENICES, THE PLEIADES AND PRAESEPE

G. CAYREL de STROBEL, R. CAYREL, Y. LEBRETON
Observatoire de Paris, DASGAL/URA CNRS 335, 92195 MEUDON Cedex, France

After having studied in great detail the observational HR diagram (log T_{eff}, M_{bol}) composed by 40 main sequence stars of the Hyades (Perryman et al.,1997, A&A., in press), we have tried to apply the same method to the observational main sequences of the three next nearest open clusters: Coma Berenices, the Pleiades, and Praesepe. This method consists in comparing the observational main sequence of the clusters with a grid of theoretical ZAMSs. The stars composing the observational main sequences had to have reliable absolute bolometric magnitudes, coming all from individual Hipparcos parallaxes, precise bolometric corrections, effective temperatures and metal abundances from high resolution detailed spectroscopic analyses. If we assume, following the work by Fernandez et al. (1996, A&A,311,127), that the mixing-lenth parameter is solar, the position of a theoretical ZAMS, in the (log T_{eff}, M_{bol}) plane, computed with given input physics, only depends on two free parameters: the He content Y by mass, and the metallicity Z by mass. If effective temperature and metallicity of the constituting stars of the 4 clusters are previously known by means of detailed analyses, one can deduce their helium abundances by means of an appropriate grid of theoretical ZAMS's. The comparison between the empirical (log T_{eff}, M_{bol}) main sequence of the Hyades and the computed ZAMS corresponding to the observed metallicity Z of the Hyades (Z= 0.0240 ± 0.0085) gives a He abundance for the Hyades, Y= O.26 ± 0.02. Our interpretation, concerning the observational position of the main sequence of the three nearest clusters after the Hyades, is still under way and appears to be greatly more difficult than for the Hyades. For the moment we can say that:

- The 15 dwarfs analysed in detailed in Coma have a solar metallicity: [Fe/H] = -0.05 ± 0.06. However,their observational main sequence fit better with the Hyades ZAMS.
- The mean metallicity of 13 Pleiades dwarfs analysed in detail is solar. A metal deficient and He normal ZAMS would fit better. But, a warning for absorption in the Pleiades has to be recalled.
- The upper main sequence of Praesepe, (the more distant cluster: 180 pc) composed by 11 stars, analysed in detail, is the one which has the best fit with the Hyades ZAMS. The deduced 'turn-off age' of the cluster is slightly higher than that of the Hyades: 0.8 Gyr instead of 0.63 Gyr.

J. Andersen (ed.), Highlights of Astronomy, Volume 11A, 565.

THE CONTRIBUTION OF HIPPARCOS TO THE STUDY OF THE STELLAR METAL-RICH POPULATION IN THE SOLAR NEIGHBOURHOOD

G. CAYREL de STROBEL[1], C. SOUBIRAN[2], Y. LEBRETON[1]
[1] *Observatoire de Paris, DASGAL/URA CNRS 335, 92195 Meudon Cedex, France*
[2] *Observatoire de Bordeaux, BP 89, 33270 Floirac, France*

The '1996 Edition' of the Catalogue of [Fe/H] determinations by Cayrel de Strobel et al. (1997, A&A S 124,1) and two recent papers by Castro et al. (1997, AJ Vol.114, N.1) and by Feltzing and Gustafsson (A&A in press) have made possible to increase in the theoretical HR diagram (log T_{eff}, M_{bol}) the number of SMR stars.

The SMR Stars of this new enlarged sample had to have reliable absolute magnitudes, coming all from Hipparcos parallaxes, precise bolometric corrections, effective temperatures and metal abundances from high resolution detailed spectroscopic analyses.

With the help of an appropriate grid of isochrones computed by Lebreton (1997, Perryman et al. A&A, in press), 'turn-off ages' could then be attributed to the slightly evolved stars (subgiants) of the sample. The (log T_{eff}, M_{bol}) diagram constituted by the new sample of SMR stars, shows that the conclusions in a former paper by Cayrel de Strobel (1987, A&AJ 8,141) remain valid: the SMR stars are old stars in spite of their higher than solar metallicity. The result, that metal-rich stars were in the mean old stars, was interpreted in the 1987 paper as due to a more chemical uniformity of the nowadays interstellar medium of the Galaxy with respect of the older much more active interstellar medium.

THE ABSOLUTE MAGNITUDE OF THE EARLY-TYPE MK STANDARDS FROM HIPPARCOS PARALLAXES

C. JASCHEK [1], A.E. GÓMEZ[2]
[1] *Teso de la Feria, 39, E37008 Salamanca, Spain*
[2] *Observatoire de Paris, DASGAL/CNRS URA 335, 92195 Meudon Cedex, France*

We have analysed the standards of the MK system in the B0-F5 spectral region with the help of Hipparcos parallaxes, using only stars for which the error on the absolute magnitude is ≤ 0.3 mag. The sample stars (about one hundred) were scrutinized for companions and for interstellar extinction. We find that the main sequence is a wide band and that, although in general giants and dwarfs have different absolute magnitudes, the separation between luminosity class V and III is not clear. We conclude that there is no strict relation between luminosity class and absolute magnitude. The relation is only a statistical one and has a large intrinsic dispersion.

We have analysed similarly the system of standards defined by Garrison and Gray (1994) separating low and high rotational velocity standards. We find similar effects as in the original MK system.

References

Garrison, R.F., Gray, R.O. (1994) *AJ*, **Vol.104**, pp. 1556–1564

J. Andersen (ed.), Highlights of Astronomy, Volume 11A, 566.

LUMINOSITY OF δ SCUTI STARS AFTER HIPPARCOS SATELLITE

E. ANTONELLO, L. MANTEGAZZA, E. PORETTI
Osservatorio Astronomico di Brera, Via E. Bianchi 46, 22055 Merate, Italy

The absolute magnitudes of δ Scuti stars derived from parallaxes measured by the Hipparcos satellite were compared with the previous estimates based on photometric $uvby\beta$ indices, and significant differences were found which are related to photometric effects of metallicity and rotational velocity. A reliable calibration of M_V in terms of the photometric indices shall include an estimate of these effects. It is important also to take into account the possible presence of unresolved close companions in order to fully exploit the accuracy of M_V of nearby stars derived from the trigonometric parallaxes.

The M_V of few bright SX Phe stars support the period–luminosity relation obtained with ground based observations of globular clusters, while it does not seem to confirm the empirical dependence of this relation on the metallicity.

Some high amplitude δ Scuti stars with intermediate or normal metallicity and small and uncertain parallax have apparently a very low luminosity. Simulations of M_V determinations from observed parallaxes based on the discussion of observational errors by Lutz and Kelker (1973) have shown that the low luminosity could be an effect related to these errors.

References

Lutz, T.E. and Kelker, D.H. (1973) *Publ.Astr.Soc.Pacific*, **Vol. no. 85**, pp. 573–578.

THE DISTANCES AND ABSOLUTE MAGNITUDES OF SOME WELL-KNOWN RED VARIABLES

ROBERT F. WING
Astronomy Department, Ohio State University, Columbus OH, USA

Distances and absolute magnitudes have been obtained from an early distribution of Hipparcos data for a number of well-known late-type variables. These data were requested in 1981 in order to ensure that these important stars were not excluded from the Input Catalogue on account of their variability or very red color.

Measurements were requested for 20 Mira variables, and 14 of these were successfully observed. Of these, the nearest to the Sun were found to be R Leo (at 102 pc), χ Cyg (106), R Cas (107), R Car (128), T Cep (210), and R Aql (211), with formal errors ranging from ±12 to ±57 pc. Mira itself was not considered because the presence of a companion causes an 11-month wobble in its photocenter which is likely to lead to a spurious parallax. The absolute visual magnitudes of Miras at mean maximum range approximately from −2.5 to +2.0 with a clear dependence on spectral type (cooler stars being fainter). This trend is reversed for absolute magnitudes in the near infrared.

Results are also given for 10 small-amplitude variables of late type. The nearest of these are L_2 Pup (M5 III, 61 pc) and R Dor (M8e, 62 pc). The symbiotic star CH Cyg lies at a rather uncertain distance of 268 ± 65 pc.

Data were obtained for 23 carbon stars, including the brightest representatives of each of several subgroups. The nearest of these was found to be the SRb variable U Hya, at 162 ± 20 pc. The stars 19 Psc, Y CVn, W Ori, and HD 201626 were all found to be somewhat more than 200 pc away.

J. Andersen (ed.), Highlights of Astronomy, Volume 11A, 567.

THE DISTANCE, ABSOLUTE MAGNITUDE AND SPACE MOTION OF ALPHA ORIONIS

R. F. WING[1], E. F. GUINAN[2]
[1] *Ohio State University, Columbus OH, USA*
[2] *Villanova University, Villanova PA, USA*

Alpha Orionis (Betelgeuse) is the brightest, and presumably the nearest, of the M-type supergiants. It is also one of the least reddened. These features make it important for the calibration of the absolute magnitudes and intrinsic colors of M supergiants. However, its distance has been difficult to establish, and there has been considerable debate concerning its reddening.

The Hipparcos parallax for α Ori is 7.63 ± 1.64 mas, corresponding to a distance of 131 ± 30 pc and a true distance modulus $(m-M)_o$ of 5.59 ± 0.48 mag. If we adopt $<V>$ = 0.50 and $E(B-V)$ = 0.05, the mean absolute visual magnitude becomes $<M_V>$ = –5.24, a result consistent with a luminosity classification of Iab.

The CN strength of α Ori measured in the near infrared also indicates luminosity class Iab. Furthermore, the star's absolute magnitude at 1.04 μm, for the Hipparcos distance, is –8.34, in agreement with the M2 Iab supergiants in h & χ Per for a distance modulus of $(m-M)_o = 12.0$.

From the Hipparcos proper-motion components and distance and a published radial velocity, we derive a space motion of 27.8 km s^{-1} with components U = +19.7, V = –7.3, and W = –1.5 km s^{-1} with respect to the Sun. This motion is not very different from the local standard of rest. It seems unlikely that α Ori, at a distance only one-third that of the Trapezium, is physically associated with the Orion Complex.

At 131 pc, the observed optical angular diameter of α Ori, 0.055 arcsec, translates to a linear diameter of 7.2 AU. Centered on the Sun, its photosphere would be in the asteroid belt. Its angular diameter measured by the *Hubble Space Telescope* in the UV (corresponding to the size of the chromosphere) is a factor of 2 larger. Surface features detected interferometrically are believed to be ~ 10% of the diameter in extent, corresponding to 0.7 AU or 10^8 km.

ON THE MASS-LUMINOSITY RELATION

P. LAMPENS[1], J. KOVALEVSKY[2], M. FROESCHLÉ[2], G. RUYMAEKERS[1]
[1] *Royal Observatory of Belgium, Ringlaan 3, B-1180 Brussels, Belgium*
[2] *Observatoire de la Côte d'Azur, CERGA, UMR CNRS 6527, 06130 Grasse, France*

The new **Hipparcos** parallaxes and photometry are used to determine individual masses and absolute (bolometric) magnitudes for the components of nearby visual binaries with good to very good orbits. The impact on the mass-luminosity relation (MLR) in the range $0 < M_{Bol} < +7.5$ mag is then evaluated.

We selected 335 visual binaries within 50 pc ($\sigma_\pi/\pi < 10\%$) for a full error analysis of their orbits by computing the covariance matrix of the orbital elements with Pourbaix' (1994) algorithm. Using ΔH_p and π, we estimated fractional and component masses as well as absolute magnitudes with their respective errors: 52 binary systems have *relative* mass errors smaller than 15%. Lutz and Kelker (1973) corrections have been applied to both data types. A new relation $BC(H_p)$ as a function of T_{eff} was obtained for the conversion to bolometric magnitudes. A doubly weighted linear regression model was applied next (Babu and Feigelson, 1996): we derived a "mean" slope 3.82 ± 0.07 and zero point 4.94 ± 0.03 for the MLR, assuming a linear relationship.

Conclusions: a) the improvement of the data on masses by **Hipparcos** is largely quantitative; b) not all systems agree: small fluctuations from a "mean" MLR are found as expected from evolutionary or abundance effects; c) the break in the slope of the MLR near $M_{Bol} = +7$ cannot be assessed due to a lack of low-mass binaries in our sample.

References

Babu G.J., Feigelson E.D. (1996), *"Astrostatistics"*, ed. Chapmann and Hall, p. 127.
Lutz T.E., Kelker D.H. (1973), *PASP* **85**, 573.
Pourbaix D. (1994), *A&A* **290**, 6.

J. Andersen (ed.), Highlights of Astronomy, Volume 11A, 568.

ANALYSIS OF THE HIPPARCOS SAMPLE OF ECLIPSING BINARIES

PARALLAXES AND MULTIPLICITY

E. OBLAK[1], M. KURPINSKA-WINIARSKA[2], T. KUNDERA[2]
AND
S. ZOLA[2], T.Z. DWORAK[3]
[1] *Observatoire de Besançon, 25010 Besançon Cedex, France*
[2] *Cracow Observatorium, Cracow, Poland*
[3] *Academy of Mining and Metallurgy, Cracow, Poland*

The Hipparcos astrometric satellite observed 993 known eclipsing binaries. From 1048 eclipsing binaries of the Brancewicz and Dworak 's catalogue (1980), there are only 368 observed by Hipparcos, as the others are too faint.

On the basis of these common stars, we have undertaken the comparison of the satellite parallaxes with the photometric ones. We obtain a good agreement between the two systems with a mean dispersion of 5 mas. The greatest dispersion is for the goup of eclipsing binaries of the EW type. The eclipsing binaries of this group have a mass-luminosity relation different from those of the main sequence stars.

An inadequate knowledge of the component spectra of an eclipsing binary seems to be one of the most important cause of discrepancies. For some brightest stars, the new Hipparcos trigonometric parallaxes differ very significantly from those previously known from groundbased measurements. It seems that the presence of a close visual components causes statistically an underestimation of photometric or overestimation of Hipparcos parallaxes.

172 Hipparcos eclipsing binaries have a close visual component, generally less than 10 arcseconds apart. The list of these stars for various stellar groups and a brief analysis are presented.

References

Brancewicz H.K., Dworak T.Z. (1980), *'A Catalogue of Parameters for Eclipsing Binaries'*, Acta Astron., **30**, N° 4.

J. Andersen (ed.), Highlights of Astronomy, Volume 11A, 569.

TOWARDS AN IMPROVED MODEL OF THE GALAXY

JOHAN HOLMBERG, LENNART LINDEGREN
Lund Observatory
Box 43, SE-22100 Lund, Sweden

AND

CHRIS FLYNN
Tuorla Observatory
Väisäläntie 20, FI-21500 Piikkiö Finland

We use the Hipparcos survey to derive an improved model of the local galactic structure. The availability of parallaxes for all the stars permits direct determination of stellar distributions, eliminating the basic indeterminacy of classical methods based on star counts. Hipparcos gives for the first time a truly three-dimensional view of the solar vicinity, and a complete, homogeneous and highly accurate set of magnitudes and colours. This means that new techniques can be applied in the treatment of the data which place strong constraints on a model that tries to describe the local Galactic structure. Here we investigate how well a static model of low complexity can describe the Hipparcos observations.

The interpretation of the Hipparcos data is complicated by various observational errors and selection effects that are hard to treat correctly. We do not try to correct the data, but instead use a model and subject this model to the same observational errors and selection effects. A model catalogue is created that can be compared with the observed catalogue directly in the observational domain, thereby eliminating the effects from various biases.

Many features in the HR diagram are for the first time seen in field stars thanks to Hipparcos, such as the slanted red giant clump, previously seen in rich old open clusters such as Berkeley 18. This and other features of the observed HR diagram are well reproduced by the model thanks to the rather detailed modelling of the joint $M_V/B-V$ distribution. Actually, separate distributions were derived for the three different components, disk, thick disk and halo, using the kinematic characteristics of the components to discriminate between them.

For galaxies other than the Milky Way seen edge-on, the vertical distribution of light has a distribution function that has a form between the two ones usually employed: the exponential and the sech^2 (isothermal). We have found the same to be true for our own Galaxy as observed by Hipparcos. From a family of density laws for the vertical mass with with the form: $\rho(Z) = \rho(0)\ \mathrm{sech}^{2/n}(nZ/2Z_e)$, where $\rho(0)$ is the density in the plane and Z_e the exponential scale height at large Z, we find $n = 3$ to give the best representation of the stellar distribution in the Hipparcos catalogue. The limiting cases are the exponential for $(n \to \infty)$ and the isothermal for $(n = 1)$.

Most studies agree that the Sun is not exactly at the plane of the Galaxy, but some distance above it. Before Hipparcos measured values of $Z_\odot$ fell in the interval 10–20 pc (mostly studies of gas, dust and young stars) and 20–40 pc (mostly optical stars counts). Hipparcos finally solves the question in a very convincing way by placing the Sun at $Z_\odot \simeq 8 \pm 4$ pc.

The Hipparcos survey was intended to be a complete sample of some 50 000 stars, limited by apparent magnitude and colour. In reality, the faint part of the survey is incomplete, and the completeness factor is depending on many variables that are hard to reproduce in a model. Therefore only the Hipparcos stars brighter then $V = 7$ were used in the construction of the the model. The much more complete and homogeneous sample offered by the Tycho Catalogue can be used to constrain the model for fainter stars down to about $V = 10$. Here the colour distribution of the stars is still very well represented by the model apart from a narrow region around $B - V = 0.45$ where the model underpredicts the counts and a clear age dependence of the scaleheight seems to be needed in the MS turnoff region.

J. Andersen (ed.), Highlights of Astronomy, Volume 11A, 570.

THE METALLICITY DISTRIBUTION OF LATE TYPE DWARFS

M. HAYWOOD, J. PALASI, A. GÓMEZ, L. MEILLON
DASGAL, URA 335 du CNRS, Observatoire de Paris - Meudon, France

The Hipparcos catalogue provides an accurate and extensive sampling of the solar neighbourhood HR diagram. The morphology of this diagram depends on selection criteria of the catalogue such as the limiting magnitude, angular separation and on the characteristics of the stellar populations near the sun (space density, metallicity, star formation rate, etc). Since the Hipparcos data are so accurate, one needs to model precisely the different selection bias and, at the same time, parametrize models of the galactic stellar populations with sufficient flexibility that as much information as possible can be grasped from the catalogue.

Comparisons between our model and the Hipparcos catalogue will be presented elsewhere. Since the quantity of information contained in the Hipparcos catalogue is so important, models ought to be complex, and external contraints, obtained prior to any general comparison with the model, are welcome.

A major factor that influences the distribution of the stars in the HR diagram is the metallicity. For the late type stars, the metallicity distribution can be best studied by re-analysing a volume-limited sample of stars from the catalogue.

Completness tests show that within a 25 pc sphere, the Hipparcos catalogue is approximately complete to M_v=8. However, available photometric metallicity indicator are calibrated down to early K type dwarfs only. Stars have been selected bluer than the limit B-V=1.05, and metallicities were calculated using the Geneva, after rejection of suspected binaries. Long lived stars were found by comparisons with adequate isochrones, and we endded with a sample containing 243 stars. The metallicity distribution of these stars shows two main differences with previous determinations. The first one is the relatively low amount of metal deficient stars, with only 6.5% of the sample having [Fe/H]<-0.5. This is compatible with the amount of thick disc found in deeper surveys of the Galaxy, but is to be compared with the 20 percents found in former studies of the G-dwarf problem. The second difference concerns the metal-rich end of the distribution. There are 14 stars with [Fe/H]>0.2 (none in Wyse & Gilmore (1995)), for instance).

Around solar metallicity (-0.15,+0.15 dex), stars have a rather cold kinematic behavior, with vertical velocity dispersion around 13-15 km.s^{-1}. Outside this metallicity interval, the dispersion is rising to values of the order of 25-30 km.s^{-1}.

More details about this study are published in the proceedings of the symposium held in Venice, may 1997, and in a paper in preparation.

References

Haywood, M., Palasi, J., Gómez, A., Meillon, L. (1997) *Hipparcos Venice'97*, p 489
Wyse, R., F., G, & Gilmore G. (1995) *AJ* **110**, 2771

J. Andersen (ed.), Highlights of Astronomy, Volume 11A, 571.

ABSOLUTE PROPER MOTIONS OF BULGE GIANTS IN THE HIPPARCOS SYSTEM

D. MINNITI[1], M.G.LATTANZI[2], J.J. CLARIA[3], G. MASSONE[2], R. CASALEGNO[2]
[1] *Lawrence Livermore National Laboratory, Livermore, CA, USA*
[2] *Osservatorio Astronomico Torino, Pino T.se, TO, ITALY*
[3] *Observatorio Astronomico, Cordoba, ARGENTINA*

Three windows towards the Galactic bulge, located at $(l,\ b) = (1.6, -2.8)$, $(5.2, -3.5)$, and $(4.2, -4.8)$, were identified from deep astrometric plates. Deep BVRI CCD photometry of these fields reaching $V = 18.5$ was obtained at the ESO Dutch telescope. The color-color and color magnitude diagrams of these fields allow us to select different bulge tracers, including blue horizontal branch stars, red clump giants, and RGB tip stars, as well as disk main-sequence stars. A first catalogue of some 2000 red clump giants was constructed, and measurement of their proper motions in under way. We are deriving absolute proper motions in the Hipparcos system using a two step procedure. First, plates from the Cordoba archive of the Astrographic Catalogue (epochs 1903-1913) are used in combination with recent plates from the ESO GPO telescope to tie a set of intermediate magnitude field stars in the Hipparcos system of positions and proper motions. Second, deeper intermediate epoch (1958) and final epoch CCD images (epoch 1996.5) are used to determine the absolute proper motions of the fainter bulge stars. The average uncertainty in the *absolute* proper motion of a single giant, based on a preliminary sample of few tens of stars, is 0".006 yr^{-1}. This relatively small error, in combination with the sample size, would allow us to study the shape of the velocity ellipsoid for the different Galactic components present in the inner regions as function of Galactocentric distance, and also to measure an accurate distance to the Galactic center and the mass of the bulge. In addition, we identified about 20 faint objects which are potential members of the Sgr dwarf, based on their location in the color-magnitude diagrams, and are also measuring their proper motions.

Other windows with measured proper motions are located along the Galactic minor axis, including Baade's window at $l,\ b = (1, -4)$ (Spaenhauer, A., Jones, B. F., & Whitford, A. E., 1992, *AJ*, **103**, 297), and the Plaut field at $l,\ b = (0, -8)$ (Mendez, R. et al., 1997, in 4th ESO/CTIO Workshop on "The Galactic Center", *in press*). Our proper motions are complementary, probing the kinematics off the Galactic minor axis, where the signature of rotation should be evident. The fields chosen for this study overlap fields that the MACHO project is following since 1993. Their variable star database would provide with excellent tracers of different populations. The 3-D motions of these different Galactic components would be measured for the first time. These include RR Lyrae tracers of the inner metal-poor halo, Miras, LPVs and delta Scuti stars tracers of the metal-rich bulge, and eclipsing binaries tracers of the Galactic disk (Minniti, D., et al., 1996, in IAP Coll. on "Variable Stars and the Astrophysical Returns of Microlensing Surveys", ed. R. Ferlet, p. 257; Alcock, C., et al., 1997, *astro-ph/9707311*).

J. Andersen (ed.), Highlights of Astronomy, Volume 11A, 572.

THE GALACTIC WARP SIGNATURE AND MOVING GROUPS

R.L. SMART, R. DRIMMEL, M.G. LATTANZI
Osservatorio Astronomico Di Torino
AND
J.J. BINNEY
Oxford University

In Smart et al (Hipparcos Venice Symposium 1997 ESA SP) we showed that the kinematic signature of the Galactic Warp was very different to that predicted both from the spatial warp parameters and from a simple discussion of HI observations. It was suggested that the kinematic signature may actually be reflecting the signature of one or more moving groups, here we test that interpretation. We split the data into three longitude bins, [70-130], [150-230] and a bin that overlaps the outside of these two bins [70-100,200-230]. We then compare a plot of velocity vs radius, figure 1, for all bins and the full data set. If the signature is due to a moving group we would expect one of the bins to reflect a non moving group trend. While the details of the correlations change for each bin, the overall trend of a decrease in velocity with radius is the same for all three bins and the combined data. A consideration of the geometry here will show that for a single moving group to cause such a velocity signature in all three bins, it would have to be approximately 5kpc in extent.

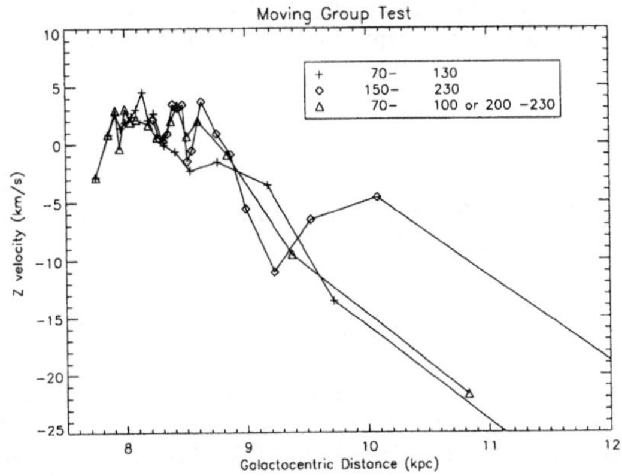

Figure 1. *Velocity vs Radius for OB stars split into three longitude bins as indicated in the legend.*

J. Andersen (ed.), Highlights of Astronomy, Volume 11A, 573.

KINEMATICS OF DISK STARS IN THE SOLAR NEIGHBOURHOOD

A.E. GÓMEZ[1], S. GRENIER[1], S. UDRY[2], M. HAYWOOD[1]
AND
V. SABAS[1], L. MEILLON[1], F. ROYER[1], Y. LEBRETON[1]
[1] *Observatoire de Paris, D.A.S.G.A.L., CNRS URA 335, 92195 Meudon Cedex, France*
[2] *Observatoire de Genève, 1290, Sauverny, Switzerland*

Using Hipparcos parallaxes and proper motions together with radial velocity data and individual ages estimated from isochones, the velocity ellipsoid has been determined as a function of age. On the basis of the available kinematic data two different samples were considered: a first one (7789 stars) for which only tangential velocities were calculated and a second one containing 3104 stars with available U, V and W velocity components and total velocities $\leq$ 65 km.s^{-1}.

The main conclusions are: -Mixing is not complete at about 0.8-1 Gyr. -The shape of the velocity ellipsoid changes with time getting rounder from $\sigma_U/\sigma_V/\sigma_W = 1/0.63/0.42 \pm 0.04$ at about 1 Gyr to $1/0.7/0.62 \pm 0.04$ at 4-5 Gyr. -The age-velocity-dispersion relation (from the sample with kinematical selection) rises to a maximum, thereafter remaining roughly constant; there is no dynamically significant evolution of the disk after about 4-5 Gyr. -Among the stars with solar metallicities and log(age) $>$ 9.8 two groups are identified: one has typical thin disk characteristics, the other is older than 10 Gyr and lags the LSR at about 40 km.s^{-1}. -The variation of the tangential velocity with age (without selection on the tangential velocity) shows a discontinuity at about 10 Gyr, which may be attributed to stars typically of the thick disk populations for ages $>$ 10 Gyr.

ON THE CHARACTERISTICS OF THE VELOCITY FIELD OF YOUNG STARS IN THE SOLAR NEIGHBOURHOOD

J. TORRA[1], A.E. GÓMEZ[2], F. FIGUERAS[1], F. COMERÓN[3], S. GRENIER[2]
AND
V. SABAS[2], B. CHEN[1], M.O. MENNESSIER[4], R. ASIAIN[1], M. MESTRES[1]
[1] *Universitat de Barcelona, Diagonal, 647, 08028 Barcelona, Spain*
[2] *Observatoire de Meudon, DASGAL/URA CNRS 335, 92195 Meudon Cedex, France*
[3] *European Southern Observatory, Garching bei Munchen, Germany*
[4] *Universite de Montpellier II, Pl. E. Bataillon, 34095 Montpellier Cedex 5, France*

The analysis of stellar kinematics of the local system of young stars from Hipparcos data offers an excellent opportunity to revise models accounting for the origin of the observed deviations from circular motions, and to give new hints on the history of star formation and the dynamical processes involved in the evolution of our Galaxy. We have complemented the astrometric Hipparcos data with a careful compilation of available radial velocities and Strömgren photometry, information which enables us to obtain reliable space velocities and individual ages for a large number of stars. The velocity field has been studied by means of the classical first order approach, interpreting the trends observed in the Oort's contants in terms of the local expansion of the Gould's Belt (stars with ages less than 30-50 Myr, at heliocentric distances $\rho < 400 - 600$ pc). The fundamental geometric parameters of such structure have been determined. Several statistical methods have been applied to identify moving groups in the 4-dimensional space of (U,V,W,age). Our results definitively confirm their existence with mean velocity dispersions of the order of 5-7 km/s. The obtained results will be published in Astronomy & Astrophysics.

J. Andersen (ed.), Highlights of Astronomy, Volume 11A, 574.

POPULATIONS AMONG HIGH-VELOCITY EARLY-TYPE STARS

F. ROYER, A.E. GÓMEZ
DASGAL URA 335, Observatoire de Paris-Meudon, F-92195 Meudon

As far as the standard picture of galactic evolution and structure is concerned, early-type stars belong to disk populations. However, the existence of a population of early-type stars with near-solar metallicities and main-sequence surface gravities, together with high velocities and/or standing at large distances away from the plane, is a well known anomaly to this picture.

In order to provide an insight into the different populations of early-type stars, a sample of high-velocity stars has been gathered from the Hipparcos catalogue. We have selected the stars according to $B - V < 0.4$, and a tangential velocity larger than 80 km/s. The sample is composed of 101 stars.

The high-velocity early-type stars in the solar neighbourhood are a mixture of different stellar populations : blue horizontal halo stars passing near the sun and high-velocity disk stars. We have decomposed into gaussians the V and W-distributions of our sample, to isolate the disk from the halo populations. The U-distribution of high-velocity disk stars is clearly asymmetric. The selection bias over the tangential velocities creates a depletion of low U values. Moreover, two thirds of the sample have $U < 0$ (toward the galactic anticentre, $l = 180°$). This cannot be explained by the distribution tail of disk stars or stochastic events such as runaway stars, which privilege no direction.

A SEARCH FOR STARS PASSING CLOSE TO THE SUN

J. GARCIA-SANCHEZ, R.A. PRESTON, D.L. JONES, P.R. WEISSMAN
JPL, California Institute of Technology, Pasadena, CA 91109, USA
J.-F. LESTRADE
Observatoire de Paris-Meudon, F-92195 Meudon Principal Cedex, France
AND
D.W. LATHAM AND R.P. STEFANIK
Harvard-Smithsonian CfA, Cambridge, MA 02138, USA

We have combined radial velocity measurements with Hipparcos proper motion and parallax data for 552 nearby stars of a selected sample of 1208 stars to find which stars may have passed (or will pass) close enough to the Sun to perturb the cometary Oort cloud. We find that the rate of close stellar approaches within a distance D (in pc) from the Sun is N = 4.3 $D^{2.02}$ Myr^{-1}, in excellent agreement with the predicted functional form of simple stellar dynamics models though the number of stars is less than predicted. However, we consider this a lower limit because there is evidence for observational incompleteness in our data set. One star (Gliese 710) is found with a future closest approach of less than 0.4 pc, and several stars come within about 1 pc during about a ± 10 Myr interval. We have performed dynamical simulations which show that none of the passing stars in this sample can perturb the Oort cloud sufficiently to create a substantial increase in the long-period comet flux at the Earth's orbit.

J. Andersen (ed.), Highlights of Astronomy, Volume 11A, 575.

HIPPARCOS AND PRIMARY DISTANCE SCALE INDICATORS

C. TURON
DASGAL, URA CNRS 335, Observatoire de Paris, 92195, Meudon Cedex, France.

1. Introduction

The Hipparcos data are providing a dramatic increase, qualitatively and quantitatively, of the basic available distance information. For example, the numbers of stars for which a relative accuracy better than 10 % is available from Hipparcos data and from ground-based data are respectively 22 396 and about 1 000. Moreover, the range of spectral types and luminosity classes for which precise parallaxes are available is considerably enlarged, including many stars in open clusters and a small number of Cepheids and RR Lyrae. The bottom of the main sequence is populated down to absolute Hp magnitude 14, including a number of subdwarf stars essential to derive globular clusters distances and ages (Pont *et al.*, 1997a).

Finally, the Hipparcos data show how difficult are the calibration of photometric distances and the transformation of relative trigonometric parallaxes to absolute parallaxes. This is illustrated by the comparison of distances given in the last edition of the Catalogue of Nearby Stars (CNS3, Gliese & Jahreiß 1991), which is the best available compilation of stars said to be closer than 25 pc from ground-based data. About a third of them are found by Hipparcos to be (much) further than this limit (Perryman *et al.*, 1995). A second example is given by the study of (Binney *et al.*, 1997).

2. Open clusters

Open clusters are widely used for distance determination by main sequence fitting. Hipparcos offers for the first time the possibility of thoroughly studying the three-dimensional structure and the kinematics of the closest cluster, the Hyades, giving a mean distance modulus of 3.33 $\pm$0.01 mag for the stars observed by Hipparcos within 10 pc of the cluster center (Perryman *et al.*, 1997a), and to precisely place the main sequences of about 15 open clusters of various ages and chemical compositions in the HR diagram (Mermilliod *et al.*, 1997), (Robichon *etal.*, 1997), (van Leeuwen & Hansen Ruiz, 1997). The surprise is that the positions of the different sequences cannot be explained by differences in metallicities, and that a scatter of Helium abondance of about 0.05 would be necessary to explain these discrepancies. A further use of the main-sequence fitting method requires the understanding of this problem.

3. Pulsating variables

Pulsating variables are some of the more powerful objects for distance determinations, after proper calibration of period-(colour-metallicity)-luminosity relations. The main difficulty is to obtain a representative sampling of the stars considered, covering the whole period, colour, and metallicity ranges for each type of objects. A few trigonometric parallaxes, good proper motions and improved cluster membership for Cepheids are obtained from Hipparcos data. Various authors have presented period-luminosity relations: (Feast & Catchpole, 1997), (Feast & Whitelock, 1997), (Paturel *et al.*, 1997), (Szabados, 1997), (Madore & Freedman, 1997) and (Luri *et al.*, 1997) for Cepheids; (Fernley *et al.*, 1997), (Gratton *et al.*, 1997), (Reid, 1997), (Tsujimoto *etal.*, 1997), (Feast & Whitelock, 1997), (Chaboyer *et al.*, 1997), and (Luri *et al.*, 1997) for RR Lyrae. There is still a scatter in the results, easily explained by the small number of stars used in each study.

J. Andersen (ed.), Highlights of Astronomy, Volume 11A, 576–577.

More photometric observations are strongly needed to better evaluate the intrinsic scatter in period-luminosity relations, the interstellar extinction, and the metallicities. However, only a project such as Gaia (Lindegren & Perryman, 1997), (Perryman *et al.*, 1997b) would bring extremely accurate determinations of absolute magnitudes for extensive samples of pulsating variable stars and allow a comprehensive calibration of period-luminosity relations for all galactic (and even the brightest Magellanic) variables of the various stellar populations (both disk and halo).

4. Results and conclusion

The distance of the Large Magellanic Cloud and the age of globular clusters have been derived by various authors, from the Hipparcos data of several types of objects, and using many other ground-based observations. The results are given below. Most results are converging to a LMC distance modulus of 18.5-18.7, but the only paper with no discrepancy between the distance obtained from RR Lyrae and from Cepheids (Luri *et al.*, 1997) gives about 18.3. For the age of globular clusters, there is some spread in the results, but most are converging to 12-14 Gyr. These results go towards solving the previous discrepancy between the age of the Universe, and the age of the oldest objects.

To go further, many more very accurate distances are needed, but also many more accurate photometric and spectroscopic observations, and more detailed theoretical models.

LMC distance modulus		**Globular cluster age**		
Objects/Authors	$(M\text{-}m)_0$	Objects/Authors	Age (Gyr)	
from Cepheids		**from subdwarfs**		
(Feast & Catchpole, 1997)	18.70 ± 0.10	(Reid, 1997)	$12 - 13$	
(Paturel *et al.*, 1997)	18.7	(Gratton *et al.*, 1997)	11.8	$^{+2.1}_{-2.5}$
(Madore & Freedman, 1997)	18.57 ± 0.11			
(Luri *et al.*, 1997) (1)	18.29 ± 0.17	(Pont *et al.*, 1997b)	14	± 1
(Luri *et al.*, 1997) (2)	18.21 ± 0.20	(Cayrel *et al.*, 1997)	$14 - 15$	
from RR Lyrae		**various methods + RR Lyrae**		
(Fernley *et al.*, 1997)	18.31 ± 0.10	(Chaboyer *et al.*, 1997)	11.78	± 1.4
(Luri *et al.*, 1997)	18.37 ± 0.23			
from RR Lyrae, via subdwarfs		**via RR Lyrae**		
(Reid, 1997)	18.65	(Fernley *et al.*, 1997)	15.8	
(Gratton *et al.*, 1997)	18.60 ± 0.07	using (Chaboyer *et al.*, 1997) Mv for RR Lyrae		
from Mirae		**via Cepheids**		
(van Leeuwen *et al.*, 1997)	18.54 ± 0.18	(Feast & Whitelock, 1997)	11	
(Whitelock *etal.*, 1997)	18.60 ± 0.18			

References

Binney, J., Dehnen, W., Houk, N., Murray, C.A., Penston, M.J., (1997), *Hipparcos Venice '97*, ESA SP-402, p. 473

Cayrel, R., Lebreton, Y., Perrin, M.N., Turon, C. (1997), *Hipparcos Venice '97*, ESA SP-402, p. 219

Chaboyer, B., Demarque, P., Kernan, P.J., Krauss, L.M. (1997), *ApJ*, astro-p/9706128

Feast, M.W., Catchpole, R.M., (1997), *MNRAS*, **286**, L1

Feast, M.W., Whitelock, P.A., (1997), *Hipparcos Venice '97*, ESA SP-402, p. 625

Fernley, J., Barnes, T.G., Skillen, I., *et al.*, (1997), *Hipparcos Venice '97*, ESA SP-402, p. 635

Gliese, W., Jahreiß, H. (1991), Astron. Rechen-Institut, Heidelberg

Gratton, R.G., Fusi Pecci, F., Carretta, E., Clementini, G., et al. (1997), *Hipparcos Venice '97*, ESA SP-402, p. 651

van Leeuwen, F., Hansen Ruiz, C.S. (1997), *Hipparcos Venice '97*, ESA SP-402, p. 689

van Leeuwen, F., Feast, M.W., Whitelock, P.A., Yudin, B, (1997), *MNRAS*, **287**, 955

Lindegren, L., Perryman, M.A.C. (1997), *Hipparcos Venice '97*, ESA SP-402, p. 799

Luri, X., Gómez, A.E., Torra, J., Figueras, F., Mennessier, M.O. (1997), *Astron. Astrophys. Letters*, submitted

Madore, B.F., Freedman, W.L. (1997), *ApJ Letters*, in press

Mermilliod, J.C., Turon, C., Robichon, N. et al. (1997), *Hipparcos Venice '97*, ESA SP-402, p. 643

Paturel, G., Lanoix, P., Garnier, R., *et al.*, (1997), *Hipparcos Venice '97*, ESA SP-402, p. 629

Perryman, M.A.C., Brown, A.G.A., Lebreton, Y. et al. (1997a), *Astron. Astrophys.*, in press

Perryman, M.A.C., Lindegren, L., Turon, C. (1997b), *Hipparcos Venice '97*, ESA SP-402, p. 743

Perryman, M.A.C., Lindegren, L., Kovalevsky, J., Turon, C., Høg, E., et al. (1995), *Astron. Astrophys.*, **304**, 69

Pont, F., Mayor, M., Turon, C. (1997a), *IAU Symp. 189*, T.R. Bedding et al. (eds.), p. 433

Pont, F., Charbonnel, C., Lebreton, Y. et al. (1997b), *Hipparcos Venice '97*, ESA SP-402, p. 699

Reid, N. (1997), *Astron. J.*, **114**, 161

Robichon, N., Arenou, F., Turon, C., et al. (1997), *Highlights in Astronomy*, **11**, in press

Szabados, L. (1997), *Hipparcos Venice '97*, ESA SP-402, p. 661

Tsujimoto, T., Miyamoto, M., Yoshii, Y., (1997), *Hipparcos Venice '97*, ESA SP-402, p. 639

Whitelock, P.A., van Leeuwen, F., Feast, M.W. (1997), *Hipparcos Venice '97*, ESA SP-402, p. 213

THE HYADES: DISTANCE, STRUCTURE AND DYNAMICS

A.G.A. BROWN
Sterrewacht Leiden, Postbus 9513, 2300 RA Leiden, The Netherlands

AND

M.A.C. PERRYMAN
Astrophysics Division, European Space Agency, ESTEC, Noordwijk 2200AG, The Netherlands

The Hipparcos parallaxes and proper motions together provide a consistent picture of the Hyades distance, structure and dynamics. They yield a cluster convergent point motion consistent with the individual trigonometric parallaxes, and together explain the large distance modulus derived from the most recent ground-based proper motion investigations as originating from differences in the space velocity and small systematic effects in the ground-based proper motions. Conversely, the smaller distance modulus generally derived from a variety of ground-based trigonometric parallax programmes originates from systematically larger values of the ground-based parallaxes. Although recent distance determinations to individual objects in the cluster, most notably the results of Torres et al. (1997a,b,c), are in excellent agreement with the Hipparcos trigonometric parallaxes, extrapolation to a mean cluster distance is again affected by systematic effects in the ground-based proper motion studies.

The combination of the Hipparcos astrometry with radial velocity measurements from ground-based programmes provides three-dimensional velocities allowing candidate membership selection to be based on 3-dimensional positional and kinematical criteria. A number of new cluster members have been found within 20 pc of the cluster centre. No evidence for systematic internal velocity structure is found; rather, the results are fully consistent with a uniform cluster space motion with an internal velocity dispersion of about 0.3 km s^{-1}. Spatial distribution, mass segregation, and binary distributions are consistent with N-body simulations.

The cluster has a tidal radius of $\sim$ 10 pc. Outside this region, the stellar distribution is elongated along the direction of the Galactic centre and anti-centre, and is slightly flattened in the direction perpendicular to the Galactic plane. Inside this sphere, the cluster has spherical symmetry with a core radius of $r_c \simeq 2.7$ pc, and a half-mass radius of $\sim$ 6 pc. The presence of objects closely linked kinematically with the cluster core, but well beyond the tidal radius, probably originates from dynamical interactions in the centre combined with diffusion beyond the Lagrangian points.

With the caveat that the importance of the Hipparcos results is to provide individual distances to cluster members, rather than a mean cluster centre of mass (a concept meaningful only in the restricted context of the cluster members contained in the Hipparcos Catalogue), our estimated distance to the observed centre of mass for the objects within 10 pc of the cluster centre is 46.34 ± 0.27 pc, corresponding to a distance modulus $m - M = 3.33 \pm 0.01$ mag. This mean distance is only marginally modified (by about 0.4 pc) for the derived centre of mass for Hipparcos objects within $r < 20$ pc of the cluster centre. A possible bias in the cluster distance due to the Lutz-Kelker effect is small and of the same size as the error in the mean distance.

For full details on these results we refer to Perryman *et al.* (1997).

References

Perryman, M.A.C., Brown, A.G.A., Lebreton, Y., Gómez, A., Turon, C., Cayrel, G., Mermilliod, J.-C., Robichon, N., Kovalevsky, J., Crifo, F., 1997, *Astron. & Astroph.*, in press

Torres, G., Stefanik, R.P., Latham, D.W., 1997a, *Astroph. Journ.* **474**, 256

Torres, G., Stefanik, R.P., Latham, D.W., 1997b, *Astroph. Journ.* **479**, 268

Torres, G., Stefanik, R.P., Latham, D.W., 1997c, *Astroph. Journ.* **485**, 167

J. Andersen (ed.), Highlights of Astronomy, Volume 11A, 578.

NEARBY OPEN CLUSTERS AND HR DIAGRAM CALIBRATION

N. ROBICHON[1], F. ARENOU[1], C. TURON[1], Y. LEBRETON[1], J.-C. MERMILLIOD[2]
[1] *DASGAL - CNRS URA 335*
Observatoire de Paris, 5, place J. JANSSEN, F92195 Meudon cedex
[2] *Institut d'astronomie, Université de Lausanne*
CH1290 Chavannes-des-Bois

13 clusters have been analysed (Coma Ber, the Pleiades, IC 2602, IC 2391, Praesepe, Alpha Per, NGC 6475, Blanco 1, Stock 2, NGC 6633, IC 4756, NGC 2516 and NGC 3532). Cluster members have been selected using Hipparcos proper motions and parallaxes. The mean cluster parallaxes have been determined using Hipparcos intermediate data (abscissae on the Reference Great Circles) taking into account star to star correlations. Both selection of members and computation of cluster mean parallaxes are described in Mermilliod *et al.*, 1997 and Robichon *et al.*, 1997.

For each cluster, a selection of fainter stars in the 'Base des Amas' (Mermilliod, 1988), based on photometric data as well as radial velocities and proper motions when available, has been done. From this selection and the computed mean parallaxes, the location of the 13 cluster sequences in the observational HR diagram is derived very accurately, at least for the closest clusters for which good photometry and small errors on the mean parallaxes are available. Relative positions of cluster sequences cannot only be explained by differences between cluster metallicities. For example, the Pleiades sequence is 0.5 magnitude bellow the Praesepe sequence whereas the adopted metallicities (obtained spectroscopically i.e. whithout any assumption on the helium content) are respectively -0.034 and +0.10.

Theoretical ZAMS (Lebreton, 1997) computed for each cluster according to its metallicity and a grid of Helium content has been achieved in order to estimate roughly the helium abundance of each cluster by comparison with the observed sequences. The scatter of the helium abundance Y from the less abundant cluster (Praesepe: $Y \simeq 0.25$) to the richest one (Pleiades: $Y \simeq 0.30$) is about 0.05.

These results are crucial for any further calibration of photometric parallaxes as well as for constraining theoretical models of internal structure and stellar evolution. Such a scatter in helium content is difficult to explain and is to be confirmed with more accurate cluster metallicities and (T_{eff}, colour) relations. Furthermore, for the majority of the 13 clusters, accurate photometric data are lacking. In particular narrow band photometry such as Strömgren photometry would be very useful.

References

Lebreton, Y. (1997), in preparation

Mermilliod, J.-C., Turon, C., Robichon, N., Arenou, F. and Lebreton, Y. (1997), *Hipparcos Venice'97*, **ESA SP-402, p. 643**

Mermilliod, J.-C. (1988), *Bull. Inf. Centre Données Stellaires*, **Vol. no. 35, pp. 77-91**

Robichon, N., Arenou, F., Turon, C., Mermilliod, J.-C. and Lebreton, Y. (1997), *Hipparcos Venice'97*, **ESA SP-402, p. 567**

J. Andersen (ed.), Highlights of Astronomy, Volume 11A, 579.

THE ABSOLUTE MAGNITUDE OF RR LYRAE STARS

T. TSUJIMOTO[1], M. MIYAMOTO[1], Y. YOSHII[2]
[1] *National Astronomical Observatory, Mitaka, Tokyo 181, Japan*
[2] *Institute of Astronomy, University of Tokyo, Mitaka, Tokyo 181, Japan*

The present determination of the absolute magnitude M_V(RR) of RR Lyrae stars is twofold, relying upon Hipparcos proper motions and trigonometric parallaxes separately. First, applying the statistical parallax method to the proper motions, we find $< M_V(\mathrm{RR}) >= 0.69 \pm 0.10$ for 99 halo RR Lyraes with <[Fe/H]> = −1.58. Second, applying the Lutz-Kelker correction to the RR Lyrae HIP95497 with the most accurately measured parallax, we obtain M_V(RR) = 0.57–0.74 at [Fe/H]=−1.6. Furthermore, allowing full use of low accuracy and negative parallaxes as well for 125 RR Lyraes with $-2.49 \leq$[Fe/H]≤ 0.07, the maximum likelihood estimation yields the relation, M_V(RR) = (0.59±0.37)+(0.20±0.63)([Fe/H]+1.60), which formally agrees with the recent preferred relation. The same estimation yields again $M_V(\mathrm{RR}) = 0.65 \pm 0.33$ for the 99 halo RR Lyraes. Although the formal errors in the latter two estimates are rather large, all of the four results suggest the fainter absolute magnitude, M_V(RR)=0.6–0.7 at [Fe/H]=−1.6. The present results still provide the lower limit on the age of the universe which is inconsistent with a flat, matter-dominated universe and current estimates of the Hubble constant.

SUGGESTION TO REVISE THE HIPPARCOS-PARALLAX OF CEPHEIDS BELONGING TO BINARY SYSTEMS

L. SZABADOS
Konkoly Observatory, P.O. Box 67, H-1525, Budapest XII, Hungary

It has been suspected (Szabados 1997) that the apparent orbit of previously unresolved binaries involving a Cepheid component falsifies the Hipparcos-parallaxes deduced for these Cepheids. Since the incidence of binaries among Cepheids exceeds 50 per cent, this adverse effect on the parallax determination is a major error source.

An obvious remedy is to repeat the determination of the five astrometric parameters for the Cepheids, members in binary systems, taking into account the orbital elements known from spectroscopic studies. This will result in more precise parallaxes and cosmic distance scale. Moreover, these Cepheids are ideal targets for the future microarcsec astrometric missions.

This study was partly supported by OTKA-grants T 014852 and T 022946.

References

Szabados L., 1997, in Proc. Conf. Hipparcos Venice'97; ESA SP-402, p. 631

J. Andersen (ed.), Highlights of Astronomy, Volume 11A, 580.

MICROARCSEC ASTROMETRY: THE GAIA MISSION

L. LINDEGREN
Lund Observatory
Box 43, SE-22100 Lund, Sweden

AND

M.A.C. PERRYMAN
Astrophysics Division, ESTEC
2200AG Noordwijk, The Netherlands

1. The GAIA Concept

The Hipparcos mission demonstrated the efficiency of space astrometry (in terms of number of objects, accuracy, and uniformity of results) and the fact that a relatively small instrument can have a very large scientific potential in the area of astrometry. However, Hipparcos could probe less than 0.1 per cent of the volume of the Galaxy by direct distance measurements. Using a larger instrument and more efficient detectors, it is now technically feasible to increase the efficiency of a space astrometry mission by several orders of magnitude, thus encompassing a large part of the Galaxy within its horizon for accurate determination of parallaxes and transverse velocities. Such a mission will have immediate and profound impact in the areas of the physics and evolution of individual stars and of the Galaxy as a whole.

The Global Astrometric Interferometer for Astrophysics (GAIA) is studied as a possible Cornerstone Mission within the European Space Agency's scientific programme Horizon 2000 Plus. Consisting of up to four optical Fizeau interferometers contained in a single payload and operated by continuous scanning, its target accuracy is about 10 microarcsec or better for the positions, parallaxes and annual proper motions of some 50 million objects down to $V = 15$ mag, with many more fainter objects measured at a correspondingly reduced accuracy. Direct recording of the interference fringes by CCDs operated in drift-scan mode will give a limiting magnitude of $V \sim 20$ and capability to observe in crowded regions. Real-time detection of all point sources to well-defined limits may eliminate the need for an input catalogue and result in the observation of up to 10^9 stars, quasars and minor planets. A mission lifetime of 5 years is targeted.

Each Fizeau interferometer has a modest (< 3 m) baseline, a large ($\sim 1°$) field of view, a wide coverage of visual to near-IR wavelengths ($\lambda \sim 400$–1000 nm), and a collecting area of about 0.5 m^2. The field of view is covered by a mosaic of CCDs simultaneously recording the fringe patterns of thousands of stars. The interferometers point in different directions perpendicular to the common axis of rotation; this enables building up a rigid reference frame of positions and proper motions, as well as the determination of absolute parallaxes. The faint limiting magnitude allows a direct link to the extragalactic reference frame, making the stellar proper motions inertial. The required short-term stability of the interferometer baselines may be maintained by laser metrology, by means of beam combiners and passive control, or through a combination of both. Some of the CCDs will be equipped with colour filters for wideband (e.g. UBVRI) and intermediate-band photometry (e.g. Strömgren/Vilnius system) of all observed objects. The possibility to include a special radial-velocity instrument, aiming at a few km/s accuracy at 15th magnitude, is being considered. A more complete description of the GAIA concept is found in Lindegren & Perryman (1997).

2. Scientific Objectives

Astrometric measurements provide model-independent estimates of basic geometric and kinematic properties of astronomical sources, in particular stellar distances and transverse velocities. The projected accuracy of 10 μas (microarcsec) for the positions and parallaxes, and 10 μas/yr for the

J. Andersen (ed.), Highlights of Astronomy, Volume 11A, 581–582.

proper motions, for stars down to $V \sim 15$, corresponds to a horizon of ~ 10 kpc for *individual* stellar distances and many times larger for statistical uses. The proper motion accuracy corresponds to transverse velocity errors of 1 km/s at 20 kpc (for known distances). Such astrometric accuracies (or better) would be obtained for tens of millions of stars to well-defined magnitude limits, and would be accompanied by accurate multi-colour, multi-epoch, sub-millimagnitude photometric measurements. The following is a brief summary of some important astrophysical applications of such data (for additional details, see Perryman et al. 1997).

Physics and Evolution of Stars. GAIA will provide extremely accurate (< 1 per cent) distances for stars out to 1 kpc distance, encompassing the entire luminosity range from $M_V = -10$ to $+20$, including all population types (also rare ones) represented in that volume, as well as numerous open clusters and associations. Detailed comparison with theoretical models will profoundly affect our understanding of stellar atmospheres, structure and evolution; indirectly, through the much improved determinations of stellar ages and chemical compositions, this will generate new information on the formation and evolution of the Galaxy.

Stellar Luminosities and the Distance Scale. The trigonometric parallaxes will provide direct and accurate luminosity calibration of an extremely wide variety of stellar objects, including evolutionary interesting objects such as massive stars, novae and planetary nebulae, and primary distance indicators such as Cepheids and RR Lyrae variables. For the first time this will establish an extensive network of distance measurements throughout a significant part of our Galaxy, including the central region, the bulge, spiral arms, and the halo. This network will even touch on the nearest companion galaxies, in particular the Magellanic Clouds, allowing the effects of the different formation histories of these systems to be studied.

Dynamics of Stellar Systems. At the sensitivity level of GAIA some 25 per cent of all stars will be resolved into components or otherwise manifest themselves as non-single (e.g. astrometric binaries). Many clusters, including a few globular clusters, will be available for detailed dynamical analysis based on the individual proper motions. Details of mass segregation, binary frequency, internal velocity structure, evaporation and tidal distortion can be studied for these objects.

Galactic Dynamics. The huge number of stars, the high accuracy for the parallaxes and absolute proper motions, and the faint limiting magnitude of GAIA would totally revolutionise dynamical studies of our Galaxy. Among the many questions that can be addressed with such data are: the nature of the spiral arms; the origin and history of the galactic bulge; the large-scale, three-dimensional distribution of mass throughout the Galaxy, including dark matter in the disk and halo, and the total mass of the Galaxy; traces of galactic mergers and other important events in the history of the Galaxy, detectable by accurate mapping of a large volume of phase space.

Planetary Systems. GAIA will be able to screen hundreds of thousands of the bright nearby (< 100 pc) stars for possible astrometric signatures of planetary and brown dwarf companions. Jupiter-mass planets with periods up to 5 years can be detected out to some 50 pc and less massive companions to proportionally smaller distances.

General Relativity. From a global analysis of the light bending by the Sun, the PPN parameter γ, equal to unity in General Relativity, could be determined to about 10^{-6}, a precision level where differences among alternative theories of gravitation may be discernible, and where higher-order effects of the solar system metric start to become apparent.

References

Lindegren, L. and Perryman, M.A.C. (1997) GAIA: Global Astrometric Interferometer for Astrophysics, Proc. *HIPPARCOS Venice '97*, ESA SP–402, pp. 799–802

Perryman, M.A.C., Lindegren, L. and Turon, C. (1997) The Scientific Goals of the GAIA Mission, Proc. *HIPPARCOS Venice '97*, ESA SP–402, pp. 743–748

DIVA – A SMALL SATELLITE FOR GLOBAL ASTROMETRY AND PHOTOMETRY

S. RÖSER[1], U. BASTIAN[1], K.S. DE BOER[2], E. HØG[3], E. SCHILBACH[4]
[1] *Astronomisches Rechen-Institut, Heidelberg*
[2] *Sternwarte Universität Bonn*
[3] *Copenhagen University Observatory, Copenhagen*
[4] *Astrophysikalisches Institut Potsdam*

AND

CH. DE VEGT[5], S. WAGNER[6]
[5] *Hamburger Sternwarte, Hamburg*
[6] *Landessternwarte Heidelberg*

DIVA (Double Interferometer for Visual Astrometry) is a Fizeau interferometer on a small satellite. It will perform astrometric and photometric observations of at least 4 million stars. A launch in 2002 and a minimum mission length of 24 months are aimed at. A detailed description of the experiment can be obtained from the DIVA homepage at `http://www.aip.de:8080/~dso/diva`. An overview is given by Röser *et al.*, 1997.

The limiting magnitude of DIVA is about V = 15 for spectral types earlier than M0, but drops to about V = 17.5 for stars later than M5. Table 1 gives a short overview on DIVA's performance. DIVA will carry out a sky survey complete to V = 12.5. For the first time this survey will comprise precise photometry in at least 8 bands in the wavelength range from 400 to 1000 nm.

TABLE 1. Astrometric and photometric performance of DIVA after two years of mission. The photometric accuracy is given for 8 photometric bands in the range from 400 to 1000nm.

V (mag)	$\sigma_{\alpha,\delta,\pi}$ (mas)	$\sigma_{\mu_{\alpha,\delta}}$ (mas/y)	$\sigma_{phot.}$ (milli-mag)
≤ 8	0.15	0.2	1 - 3
10	0.35	0.5	3 - 8
12	1.1	1.5	7 - 25

DIVA will improve parallaxes by a factor of 3 compared to Hipparcos; proper motions by at least a factor of 2 and, in combination with the Hipparcos observations, by a factor of 10 for Hipparcos stars. At least 30 times as many stars as Hipparcos will be observed, and doing this DIVA will fill the gap in observations between Hipparcos and GAIA. DIVA's combined astrometric and photometric measurements of high precision will have important impacts on astronomy and astrophysics in the next decade.

References

Röser, S., Bastian, U., de Boer, K.S., Høg, E., Röser, H.P, Schalinski, C., Schilbach, E., de Vegt, Ch., sWagner, S. (1997) DIVA - Towards Microarcsecond Global Astrometry, *ESA-SP-402*, (in press)

J. Andersen (ed.), Highlights of Astronomy, Volume 11A, 583.

ASTROMETRIC AND PHOTOMETRIC UTILIZATION OF DISPERSED FRINGES FROM A SPACE INTERFEROMETRY MISSION

R.-D. SCHOLZ[1], S. HIRTE[1], U. BASTIAN[2], S. RÖSER[2]
[1] *Astrophysikalisches Institut, An der Sternwarte 16, 14482 Potsdam, Germany*
[2] *Astronomisches Rechen-Institut, Mönchhofstraße 12-14, 69120 Heidelberg, Germany*

Dispersed interferometer fringes as a new type of signal are currently under discussion for several space interferometry mission projects (GAIA, FAME, DIVA). The dispersed fringes are directly recorded by CCD mosaics which are operated in drift-scan mode on board a continuously rotating satellite. They allow combined astrometric (perpendicular to the fringes, i.e. in the direction of the drift scan) and spectro-photometric measurements (in the direction of the dispersion of the fringes, i.e. perpendicular to the drift scan). We have simulated polychromatic dispersed fringe patterns of stars with known spectra. The specific instrument parameters used for the simulation are those envisaged for the DIVA instrument, with realistic assumptions on various noise contributions. The results can be scaled to GAIA and FAME instrument parameters. We have investigated simple algorithms which can be implemented in an on-board data reduction and used for the real-time attitude determination with bright and single stars. On the preliminary basis of 100 simulations per star of a given magnitude and spectral type, we have obtained first estimates of the achievable accuracy of the astrometric measurements and of the location of the photometric zero point λ_0.

THE INSTRUMENT OF THE ASTROMETRIC MISSION DIVA

S.J. WAGNER [1], W. SEIFERT [1], H. MANDEL [1], U. BASTIAN [2], S. RÖSER[2]
[1] *Landessternwarte, Königstuhl, 69117 Heidelberg, Germany*
[2] *Astronomisches Rechen-Institut, Heidelberg, Germany*
AND
the DIVA team
see http://www.aip.de:8080/~dso/diva

As a follow-up to the most successful Hipparcos mission we have suggested a mini-satellite. The instrument DIVA is a pair of two small Fizeau interferometers with a fixed angular separation scanning the entire sky in a Hipparcos-type fashion. Despite its small apertures the very efficient optical design and the use of CCD detectors will allow a significant increase in performance compared to the Hipparcos mission. Introducing objective prisms in front of the beam combiners will generate dispersed fringes which will allow a more accurate astrometry and much more detailed spectro-photometric information than filtered, undispered fringes. A complete optical system was designed which is suitable to achieve the desired goal and may be expanded to the size of GAIA. Likewise, several options exists to improve the capabilities and the depth of the DIVA mission.

J. Andersen (ed.), Highlights of Astronomy, Volume 11A, 584.

STRUVE – SPACE ASTROMETRY AND PHOTOMETRY PROJECT

A.E. ILIN, A.G. BUTKEVICH, M.S. CHUBEY, D.I. GORSHANOV
I.I. KANAYEV, T.R. KIRIAN, I.M. KOPYLOV, V.N. YERSHOV
Pulkovo Observatory, Russian Academy of Sciences, St.Petersburg 196140, Russia

The Space Astrometry and Photometry Project Struve is being designed at the Pulkovo Observatory in cooperation with other Russian space institutes. It is expected to be launched before 2010 with a duration of mission of at least 36 months.

The main objectives are to extend at milliarcsecond accuracy the Hipparcos satellite reference system to fainter objects including quasars and to get a second epoch for Hipparcos stars. The project Struve, along with the recently suggested project DIVA, will fill the intermediate place between Hipparcos and microarcsecond astrometry. Unlike DIVA, we propose far more extensive astrometric and photometric surveys.

We expect an Output Catalogue of 20 million stars (density of about 500 stars per square degree). A sky survey will be complete down to $V = 14$ (about 15 million stars), and selected objects down to $V = 19.5$ will be observed within a special program. The proper motions of the Hipparcos stars will be determined with an accuracy of about 0.1 mas/yr. The mean accuracy of star positions in the output catalogue is expected to be 0.6 mas which could be achieved by proper design of the satellite (symmetry, smooth rotation etc.), optics and the micrometer. A properly designed micrometer (with CCD arrays, special processors for image processing and the compression of the data flux to the ground station) will give the possibility of observing all objects of the sky down to a definite limiting magnitude.

ON THE POSSIBILITY OF IDENTIFICATION OF THE MOVING CELESTIAL OBJECTS OBSERVED WITH THE SPACE ASTROMETRIC TELESCOPE

O.P. BYKOV
Pulkovo Astronomical Observatory, Russia

In connection with the creation of the scientific grounds of the Russian Project named 'Struve' Space Astrometric System, the main principles of classification and identification of any celestial moving object observed with this System were formulated and basic algorithms were elaborated. These algorithms are invariable for a short or long observational arc, for a known or unknown heavenly body and for an artificial or natural celestial object.

For classification of observed celestial bodies the angular velocities of motion are used. These values are calculated from the statistical treatment of the 5 nearest spherical celestial body positions with a time interval between them near 7 seconds.

For an identification of well known sky objects ordinary procedure may be used by means of traditional ephemeris calculations. For an identification of unknown natural celestial bodies a special algorithm was developed. It uses a calculated angular velocity of moving object from one scan to another close scan of observations. Then, having several accurate positions of fixed object during 5-10 hours per day, we can determine an initial elliptical object's orbit by the Apparent Motion Parameters Method created at Pulkovo Observatory. It deals with a position of object, its angular velocity and acceleration, position's angle and a curvature of trajectory on a short observational arc. These last four quantities are named the Apparent Motion Parameters.

In spite of a preliminary character of the AMP-method orbits we can identify an observed object through a large interval of time, for example over 2-4 months after orbit determination. In this problem the observed and calculated angular velocities of the object's motion are very useful. They are new and important ephemeris parameters at the epoch of Space Telescope astrometric positional observations. Examples of simulations of orbit determinations will be presented.

J. Andersen (ed.), Highlights of Astronomy, Volume 11A, 585.

CONCLUDING REMARKS

M.A.C. PERRYMAN
Astrophysics Division, Space Science Department
ESA-ESTEC, 2200AG Noordwijk, The Netherlands

Seventeen years ago, the Hipparcos project was accepted within ESA's scientific programme, as the first space mission specifically devoted to the accurate measurement of stellar positions. As the result of an unprecedented and productive collaboration between ESA and the European astronomical community, the Hipparcos project has reached a decisive milestone.

The Hipparcos and Tycho Catalogues were completed in August 1996, and made widely available in June 1997, when they were published by ESA as the 17-volume publication ESA SP–1136. Volume 17 includes all data files, and documentation, in the form of six ASCII CD-ROMs. The catalogues can be obtained from the ESA Publications Division, with full details (and an order form) given on the Hipparcos www page, http://astro.estec.esa.nl/Hipparcos. The full catalogues are also available via ftp from the CDS, Strasbourg, where the catalogue data are also linked to the cross-identifications in the SIMBAD data base, and through the VizieR relational data base queries.

The Hipparcos www page also supports remote interrogation of the catalogues, either by object (HIP or TYC) identifier, or via a search range in right ascension and declination. Active links allow the user to follow selected HIP or TYC identifiers, entering the associated annexes (Variability Annexes, and the Double and Multiple Systems Annex) or inspecting the light-curve (in PDF form) when provided. A more comprehensive search and interrogation facility will become available at the end of 1997 in the form of a CD-ROM (Celestia 2000) with access and display software prepared for IBM PC and compatibles running under Windows 3.1, Windows 95, or Windows NT.

ESA has distributed an order form for the catalogues to most major astronomical institutes, observatories and libraries; and has made the 17-volume catalogue available free of charge to *bona fide* institutes for whom the purchase cost of the catalogues would likely be prohibitive.

The following are representative of some of the general questions that have been asked about the Hipparcos and Tycho Catalogues, addressed here in order to help orientate prospective catalogue users:

(a) are new versions of the Hipparcos and Tycho Catalogues expected? The answer is no; the three years between the end of satellite operations (mid-1993) and the catalogue publication (June 1996) was devoted to a full calibration, reduction, and documentation of the catalogues. No further global reductions of the data are planned or expected. Intermediate astrometric data, and photometric measurements at each epoch of observation, are published as part of the ASCII CD-ROM set, and these will allow users to undertake detailed investigations of individual objects – for example, for new variability period searches, or for revised orbital element determinations in conjunction with ground-based observations.

(b) What kind of errors been found already in the published catalogues? A full list of reported errors is maintained on the Hipparcos www page. Presently (September 1997) this list is very short, and restricted to minor editorial-type corrections.

(c) Can users trust the data? Our short answer is yes! A very considerable effort was devoted to the calibration and validation of the catalogue data (and the documentation of these tasks). The catalogue compilers are presently confident about the quality of the published data. Users should take fully into account the published standard errors and correlations, and the various notes and qualifiers related to the catalogue data. There is no present evidence that the astrometric data are affected by quantified systematic errors, or that the accompanying standard errors do not adequately reflect the true (external) errors.

J. Andersen (ed.), Highlights of Astronomy, Volume 11A, 586–587.

Accompanying the release of the Hipparcos and Tycho Catalogues, a major Symposium was organised in Venice in May. This represented the end of the formal involvement in the Hipparcos Project by ESA and the four scientific consortia: INCA (led by Catherine Turon, Meudon) responsible for the Hipparcos Input Catalogue; NDAC (led by Lennart Lindegren, Lund) and FAST (led by Jean Kovalevsky, CERGA), jointly responsible for the Hipparcos Catalogue; and TDAC (led by Erik Høg, Copenhagen) responsible for the Tycho Catalogue.

At the Venice Symposium roughly 200 oral and poster contributions were presented, communicating early results of investigations that had been undertaken during the one-year proprietary period following the catalogue completion. The resulting Proceedings, amounting to about 1000 pages, has been published as ESA SP–402 in September 1997.

These early analyses addressed a wide variety of subjects: in stellar physics, Galactic structure and evolution, reference frames, age and distance scale determinations, photometry and variability, studies of solar system objects, etc. In the three months between the Venice Symposium and the IAU Joint Discussion in Kyoto, progress has been evident on a number of fronts.

First, the Venice Symposium drew attention to some of the subtle considerations that needed to be accounted for in the detailed interpretation of the Hipparcos parallaxes (in particular) – especially in the context of the transformation of parallaxes to distances and the resulting consequences (e.g. the 'Lutz-Kelker' type effects), and in the correct use of the published correlations between astrometric parameters. Without due care, ignoring these effects may lead to incorrect astrophysical conclusions, and incorrect inferences about the data quality.

Second, a number of preliminary results presented at the Venice Symposium created some puzzlement: for example, the distance of the Pleiades was significantly different from the pre-Hipparcos value; similarly, comparisons between the pre- and post-Hipparcos distance estimates of a number of open clusters revealed some surprising results; and the various attempts to determine ages of certain Galactic tracers on the basis of stellar evolutionary models led to rather different age estimates, and to different conclusions about what these different estimates might mean.

In the short interval between these two meetings, there appears to have been already a significant increase in the confidence ascribed to the Hipparcos astrometry. In particular it does appear that investigators are now placing more confidence in the published data, and therefore in the discrepancies that these results are revealing. In parallel, different groups are starting to probe the implications that previous models of internal stellar structure and evolution might indeed have been too simplistic, and therefore do not adequately fit the new results.

These trends were evident in a number of areas presented at the Venice Symposium and again at the present Joint Discussion. Over the next few months, the Hipparcos implications about the distance and age scales, distribution of stars and hidden mass in the disk, Galactic rotation, and many other areas of astrophysics, should become much clearer.

The Hipparcos satellite data has already had a significant impact in a number of areas, and the early literature has already brought the relevance of space-based astrometric measurements more clearly into discussions of a wide range of astronomical disciplines. For the future, the value of significantly improved astrometric measurements is becoming ever more evident. Within ESA, plans for a 'super-Hipparcos' – a global astrometric interferometer which might reach 10 microarcsec accuracy at 15 mag, and which could allow the measurement of as many as 1 billion stars down to about 20 mag – are progressing. Distances accurate to 10 per cent could be measured out to 10 kpc, and proper motions corresponding to 1 km/s accuracy in the velocity vector would be possible out to 20 kpc. Some of the astrophysical problems that could be addressed with such an instrument have also been raised at this Joint Discussion.

As one major astrometric space programme is successfully concluded, the prospects for greatly enhancing our understanding of our Galaxy with improved positional measurements appear highly promising. At this level of accuracy, everything in our Galaxy would be seen to move. In years to come, the present gap in our knowledge of distances and motions within the Galaxy might seem surprising. But Hipparcos has certainly made a huge step towards acquiring this fundamental understanding.

AUTHOR INDEX

GPSR Compliance

The European Union's (EU) General Product Safety Regulation (GPSR) is a set of rules that requires consumer products to be safe and our obligations to ensure this.

If you have any concerns about our products, you can contact us on ProductSafety@springernature.com

In case Publisher is established outside the EU, the EU authorized representative is:

Springer Nature Customer Service Center GmbH
Europaplatz 3
69115 Heidelberg, Germany

Batch number: 10577874

Printed by Printforce, the Netherlands